The New Physics

The New Physics

Edited by
Paul Davies

Professor of Theoretical Physics
The University of Adelaide

CAMBRIDGE
UNIVERSITY PRESS

Published by the Press Syndicate of the University of Cambridge
The Pitt Building, Trumpington Street, Cambridge CB2 1RP
40 West 20th Street, New York, NY 10011–4211, USA
10 Stamford Road, Oakleigh, Victoria 3166, Australia

First published 1989
Reprinted 1990
First paperback edition 1992

Printed in Great Britain by
Butler & Tanner Ltd, Frome and London

British Library Cataloguing in publication data

The New Physics
1. Physics
I. Davies, P.C.W.
530   QC21.2

Library of Congress cataloguing in publication data

The New Physics/edited by Paul Davies.
p.   cm.
ISBN 0 521 30420 2
1. Physics.   Davies, P.C.W.
QC21.2.N49   1989
530—dc19

ISBN 0 521 30420 2 hardback
ISBN 0 521 43831 4 paperback

Contents

Contributors

Alastair D. Bruce
Department of Physics, James Clerk Maxwell Building, University of Edinburgh, Mayfield Road, Edinburgh EH9 3JZ, UK.

Frank Close
Nielsen Physics Laboratory, University of Tennessee, Knoxville, TN 37996–1200, USA and Rutherford Appleton Laboratory, Chilton, Didcot, OX11 0QX, UK.

Paul C.W. Davies
Department of Physics and Mathematical Physics, The University of Adelaide, GPO Box 498, Adelaide, South Australia 5001, Australia.

Joseph Ford
School of Physics, Georgia Institute of Technology, Atlanta, GA 30332, USA.

Howard M. Georgi
Department of Physics, Harvard University, Cambridge, MA 02138, USA.

Alan H. Guth
Department of Physics, Massachusetts Institute of Technology, Cambridge, MA 02139, USA.

Stephen W. Hawking
Department of Applied Mathematics and Theoretical Physics, University of Cambridge, Silver Street, Cambridge CB3 9EW, UK.

Chris. J. Isham
The Blackett Laboratory, Imperial College of Science and Technology, Prince Consort Road, London SW7 2BZ, UK.

Peter L. Knight
Optics Section, The Blackett Laboratory, Imperial College of Science and Technology, Prince Consort Road, London SW7 2BZ, UK.

Anthony J. Leggett
Department of Physics, University of Illinois at Urbana-Champaign, 1110 W. Green Street, Urbana, IL 61801, USA.

Malcolm S. Longair
Cavendish Laboratory, University of Cambridge, Madingley Road, Cambridge CB3 0HE, UK.

Gregoire Nicolis
Service de Chimie Physique, Code Postal 231, Campus Plaine ULB, Boulevard du Triomphe, 1050 Bruxelles, Belgium.

Abdus Salam
Imperial College of Science and Technology, Prince Consort Road, London SW7 2BZ, UK, and International Centre for Theoretical Physics, POB 586, Miramare, Strada Costiera 11, 34100 Trieste, Italy.

Abner Shimony
Department of Physics, Boston University, 509 Commonwealth Avenue, Boston, MA 02215, USA.

Paul Steinhardt
Department of Physics, University of Pennsylvania, 209 South 33rd Street, PA 19104, USA.

John C. Taylor
Department of Applied Mathematics and Theoretical Physics, University of Cambridge, Silver Street, Cambridge CB3 9EW, UK.

David J. Thouless
Department of Physics, FM–15, University of Washington, Seattle, WA 98195, USA.

David J. Wallace
Department of Physics, James Clerk Maxwell Building, University of Edinburgh, Mayfield Road, Edinburgh EH9 3JZ, UK.

Clifford M. Will
Department of Physics, Washington University, St Louis, MO 63130, USA.

Preface

Many elderly scientists look back nostalgically at the first thirty years of this century, and refer to it as the Golden Age of physics. Historians, however, may come to regard those years as the dawning of 'the New Physics'. The events which the quantum and relativity theories set in train are only now fully impinging on science, and many physicists believe that the Golden Age was only the beginning of the revolution.

Any attempt to take a 'snapshot' of the frontiers of fundamental research at a time when physics is developing so rapidly is bound to be risky. The intention of *The New Physics* is to communicate something of the essential flavour of the latest ideas without descending into a lot of technicalities. No doubt some of the material will date rather rapidly; that is inevitable. But most of what is contained in this volume will have enduring value.

Our contributors are not only internationally recognised expositors, they are also very distinguished scientists in their own right who have made significant contributions to the topics they discuss. They have written at a level which generally corresponds to *Scientific American* or *New Scientist*, although in places a more advanced treatment has been unavoidable. All the contributors have endeavoured to keep mathematics to a minimum, and to avoid excessive technical jargon. Such mathematics as does appear is either elementary, or confined to specific sections. A glossary is provided to help with the technical terms.

I hope that this attention to level will make the majority of the essays accessible to the non-specialist. Editorial interference has been kept to a minimum, so the essays retain the individual styles (and conventions) of their authors. I hope this imparts an informal, chatty flavour to the book, which I regard as less an encyclopedia and more a collection of personal perspectives on this most challenging of subjects.

Paul Davies

1 The New Physics: a synthesis

Paul Davies

1.1 What is the 'New Physics'?

It is fashionable to suppose that science advances in revolutionary leaps. Physicists often talk of the imminence of the 'third revolution' in physics. The first revolution is identified with the work of Galileo, Newton and their contemporaries, for it was in the seventeenth century that the foundations were laid for the systematic study of matter, force and motion by what we would today call the scientific method. The second revolution occurred around the turn of this century, with the theory of relativity, the quantum theory and the discovery of radioactivity.

For many non-physicists, the quantum and relativity theories are the 'New Physics'. Their consequences pervade much of twentieth-century science and technology. Although these theories are now many decades old it is only recently that their full importance is becoming apparent to people outside the field.

While it is true that the quantum and relativity theories form the framework of much of present-day physics, there is a growing belief that major new conceptual and practical advances are currently taking place that go beyond the mere application of these theories. Unlike in the case of the first two revolutions, where specific developments in a relatively well-defined area led to an explosion of new ideas, the 'third revolution' is taking place across a broad front. This wide sweep is reflected in the contents of *The New Physics*, which covers topics as diverse as black holes, subatomic particles, novel materials and self-organising chemical reactions.

Physics is the most pretentious of the sciences, for it purports to address all of physical reality. The physicist may confess ignorance about a particular system – a snowflake, a living organism, a weather pattern – but he will never concede that it lies outside the domain of physics in principle. The physicist believes that the laws of physics, plus a knowledge of the relevant boundary conditions, initial conditions and constraints, are sufficient to explain, in principle, every phenomenon in the universe. Thus the entire universe, from the smallest fragment of matter to the largest assemblage of galaxies, becomes the physicist's domain – a vast natural laboratory for the interplay of lawful forces.

1.2 The small

The investigation of matter on ever smaller scales of size is motivated in large part by the age-old quest for the ultimate building blocks of the physical world. Ever since the atomic theory of Leuccipus and Democritus in the fifth century BC, the idea that all material substance is composed of a small number of truly elementary particles bound together has been a compelling one. What we today call atoms are not, of course, elementary particles at all, but composite entities with an exceedingly complicated internal structure. Nor is atomic matter the only sort that can exist. With their giant particle accelerators, physicists have produced in high-energy collisions a multitude of short-lived particles.

When I was a student in the 1960s particle physics was in a mess. More and more particles were being discovered and theorists were less and less sure about what they all were, how they related to one another or whether any sort of systematic understanding of their properties would ever be possible. The four fundamental forces of nature that act on these particles – electromagnetism, gravitation, and the weak and strong nuclear forces – were also ill-understood at the quantum (microscopic) level. Only one of these forces, electromagnetism, had a consistent theoretical description. The weak force could not be properly understood, and many calculations of its effects gave manifest nonsense. The theory had little predictive power. The strong force appeared to be not a single force at all, but a complex tangle of perplexing interactions that seemed to have no simple underlying form. Gravitation was dismissed as irrelevant to particle physics, and the most strenuous attempts at providing it with a quantum description gave mathematical rubbish for almost all predictions.

Today all this has changed. The change can be traced in large part to two crucial developments that both took place in the 1960s, but reached fruition only in the 1970s.

The first of these developments was the quark theory. The basic idea is simple. Particles of matter can be grouped into two classes. One class consists of heavy, strongly interacting particles called hadrons, which includes the neutron, the proton, and other particles which feel the strong nuclear force. The other class, known collectively as leptons, consists of the electron, the neutrino and other particles that are generally light and interact only weakly (i.e. they do not feel the strong force at all). According to the quark theory leptons are elementary, but hadrons are not. Hadrons are made of quarks. These quarks (which *are* treated as elementary) combine together either in triplets, to form the so-called baryons, or in pairs to form mesons. The quarks and leptons, which probably number only a handful in their different types, could well be the ultimate building blocks of all matter. If so, then the two-and-a-half millenia quest would be at an end.

Particle physics is probably the most spectacular branch of the New Physics, and the one most readily identified by the layman. Particle accelerators are gargantuan machines requiring large teams of scientists and engineers to operate them. Accelerator laboratories are prestige institutions, often involving a degree of international collaboration that is unusual in science. A particle accelerator can be regarded as a giant microscope for peering into the innermost recesses of matter, an awesome complement to the giant telescopes that probe the edges of the universe.

Several essays in this volume deal in one way or another with particle physics. Frank Close reviews the latest developments concerning quarks and gluons – the 'messenger' particles responsible for binding the quarks together. Most particle accelerator experiments involve the high energy collision of two hadrons. This represents a complicated interaction between several quarks, and the data from collision experiments involves a lot of difficult analysis. It is clear that our understanding of hadronic matter is still in its infancy, comparable perhaps to the state of knowledge of atomic physics and chemistry fifty years ago.

Subatomic particle physics is important for more than its role in the search for the ultimate building blocks of matter. It is also the great testing ground for both the special theory of relativity and quantum mechanics. A proper union of these two theories leads to the subject of relativistic quantum field theory, the jewel in the crown of the New Physics. Relativistic quantum field theory is the starting point for almost all current attempts to provide a fundamental description of subatomic particles. It is not an easy subject to understand, but it is so central to modern thinking about the microcosmos that I regard its emphasis in this volume as mandatory.

When the quantum theory is properly applied to the electromagnetic field, it leads to quantum electrodynamics (QED), a consistent theory of electrons, positrons and photons in interaction. Central to this theory is the idea that electrons and positrons interact by exchanging photons. Thus the photon can be regarded as a 'messenger' particle, conveying the electromagnetic force between particles of matter.

Unquestionably QED is quantitatively the most successful relativistic quantum field theory. Using QED, subtle physical effects have been predicted and confirmed by experiment to ten-figure accuracy. The key to the success of QED is the fact that the electromagnetic field possesses an abstract, but powerful, sort of symmetry known as *gauge symmetry*. The most familiar example of this gauge symmetry is the fact that the electric potential is not itself an observable quantity, only the potential *difference*. Thus one is free to 're-gauge' the zero of electric potential without affecting the physics of an electrical system.

It is now generally believed that all the fundamental forces of nature – gravity, the weak and strong nuclear forces, as well as electromagnetism – possess gauge symmetries, albeit of a more complicated kind, and that these forces are transmitted by the exchange of 'gauge particles'. The crucial topic of gauge theories thus deserves a chapter of its own, and this is duly provided by John Taylor.

We are here touching upon the second great theoretical development that took place in the 1960s. Some far-sighted theorists spotted that the success of QED was based on the crucial gauge symmetry, and that the weak force might possess a form of gauge symmetry too, but one which is hidden from us. Abdus Salam and Steven Weinberg, basing their work on the ideas of Sheldon Glashow, spotted how nature performs the trick of hiding the gauge symmetry of the weak force. This enabled them to reformulate the theory of the weak force in such a way that it could be amalgamated with the electromagnetic force to produce a consistent theory of an integrated *electroweak* force. Just as James Clerk Maxwell had shown that electricity and magnetism were really two aspects of a single electromagnetic force, so Salam and Weinberg showed that the electromagnetic and weak forces were not independent, but part of a more embracing scheme.

The new electroweak theory made a very specific prediction. There should exist in nature three hitherto undetected particles, called W^+, W^- and Z, that convey the weak force between particles of matter. These new particles go alongside the photon that conveys the electromagnetic force. The highpoint of the theory came in 1983 when the W and Z particles were found in the proton–antiproton collider at CERN, Europe's large accelerator laboratory near Geneva. This was the New Physics at its best.

Building upon the success of the electroweak theory, theorists began extending gauge theories to incorporate the strong force too. Whereas the strong force between hadrons manifests the hideous complexity I have already described, the force between individual quarks *within* a hadron is basically very simple. A gauge theory of the interquark force was developed, in which gluons play the role of the 'messenger' particles that get exchanged between quarks. It was called QCD, which stands for 'quantum chromodynamics'. It is a close analogue of quantum electrodynamics, with gluons in place of photons and a new quality – a strong force 'charge' – playing the role of electric charge. This new quality is whimsically called 'colour' – hence *chromo*dynamics.

With a gauge field description of the strong force at hand, the way was open for an amalgamation of the strong force with the similarly formulated electroweak gauge theory that Salam and Weinberg had proposed. These so-called *grand unified theories*, or GUTs for short, have led to much excitement. They offer the possibility of being tested through two unusual phenomena. One is the existence of magnetic monopoles, the other is the very weak but immensely significant decay of the proton. At the time of writing both these have been reported, but the reports have been generally discounted. The future of GUTs thus remains open.

Some of these exhilarating developments are taken up in John Taylor's chapter. More details are given by Howard Georgi, one of the originators of GUTs. He relates a fascinating personal perspective of GUTs, and also provides an essay on the more refined application of quantum field theory.

With promising theories of three out of the four of nature's forces 'in the bag' the conspicuous odd man out is gravitation. Gravitation was the first of nature's forces to receive a systematic mathematical description (by Newton), but it continues to resist attempts to provide it with a quantum field description, in spite of its gauge nature. Direct attempts to quantise gravity in analogy with QED soon run into in-superable mathematical problems associated with the appear-ance of infinite terms in the equations. These 'divergences' have plagued all quantum field theories over the years, but the gauge nature of the other forces enables the divergences in their theories to be circumvented.

Many ingenious alternative approaches to quantising gravity have been made, and these are described in some detail in Chris Isham's chapter. So long as gravity remains an unquantised force there exists a devastating inconsistency at the heart of physics. Although quantum effects of gravitation are unlikely to have any detectable results in particle physics (and precious few elsewhere, save possibly in early-universe cosmology), nevertheless it is vital that a consistent quantum description be found, otherwise gravitation cannot be properly interfaced with the rest of physics.

Today, most theorists are pinning their hopes on a theory that will simultaneously unite gravitation with the other three forces of nature in a superunified theory, and provide a consistent quantum description of all four forces. These 'Theories of Everything' are currently occupying the attention of a small army of physicists. The most promising theory at the time of writing goes by the name of *superstrings*. It takes as its starting point the idea that the world is built, not from particles, but loops of string inhabiting a ten-dimensional spacetime. Abdus Salam, in his overview of particle physics, gives the latest thinking on the Theories of Everything, including superstrings,

as well as summarising the status of the particle physics experimental programme.

1.3 The large

Astronomy has always been a rich source of phenomena for the physicist, and modern astrophysics provides a whole plethora of exotic processes that can throw light on almost every branch of physics. Newton appealed to astronomy, in the form of the orbital analysis of the solar system, to test his new theory of gravitation. Three-hundred years on scientists are still using the motions of the planets, as well as those of less familiar astronomical objects such as black holes and neutron stars, as a gravitational laboratory.

The currently accepted theory of gravitation is Einstein's general theory of relativity, published in 1915. The theory was applied to three so-called classical tests: the precession of the perihelion of the planet Mercury, the bending of light by the sun and the red-shift of light escaping from a massive body such as a star. With the enormous advances that have occurred in telescope design, in modern electronics and instrumentation and in spaceflight, there are now a whole range of tests of general relativity to supplement the three classical ones. Twenty years ago, gravitation was a backwater of physics, confined almost exclusively to theory. Now it is a major experimental enterprise.

No volume on the New Physics would be complete without an essay on the burgeoning field of gravitational tests. Clifford Will provides a comprehensive summary of all the modern lines of research, ranging from laboratory and ground-based experi-ments, through lunar and solar system tests, to observations of far-flung astronomical objects. He also discusses the prospects for detecting gravitational radiation – gravity waves – from space. This promises to be among the New Physics of the early twenty-first century, and would open up a powerful new window on the universe.

Astronomical objects are not, of course, merely gravitating masses. They engage in a whole range of physical processes, from nuclear reactions to superconductivity. The modern astronomer must be prepared to model X-rays from material accreting onto black holes, neutrino processes and nuclear reactions in supernova explosions, magnetic and plasma effects near rotating neutron stars, complex electromagnetic and gravitational processes near violently erupting objects such as quasars and disturbed galaxies, and much more. The New Astrophysics is not so much a branch of the New Physics as a subject in its own right, and Malcolm Longair takes us on a mammoth tour through all of modern astrophysics, touching upon the vast range of different astronomical objects currently being studied, and the physics that goes with them.

When applied to the very largest length scales, physics becomes cosmology – the study of the overall structure and evolution of the universe. For three centuries cosmology was merely a speculative and unimportant branch of the theory of gravitation. Two major discoveries have transformed the subject, however. The first happened in the late 1920s when the astronomer Edwin Hubble found that the universe is expanding, and by implication must have originated a finite time ago in an explosion popularly called the big bang. The second occurred in 1965 when two radio engineers accidentally stumbled on a cosmic background of heat radiation, thought to be the last fading remnant of the primeval heat that accompanied the big bang. By extrapolation backwards in time, one infers that at moments successively closer to the initial explosion the temperature rises without limit. Thus the primeval universe can be viewed as a vast high energy physics laboratory in which all the processes of interest to particle physicists (and more) must have occurred, if fleetingly. Conversely, particle accelerators can be viewed as simulators of the conditions which prevailed a split second after the creation.

The use of cosmology as the testing ground for high energy particle physics represents one of the most pleasing confluences of science; it marries the very small with the very large. No longer are cosmologists concerned with gravitation alone. They must now attend to the other three forces as they occur in modern quantum field theory and particle physics.

Indeed, perhaps the most significant advance in cosmological ideas in the last decade came out of the application of GUTs to the early universe. As remarked above, one of the predictions of GUTs is the existence of magnetic monopoles. Calculations suggest that such monopoles would have been produced in huge proliferation in the big bang. Yet direct searches for monopoles have proved negative. Where are they?

An explanation for the lack of cosmic monopoles was proposed in a bold conjecture by Alan Guth. In gauge field theories such as GUTs the nature of the vacuum state, as interpreted quantum mechanically, is very strange. It is possible for the vacuum to become excited. Though the excited vacuum remains devoid of particles, it possesses enormous energy and pressure. The pressure, however, has the peculiar feature that it is *negative*. In Einstein's general theory of relativity energy and pressure, along with mass, both gravitate. It turns out in the case of the false vacuum that the pressure contributes three times as much as the energy, and being negative the effect is to cause *antigravity*, rather than gravity (strictly, it causes a cosmological constant term to appear in the gravitational field equations).

Guth reasoned that if the universe originated in an excited vacuum state, it need not have begun with a bang at all, but a mere quantum whimper. The effect of the 'antigravity' would have been to cause the universe to embark upon a phase of exponentially rapid expansion – which he dubbed *inflation* – with an e-folding time of 10^{-35} s or so. After only a few dozen e-folding times the whimper would have become a bang, and any monopoles present prior to inflation would have been diluted in density by the huge distension in length scales to become utterly insignificant. So, not only did Guth's inflationary universe scenario explain the absence of magnetic monopoles in the universe, it also accounted for the origin of the big bang and why it had the precise form it did.

Inflation has transformed modern cosmology, and offers the hope that many of the very special large scale features of the universe, such as its remarkable smoothness, emerge automatically from an inflationary phase, rather than requiring contrived and highly special initial conditions for their explanation. In his essay with Paul Steinhardt, Alan Guth explains the virtues of inflation and the very natural way in which it accounts for cosmological features that previously had to be inserted 'by hand'.

The application of quantum field theory to the universe as a whole has spawned the subject of 'quantum cosmology', perhaps the weirdest branch of the New Physics. Although a consistent theory of quantum gravity remains elusive, some theorists have boldly gone ahead and applied quantum mechanics to simplified model universes anyway. The results are far from absurd. Among the more intriguing ideas to emerge from quantum cosmology is the possibility of explaining the cosmological initial conditions. In one language, it is said that quantum mechanics can make sense of the idea that the universe appears spontaneously from nothing. Another language says that the origin as such is abolished by quantum mechanics. Either way, it is possible to formulate very natural mathematical statements that fix the quantum state of the universe. Moreover, this state corresponds to something very close to what is actually observed (in a crude, overall sense) to emerge from the big bang. In his provocative essay, Stephen Hawking argues that quantum cosmology has removed the need to impose special initial conditions on the universe; indeed, there is no 'origin' to the universe at all, in spite of the fact that time is finite in the past!

1.4 The complex

There are really three ultimate frontiers of physics: the very small, the very large and the very complex. It is only comparatively recently that complex systems have received systematic study as a physical science. In large part this is due to the advent of fast electronic computers, enabling scientists to model systems that have no simple analytical description.

Complex systems cease to be merely complicated when they display coherent behaviour involving the collective organisation of vast numbers of degrees of freedom. It is one of the universal miracles of nature that huge assemblages of particles, subject only to the blind forces of nature, are nevertheless capable of organising themselves into patterns of cooperative activity.

Perhaps the most spectacular example of the spontaneous appearance of ordered behaviour in a macroscopic system is superfluidity and superconductivity. This is a field that is in some sense the opposite of high energy particle physics, for superfluidity takes place at temperatures close to absolute zero. Here the challenge is to take energy *out* of the system.

Superfluidity is an essentially quantum phenomenon that involves the coherent behaviour of enormous numbers of atomic particles. It can result in bizarre phenomena such as the completely frictionless flow of a liquid or the completely resistanceless flow of electric current. The recent announcement of ceramic superconductors that operate at temperatures as high as 90 K and above has propelled the subject to public fame, and there are real prospects that much cheaper superconductors will soon be used in a wide range of technological devices.

Anthony Leggett surveys the subject of low temperature physics in his essay. He explains how quantum liquids come about and explores their fascinating and peculiar properties. He then explains how electrons can pair together so that their fermionic properties are suppressed to produce a superconducting state. This is a part of the New Physics at the forefront of basic research that is likely to have a big impact on our lives in the years ahead.

The same can be said of quantum optics, which after superfluidity involves the best-known example of spontaneous self-organisation: the laser. Shortly after the first lasers were made they were described as an invention looking for a use. Today, lasers have found a multitude of uses, ranging from burglar alarms to telecommunications.

Lasers are, however, much more than mere gadgets. They offer a means to explore some very fundamental aspects of quantum electrodynamics. In an ordinary lamp, photons are emitted independently from each atom. The wave pulses overlap at random to produce continuous light. But if the system is driven far from equilibrium, it can suddenly undergo a transition to the lasing mode, wherein all the atoms cooperate and emit their photons precisely in synchronism, producing a giant coherent wave train in which all the individual wavelets are exactly in step. As Peter Knight explains in his essay, this exceedingly ordered form of light can reveal an amazing amount of information about the interaction of photons with matter.

The transitions to superfluidity and the lasing mode are two examples of what physicists call phase transitions. Many more are familiar from daily life, such as the transition of ice to liquid water, or water to vapour, or the onset of ferromagnetism in iron when it cools below the Curie temperature. The existence of more than one phase for a physical system is a quite general property.

Less familiar is the fact that, if the parameters constraining the system are chosen correctly, the two phases can become indistinguishable. This happens, for example, at high pressure, when the distinction between a gas and a liquid disappears. The point in parameter space at which such a phenomenon occurs is known as a critical point. Critical point phenomena are a classic example of complexity yielding to deep physical insight. As the critical point is approached a system will adopt a seemingly random and unpredictable configuration, in which regions of one phase interpenetrate another in a complicated manner, these regions fluctuating on ever-larger length scales – a hopeless system to understand, one might think. Nevertheless, work by Kenneth Wilson and others has demonstrated that even processes as complex as critical point phenomena have an underlying mathematical order.

In their essay on critical phenomena, Alastair Bruce and David Wallace describe how certain fundamental scaling properties exist that make the behaviour of these phenomena predictable in a limited fashion. There is a fascinating universality apparent in this scaling behaviour that can be understood in terms of the so-called renormalisation group – a mathematical concept that also finds application in quantum field theory and particle physics. Here, then, is another example of deep connections between hitherto disparate branches of physics.

The unusual propensity for matter and energy to self-organise into coherent structures and patterns is only very recently becoming appreciated by physicists. Partly this is because of the longstanding emphasis that physicists have given to linear systems. Self-organisation (and the related subject of chaos – see below) are essentially nonlinear in nature. As a result they are harder to understand, but they possess a much richer variety of behaviour. Of course, biologists have long studied self-organisation and pattern formation. Today, however, physicists and chemists are joining in, and self-organisation has become a distinctive branch of the New Physics.

Spontaneous self-organisation in simple physical and chemical systems is as striking as it is easy to demonstrate. To give a simple example, if a layer of liquid is heated from below, a critical temperature gradient is reached at which the liquid suddenly begins to convect. The formerly featureless fluid abruptly organises itself into a coherent pattern of convection cells or rolls. In the convecting mode, vast numbers of molecules move in unison as though to some unseen command.

Processes such as these have been brought to fame by the work of Ilya Prigogine and his co-workers, who have developed a whole science of far-from-equilibrium thermodynamics. Gregoire Nicolis, in a review of self-organising systems, describes his work with Prigogine, and shows how many systems, when forced away from equilibrium, can abruptly leap into new and more ordered phases. This 'order out of chaos' seems to fly in the face of the hallowed second law of thermodynamics. In fact there is no contradiction with the letter of the second law, because self-organising systems are always open to their environment, and so can export excess entropy. Nevertheless, self-organisation certainly challenges the spirit of the second law, as well as the prevalent world view that goes with it, based as it is on the idea that the universe is running down amid spiralling entropy. Prigogine and his colleagues believe they have initiated nothing less than a fundamental paradigm shift.

In addition to matter's innate ability to self-organise out of chaos, man is able to artificially order matter. Some of the most spectacular advances made in this direction have been in materials science. Solid state physics has a reputation for dealing with ever-smaller structures, and the pace of technological advance is now such that materials can be grown atomic layer by atomic layer. Physicists can literally 'order up' totally new material structures, and then proceed to explore their electronic properties.

Among the many new materials currently being produced and investigated are those that involve single layers of atoms deposited in some crystal lattice. From the electronic point of view these systems can often be regarded as two dimensional, because current flow transverse to the layer may be inhibited. As a result a whole new field of quantum physics has appeared – the physics of low-dimensional structures. New phenomena, many with obvious practical applications, are being discovered almost daily. This exciting and fast-moving field is reviewed for us by David Thouless.

Complexity is usually associated with many degrees of freedom. We can understand why 10^{23} atoms can behave in a complicated way, and we are beginning to understand how they might organise themselves to behave in a cooperative way under some circumstances. But what is quite unexpected intuitively is that even very simple systems, perhaps with only one or two degrees of freedom, can behave in a fashion which is in some sense infinitely complex. Such is the conclusion of those who work on the subject of 'deterministic chaos'.

Imagine a pendulum free to swing both north–south and east–west – a conical pendulum. We neglect quantum effects, so this is a strictly deterministic system; the motion for all time is uniquely determined by the initial conditions. If the pendulum is driven just below its natural frequency, it settles down to a predictable pattern of motion – stable closed loops. Now increase the driving frequency a bit. Something extraordinary happens. The pendulum begins to gyrate erratically. Sometimes it swings clockwise, sometimes anticlockwise, sometimes in a line. Observation of a few swings cannot tell you a jot about how the pendulum will be moving five minutes later. Predictability has gone. The system is said to be 'chaotic'.

The idea that a system can be both deterministic yet unpredictable is still rather a novelty. (We are dealing with something quite different from quantum uncertainty here, which is based on *indeterminism*.) The reason can be traced to the system's extreme sensitivity to initial conditions. In a more familiar predictable system of the sort studied in elementary mechanics, small errors in input description propagate to small errors in output. In a chaotic system the errors grow exponentially with time, so that the slightest error in input soon leads to complete loss of predictive power.

Chaotic behaviour has been found in an astonishingly wide range of systems. Some famous examples include turbulent fluids, weather systems, geomagnetic reversals, cardiac fibrillations, dripping taps, insect populations, lasers, electrical circuits and chemical reactions. Remarkably, there are some universal features whenever certain forms of chaos occur. A few years ago Mitchell Feigenbaum discovered that in one rather common route from non-chaotic to chaotic behaviour certain universal scaling relations exist (cf. the renormalisation group and critical phenomena). Thus, although chaos represents the breakdown of predictive science, there nevertheless remains an underlying mathematical order. In fact, chaos and self-organisation turn out to be deeply related. One finds, for example, that systems which undergo self-organising transitions also tend to undergo transitions to chaos. (Chaotic behaviour is observed in convecting fluids and lasers, for instance.)

Joseph Ford, in his essay, shows that chaotic systems behave so erratically that they can be considered as truly random in a fundamental sense. He argues that chaos provides the 'missing link' between the laws of physics, so familiar to the scientist, and the laws of chance, equally familiar to the gambler. But a paradox lurks, he believes, just around the corner. When a system which is chaotic according to classical mechanics is given a quantum mechanical description, chaos seems to go away. How can we reconcile the classical world as the large scale limit of the quantum world if classical systems are chaotic but quantum systems are not?

Joining the classical to the quantum world is a deep and longstanding enigma, chaos or no chaos. It forms the subject of the chapter by Abner Shimony. In one sense the interpretational problems of quantum mechanics (which were already apparent to Einstein and Bohr in the early 1930s) constitute the culmination of our investigation of the complex, for these problems address the issue of the observer – surely the most complex of all physical systems.

At rock bottom, quantum mechanics provides a highly successful procedure for predicting the results of observations on microsystems, but when we ask what actually happens when an observation takes place we get nonsense! Attempts to break out of this paradox range from the bizarre, such as the many universes interpretation of Hugh Everett, to the mystical ideas of John von Neumann and Eugene Wigner, who invoke the observer's consciousness. After half a century of argument, the quantum observation debate remains as lively as ever. The problems of the physics of the very small and very large are formidable, but it may be that this frontier – the interface of mind and matter – will turn out to be the most challenging legacy of the New Physics.

2 The renaissance of general relativity

Clifford Will

2.1 Back in the mainstream again

During the two decades 1960–80, the subject of general relativity experienced a rebirth. Despite its enormous influence on scientific thought in its early years, by the late 1950s general relativity had become a sterile, formalistic subject, cut off from the mainstream of physics. It was thought to have very little observational contact, outside of cosmology and a few tests. It was believed to be an extremely difficult subject to learn and comprehend. It was also viewed as a field that was full of ambiguities and unanswerable questions.

Yet by 1970, general relativity had become one of the most active and exciting branches of physics. It took on new roles both as an important theoretical tool of the astrophysicist, and as a new arena for the elementary-particle physicist. Alternative theories challenged its validity as never before, while new experimental tests verified its predictions in unheard-of ways, and to remarkable levels of precision. New fields of study were created, such as 'black-hole physics' and 'gravitational-wave astronomy', that brought together the efforts of theorists and experimentalists. One of the most remarkable and important aspects of this renaissance of relativity was the degree to which experiment and observation motivated and complemented theoretical advances.

It was not always so.

In deriving general relativity during late 1915, Einstein himself was not particularly motivated by a desire to account for observational results. Instead, he was driven by purely theoretical criteria of elegance and simplicity. His goal was to produce a theory of gravitation that incorporated in a natural way both his 1905 special theory of relativity that dealt with physics in inertial frames, and the principle of equivalence, the proposal that physics in a frame falling freely in a gravitational field was in some sense equivalent to physics in an inertial frame.

Once the theory was formulated, however, he did try to confront it with experiment. This confrontation was based on what came to be known as the 'three classical tests'. One of these tests was an immediate success – the explanation of the anomalous advance in the perihelion of Mercury of 43 arcseconds per century, a problem that had bedeviled celestial mechanicians of the latter part of the nineteenth century. The next 'classical test', the deflection of light by the Sun, was such a success that it produced what today would be called a 'media event'. The measurements of the deflection, amounting to 1.75 arcseconds for a ray that grazes the Sun, by two teams of British astronomers in 1919, made Einstein an instant international celebrity. However, these measurements were not all that accurate, and subsequent attempts to measure the deflection of light weren't much better. The third 'classical test', actually proposed by Einstein in 1907, was the gravitational redshift of light, but it was a test that remained unfulfilled until 1960, by which time it was no longer viewed as a true test of general relativity.

Thus, by the late 1950s, it could be argued that the validity of general relativity rested upon one test of moderate precision (the perihelion shift, approximately 1%), one test of low precision (the deflection of light, no worse than 50%), and one unsuccessful test that was not a real test anyway (the gravitational redshift).

Cosmology was the other area where general relativity was believed to have observational relevance. Although the general relativistic picture of the expansion of the universe from a 'big bang' was compatible with observations, there were problems. As late as the middle 1950s the measured values of the universal expansion rate implied that the universe was younger than the Earth! This problem was partly responsible for the rise in popularity of the steady-state cosmology of Hermann Bondi, Thomas Gold and Fred Hoyle, which avoided the big bang by allowing the continuous creation of matter to produce the observed expansion. Even though this 'age' problem had been resolved by 1960, cosmological observations were still in their infancy, and could not distinguish between various alternative models.

Largely because of this paucity of experimental or observational contact, the science of general relativity became stagnant. Moreover, there seemed to be as many unresolved issues of principle in the theory as there were answers; issues such as how should gravitational energy be defined, is gravitational radiation a physically observable phenomenon, what is the meaning of the singularity at the Schwarzschild radius, what are the correct equations of motion for finite bodies, and so on. As late as 1962, the now renowned general relativist and astrophysicist Kip Thorne was advised not to go into this field for his PhD because it had so little connection with the rest of physics and astronomy.

The turning point for general relativity came during the academic year 1959–60, although it is unlikely that anyone at the time recognised it as such.

During that year, a number of scientific events occurred that foretold a new era for general relativity, in which the theory not only would play an important role in physics and astronomy, but also would have its validity put to the test as never before.

The first event was the successful recording of a radar echo from the planet Venus, in data taken on September 14, 1959 (although the analysis of the data was done in 1961). This opened up the solar system as a laboratory for high-precision tests of general relativity. The rapid development of the interplanetary space programme during the 1960s made radar ranging to planets and artificial satellites an indispensable tool for probing relativistic gravitational effects. The second event occurred on March 6, 1960, when the journal *Physical Review Letters* received a paper by Robert Pound and Glen Rebka, Jr, reporting the first successful laboratory measurement of the gravitational redshift of light. Besides verifying this important effect, the experiment demonstrated the powerful use of quantum technology in new devices that would be important both in high-precision gravitational experiments and in astronomical observations. That summer, in the third event, the journal *Annals of Physics* published a paper by Roger Penrose on a 'spinor' approach to general relativity, that was the beginning of a new line of attack that used elegant techniques of pure mathematics to streamline calculations in the theory and to help elucidate its physical consequences. During the same period, the finishing touches were being put on a new theory of gravity by Carl Brans, who was a student of Robert Dicke. The Brans–Dicke theory provided a viable alternative to general relativity and accentuated the need for high-precision experiments to distinguish between competing theories, and for new theoretical work to determine further the observable consequences of general relativity. The fifth event occurred just over a year after the Venus radar measurement. On September 26, 1960, astronomers at Mount Palomar detected an unusual, starlike object at the precise location of the radio source 3C48. The name 'quasar' was soon applied to it and to others like it. This and subsequent astronomical discoveries such as pulsars, the cosmic microwave background radiation, and systems possibly containing black holes, demonstrated that general relativity would have important applications in astrophysical situations.

After this, the pace of research in general relativity and in a new field called 'relativistic astrophysics' began to accelerate.

New advances, both theoretical and observational, came at an ever increasing rate, including the discovery of the microwave background; the theoretical analysis of the synthesis of helium in the big bang; observations of pulsars and of black-hole candidates; the development of the theory of relativistic stars and black holes; the theoretical study of gravitational radiation and the beginning of an experimental program to detect it; improved versions of old tests of general relativity, and brand new tests, discovered after 1959; the discovery of the binary pulsar; the analysis of quantum effects outside black holes and of black-hole evaporation; the discovery of a gravitational lens; and the beginnings of a unification of gravitation theory with the other interactions and with quantum mechanics.

During the two decades following 1960, general relativity rejoined the world of physics and astronomy. Research in relativity took on an increasingly interdisciplinary flavour, spanning such subjects as celestial mechanics, pure mathematics, experimental physics, quantum mechanics, observational astronomy, particle physics, theoretical astrophysics, and so on.

Before we examine the activities and advances of the new general relativity, let us first look at the empirical basis for general relativity, the reason for believing in it in the first place. Here, too, there was a renaissance.

2.2   Curved spacetime: the straight story

The principle of equivalence has historically played an important role in the development of gravitation theory. Isaac Newton regarded this principle as such a cornerstone of mechanics that he devoted the opening paragraph of his 1687 *Philosophiae Naturalis Principia Mathematica* to a detailed discussion of it. He also made the principle an empirical issue by carrying out experiments using pendula to verify it with modest precision. In 1907, Einstein used the principle as a basic element of general relativity. Yet it is only relatively recently that we have arrived at a deeper understanding of the principle of equivalence for gravitation theory. Largely through the work of Robert Dicke around 1960, we have come to see that the principle of equivalence is the foundation, not of general relativity itself, but of the broader idea that spacetime is curved.

One elementary equivalence principle is the kind Newton had in mind, and is known as the weak equivalence principle. It states that the trajectory of a freely falling body (one not acted upon by such forces as electromagnetism) is independent of its internal structure and composition. In the simplest case of dropping two different bodies in a gravitational field, the weak equivalence principle states that the bodies fall with the same acceleration.

A much more powerful and far-reaching equivalence principle is known as the Einstein equivalence principle (EEP). It states (i) the weak equivalence principle is valid, (ii) the outcome of any local non-gravitational experiment is independent of the velocity of the freely falling reference frame in which it is performed, and (iii) the outcome of any local non-gravitational experiment is independent of where and when in the universe it is performed. The second piece of EEP is called local Lorentz invariance, and the third piece is called local position invariance.

For example, a measurement of the electric force between two charged bodies is a local non-gravitational experiment; a measurement of the gravitational force between two bodies (Cavendish experiment) is not.

More so than the weak principle, the Einstein equivalence principle is the heart and soul of gravitational theory, for it is possible to argue convincingly that if EEP is valid, then gravitation must be a 'curved spacetime' phenomenon; in other words, the effects of gravity must be equivalent to the effects of living in a curved spacetime. A key ingredient of curved spacetime is a mathematical variable known as the 'metric'. The metric determines the geometrical relations between events, such as the distance between two spatial locations at a common time, the time between two events at a common location, and a generalised 'distance' between two events at different locations and different times. Special relativity can be described in terms of spacetime with a metric, but, in this case, the metric, known as the Minkowski metric, is that of a 'flat' spacetime. As a consequence of this argument, the only theories of gravity that can embody EEP are those that satisfy the postulates of 'metric theories of gravity', which are (i) spacetime is endowed with a metric, (ii) the trajectories of freely falling bodies are geodesics ('straightest lines') of that metric, and (iii) in local (i.e. small) freely falling reference frames, the non-gravitational laws of physics are those written in the language of special relativity. The argument that leads to this conclusion simply notes that, if EEP is valid, then in local freely falling frames the laws governing experiments must be independent of the velocity of the frame (this property is known as 'Lorentz invariance'), with constant values for the various atomic constants (in order to be independent of location). The only laws we know of that fulfil this are those that are compatible with special relativity, such as Maxwell's equations of electromagnetism. Furthermore, in local freely falling frames, test bodies appear to be unaccelerated; in other words, they move on straight lines, but such 'locally straight' lines simply correspond to 'geodesics' in a curved spacetime (see figure 2.1).

General relativity is a metric theory of gravity, but then so are many others, including the Brans–Dicke theory. So the notion of curved spacetime is a very general and fundamental one.

Because EEP is so central to this conclusion about the nature of gravity and spacetime, let us first examine the experimental evidence that supports it.

A direct test of the weak equivalence principle is the comparison of the acceleration of two laboratory-sized bodies of different composition in an external gravitational field. If the

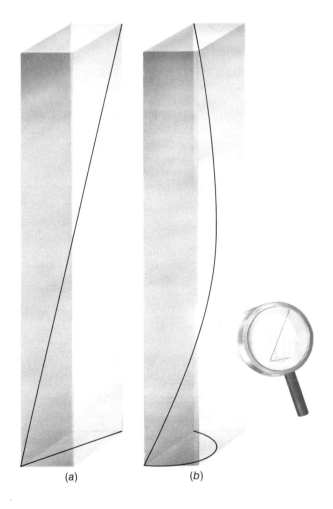

(a)   *(b)*

Figure 2.1. Flat spacetime and curved spacetime. Shown are 'spacetime diagrams', with two spatial directions plotted along the base, and time plotted along the vertical axis. Time is to be plotted using the same units as distance, so one second of time converts through the speed of light to 300 million metres; in order to fit on the page, the scale of the spacetime diagrams has been severely compressed in the vertical direction. (*a*) Flat spacetime. A freely moving body moves along a straight line in spacetime, as well as in space (the curve projected onto the base of the diagram). (*b*) Curved spacetime. A body moves under the influence of a gravitational field directed towards the left of the page. Its curve in space is a parabola (base of the diagram), while its curve in spacetime is a geodesic that appears 'straighter' because of the scale of the time axis. In the spacetime diagram of a freely falling frame, a portion of the curve is seen to be almost perfectly straight, in agreement with that of flat spacetime.

principle were violated, then the accelerations of different bodies would differ. Newton's pendulum experiments tested the weak equivalence principle, but by far the most precise results come from experiments performed by Dicke and his colleagues at Princeton University in the early 1960s and by Vladimir Braginsky and colleagues at Moscow State University in the early 1970s. These experiments adopted the classic method first used by Baron R. von Eötvös around the turn of the century, in which balls of different material (aluminium and platinum, for instance) are attached to opposite ends of a rod that is suspended horizontally from its centre by a fibre. If the balls accelerate differently toward a distant body, in this case the Sun, the rod will try to rotate about the fibre axis in one sense during the day, when the Sun is in the sky, and in the opposite sense at night, when the Sun is 'underneath'. No such rotations were seen to the limits of accuracy of the measurements, with the result that different materials fall with the same acceleration to one part in 10^{12}.

Local Lorentz invariance is tested in a sense every time that special relativity is validated in high-energy physics experiments. However, such experiments are rarely 'clean' tests, because they often involve complicated effects of the weak and strong interactions. But there is one experiment that can be interpreted as a 'clean' test of local Lorentz invariance, and a very accurate one at that. It is the Hughes–Drever experiment, performed in the period 1959–60 independently by Vernon Hughes and collaborators at Yale University, and by Ron Drever at Glasgow University. In the Glasgow version, the experiment examined the ground state of the lithium-7 (^{7}Li) nucleus in an external magnetic field. The state has total angular momentum quantum number $\frac{3}{2}$, and thus is split into four equally spaced levels by the magnetic field. When the nucleus undergoes a transition between a pair of adjacent levels, the photon emitted has the same energy or frequency, no matter which pair of levels was involved. The result is a single narrow spectral line. Any external perturbation of the nucleus that is associated with a preferred direction in space, such as the motion of the Earth relative to the mean rest frame of the universe, will destroy the equality of the energy spacing between the four levels, since the nuclear wave functions of the four levels have different spatial dependences relative to the magnetic field. This will split the spectral line of the emitted photons. Using nuclear magnetic resonance techniques, the experiments set a limit on the separation or spread in frequency of the line that corresponded to a limit on any anisotropy or directional dependence in the energy of the nucleus at the level of a part in 10^{23}. A recent experiment at the US National Bureau of Standards using beryllium improved this limit by a factor of several hundred. These and many other experiments confirm local Lorentz invariance to high precision.

One of the main tests of local position invariance is the gravitational redshift experiment, which tests the existence of spatial dependence in the locally measured rates of atomic frequency standards, or clocks.

Einstein viewed the gravitational redshift as a test of general relativity, but we now regard it as a more basic test of EEP and of the existence of curved spacetime. A typical gravitational redshift experiment measures the frequency shift between two identical clocks, placed at rest at different heights in a gravitational field. To derive the frequency shift from EEP, one argues that if the frequency of a given type of atomic clock is the same when measured in a local, momentarily comoving, freely falling frame, independent of the location of that frame, then the comparison of the frequencies of two clocks at rest at different locations comes down to a comparison of the velocities of two local freely falling frames: one at rest with respect to one clock at the moment it emits its signal, the other at rest with respect to the other clock at the moment it receives the signal. From this viewpoint, the frequency shift is then a consequence of the first-order Doppler shift between the frames, created by the velocity of fall that the first frame has picked up relative to the second during the time of transit of the connecting signal. The result is a shift in the frequency Δv, given by

$$\Delta v/v = \Delta U/c^2,$$

where U is the Newtonian gravitational potential, and c is the speed of light. For small separations h between emitter and receiver, $\Delta v/v = gh/c^2$, where g is the local gravitational acceleration. If the receiver is at a lower height than the emitter, the received signal is shifted to higher frequencies ('blueshift'), while if the receiver is higher the signal is shifted to lower frequencies ('redshift'). The generic name for the effect is 'gravitational redshift'.

The first and most famous high-precision redshift measurement was the series of Pound–Rebka–Snider experiments of 1960–65, that measured the frequency shift of gamma-ray photons from the decay of iron-57 (^{57}Fe) as they ascended or descended the Jefferson Physical Laboratory tower at Harvard University. The high accuracy achieved, 1%, was obtained by making use of the Mössbauer effect to produce a narrow resonance line whose shift could be accurately determined. Other experiments since 1960 have measured the shift of atomic spectral lines in the Sun's gravitational field, and the changes in rate of atomic clocks transported aloft on aircraft, rockets and satellites.

But the most precise gravitational redshift experiment performed to date was a rocket experiment carried out in June 1976 by Robert Vessot and Martin Levine of Harvard University, in collaboration with NASA. In this experiment a hydrogen maser atomic clock was flown on a Scout D rocket to an altitude of about 10 000 km, and its frequency compared to a similar clock on the ground. The experiment took advantage of the extraordinary frequency stability of hydrogen maser clocks by monitoring the relative frequency shifts as a function of altitude. A sophisticated data acquisition scheme accurately eliminated all effects of the first-order Doppler shift caused by the rocket's motion relative to the ground, while tracking data were used to determine the payload's location and velocity at all times, in order to evaluate the potential difference U, and to correct for the special relativistic time dilation of the moving clock. The result was a confirmation of the prediction to seven parts in 10^5.

The wide range of empirical evidence supporting the Einstein equivalence principle has convinced many theorists that spacetime is curved, and that only theories compatible with the metric postulates described above have a hope of being viable. All metric theories treat gravitation as a manifestation of curved spacetime, and the response of all particles and non-gravitational fields, such as electromagnetism, to gravity is universally described by the equations of curved spacetime. Mathematicians refer to such a spacetime as a 'four-dimensional Riemannian geometry'.

What then is general relativity, and how does it differ from other metric theories of gravity? General relativity provides a set of equations, called Einstein's equations, or the Einstein field equations, that determine *how much* spacetime metric or curvature is generated by a given distribution of matter. Once the spacetime metric is known, the equivalence principle determines how matter and fields respond to it.

Different, alternative metric theories of gravity therefore provide different equations for determining the amount of metric or curvature. The most common way that this is done is to introduce additional gravitational fields that mediate the way in which the metric is produced by matter. The metric is known as a 'tensor' field: to each point in spacetime it assigns ten numbers or values (called 'components'); other kinds of fields in spacetime include 'vector' fields, which assign four numbers to each point, or 'scalar' fields, which assign only one number to each point. For instance, the Brans–Dicke theory introduces a scalar field, in addition to the metric, and gives equations that determine how much scalar field is generated by matter, then how much metric is generated by matter in combination with the scalar field. A 'bimetric' theory devised in the 1970s by Nathan Rosen introduced in addition to the metric a tensor field that corresponded to the metric of a flat spacetime, hence the name 'bimetric'. Other theories introduce other kinds of fields in addition to the metric, but in all cases, once the metric is determined, matter responds only to it. Thus the metric and the equations of motion for matter are the primary theoretical entities for describing observable effects, and all that distinguishes one metric theory from another is the particular way in which matter and possibly other gravitational fields generate the metric.

This point of view has led to a very useful and powerful technique for analysing experimental tests of general relativity. It exploits the fact that the comparison of metric theories of gravity with each other and with experiment becomes particularly simple when one takes the slow-motion weak-field limit, which is appropriate for systems in which the square of characteristic speeds and the characteristic gravitational potential are small compared to c^2. This approximation, known as the 'post-Newtonian limit' because it takes into account the

first corrections beyond Newtonian gravitation, is sufficiently accurate to encompass most tests of gravitation theory that can be performed in the solar system. It turns out that, in this limit, the spacetime metric predicted by nearly every metric theory of gravity has the same structure, and can be categorised by the values of a set of ten dimensionless parameters (that this number is the same as the number of components of the metric is purely coincidental). In the special case of theories of gravity which possess conservation laws for energy and momentum or are based on an action principle (which includes most interesting theories), five parameters are sufficient. Each of these parameters has a qualitative physical interpretation, as shown in table 2.1. For illustration purposes, the table lists the values of the parameters in general relativity, the Brans–Dicke theory, and the Rosen bimetric theory.

This framework is known as the parametrised post-Newtonian (PPN) formalism, and the parameters are called PPN parameters. Although the astronomer and relativist Sir Arthur Eddington made use of a primitive version of such a formalism in the 1920s, the first sophisticated and powerful

Table 2.1. *The parametrised post-Newtonian (PPN) parameters and their significance*

Parameter	What it measures, relative to general relativity	Value in general relativity	Value in Brans–Dicke theory[a]	Value in Rosen bimetric theory[b]
γ	How much space curvature does mass produce?	1	$\dfrac{1+\omega}{2+\omega}$	1
β	How non-linear is gravity?	1	1	1
ξ	Are there gravitational effects due to location?	0	0	0
α_1 } α_2 }	Are there gravitational effects due to velocity through the universe?	0 0	0 0	0 α

[a]Brans–Dicke theory has an adjustable 'coupling constant' ω whose value can range from $-\frac{3}{2}$ to $+\infty$. The larger the value of ω the less the importance of the scalar field. In the limit as ω tends to ∞, Brans–Dicke theory becomes identical with general relativity.
[b]In the Rosen bimetric theory, the coefficient α depends on the background cosmological model. Although, when $\alpha = 0$, the PPN parameters are the same as those of general relativity, the theory continues to make different predictions in other areas, such as gravitational radiation.

version was developed by Kenneth Nordtvedt Jr in the 1960s. Over forty theories of gravity have had their PPN parameters calculated by various theorists.

However, the PPN formalism is perhaps most useful for analysing solar system experiments to test metric theories, and it is to these that we now turn.

2.3   Putting general relativity to the test

The attempt to verify general relativity experimentally has always been a central part of the subject, from the moment Einstein first calculated the observable consequences of the theory. However, it has only been during the period of the renaissance of general relativity that technological advances have permitted a vigorous and systematic confrontation between general relativity and experiment. Assisting in this effort was the general theoretical viewpoint discussed in the previous section that compared and contrasted general relativity with other metric theories of gravity in an unbiased way. For example, using a framework such as the PPN formalism, it became possible to regard solar system tests of post-Newtonian effects simply as measurements of the 'correct' values of the PPN parameters. One could then determine which theory best fitted the measured values of the parameters, within the measurement error.

It is useful and instructive to separate solar system experiments into several classes, including the 'classical tests' and 'tests of the strong equivalence principle'.

The classical tests

Three solar system experiments: the deflection of light, the time delay of light, and the perihelion shift of Mercury, can be called the three 'classical tests'. This terminology differs from the conventional usage. Traditionally, the term 'classical tests' has referred to the gravitational redshift experiment, the deflection of light, and the perihelion shift of Mercury. The reason is historical. These were the first three testable effects of general relativity that Einstein calculated. However, we have already seen that the gravitational redshift experiment is really not a test of general relativity, rather it is a test of the Einstein equivalence principle, upon which general relativity and every other metric theory of gravity are founded. Another way of saying this is that every metric theory of gravity automatically predicts the correct redshift. For this reason it has become conventional to drop the redshift as a classical test of general relativity (it is still an important test of EEP, of course).

However, we can immediately replace it with an experiment that is just as important as the other two, the time delay of light. This effect is closely related to the deflection of light, as one might expect, since any physical mechanism in electromagnetism (refraction, dispersion, gravity) that bends light can also be expected to delay it. In fact, it is a bit of a mystery

why Einstein did not discover this effect. It was not discovered as a theoretical prediction until 1964, by Irwin Shapiro, then at MIT. The simplest explanation seems to be that Shapiro anticipated that the space technology of the 1960s and 70s would make feasible a measurement of the effect, a delay of around $200\,\mu s$ in a round trip radar signal to Mars. No such technology was known to Einstein. He was aware only of the known problem of Mercury's excess perihelion motion, and of the potential ability to measure the deflection of starlight by the Sun. Nevertheless, despite its late arrival, the time delay deserves a place in the triumvirate of classical tests, not the least because it has given one of the most precise tests of general relativity to date.

Let us begin with a test that made Einstein's name a household word, the deflection of light. According to the PPN formalism, a light ray which passes the Sun at a distance d is deflected by an angle

$$\Delta\theta = \tfrac{1}{2}(1+\gamma)1\,''.75/d,$$

where d is measured in units of the solar radius (see figure 2.2) and the notation $''$ denotes seconds of arc. Notice the dependence on the coefficient $\tfrac{1}{2}(1+\gamma)$ which has the value unity in general relativity. The expression for $\Delta\theta$ has a simple physical interpretation. The '$\tfrac{1}{2}$' part of the deflection, corresponding to the first term in the coefficient $\tfrac{1}{2}(1+\gamma)$, can be calculated by appealing only to the principle of equivalence. The idea is to determine what observers in a sequence of free falling frames all along the trajectory of the photon would find for the deflection from one frame to the next. It turns out that the net deflection over all the frames is $0''.875/d$. The same result can be obtained by calculating the deflection of a particle using ordinary Newtonian theory, and then take the limit in which the particle's velocity becomes that of light. But the result of both versions is the deflection of light only relative to local straight lines, as defined for example by rigid rods laid end-to-end. However, because of space curvature, local straight lines are themselves bent relative to straight lines far from the Sun by an amount that yields the remaining factor '$\gamma/2$'. The first factor, $\tfrac{1}{2}$, is a consequence of the equivalence principle and holds in any metric theory, the second factor, $\gamma/2$, varies from theory to theory.

The prediction of the bending of light by the Sun was one of the great successes of general relativity. Eddington and Andrew Crommelin's confirmation of the bending of optical starlight during a total solar eclipse in the first months following World War I helped make Einstein a celebrity. However, the experiments of Eddington and his co-workers had only 30% accuracy, and succeeding eclipse experiments were not much better; the results were scattered between one half and twice the Einstein value, and the accuracies were low.

However, the development of long-baseline radio interferometry produced a method for greatly improved determinations of the deflection of light. Radio interferometry is a technique of combining widely separated radio telescopes in

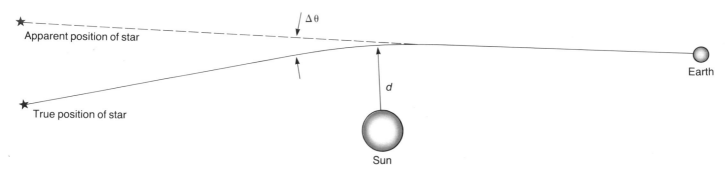

Figure 2.2. Deflection of light by the Sun through an angle $\Delta\theta$.
The distance of closest approach of the ray to the Sun is d.

such a way that the direction of a source of radio waves can be measured by determining the difference in phase of the signal received at the different telescopes. Modern interferometry has the ability to measure angular separations and changes in angles as small as a few hundred microarcseconds. Coupled with this technological advance is a series of heavenly coincidences: each year groups of strong quasars pass near the Sun (as seen from the Earth), including the group 3C273, 3C279 and 3C48, and the group $0111+02$, $0119+11$ and $0116+08$. The idea is to measure the differential deflection of radio waves from one quasar relative to those from another as they pass near the Sun. A number of measurements of this kind occurred almost annually over the period 1969–75, yielding an accurate determination of the coefficient $\frac{1}{2}(1+\gamma)$. The results are shown in figure 2.3. Recent improvements in very long baseline interferometry (VLBI) have made it necessary to take the deflection of light into account over the entire celestial sphere. For a source at 90° from the Sun, for instance, the deflection is only a milliarcsecond, but it is still detectable. Results of a recent analysis of such data are also shown in figure 2.3.

The 1979 discovery of the 'double' quasar Q0957+561 converted the deflection of light from a test of relativity to a useful tool in astronomy and cosmology. The double quasar was interpreted as a multiple image of a single quasar caused by the gravitational lensing effect of a galaxy or a cluster of galaxies along the line of sight between us and the quasar. It turns out that the number and characteristics of the images in such lensed systems can be used as a probe of the mass distribution of the lensing galaxy or cluster of galaxies, and may even be useful for refining the distance to the quasar.

Closely related to light deflection is the Shapiro time delay, a retardation of light signals that pass near the Sun. For a signal that passes the Sun at a distance d on a round trip from Earth to

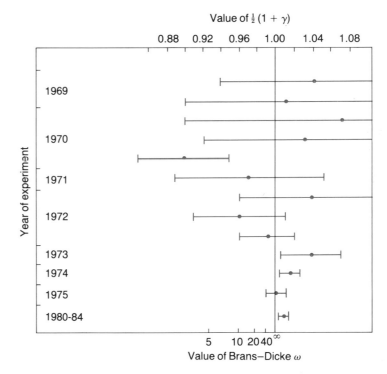

Figure 2.3. Results of radio-wave deflection experiments. Between 1969 and 1975, experiments measured the deflection of light from specific groups of quasars. The 1980–84 measurements were surveys of many quasars and radio sources, in which the deflection had to be taken into account over the entire celestial sphere. Plotted are the results, with measurement errors, for the value of the coefficient $\frac{1}{2}(1+\gamma)$, whose value is unity in general relativity. For comparison, the lower scale shows the corresponding prediction of the coefficient in Brans–Dicke theory, for various values of the 'coupling constant' ω. When ω is infinite, the two theories are indistinguishable.

Mars at superior conjunction (when Mars is on the far side of the Sun), for example, the round trip travel time is increased over what Newtonian theory would give by an amount

$$\Delta t = \tfrac{1}{2}(1+\gamma)250(1-0.16\ln d)\mu s,$$

where d is again in solar radii (see figure 2.4). The close connection between this effect and the deflection of light is reflected in the fact that the coefficient is the same, the factor $\tfrac{1}{2}$ coming from the principle of equivalence, the factor $\gamma/2$ coming from the 'stretching' of space near the Sun.

Figure 2.4. The time delay of light. A target such as a planet or a spacecraft moves from left to right on the far side of the Sun while periodic radar tracking signals are sent to it from Earth. As the signals pass the Sun at closer range, they suffer an additional delay of up to several hundred microseconds, over and above the expected round trip travel time, which for Mars would be around 42 minutes. Shown in the lower half of the figure is a schematic (and exaggerated) plot of the observed round-trip travel time as a function of time, showing the excess delay for rays that pass near the Sun.

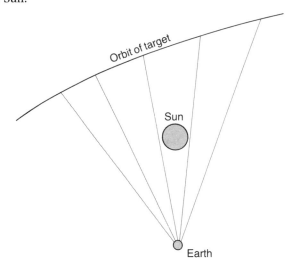

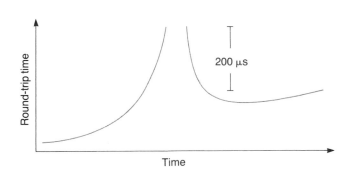

In the two decades following Shapiro's discovery of this effect, several high-precision measurements have been made using radar-ranging techniques that evolved from the Venus echo work of 1959–60. Three types of targets were employed: planets such as Mercury and Venus, used as passive reflectors of the radar signals; spacecraft such as *Mariners* 6 and 7, used as active retransmitters of the signals; and combinations of planets and spacecraft, known as 'anchored spacecraft', such as the *Mariner* 9 Mars orbiter and the 1976 *Viking* Mars landers and orbiters. The *Viking* experiments produced dramatic improvements in the determination of the time delay, because anchoring the spacecraft reduced errors due to random fluctuations in their orbits (planets are very imperturbable), and because noise introduced into the tracking signal by the rough planetary topography and poor planetary reflectivity is removed by the use of transponding spacecraft. The results are summarised in figure 2.5.

Whereas these two classical tests have undergone enormous improvements since 1960, the third classical test, the perihelion shift of Mercury, is in some sense in worse shape now than it was in 1960. Originally, the explanation of the anomalous perihelion shift of Mercury's orbit was another of the triumphs of general relativity. This had been an unsolved problem in celestial mechanics for over half a century, since the announcement by Le Verrier in 1859 that, after the perturbing effects of the planets on Mercury's orbit had been accounted for, there remained in the data an unexplained advance in the perihelion of Mercury (figure 2.6). The modern value for this discrepancy is about 43 arcseconds per century. Many *ad hoc* proposals were made in an attempt to account for this excess, including, among others, the existence of a new planet Vulcan near the Sun, a ring of planetoids, a solar oblateness, and a deviation from the inverse square law of gravitation. Although these proposals could be made to account for the perihelion advance of Mercury, they either involved objects that were detectable by direct optical observation yet were not observed, or predicted perturbations on the other planets that were inconsistent with observations. Thus they were doomed to failure. General relativity accounted for the anomalous shift in a natural way, without disturbing the agreement with other planetary observations. However, since 1966 there has been considerable controversy over whether the perihelion shift is a confirmation or a refutation of general relativity because of the apparent existence of a solar oblateness that could modify the external Newtonian gravitational field of the Sun and contribute a portion of the observed perihelion shift.

The predicted advance rate, including both relativistic PPN contributions and the Newtonian contribution due to a possible oblateness, is given, in arcseconds per century, by

$$d\omega/dt = 42\overset{''}{.}98\lambda,$$

where

$$\lambda \equiv \tfrac{1}{3}(2+2\gamma-\beta)+0.0003(J_2/10^{-7}).$$

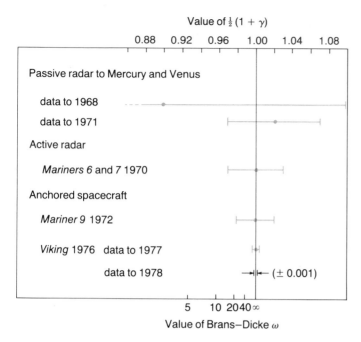

Value of $\frac{1}{2}(1 + \gamma)$

Passive radar to Mercury and Venus

data to 1968

data to 1971

Active radar

Mariners 6 and *7* 1970

Anchored spacecraft

Mariner 9 1972

Viking 1976   data to 1977

data to 1978 ($\pm$ 0.001)

Value of Brans–Dicke ω

Figure 2.5. Results of time delay measurements. The layout is the same as for figure 2.3. The *Viking* measurements place a lower limit of 500 on the Brans–Dicke coupling constant.

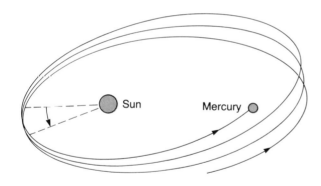

Figure 2.6. Perihelion advance of Mercury. The point of closest approach of Mercury to the Sun, the perihelion, advances by about 574″ per century. Of this, 531″ are caused by gravitational perturbations from the other planets, primarily Venus, Earth and Jupiter. The difference, 43″ per century, was accounted for by general relativity.

The first term in the coefficient λ is the 'classical' relativistic perihelion shift contribution, which depends on the PPN parameters γ and β. In general relativity, this term is unity (see table 2.1). The second term depends on the amount of the Sun's flattening or oblateness, which is measured by the dimensionless parameter J_2. For a Sun that rotates uniformly with its observed surface angular velocity, so that the oblateness is

produced by centrifugal flattening, J_2 is estimated to be 10^{-7}, so that in such a case its contribution to λ would be very small.

Now, the measured shift is known accurately: after the perturbing effects of the other planets have been accounted for, the excess perihelion shift is known to about 0.5% from radar observations of Mercury since 1966, with the result that $\lambda = 1.003 \pm 0.005$. If J_2 were indeed as small as 10^{-7} this would be in complete agreement with general relativity. However, measurements of the visual solar oblateness by Dicke and Mark Goldenberg in 1966 were interpreted as corresponding to a value $J_2 = (2.5 \pm 0.2) \times 10^{-5}$, which would contribute an anomalous 3″ per century to the shift. Coincidentally, such a result would then agree with the 40″ per century relativistic prediction of Brans–Dicke theory, assuming a value of the coupling constant (see table 2.1) of about 7, a value preferred by Dicke on other grounds. Later measurements by Henry Hill and colleagues yielded an upper limit 5×10^{-6}. More recently, values of J_2 ranging between 6×10^{-6} and 10^{-7} have been inferred from models of the Sun with rapidly rotating cores, constructed to be compatible with the patterns of discrete oscillation frequencies of the Sun that have been observed since 1976. Values of J_2 larger than 2×10^{-6} would produce a disagreement with general relativity at least at the level of one standard deviation.

Thus there remains some uncertainty in the interpretation of perihelion shift measurements as tests of general relativity. An unambiguous measurement of J_2 through direct study of the Sun's gravitational field over a large range of distances could be provided by a space mission that was at one time under study by NASA. Known as *Starprobe*, it was a spacecraft that would approach the Sun to within four solar radii. Feasibility studies indicated that J_2 could be measured to an accuracy of 10% of its conventional value of 10^{-7}. Unfortunately, this mission is not a part of NASA's current plans.

Tests of the strong equivalence principle

There is a class of experimental tests of metric theories of gravity that is analogous to tests of the Einstein equivalence principle. One can formulate a 'strong' equivalence principle (SEP), in which not only laboratory-sized bodies, but also planetary and stellar bodies with their own internal gravitational binding energy, should all fall with the same acceleration in an external gravitational field. Furthermore, in a freely falling frame large enough to contain, say, a star or a planet, the laws of gravity as well as the laws of non-gravitational physics should behave independently of the velocity and location of the frame relative to other matter in the universe. This is a much stronger principle than EEP, and to our knowledge it is obeyed only by general relativity: every other known metric theory of gravity violates it at some level. Thus, tests of the strong equivalence principle are important tests of general relativity.

The possibility that gravitationally bound bodies could fall with different accelerations according to different metric theories of gravity was pointed out by Nordtvedt in 1968. The difference in acceleration depends upon the gravitational binding energy of the bodies, and upon the coefficient $4\beta - \gamma - 3 - (10/3)\xi - \alpha_1 + (2/3)\alpha_2$, whose value is zero in general relativity (see table 2.1), but non-zero, for instance, in the Brans–Dicke theory. The best test for the existence of this 'Nordtvedt effect' is in the Earth–Moon system, a kind of planetary Eötvös experiment, in which one looks for the difference in acceleration of the Earth and the Moon toward the Sun. If such an effect existed, it would cause an apparent elongation of the lunar orbit oriented along the Earth–Sun line, of amplitude

$$(4\beta - \gamma - 3 - \tfrac{10}{3}\xi - \alpha_1 + \tfrac{2}{3}\alpha_2) \times 9.2 \text{ m}.$$

Technological advances in lasers, coupled with the US space programme, made such an Eötvös experiment possible. In July 1969, *Apollo* astronaut Neil Armstrong deployed on the Moon a laser retroreflector, a device designed to reflect a laser beam back to Earth, and later *Apollo* and Soviet unmanned missions deployed others. A month later, the first successful acquisition of a reflected laser signal was made, and since then the Lunar Laser Ranging Program has made regular measurements of the round-trip travel times of laser pulses between such places as the McDonald Observatory in Texas and the lunar reflectors, with accuracies of 1 ns, or 30 cm in distance to the Moon. Analysis of the data has yielded no evidence for the Nordtvedt effect, down to a level of better than 30 cm, showing that the combination of PPN parameters $4\beta - \gamma - 3 - (10/3)\xi - \alpha_1 + (2/3)\alpha_2$ must be zero within a few parts in 100, in agreement with SEP and general relativity. Another way of looking at this result is to say that the Earth and the Moon fall with the same acceleration to a few parts in 10^{12}.

Another important consequence of the strong equivalence principle is that the gravitational constant G should be a true constant, independent of the surrounding environment. There is good evidence from geophysical observations of tides of the solid Earth and of the Earth's rotation rate that the gravitational constant is independent of the velocity of the Earth relative to the mean rest frame of the universe, and is independent of any particular direction associated with a nearby concentration of matter, such as that at the galactic centre. The evidence is somewhat weaker that the gravitational constant is independent of the mean matter density of the universe, which decreases on a time scale associated with the expansion of the universe (five parts in 10^{10} per year). From analyses of passages of the moon in front of stars (lunar occultations), planetary and spacecraft radar ranging, and lunar laser ranging, about all that can be said is that, if G varies on a cosmic time scale, it does so no faster than several parts in 10^{11} per year. While some investigators, chiefly involved in lunar occultation studies, once claimed to have seen statistically significant variations in G at about the above rate, other groups claimed that the data is consistent with constant G,

within measurement errors. Continued analysis of radar data from *Viking*, or laser observations of a future orbiter of Mercury, could reduce the uncertainty to a part in 10^{12} per year.

The frontiers of experimental gravitation

Despite the success of general relativity in confronting the experiments described in the previous sections, the subject of experimental gravitation is far from being a closed book. Work continues to improve many of the measurements, for example, by further analysis of *Viking* radar data to improve the determinations both of PPN parameters and possibly of J_2, and to improve the limits on a cosmological variation in the gravitational constant. Other experiments are underway or are planned that will measure effects that have not been seen before.

One of these is the Stanford Relativity Gyroscope Experiment, that has been under development since Leonard Schiff proposed a new test of general relativity using gyroscopes in 1960. The goal of the experiment is to measure the precessions of a set of orbiting gyroscopes that result from two effects, the curvature of space around the Earth (net effect $\sim 7''$ per year) and the 'dragging of inertial frames' by the rotation of the Earth (net effect $\sim 0''.05$ per year). The gyroscopes are 4 cm diameter quartz spheres coated with a layer of superconducting niobium; at liquid helium temperatures the sphere develops a magnetic moment parallel to its spin axis whose direction can then be determined by super-precise magnetometers. The precession of the gyroscope axes will be measured relative to the optical axis of a telescope fixed on a distant star (Rigel). The entire system will be in a drag-compensated satellite. Current plans call for a proving flight to test the components on a 1991 Space Shuttle mission. If all goes as planned, an operational flight could follow in a few years.

The possibility of measuring the second-order, or post-post-Newtonian contributions to solar system relativistic effects is being studied by several groups. The ideas include a precision optical interferometer in space (POINTS) with microarcsecond accuracy, to measure the second-order contributions to the deflection of light, and the use of ultra-stable hydrogen maser clocks on a *Starprobe*-type mission to measure the second-order part of the gravitational redshift.

During the past twenty-five years, an intensive theoretical and experimental effort has gone into testing the predictions of general relativity and of other theories of gravitation in many different arenas, and to high precision. General relativity has passed every test, while numerous theories have fallen by the wayside. Although many opportunities remain for further testing of gravitational theory, we can be sufficiently secure about the empirical underpinnings of general relativity to assume now that it is right, and to use it as a practical tool in physics and astronomy.

2.4 Gravitational waves: new ripples on an old pond

The subject of gravitational radiation is almost as old as general relativity itself. By 1916, Einstein had succeeded in showing that the field equations of general relativity admitted wavelike solutions analogous to those of electromagnetic theory. For example, a dumb-bell rotating about an axis passing at right angles through its handle will emit gravitational waves that travel at the speed of light. But Einstein also found that the waves have a very important property: they carry energy away from the rotating dumb-bell, just as electromagnetic waves carry energy away from a light source. He even derived a formula to determine the rate at which energy would be lost from a system such as a rotating dumb-bell, as a consequence of the emission of gravitational waves. As it turned out, the assumptions that he made to simplify the calculation were not completely rigorous, and he also made a trivial mathematical error that made his answer a factor two too large, but the basic analysis was correct. (The error was first pointed out by Eddington.)

That was about all that was heard on the subject for over forty years. One reason was that the effects associated with gravitational waves were extremely tiny, unlikely (it was thought) ever to be of experimental or observational interest. Another reason was that, for a long time, there was disagreement over whether the waves were 'real', or whether they were some artifact of the theory's inherent freedom to express the equations in any coordinate system or reference frame, and would therefore not have observable consequences.

But by 1960, the beginning of the relativity renaissance, two developments sparked the resurrection of the idea of gravitational radiation. One was the work of the group of relativists headed by Hermann Bondi that demonstrated in invariant, coordinate-free terms that gravitational radiation *was* a physically observable phenomenon, that it carried energy and angular momentum away from systems, and that the mass of systems that radiate gravitational waves must decrease. The second development was the decision by Joseph Weber of the University of Maryland to begin to build detectors for gravitational waves.

Since that time, gravitational radiation has been one of the most active and challenging branches of the new general relativity, involving both theorists and experimentalists (and, as we shall see, radio astronomers). The subject divides naturally into two parts; one part dealing with the waves that impinge on the Earth, asking how can they be detected and what is the nature of their source, and the other part dealing with the effect on the source itself of the energy lost through the emission of gravitational waves. Let us consider the second part first, since it has yielded the first evidence that gravitational waves actually exist.

The binary pulsar: gravitational waves exist!

The summer of 1974 was a memorable one for Russell Hulse and Joseph Taylor, and an important one for the subject of gravitational radiation. During a systematic search for new pulsars using the Arecibo radio telescope in Puerto Rico, Hulse and Taylor discovered PSR1913+16, colloquially known as the 'binary pulsar'. The pulsar proved to be a member of a close binary system with an as yet unseen companion. This would have been only a mild curiosity (of 300 radio pulsars, only three are known to be in binary systems), were it not for two important properties of the system. The orbit is so close, with an average separation of the order of a solar radius, orbital velocities up to $300 \, \text{km s}^{-1}$, and an orbital period of only eight hours, that relativistic effects such as the periastron shift (the binary-system analogue of Mercury's perihelion shift) can be significant. Furthermore, the pulsar appears to be one of the most stable clocks in the universe, its pulse period of 59 ms drifting by only 0.25 ns, or four parts in 10^9, per year (only the millisecond pulsar PSR1937+214 discovered in 1982 is more stable).

These two circumstances made the binary pulsar an exciting new laboratory outside the solar system for studying relativistic effects. By measuring arrival times of individual groups of radio pulses at Earth to accuracies of $50 \, \mu\text{s}$, the observers were able to determine the motion of the pulsar about its invisible companion (the system is a special kind of what astronomers call a 'single-line spectroscopic binary'), and thereby measure many of the important orbital elements with an accuracy that boggles the mind. For instance the intrinsic pulse period (referred to September 1, 1974) is 0.059029995271 s, the rate of change of the pulse period is 0.273 ns per year, the eccentricity of the orbit is 0.617127, the orbital period is 27 906.98163 s, and so on.

This accuracy made possible the measurement of several relativistic effects. The first was the periastron shift: the measured value was 4.2263 degrees per year (compare with Mercury's 43″ per century!). According to general relativity, the predicted shift depends on known orbital parameters such as the period and the eccentricity, and on the total mass of the system, which is unknown. We are assuming here that the companion does not have a significant oblateness, in order not to generate an additional contribution to the periastron shift like that produced by an oblateness of the Sun. This assumption will be valid if, as seems likely, the companion is a neutron star, a black hole, or a slowly rotating white dwarf. Now, the observation of the periastron shift cannot be used to test general relativity because of the unknown total mass; instead, the tables can be turned, and general relativity can be used as a tool to determine the total mass of the system. The result turned out to be 2.8275 solar masses.

Another relativistic effect that was observed, although with more difficulty, was the combined effect of the gravitational redshift of the pulsar signal, due to the companion's gravitational potential, and of the special relativistic time dilation due to the pulsar's orbital motion. The observed effect was a periodic variation in the arrival times of pulses, with an amplitude of 4.38 ms. By comparing this observation with the prediction, one obtains another, independent, piece of information about the masses, which makes it possible to determine them separately. The results, in solar masses, were 1.42 ± 0.03 for the pulsar, and 1.40 ± 0.03 for the companion. For the first time general relativity was used as a direct tool for an astrophysical measurement – the weighing of a pulsar!

But the biggest payoff of the binary pulsar had to do with gravitational radiation. For systems with characteristic velocities small compared to the speed of light, and with characteristic gravitational fields that are weak (in an appropriate sense), the generation of gravitational radiation within general relativity is described by an approximation known as the quadrupole formalism. The dominant contribution to gravitational radiation emitted to infinity comes from variations in the quadrupole moment or moment of inertia for the dynamical system, and the rate of energy loss is given by a simple formula, called the 'quadrupole energy-flux formula', which can be represented schematically by

$$\frac{\mathrm{d}E}{\mathrm{d}t} \sim -\frac{1}{5}\frac{G}{c^5}(\dddot{I})^2,$$

where I represents a three-dimensional tensor or a matrix related to the moment of inertia of the system, dots denote derivatives with respect to time, and the superscript 2 represents a kind of square of this matrix. This is indeed just the formula that Einstein derived in 1916 for emission from a rotating rod (as corrected by Eddington).

The loss of energy to infinity must show up within the system as a loss of orbital energy, whose consequence is that the orbit spirals in, and the orbital period shortens. In general relativity, with the measured values of the orbital elements, and with the inferred values for the masses of the two bodies, the quadrupole formula makes a definite prediction for the rate of decrease of the orbital period: about $75\,\mu s$ per year, or more precisely $(2.403 \pm 0.002) \times 10^{-12}\,\mathrm{s}\,\mathrm{s}^{-1}$. Shortly after the discovery of the binary pulsar in 1974, it was thought that measurement of such a small effect would require ten to fifteen years of data, but through brilliant efforts to improve electronic data acquisition techniques at the telescope and to refine the data analysis, Taylor and his team were able to do it in just over four years, in time to open the 1979 Einstein centenary year. Their initial result, announced in December 1978, agreed with the prediction with 20% uncertainties, but subsequent data to 1983 have led to the improved value of $(2.40 \pm 0.09) \times 10^{-12}\,\mathrm{s}\,\mathrm{s}^{-1}$ in complete agreement with the quadrupole prediction of general relativity. No other plausible source of orbital period decrease has been proposed which could account for all or part of the observed decrease. One interesting by-product of this was the knocking down of the Rosen bimetric theory of gravity, which

hitherto was in agreement with solar system experiments. The theory turned out to make radically different predictions for gravitational wave energy loss than general relativity, and was in severe disagreement with the observations. The recent, post-1980 data on the binary pulsar has become so good that it has been possible to detect even smaller relativistic, post-Newtonian effects, such as the Shapiro time delay of the pulsar signal as it passes by the companion, the first time that this effect has been seen outside the solar system.

The discovery of the binary pulsar and the possibility of detecting the energy loss due to gravitational radiation brought to the forefront a controversy that had been simmering in the background of gravitational-wave theory since the late 1950s. A number of theorists began to argue forcefully that the quadrupole formula was not to be trusted, because the purported derivations of it from general relativity were severely flawed. Other theorists argued that it was just plain wrong. The binary pulsar made this controversy more than just a formal, technical question, because now it was observationally relevant.

In order to understand the origins of this controversy and its ultimate resolution, it is useful to compare and contrast gravitational waves with electromagnetic waves. This comparison is simplest when one restricts attention to systems in which the characteristic velocities v, are small compared to the speed of light c. In this case, the radiation can be described by a multipole decomposition. For the electromagnetic radiation from a system of charges, it is well known that the lowest non-vanishing component is electric dipole, with magnetic dipole, quadrupole and higher moments typically smaller in their contribution, because they depend on higher powers of (v/c). There is no monopole moment contribution to radiation because of the conservation of charge; in other words, the monopole or Coulomb electric field is always static. Thus the rate of electromagnetic energy radiated to infinity is determined by the square of the second time derivative of the electric dipole moment, the so-called Larmor formula.

By a similar argument, one can see that the lowest multipole contribution to gravitational radiation is quadrupole. There is no monopole contribution in general relativity because of the conservation of mass–energy, and there is no dipole contribution because of the conservation of momentum and angular momentum. In other words, the external monopole and dipole gravitational fields are static. It is not surprising, then, that the rate of energy radiated in gravitational waves should be given by a formula similar to the Larmor formula, only in this case determined by the square of the *third* time derivative of the mass quadrupole moment or moment of inertia, as given schematically in the formula for $\mathrm{d}E/\mathrm{d}t$ above.

Just as in electromagnetic radiation, the loss of gravitational radiation to infinity must manifest itself within the radiating system in forces, called 'radiation-reaction forces' that damp the motions within the system, reducing its internal energy by an amount that balances that radiated away. An heuristic argument leads to a reaction force that is consistent with the energy-loss rate, given by a simple formula involving five time derivatives of the moment of inertia of the system.

However, these statements are only heuristic, they do not amount to a derivation of either the energy-flux or the radiation-reaction quadrupole formula. What right do we have to believe the two formulae discussed above? The problem is, these formulae do not come from any exact solution of Einstein's equations. In fact, there is no known exact solution of Einstein's equations that contains realistic dynamical material sources, together with gravitational radiation propagating to infinity and radiation reaction experienced by the sources, despite concerted searches by theorists over the past seventy years. Because of this, our knowledge about gravitational radiation and radiation reaction is based upon approximations of Einstein's equations.

The controversy over the validity of the quadrupole formulae centred around how Einstein's equations should best be approximated in order to obtain accurate, reliable formulae for energy-flux and radiation reaction that bear some relation to reality.

One of the issues involved the original derivation of the quadrupole energy-flux formula by Einstein for gravitational radiation from a system whose internal structure and dynamics is dominated by non-gravitational forces, such as a dumb-bell. Such systems are called 'non-self-gravitational'. The question was, does the formula apply equally well to a system whose dynamics is dominated by internal gravitational forces, a so-called 'self-gravitational' system, such as a binary-star system? In other words, do freely falling bodies such as orbiting stars (as opposed to dumb-bells constrained by solid-state or electromagnetic forces) radiate gravitational waves and experience radiation? The answer turned out to be yes; in fact the same formula applies to both the dumb-bell and the binary-star cases. The difficulty in arriving at this conclusion was caused by the fact that Einstein's equations are non-linear, in other words by the fact that gravity can act as a source for more gravity. Because of this, it is usual to solve the equations by a sequence of successive approximations, provided the gravitational fields are weak. For the case of a dumb-bell as the source, the first approximation, usually called 'linearised general relativity', was sufficient to determine both the radiation flux and the radiation reaction. But for a binary-star source, the situation was more subtle: a second approximation was required to get the flux, while a *third* approximation was required to get the reaction force. Purely by coincidence, the final formulae were the same for both the non-self-gravitational and the self-gravitational cases.

Another factor in the controversy was the existence in the literature of counterexamples to the quadrupole reaction formula, derivations whose result differed from the quadrupole formula either in the numerical coefficient or in functional form. For example, a 1965 derivation gave a positive sign for the reaction force, corresponding to anti-damping, rather than to damping of the motion of the source! In the end it was found that most of these purported counterexamples arose because their authors simply failed to iterate or approximate Einstein's equations to the required third level.

A third factor in the controversy was really a question of applied mathematics, and had to do with two features of gravitational radiation. First, since the theory is non-linear, gravity can generate gravitational radiation, just as can the motion of the source, but the gravitational field extends all the way to infinity – it is not localised as are the material sources. A consequence of this was the appearance in many formulae of integrals that diverged or became infinite. This problem does not arise in the linear electromagnetic theory, where the only source of radiation is the localised charge. Secondly, the problem of radiation reaction is really two problems involving very different approximations. In the regions far from the source, the so-called far zone, the solutions are wavelike, and the approximation is built around the fact that the wavelength is much smaller than the region. But in the near zone, within the source, the approximation builds upon the fact that the wavelength is much larger than the size of the zone. This problem occurs in electromagnetism, fluid dynamics, indeed in any problem involving radiation of waves by sources. Most previous attempts to solve these problems encountered difficulties, such as divergent integrals, which made it impossible to make error estimates to gauge how close the approximation was to a 'true' solution of general relativity.

But by the early 1980s most of the problems of this kind were under control, through the application of mathematical techniques of 'matched asymptotic expansions', and other asymptotic methods. These methods treated the near zone and the far zone as separate approximation problems, then matched the two solutions asymptotically in an intermediate overlap zone. The results were much more satisfactory, and yielded reliable approximations to general relativity. The final answers in all cases were the quadrupole formulae for energy flux and radiation reaction. At present, there is nearly unanimous agreement with the quadrupole approximation as the basis for understanding gravitational radiation from slow-motion systems, such as the binary pulsar. It has even been shown that, in the binary pulsar, the quadrupole formula is a valid approximation despite the fact that at least one of the two bodies has highly relativistic internal gravitational fields. It is interesting to notice how, in a subject historically characterised by an absence of empirical contact, an observation, that of the binary pulsar, had such a direct impact on theoretical progress.

The search for gravitational radiation

The binary pulsar has confirmed that gravitational waves exist, and that the quadrupole formula for energy loss is in accord with the observations, but the goal of actually detecting gravitational waves themselves has not yet been attained. Once that happens, however, a new field which relativists like to call

Gravitational-Wave Astronomy, will open up. Because of the binary pulsar, the detection of gravitational radiation as a confirmation of a prediction of general relativity has taken a back seat to the potential use of gravitational radiation as a probe of highly dynamical and violent events in the universe.

To see how this might come about, and to understand some of the difficulties involved, let us describe the effect a passing gravitational wave has on matter, again comparing and contrasting it with that of an electromagnetic wave. An electromagnetic wave consists of an oscillating electric field with a magnetic field perpendicular to it, and both perpendicular to the direction of propagation of the wave. In vacuum, the wave pattern propagates with the speed of light. When the wave passes a charged particle, such as an electron in an antenna, the electric field causes the particle to move, creating an electric current that can be picked up and transformed, say into a television image.

A gravitational wave also consists of an oscillating field that is perpendicular to the direction of propagation, and that propagates with the speed of light. However, the nature of the force field is rather different from the electromagnetic force field. Because of the equivalence principle, the force field cannot just cause an individual particle to move, since a freely falling particle feels no gravitational forces in its own reference frame, no matter what the forces are doing. However, if the force field varies in strength from place to place, it can cause a pair of particles to move relative to each other. Such a force field is called a tidal gravitational force field. Specifically, the pattern of tidal forces from a gravitational wave is illustrated in figure 2.7, in which a circular ring of particles is placed perpendicular to the incident direction of the wave. On each half cycle, the wave distorts the circular shape alternately into ellipsoidal shapes perpendicular to each other. There is no motion of the particles parallel to the direction of propagation of the wave. Just as electromagnetic radiation has a second independent polarisation, with the electric field rotated by 90° relative to the first, so too does gravitational radiation have a second polarisation. But, in this case, the pattern of ellipsoidal shapes produced by the second polarisation is rotated by 45° relative to that of the first, as shown in figure 2.7.

While the source of electromagnetic waves is moving charges, the source of gravitational waves is moving masses. But only a particular kind of mass motion is applicable: motion in which the source changes its shape, in other words in which the dynamics are non-spherical. For example, in the slow-motion limit, as we have already seen, a varying quadrupole moment is required to get radiation. The kinds of sources that might produce gravitational waves therefore include a star that vibrates or collapses in a non-spherical manner (purely spherical vibration or collapse produces exactly zero radiation), a binary star system, a black hole swallowing a star, two stars or black holes colliding or flying past each other, and so on. Laboratory-sized objects such as rotating dumb-bells can also

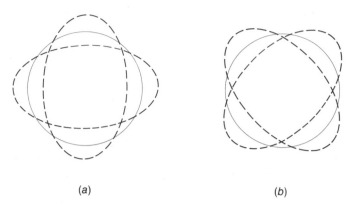

(a)   (b)

Figure 2.7. Effect of a gravitational wave on a ring of particles. The wave is propagating out of the page. In (a), the initially circular ring is distorted into an elliptical shape after one-quarter of a cycle of the wave. After half a cycle the ring returns to its circular shape. After three-quarters of a cycle, the ring is distorted into the perpendicular elliptical shape. After one complete cycle of the wave, the ring returns to its initial shape. In (b), the second independent polarisation state of the wave causes the same sequence of distortions, with the pattern rotated by 45° relative to that in (a). A general wave causes a superposition of the two distortions.

generate gravitational waves, but they are too weak ever to be of interest.

The previous sentence illustrates the discouraging contrast between electromagnetic radiation and gravitational radiation, and that is their relative strength. While electromagnetic radiation is strong enough to have many practical (and impractical) applications, gravitational radiation is extraordinarily weak. One of the strongest imagined sources of gravitational waves, a rotating star that collapses to form a black hole in our galaxy, will produce a field of tidal forces that will cause two masses separated by one metre to move together and apart by only one-hundredth the diameter of an atomic nucleus. For sources farther away, the effect will be even weaker, because the size of the radiating tidal force field falls off as the inverse distance to the source.

The most useful way to characterise the strength of a gravitational wave is by a quantity that gravitational-wave astronomers call h, which measures the strain, or the change in separation divided by the separation between two masses, such as two on opposite sides of one of the rings in figure 2.7. This quantity h is also the deviation in the spacetime metric from its nominal form, produced by the gravitational wave. It is also sometimes called the gravitational waveform. One of the central activities of theorists is to attempt to calculate or estimate the size of the waveform from various proposed sources of gravitational waves. Also of interest is the frequency spectrum contained in h, whether the wave is emitted in a 'burst', or whether it is continuous, and how often such sources occur in the universe. This activity is a blend of pure general relativity, which provides the basic equations, large-scale numerical computations, which are often necessary, especially in some of the collapse and collision sources, and astrophysics, which is needed to determine when, where and how often sources might occur.

Figures 2.8 and 2.9 display some of the theorists' current best guesses concerning the expected waveform sizes and frequency bands of a variety of sources. One source of 'burst' or impulsive gravitational waves is a supernova. Estimates of the size of h from a supernova, say at the centre of our Galaxy, vary widely, from 10^{-18} on down, depending on the amount of rotation the initial star has, on the equation of state of the collapsing matter, and on the detailed dynamics of the collapse. Unfortunately, the last known supernova in our Galaxy was in 1604, so in order to get a decent event rate, one has to be able to see supernovae as far away as the Virgo cluster of galaxies, but then, because of the distance factor, the size of h from the strongest supernova is down from 10^{-18} to 10^{-22}. Neutron-star binary systems that have lost orbital energy to gravitational radiation emission, and are in the final stage of spiralling in until they coalesce, can be sources of strong bursts of gravitational waves in the 10 to 1000 Hz band. Another hypothetical source of lower-frequency bursts is the formation of supermassive black holes (between 100 and 10^7 solar masses) at various early epochs in the evolution of the universe.

Proposed sources of periodic or continuous gravitational waves include known close binary systems, including the binary pulsar (whose waves, unfortunately, are too weak to be detected directly in the foreseeable future), postulated close white-dwarf binaries, and pulsars with appropriately distorted shapes.

Another important type of gravitational radiation is a possible stochastic, or 'noisy' background from the super-imposed waves from all sources that have ever existed, and possibly from exotic processes that might have occurred in the very early universe, such as the formation of cosmic 'strings' (for further details on such processes in the early universe, see the chapter by Guth and Steinhardt).

A glance at figures 2.8 and 2.9 indicates that, to have a hope of seeing gravitational waves from interesting and reasonably abundant sources, one must be able to detect a strain or h of 10^{-22}. The goal of gravitational wave antenna builders is to do exactly this.

The first antenna builder, and the pioneer of gravitational-wave astronomy was Joseph Weber. Around 1958, Weber began working on the problem of how to detect the tiny tidal forces associated with a gravitational wave, first doing the theoretical calculations to determine just what the effects of a passing wave would be, and then building an apparatus. By 1965 he had put a simple detector into operation.

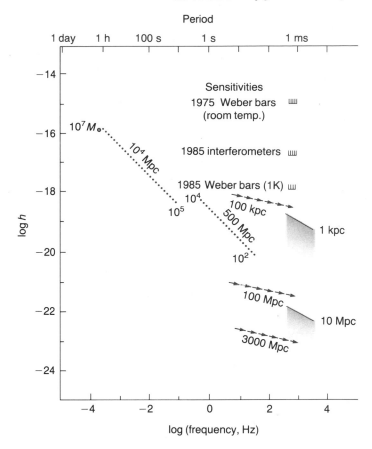

Figure 2.8. Sources of 'burst' gravitational waves. Shown are estimated waveform amplitudes h, plotted on a logarithmic scale, versus the frequency or period of oscillation of the waves. One parsec (pc) equals about three light years. Expected amplitudes from supernovae in our Galaxy (at 1 kpc) and out to the Virgo cluster (10 Mpc) are shown; the large uncertainties in these estimates are illustrated by the shaded regions. Neutron-star binary systems in the final stage of coalescing are shown, for various distances. Formation of very massive black holes from 100 to 10 million solar masses ($M_\odot$), at distances ranging to the edge of the visible universe, could produce sizable amplitudes at lower frequencies. Also shown are the claimed sensitivities of Weber bars circa 1975 (room temperature), and of current interferometers and low-temperature Weber bars. Sources: →▸→ neutron-star binary (distance); ⊔⊔⊔ supernova (distance); ········ black hole formation (mass, distance). (After Wilkinson, D., ed. (1985). *Survey of Gravitation, Cosmology and Cosmic Ray Physics*. National Academy Press.)

Weber's basic concept for a detector is still in use today. It consists of a freely suspended solid cylinder several metres in length, weighing a few tonnes. For reasons of cost, his and many later detectors were made of aluminium. When a gravitational wave passes the cylinder in a direction perpendicular to its axis, the tidal force acts on the distributed

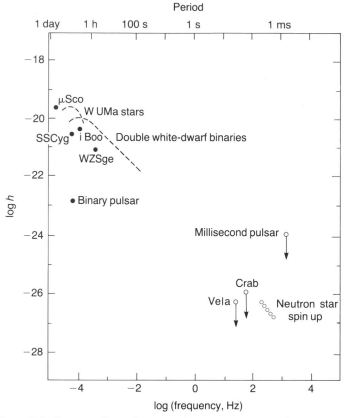

Figure 2.9. Sources of steady or continuous gravitational waves. Shown are expected amplitudes from known binary systems, and from hypothetical binary systems, such as double white-dwarf binaries. If pulsars have distorted shapes, caused, for example, by strong magnetic fields, they could emit gravitational waves of sizes shown for the Crab, Vela, and millisecond pulsars. The main uncertainty is the size or even existence of appropriate distortions. Neutron-star spin up via accretion could lead to an unstable oscillation that produces gravitational radiation (see Section 2.5 for discussion). – – – – Background binaries; • individual binary systems; ○ distorted pulsars. (After Wilkinson, D., ed. (1985). *Survey of Gravitation, Cosmology and Cosmic Ray Physics.* National Academy Press.)

material in the bar, but instead of moving freely as in the case of the rings in figure 2.7, the material oscillates in one of the characteristic end-to-end resonant normal modes of the bar, with a strain amplitude of order h. The resonant normal-mode frequencies for bars of the size of those of Weber are typically in the kilohertz band. The oscillation of the bar continues until internal friction damps it out. It can be detected by attaching strain-measuring devices around the middle of the cylinder (Weber's original method) or by attaching various kinds of motion sensors to the end faces.

Between the period 1968 and 1975, Weber announced the detection of coincident events in similar detectors spaced 1000 km apart, one in Maryland and the other at Argonne National Laboratory near Chicago. Since the bars are constantly in a state of excitation because of thermal and environmental effects, coincident detection in widely separated bars would presumably increase the likelihood of identifying an event of extraterrestrial origin. Weber reported coincidence rates that were apparently much larger than could be explained by random chance. In addition, after 1970, he reported a significant correlation between high rates of coincidences and the time at which the bars were oriented to be most sensitive to waves from the galactic centre. These reports caused a significant stir, both in the relativity and astrophysics communities, and in the popular press.

There were two problems, however. The observed events occurred with a strain amplitude and at a rate (around three times per day) that were much too large to be explained by the current understanding of sources of waves. The second problem was more telling, however. The main body of Weber's coincidence results was reported between 1968 and 1975. But, by 1970, independent groups had built their own detectors with sensitivities equal to or better than Weber's, yet between 1970 and 1975 none of these groups saw any unusual disturbances over and above the inevitable thermal and environmental noise.

Weber's reported detections are now generally regarded as a false alarm, although there is still no good explanation for the coincidences if they were not gravitational waves. Nevertheless, Weber's experiments did initiate the program of gravitational-wave detection, and inspired other groups to build better detectors. One of the main improvements has been to cool the entire bar and associated sensing devices to temperatures a few degrees above absolute zero. At a finite temperature, any mechanical object such as a Weber bar is in a constant state of thermal agitation that produces essentially random oscillations of the bar that, at room temperature, are orders of magnitude larger than the expected gravitational-wave excitations. The effects of thermal agitations are reduced by going to low temperatures, and by employing some sophisticated data analysis techniques that allow detection of externally produced excitations even smaller than the thermal ones. One of the most sensitive low-temperature bar detectors at present is at Stanford University, said to operate at a strain sensitivity of 10^{-18}. Temperatures in the millidegree range are planned for the future.

A dozen other laboratories around the world (figure 2.10) are engaged in building and improving upon the basic 'Weber bar' detector, some using bigger bars, some using smaller bars, some working at room temperature, and some working near zero degrees absolute, some using aluminium, some using other materials such as sapphire that might have reduced internal damping following the gravitational-wave excitations. The goal in all cases is to reduce noise from thermal, electrical and

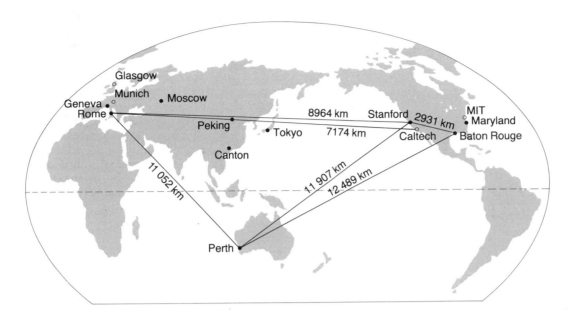

Figure 2.10. World-wide network of gravitational-wave detectors. Closed circles represent Weber bars, and open circles represent interferometers. By taking into account the time delay between receipt of waves from some source at the various 'observatories' researchers can obtain information about the direction to the source. (From Shapiro, S. L., Stark, R. F. and Teukolsky, S. A. (1985). *American Scientist* 73, 255.)

environmental sources as much as possible, in order to reach the 10^{-22} strain level.

The second important type of detector is the laser interferometer, currently being developed at such places as Munich, Glasgow, Caltech and MIT (figure 2.10). Schematically, a laser interferometer takes light from a laser, splits it into two beams travelling at right angles to one another, reflects the beams from two mirrors that return the beams to the starting point, and combines them in such a way that they interfere with each other. A passing gravitational wave will change the length of one arm of the apparatus relative to the other, and cause the interference pattern to vary. Unlike the Weber bar, a laser interferometer is not a resonant detector, because the mirrors are typically suspended in such a way that they can move effectively freely in response to the wave. Therefore they are sensitive to a range of frequencies, between 100 Hz and 10 kHz. One way to achieve higher sensitivity is to increase the path lengths of the arms, and a number of research groups, including a Caltech–MIT collaboration, as well as groups in Europe, are planning to do this. The Caltech–MIT proposal is to build two independent interferometers, each with arm lengths of 4 km, and separated by 4300 km, in order to run in coincidence. Such a device might initially be sensitive to bursts at the 10^{-20} level, and to continuous waves at the 10^{-24} level.

Another important method for searching for gravitational waves, in this case of very low frequency (10^{-6} to 10^{-2} Hz) is to look for fluctuations that they would cause in the frequencies of radio signals from sources ranging from interplanetary spacecraft such as *Voyager*, to ultra-stable pulsars, such as the millisecond pulsar. Measurements using *Voyager* tracking data in 1979 set an upper limit of 10^{-14} on strains produced by low-frequency waves.

The 'first generation' detectors of Weber probably were sensitive to strains at the 10^{-15} level, while the current 'second generation', low-temperature bars and interferometers claim to operate at the 10^{-18} level. Much improvement remains before the 10^{-22} level is reached. So the programme started by Weber a quarter of a century ago is far from over, but, when it achieves its first successful detection, an exciting new kind of astronomy will begin.

2.5   Neutron stars: relativity's limiting role

For the most part, the effects of general relativity in stars are very small. The characteristic measure of the size of relativistic effects in bodies is the quantity GM/Rc^2, where M and R are the mass and radius of the body. For the Sun, this factor is 10^{-5}, while for a white-dwarf star it is 10^{-3}, both small corrections to the overall structure. One of the few stellar situations in which general relativity was found to have a qualitative effect was supermassive stars, with masses of between 1000 and 10^9 solar masses, once proposed as possible models for quasars. It turned out, however, that these stars were on the dividing line between stability and instability to collapse according to Newtonian theory, and so the general relativistic effects, though only of order 10^{-4}, were enough to make such stars unstable, and make them unattractive as candidates for quasars.

The one material stellar object for which general relativity is important is the neutron star. (The black hole is, of course, another object in which general relativity is crucial, but since the matter in a black hole has fallen across the 'event horizon', it is not a 'material' body in the normal sense. Black holes will be described in the next section.)

The factor GM/Rc^2 for a neutron star can be as large as 0.3, so general relativity can introduce sizable corrections to the structure of the body. Furthermore, general relativity introduces important upper bounds on the masses and rotation rates of neutron stars that are observationally relevant.

Neutron stars were first suggested as theoretical possibilities by Walter Baade and Fritz Zwicky in the 1930s as the end product of the gravitational collapse in the interior of a supernova explosion. The collapsed star is so highly condensed by gravitational forces that atomic electrons are crushed together with the nuclear protons to form neutrons, and the density of the matter is raised above nuclear densities $(4 \times 10^{11} \mathrm{~g~cm}^{-3})$. Under such conditions, the neutrons become 'degenerate', a state in which their behaviour is governed by the Pauli exclusion principle of quantum mechanics which prevents any two neutrons from occupying the same region of space in the same state (the same principle applied to electrons in atoms leads to the shell structure evident in the periodic table of the elements). A typical neutron star model has a mass of one solar mass, and a radius of about 10 km.

However, neutron stars remained just theoretical possibilities until the discovery of pulsars in 1967 and their subsequent interpretation as rotating neutron stars. Since that time, much effort has been directed toward calculating detailed neutron-star models, with particular interest in masses, moments of inertia, internal structure, and oscillations. These quantities are important in understanding both the steady changes and the discontinuous jumps ('glitches') in the observed periods of pulses from pulsars. The principal uncertainty in these calculations is not general relativity; the equations of relativistic stellar structure are well known and relatively simple to implement on the computer. Rather the uncertainty is in the equation of state of matter above nuclear densities, for which there is virtually no experimental information. In fact, some theorists regard pulsar observations and neutron-star models as a way to test various candidate equations of state.

One area in which general relativity plays a major role is in the maximum mass of a neutron star, a quantity that turns out to be vital in the search for black holes. Ever since the work of

Subrahmanyan Chandrasekhar and Lev Landau in the late 1930s, it has been known that stars in which the pressure needed to resist gravity comes from quantum-mechanical degeneracy have a maximum possible mass. The reason is that, as one considers more massive degenerate configurations, the degenerate particles become relativistic (speeds near that of light), and the pressure they can exert is limited in such a way that a point can be reached where the pressure is no longer sufficient to counteract gravity, and no equilibrium configuration is possible. If the degenerate particles are electrons, as they are in a white-dwarf star, the maximum mass turns out to be about 1.4 solar masses, depending somewhat on the composition of the star, while, if the degenerate particles are neutrons, the maximum mass is about 5.7 solar masses. These are Newtonian estimates. General relativity has a negligible effect on the white-dwarf maximum mass, but it tends to lower the neutron-star maximum mass, for two reasons: general relativistic gravitational forces inside stars are stronger than Newtonian forces (other variables being equal), so that the degeneracy pressure forces are overwhelmed for less massive configurations; and the observed mass is less than the total mass of all the neutrons in the star, because of the sizable negative gravitational binding energy.

As usual, the actual values of the maximum mass are sensitive to the equation of state for nuclear matter, but for a wide range of models the values for the maximum mass lie between 1.5 and 2.7 solar masses. Even if one takes the most extreme possible equation of state that doesn't violate all our cherished beliefs, namely one in which the speed of sound is equal to the speed of light, the maximum mass is always less than 3.6 solar masses. Thus general relativity can reduce the maximum neutron star mass by as much as a factor of two over the Newtonian value.

General relativistic effects also limit the rotation rate of neutron stars. In 1970, Chandrasekhar was the first to point out that, if a uniform-density Newtonian star rotates more rapidly than some critical value, the possibility of the emission of gravitational radiation can induce an unstable, non-axisymmetric mode of oscillation. The gravitational radiation emitted by this oscillation takes angular momentum away from the star, thereby reducing its rotation rate back below the critical value. Later workers confirmed this result for realistic, general relativistic stellar models. A consequence of this instability is that there should be a lower bound on the rotation period of a neutron star. The actual value of the minimum period is sensitive to the equation of state and to the mass of the model, ranging from 1.5 to 0.4 ms. These values are very close to the period of the millisecond pulsar PSR1937+214, discovered in 1982, whose period is 1.55 ms. The discovery of pulsars with shorter periods could begin to constrain the possible viable equations of state.

It is interesting to note that rotation was once believed to be a way to increase the maximum mass of a neutron star, because centrifugal forces would help counter the gravitational forces and permit more massive models. However, the discovery of a maximum rotation rate limits the possibilities, and numerical calculations show that the maximum mass is increased by at most 20% by rotation. As we will see shortly, this will have important consequences for the identification of black holes.

Other aspects of neutron stars, such as the pulsar emission mechanism, accretion onto neutron stars, the nature of the matter inside a neutron star, are more in the astrophysical realm, and for a discussion of these issues the reader is referred to Chapter 6 by Longair.

2.6 The new black hole physics

One of the most important and exciting aspects of the relativity renaissance is the study and search for black holes. The subject began in the early 1960s as astrophysicists looked for explanations of the quasars, warmed up in the early 1970s following the possible detection of a black hole in an X-ray source, became white hot in the middle 1970s when black holes were found theoretically to evaporate, and continues today as one of the most active branches of general relativity.

However, the first glimmerings of the black hole idea date back to the eighteenth century, in the writings of a British amateur astronomer, the Reverend John Michell. Reasoning on the basis of the corpuscular theory, that light would be attracted by gravity in the same way as ordinary matter is attracted, he noted that light emitted from the surface of a body such as the Earth or the Sun would be reduced in velocity by the time it reached great distances. How large would a body of the same density as the Sun have to be in order that light emitted from it would be stopped and pulled back before reaching infinity? The answer he obtained was 500 times the diameter of the Sun. Light could therefore never escape from such a body. Fifteen years later, the French mathematician Pierre Simon Laplace performed a similar calculation. In today's language, such an object would be a supermassive black hole of about 100 million solar masses.

The second stab at black hole physics occurred in 1939. That year, J. Robert Oppenheimer and Hartland Snyder published a paper that described what happens to a star that has exhausted the thermonuclear fuel necessary to produce the heat and pressure that support it against gravity. According to their calculations, the star begins to collapse, and if it is massive enough not to be halted at the white-dwarf or neutron-star stage by the degeneracy pressure of electrons or neutrons, it continues to collapse until the radius of the star approaches a value called the gravitational radius or Schwarzschild radius. This radius has a value given by $2GM/c^2$, where M is the mass of the star. For a body of one solar mass, the gravitational radius is about 3 km, for a body of the mass of the Earth, it is about 9 mm. An observer sitting on the surface of the star sees the collapse continue to smaller and smaller radii, until both star and observer reach the origin, with consequences too horrible to describe in detail. On the other hand, an observer at great distances observes the collapse to slow down as the radius approaches the gravitational radius, a consequence of the gravitational redshift of the signals sent outward. However, in this case, the redshifting or slowing down becomes so extreme that the star appears almost to stop, and to hover just at the gravitational radius. The distant observer never sees any signals emitted by the falling observer once the latter is inside the gravitational radius. The calculations showed that any signal emitted inside can never escape the sphere bounded by the gravitational radius.

The idea that there was something unusual about this gravitational radius did not originate with Oppenheimer and Snyder; it dates back almost to the inception of general relativity itself. Within two months of the publication of the final form of the theory, the German astronomer Karl Schwarzschild had obtained a pair of rigorous, exact solutions to the field equations, the first corresponding to an ideal body consisting of a mass point at the origin, the other corresponding to a spherical body of finite extent. In the first solution (conventionally called the 'Schwarzschild solution'), the gravitational radius appeared to be a pathological place because part of the spacetime metric became infinite there. In the second solution, the gravitational radius lay inside the star, but, because the presence of the material making up the star altered the interior solution, the gravitational radius was a perfectly ordinary place. The second solution applied to any stationary, spherical body, and is the solution sometimes used, for example, in modelling non-rotating neutron stars. The pathology of the first Schwarzschild solution led most relativists, Einstein included, to believe that it would be impossible for any body to be so small that the gravitational radius would lie outside it, thereby revealing this pathological surface to the external world.

Despite the explicit demonstration by Oppenheimer and Snyder of a solution in which precisely such a thing would occur, their result was not taken particularly seriously, and the subject lay dormant for another two decades. The real revival of black hole physics coincided with the renaissance of general relativity brought about in the 1960s, and was due to two things. The first, as I point out above, was the discovery of quasars. To understand the enormous energy output of these objects, theorists turned to the strong gravitational fields of superdense objects, and what better objects to consider than the collapsing objects of Oppenheimer and Snyder, whose endpoints were the 'pathological' Schwarzschild solution? The second contributor to the revival was the discovery in 1963 of a new exact solution of Einstein's equations by Roy Kerr.

The Kerr solution turned out to be the solution for a rotating black hole, with the Schwarzschild solution being just the special case corresponding to no rotation. Egged on by the problem of the quasars, relativistic astrophysicists spent the next ten years proving many important features of the Schwarzschild and Kerr solutions, and of black holes in general.

The pathological behaviour of the Schwarzschild solution at the gravitational radius (and a similar behaviour of the Kerr solution) proved to be purely a product of an inappropriate choice of mathematical variables. This did not alter the fact that the gravitational radius was special, however. The name 'event horizon' was attached to the surface corresponding to this radius, because it is a boundary for communication, just as the Earth's horizon is a boundary for our vision. An observer inside the event horizon cannot communicate with an observer outside by any means whatsoever, even by sending light signals. Nothing, not even light, can escape from the interior. On the other hand, light, matter, physicists, are all free to cross the event horizon going inward (figure 2.11). This one-way property of the event horizon, allowing nothing to emerge, led John Wheeler, one of the fathers of the new relativity, to coin the term 'black hole', during a 1967 conference in New York.

To an observer outside the horizon, the only feature of the black hole itself that is detectable is its gravitational field. Any matter or radiation that remains outside the horizon, of course, is detectable. Far away from the black hole, this gravitational field is indistinguishable from the gravitational field of any object of the same mass and angular momentum, such as a star. However, to an observer close to the horizon, things can be very unusual. In Newtonian gravitational theory any particle that approaches a point mass with a small amount of angular momentum (i.e. not head on) automatically misses the body and continues back to large distances with its direction deflected. But for a Schwarzschild black hole, for example, if the angular momentum of the particle is too low for its initial energy, it can be captured by the strong gravitational field near the hole, and can be pulled in across the horizon (see figure 2.12). The deflection of light can be so large that light can move on circular orbits, just outside the horizon (at $3GM/c^2$ for the Schwarzschild hole), and light can also be captured by the hole.

For the Kerr solution, rotation of the black hole produces the same dragging of inertial frames effect whose detection is the goal of the Stanford gyroscope experiment (in the more mundane terrestrial setting), but if the observer goes close enough to the horizon near the equator, and attempts to hover without falling inwards, the dragging of spacetime becomes so strong that it is impossible for him to avoid being dragged around bodily with the rotation of the hole, no matter how hard he blasts his rockets in the reverse direction to try to avoid it. The region inside which the dragging effect has this property is called the ergosphere (see figure 2.13). This property of the ergosphere was exploited by Roger Penrose to devise a remarkable process that bears his name. In this Penrose process, a particle is injected into the ergosphere and is broken into two, one piece going across the horizon in a direction opposite to that of the rotation (going against the rotation is possible as long as the particle approaches the hole), the other being ejected from the ergosphere and returning to infinity. As a consequence, the total mass-energy of the particle returning to infinity can be larger than that of the initial ingoing particle, and the mass and angular momentum of the black hole can be

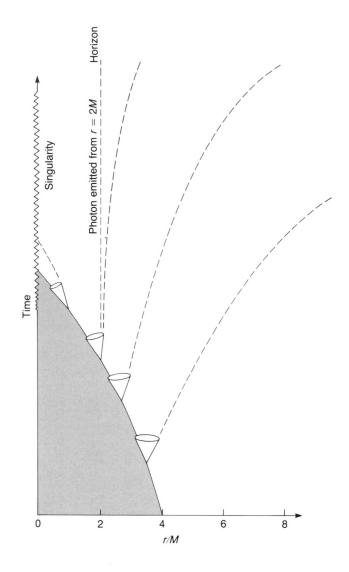

Figure 2.11. Spherical gravitational collapse to a black hole. Radial distance is plotted horizontally, a variable that measures time is plotted vertically. Here 'M' is shorthand for GM/c^2 which has units of length. The shaded region is the interior of the collapsing star. Representative 'light cones' are shown. Light cones determine the possible trajectories of light and particles in spacetime. Outgoing light rays move along the outward-pointing edge of the cone, while ingoing light rays move along the inward-pointing edge. In the absence of gravity, light rays would move along 45° lines in this diagram. Since particles must move more slowly than light, their trajectories must lie within the opening of the cone. A distant observer sees light rays from an observer falling in on the surface of the star, but as the radius of the star approaches $2M$, the light cones are 'tipped' by spacetime curvature so that the rays take longer and longer to reach the distant observer, who sees the process of collapse slow down. Just outside $r = 2M$, the outgoing photon barely escapes, while a photon emitted at $r = 2M$ hovers there, becoming a generator of the event horizon. Inside $r = 2M$, the light cones are tipped over, so that even the 'outgoing' photons actually travel to smaller radii. Everything inside the horizon is forced to reach $r = 0$, where a spacetime singularity resides.

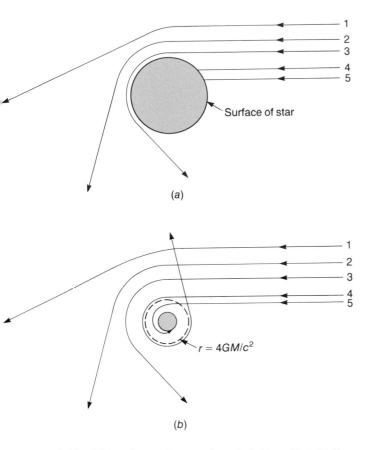

(a)

(b)

Figure 2.12. Orbits of particles sent from infinity with initially small velocities toward a star and a black hole. (*a*) For a star, particles 1, 2 and 3 are deflected and return to infinity, while particles 4 and 5 hit the stellar surface. (*b*) For a black hole of the same mass, particles 1, 2 and 3 experience the same orbits as in (*a*). In general relativity, the spacetime outside a spherical star is identical to the spacetime outside a black hole of the same mass. Particle 4 comes close to a critical 'capture radius' given by $4GM/c^2$, and can circle the hole many times before escaping to infinity. Particle 5 crosses the critical radius and continues across the horizon (if particle 5 had the right rockets, it could still escape from the capture region before crossing the horizon). For photons, the critical capture radius is at $3GM/c^2$.

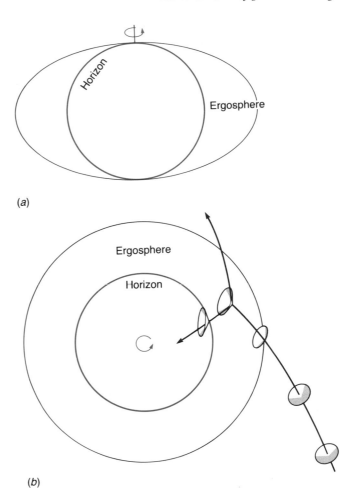

(a)

(b)

Figure 2.13. The Kerr black hole and the Penrose process. (*a*) Side view of the black hole, showing the horizon and the ergosphere. (*b*) Top view of the black hole. A particle enters the ergosphere, splits into two particles, one going across the horizon against the rotation direction, the other leaving the ergosphere, and reaching infinity with more mass-energy than the initial particle had. The hole's mass and angular momentum are reduced. The dragging of inertial frames is illustrated schematically by light cones. Time is imagined to be plotted out of the page. The cones are not only tipped inward toward the hole, as they were in figure 2.11, they are also tipped in the direction of rotation of the hole. Outside the ergosphere, an observer can move more or less freely in any direction. At the boundary of the ergosphere, the observer can remain at rest relative to the page only if he travels against the rotation of the hole at the speed of light, otherwise he is free to move inward with or against the rotation or outward with the rotation at ordinary speeds. He can stay on the boundary as long as he moves around with the rotation. Inside the ergosphere, the light cone is tipped so far that he cannot stay 'at rest'; he must move inward, or outward, or travel around with the rotation of the hole. At the horizon, the observer must move inward. Only an outgoing light ray can hover on the horizon, but, because of the rotation and the tipping of the light cones, the ray circles the horizon with the rotation of the hole.

decreased. In this way, energy can be extracted from the hole, at the expense of its rotational kinetic energy. One of the unusual and suggestive consequences of the Penrose process is that a certain quantity made up of the mass and angular momentum of the Kerr black hole always increases (or stays the same), no matter how the mass and angular momentum of the hole separately change. This quantity was called the irreducible mass, and its square turned out to be proportional to the surface area of the horizon. We will see that the area plays an important role in black hole dynamics.

It turns out that there are other black hole solutions corresponding to the case when the hole has a net charge, known as the Reissner–Nordstrom solution for the non-rotating charged case, and the Kerr–Newman solution for the rotating charged case. However, there is no evidence in astronomy for bodies carrying a significant net charge, because, it is believed, such bodies would be rapidly neutralised by charge separation of the interstellar ionised gas. The same effect applies to charged black holes, so for the most part they have not played an important role in black hole astrophysics, although they do have very interesting and remarkable properties in their own right. However, for the rest of this article, we shall ignore them.

One question that began to concern black hole theorists in the middle 1960s was whether or not the Kerr and Schwarzschild solutions were the only (neutral) black holes in vacuum (in other words without stars or gas outside), or whether there was a whole class of solutions just as there is for stars. As early as 1967, Werner Israel had proven that Schwarzschild was the unique non-rotating vacuum black hole solution.

During the next several years, a variety of calculations of gravitational collapse with perturbations gave support to a theorem that came to be stated as 'black holes have no hair'. The idea was this: if you consider a spherical collapse of matter, which would normally produce a Schwarzschild black hole, but perturb it slightly, by introducing for example some small angular momentum, or some slight mass anisotropy (such as a quadrupole moment), or a long-range field corresponding to a spinless particle, or a neutrino field, or a field corresponding to a strongly interacting particle, then, as the collapse proceeds and a horizon forms, all the external effects of the perturbations are radiated away to infinity and disappear after sufficiently long times, with the exception of the effect of the rotation! In other words whatever additional characteristics, or 'hair', one tries to attach to the collapsing matter, they all vanish as far as the external spacetime is concerned, except for those associated with the conserved quantities of general relativity, namely the mass and angular momentum. (Since charge is also conserved, introducing an electromagnetic perturbation led to one of the charged black holes.) The outcome of these perturbed collapse calculations was either Schwarzschild or a Kerr solution with a small amount of rotation. There was no way for any external observer to determine, for example, the total number of baryons or leptons in a black hole, since all evidence of the fields associated with these variables was gone. Similarly, while the Earth has an equatorial bulge and associated gravitational quadrupole, octopole, and higher multipole fields determined by its rotation and by its internal constitution (strength of the crust, depth of the mantle, etc.), the gravitational multipole fields of the Kerr black hole are determined uniquely by its mass and angular momentum. This is the sense in which black holes have no hair; they are uniquely characterised by their mass and angular momentum. But these calculations involved only small

perturbations away from Schwarzschild. It was 1975 before rigorous theorems were proved that in general the Kerr solution was the unique vacuum black hole solution, with Schwarzschild the limiting case of no angular momentum.

The fact that neutral black holes are uniquely Kerr suggests that there is something fundamental that separates black holes from other gravitating systems. Obviously that something is the event horizon. But the issue goes even deeper than this simple observation, for it became possible to learn much more about black holes than could be gained from an examination of the Kerr solution alone, by studying the event horizon itself. The horizon is a boundary between two regions, one that can communicate with observers at infinity, and one that cannot. This simple definition implies for example that the horizon is generated by 'outgoing' light rays that in some sense are 'hovering' without going anywhere, since any outgoing light ray just outside the horizon escapes to infinity, and any apparently outgoing light ray inside the horizon is actually pulled inward.

Using such elementary definitions, together with general properties of Einstein's equations, Penrose, Stephen Hawking, Robert Geroch and others proved a variety of powerful theorems about black holes, theorems that applied to more general situations than just Schwarzschild and Kerr. For example, the 1969 Hawking–Penrose theorem stated essentially that, inside a horizon, there must exist a singularity of spacetime at which the path or world line of an observer who hits it must terminate, and physics as we know it must break down. In the Schwarzschild solution, *any* observer who crosses the horizon ends up at a singularity, while in Kerr the singularity is there but can be avoided by clever choice of inward direction. More generally, the theorem says that any spacetime containing a horizon also contains a singularity inside.

Another important theorem was Hawking's 1972 theorem that the area of the horizon can never decrease, in other words it must be constant, as in a stationary black hole, or it must increase with time, as in any dynamical situation. A heuristic way to see this is to imagine a particle falling inward across a horizon. Since the particle has its own gravitational field, it bends light around it, causing light rays to converge. Rays initially on the horizon then begin to converge; but if 'outgoing' rays actually begin to converge, then they must now be in the interior of the horizon. But then there must exist outgoing rays initially just outside the horizon (about to start their trips to infinity) that are made to converge by the passing particle just enough to become new horizon generators. But because these rays were originally outside the horizon, the total surface area of the horizon must be greater than it was before. This is a gross oversimplification of the actual theorem, but it applies to any situation involving black holes, such as a black hole embedded in a gas, or two black holes colliding with each other.

Similar general 'horizon' techniques made it possible to prove that a quantity called the 'surface gravity', a renormalised value of the acceleration required for an observer to hover just at the horizon (the actual acceleration is infinite), is constant over the horizon, for a stationary situation. This is to be contrasted with the surface gravity of the Earth, which varies over the Earth's surface because of the centrifugal force.

The idea that there should exist singularities of spacetime, places where general relativity and all the other laws of physics should break down, is very disturbing. Physicists are uncomfortable with the possibility of being unable to predict the future from some given initial data, but this was precisely what a singularity would do, because the physics that emerges from a singularity is not under any control. However, as long as the singularity is hidden behind an event horizon, we are safe, because whatever bizarre physics it produces can never be sensed by an external observer. The Hawking–Penrose theorem established that a singularity existed inside any horizon. What about the converse? Is *every* singularity hidden behind a horizon? This question is called the *Cosmic Censorship Hypothesis*, and if true would protect physicists from all spacetime singularities. (Actually, we are excluding the singularity of the kind that may have occurred at the big bang. This is of a different type than those associated with black holes, and is considered more acceptable since it started everything in the universe off 'with a bang'. Physicists did not exist before this singularity to worry about whether it would destroy predictability. Nevertheless, this kind of singularity is sufficiently worrisome that one of the motivations for developing a quantum theory of gravity is to find a way around it. For more on this, see Chapter 5 by Isham.)

Unfortunately, there is no convincing proof of the Cosmic Censorship Hypothesis. There is not even general agreement on how to formulate the vague notion of censorship into terms that can be translated into mathematics the way the horizon theorems were translated. The hypothesis may be true only for 'reasonable' physical situations that might occur in the universe, such as realistic gravitational collapse of a star to form a black hole (with the singularity safely nestled inside the horizon), yet it might not be true for special or contrived processes. In fact, counterexamples of the latter type have recently been devised.

One property of Kerr black holes that is at least consistent with Cosmic Censorship has to do with the maximum rotation rate of Kerr. For a Kerr black hole of a given mass, there is a maximum rotation rate. If the rate is larger than this value, the solution of Einstein's equations is no longer a black hole, but is a 'naked' singularity, visible from the outside world. Now suppose one tried to create such a naked singularity from a Kerr black hole by injecting particles down the hole so as to speed up its rotation above the critical value. It turns out that the critical value of the rotation rate depends on the mass of the hole, and as you inject particles into the hole, its mass increases in just such a way that the critical rotation rate always stays a bit larger than the actual rotation rate achieved. This, then, is Cosmic Censorship in action. However, it is only a special

example, and one of the most important unsolved issues in classical general relativity is the validity (and even the meaning) of the Cosmic Censorship Hypothesis.

By 1972, enough was known about both the specific Schwarzschild and Kerr black holes, and about horizons in general, that it became possible for James Bardeen, Brandon Carter and Hawking to codify many of their properties into a set of laws that were very suggestive. The 'four laws of black hole mechanics' that they listed were:

The zeroth law. In a stationary situation, or equilibrium, the surface gravity κ of a black hole is constant over the horizon.

The first law. In a transformation from one state to a nearby state, the energy of the system changes by

$$\Delta E = \frac{c^2}{G} \frac{\kappa \Delta A}{8\pi} + W,$$

where A is the surface area of the horizon and W is the total of any work done in changing the rotation of the black hole, and any work done on any matter that may be outside the black hole.

The second law. During any process in an isolated system, the total area of all horizons is non-decreasing.

The third law. It is impossible by any finite sequence of steps to bring the surface gravity to the value zero (for the Kerr solution with the critical rotation rate – a naked singularity – κ is zero).

These laws are in exact parallel with the laws of thermodynamics, with κ playing the role of temperature, and the area playing the role of the entropy. In 1972, Jacob Bekenstein suggested that, because of quantum mechanics, this was more than just a coincidence, that black holes could indeed have a real entropy. His argument was based on the notion that black holes have no hair. When a particle falls down a black hole, information (hair) is lost, but, according to the information theory approach to thermodynamics, missing information is equivalent to entropy. Therefore, there must be an entropy associated with a black hole. Bekenstein went on to show that if the entropy were proportional to the area, as suggested by the four laws, the proportionality constant could be estimated by noticing that the Heisenberg uncertainty principle applied to the injection of a particle into a black hole provided a minimum amount by which the area of the hole could be increased during any injection. By equating this to the entropy gained by the hole upon loss of information about the injected particle, he arrived at a value that was surprisingly close to the correct proportionality constant. By 1974, Hawking had applied the mathematics of quantum field theory to black hole spacetimes, and proved that black holes could evaporate by creation of particle–antiparticle pairs from the vacuum near the horizon. The emitted particles were found to have a thermal, or black-body spectrum with a precise temperature given by $T = \hbar\kappa/2\pi kc$, where $\hbar$ is Planck's constant and k is Boltzmann's constant. The associated entropy was indeed proportional to the area of the horizon, and Hawking's calculation gave the precise proportionality constant. The entropy S was given by

$$S = \frac{1}{4}\frac{kc^3}{G\hbar}A.$$

For a Schwarzschild hole, the area and hence the entropy are proportional to M^2, and the temperature is inversely proportional to M. Because the black hole can emit particles with a thermal spectrum at a definite temperature, and can of course absorb such particles, it can be in thermal equilibrium with external systems. Thus the four laws of black hole mechanics are truly thermodynamic laws. It is now possible to state, for example, a generalised second law of thermodynamics: in any process in a closed system the total entropy of matter *and black holes* can never decrease. This merging of the laws of gravitation, quantum mechanics and thermodynamics is one of the stunning achievements of modern theoretical physics. For further details, see Chapter 5 by Isham.

This emission or evaporation process will ultimately cause a black hole to decay away entirely. However, for most astrophysical purposes, it is unimportant. The reason is that the temperature of a solar mass black hole is less than a microkelvin, and its lifetime against the evaporation is greater than 10^{70} years. So, to all intents and purposes, black holes formed by stellar collapse do not evaporate. However, Hawking and others suggested the possibility that mini-black holes were created in the big bang. For mini-black holes of mass about 10^{12} kg, the lifetime would be about 20 billion years, so they would be in the final stages of evaporation today. Because their masses would now be very small, their temperatures and evporation rates would be very high, so they would emit a final burst of high-energy gamma rays and elementary particles that could be detectable. Such bursts have not been detected to date, a result that sets a stringent upper limit on the possible number of such mini holes.

Although a great deal is known about black holes in theory, rather less is known about them observationally. There are several instances in which the evidence for the existence of black holes is impressive, but in all cases it is indirect. Because black holes have no hair, it is difficult to obtain critical pieces of evidence (so-called 'smoking guns') as one could, for example, in the case of pulsars. For instance, in the binary X-ray source Cygnus X1, the source of the X-rays is believed to be a compact object with a mass larger than about six solar masses, too large to be either a white dwarf or a neutron star (this is where the neutron-star mass limits are crucial). Thus the black hole model wins by default. Other models for Cygnus X1 are possible, but implausible. Similarly, there is evidence in certain galactic nuclei, such as M87, and possibly even our own, of collapsed objects of between 10^2 and 10^8 solar masses. Again, the black hole model is consistent with the observations, but is not necessarily required. Accretion of matter onto supermassive rotating black holes may produce the jets of outflowing matter that are observed in many quasars, and active galactic nuclei. These and other astrophysical processes that might aid in the detection of black holes can be found in the chapter by Longair.

Perhaps the 'smoking gun' for a black hole will be found when gravitational radiation detectors pick up the gravitational waves emitted from the normal-mode oscillations of a black hole formed during a supernova collapse. Until then, however, we must be content with indirect evidence that they exist.

2.7 Cosmology: the new respectability

Until the 1960s general relativity's record in cosmology was decidedly mixed. Over the years since 1917, when Einstein first calculated a cosmological model using general relativity, the theory has bounced from failure to success, back to failure, and finally, in the middle 1960s, to success.

The first cosmological failure of general relativity was in a sense a failure of will for Einstein. In 1917, he applied general relativity to the problem of the universe at large. This in itself was a bold step, because in 1917 it was not yet known whether there was anything but a void outside our own Milky Way Galaxy. For example, the Andromeda Galaxy, then called a 'nebula', was still believed to lie inside the Milky Way. Yet Einstein assumed that the universe could be idealised as a homogeneous distribution of matter, with a density that was the same everywhere. Eventually, this assumption was seen to be valid, at least as a first approximation. For a variety of philosophical reasons, he chose a model for the universe that was closed, in the sense that any observer setting off on a 'straight' line (a geodesic) would ultimately return to his starting place, as a consequence of the curvature of space. The model was finite, yet unbounded, in the same sense that a two-dimensional analogue, the surface of a balloon, is finite in area, yet has no boundaries within the surface. The final assumption he made was reasonable enough: the model should be static, unchanging in time. This certainly fitted the observational situation in 1917.

But to his horror he discovered that the theory did not admit any such solutions. The only solutions allowed were either expanding or contracting. In order to get the needed static solutions, he had to modify the original equations of the theory by adding a term called the 'cosmological term'. With the modified equations, he found that he could obtain static models for the universe.

He later referred to this as 'the biggest blunder' of his scientific life, for the discovery by Edwin Hubble in 1929 that the universe was indeed expanding showed that the cosmological term was unnecessary. In 1931, Einstein recommended that it be dropped, and that the original field equations be restored. In this roundabout way, the expanding universe was converted from a failure to a success for the theory. It is amusing to speculate what the course of general relativity and cosmology might have been if Einstein had stuck with the theory as he had developed it, predicted in 1917 that the universe *must* be evolving with time, and then sat back to await confirmation by observations.

With the cosmological term banished, general relativistic cosmology gave a decent accounting of the expansion of the universe from an initial 'big bang', although there was much debate, both philosophical and technical, over the nature and significance of a closed universe. However, by the late 1940s the theory appeared to be in jeopardy again, because the rate of expansion that astronomers had established implied an age of the universe of 2 billion years. There was only one problem. By examining radioactive elements in rocks, geologists had determined that the age of the Earth was at least 3.5 billion years, older than the universe itself.

This was an embarrassment, and it was partly responsible for the rise and popularity during the 1950s of the steady-state theory of the universe devised by Fred Hoyle, Hermann Bondi and Thomas Gold. The steady-state theory got around the embarrassment by postulating that the universe has existed and has been expanding forever (and is therefore quite a bit older than the Earth), and then squared the expansion with the idea of a steady state by postulating the continuous creation of matter in the void between the existing matter.

General relativity started to recover from this failure in the late 1950s. Astronomers began to find serious inadequacies in the methods that had been employed to determine the distances to the galaxies used to study the expansion of the universe. The new distance determinations put these galaxies much farther away than had previously been thought. The effect of this was to increase the time the galaxies would have been expanding since the big bang. The age of the universe implied by these new measurements was then in a much more comfortable range of larger than 10 billion years.

The real cosmological successes for general relativity came in the 1960s. First was the discovery of the cosmic background radiation in 1965, by Arno Penzias and Robert Wilson of Bell Telephone Laboratories. This radiation is the remains of the hot electromagnetic black-body radiation that once dominated the universe in its earlier phase, now cooled to three degrees above absolute zero by the subsequent expansion of the universe. Second came calculations of the amount of helium that would be synthesised from hydrogen in the very early universe, around 1000 s after the big bang. The amounts, approximately 25% by weight, were in agreement with the abundances of helium observed in stars and in interstellar space. This was an important confirmation of the hot big bang picture, because the amount of helium believed to be produced by fusion in the interiors of stars was woefully inadequate to explain the observed abundances. These two results together with other observations spelled the death knell of the steady-state theory. Since the universe in that model is in a steady state, it was

always as cold as it is today, while to generate the background radiation and the helium a hot universe seems to be necessary.

Today, the general realtivistic hot big bang model of the universe has broad acceptance, and cosmologists now focus their attention on more detailed issues, such as how galaxies and other large-scale structures formed out of the hot primordial soup, and on what the universe might have been like earlier than 1000 s, all the way back to 10^{-36} s (and some brave cosmologists are going back even further) when the laws of elementary-particle physics played a major role in the evolution of the universe. This 'new cosmology' has become such an active and exciting field, bringing together general relativists, particle physicists and astrophysicists, that the whole of Chapter 3 by Guth and Steinhardt has been devoted to it. Let us return to the world of general relativity.

2.8   General relativity, pure and simple

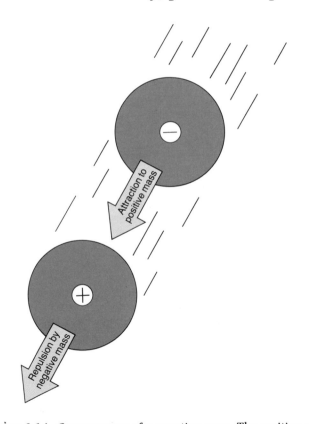

Figure 2.14. Consequences of a negative mass. The positive-mass body attracts the negative-mass body, while the negative-mass body repels the positive-mass body. The pair accelerates itself off, without any outside help or use of propulsion. Such a situation would revolutionise space travel. In electromagnetism, charges either attract each other or repel each other, never one of each, so pairs of charges do not experience this effect.

The renaissance of general relativity had its counterpart in that part of the subject concerned with somewhat more formal, mathematical issues. Although the observational implications of work in 'pure general relativity' were not immediately obvious, great progress was made in understanding the structure of Einstein's equations, and the nature of broad classes of solutions of Einstein's theory. It should be pointed out, of course, that it was precisely this kind of pure mathematical research that led to the discovery of the Kerr metric, so the potential observational impacts of pure general relativity should not be ignored.

One fruitful line of research that became very active during the 1970s and 1980s was the use of new techniques to search for exact solutions of Einstein's equations. For example, the use of complex, as opposed to real, versions of spacetime allowed one to generate new solutions from old solutions. For instance, the Kerr solution could be generated from the Schwarzschild by a transformation involving complex variables. One particularly rich area for generating solutions was the class of stationary, axially symmetric solutions that are asymptotically flat (vanishing gravitational fields far away), with either no matter present (vacuum solutions) or only electromagnetic fields present ('electro-vac' solutions). In this case, the Einstein equations can be cast into the form of a set of complex, non-linear differential equations, and a broad class of solutions to these equations could be generated using 'inverse scattering' methods or 'soliton' methods for solving non-linear evolution equations such as the Sine–Gordon or the Korteweg de Vries equations, that have recently become popular in such diverse fields as solid-state physics, fluid mechanics, and quantum field theory. The physical relevance of all these new solutions has not yet been systematically explored.

One of the factors that led to renewed interest in gravitational radiation was the rigorous proof that gravitational radiation was a physically observable phenomenon, carrying energy and angular momentum to infinity. These proofs used extremely elegant techniques for analysing the nature of Einstein's equations far from any localised source. This study of the asymptotic structure of spacetime, or 'asymptopia' as it came to be called, focused on different ways in which one could go 'far' from a system. One could imagine a limit in which the distance from the source becomes infinite, while the instant of time in question remains finite. This was called 'spacelike' infinity. Or one could imagine going to infinite distance along the trajectory traced by a light ray. This was called 'lightlike' or 'null' infinity. The third limit, which is less useful for the study of asymptopia, is to follow the paths of all observers travelling at less than the speed of light; this limit is called 'timelike' infinity. Beginning with the work of the school of Bondi in the late 1950s, the study of asymptotic null infinity became an active field, stimulated by the work of Ezra Newman and Penrose who developed an elegant formalism tailor-made for studying the properties of spacetime near null infinity. Using these techniques, one could obtain rigorous definitions for the mass and angular momentum of an isolated system, study the properties of outgoing gravitational radiation from such a system, and analyse the asymptotic symmetry groups associated with gravitating systems. More recently, it has become possible to make similar rigorous definitions for mass and angular momentum at spacelike infinity. Other, more exotic 'infinities' were developed to simplify further the study of asymptopia, one example being a complex space that could be defined at null infinity, called H-space, or 'Heaven'. The properties of this space were so beautiful that its developers regarded the second name as more appropriate.

The reason that such asymptotic methods were needed to define such quantities as mass rigorously was that, in general relativity, the concept of a local density of gravitational energy is meaningless, except under very special circumstances. This is because of the equivalence principle. In a freely falling frame, gravity disappears, so that what one thought was gravitational energy density in one frame is nothing in another frame. This is what makes it impossible to determine the mass of a gravitating body by simply adding up the contributions of particles and non-gravitational fields and the contribution of the gravitational field. The only meaningful definition of mass is a global one, namely the quantity that determines the asymptotic gravitational field, as probed for example by orbiting particles. Nevertheless, it had long been known, both in Newtonian theory and in general relativity, that the effect of 'gravitational energy' was negative, in other words the total asymptotic mass of a star was in general less than the sum of the masses of the particles making it up, because of the negative gravitational binding energy.

But if such a body became more and more compact, so that the gravitational binding energy became more and more negative, the possibility arose that the total mass of such a body could become negative! This would be a very bizarre situation, because such a body would exert antigravity: if brought close to a normal body, it would repel the normal body, while the normal body would attract it, and the pair would accelerate itself away (figure 2.14). In the middle 1960s the 'Positive Energy Theorem' was posed, which stated that such a thing could never happen, that the total asymptotically determined mass of any isolated body in general relativity must be non-negative. Physically, one expected that any body that came close to violating the positive energy theorem would be unstable and would collapse to form a positive-mass black hole. Ironically, although relativists debated and discussed this conjecture for almost fifteen years, the actual first proofs were not found by relativists at all. In 1979, the theorem was first proved by two pure mathematicians using sophisticated techniques from differential geometry, and in 1981 a different proof was given by a particle physicist using an argument motivated by a new, unified theory of particle interactions called supergravity. It was only after the fact that relativists were able to generate proofs of the theorem based on more familiar approaches.

Another area that has become a major activity in general relativity, as it has in many other fields, is the use of computers. The use of computers in general relativity divides itself into two rather different branches: algebraic computing, and numerical computing. The first branch remains largely within the realm of pure general relativity, while the second is more concerned with the real world.

Because of the complexity of the language of curved spacetime – the metric has ten components or variables, the gravitational force field or 'connection coefficients' have forty, the 'Riemann' tensor that measures spacetime curvature has twenty, and so on – algebraic manipulations can be tedious and subject to error. For example, to calculate the Riemann tensor of one well-known metric takes six months by hand, while on a modern algebraic computing system it takes thirty seconds. Since the early development in the late 1960s of FORMAC-based programs designed specifically for gravitational algebra, many different approaches have been taken, some based on high level computer languages for algebraic manipulation, such as LISP, others based on languages such as FORTRAN, FORMAC, or BCPL (the forerunner of the C language). Some workers use systems dedicated to gravitational algebra, while others use general-purpose systems such as MACSYMA and REDUCE. Algebraic manipulations have many important uses: calculating all the relevant tensors from a known metric, generating or finding solutions from the field equations, determining whether two metrics are physically different, or whether they are physically equivalent but expressed in different coordinate variables, and so on. Such algebraic techniques have also become important in quantum gravity and supergravity (see Chapter 5 by Isham for examples).

Numerical relativity is somewhat more directly concerned with the realistic predictions of general relativity in complicated situations. One of the most significant advances in the past fifteen years has been the use of powerful two-dimensional hydrodynamics codes in relativistic calculations. For example, a classical numerical calculation of the head-on collision of two vacuum black holes by Larry Smarr and colleagues in the early 1970s required sophisticated numerical integration in two spatial plus one time dimension (the problem was axially symmetric). Similarly, the fully relativistic collapse of a rotating star (the rotating version of the Oppenheimer–Snyder collapse) has only recently been amenable to solution with reasonable confidence, with the use of giant two-dimensional hydrodynamic codes. As relativists turn their attention to more astrophysically realistic relativistic problems, such as collisions between stars and black holes, accretion of matter onto and ejection of jets from black holes, and so on, two-dimensional and possibly three-dimensional hydro codes and the use of supercomputer facilities will play an increasing important role in research in the new general relativity.

2.9   A tool for physics and everyday life

One of the outgrowths of the renaissance of general relativity that occurred between 1960 and 1980 has been a change in attitude about the importance and use of the theory. Its importance as a fundamental theory of the nature of spacetime and gravitation has not been diminished in the least; if anything it has been enhanced by the flowering of research in the subject that has taken place. Its importance as a foundation for other theories of physics has been strengthened by current searches for unified and grand unified quantum theories of nature that incorporate gravity along with the other interactions.

But the real change in attitude about general relativity has been in its use as a tool in the real world. In astrophysical situations, we have already seen several examples of this. The general relativistic bending of light in gravitational lenses can help astrophysicists probe the structure of galaxies. General relativistic effects in the binary pulsar gave a high-precision determination of the mass of a pulsar. Had the result been very different from 1.4 solar masses, it could have affected our understanding of supernovae in close binary systems. Neutron-star mass limits from general relativity are important in the observational search for black holes. Finally, gravitational radiation may one day provide a completely new tool for exploring and examining the universe.

Relativity even plays a role in everyday life. For example, the gravitational redshift effect on clocks *must* be taken into account in satellite-based navigation systems, such as the US Global Positioning System, in order to achieve the required positional accuracy of a few metres or time transfer accuracy of a few nanoseconds.

To general relativists, always eager to find practical consequences of their subject, these have been very welcome developments!

Further reading

Davies, P. C. W. (1980). *The Search for Gravity Waves*. Cambridge University Press.
This book presents a discussion of the nature of gravitational waves, how they are generated, how they can be detected, and the status of the search as of 1980.

Greenstein, G. (1984). *Frozen Star: Of Pulsars, Black Holes and the Fate of Stars*. Freundlich Books, New York.
This is an award-winning account of nature and discovery of compact or collapsed objects in the universe.

Pais, A. (1980). '*Subtle is the Lord . . .': The Science and Life of Albert Einstein*. Oxford University Press, New York.
A beautiful biography of Einstein the man and of Einstein's physics. The discussion of the physics is quite technical in places.

Schutz, B. F. (1985). *A First Course in General Relativity*. Cambridge University Press.
An undergraduate-level textbook on general relativity.

Will, C. M. (1986). *Was Einstein Right? Putting General Relativity to the Test*. Basic Books, New York.
Award-winning discussion of experiments carried out since 1960 to verify general relativity.

3 The inflationary universe*

Alan Guth and Paul Steinhardt

3.1 Introduction

In the past few years certain flaws in the standard big-bang theory of cosmology have led to the development of a new model of the very early history of the universe. The model, known as the inflationary universe, agrees precisely with the generally accepted description of the observed universe for all times after the first 10^{-30} s. For this first fraction of a second, however, the story is dramatically different. According to the inflationary model, the universe had a brief period of extraordinarily rapid expansion, or 'inflation', during which its diameter increased by a factor at least 10^{25} times larger (and perhaps much larger still) than had been previously thought. In the course of this stupendous growth spurt all the matter and energy in the universe could have been created from virtually nothing. The inflationary process also has important implications for the present universe. If the new model is correct, the observed universe is only a minute fraction of the vast universe in which we live.

The inflationary model has many features in common with the standard big-bang model, which over the past twenty years has become the accepted description of the evolution of our universe. In both models the universe began between ten and twenty billion years ago as a primeval fireball of extreme density and temperature, and it has been expanding and cooling ever since. This picture has been successful in explaining many aspects of the observed universe, including the redshifting of the light from distant galaxies, the cosmic microwave background radiation, and the primordial abundances of the lightest elements. Many of these topics are discussed in Chapter 6 by Longair. These predictions depend only on the behaviour of the models for times after the first second, when the two models coincide.

Until the last decade there were few if any serious attempts to describe the universe during its first second. The temperature during this period is believed to have been hotter than ten billion degrees kelvin (10^{10} K), and very little was known about the behaviour of matter under these conditions. However, using recent developments in the theory of elementary particles, cosmologists are now attempting to understand the history of the universe back to 10^{-43} s after its beginning. (At even earlier times the energy density would have been so great that Einstein's theory of general relativity would have to be replaced by a quantum theory of gravity, which so far does not exist. Chris Isham discusses the problems of quantum gravity in Chapter 5.) When the standard big-bang model is extended to these earlier times, various problems arise. First, it becomes clear that the model requires a number of very stringent assumptions about the initial conditions of the universe. Since the model does not explain why this special set of initial conditions came to exist, one is led to believe that the model is incomplete – a theory which included a dynamical explanation for these conditions would certainly be more convincing. In addition, most of the new theories of elementary particles imply that exotic particles called magnetic monopoles (each of which corresponds to an isolated north or south magnetic pole) would be produced in the early universe. In the standard big-bang model, the number of monopoles would be so great that their mass would dramatically alter the evolution of the universe, with results which are clearly inconsistent with observations.

The inflationary universe model was developed to overcome these problems. The dynamics that govern the period of inflation have a very attractive feature: from almost any set of initial conditions the universe evolves to precisely the situation that had to be postulated as the initial state in the standard model. Moreover, the predicted density of magnetic monopoles becomes small enough to be consistent with the fact that they have not been observed. In the context of the recent developments in elementary particle theory, the inflationary model seems to be a simple and natural solution to many of the problems of the standard big-bang picture.

---

* This article is an update and elaboration of 'The inflationary universe', by Alan H. Guth and Paul J. Steinhardt, *Scientific American*, May 1984, pp. 116–28.

To understand the motivation of the inflationary universe model in detail, one must first understand the successes and failures of the standard big-bang model. The standard model is based on several assumptions. First, it is assumed that the fundamental laws of physics do not change with time, and that the effects of gravitation are correctly described by Einstein's theory of general relativity, considered in detail in Chapter 2 by Will. It is also assumed that the early universe was filled with an expanding, intensely hot gas of elementary particles in thermal equilibrium. In addition, one assumes that the universe is homogeneous on large scales – this means that if any physical quantity (energy density, for example, or the number density of electrons) is averaged over large regions, then these average values are uniform from place to place. Implicit in the assumption of homogeneity is the notion that the universe has no centre and no edge; the matter in the universe fills all of space, with the matter and space expanding together at the same rate. The universe is assumed to have been homogeneous from the start, and has therefore remained homogeneous as it evolved. Finally, it is assumed that any changes in the state of the matter and radiation have been so smooth that they have had a negligible effect on the thermal history of the universe. The violation of the last assumption is a key to the inflationary universe model.

Given these simple assumptions, the temperature, energy density and size of the universe can be traced over its history, as is shown in figure 3.1. The figure also shows the corresponding curves for the inflationary universe model, to be explained later in the article.

The big-bang model leads to three important, experimentally testable predictions. First, the model predicts that, as the universe expands, the galaxies recede from one another with a velocity proportional to the distance between them. In the 1920s Edwin P. Hubble inferred just such an expansion law from his study of the red-shifts of distant galaxies. Second, the big-bang model predicts that there should be a background of microwave radiation bathing the universe as a remnant of the intense heat of its origin. According to the model, the universe became transparent to this radiation several hundred thousand years after the big bang. Ever since then the matter has been clumping into stars, galaxies, and the like, but the radiation has simply continued to expand and red-shift, and in effect to cool. This prediction was confirmed in 1964 when Arno A. Penzias and Robert W. Wilson of the Bell Telephone Laboratories discovered a background of microwave radiation, received uniformly from all directions in the sky with an effective temperature of about 3 K. Third, the model leads to successful predictions for the synthesis of light atomic nuclei from protons and neutrons during the first minutes after the big bang. Successful predictions can be obtained in this way for the abundances of ^{4}He, ^{2}H, ^{3}H and ^{7}Li. (Heavier nuclei are thought to have been produced much later in the interior of stars. For details see Longair's Chapter 6.)

The successes of the big-bang picture all pertain to the behaviour of the model a second or more after the big bang. While the successes of the model have been energetically pursued, various problems of the standard big-bang picture have also emerged. These problems, however, all involve times when the universe was much less than a second old.

One set of problems concerns the very special conditions that are required by the model as the universe emerged from the big bang. The first problem is the difficulty of explaining the large-scale uniformity of the observed universe. In the standard model the universe evolves so quickly that it is impossible for this uniformity to be created by any physical process. This impossibility is a consequence of the fact that no information or physical process can propagate faster than a light signal. At any given time there is a maximum distance, known as the horizon distance, that a light signal could have travelled since the beginning of the universe. The variation of the horizon distance with time is shown in figure 3.1(c). As can be seen, the horizon distance in the standard model has been much smaller than the radius of the observed universe for most of its history, which makes it difficult to comprehend how the large-scale uniformity of the universe came to be.

The microwave background provides a striking illustration of this puzzle. On the one hand, observations of the microwave background radiation show that the temperature is extraordinarily uniform across all directions in the sky (to at least one part in 10 000), suggesting that the sources of the radiation were in thermal equilibrium. Yet, if the universe expanded according to the standard model, calculations show that the sources were as much as ninety horizon distances apart at the time the radiation was emitted, in which case there is no physical process by which the sources could have reached equilibrium. The situation is illustrated in figure 3.2.

The puzzle of explaining why the universe appears to be uniform over distances that are large compared with the horizon distance is known as the horizon problem. It is not a genuine inconsistency of the standard model – if the uniformity is assumed in the initial conditions, then the universe will evolve uniformly. The problem is that one of the most salient features of the observed universe – its large-scale uniformity – cannot be explained by the standard model; it must be assumed as an initial condition.

Even with the assumption of large-scale uniformity, the standard big-bang model requires yet another assumption to explain the nonuniformity which is observed on smaller scales. To account for the clumping of matter into galaxies, clusters of galaxies, superclusters of clusters, and so on, a spectrum of primordial inhomogeneities must be assumed as part of the initial conditions. The fact that the spectrum of inhomogeneities has no explanation is a drawback in itself, but the problem becomes even more pronounced when the model is extended to 10^{-43} s after the big bang. The incipient clumps of matter develop rapidly with time as a result of their gravi-

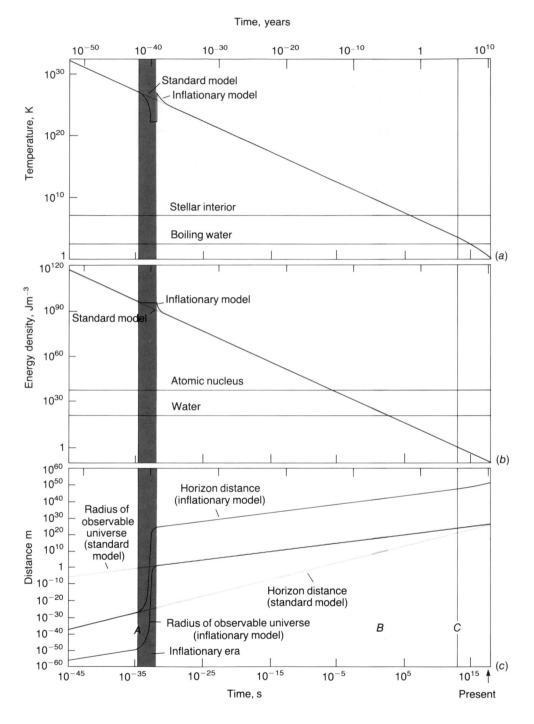

Time, years

10^{-50} 10^{-40} 10^{-30} 10^{-20} 10^{-10} 1 10^{10}

10^{30}

Standard model

Inflationary model

10^{20}

10^{10}

Stellar interior

Boiling water

1   (a)

Temperature, K

10^{120}

Inflationary model

10^{90}

Standard model

10^{60}

Atomic nucleus

10^{30}

Water

1   (b)

Energy density, Jm^{-3}

10^{60}
10^{50}
10^{40} Horizon distance
10^{30} (inflationary model)
Radius of
10^{20} observable
universe
10^{10} (standard
model)
1
10^{-10}
10^{-20}
Horizon distance
10^{-30} (standard model)
10^{-40} Radius of observable universe   B   C
A   (inflationary model)
10^{-50} Inflationary era
10^{-60} (c)

10^{-45} 10^{-35} 10^{-25} 10^{-15} 10^{-5} 10^{5} 10^{15}

Distance m

Time, s   Present

tational self-attraction, and so a model that begins at a very early time must begin with only a very small degree of inhomogeneity. To begin at 10^{-43} s, the matter must start in a peculiar state of extraordinary but not quite perfect uniformity. A normal gas in thermal equilibrium would be far too inhomogeneous, owing to the random motion of the particles

of which it is composed. This peculiarity of the initial state of matter required by the standard model is called the smoothness problem.

Another problem of the standard model concerns the energy density of the universe. According to general relativity, the space of the universe can in principle be curved, and the nature

Figure 3.1. Comparison of standard and inflationary models of the universe. The standard big-bang model of the universe is represented by the grey curves in this set of graphs, showing how several properties of the observed universe could have changed with time starting at 10^{-45} s after the big bang. The black curves represent the inflationary model, which is coincident with the standard model for all times after 10^{-30} s. For comparison the graph of temperature (*a*) also includes an indication of the boiling point of water (373 K) and the temperature at the centre of a typical star (10^7 K). Similarly, the graph of energy density (*b*) indicates the energy density of water (10^{20} J m^{-3}) and of an atomic nucleus (10^{35} J m^{-3}). On the graph of spatial dimensions (*c*) each cosmological model is represented by two curves. One curve shows the region of space that evolves to become the observed universe and the other shows the horizon distance: the total distance a light signal could have travelled since the beginning of the universe. On the time axis several significant events are marked. *A* indicates the time of the phase transition predicted in the standard big-bang model by grand unified theories of the interactions of elementary particles. A key feature of the inflationary model is the prolongation of the phase transition, which extends through a period called the inflationary era (dark grey band); during this era the universe expands by an extraordinary factor – the figure shows a factor of 10^{50}, but the actual number could even be much larger than that. Meanwhile the temperature plunges, but it is stabilised at about 10^{22} K by quantum effects which arise in the context of general relativity. The light grey band labelled *B* indicates the period when the lightest atomic nuclei were synthesised, and *C* indicates the time when the universe became transparent to electromagnetic radiation.

Figure 3.2. The horizon problem. The horizon problem is a serious drawback to the standard big-bang theory. In this three-dimensional space-time diagram the scales have been drawn in a nonlinear way so that the trajectory of a light pulse is represented by a line at 45° to the vertical axis. Our position in space and time is indicated by the point *A*. Since no signal can travel faster than the speed of light, we can receive signals only from the large cone, which is called our past light cone. Events outside the past light cone of a given point cannot influence an event at that point in any way. The grey horizontal plane shows the time at which the microwave background radiation was released. Radiation that is now reaching us from opposite directions was released at points *B* and *C*, and since then it has travelled along our past light cone to point *A*. The past light cone of point *B* has no intersection with the past light cone of point *C*, and therefore the two points were not subject to any common influences. The horizon problem is the difficulty of explaining how the radiation received from the two opposite directions came to be at the same temperature. In the standard model the large-scale uniformity of temperature evident in the microwave background radiation must be assumed as an initial condition of the universe. (Note that if the diagram were drawn accurately, the grey plane would be about ten times closer to the bottom.)

of the curvature depends on the energy density. Today the energy density of the universe is believed to be composed primarily of the 'rest energy' of its particles, with each particle of mass M contributing to the energy E according to $E = Mc^2$, where c denotes the speed of light. Until about 10 000 years after the big bang, however, the main contribution to the energy density is believed to have come from the intense glow of radiation that existed in the heat of the primordial gas. In any case, if the energy density exceeds a certain critical value, which depends on the expansion rate, the universe is said to be closed: space curves back on itself to form a finite volume with no boundary. (A familiar analogy is the surface of a sphere, which is finite in area and has no boundary.) If the energy density is less than the critical density, the universe is open: space curves but does not turn back on itself, and in the simplest formulation the volume is infinite. If the energy density is just equal to the critical density, the universe is flat: space is described by the familiar Euclidean geometry (again with infinite volume). These possibilities are summarised in figure 3.3, and are also considered in Chapter 4 by Hawking.

The ratio of the energy density of the universe to the critical density is a quantity cosmologists designate by the capital Greek letter Ω (omega). The value $\Omega = 1$ (corresponding to a flat universe) represents a state of unstable equilibrium. If Ω was ever exactly equal to one, it would remain exactly equal to one forever. But if Ω differed slightly from one an instant after the big bang, then the deviation from one would grow rapidly with time. Given this instability, it is surprising that Ω is measured today as being between 0.1 and 2. (Cosmologists are still not sure whether the universe is open, closed or flat.) In order for Ω to be in this rather narrow range today, its value a second after the big bang had to equal one to within one part in 10^{15}. The standard model offers no explanation of why Ω began so close to one, but merely assumes the fact as an initial condition. This shortcoming of the standard model, called the flatness problem, was first pointed out in 1979 by Robert H. Dicke and P. James E. Peebles of Princeton University.

It is interesting to note that the value of Ω is also related to the ultimate fate of the universe. If $\Omega > 1$ then the gravitational self-attraction is sufficient to eventually halt the expansion and cause the universe to recollapse. If $\Omega \leqslant 1$, the expansion of the universe will continue forever. A graph of this behaviour is shown in figure 3.4. These standard predictions can be altered, however, if a nonzero 'cosmological constant' is included in the equations of general relativity, as will be discussed later. For the benefit of those readers with a background in Newtonian physics, a derivation of the critical energy density is given in box 3.1 on page 54. The equations of evolution for a universe evolving according to Newtonian mechanics are given in box 3.2 on page 56.

3.2   Grand unified theories

The successes and drawbacks of the big-bang model that we have considered so far involve cosmology, astrophysics and nuclear physics. As the big-bang model is traced backward in time, however, one reaches an epoch for which these branches of physics are no longer adequate. In this epoch all matter is so hot that it is decomposed into its elementary particle constituents. In their attempts to understand this epoch, cosmologists have made use of recent progress in the theory of elementary particles. Indeed, one of the important developments of the past decade has been the fusing together of interests in particle physics, astrophysics and cosmology. The result for the big-bang model appears to be at least one more success and at least one more failure.

Elementary particle physicists use the word 'interaction' to refer to any process that elementary particles can undergo, whether it involves scattering, decay, particle annihilation, or particle creation. All of the known interactions of nature are divided into four types. From the weakest to the strongest, these interactions are gravitation, the weak interactions, electromagnetism, and the strong interactions. The force of gravity appears to be strong in our everyday lives because it is long-range and universally attractive – thus we are accustomed to feeling the force that acts between all the particles in the Earth and all the particles in our own bodies. The force of gravity acting between two elementary particles, however, is so weak that it has never been detected. The weak interactions have a range that is roughly one-hundred times smaller than the size of an atomic nucleus, and they are seen primarily in the radioactive decay of many kinds of nuclei. The weak interactions are also responsible for the scattering of particles called neutrinos, a type of experiment that is now routinely carried out at high energy accelerator laboratories. Electromagnetism includes both electric and magnetic forces, and is responsible for holding the electrons of an atom to the nucleus. Light waves, radio waves and X-rays are also electromagnetic phenomena. The strong interactions have a range of about the size of an atomic nucleus, and they account for the force which binds the protons and neutrons inside a nucleus. They also account for the tremendous energy release of a hydrogen bomb, as well as the interactions of many short-lived particles that are investigated in particle accelerator experiments.

Type of universe	Ratio of energy density to critical density (Ω)	Spatial geometry	Volume	Temporal evolution
Closed	> 1	Positive curvature (spherical)	Finite	Expands and recollapses
Open	< 1	Negative curvature (hyperbolic)	Infinite	Expands forever
Flat	1	Zero curvature (Euclidean)	Infinite	Expands forever, but expansion rate approaches zero

Figure 3.3. The geometry of the universe. Three possible geometries of the universe, classified as closed, open, and flat, can arise from the standard big-bang model. Under the usual assumption that the equations of general relativity are not modified by the introduction of a nonzero cosmological constant, the distinction between the different geometries depends on the quantity designated Ω, the ratio of the energy density of the universe to the critical density. The value of the critical density depends in turn on the rate of expansion of the universe. The value of Ω today is known to lie between 0.1 and 2, which implies that its value a second after the big bang was equal to one to within one part in 10^{15}. The failure of the standard big-bang model to explain why Ω began so close to one is called the flatness problem.

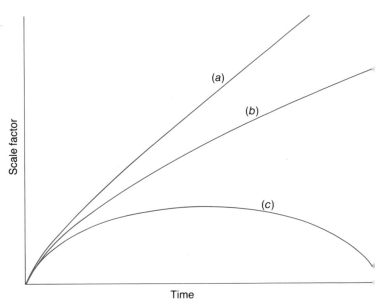

Figure 3.4. Evolution of a Newtonian universe. The time evolution of a cosmological model evolving according to Newtonian mechanics is shown for three types of solutions. The evolution is described by a scale factor which characterises the expansion of the universe – when the scale factor doubles, it means that all distances in the universe have doubled. (*a*) represents the evolution of an open universe, with a mass density parameter $\Omega < 1$; (*c*) shows the behaviour of a closed universe ($\Omega > 1$); and (*b*) shows the behaviour of a flat universe. All the curves are drawn under the usual assumption that the cosmological constant is zero.

The strong, the weak and the electromagnetic interactions appear to be accurately described by theories developed during the early 1970s. The strong interactions are described by quantum chromodynamics, or QCD, a theory based on the hypothesis that all strongly interacting particles are composed of quarks. The theory provides a detailed description of the interactions that bind the quarks into the observed particles, and the residual effect of these quark interactions can account for the observed interactions of the particles. Although our ability to extract quantitative predictions from QCD is severely limited by shortcomings in our calculational techniques, there is nonetheless convincing evidence that QCD correctly describes the strong interactions over the full range of available energies. The weak and electromagnetic interactions are successfully described by the unified electroweak theory, also known as the Glashow–Weinberg–Salam model (named for Sheldon Lee Glashow of Harvard University, Steven Weinberg of the University of Texas at Austin, and Abdus Salam of the International Center for Theoretical Physics in Trieste, who shared the 1979 Nobel Prize in Physics for this work). Known calculational techniques are very effective in extracting predictions from this theory, owing to the inherent weakness of the interactions being described. Quantum chromodynamics and the unified electroweak theory, when taken together, are called the standard model of elementary particle interactions. While a quantum theory of gravity remains to be developed, we nonetheless believe that general relativity provides an adequate description of gravity at the classical level. Since the gravitational interactions of individual elementary particles are negligible at the energies presently available, the classical theory of general relativity is sufficient to describe all the observed properties of gravity. Thus, elementary particle physics is in a state of unprecedented success – all feasible experiments can be described by the standard model of particle interactions and/or the theory of general relativity.

In this atmosphere of enormous success, physicists have attempted to extend our understanding of particle interactions. One of the major recent developments in the theory of elementary particles has been the notion of grand unified theories, the first successful version of which was proposed in 1974 by Sheldon Glashow and Howard M. Georgi, also of Harvard University. This is recounted by Georgi in his article on GUTs in this volume (Chapter 15). The theories are difficult to verify experimentally, because their most distinctive predictions can be tested only by doing experiments at very high energies. The characteristic energy scale of these theories is about 10^{14} GeV (1 GeV = one billion electron volts), which is approximately the energy necessary to light a 100 W light bulb for one minute. While this amount of energy is not much by the standard of an electric power company, the idea of accelerating a single elementary particle to this energy is extraordinary – it is about eleven orders of magnitude beyond the reach of the largest existing accelerators. Nevertheless, grand unified theories have some experimental support. For example, they

successfully predict a relationship involving the strengths of certain weak and electromagnetic processes, a relationship which has no reason to hold in the standard model of particle interactions. They also explain the well-known fact that the magnitude of the charge of an electron is equal to that of a proton – in the standard model this equality appears to be an extraordinary coincidence. Finally, grand unified theories so elegantly unify our understanding of elementary particle interactions that many physicists find them extremely attractive.

More recent developments in elementary particle theory, such as supergravity and superstrings (discussed in Chapters 5 and 18 by Isham and Salam, respectively), have carried the thoughts of theoretical physicists to even higher energies. The characteristic energy scale for these theories is known as the Planck energy, about 10^{19} GeV, which is the energy at which the gravitational force on an individual elementary particle becomes comparable in strength to its other interactions. Indeed, the primary motivation of supergravity and superstrings has been the unification of gravity with the other elementary particle interactions, with the hope of developing an acceptable quantum theory of gravity. At energies well below the Planck scale, these theories are expected to closely approximate the behaviour of grand unified theories. Although these theories clearly represent a major step in the development of theoretical physics, our understanding of their implications for cosmology is still very incomplete. Since the energy scales relevant to most inflationary models are much less than the Planck scale, we will discuss inflation solely in the context of grand unified theories.

The basic idea of a grand unified theory is that what are perceived to be three independent forces – the strong, the weak and the electromagnetic – are actually components of a single unified force. In the theory each component of the force is related to the others by an underlying symmetry, so that all the components are described by a single force law. The same symmetry implies that, at the fundamental level, there is no distinction between electrons, the weakly interacting massless particles known as neutrinos and the quarks, which bind strongly together to make up protons and neutrons. Experimentally, on the other hand, electrons, neutrinos and quarks behave very differently, and the strong, weak and electromagnetic forces are very different in both strength and character. Thus, the theory is constructed so that the symmetry is 'spontaneously broken' in the present universe. Grand unified theories were not the first to make use of spontaneous symmetry breaking – the mechanism had previously been used in the unified electroweak theory.

A spontaneously broken symmetry is one that is present in the underlying theory describing a system, but that is hidden in the equilibrium state of the system. For example, the physical laws that describe a liquid are rotationally symmetric, and the liquid state exhibits rotational symmetry: the distribution of molecules looks the same no matter how the liquid is turned.

When the liquid freezes into a crystal, however, the atoms arrange themselves along crystallographic axes and the rotational symmetry is broken. We call the high temperature phase the symmetric phase and the low temperature phase the broken symmetry phase. One would expect that if the temperature of a system in a broken-symmetry state were raised, it could undergo a 'phase transition' to a state in which the symmetry is restored, just as a crystal can melt into a liquid. Grand unified theories predict such a transition at a critical temperature of roughly 10^{27} K, which corresponds to a mean thermal energy per particle of about 10^{14} GeV.

The analogy between crystals and grand unified theories can be made more illustrative by whimsically imagining a world of intelligent creatures living inside a crystal. Let us assume that these creatures can somehow move about and carry on the task of scientific investigation, but that they cannot muster enough energy to come close to melting the crystal in which they live. In this world the crystal would not be considered an object, but instead the crystalline structure would be taken as a fundamental property of space. A physics book would make no mention of rotational symmetry, but would instead contain a chapter discussing the properties of space and its primary axes. The crystalline structure would, for example, affect the propagation of light through the medium, and for many crystals the speed of light would vary with direction. Thus, a table of physical constants in the crystal world might list several speeds of light, one for each primary axis. If grand unification is correct, then the world of our experience is similar to this crystal world; our tabulation of the different properties of the strong, weak and electromagnetic interactions is analogous to the tabulation of the different speeds of light.

One direct result of combining the big-bang model with grand unified theories is the successful understanding of the excess of matter over antimatter in the universe. The result hinges on a novel property of the grand unified theories concerning a principle known as conservation of baryon number. The baryon number of a system is defined as the total number of quarks which it contains, minus the number of antiquarks, all divided by three. Since protons and neutrons are each composed of three quarks, each such particle contributes one unit to the total baryon number; antiprotons and antineutrons each contribute negative one unit. (Other elementary particles possessing nonzero baryon numbers, such as the delta, lambda, etc., are heavier and very short lived. Quarks are believed not to exist as free particles, provided that the temperature is less than about 10^{12} K. Instead they are bound in particles of integer baryon number.) All physical processes observed up to now obey the principle of baryon number conservation – the total baryon number of an isolated system cannot be changed. This principle implies, for example, that the proton must be absolutely stable; because it is the lightest baryon, it cannot decay into another particle without changing the total baryon number. Experimentally the lifetime of the proton is now known to exceed 10^{32} years.

Grand unified theories, however, imply that baryon number is not exactly conserved. At low temperature, in the broken-symmetry phase, the conservation law is an excellent approximation, and the observed limit on the proton lifetime is consistent with many versions of grand unified theories. At temperatures of order 10^{27} K and higher, however, processes that change the baryon number of a system of particles are expected to be quite common.

Cosmologists have wondered for many years whether all the stars, galaxies and dust observed in the universe are composed of matter, or whether some of the distant objects might perhaps be formed from antimatter. The total baryon number depends crucially on this question, since the nuclear particles of matter are baryons, while those of antimatter are antibaryons. It seems clear that all the observed objects within our own supercluster are made of matter – if there were patches of antimatter at these distances and densities, observable radiation would be emitted by the matter–antimatter annihilation occurring at the boundaries. Beyond our own supercluster, on the other hand, the annihilation radiation would be too weak to observe, and there is no other direct observational evidence concerning this question. Thus the issue cannot be definitively settled, but there is a strong consensus that the universe is probably made entirely of matter. The belief is motivated primarily by the absence of any known mechanism that could have separated the matter from the antimatter over such large distances. If this belief is valid, then the net baryon number of the observed universe is about 10^{78}.

Before the advent of grand unified theories, when baryon number was thought to be conserved, this net baryon number had to be postulated as yet another initial condition of the universe. In the context of grand unified theories, however, the observed excess of matter over antimatter can be produced naturally by elementary-particle interactions at temperatures just below the critical temperature of the phase transition. Grand unified theories contain two ingredients that make this possible. First, as discussed above, grand unified theories allow for baryon nonconserving processes. Second, grand unified theories incorporate a small but important asymmetry in the behaviour of matter versus antimatter. This asymmetry was first discovered experimentally in 1964 by Val L. Fitch of Princeton University and James W. Cronin of the University of Chicago (who shared the 1980 Nobel Prize in Physics for this discovery). The matter–antimatter asymmetry is necessary to explain the baryon excess, since it can provide a tendency for the baryon nonconserving processes to produce more matter than antimatter. The grand unified theories depend on too many arbitrary parameters for a quantitative prediction of baryon production, but the observed matter excess can be produced with a reasonable choice of values for these parameters.

A serious problem that results from combining grand unified theories with the big-bang picture is that a large number of 'defects' are generally formed during the transition from the symmetric phase to the broken-symmetry phase. The defects are created when regions of symmetric phase undergo a transition to different broken-symmetry states. In an analogous situation, when a liquid crystallises, different regions may begin to crystallise with different orientations of the crystallographic axes. The 'domains' of different crystal orientation grow and coalesce, and it is energetically favourable for them to smooth the misalignment along their boundaries. The smoothing is often imperfect, however, and localised defects remain.

In the grand unified theories there are serious cosmological problems associated with pointlike defects, which correspond to magnetic monopoles, and surfacelike defects called domain walls. Both are expected to be extremely stable and extremely massive – the monopole would be about 10^{16} times as heavy as the proton. (Stringlike defects called cosmic strings are also predicted by some grand unified theories, but they do not cause cosmological problems. In fact, we will later mention how the presence of cosmic strings may play an important role in galaxy formation.) A domain of correlated broken-symmetry phase cannot be much larger than the horizon distance at that time, so the minimum number of defects created during the transition can be estimated. The result is that there would be so many defects after the transition that their mass would dominate the energy density of the universe by a large factor. The gravitational attraction would slow the expansion of the universe so rapidly that the present expansion rate would be reached after only 30 000 years, rather than ten to twenty billion years. All the successful predictions of the big-bang model would be lost. Thus any successful union of grand unified theories and the big-bang picture must incorporate some mechanism to drastically suppress the production of magnetic monopoles and domain walls.

3.3 Spontaneous symmetry breaking and the Higgs mechanism

The inflationary universe model appears to provide a simple and elegant solution to these problems. Before the model can be described, however, we must first explain a few more of the details of symmetry breaking and phase transitions in grand unified theories.

Grand unified theories, like all modern particle theories except those based on superstrings, are examples of quantum field theories – to understand their structure one must understand the relationship between particles and fields. As an example we can consider the best-known field theory, the one that describes electromagnetism. According to the classical (i.e., nonquantum) theory of electromagnetism developed by James Clerk Maxwell in the 1860s, electric and magnetic fields have a well-defined set of equations. Early in the twentieth century, however, Maxwell's theory was modified in order to achieve consistency with the quantum theory which was developed to explain the behaviour of atoms and molecules. In the classical theory it is possible to increase the energy of an electromagnetic field by any amount, but in the quantum theory the increases in energy can come only in discrete lumps, or quanta, which in this case are called photons. The photons have both wavelike and particlelike properties, but in the lexicon of modern physics they are usually called particles. In general the formulation of a quantum field theory begins with a classical theory of fields, and it becomes a theory of particles when the rules of the quantum theory are applied.

As we have already mentioned, an essential ingredient of grand unified theories is the phenomenon of spontaneous symmetry breaking. The detailed mechanism of spontaneous symmetry breaking in grand unified theories is simpler in many ways than the analogous mechanism in crystals. The formation of crystalline structure is thought to be the result of the complicated interplay between large numbers of electrons and atomic nuclei, interacting via electromagnetic forces and moving according to the laws of quantum theory. In grand unified theories, on the other hand, a mechanism is used in which a set of fields is added for the specific purpose of spontaneously breaking the symmetry. These fields are known as Higgs fields (after Peter W. Higgs of the University of Edinburgh), and the spontaneous symmetry breaking mechanism, which occurs in a variety of particle physics theories, is known as the Higgs mechanism. The fields are 'scalar', which means that unlike the electric and magnetic fields they do not point in any direction – each Higgs field is specified by giving its value, a single number, at each point in space.

The theory is formulated so that the grand unified symmetry is unbroken when all the Higgs fields have a value of zero (just as rotational symmetry is unbroken by the distribution of molecules in a liquid state). The grand unified symmetry is spontaneously broken, by contrast, if one or more of the Higgs fields acquire a nonzero value. Furthermore, it is possible to formulate the theory in such a way that the state in which all the Higgs fields vanish is *not* the state of lowest possible energy density. Although it seems counterintuitive that energy should be required in order for the value of the Higgs fields to be zero, particle physicists find that this property can be incorporated into quantum field theories without causing any inconsistencies. The lowest possible energy density is therefore attained in a state for which one or more Higgs fields have nonzero values. Such a state is related by the grand unified symmetry to an infinite number of other states, in which the Higgs fields differ from those of the first in a manner prescribed by the symmetry property. (In the crystal analogy, any state with a crystalline structure is related by the rotational symmetry to an infinite number of other states which are distinguished by the orientations of the crystallographic axes.) The symmetry implies that each of these states has the same energy density as the first, and hence the state of lowest possible energy density is not unique. A state belonging to the class of states with the lowest possible energy density is called a 'true vacuum'. Empty space is assumed to consist of one such true vacuum state. A typical plot showing how the energy density depends on the value of the Higgs fields is displayed in figure 3.5.

The oscillation of the Higgs fields about their true vacuum values is interpreted in the quantum version of the theory as the presence of Higgs particles. These particles would be very short lived, so we would not expect to observe relic Higgs particles remaining from the big bang. Nor would the Higgs particles be produced in accelerator experiments, since their masses are expected to correspond to energies of order 10^{14} GeV. Thus, the Higgs particles associated with grand unified theories cannot be observed by any method available now or in the foreseeable future.

The Higgs mechanism is also used in the unified electroweak theory, which contains a set of Higgs fields which are distinct from those that are hypothesised to break the grand unified symmetry. The electroweak Higgs fields are associated with an electrically neutral particle, with a mass corresponding to an energy of the order of 100 GeV. The precise value of the electroweak Higgs mass, however, is a free parameter of the theory which has not yet been measured. In contrast to the grand unified Higgs particles, the electroweak Higgs particle could conceivably be detected in upcoming accelerator experiments.

In a region at zero temperature, the Higgs fields would assume values which give the lowest possible energy density. That is, the values of the fields would correspond to one of the true vacuum states, and the symmetry would be broken. As the temperature of the region is raised, the Higgs fields would begin to fluctuate about their zero-temperature values. These fluctuations would correspond, in the particle interpretation, to the presence of Higgs particles that were created by collisions in the hot gas. Provided that the temperature is not too high, these fluctuations would centre on the original zero-temperature values and the system would remain in a broken symmetry phase. Once the temperature exceeds a certain critical value, however, the fluctuations in the Higgs fields would become so large that all evidence of the initial values would be lost – each field would oscillate about an equilibrium value of zero, and the system would therefore be in a symmetric phase. In a typical grand unified theory this phase transition from the low temperature broken symmetry phase to the high temperature symmetric phase occurs at a critical temperature of about 10^{27} K.

Note that the energy density of the true vacuum states is shown as zero in figure 3.5. The zero point of energy can be chosen arbitrarily whenever the effects of gravity are negligible, and in these cases it is conventional to *define* the energy density of the true vacuum states to be zero. Gravity, however, is not

Figure 3.5. Energy density of the Higgs fields. This three-dimensional diagram illustrates a typical form for the energy density of a pair of Higgs fields. These fields are members of a special set of fields postulated in grand unified theories to account for spontaneous symmetry breaking. The surface, shown in cross section, is rotationally symmetric about a vertical axis; this symmetry of the diagram reflects a symmetry of the theory relating the behaviour of one Higgs field to that of the other. The surface contains exactly one point – the intersection with the vertical axis – which remains undisplaced by a symmetry rotation.

This point corresponds to a state in which each Higgs field has a value of zero. In the absence of thermal excitations this state of unbroken symmetry is known as the false vacuum; it would have an energy density of about 10^{93} J m^{-3}, or some 10^{58} times the energy density of an atomic nucleus. The symmetry is broken whenever one or both of the Higgs fields acquire nonzero values. The theory has been formulated in such a way that the states of lowest possible energy density, known as the true vacuum states, are states of broken symmetry, forming a circle in the horizontal plane at the bottom of the diagram.

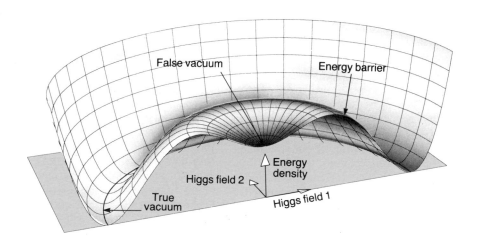

negligible in cosmology – a nonzero value of the vacuum energy density corresponds to what is usually called a cosmological constant, and alters the dynamics of the cosmic expansion. (The mathematics of this alteration is shown in boxes 3.1, 3.2 and 3.4.) In fact, astrophysical observations imply that the magnitude of the vacuum energy density is less than 1.5×10^{-9} J m^{-3}, including the possibility that the vacuum energy density is identically zero. Figure 3.5, on the other hand, is drawn on the scale of 10^{93} J m^{-3}, so the vacuum energy density is certainly negligible on the scale of this figure.

Grand unified theories can easily accommodate a value of zero for the true vacuum energy density, but there appears to be no reason to expect the value to be zero. While a value of zero might have been expected for the symmetric vacuum in which the Higgs fields have zero magnitude, the limits obtained from astronomical observations refer instead to the broken-symmetry state that describes our present universe, a state with large, nonzero values for the Higgs fields. Our inability to explain the extreme smallness of the vacuum energy density, or equivalently the cosmological constant, is regarded by many particle theorists as one of the most important problems in physics. This 'cosmological constant problem' seems to indicate that even a state as simple as the vacuum has properties that we do not yet understand.

3.4   The original inflationary universe model

We have now assembled enough background information to describe the inflationary model of the universe, beginning with the form in which it was first proposed by one of us (Guth) in 1980. Any cosmological model must begin with some assumptions about the initial conditions, but for the inflationary model the initial conditions can be rather arbitrary. One assumes that the early universe included at least some regions of gas that were hot compared with the critical temperature of the phase transition and that were also expanding. In such a hot region the Higgs fields would each have a value of zero. As the expansion caused the temperature to fall it would become thermodynamically favourable for one or more of the Higgs fields to develop a nonzero value, bringing the system to its broken-symmetry phase.

For some values of the unknown parameters of the grand unified theories this phase transition would occur very quickly compared with the cooling rate, in which case the effect on cosmological evolution would be minimal. The inflationary universe model, however, assumes that the parameters lie in a range for which the phase transition is very slow compared to the cooling. As a result the system could cool to well below 10^{27} K with the values of the Higgs fields remaining at zero. This phenomenon, known as supercooling, is quite common in condensed-matter physics; water, for example, can be super-cooled to more than 20 K below its freezing point, and glasses are formed by rapidly supercooling a liquid to a temperature well below its freezing point.

As the region of gas continued to supercool, it would approach a peculiar state of matter known as a false vacuum. This state of matter has never been observed, but it has properties that are unambiguously predicted by quantum field theory. The temperature, and hence the thermal component of the energy density, would rapidly decrease and the energy density of the state would be concentrated entirely in the Higgs fields. A zero value for the Higgs fields implies a large energy density for the false vacuum, as can be seen in figure 3.5. In the classical form of the theory such a state would be absolutely stable, even though it would not be the state of lowest possible energy density. States with a lower energy density would be separated from the false vacuum by an intervening energy barrier, and there would be no energy available to take the Higgs fields over the barrier.

In the quantum version of the theory, however, the false vacuum is not absolutely stable. Under the rules of quantum theory all the fields would be continually fluctuating. As was first described by Sidney Coleman of Harvard University, a quantum fluctuation would occasionally cause the Higgs fields in a small region of space to 'tunnel' through the energy barrier, nucleating a 'bubble' of the broken-symmetry phase. The bubble would then start to grow at a speed that would rapidly approach the speed of light, converting the false vacuum into the broken-symmetry phase. The rate at which bubbles form depends sensitively on the unknown parameters of the grand unified theory, and therefore cannot be predicted. In the inflationary model it is assumed that the rate would be extremely low.

The most peculiar property of the false vacuum is probably its pressure, which is both large and negative. To understand why, consider a bubble of true vacuum which is surrounded by a region of false vacuum. The true vacuum bubble would grow and thereby displace the false vacuum, a process which is favoured energetically because the true vacuum has a lower energy density than the false vacuum. The growth also indicates, however, that the pressure of the true vacuum must be higher than the pressure of the false vacuum, forcing the bubble wall to expand outward. Because the pressure of the true vacuum is zero, the pressure of the false vacuum must be negative. A more detailed argument, described in box 3.3 on page 57, shows that the pressure of the false vacuum is equal to the negative of its energy density (when the two quantities are measured in the same units).

The negative pressure would not result in mechanical forces within the false vacuum, because mechanical forces arise only from differences in pressure. Nevertheless, there would be gravitational effects. Under ordinary circumstances the expansion of the region of gas would be slowed by the mutual gravitational attraction of the matter within it. In Newtonian physics that attraction is proportional to the mass density, which in relativistic theories is equal to the energy density divided by the square of the speed of light. The mathematics of a universe evolving according to Newtonian gravity was described in box 3.2. According to general relativity, however, the pressure also contributes to the attraction; to be specific, the gravitational force is proportional to the energy density plus three times the pressure. This relationship can also be obtained by applying relativistic corrections to the Newtonian equations, as is shown in box 3.4 (page 58). For the false vacuum the contribution made by the pressure would therefore overwhelm the energy-density contribution and would have the opposite sign. Hence, the bizarre notion of negative pressure leads to the even more bizarre effect of a gravitational force that is effectively repulsive. As a result the expansion of the region would be accelerated – the region would grow exponentially, doubling in diameter during each interval of about 10^{-34} s. A derivation of the exponential expansion law is given in box 3.5 (page 59).

This period of accelerated expansion is called the inflationary era, and it is the key element of the inflationary model of the universe. According to the model, the inflationary era continued for 10^{-32} s or longer, and during this period the diameter of the universe increased by a tremendous factor. The factor must have been at least 10^{25} in order for the cosmological problems discussed in Section 3.1 to be avoided. The expected value of the factor, however, depends on the highly uncertain details of the underlying particle theory. We will use a factor of 10^{50} as an example, but the actual number could be much larger still. It is assumed that after this colossal expansion the transition to the broken-symmetry phase finally took place. The energy density of the false vacuum was then released, resulting in a tremendous amount of particle production. The region was reheated to a temperature of almost 10^{27} K. (In the language of thermodynamics the energy released is called the latent heat; it is analogous to the energy released when water freezes.) From this point on, the region would continue to expand and cool at the rate described by the standard big-bang model. A volume the size of the observable universe would lie well within such a region.

The horizon problem is avoided in a straightforward way. In the inflationary model the observed universe evolves from a region that is much smaller in diameter (by a factor of 10^{50} in our example) than the corresponding region in the standard model. Before inflation begins the region is much smaller than the horizon distance, and it has time to homogenise and reach thermal equilibrium. This small homogeneous region is then inflated to become large enough to encompass the observed universe. Thus, the sources of the microwave background radiation arriving today from all directions in the sky were once in close contact; they had time to reach a common temperature before the inflationary era began.

The flatness problem is also evaded in a simple and natural way. The equations describing the evolution of the universe during the inflationary era are different from those for the standard model, and it turns out that the ratio Ω is driven rapidly toward one, no matter what value it had before inflation. This behaviour is most easily understood by recalling that a value of $\Omega = 1$ corresponds to a space that is geometrically flat. The rapid expansion causes the space to become flatter just as the surface of a balloon becomes flatter when it is inflated, as illustrated in figure 3.6. The mathematical description of this process is given in box 3.6 on page 59. The mechanism driving Ω towards one is so effective that one is led to an almost rigorous prediction: the value of Ω today should be very accurately equal to one. Many astronomers (although not all) think a value of one is consistent with current observations, but a more reliable determination of Ω would provide a crucial test of the inflationary model.

Figure 3.6. The flatness problem. The solution of the flatness problem is illustrated by this sequence of perspective drawings of an inflating sphere. The illustration shows how a flat spatial geometry (which corresponds to $\Omega = 1$) can be produced by the inflationary scenario in a simple and natural way. In each successive frame the sphere is inflated by a factor of three (while the number of grid lines on the surface is increased by the same factor). The curvature of the surface quickly becomes undetectable on the scale of the illustration.

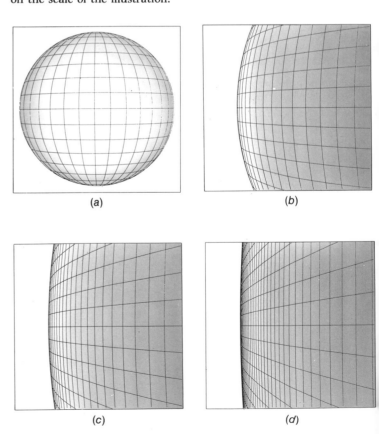

(a)   (b)

(c)   (d)

In spite of these successes, the original form of the inflationary model had a crucial flaw: under the circumstances described, the phase transition itself would create inhomogeneities much more extreme than those observed today. As we have already described, the phase transition would take place by the random nucleation of bubbles of the new phase. Erick Weinberg of Columbia University and one of us (Guth) have shown that the bubbles would always remain in finite clusters disconnected from one another, and that each cluster would (with overwhelming probability) be dominated by a single largest bubble, as is illustrated in figure 3.7. Almost all the energy in the cluster would be initially concentrated in the surface of the largest bubble, and there is no apparent mechanism to redistribute the energy uniformly. Such a configuration bears no resemblance to the observed universe.

3.5 The new inflationary universe

For almost two years after the invention of the inflationary universe model it remained a tantalising but clearly imperfect solution to a number of important cosmological problems. Near the end of 1981, however, a new approach was developed by A. D. Linde of the P. N. Lebedev Physical Institute in Moscow and independently by Andreas Albrecht and one of us (Steinhardt) of the University of Pennsylvania. This approach, known as the new inflationary universe, avoids all the problems of the original model while maintaining all its successes.

The key to the new approach is to consider a special form of the relation that describes the energy density of the Higgs fields (see figure 3.8). Quantum field theories with energy-density relations of this type were first studied by Sidney Coleman and Erick J. Weinberg. In contrast to the more typical case shown in figure 3.5, there is no energy barrier separating the false vacuum from the states of lower energy; instead the false vacuum lies at the top of a rather flat plateau. In the context of grand unified theories such an energy-density relation is achieved by a special choice of parameters. As we shall explain below, this energy-density relation leads to a special type of phase transition that is sometimes called a slow-rollover transition.

The scenario begins just as it does in the original inflationary model. Again one assumes that the early universe had regions that were hotter than about 10^{27} K and were also expanding. In these regions thermal fluctuations would drive the equilibrium value of the Higgs fields to zero and the symmetry would be unbroken. As the temperature fell it would become thermodynamically favourable for the system to undergo a phase transition in which at least one of the Higgs fields acquired a nonzero value, resulting in a broken-symmetry phase. As in the previous case, however, the rate of this phase transition would be extremely low compared with the rate of cooling. The system would supercool to a negligible temperature with the Higgs fields remaining at zero, and the resulting state would again be considered a false vacuum.

The important difference in the new approach is the way in which the phase transition would take place. Quantum fluctuations or small residual thermal fluctuations would cause the Higgs fields to deviate from zero. In the absence of an energy barrier the value of the Higgs fields would begin to increase steadily; the rate of increase would be much like that of a ball rolling down a hill of the same shape as the curve in figure 3.8, under the influence of a frictional drag force. Since the energy-density curve is almost flat near the point where the Higgs fields vanish, the early stage of the evolution would be very slow. As long as the Higgs fields remain close to zero, the energy density would be almost the same as it is in the false vacuum. As in the original scenario, the region would undergo accelerated expansion, doubling in diameter every 10^{-34} s or so. Now, however, the expansion would cease to accelerate when the value of the Higgs fields reached the steeper part of the curve. By computing the time required for the Higgs fields to evolve, the amount of inflation can be estimated. An expansion factor of 10^{50} or more is quite plausible, but the actual factor depends on the details of the particle theory one adopts.

The variation of the Higgs fields with time was discussed in the previous paragraph, but there is also a variation with position. Recall that there are actually many different broken-symmetry states, just as there are many possible orientations for the axes of a crystal. There are a number of Higgs fields, and the various broken-symmetry states are distinguished by the combination of Higgs fields that acquire nonzero values. Since the fluctuations that drive the Higgs fields from zero are random, different regions of the primordial universe would be driven toward different broken-symmetry states. The Higgs fields would, however, be approximately homogeneous over small regions. The distance across a typical homogeneous region is called the 'correlation length'. The correlation length is expected to be approximately equal to the horizon distance, which at the start of the phase transition would be about 10^{-26} m. Once the Higgs fields within such a homogeneous region began to deviate from zero in a definite combination, the region would evolve toward one of the stable broken symmetry states and would inflate by a factor of 10^{50} or more. The size of the region after inflation would then be greater than 10^{24} m. The entire observable universe, which at that time would be only about 10 cm across, would be able to fit deep inside such a region.

Note that the homogeneous regions of the slow-rollover transition are rather different from the bubbles that were discussed in the original model. First, unlike the bubbles, the homogeneous regions of the slow-rollover transition would tend not to be spherical. Second, whereas the bubbles are separated from each other by false vacuum at the time of formation, the homogeneous regions tend to be juxtaposed.

(a)

(b)

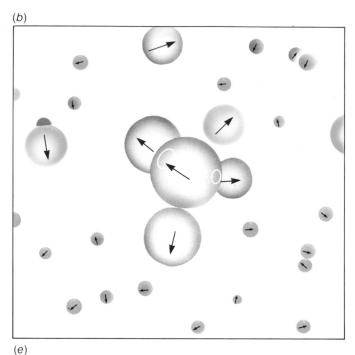

(d)

(e)

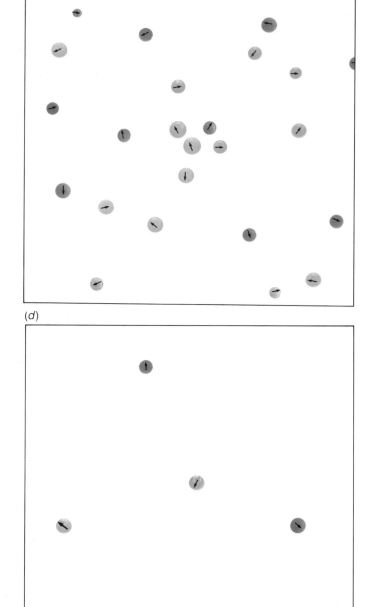

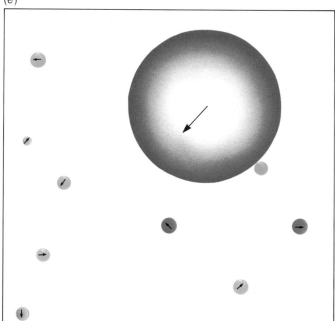

Figure 3.7. Bubble formation: standard model vs. original
inflationary model. Expanding bubbles of broken-symmetry phase
form in an expanding region of symmetric phase in the two highly
schematic time sequences (*a–c*) and (*d–f*). The sequence
representing the standard big-bang model (*a–c*) covers a much
shorter time span than the sequence representing the original form
of the inflationary model (*d–f*). In both cases the Higgs fields have
a value of zero in the region outside the bubbles, while at least one
Higgs field has a nonzero value inside each bubble. In a grand
unified theory the broken-symmetry states can in general be

distinguished by parameters of two kinds: discrete and continuous.
Here each bubble is labelled in two ways: by a colour (blue or red)
to indicate the discrete parameter and by an internal black arrow
to indicate the value of the continuous parameter. In the standard
model the bubbles would quickly coalesce and complete the
transition from the symmetric phase to the broken-symmetry
phase. A surfacelike defect called a domain wall would form at any
boundary between regions with different values of the discrete
parameter (purple areas). Within a region of uniform colour a
pointlike defect called a magnetic monopole would form at a centre

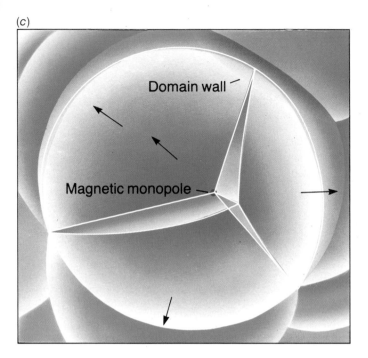

(c)

Domain wall

Magnetic monopole

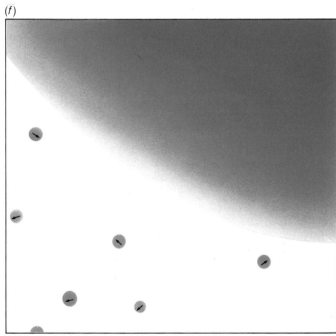

(f)

created by the intersection of many bubbles whenever the arrow representing the continuous parameter points everywhere away from the centre. In the original form of the inflationary model the rapid expansion of the false vacuum, or symmetric phase, region would keep the bubbles from ever coalescing into an infinite cluster. Either hypothetical situation has consequences that are contrary to observation; the new inflationary model was developed to avoid both of them.

Furthermore, at the time of formation the bubbles would be bounded by sharply defined bubble walls, while the boundaries between homogeneous regions would be diffuse. Finally, in the case of bubbles false vacuum is converted into the broken symmetry phase at the bubble boundary as it moves outward. For the homogeneous regions, on the other hand, a gradual conversion to the broken-symmetry phase occurs nearly uniformly throughout the entire region.

As the region underwent this enormous inflation, any density of particles that might have been present initially would be diluted to virtually zero. The energy content of the region would then consist entirely of the energy stored in the Higgs field. How could this energy be released? Once the Higgs field evolved away from the flat of the energy-density curve, it would start to oscillate rapidly about the true-vacuum value. Drawing once more on the relation between particles and fields implied by quantum field theory, this situation can also be described as a state with a high density of Higgs particles (just as a state with an oscillating electromagnetic field can be described as a state with a high density of photons). The Higgs particles, however, would be unstable: they would rapidly decay to lighter particles, which would interact with one another and possibly undergo subsequent decays. The system would quickly become a hot gas of elementary particles in thermal equilibrium, just as was postulated in the initial conditions for the standard model. The reheating temperature is calculable and is typically a factor of between two and ten below the critical temperature of the phase transition. From this point on, the scenario coincides with that of the standard big-bang model, and so all the successes of the standard model are retained.

Note that the crucial flaw of the original inflationary model is deftly avoided. As in the original inflationary scenario, the phase transition is driven by random processes – in this case it is the random fluctuations of the Higgs fields. In the new inflationary model, however, the transition occurs gradually, so that significant inflation can take place after the phase transition is already well underway, and after the random choices have been made. Thus each homogeneous region inflates in the course of its evolution toward the true vacuum, producing a vast region of uniformity within which the observable universe can easily fit.

Since the reheating temperature is near the critical temperature of the grand unified theory phase transition, the matter–antimatter asymmetry could be produced by particle interactions just after the phase transition. The production mechanism is the same as the one predicted by grand unified theories for the standard big-bang model. In contrast to the standard model, however, the inflationary model does not allow the possibility of assuming the observed net baryon number of the universe as an initial condition; the subsequent inflation would dilute any initial baryon-number density to an imperceptible level. Thus the validity of the inflationary model depends crucially on the viability of particle theories, such as the grand unified theories, in which baryon number is not conserved.

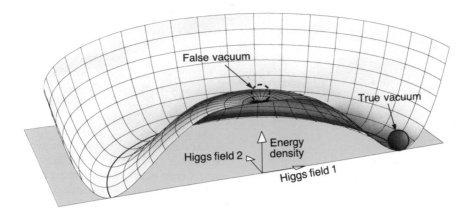

Figure 3.8. Higgs fields in the new inflationary universe. The energy density of a pair of Higgs fields is shown for the special form of the energy density that is required for the new inflationary universe model. The evolution of the Higgs fields in the early universe can be traced by imagining the analogy of a ball rolling on the surface. The ball's distance from the central axis represents the combined values of the Higgs fields, and its height above the horizontal surface represents the energy density of the universe. When the Higgs fields both have a value of zero, the ball is poised at the axis of symmetry; this is the false vacuum state. When the Higgs fields have a value that corresponds to the lowest possible energy density, the ball is lying somewhere in the circular trough that defines the broken-symmetry, or true-vacuum, states. This diagram differs from figure 3.5 in that the false vacuum is not surrounded by an energy barrier, but is instead at the top of a rather flat plateau. Under these circumstances the transition from the false vacuum to the broken-symmetry phase occurs by means of a slow-rollover mechanism: the Higgs fields are pushed from the initial value of zero by thermal or quantum fluctuations, and they proceed toward their true-vacuum values much as a ball would roll off a plateau of the same shape. The accelerated expansion of the universe takes place during the early stages of the rollover, while the energy density remains roughly constant. A single homogeneous region of broken-symmetry phase could then grow large enough to encompass the entire observable universe. When the Higgs fields reach the bottom of the trough, they would oscillate about the circle of true vacuum values, causing a reheating of the universe.

One can now grasp the solutions to the cosmological problems discussed above. The horizon and flatness problems are resolved by the same mechanisms as in the original inflationary universe model. In the new inflationary scenario the problem of monopoles and domain walls can also be solved. Such defects would form along the boundaries separating homogeneous regions, but these regions would have been inflated to such an enormous size that the defects would typically lie far beyond any observable distance. (A few defects might be generated by thermal effects after the transition, but they are expected to be negligible in number.)

The evolution of the universe in the new inflationary model, as compared with the standard model, is illustrated schematically in figure 3.9.

Thus with a few simple ideas the improved inflationary model of the universe leads to a successful resolution of several major problems that plague the standard big-bang picture: the horizon, flatness, magnetic-monopole and domain-wall problems. Unfortunately the necessary slow-rollover transition requires the fine tuning of parameters; i.e. calculations yield reasonable predictions only if the parameters are assigned values in a narrow range. Most theorists (including both of us) regard such fine tuning as implausible. The consequences of the scenario are so successful, however, that we are encouraged to go on in the hope that we may discover realistic theories in which such a slow-rollover transition occurs without fine tuning.

The successes already discussed offer persuasive evidence in favour of the new inflationary model. Moreover, it was subsequently discovered that the model may also resolve an additional cosmological problem not even considered at the time the model was developed: the smoothness problem. The generation of density inhomogeneities in the new inflationary universe was addressed in the summer of 1982 at the Nuffield Workshop on the Very Early Universe by a number of theorists, including James M. Bardeen of the University of Washington, Stephen W. Hawking of Cambridge University, So-Young Pi of Boston University, Michael S. Turner of the University of Chicago, A. A. Starobinsky of the L. D. Landau Institute of Theoretical Physics in Moscow, and the two of us. It was found that the new inflationary model, unlike any previous cosmological model, leads to a definite prediction for the spectrum of inhomogeneities. Basically the process of inflation first smoothes out any primordial inhomogeneities that might have been present in the initial conditions. Then in the course of the phase transition inhomogeneities are generated by the quantum fluctuations of the Higgs field in a way that is completely determined by the underlying physics. The inhomogeneities are created on a very small scale of length, where quantum phenomena are important, and they are then enlarged to an astronomical scale by the process of inflation.

The predicted shape for the spectrum of inhomogeneities is essentially scale-invariant; that is, the magnitude of the inhomogeneities is approximately equal on all length scales of astrophysical significance. While the precise shape of the spectrum depends on the details of the underlying grand unified theory, the approximate scale-invariance holds in almost all cases. It turns out that a scale-invariant spectrum was proposed in the early 1970s as a phenomenological model for galaxy formation by Edward R. Harrison at the University of Massachusetts at Amherst and Yakov B. Zel'dovich of the Institute of Physical Problems in Moscow, working independently. The details of galaxy formation are complex and still not well understood, but for years many cosmologists have thought that a scale-invariant spectrum of inhomogeneities is precisely what is needed to explain how the present structure of galaxies and galactic clusters evolved.

The new inflationary model also predicts the magnitude of the density inhomogeneities, but the prediction depends sensitively on the details of the underlying particle theory. The magnitude that results from the simplest grand unified theory is far too large to be consistent with the observed uniformity of the cosmic microwave background. It should be pointed out, though, that the same grand unified theory also predicts a lifetime for the proton that is inconsistent with the limit set by recent experiments. These problems are surmountable. For example, one can construct more complicated grand unified theories that result in density inhomogeneities of the desired magnitude (as well as a value for the proton lifetime that is consistent with experiment). In these models the field which drives the inflation is generally not one of the usual Higgs fields, which seem to lead unavoidably to density inhomogeneities which are much too large. Instead one postulates the existence of another scalar field with an energy-density function similar to that of the Higgs fields, but which interacts much more weakly with the other fields in the theory.

Thus, many investigators believe that, with the development of the correct particle theory, the new inflationary model would lead inevitably to a correct prediction for the density inhomogeneities. The most recent developments in elementary particle physics and in the theory of galaxy formation, however, suggest that it will be some time before we will know if these beliefs are correct.

Particle physicists are now actively considering a radically new kind of particle theory, known as a 'superstring theory', that is supposed to provide the fundamental quantum theory of *all* the elementary particle forces, including gravity. Superstring theories represent a dramatic departure from conventional particle theories in that they are *not* field theories – particles correspond to quantum excitations of ultramicroscopic strings (length $\simeq 10^{-35}$ m) instead of quantum excitations of fields. Furthermore, according to the theory, the universe has nine spatial dimensions. Early in the history of the universe, when the temperature cooled below 10^{32} K, all spatial dimensions but the three we know today stopped expanding and remained curled up with an unobservably small radius. As bizarre as the theory may sound, the superstring theory has been shown to possess a number of unique properties crucial to a unified quantum theory of particle forces and, as a result, has totally captured the attention of many of the leading theorists.

Because superstring theories are not field theories, very little is known at present about their predictions for temperatures below 10^{32} K (when the inflationary transition is supposed to have taken place). However, it is generally believed that, once the three familiar spatial dimensions have become large compared to the extra dimensions, the superstring theories closely approximate quantum field theories with a kind of symmetry called 'supersymmetry' (whence the 'super' in superstring). Supersymmetry relates the properties of particles with integer angular momentum to those of particles with half-integer angular momentum; it thereby highly constrains the form of the theory. A tantalising property of models incorporating supersymmetry is that many of them contain fields of the type needed to drive inflation, and they frequently give slow-rollover phase transitions with little or no tuning of parameters. Thus, it will be very interesting to follow the developments in superstring theories to see if these very special and highly constrained particle physics models eventually give rise to an inflationary phase transition and to the correct magnitude for the density inhomogeneities.

At the same time, many issues related to the understanding of galaxy formation have been raised. One issue concerns the mass density of the universe: while inflationary cosmology predicts a critical mass density ($\Omega = 1$), it is known that ordinary matter accounts for much less than this value. Here the term 'ordinary matter' refers to matter composed of protons, neutrons and electrons, of which the first two form the dominant component of the mass. Visible forms of ordinary matter, mainly in stars, account for less than 1% of the critical density. Although significant amounts of ordinary matter might be hidden in other forms that are difficult to observe (e.g. intergalactic dust, rocks or brown dwarfs), the successful predictions of nucleosynthesis require that the total mass in baryons (i.e. protons and neutrons) is less than 15% of the critical density. Therefore, if inflationary cosmology is correct, the mass density of the universe must be dominated by some form of dark (i.e. invisible) nonbaryonic matter (see Chapter 6 by Longair). This proposal may seem radical, but recent developments in astrophysics and particle physics tend to make it acceptable. Observationally, there is now clear evidence that a galaxy is surrounded by a large 'halo' of dark matter. The existence of these halos is inferred by observing the velocity of matter on the outskirts of the galaxy. Assuming that the matter is moving in gravitationally stable orbits, one finds that the observed velocity distribution implies that the total mass of the halo is much larger than the visible mass of the galaxy.

(a)

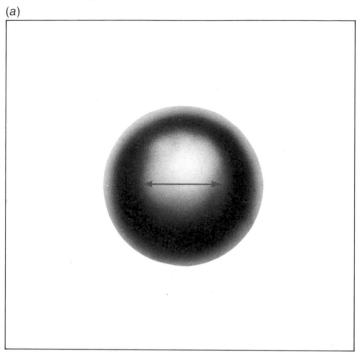

(b)

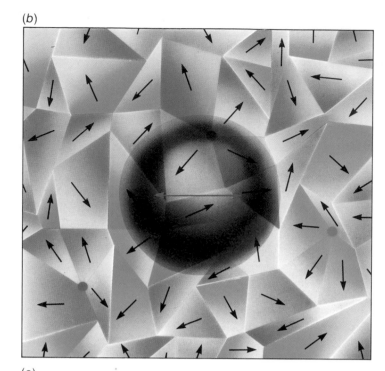

(d)

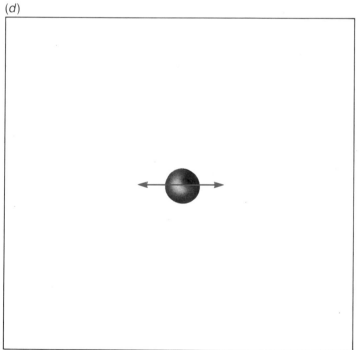

(e)

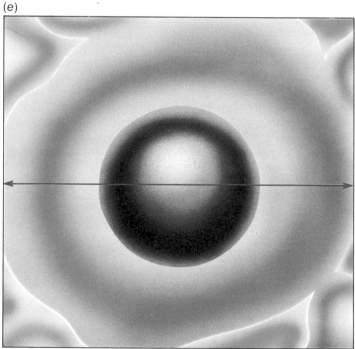

Figure 3.9. Cosmic evolution: standard model vs. new inflationary model. The new inflationary model deftly evades the horizon, magnetic-monopole and domain-wall problems. In the two series of drawings, representing the standard big-bang model (*a–c*) and the new inflationary model (*d–f*), the grey sphere corresponds to the region of space that evolved to become the observed universe and the two-headed green arrow represents the horizon distance. (The relative scales shown here are suggestive only; the actual scales differ by factors that are too extreme to depict.) Three evolutionary stages are shown for each scenario; just before the phase transition (*a* and *d*), just after the phase transition (*b* and *e*) and today (*c* and *f*). In the standard model, the horizon distance is always smaller than the grey sphere, making the large-scale uniformity of the observed universe puzzling. Since in the standard model a domain of broken-symmetry phase created in the phase transition would have a radius comparable to the horizon distance, many monopoles and domain walls would be present in the observed universe. In the new inflationary model the horizon distance is

(c)

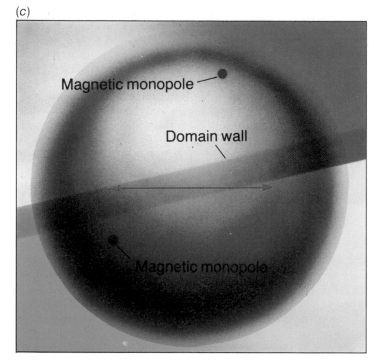

Magnetic monopole

Domain wall

Magnetic monopole

(f)

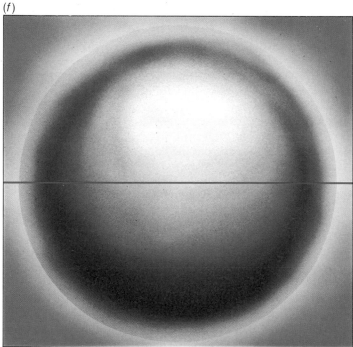

always much larger than the grey sphere, and so the observed universe is expected to be uniform on a large scale and to have few, if any, monopoles and domain walls. Just before the phase transition the grey sphere in the inflationary model is much smaller than it is in the standard model; during the phase transition the grey sphere in the inflationary model expands by a factor of 10^{50} or more in radius to match the size of the corresponding sphere in the standard model.

Extrapolating the mass in halos to the scale of the entire universe, astrophysicists conclude that the dark matter in halos adds up to at least 10% of the critical density. While this lower bound is consistent with the halo being some invisible form of baryons, the existence of halos opens the possibility that there are large numbers of weakly interacting, nonbaryonic, invisible particles occupying the halos. These new particles could well dominate the universe and account for the missing mass necessary for the universe to have a critical density. The plausibility of this line of argument is enhanced by the fact that many particle theories predict the existence of weakly interacting massive particles that could account for a large dark matter contribution to the mass density. Thus, although a critical mass density has certainly not been found, it appears that the possibility is by no means excluded.

The presence of dark matter is critical to the galaxy formation issue because inhomogeneities in the density of dark matter may be important in the galaxy formation process. Primordial inhomogeneities can develop into galaxies as the matter clumps under the influence of gravity, but for ordinary matter this process cannot begin until several hundred thousand years after the big bang. (At earlier times the matter was ionised by the intense heat, and the ionised matter interacted so strongly with the high pressure gas of hot photons that it was prevented from clumping.) Given the smoothness of the early universe as indicated by the uniformity of the microwave background radiation, it is difficult to understand how ordinary matter could have clumped fast enough to account for the distribution of galaxies observed today. Weakly interacting dark matter, on the other hand, would begin to clump at a much earlier time, so inhomogeneities in its density may have grown into significant clumps by the time that the ordinary matter ceased to be ionised. The ordinary matter would then be rapidly attracted to the clumps of dark matter, in time to form the galaxies we observe. Although there is continuing debate as to the precise nature of the dark matter, the dark matter picture has emerged as one of the leading approaches to explaining galaxy formation at the present time.

While cosmologists have investigated the evolution of inhomogeneities created by inflation, a competing model has arisen in which the density inhomogeneities are provided by 'cosmic strings' (not related to superstrings). According to the cosmic string picture, the inhomogeneities that lead to galaxy formation were provided by linelike defects, known as 'strings', that can be produced in a transition from a symmetric to a broken-symmetry phase. As their name suggests, strings are very thin, spaghetti-like objects that can form infinite curves or closed loops of astrophysical size. In the simplest theories a string can never end, but in some theories a string can terminate at a magnetic monopole. Unlike monopoles or domain walls, these defects can form loops, contract, and transform their energy into radiation fast enough so that they do not come to dominate the energy density of the universe and thereby destroy the successful predictions of hot big-bang cosmology. At the same time, they can remain long enough to provide the density inhomogeneities that are needed to seed the formation of galaxies.

Cosmic strings occur only in some particle theories, and there is still much uncertainty about the detailed predictions of the cosmic string picture. There are nevertheless some preliminary indications that the cosmic string picture will lead to successful predictions for the distribution of galaxies, and so the model is presently a subject of active research among cosmologists.

While cosmic strings may provide the inhomogeneities needed for galaxy formation, the cosmic string picture is not intended to be a replacement for inflationary cosmology. The cosmic string picture requires some additional mechanism to smooth and flatten the universe and to suppress magnetic monopoles, and an inflationary epoch is still the only known approach for accomplishing these feats. If the cosmic string picture is correct, however, the inflationary transition must be adjusted to yield inhomogeneities of an even smaller magnitude so that they will not compete with those formed by cosmic strings. Furthermore, the cosmic strings must form after inflation, so that the density of strings is not diluted to a negligible value.

Thus, the inflationary universe model is an economical theory that accounts for many features of our observable universe which had no explanation in the standard hot big-bang model. It is very attractive, on the basis of simplicity, to assume that the inflationary phase transition even accounts for the primordial inhomogeneities necessary for galaxy production. However, even if cosmic strings or some other excitations are found to play this role, some sort of inflationary epoch is still necessary to set the stage.

3.6   The origin and very large scale structure of the universe

The beauty of the inflationary model is that the evolution of the universe becomes almost independent of the details of the initial conditions, about which little if anything is known. It follows, however, that if the inflationary model is correct, it will be difficult for anyone to ever discover observable consequences of the conditions existing before the inflationary phase transition. Similarly, the vast distance scales created by inflation would make it essentially impossible to observe the structure of the universe as a whole. Nevertheless, one can still discuss these issues, and a number of remarkable scenarios seem possible.

The simplest possibility for the very early universe is that it actually began with a big bang, expanded rather uniformly until it cooled to the critical temperature of the phase transition, and then proceeded according to the inflationary scenario. Extrapolating the big-bang model back to zero time brings the universe to a cosmological singularity, a condition of infinite temperature and density in which the known laws of physics must certainly break down. The instant of creation remains unexplained. In this picture, however, the plausibility of the inflationary phase transition can be questioned. In order to produce density inhomogeneities of sufficiently small magnitude, the field responsible for the inflationary transition must be very weakly interacting. As a result, the interactions would be too weak to bring the field into thermal equilibrium prior to the time of the expected phase transition. It has been suggested that gravitational forces may have been much stronger prior to the inflationary transition and may have driven the field into thermal equilibrium, but there is at present no well-defined theory to support this speculation. It has also been suggested that thermal equilibrium is not necessary, but instead there is a wide class of initial conditions for which the field will undergo what is effectively a slow-rollover transition. In any case, some sort of a 'pre-inflationary scenario' appears to be required.

A second possibility is that the universe began (again without explanation) in a random, chaotic state. The matter and temperature distributions would be nonuniform, with some parts expanding and other parts contracting. In this scenario, certain small regions in which the conditions happened to be appropriate would undergo inflation, evolving into huge regions easily capable of encompassing the observable universe. Outside these regions the chaos would remain. The chaos would gradually creep into the regions that had inflated, but only on a time scale many orders of magnitude larger than the current age of the universe.

One version of this approach is the 'chaotic inflation' model developed by Linde and his collaborators. In this model, the chaotic conditions in the early universe produce large fluctuations in the values of the Higgs fields. In contrast to the forms of inflation discussed above, in this model inflation takes place in those regions where the initial value of one or more Higgs fields is very *large*. These large values of the Higgs fields correspond to points high up on the energy density surface shown in figure 3.8, far from the flat plateau region associated with the slow Higgs field evolution described above. In fact, an important feature of this model is the lack of any need for a flat plateau in the energy density curve. Although the surface is very steep for large values of the Higgs fields, the large values of the energy density cause the inflation to be very rapid. Linde has shown that, for sufficiently large values of the Higgs fields, the evolution of the fields back to the energy density minimum will be slow enough to produce substantial inflation. In fact, under these circumstances quantum fluctuations in the Higgs fields would be so large that there would always be regions in which the value of the Higgs fields would remain large. As the Higgs fields relaxed to the energy density minimum in various randomly chosen regions, a complicated network of mini-universes would be produced. This process can continue forever, leading to what Linde calls the 'eternally existing self-reproducing chaotic inflationary universe'. While this model provides somewhat greater freedom in the choice of the energy density curve, it still requires a scalar field which interacts much more weakly than the Higgs fields which appear in grand unified theories. Furthermore, some physicists question the plausibility of the initial conditions that the model assumes. The concept of a chaotic initial condition is somewhat vague, so it is difficult to assess the probability of finding a sufficiently large region in which the scalar field has a large and sufficiently uniform value.

Recently, there has been some serious speculation that the actual creation of the universe is describable by physical laws. In this view the universe would originate as a quantum fluctuation, starting from absolutely nothing. The idea was first proposed by Edward P. Tryon of Hunter College of the City University of New York in 1973, and it was put forward again in the context of the inflationary model by Alexander Vilenkin of Tufts University in 1982. In this context 'nothing' might refer to empty space, but Vilenkin uses it to describe a state devoid of space, time and matter. In a similar vein, James B. Hartle of the University of California at Santa Barbara, working with Stephen Hawking, has recently proposed a theory in which the universe exists in a quantum state that is uniquely determined by the fundamental laws of physics, bypassing the issue of initial conditions. Calculations in this model show that the universe would spontaneously enter an era of inflation, so that successes of the inflationary model can be incorporated. Of course, a quantum description of the structure of space-time can be discussed only in the context of quantum gravity, so these ideas must be considered highly speculative until a working theory of quantum gravity is formulated (see the chapter by Hawking). Nevertheless, it is fascinating to contemplate that physical laws may determine not only how a given state evolves in time, but they may also remove the need for assumptions about the initial conditions for our universe.

As for the structure of the universe as a whole, the inflationary model allows for several possibilities, depending on the details of the Higgs field energy diagram. (In all cases the observable universe is a very small fraction of the universe as a whole; it is likely that we are living in a homogeneous region with a size of 10^{35} light years or more.) One possibility is that the homogeneous regions meet one another and form sharply defined domains which fill all space. The domains are then separated by domain walls, and in the interior of each wall is the symmetric phase of the grand unified theory. Protons or neutrons passing through such a wall would decay instantly. Domain walls would tend to straighten with time. After 10^{35} years or more smaller domains (possibly even our own) would disappear and larger domains would grow.

Alternatively, some versions of grand unified theories do not allow for the formation of sharp domain walls. In these theories it is possible for different broken-symmetry states to merge smoothly into each other. If one moved over distances comparable to the correlation length, there would be gradual (and imperceptible) variations in the density and velocity of matter, and one would also find an occasional magnetic monopole.

A quite different possibility would result if the energy density of the Higgs fields were described by a curve such as the one in figure 3.10. As in the other two cases, regions of space would supercool into the false-vacuum state and undergo accelerated expansion. As in the original inflationary model, the false-vacuum state would decay by the mechanism of random bubble formation: quantum fluctuations would cause at least one of the Higgs fields in a small region of space to tunnel through the energy barrier, to the value marked A in figure 3.10. In contrast

Figure 3.10. A variant of the new inflationary universe. In a variant of the new inflationary model the false vacuum is surrounded by a small energy barrier. As in the original inflationary model, the false vacuum decays by the random formation of bubbles, created by the tunneling of the Higgs fields through the energy barrier. Because the energy barrier is small in this case, the Higgs fields tunnel only as far as the circle labelled A. Since the slope is quite flat at A, the Higgs fields evolve very slowly toward their true-vacuum values. The accelerated expansion of the universe continues as long as the Higgs fields remain near A, and a single bubble could grow large enough to encompass the observable universe.

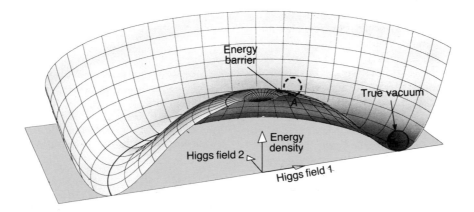

to the original inflationary scenario, the Higgs field would then evolve very slowly (because of the flatness of the curve near A) to its true-vacuum value. The accelerated expansion would continue, and the single bubble would become large enough to encompass the observed universe. If the rate of bubble formation were low, bubble collisions would be rare. The fraction of space filled with bubbles would become closer to one as the system evolved, but space would be expanding so fast that the volume remaining in the false-vacuum state would increase with time. Bubble universes would continue to form forever, and there would be no way of knowing how much time had elapsed before our bubble was formed. This picture is much like the old steady-state cosmological model on the very large scale, and yet the interior of each bubble would evolve according to the big-bang model, improved by inflation.

3.7 Conserved quantities in the universe

From a historical point of view probably the most revolutionary aspect of the inflationary model is the notion that all the matter and energy in the observable universe may have emerged from almost nothing. This claim stands in marked contrast to centuries of scientific tradition in which it was believed that nothing can come from nothing. The tradition, dating back at least as far as the Greek philosopher Parmenides in the fifth century BC, has manifested itself in modern times in the formulation of a number of conservation laws – laws which state that certain physical quantities cannot be changed by any physical process. A decade or so ago the list of quantities thought to be conserved included energy, linear momentum, angular momentum, electric charge and baryon number.

Since the observed universe apparently has a huge baryon number and a huge energy, the idea of creation from nothing has seemed totally untenable to all but a few theorists. (The other conservation laws mentioned above present no such problems: the total electric charge and the angular momentum of the observed universe have values consistent with zero, and the total linear momentum depends on the velocity of the observer and so cannot be defined in absolute terms.) With the advent of grand unified theories, however, it now appears quite plausible that baryon number is not conserved. Hence only the conservation of energy requires further discussion.

The total energy of any system can be divided into a gravitational part and a nongravitational part. The gravitational part (that is, the energy of the gravitation field itself) is negligible under laboratory conditions, but cosmologically it can be quite important. The nongravitational part is not by itself conserved; in the standard big-bang model it decreases drastically as the early universe expands, and the rate of energy loss is proportional to the pressure of the hot gas. During the era of inflation, on the other hand, the region of interest is filled with a false vacuum that has a large negative pressure. In this case the nongravitational energy increases drastically. Essen-

tially all the nongravitational energy of the universe is created as the false vacuum undergoes its accelerated expansion. This energy is released when the phase transition takes place, and it eventually evolves to become everything that we see, including the stars, the planets, and even ourselves. Thus, the inflationary model offers what is apparently the first plausible scientific explanation for the creation of essentially all the matter and energy in the observable universe.

Under these circumstances the gravitational part of the energy is somewhat ill-defined, but crudely speaking one can say that the gravitational energy is negative, and that it precisely cancels the nongravitational energy. The total energy is then zero and is consistent with the evolution of the universe from nothing.

If grand unified theories are correct in their prediction that baryon number is not conserved, there is no known conservation law that prevents the observed universe from evolving out of nothing. The inflationary universe model provides a possible mechanism by which the observed universe could have evolved from an infinitesimal region. It is then tempting to go one step further and speculate that the entire universe evolved from literally nothing. The recent developments in cosmology strongly suggest that the universe may be the ultimate free lunch.

Box 3.1 The critical energy density

Under the usual assumption that the value of Einstein's cosmological constant is zero, the critical energy density is defined to be that energy density which would be just sufficient to eventually halt the expansion of the universe. The value of this critical density is therefore determined by the expansion rate of the universe. Here we will derive the value in the context of Newtonian mechanics, but the answer that we will find will agree exactly with the answer implied by Einstein's general relativity.

According to Hubble's law, each distant galaxy is receding from us with a velocity proportional to its distance. If we let v denote the recession velocity and r denote the distance to the galaxy, then Hubble's law can be written as

$$v = Hr, \qquad (3.1)$$

where the expansion rate of the universe is characterised by the quantity H, known as Hubble's constant. (H is called a 'constant' because it does not vary with position – however, its value does vary as the universe evolves.)

For the purpose of this derivation, we assume that the universe has a perfectly uniform density of matter, with each particle moving according to equation (3.1). We consider the future trajectory of a test particle of mass m which is now separated from us by a distance r. To find the force on the particle, we can divide the mass in the universe into spherical shells centred on us, as shown in figure 3.11.

The particle lies outside those shells with radii less than r. According to Newtonian mechanics these shells will produce a gravitational attraction on the test particle which is equal to the

Box 3.1 continued

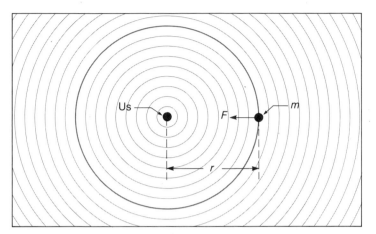

Figure 3.11.

attraction that would be produced by a point mass of magnitude M, located at the centre, where M is the total mass within the radius r. The force on the test particle due to these shells is therefore directed towards us, with a magnitude given by

$$F = \frac{GMm}{r^2},$$

(3.2)

where G is Newton's constant. M can be related to the mass density ρ by

$$M = \frac{4\pi}{3}r^3\rho.$$

(3.3)

Newtonian mechanics implies that a spherical shell of matter produces no net force on a particle inside the shell, so the shells with radii greater than r will make no contribution to the force on the test particle.

As the model universe undergoes uniform expansion, the mass density ρ and the distance to the test particle r will both vary. The quantity M, however, which represents the total mass of the matter whose distance from us is less than that of the test particle, will not vary. Thus, equation (3.2) implies that the **trajectory of the test particle will be the same as if it were moving in the gravitational field of a fixed point mass M at our location.**

The motion of a test particle in the field of a fixed point mass can be understood by considering the gravitational potential energy, given by

$$E_{\text{pot}} = -\frac{GMm}{r}.$$

(3.4)

The total energy is then obtained by adding this to the kinetic energy, giving

$$E_{\text{tot}} = \tfrac{1}{2}mv^2 - \frac{GMm}{r}.$$

(3.5)

The total energy is conserved, and at infinite values of r it is purely kinetic, since the potential term vanishes in this limit. Thus the test particle can escape to infinite distance only if E_{tot} is positive, which means that

$$v > \left(\frac{2GM}{r}\right)^{\frac{1}{2}}.$$

(3.6)

(One could also use equation (3.6) to calculate the escape velocity from the surface of a planet, in which case one would set M equal to the mass of the planet, and r equal to its radius.) Using equations (3.1) and (3.3), equation (3.5) can be rewritten as

$$E_{\text{tot}} = \tfrac{1}{2}mr^2\left[H^2 - \frac{8\pi}{3}G\rho\right].$$

(3.7)

The test particle can escape to infinite distance only when this quantity is positive, which occurs when $\rho < \rho_c$, with

$$\rho_c = \frac{3H^2}{8\pi G}.$$

(3.8)

The quantity ρ_c is therefore the critical mass density. According to special relativity the critical energy density is obtained by multiplying by c^2, the square of the speed of light.

Numerically, the value of H today is believed to lie between 15 and 30 km s^{-1} per million light-years, which upon conversion of units becomes a number in the range of $(1.5–3) \times 10^{-18}$ s^{-1}. Using $G = 6.67 \times 10^{-11}$ N m^2 kg^{-2}, one finds a critical mass density in the range of $(4–16) \times 10^{-27}$ kg m^{-3}. This corresponds to only about $2\tfrac{1}{2}$ to 10 hydrogen atoms per cubic metre. For comparison, an extremely good vacuum of 10^{-9} N m^{-2} at 300 K contains about 2×10^{11} molecules per cubic metre. The mean mass density of the universe is within an order of magnitude of the critical density, and is therefore extraordinarily low.

If Einstein's cosmological constant had a nonzero value, then the calculation would have to be modified. The cosmological constant is denoted by the capital Greek letter lambda (Λ), and it is bounded empirically by

$$|\Lambda| < 3 \times 10^{-52} \text{ m}^{-2}.$$

(3.9)

The effect on this problem is to modify equation (3.4) for the potential energy of the test particle to read

$$E_{\text{pot}} = -\frac{GMm}{r} - \tfrac{1}{6}\Lambda mc^2 r^2.$$

(3.10)

The additional term corresponds to an outward force on the test particle, independent of the mass M at the centre. The potential energy function (3.10) is strongly negative at small r and at large r, and has a maximum at some intermediate value of r. The test particle will escape to infinity (and the universe will expand forever) only if its energy is high enough to allow it to cross the maximum of the potential energy function.

The reader has probably noticed that the above derivation has a distinctly geocentric tone: our position is the centre of the picture, all velocities are defined relative to us, and all forces are calculated by considering spherical shells centred on our position. Since the geocentric viewpoint has been strongly disfavoured since the work of Copernicus, it is important to note that this viewpoint was used only for convenience of discussion – it is not an intrinsic feature of the mathematical model. It can be shown in fact that all the particles in this calculation are equivalent. If an observer following any particle were to measure all positions, velocities and accelerations relative to himself, he would construct a picture centred on himself which would be identical to the one that we drew, centred on ourselves.

Box 3.2   Evolution of a Newtonian universe

The standard big-bang model pictures the universe as a homogeneous distribution of mass which is undergoing uniform expansion. Uniform expansion can be described in terms of a hypothetical three-dimensional map of the universe. The map could be constructed, for example, as a block of rigid transparent lucite; each object in the universe (e.g. stars, dust clouds or galaxies) could be represented by embedding a small coloured ball in the lucite. Uniform expansion means that the map remains accurate for all time, except for the fact that the scale must be increased as the universe expands. At one time a centimetre on the map might correspond to a million light-years, and at a later time a centimetre might correspond to one and a half million light-years. Let s denote the distance on the map between two arbitrary coloured balls, and let $r(t)$ denote the distance, measured in the real universe, between the objects which the two balls represent. Note that $r(t)$ grows with time t, while s is time-independent. These two quantities are related by

$$r(t) = R(t)s, \qquad (3.11)$$

where $R(t)$ is the scale factor of the map. In this box we will derive the Newtonian evolution equation for this scale factor, using the equation of energy conservation which was given in box 3.1.

Hubble's law follows as an immediate consequence of the assumption of uniform expansion, as can be seen by considering the rate of change of each side of equation (3.11). Using $\dot{r}$ to denote the rate of change of r, and using a similar notation for other quantities, one has

$$\dot{r}(t) = \dot{R}(t)s = \left[\frac{\dot{R}(t)}{R(t)}\right][R(t)s] = \left[\frac{\dot{R}(t)}{R(t)}\right]r(t). \qquad (3.12)$$

This equation agrees exactly with Hubble's law, as expressed in equation (3.1), with the (time-dependent) Hubble 'constant' identified as

$$H(t) = \frac{\dot{R}(t)}{R(t)}. \qquad (3.13)$$

Now consider again a test particle of mass m which is presently separated from us by a distance r, as described in box 3.1. The total energy of such a particle is given by equation (3.7). Using equation (3.11) to eliminate r, the total energy can be re-expressed as

$$E_{\text{tot}} = \tfrac{1}{2}ms^2R^2(t)\left[H^2(t) - \frac{8\pi}{3}G\rho(t)\right]. \qquad (3.14)$$

It is both useful and conventional to define a parameter k by

$$k \equiv \frac{2E_{\text{tot}}}{mc^2s^2}. \qquad (3.15)$$

Since none of the quantities on the right-hand side depend on time, neither does k. Using equation (3.14) one has

$$k = \frac{1}{c^2}R^2(t)\left[\frac{8\pi}{3}G\rho(t) - H^2(t)\right], \qquad (3.16)$$

which shows that k has the same value for all test particles, independent of their mass or position.

If $k > 0$ then $E_{\text{tot}} < 0$, implying that the universe will ultimately recollapse under the influence of gravity; $k < 0$, on the other hand, implies that the universe will expand forever. The numerical value of k, however, does not convey any information about the universe. This can be seen from the definition (3.15): the quantities on the right-hand side depend only on the properties of the universe and/or the test particle, except that s depends on the scale of the three-dimensional map. By changing either the physical size of the map or equivalently the size of the units with which it is calibrated, the magnitude of k can be rescaled by an arbitrary factor, although the sign is unchanged. It is conventional to use this freedom to restrict the possible values of k to be either $+1$, -1, or 0. The scale of the map is then fixed for $k = \pm 1$, but remains arbitrary for the case $k = 0$.

The equation that governs the time dependence of $R(t)$ can now be obtained from equation (3.16), using equation (3.13) to eliminate the Hubble constant:

$$\left[\frac{\dot{R}(t)}{R(t)}\right]^2 = \frac{8\pi}{3}G\rho(t) - \frac{kc^2}{R^2(t)}. \qquad (3.17)$$

To solve equation (3.17) one needs the relation between the time dependence of $\rho(t)$ and that of $R(t)$, which can be found by using equations (3.3) and (3.11):

$$\rho(t) = \frac{3M}{4\pi s^3 R^3(t)} \propto \frac{1}{R^3(t)}. \qquad (3.18)$$

Equations (3.17) and (3.18) completely determine the time dependence of $R(t)$. The simplest case is $k = 0$, for which the equation takes the form

$$\dot{R}^2(t) = \frac{\text{constant}}{R(t)}. \qquad (3.19)$$

The reader with a basic knowledge of calculus can verify that this equation is solved by

$$R(t) \propto t^{2/3}. \qquad (3.20)$$

The constant of proportionality in the above equation has no physical relevance, since it depends on the scale of the three-dimensional map. As discussed above, this scale remains arbitrary if $k = 0$. The Hubble constant is determined by equation (3.13), which gives.

$$H(t) = \frac{2}{3t}. \qquad (3.21)$$

The evolution for the cases $k = \pm 1$ is given by a more complicated set of formulas, and is shown graphically in figure 3.4.

The dynamics of the universe would be modified if Einstein's cosmological constant had a nonzero value, in which case equation (3.10) would replace equation (3.4). Keeping the definition (3.15), one finds that the result (3.17) must be modified to become

$$\left[\frac{\dot{R}(t)}{R(t)}\right]^2 = \frac{8\pi}{3}G\rho(t) + \tfrac{1}{3}\Lambda c^2 - \frac{kc^2}{R^2(t)}. \tag{3.22}$$

The form of this equation can be simplified by defining

$$\rho_{\mathrm{vac}} \equiv \frac{\Lambda c^2}{8\pi G}. \tag{3.23}$$

and

$$\rho_{\mathrm{eff}}(t) \equiv \rho(t) + \rho_{\mathrm{vac}}. \tag{3.24}$$

With this notation equation (3.22) can be written as

$$\left[\frac{\dot{R}(t)}{R(t)}\right]^2 = \frac{8\pi}{3}G\rho_{\mathrm{eff}}(t) - \frac{kc^2}{R^2(t)}. \tag{3.25}$$

which has a form identical to that of equation (3.17). Thus the term ρ_{vac} enters this equation exactly as if it were a contribution to the mass density. The existence of a cosmological constant is therefore interpreted as a fixed mass density of the vacuum given by ρ_{vac}. The bound on Λ given in equation (3.9) implies that $|\rho_{\mathrm{vac}}| < 1.6 \times 10^{-26} \mathrm{kg\, m^{-3}}$.

Although the derivation presented here is based on Newtonian mechanics, the results agree exactly with those obtained from general relativity under the assumption that the pressure of the matter in the universe is negligible (i.e. pressure $\ll \rho c^2$). It is believed that this assumption has been valid for our universe since about 10 000 years after the big bang.

Box 3.3   **Pressure of the false vacuum**

The energy density of the false vacuum can be attributed entirely to the Higgs fields. Furthermore, as can be seen in figure 3.5, any deviation of the Higgs fields from their false-vacuum values would require an increase in energy density. This fact implies that the false vacuum has a peculiar property which distinguishes it from all familiar materials: when the volume of a region of false vacuum is increased, the energy density cannot decrease, but instead remains at a constant value. The pressure of the false vacuum can be deduced from this property. (Note that we are considering changes of volume that happen so quickly that the probability of bubble nucleation is negligible.)

Imagine a hypothetical situation in which a cylindrical chamber is filled with false vacuum. Although the enormous energy requirement makes this situation totally unrealistic, we can still use it for a 'thought experiment'. Suppose that the chamber is separated by a movable piston from an exterior region of true vacuum, which has zero energy density and zero pressure. If the piston is moved outward a distance Δl, then the volume of the chamber (and hence the volume of the false-vacuum region) is increased by

$$\Delta V = A\Delta l, \tag{3.26}$$

where A is the cross sectional area of the piston (see figure 3.12). The energy density of the false vacuum remains fixed at a value which we denote by u_{f}, so the energy of the system is increased by

$$\Delta E = u_{\mathrm{f}}\Delta V. \tag{3.27}$$

Since the energy has increased, it follows that work must have been done in moving the piston – the 'agent' which moved the piston must have been pulling against a force. Since the outside pressure is zero, the pressure p_{f} of the false vacuum must be negative. The outward force that the agent must apply to the piston to balance the inward force due to the pressure difference is then given by

$$F = -p_{\mathrm{f}}A. \tag{3.28}$$

The energy supplied by the agent is equal to the force times the distance, so

$$\Delta E = F\Delta l = -p_{\mathrm{f}}\Delta V. \tag{3.29}$$

By comparing equations (3.27) and (3.29), one sees that

$$p_{\mathrm{f}} = -u_{\mathrm{f}}. \tag{3.30}$$

Thus the pressure of the false vacuum is negative, with a magnitude which is equal to the energy density.

A typical number for the energy density (or pressure) can be estimated without discussion of details by using the characteristic energy scale of the grand unified theories, $E_{\mathrm{GUT}} \simeq 10^{14}\,\mathrm{GeV} \simeq 10^4\,\mathrm{J}$. Allowing E_{GUT} and the fundamental constants $c = 3.00 \times 10^8\,\mathrm{m\,s^{-1}}$ (speed of light) and $\hbar = 1.05 \times 10^{-34}\,\mathrm{J\,s}$ (Planck's constant), a quantity with the units of energy density can be constructed as

$$u_{\mathrm{f}} \simeq \frac{E_{\mathrm{GUT}}^4}{(\hbar c)^3} \simeq 10^{92}\,\mathrm{J\,m^{-3}}. \tag{3.31}$$

Note that $1\,\mathrm{J\,m^{-3}} = 1\,\mathrm{N\,m^{-2}}$, so that energy density and pressure can be measured in the same units.

Figure 3.12.

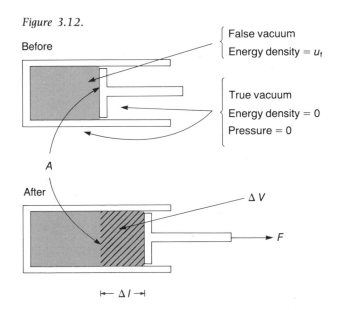

Before

False vacuum
Energy density = u_{f}

True vacuum
Energy density = 0
Pressure = 0

A

After

ΔV

F

$\vdash \Delta l \dashv$

Box 3.4   The deceleration of the universe

It is useful, for both conceptual and calculational purposes, to derive a formula for $\ddot{R}$, which describes the deceleration of the cosmic expansion.

In the context of Newtonian mechanics, the desired equation can be obtained directly from equations (3.17) and (3.18), using some elementary calculus. First, equation (3.18) is differentiated to obtain

$$\dot{\rho}(t) = -3\frac{\dot{R}(t)}{R(t)}\rho(t). \tag{3.32}$$

One then differentiates equation (3.17), after first multiplying both sides by $R^2(t)$. Using equation (3.32) to eliminate $\dot{\rho}(t)$, one finds

$$\ddot{R}(t) = -\frac{4\pi}{3}G\rho(t). \tag{3.33}$$

The rate of slowing down of the cosmic expansion is proportional to the mass density, as expected.

The standard relativistic description of cosmology is formulated in the context of general relativity, which is mathematically too complex to present here. It can be shown, however, that equation (3.17) is valid in general relativity as well as in Newtonian mechanics, even when the pressure is not negligible. Equation (3.18), on the other hand, requires modification – but these modifications can be derived by using energy conservation and the relativistic mass–energy relationship, without recourse to the complete formalism of general relativity.

In a relativistic theory the energy density $u(t)$ is related to the mass density $\rho(t)$ by

$$u(t) = \rho(t)c^2. \tag{3.34}$$

As the universe expands, the variation of $u(t)$ can be found by accounting for the total energy contained within a region corresponding to a fixed volume V_{map} of the three-dimensional map introduced in box 3.2. The corresponding volume of the physical universe is given by

$$V_{\text{phys}}(t) = R^3(t)V_{\text{map}}, \tag{3.35}$$

and the total energy in this region is given by

$$E(t) = u(t)V_{\text{phys}}(t) = u(t)V_{\text{map}}R^3(t). \tag{3.36}$$

Remembering that V_{map} is time independent, one can differentiate the above equation to obtain

$$\dot{E}(t) = \dot{u}(t)V_{\text{phys}}(t) + 3\frac{\dot{R}(t)}{R(t)}u(t)V_{\text{phys}}(t). \tag{3.37}$$

As the universe expands, the energy E is not expected to remain constant. The region contains a gas which is expanding under pressure, and such a gas does work against the external environment. In this case the energy is converted into gravitational potential energy. When the volume increases by an infinitesimal amount ΔV, then the energy remaining in the gas decreases by an amount $p\Delta V$, where p denotes the pressure. (This result was derived for a special geometry in equation (3.29), but it can be shown to hold for an arbitrary geometry.) Therefore

$$\dot{E}(t) = -p(t)\dot{V}_{\text{phys}}(t) = -3\frac{\dot{R}(t)}{R(t)}p(t)V_{\text{phys}}(t), \tag{3.38}$$

where equation (3.35) was used to obtain the second equality. Combining equations (3.37) and (3.38), one finds

$$\dot{u}(t) = -3\frac{\dot{R}(t)}{R(t)}[u(t) + p(t)]. \tag{3.39}$$

The calculation of $\ddot{R}(t)$ can now be carried out as before, except that equation (3.39) is used instead of equation (3.32), and equation (3.34) is used to relate the mass density to the energy density. The result is

$$\ddot{R}(t) = -\frac{4\pi G}{3c^2}[u(t) + 3p(t)]R(t). \tag{3.40}$$

In the present universe the second term in square brackets, $3p(t)$, is a very small relativistic correction to the first term, $u(t)$. During the first 10 000 years of the history of the universe, however, when the energy density was dominated by radiation resulting from the intense heat of the big bang, this term was equal to the first. During the inflationary era, on the other hand, with the pressure equal to the negative of the energy density, this term would dominate over the first and result in a change of sign for $\ddot{R}(t)$: the expansion of the universe would be accelerated.

If the derivation of equation (3.40) is generalised to allow for the possibility of a nonzero cosmological constant, one finds

$$\ddot{R}(t) = -\frac{4\pi G}{3c^2}[u(t) + 3p(t)]R(t) + \tfrac{1}{3}\Lambda c^2 R(t). \tag{3.41}$$

It was pointed out in box 3.2 that the cosmological constant can be interpreted as a vacuum mass density given by $\Lambda c^2/8\pi G$, which is equivalent to a vacuum energy density

$$u_{\text{vac}} = \rho_{\text{vac}}c^2 = \frac{\Lambda c^4}{8\pi G}. \tag{3.42}$$

Equation (3.41) can be simplified by defining

$$p_{\text{vac}} \equiv -u_{\text{vac}} = -\frac{\Lambda c^4}{8\pi G}, \tag{3.43}$$

which leads to

$$\ddot{R}(t) = -\frac{4\pi G}{3c^2}[u_{\text{eff}}(t) + 3p_{\text{eff}}(t)]R(t), \tag{3.44}$$

where

$$u_{\text{eff}}(t) = u(t) + u_{\text{vac}},$$
$$p_{\text{eff}}(t) = p(t) + p_{\text{vac}}. \tag{3.45}$$

Since p_{vac} enters the equations exactly as if it were a contribution to the pressure, it is interpreted as the pressure of the vacuum. Thus the true vacuum, like the false vacuum, has a pressure which is equal to the negative of the energy density. In the case of the true vacuum, however, it is possible that both quantities are zero.

Box 3.5 Exponential expansion in the inflationary era

According to general relativity, the evolution of the cosmic scale factor $R(t)$ is governed by the equation

$$\ddot{R}(t) = -\frac{4\pi G}{3c^2}[u(t) + 3p(t)]R(t), \tag{3.46}$$

where $u(t)$ denotes the energy density and $p(t)$ denotes the pressure. A derivation of this equation was given in box 3.4.

As was discussed in box 3.3, the false vacuum is a state of fixed energy density u_f, with a pressure given by

$$p = -u_f. \tag{3.47}$$

Thus, for a universe filled with false vacuum equation (3.46) becomes

$$\ddot{R}(t) = \frac{8\pi G}{3c^2}u_f R(t). \tag{3.48}$$

This equation has the solution

$$R(t) \propto e^{t/\tau}, \tag{3.49}$$

where

$$\tau = \left(\frac{3c^2}{8\pi G u_f}\right)^{\frac{1}{2}}. \tag{3.50}$$

Taking equation (3.31) as an estimate of u_f, one finds that the expansion time constant τ is given numerically by

$$\tau \simeq 10^{-33} \text{ s}. \tag{3.51}$$

The most general solution to equation (3.48) contains also a decaying exponential term ($e^{-t/\tau}$) with the same time constant τ, but this term rapidly becomes negligible. Thus, the cosmic scale factor $R(t)$ rapidly approaches the behaviour of a growing exponential function. The behaviour of the Hubble constant during the inflationary era is then determined by equation (3.13), which yields

$$H(t) = \frac{1}{\tau}. \tag{3.52}$$

Box 3.6 Solution to the flatness problem

The flatness problem is avoided in the inflationary model because the ratio Ω is driven rapidly toward one during the era of inflation, regardless of its previous value. This can be seen by combining equations (3.13) and (3.17) to obtain

$$H^2(t) = \frac{8\pi}{3}G\rho(t) - \frac{kc^2}{R^2(t)}. \tag{3.53}$$

Using equation (3.8) for the critical mass density, the above equation can be rewritten as

$$\Omega(t) = 1 + \frac{kc^2}{R^2(t)H^2(t)}. \tag{3.54}$$

The second term represents the deviation of Ω from one. During an era of Newtonian evolution in a $k = 0$ universe, for example, equations (3.20) and (3.21) tell us that $R(t) \propto t^{2/3}$, and $H(t) \propto t^{-1}$. Equation (3.54) then implies that the deviation of Ω from one grows as $t^{2/3}$, exemplifying the instability which forms the basis of the flatness problem. During the inflationary era, on the other hand, we found in box 3.5 that $R(t) \propto e^{t/\tau}$, and $H(t) \simeq 1/\tau$. For this case equation (3.54) implies that the deviation of Ω from one decreases as $e^{-2t/\tau}$. So if $R(t)$ increases by a factor of 10^{50}, as in our example, then the deviation of Ω from one is suppressed during the inflationary era by a factor of 10^{100}.

If the cosmological constant were nonzero, then the prediction of inflation would have to be restated. In this context the relation $\rho_c = 3H^2/8\pi G$ is conventionally taken as the definition of ρ_c, in spite of the fact that this value of the mass density has no special significance when the cosmological constant is nonzero. Using this definition and taking equation (3.22) to replace equation (3.17), one finds that

$$\Omega(t) + \frac{\Lambda c^2}{3H^2(t)} = 1 + \frac{kc^2}{R^2(t)H^2(t)}. \tag{3.55}$$

During the inflationary era the term kc^2/R^2H^2 is enormously suppressed, so the model predicts that the relation

$$\Omega(t) + \frac{\Lambda c^2}{3H^2(t)} = 1 \tag{3.56}$$

holds to a high degree of accuracy.

Using equation (3.23) for the mass density of the vacuum, one sees that the second term on the left-hand side can be interpreted as the vacuum contribution to Ω:

$$\Omega(t) + \frac{\rho_{vac}}{\rho_c} = 1. \tag{3.57}$$

Acknowledgements

A.H.G. was supported in part through funds provided by the US Department of Energy (DOE) under contract DOE-AC02-76ERO3069, and in part by the National Aeronautics and Space Administration (NASA) under grant NAGW-553. P.J.S. was supported in part through funds provided by the US Department of Energy (DOE) under contract DOE-AC02-76ERO3071.

Further reading

References for general background
Silk, J. (1980). *The Big Bang*. W. H. Freeman and Co., NY.
Weinberg, S. (1977). *The First Three Minutes*. Basic Books, NY.

Semipopular review of inflation
Tryon, E. P. (1987). Cosmic inflation. In *The Encyclopedia of Physical Science and Technology*, vol. 3, pp. 709–43. Academic Press, NY.

Technical reviews of inflation
Blau, S. K. and Guth, A. H. (1987). Inflationary cosmology. In *300 Years of Gravitation* (S. W. Hawking and W. Israel, eds.), pp. 524–603. Cambridge University Press.
Brandenberger, R. H. (1985). *Rev. Mod. Phys.* 57, 1.
Linde, A. D. (1984). *Rep. Prog. Phys.* 47, 925.
Steinhardt, P. J. (1986). Inflationary cosmology. In *High Energy Physics, 1985. Proceedings of the Yale Theoretical Advanced Study Institute* (M. J. Bowick and F. Gürsey, eds.), vol. 2, p. 567. World Scientific, Singapore.
Turner, M. S. (1987). Cosmology and particle physics. In *Architecture of Fundamental Interactions at Short Distances* (P. Ramond and R. Stora, eds.), p. 513. North Holland, Amsterdam.

Technical books
Abbott, L. F. and So-Young Pi, eds. (1986). *Inflationary Cosmology.* World Scientific, Singapore.
Kolb, E. W., Turner, M. S., Lindley, D., Olive, K. and Seckel, D., eds. (1986). *Inner Space/Outer Space: The Interface between Cosmology and Particle Physics.* Chicago University Press.

4 The edge of spacetime

Stephen Hawking

4.1 Early cosmologies

From the dawn of civilisation people have asked questions like: 'Did the universe have a beginning in time? Will it have an end? Is the universe bounded or infinite in spatial extent?' In this article I shall outline some of the answers to these questions which are suggested by modern developments in science. Most of what I describe is now fairly generally accepted, though some of it was controversial. My final conclusion, however, is based on some very recent work on which there has not yet been time to reach a consensus.

In most of the early mythological or religious accounts the universe, or at least its human inhabitants, was created by a divine Being at some date in the fairly recent past like 4044 BC. Indeed the necessity of a 'first cause' to account for the creation of the universe was used as an argument to prove the existence of God. The Greek philosophers like Plato and Aristotle, on the other hand, did not like the thought of such direct divine intervention in the affairs of the world and so mostly preferred to believe that the universe had existed and would exist forever.

Most people in the ancient world believed that the universe was spatially bounded. In the earliest cosmologies the world was a flat plate with the sky as a pudding basin overhead. However, the Greeks realised that the world was round. They constructed an elaborate model in which the Earth was a sphere surrounded by a number of spheres which carried the Sun, the Moon and the planets. The outermost sphere carried the so-called fixed stars, which maintain the same relative positions but which appear to rotate across the sky.

This model with the Earth at the centre was adopted by the Christian Church. It had the great attraction that it left plenty of room outside the sphere of the stars for Heaven and Hell, though quite how these were situated was never clear. It remained in favour until the seventeenth century when the observations of Galileo showed that this model of the universe had to be replaced by the Copernican model in which the Earth and the other planets orbited around the Sun. Not only did this get rid of the spheres but it also showed that the 'fixed stars' must be at a very great distance because they did not show any apparent movement as the Earth went round the Sun, apart from that caused by the rotation of the Earth about its own axis. Having realised this, and having abandoned the belief that the Earth was at the centre of the universe, it was fairly natural to postulate that the stars were other suns like our own and that they were distributed roughly uniformly throughout an infinite universe.

This, however, raised a problem: according to Newton's theory of gravity, published in 1687, each star would be attracted towards every other star in the universe. Why then did the stars not all fall together to a single point? Newton himself tried to argue that this would indeed happen for a bounded collection of stars but that, if one had an infinite universe, the gravitational force on a star caused by the attraction of the stars on one side of it would be balanced by the force arising from the stars on the other side. The net force on any star would therefore be zero and so the stars could remain motionless. This argument is in fact an example of the fallacies one can fall into when one adds up an infinite number of quantities: by adding them up in different orders one can get different results. We now know that an infinite distribution of stars cannot remain motionless if they are all attracting each other; they will start to fall towards each other. The only way that one can have a static infinite universe is if the force of gravity becomes repulsive at large distances. Even then, the universe is unstable because, if the stars get slightly nearer each other, the attraction wins out over the repulsion and the stars fall together. On the other hand, if they get slightly further away from each other, the repulsion wins and they move away from each other.

There was another difficulty with the idea of an infinite unchanging universe, first noticed by the Swiss astronomer J. L. de Cheseaux and later elaborated by H. Olbers. If space is populated uniformly with stars of the same average brightness, then a simple calculation leads to the result that the total flux of starlight arriving at any given point in the universe is infinite! The reason is apparent from figure 4.2. Imagine a sequence of concentric spherical shells of equal thickness drawn around

Figure 4.1. The Ptolemaic system from Apian's *Cosmographia*
(1553). Alfonso X (1221–84), King of Castile and Leon, is said to
have remarked, after the Ptolemaic system had been explained to
him, 'If the Lord Almighty had consulted me before embarking
upon Creation, I should have recommended something simpler.'

some location, such as Earth. Because of the uniformity
assumption, the number of stars in each shell will be proportional to the volume of the shell. For a shell of radius r, the
volume is in turn proportional to r^2. On the other hand, the
inverse square law of light flux tells us that the intensity of light
from a shell that is at distance r from Earth is proportional to
$1/r^2$. The factors of r^2 thus cancel, implying that the total flux of
light at Earth due to the stars in a given shell of space is
independent of the radius of the shell. Adding together the light
from all the shells in a spatially infinite universe therefore leads
to an infinite light flux.

A more refined calculation takes into account the fact that
the nearby stars tend to block the light from the far away ones,
and instead of infinity one gets the result that the intensity of
radiation at the surface of the Earth should be comparable to
that at the surface of the Sun. Instead of the familiar dark night
sky, the heavens should be ablaze like a furnace. Clearly there is
something wrong with the static model universe, and the
problem of why the sky is dark at night became known as
Olbers' paradox.

Despite these and other difficulties, nearly everyone in the
eighteenth and nineteenth centuries believed that the universe
was essentially unchanging in time. For such a universe the
question of whether it had a beginning was metaphysical; one
could equally well believe that it had existed forever or that it
had been created in its present form a finite time ago. The belief
in a static universe still persisted in 1915 when Einstein
formulated his general theory of relativity (see Chapter 2 by
Will) which modified Newton's theory of gravity to make it
compatible with discoveries about the propagation of light. He

therefore added a so-called 'cosmological constant', which produced a repulsive force between particles at a great distance. This repulsive force could balance the normal gravitational attraction and allow a static uniform solution for the universe.

It turned out that this balancing act was unstable, and could not yield a satisfactory cosmological model. Einstein's model did, however, possess an additional interesting feature that had enduring value. This concerns its spatial geometry. A fundamental feature of Einstein's general theory of relativity is that space (strictly spacetime) can be curved. Indeed, gravitation manifests itself precisely as a distortion or curvature in the geometry of spacetime. Einstein's model universe serves as a good example of space curvature.

In flat space (i.e. when gravitation is absent) the usual rules of school geometry apply. In particular, the volume of a sphere of radius r is $\frac{4}{3}\pi r^3$. In the Einstein model universe, the averaged gravitation of all the stars causes the volume of space enclosed by a sphere of radius r to be *less* than $\frac{4}{3}\pi r^3$. A convenient way of depicting this is to use a two-dimensional analogy for three-dimensional space. The two-dimensional surface of a ball has a property similar to that of the Einstein universe. Figure 4.3 shows concentric circles drawn on the ball. In this case the *area* enclosed by the circles corresponds to the *volume* enclosed by the spherical surfaces in three dimensions. It is easy to believe by looking at the figure that the area enclosed by a circle of radius r is less than πr^2, the formula we all remember from school geometry. The reason is that the surface of the ball is curved.

The uniform inward curvature of the ball is important not only for the geometry of its surface, but also for its *topology*. Topology is that branch of mathematics which deals with the way lines, surfaces etc. connect together. The precise geometry is disregarded. Thus a topologist recognises that a doughnut is different from a potato, because the latter has no hole in it, but no distinction is made between a doughnut and a teacup (see figure 4.4).

Clearly the ball has a different topology from an infinite flat sheet – which would be the two-dimensional analogue of an infinite universe without gravitation. The surface of the ball has the property that it is finite, yet unbounded. What this means is that a flat creature living on the surface of the ball could visit every point on the surface, but nowhere encounter an edge or boundary. We might say that the curvature of the ball causes the surface to 'close up on itself'.

Something very similar prevails in the Einstein universe. Its volume is finite, and an observer could in principle explore all of space, yet there is no edge or boundary. Space curves in on itself and 'closes up' to form a three-dimensional analogue of the spherical surface. This arrangement is not easy to visualise, nor is it necessary to try. Mathematically the idea of a closed three-dimensional spherical (strictly hyperspherical) space makes perfect sense, and Einstein believed that the universe was indeed spatially finite. Time, however, was still infinite in this model.

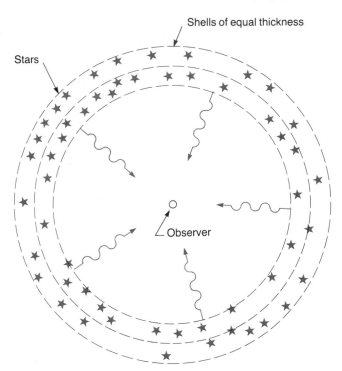

Figure 4.2. If the universe is unchanging with time and uniformly populated to infinity with stars of the same average brightness, then the flux of light received by the observer is the same from each equal-thickness shell. This leads to the conclusion that the night sky should be ablaze with light.

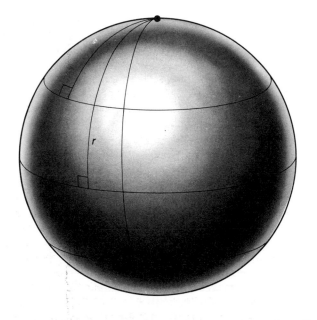

Figure 4.3. If concentric circles are drawn on the surface of a sphere, then the area enclosed is less than πr^2. Similarly the circumference is less than $2\pi r$.

4.2   Modern cosmology

Einstein's static model of the universe was one of the great missed opportunities of theoretical physics; if he had stuck to his original version of general relativity without the cosmological constant he could have predicted that the universe ought to be either expanding or collapsing. As it happened, however, it was not realised that the universe was changing with time until astronomers like Slipher and Hubble began to observe the light from other galaxies. Visible light is made up of waves, like radio waves only with a much shorter wavelength or distance between wave crests. If one passes the light through a prism it is decomposed into its constituent wavelengths or colours like a rainbow. Slipher and Hubble found the same characteristic patterns of wavelengths or colours as for the light from stars in our own galaxy but the patterns were all shifted towards the red, or longer, wavelength end of the spectrum. The only reasonable explanation of this was that the galaxies are moving away from us. In this case the distance between the wave crests would be increased. Similarly, if one observed light from a source that was moving towards us, the wave crests would be crowded up and the wavelength would be reduced. This effect, known as the Doppler shift, is used by the police to measure the speed of cars.

During the 1920s Hubble observed the remarkable fact that the red shift was greater the further the galaxy was from our own. This meant that other galaxies were moving away from us at rates that were roughly proportional to their distance from us (see figure 4.5). The universe was not static as had been previously thought, but was expanding. The rate of expansion is very low; it will take something like twenty-thousand-million years for the separation of two galaxies to double, but it completely changes the nature of the discussion about whether the universe has a beginning or an end. This is not just a metaphysical question as in the case of a static universe; as I shall describe, there may be a very real physical beginning or end to the universe.

The first realistic model of an expanding universe that was consistent with Einstein's general theory of relativity and Hubble's observations of red-shifts was proposed by the Russian physicist and mathematician Alexander Friedmann, in 1922. However, it received very little attention until similar models were discovered by other people towards the end of the 1920s. The Friedmann model and its later generalisations assumed that the universe is the same at every point in space and in every direction. This is obviously not a good approximation in our immediate neighbourhood; there are local irregularities like the Earth and the Sun and there are many more visible stars in the direction of the centre of our galaxy than in other directions. However, if we look at distant galaxies, we find they are distributed roughly uniformly throughout the universe, the same in every direction. Thus, it does seem to be a good approximation on a large scale.

The best evidence for large scale uniformity comes from

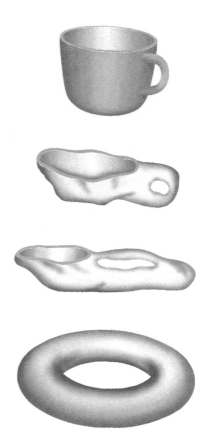

Figure 4.4. From the topologist's point of view a teacup is indistinguishable from a doughnut – both contain one hole. Thus, it is possible to continuously deform the teacup shape into the doughnut shape, without having to tear or join at any stage.

observations of the background of microwave radiation that was discovered in 1965 by Arno Penzias and Robert Wilson at the Bell Telephone Laboratories (see figure 4.6). The universe is very transparent to radio waves of a few centimetres wavelength so this radiation must have travelled to us from very great distances. Any large scale irregularities in the universe would cause the radiation reaching us from different directions to have different intensities. Yet the observed intensity is the same in every direction to a very high degree of accuracy (about one part in 10^4).

There are three kinds of generalised Friedmann models (see figure 4.7). In one of them the galaxies are moving apart sufficiently slowly that the gravitational attraction between them will eventually stop them moving apart and start them approaching each other. The universe will expand to a maximum size and then recollapse. In the second model, the galaxies are moving apart so fast that gravity can never stop them and the universe expands forever. Finally, there is a third model in which the galaxies are moving apart at just the critical rate to avoid recollapse. In principle we could determine which model corresponds to our universe by comparing the present rate of expansion with the present average mass density. The mass of the matter in the universe that we can observe directly is not enough to stop the expansion. However, we have indirect evidence that there is more mass that we cannot see; this is sometimes called the dark matter problem (see Chapter 6 by Longair). Whether this 'invisible' mass could be enough to stop the expansion eventually remains an open question.

4.3 The Big Bang singularity

In the Friedmann model which recollapses eventually, space is finite but unbounded, like in the Einstein static model. In the other two Friedmann models, which expand forever, space is infinite. Time, on the other hand, has a boundary or edge in all three models. The expansion starts from a state of infinite density called the Big Bang singularity. The best way of visualising the Big Bang is to imagine a movie of the expanding universe being played backwards in time. A given spherical region of the universe then shrinks, more and more rapidly, until its radius reaches zero. At this juncture, all the matter and energy that was contained in that spherical volume of space will be compressed into a single point, or singularity. In this idealised model, the entire observable universe is considered to have started out compressed into such a point. Moreover, in the model which recollapses, the universe eventually returns to a singularity at the end – the so-called Big Crunch.

If one takes the singularities in these idealised models seriously, then some profound conclusions follow. Because of the infinite compression of matter and energy, the curvature of

Figure 4.5. A representative velocity–distance curve showing the expansion of the universe. The lower left corner is the region surveyed by Hubble up to 1929.

Figure 4.6. Arno Penzias and Robert Wilson, shown with the equipment that accidentally detected the heat radiation from the Big Bang in 1965. Penzias and Wilson received the Nobel Prize for their important discovery in 1978.

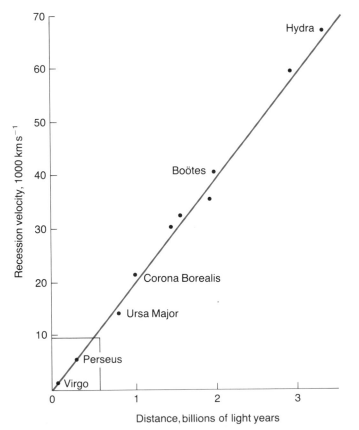

spacetime is infinite at the Friedmann singularities too. Under these circumstances the concepts of space and time cease to have any meaning. Moreover, because all present scientific theories are formulated on a spacetime background, all such theories will break down at these singularities. So, if there were events before the Big Bang, they would not enable one to predict the present state of the universe because predictability would break down at the Big Bang. Similarly, there is no way that one can determine what happened before the Big Bang from a knowledge of events after the Big Bang. This means that the existence or non-existence of events before the Big Bang is purely metaphysical; they have no consequences for the present state of the universe. One might as well apply the principle of economy, known as Occam's razor, to cut them out of the theory and say that time began at the Big Bang. Similarly, there is no way that we can predict or influence any events after the Big Crunch, so one might as well regard it as the end of time.

This beginning and possible end of time that are predicted by the Friedmann solutions are very different from earlier ideas. Prior to the Friedmann solutions, the beginning or end of time was something that had to be imposed from outside the universe; there was no necessity for a beginning or an end. In the Friedmann models, on the other hand, the beginning and end of time occur for dynamical reasons. One could still imagine the universe being created by an external agent in a state corresponding to some time after the Big Bang, but it would not have any meaning to say that it was created *before* the Big Bang. From the present rate of expansion of the universe we can estimate that the Big Bang should have occurred between ten- and twenty-thousand-million years ago.

Many people disliked the idea that time had a beginning or will have an end because it smacked of divine intervention. There were therefore a number of attempts to avoid this conclusion. One of these was the 'steady state' model of the universe proposed in 1948 by Herman Bondi, Thomas Gold and Fred Hoyle. In this model it was suggested that, as the galaxies moved further away from each other, new galaxies were formed in between out of matter that was being 'continually created'. The universe would therefore look more or less the same at all times and the density would be roughly constant. This model had the great virtue that it made definite predictions that could be tested by observations. Unfortunately, observations of radio sources by Martin Ryle and his collaborators at Cambridge in the 1950s and early 1960s showed that the number of radio sources must have been greater in the past, contradicting the steady state model.

The final nail in the coffin of the steady state model came with the discovery of the microwave background radiation by Penzias and Wilson, which appears to bathe the entire cosmos uniformly. Measurements of the spectrum of this radiation reveal it to carry the unmistakable thumbprint of thermal equilibrium, i.e. it is a black body spectrum, corresponding to a temperature a little below 3 K (see figure 4.8). This thermal radiation has a very natural explanation in the Big Bang theory

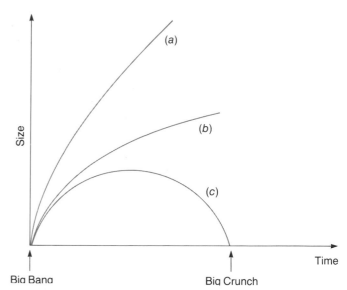

Figure 4.7. The radius of a typical region of the universe is plotted as a function of time for three distinct types of Friedmann model. In (*a*) the universe has a low density and escapes its own gravity. The rate of expansion approaches a uniform value. Curve (*b*) corresponds to the critical case in which the universe just escapes its own gravity. It continues to expand forever, but at a diminishing rate. Curve (*c*) corresponds to a high density universe. After expanding to a maximum size, it recontracts, finally obliterating itself at a Big Crunch.

– it is a relic of the primeval heat which accompanied the Big Bang. In the steady state theory, however, there was no hot dense phase in the past from which this heat radiation could issue. Its presence had no natural explanation.

Another attempt to avoid a beginning of time was the suggestion that maybe the singularity was simply a consequence of the high degree of symmetry of the Friedmann solutions. This restricted the relative motion of any two galaxies to be along the line joining them. It would therefore not be surprising if they all collided with each other at some time. However, in the real universe, the galaxies would also have some random velocities perpendicular to the line joining them. These transverse velocities might be expected to cause the galaxies to miss each other and to allow the universe to pass from a contracting phase to an expanding one without the density ever becoming infinite. Indeed, in 1963 two Russian scientists, E. Lifshitz and I. Kalatnikov, claimed that this would happen in nearly every solution of the equations of general relativity. They based this claim on the fact that all the solutions with a singularity that they constructed had to satisfy some constraint or symmetry. They later realised, however, that there was a more general class of solutions with singularities which did not have to obey any constraint or symmetry.

This showed that singularities *could* occur in general solutions of general relativity but it did not answer the question of whether they necessarily *would* occur. However, between 1965 and 1970 a number of theorems were proved which showed that any model of the universe which obeyed general relativity, satisfied one or two other reasonable assumptions and contained as much matter as we observe in the universe, must have a Big Bang singularity. The same theorems predict that there will be a singularity which will be an end of time if the whole universe recollapses. Even if the universe is expanding too fast to collapse in its entirety, we nevertheless expect some localised regions, such as massive burnt out stars, to collapse and form black holes. The theorems predict that the black holes will contain singularities which will be an end of time for anyone unfortunate or foolhardy enough to fall in.

4.4   Quantum theory to the rescue

Einstein's general theory of relativity is probably one of the two greatest intellectual achievements of the twentieth century. It is, however, incomplete because it is what is called a classical theory, that is it does not incorporate the uncertainty principle of the other great discovery of this century: quantum mechanics. The uncertainty principle states that certain pairs of quantities, such as the position and velocity of a particle, cannot be predicted simultaneously with an arbitrarily high degree of accuracy. The more accurately one predicts the position of the particle, the less accurately one will be able to predict its velocity and vice versa. Quantum mechanics was developed in the early years of this century to describe the behaviour of very small systems such as atoms or individual elementary particles. In particular there was a problem with the structure of the atom which was supposed to consist of a number of electrons orbiting around the central nucleus, like the planets around the Sun. The previous classical theory predicted that each electron would radiate light waves because of its motion. The waves would carry away energy and so would cause the electrons to spiral inwards until they collided with the nucleus. However, such behaviour is not allowed by quantum mechanics because it would violate the uncertainty principle; if an electron were to sit on the nucleus, it would have both a definite position and a definite velocity. Instead, quantum mechanics predicts that the electron does not have a definite position but that the probability of finding it is spread out over some region around the nucleus with the probability density remaining finite even at the nucleus.

The prediction of classical theory of an infinite probability density of finding the electron at the nucleus is rather similar to the prediction of classical general relativity that there should be a Big Bang singularity of infinite density. Thus one might hope that if one was able to combine general relativity and quantum mechanics into a theory of quantum gravity (see Chapter 5 by Isham) one would find that the singularities of gravitational collapse or expansion were smeared out like in the case of the collapse of the atom.

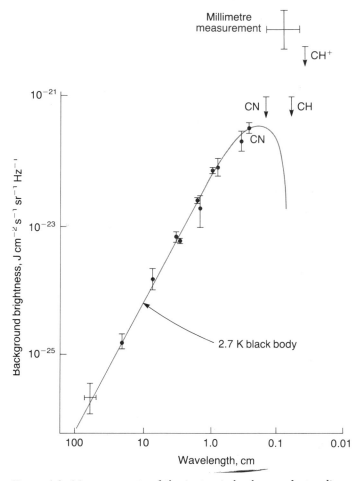

Figure 4.8. Measurements of the isotropic background at radio wavelengths, compared with a 2.7 K black body spectrum. The upper limits from interstellar molecules conflict with the millimetre measurement, which may refer to a discrete line superposed on the continuous background.

The first indication that this might be so came with the discovery that black holes, formed by the collapse of localised regions such as stars, were not completely black if one took into account the uncertainty principle of quantum mechanics. Instead, a black hole would emit particles and radiation like a hot body with a temperature which was higher the smaller the mass of the black hole. The radiation would carry away energy and so would reduce the mass of the black hole. This in turn would increase the rate of emission. It seems that, eventually, the black hole will disappear completely in a tremendous burst of radiation. All the matter that collapsed to form the black hole and any astronauts who were unlucky enough to fall into the black hole would completely disappear, at least from our region of the universe. However, the energy that corresponded to their mass by Einstein's famous equation $E = mc^2$ would survive, to be emitted by the black hole in the form of radiation. Thus the astronaut's mass–energy would be recycled to the universe. However, this would be rather a poor sort of immortality as the astronaut's subjective concept of time would almost certainly come to an end and the particles out of which he was composed would not in general be the same as the particles that were re-emitted by the black hole. Still, black hole evaporation did indicate that gravitational collapse might not lead to a complete end of time.

4.5   The problem of initial conditions

The real problem with spacetime having an edge or boundary at a singularity is that the laws of science do not determine the initial state of the universe at the singularity but only how the universe evolves thereafter. This problem would remain even if there were no singularity and time continued back indefinitely; the laws of science would not fix what the state of the universe was in the infinite past. In order to pick out one particular state of the universe from among the set of all possible states that are allowed by the laws, one has to supplement the laws by boundary conditions which say what the state of the universe was at an initial singularity or in the infinite past. Many scientists are embarrassed at talking about the boundary conditions of the universe because they feel that it verges on metaphysics or religion. After all, they might say, the universe could have started off in a completely arbitrary state. That may be so, but in that case it could also have evolved in a completely arbitrary manner. Yet all the evidence that we have suggests it evolves in a well-determined way according to certain laws. It is therefore not unreasonable to suppose that there may also be simple laws that govern the boundary conditions and determine the state of the universe.

In the classical general theory of relativity, which does not incorporate the uncertainty principle, the initial state of the universe is a point of infinite density. It is very difficult to define what the boundary conditions of the universe should be at such a singularity. However, when quantum mechanics is taken into account, there is a fresh possibility, namely that the singularity might be smeared away. The question then arises as to what shape space and time may adopt instead of the point of infinite curvature.

In investigating this point, it is necessary to take into account a curious property that quantum mechanics can bestow upon spacetime. In the theory of relativity space and time are closely linked. In fact, physicists prefer to regard space and time together as forming a four-dimensional spacetime continuum, three dimensions of space plus one of time. In spite of this intimate association, there are still physical differences between space and time. One of these refers to the measurement of the four-dimensional distance or interval between two points in spacetime. If the points have a greater separation in time than they do in space (e.g. successive moments at the same spatial location), then the square of the four-

dimensional separation is negative. By contrast, the four-dimensional separation of two points in spacetime for which the spatial separation exceeds the time separation (e.g. simultaneous events at different locations) has a positive square.

In the very early universe, when space was very compressed, the smearing effect of the uncertainty principle can change this basic distinction between space and time. It is possible for the square of the time separation to become positive under some circumstances. When this is the case, space and time lose their remaining distinction – we might say that time becomes fully spatialised – and it is then more accurate to talk, not of spacetime, but of a four-dimensional space. Calculations suggest that this state of affairs cannot be avoided when one considers the geometry of the universe during the first minute fraction of a second. The question then arises as to the geometry of the four-dimensional space which has to somehow smoothly join onto the more familiar spacetime once the quantum smearing effects subside.

One possibility is that this four-dimensional space curves around to form a closed surface, without any edge or boundary, in much the same way as the surface of a ball or the Einstein universe, but this time in *four* dimensions. In the case of the recontracting model universe, for which the three ordinary spatial dimensions are already closed into a hypersphere, this new proposal would imply that the whole of spacetime was finite and unbounded. This in turn would mean that the universe is completely self-contained and *did not require* boundary conditions. One would not have to specify the state in the infinite past and there would not be any singularities at which the laws of physics would break down. One could say that the boundary conditions of the universe are that it has no boundary.

It should be emphasised that this is simply a *proposal* for the boundary conditions of the universe. One cannot deduce them from some other principle but one can merely pick a reasonable set of boundary conditions, calculate what they predict for the present state of the universe and see if they agree with observations. The calculations are very difficult and have been carried out so far only in simple models with a high degree of symmetry. However, the results are very encouraging. They predict that the universe must have started out in a fairly smooth and uniform state. It would have undergone a period of what is called exponential or 'inflationary' expansion (see Chapter 3 by Guth and Steinhardt) during which its size would have increased by a very large factor but the density would have remained the same. The universe would then have become very hot and would have expanded to the state that we see it today, cooling as it expanded. It would be uniform and the same in every direction on very large scales but would contain local irregularities that would develop into stars and galaxies.

What happened at the beginning of the expansion of the universe? Did spacetime have an edge at the Big Bang? The answer is that, if the boundary conditions of the universe are that it has no boundary, time ceases to be well-defined in the very early universe just as the direction 'north' ceases to be well-defined at the North Pole of the Earth. Asking what happens before the Big Bang is like asking for a point one mile north of the North Pole. The quantity that we measure as time had a beginning but that does not mean spacetime has an edge, just as the surface of the Earth does not have an edge at the North Pole, or at least, so I am told; I have not been there myself.

If spacetime is indeed finite but without boundary or edge, this would have important philosophical implications. It would mean that we could describe the universe by a mathematical model which was determined completely by the laws of science alone; they would not have to be supplemented by boundary conditions. We do not yet know the precise form of the laws; at the moment we have a number of partial laws which govern the behaviour of the universe under all but the most extreme conditions. However, it seems likely that these laws are all part of some unified theory that we have yet to discover. We are making progress and there is a reasonable chance that we will discover it by the end of the century. At first sight it might appear that this would enable us to predict everything in the universe. However, our powers of prediction would be severely limited, first by the uncertainty principle, which states that certain quantities cannot be exactly predicted but only their probability distribution, and, secondly, and even more importantly, by the complexity of the equations which makes them impossible to solve in any but very simple situations. Thus we would still be a long way from omniscience.

5 Quantum gravity

Chris Isham

5.1 Introduction

Gravitational forces are a simple and rather obvious feature of daily life. It may seem surprising therefore that the construction of a quantum mechanically consistent theory of gravity remains one of the major challenges to theoretical physicists. This becomes particularly striking when it is realised that the analogous problem of electromagnetism has been tackled with considerable success. In this context, recall that the magnitude of the gravitational force between two particles with masses M and m, and separated by distance r, is

$$F_{grav} = \frac{GMm}{r^2},$$ (5.1)

where $G = 6.67 \times 10^{-11}$ Nm^2 kg^{-2} is Newton's universal gravitational constant. ('Universal' in the sense that G does not depend on the values of M, m or r.) The analogous expression for the electric force between two charges e_1 and e_2 is

$$F_{elec} = \frac{e_1 e_2}{4\pi\varepsilon_0 r^2},$$ (5.2)

where $\varepsilon_0 = 8.85 \times 10^{-12}$ F m^{-1} is the electric permittivity of free space. It should be noted that equation (5.2) has the same 'inverse square-law behaviour' as that in equation (5.1).

Quantum mechanics is usually thought to be relevant only at atomic sizes or less, and so some idea of the importance of quantum gravity might be obtained by comparing, for example, the gravitational force between the proton and the electron in a hydrogen atom with the electric force that holds the atom together. Indeed, it follows at once from equations (5.1) and (5.2) that the gravitational force is smaller by thirty-eight orders of magnitude, and so it is hard to see why there should be any problem in computing these tiny corrections and, correspondingly, why the subject should generate anything other than a passing interest. That quantum gravity in fact raises issues of the greatest mathematical and conceptual complexity rests ultimately on a remarkable feature of the simple Newtonian expression (5.1): the quantity (like M or m) that measures the ability of an object to produce (or react to) a gravitational field is the *same* as the inertial mass that appears on the right hand side of Newton's second equation of mechanical motion,

$$\text{force} = \text{acceleration} \times \text{inertial mass.}$$ (5.3)

This so-called 'equivalence principle' has the striking result that if (for the sake of simplicity) we take $M \gg m$ (so that the motion of the object with mass M can be ignored), the acceleration of the particle of mass m is given by the gravitational force

$$F_{grav} = \frac{GMm}{r^2} = m \times \text{acceleration,}$$ (5.4)

i.e.

$$\text{acceleration} = \frac{GM}{r^2}$$ (5.5)

and is thus *independent* of the value of m.

The importance and uniqueness of this result cannot be overemphasised. For example, in the analogous case of an electric force acting on an object of charge e and mass m, the acceleration is proportional to e/m, which can, and does, vary widely from one piece of matter to another. Within the framework of Newtonian gravity, the equivalence principle is a complete mystery, and its *prediction* by Einstein's general theory of relativity is rightly regarded as one of the major achievements of that modern description of gravitational phenomena. But this triumph is bought at the expense of accepting a view of the gravitational force that is radically different from that of the other classical forces of nature. As we shall discuss in Section 5.2 (see also Chapter 2 by Will), gravitational effects are regarded as arising from a *curvature* in spacetime, and it is the reconciliation of this dynamical view of spacetime with the passive role it plays in quantum theory that constitutes the primary obstruction to the creation of a satisfactory quantum theory of gravity. Indeed, so great is this disparity, and so great are the difficulties it engenders, that workers in this area have felt obliged to coin the maxim: 'What God has put asunder, let no man join together'!

It should therefore be understood that 'quantum gravity' refers to the attempts to unify, or cohere, *general relativity* and quantum theory. If gravity was nothing but the Newtonian static force of equation (5.1), the construction of a corresponding quantum theory would be a simple (and rather uninteresting) affair. However, even allowing for the complexities of Einstein's theory, it is still difficult to believe that there can be much wrong with the rough estimate made above of quantum gravitational effects at the atomic level, and it is therefore important to ask at what scale is it realistic to expect to see something of interest?

A good answer can be obtained from the simple, but never to be underrated, method of dimensional analysis. Two basic constants of nature that are presumably available in the construction of *any* quantum theory of gravity are Newton's constant G (setting the scale of gravitational forces) and Planck's constant h (setting the scale of quantum effects). A third universal constant is the speed of light c, and it is a remarkable fact that, with the aid of these three constants, we can construct a number that has the units of length:

$$L_P = \left(\frac{Gh}{c^3}\right)^{\frac{1}{2}} \simeq 10^{-35}\,\text{m}.$$

This fundamental length in nature is known as the 'Planck length', and it is at this scale that we might expect to see some effects of quantum gravity. However, before embarking on a 'do-it-yourself' kitchen sink experiment, the prudent reader should note that the sizes of an atom and nucleus are approximately 10^{-10} m and 10^{-15} m, respectively. So we are contemplating a distance that is *twenty* orders of magnitude smaller than the diameter of a proton; not perhaps the easiest regime to explore, even with the aid of the implements in a modern kitchen.

To get some idea of what *is* involved, recall that one implication of the Heisenberg uncertainty principle is that the energy required to resolve structure at a distance 'd' is inversely proportional to the value of d. Thus the energy needed to probe the Planck length is twenty orders of magnitude larger than that involved in unravelling the structure of the proton. In fact, the value is about 10^{18} GeV (L_P written in energy units) whereas the largest particle accelerators only work at around 10^3 GeV. It seems highly unlikely that a machine will ever be built with which these minute distances can be studied directly, and perhaps this is why, in the past, elementary particle physicists have tended to be rather hostile towards research in quantum gravity, regarding it as (at best) a self-indulgent hobby for general relativists who are, of course, well known 'for spending their lives with their heads permanently in the clouds'!

All this has changed dramatically in recent years, and quantum gravity has become a major branch of modern theoretical physics. One important reason for this metamorphosis is the considerable success that has been achieved in the application of quantum field theory to the strong and weak nuclear forces, and to their unification with electromagnetism. As Georgi has explained in his articles, this development of 'grand unified theories' has been accompanied by a realisation that the 'strengths' of the forces can depend on the energy at which they are probed. 'Unification' then occurs at the energy at which all the forces become of equal strength, and in the case of the strong, weak and electromagnetic forces, this happens at around 10^{15} GeV.

From the perspective of quantum gravity this is rather striking. Psychologically speaking, 10^{15} GeV is 'sufficiently close' to the Planck energy 10^{18} GeV that elementary particle physicists can now discuss the latter openly without danger of losing their membership of the club.

But there is also the genuine scientific implication that perhaps we should be looking for a unified theory of *all* of the fundamental forces, including gravity, with this unification taking place somewhere in the 10^{15}–10^{18} GeV energy range. This in turn suggests that perhaps it will be *necessary* to include gravity before a unification of the other forces can be attained. Admittedly, this is a contentious issue involving subtle questions on the precise role ascribed to quantum field theory in the description of the fundamental constituents of the physical world. But, at the very least, this new situation means that tackling quantum gravity with the methodology of elementary particle physics can be viewed with more confidence than was the case, say, fifteen years ago.

When viewed from the perspective of general relativity, the motivation for studying quantum gravity has always been clear and strong. One of the major predictions of Einstein's theory is the phenomenon of gravitational collapse in which, under a wide range of initial conditions, matter that is compressed to more than a critical density will inevitably collapse under its self-gravitational attraction until it becomes a 'point' or, more precisely, a gravitational singularity. The classical theory of general relativity cannot predict what happens beyond this stage, and this signals the collapse of the theory itself. Now 10^{-35} m may seem very small, but it is infinitely larger than 0 m (a point!), and it has long been assumed that something 'rather odd' will happen as the matter passes through the Planck length, and that the singularity will thereby be averted. Of course, the most impressive singularity of which we know is the 'big bang', and it is not surprising that workers in quantum gravity have been consistently fascinated with the idea of using 'quantum cosmology' to describe the very early universe. As discussed in Chapter 3 by Guth and Steinhardt, our understanding of the early universe has made remarkable progress recently via an analysis of the cosmological situation with ideas drawn from elementary particle physics – in particular grand unification. It would thus be most pleasing, to say the very least, if the process could be pushed back to the Planck time (defined as $L_P c \simeq 10^{-42}$ s) by incorporating quantum gravity into the unification scheme.

For many workers, the strongest motivation for studying quantum gravity has always been this expectation that something 'odd' happens at the Planck length, coupled with the belief that understanding this 'something' will involve a fundamental reappraisal of our basic concepts of the physical world, including perhaps the downfall of both general relativity and quantum theory. It must be admitted that, at both the epistemological and ontological levels, our current understanding of space and time leaves much to be desired. In a gross extrapolation from daily experience, both special and general relativity use a model for spacetime that is based on the idea of a continuum, i.e. the position of a spacetime point is uniquely specified by the values of four real numbers (the three space, and one time, coordinates in some convenient coordinate system). But the construction of a 'real' number from integers and fractions is a very abstract mathematical procedure, and there is no *a priori* reason why it should be reflected in the empirical world. Indeed, from the viewpoint of quantum theory, the idea of a spacetime point seems singularly inappropriate: by virtue of the Heisenberg uncertainty principle, an *infinite* amount of energy would be required to localise a particle at a true point; and it is therefore more than a little odd that modern quantum field theory still employs fields that are functions of such points. It has often been conjectured that the almost unavoidable mathematical problems arising in such theories (the prediction of infinite values for the probabilities of physical processes occurring, and the associated need to 'renormalise' the theory – see Chapters 15 and 16 by Georgi and Chapter 17 by Taylor) are a direct result of ignoring this internal inconsistency. Be this as it may, it is clear that quantum gravity, with its natural Planck length, raises the possibility that the continuum nature of spacetime may not hold below this length, and that a quite different model is needed.

Unfortunately, it is not easy to construct theories of physics *in vacuo*, and the absence of any definitive experimental input opens up a Pandora's box of possibilities. One common observation is that, whereas in conventional quantum theory spacetime is merely a fixed background, in general relativity it plays a more dynamic role, and perhaps this is the direction in which to look for Planck length effects. We touch here on a problem that has exercised the minds of philosophers since the dawn of Western civilisation: the dichotomy of the 'things' that make up the external world, and the 'space' and 'time' in which we encounter them. Should we follow one ancient Greek line of thought and view space and time as an *a priori* container of 'things'? Or should we reverse the ontological order and, following Leibniz, regard 'things' as the primary concept, and relegate space and time to a secondary role of expressing relations between these 'things'? Twentieth century physics has oscillated (at times uncomfortably) between these positions and, under the impact of quantum gravity studies, these oscillations threaten to become unstable!

I hope it is beginning to be clear why there is more to quantum gravity than meets the eye, and why it is such a fascinating subject. Unfortunately, it is probably also becoming clear that this is not the easiest branch of modern physics, and that it can pose singular difficulties for a would-be expositor. Any real understanding of quantum gravity requires a firm grasp of both quantum theory and general relativity, and the explication of these subjects demands full length articles in their own right. Fortunately, there are such articles in this volume, and I will restrict my preliminary remarks on these topics to highlighting those features that are of particular relevance to quantum gravity proper.

5.2   General relativity

A common starting point for discourses on general relativity is the empirically well-established 'equivalence principle' that the extent to which a piece of matter produces, or reacts to, a gravitational field is determined by its inertial mass. One consequence is the mass-independence of the acceleration of a particle in a gravitational field. Another implication was emphasised by Einstein in a famous thought experiment involving a box containing an observer who is apparently experiencing a gravitational force.

One explanation could be simply that the box and its contents are suspended over a large mass (such as the Earth) so that the gravitational force is 'genuine' (figure 5.1*a*). But it is also possible that the box is being accelerated by a rope connected to the roof, so that the force experienced by the observer is really 'nothing but' his inertial resistance to this acceleration (figure 5.1*b*). The crucial point is that, because of the equivalence principle, there is no way in which, by making observations inside the box, the observer can tell which explanation is correct.

This suggested to Einstein that any theory of gravity incorporating the equivalence principle should also include a mechanism for relating the results of measurements made by two observers who are accelerating with respect to each other, i.e. the theory should be 'covariant' with respect to such transformations of reference frame. Such a theory would necessarily go beyond the special theory of relativity in which only inertial observers (i.e. with no mutual accelerations) enjoy such a privileged status.

To understand why this involved the ideas of geometry, it is instructive to reflect on where else in mechanics mass-independent motion is encountered. One obvious case is when there is no force at all. For then Newton's second law shows that the acceleration vanishes, and the position $\mathbf{x}$ as a function of time t is of the (mass-independent) form

$$\mathbf{x} = \mathbf{b} \pm \mathbf{c}t, \qquad (5.6)$$

where $\mathbf{b}$ and $\mathbf{c}$ are constant vectors determined by the initial conditions of the particle (namely its position and velocity at $t = 0$). Now the set of points satisfying equation (5.6) (i.e. the trajectory of the particle) is a *straight* line in space – a manifestly geometrical concept. It occurred to Einstein that perhaps the

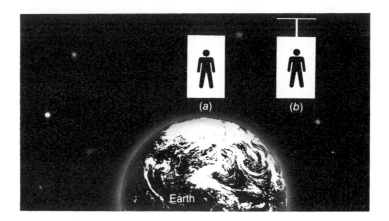

Figure 5.1. An observer in an enclosed box and experiencing an apparent gravitational field cannot tell from measurements made inside the box if this force is 'genuine' or whether it is merely an 'apparent' force coming from his inertial resistance to being accelerated via a rope attached to the outside of the box.

equivalence principle could be incorporated by postulating that, even in the presence of a gravitational field, a particle still moves on a 'straight line'. To the obvious objection that the Earth moving around the Sun is clearly *not* traversing a straight line, Einstein responded that it is, but the straight line concerned is in a *curved* space.

To appreciate the full implications of this remarkable idea, it is necessary to say a little about the way in which the curvature of a space is described. This must involve some exposure to the language of mathematics, but a reader who begins to find the exposition a little hard-going should have no reservations about skipping lightly over the remainder of this section (and much of the following one) and moving quickly to the heart of the chapter which lies in Section 5.4 onwards.

Working for the moment in two dimensions, consider the straight line drawn between two points A and B in a flat, two-dimensional space (e.g. the surface of a table), see figure 5.2. We will suppose that the coordinates of the two points are (x,y) and $(x+\delta x, y+\delta y)$, respectively, where the differences between the coordinates of the points are denoted by δx and δy since we will shortly be considering pairs of points that are 'infinitesimally' close together. Application of elementary cartesian geometry and Pythagoras's theorem, shows that the distance δs between the points A and B is related to the differences in the coordinates of the points by

$$(\delta s)^2 = (\delta x)^2 + (\delta y)^2. \qquad (5.7)$$

Now consider the analogous situation of two points on a *curved* surface which, for the sake of illustration, will be taken as the surface of a ball of radius ρ embedded in three dimensions. This 'two-dimensional' surface is thus the set of all

points (x, y, z) in three dimensions satisfying the constraint

$$x^2 + y^2 + z^2 = \rho^2 \qquad (5.8)$$

(see figure 5.3). Within any particular quadrant of the surface, the position of a point is uniquely fixed by the values of x and y (which is why the surface is said to be of dimension 2), and the value of z is determined by equation (5.8). Pythagoras's theorem still applies in the (flat) three-dimensional space, and from this perspective, the distance between the two points A and B (i.e. along the dotted line in figure 5.3) is

$$(\delta s)^2 = (\delta x)^2 + (\delta y)^2 + (\delta z)^2. \qquad (5.9)$$

For an 'infinitesimal' separation the distance between these points, viewed now as belonging to the curved surface (i.e. as measured along the curved line joining A to B), will have the same numerical value as in equation (5.9). However, since this two-dimensional surface is coordinatised with just the two numbers x and y, we must express δs in terms of the infinitesimals δx and δy alone. This can be done by noting that the infinitesimal δz that appears in equation (5.9) cannot be specified arbitrarily if both points are to lie on the surface. Rather, we must have

$$\begin{aligned} x^2 + y^2 + z^2 \\ = (x+\delta x)^2 + (y+\delta y)^2 + (z+\delta z)^2 = \rho^2. \end{aligned} \qquad (5.10)$$

Expanding the brackets and discarding the $(\delta x)^2$, $(\delta y)^2$ and $(\delta z)^2$ terms (as being 'infinitely' smaller than the terms linear in δx, δy and δz), we find

$$x\delta x + y\delta y + z\delta z = 0, \qquad (5.11)$$

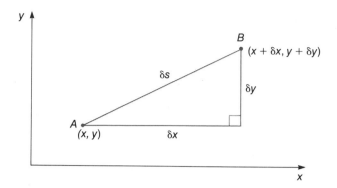

Figure 5.2. The familiar Pythagoras's theorem in a flat two-dimensional space expresses the square of the distance δs between the two points A and B as the sums of the squares of the differences δx and δy of their x and y coordinates.

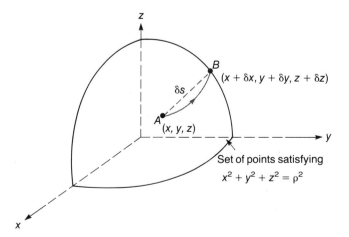

Figure 5.3. A and B are now two points on the curved two-dimensional spherical surface of a ball embedded in a flat three-dimensional space. The new 'Pythagoras' theorem for this curved space geometry is given by equation (5.12).

which can be solved for δz and then substituted in equation (5.9) to give the final expression

$$(\delta s)^2 = (\delta x)^2 + (\delta y)^2 + \frac{(x\delta x + y\delta y)^2}{(\rho^2 - x^2 - y^2)}. \qquad (5.12)$$

This is a form of Pythagoras's theorem applicable to the surface of the ball. It can be written as

$$(\delta s)^2 = g_{xx}(x,y)(\delta x)^2 + g_{yy}(x,y)(\delta y)^2 +$$
$$g_{xy}(x,y)\delta x\delta y + g_{yx}(x,y)\delta y\delta x, \qquad (5.13)$$

with

$$g_{xx}(x,y) = 1 + \frac{x^2}{(\rho^2 - x^2 - y^2)};$$

$$g_{yy}(x,y) = 1 + \frac{y^2}{(\rho^2 - x^2 - y^2)}; \qquad (5.14)$$

$$g_{xy}(x,y) = g_{yx}(x,y) = \frac{xy}{(\rho^2 - x^2 - y^2)}.$$

The three functions $g_{xx}(x,y)$, $g_{yy}(x,y)$ and $g_{xy}(x,y)$ are called the *components* of the *metric tensor*, and completely determine the curvature of the sphere. This idea generalises to give one of the major results of differential geometry: the curvature of *any* space of *any* dimension (say N) is completely reflected in the generalised Pythagoras theorem

$$(\delta s)^2 = \sum_{a,b=1}^{N} g_{ab}(x)(\delta x^a)(\delta x^b) \qquad (5.15)$$

in which $x^1, \ldots, x^N$ are a system of coordinates on this space, and the components $g_{ab}(x) = g_{ba}(x)$ can be functions of all of the coordinates. Note that:

(i) Since the metric components $g_{ab}(x)$ are *functions*, the deviation of the geometry from its flat space form

$$(\delta s)^2 = (\delta x^1)^2 + (\delta x^2)^2 + \ldots + (\delta x^N)^2 \qquad (5.16)$$

can depend on where we are in the space.

(ii) A general metric tensor has *off*-diagonal components (e.g. $g_{xy}(x,y)$ in equation 5.13). Since $g_{ab}(x) = g_{ba}(x)$, the number of independent components of this tensor is $N(N+1)/2$.

(iii) There is no unique way of selecting the coordinates that parametrise the points in the space. Different choices $x^1, \ldots, x^N$ and $y^1, \ldots, y^N$ will produce different explicit forms $g_{ab}(x)$ and $g'_{ab}(y)$ for the metric tensor. However, it is a fundamental result that these two sets are related by the transformation equation

$$g_{ab}(x) = \sum_{c,d=1}^{N} \frac{\partial y^c}{\partial x^a} \frac{\partial y^d}{\partial x^b} g'_{cd}(y(x)). \qquad (5.17)$$

Einstein's use of this mathematical machinery in his formulation of general relativity can be summarised as follows. The primary postulate of special relativity is that three-dimensional space and one-dimensional time must be taken together as a single, four-dimensional, space with the 'flat-space' Pythagoras theorem

$$(\delta s)^2 = (\delta x)^2 + (\delta y)^2 + (\delta z)^2 - (\delta t)^2, \qquad (5.18)$$

in which (x, y, z) and t are, respectively, the space and time coordinates. Note that this differs from the flat-space metric in equation (5.16) (with $N = 4$) by the presence of the minus sign in front of the $(\delta t)^2$ term. Mathematically, this corresponds to 'hyperbolic' rather than 'Euclidean' geometry, and it reflects the essential physical difference between space and time. (In this sense, the familiar remark that 'Einstein put space and time onto the same footing' is somewhat misleading.)

The central postulate of general relativity is that the gravitational field caused by the presence of matter can be described mathematically by replacing this flat 'Minkowski' metric with a general one

$$(\delta s)^2 = \sum_{a,b=1}^{4} g_{ab}(x)\delta x^a \delta x^b, \qquad (5.19)$$

in which x^1, x^2, x^3 are the spatial coordinates and x^4 points in the timelike direction. In any particular physical situation, the curvature of this metric is to be determined by solving the Einstein field equations, which are a set of ten, nonlinear second-order (hyperbolic) partial differential equations for the ten components $g_{ab}(x)$, $a,b = 1, \ldots, 4$, of the metric tensor.

A 'straight line' (usually called a geodesic) in such a curved space is defined to be a line that minimises the distance between its two endpoints, and Einstein postulated that it was on such lines that a particle experiencing the gravitational force would move. The Einstein field equations include specific contributions from both the energy and momentum of the matter and, if the theory is to be consistent, the metric in equation (5.18) must be a solution of these equations in the special case when these matter-related contributions vanish.

Fortunately this is the case, and Einstein's theory of general relativity is both internally self-consistent and aesthetically pleasing. It contains the equivalence principle via the notion of motion along geodesics and it allows for a covariant (i.e. reference frame-independent) description of all gravitational phenomena via the mathematical transformation laws in equation (5.17). It is important to note that the metric components g_{ab} in equation (5.19) depend on both spatial and time coordinates, and that, because of the latter, the gravitational field in general relativity has an intrinsically dynamical structure. Indeed, with respect to a fixed choice of time coordinate, the theory can be recast into a form in which the basic dynamical variable is the time-dependent metric tensor describing the three-dimensional physical space as it changes in time.

The experimental tests and general structure of the theory are discussed in the chapter by Will but, from the perspective of quantum gravity, two features particularly deserve emphasis.

(i) The gravitational variables $g_{ab}(x)$ are *fields* in the usual sense of being functions defined on points in spacetime. Similarly, the Einstein field equations depend only on these fields and their (finite-order) derivatives at such points. Thus the *local* structure of general relativity is broadly in line with that of other branches of physics (e.g. Maxwell's equations in electromagnetic theory), and is therefore open to the criticism discussed briefly in the Introduction. Indeed, the entire mathematical structure of general relativity rests on the notion of 'infinitesimally close' points, and is therefore particularly sensitive to any attacks on the physical validity of real numbers and the continuum.

(ii) General relativity also has *large-scale* (or *global*) features relating to the 'topology' of the spacetime manifold. Examples of two topologically inequivalent spaces are a two-sphere (discussed above) and a two-torus (the surface of an American doughnut), in the sense that, roughly speaking, no matter how much we stretch or bend the sphere, we cannot deform it into the torus without breaking the surface somewhere (thinking of it as made of some elastic material) (figure 5.4).

Einstein's field equations do not tell us anything about this global structure, and the topology of spacetime must be specified independently. On a large scale, the universe looks rather uninteresting, with a topology that is quite possibly the same as that of flat Minkowski space (i.e. it can be deformed into such; of course, the *metrics* are different – that is the whole point of general relativity). However, it has often been conjectured that one of the 'funny things happening at the Planck length' is some sort of quantum-induced change in the topology, and we shall return later to this exciting (but complicated) possibility.

Figure 5.4. No matter how much the two-dimensional sphere on the left hand side of the diagram is bent or pulled around, it cannot be smoothly deformed into the two-dimensional torus on the right hand side. This is an example of two spaces whose topologies are different.

5.3   Quantum theory

Quantum physics is a fascinating, but subtle, subject which can cause considerable difficulties on a first exposure. These stem partly from the abstract nature of the mathematics needed to describe even a simple system, and partly from the profound way in which the conceptual structure differs from that of Newtonian mechanics and common sense. In this brief exposition we shall concentrate mainly on the concept of a 'state' of the system since this illustrates the problem particularly well. However, as with the previous chapter, the mathematically shy reader should feel free to treat the text with disrespect and move quickly to the more descriptive material in Section 5.4.

In most branches of physics, the idea of a state is introduced to fulfil two distinct, but related, functions.

(i) A knowledge of the state of a system should enable us to predict the results of any appropriate measurements that may be made. The qualification 'appropriate' is necessary because what is, or is not, deemed to be measurable depends on the branch of physics under discussion. For example, if we were ambitious enough to study a gas of 10^{24} molecules from the viewpoint of precise Newtonian mechanics, then the position and velocity of each molecule would be regarded as independently measurable, and a specification of state would require giving the values of all of them. However, if the same system is analysed from the perspective of equilibrium thermodynamics, the number of relevant observables is very small (temperature, pressure, etc.) and so is the corresponding number of 'thermodynamical' states.

(ii) It is assumed above that the measurements involved are made immediately after the state has been specified. However, we would generally expect the results of measurements to change in time, and this can be incorporated into the structure by allowing the state to be time-dependent, and in fact to carry the *causal* properties of the system. In this particular context, a system is said to be 'causal' if its dynamical evolution is such that, given a precise specification of the state at some time t_0, it is possible to predict exactly what the state will be at any later time t (or what it must have been at any earlier time t to produce the given state at time t_0).

When taken together, these two properties of state enable us to pose, and answer, one of the basic questions of any branch of physics: 'If the system is prepared at some time t_0 to be in a certain state, what results will be obtained if various measurements are made on the system at a later time t?'.

A paradigmatic example of such a structure is provided by conventional Newtonian mechanics. For example, consider a particle moving in three dimensions under the influence of a force $\mathbf{F}(\mathbf{x})$, so that the equations of motion are

$$m\frac{d^2\mathbf{x}}{dt^2}(t) = \mathbf{F}(\mathbf{x}(t)), \qquad (5.20)$$

where the position of the particle is represented mathematically by the vector $\mathbf{x}$. (Notice how close the relation is between the mathematics and the physics; so close in fact that we tend to forget that mathematical three-space is only a *model* for 'physical' three-space, whatever that might be. In quantum theory the situation is different, and the relation of the mathematical structure to the physical situation is much less 'obvious'.) It might seem that the state space (i.e. the set of all states) of this system is just the set of all vectors $\mathbf{x}$, but this is not quite correct. The differential equations (5.20) are second-order in the time variable, and therefore the value of $\mathbf{x}$ at a time t is not determined uniquely by its value at time t_0; we must also specify the value of the velocity (or momentum) at that time. It follows that the causal structure can be made manifest by rewriting these equations in the first-order form

$$\frac{d\mathbf{p}}{dt}(t) = \mathbf{F}(\mathbf{x}(t)) \qquad (5.21)$$

$$\frac{d\mathbf{x}}{dt}(t) = \frac{1}{m}\mathbf{p}(t) \qquad (5.22)$$

in which the position $\mathbf{x}$ and the momentum $\mathbf{p}$ are regarded as independent variables. Substituting equation (5.22) in equation (5.21) will of course simply reproduce the second-order equations (5.20). However, considered in their own right, the pair of equations (5.21) and (5.22) are first-order in time, and so the causality requirement is satisfied in the sense that the values of $\mathbf{x}(t)$ and $\mathbf{p}(t)$ can be determined once their values at some other time t_0 are specified. Thus, from the viewpoint of causality, it is consistent to represent the states of the system by all pairs of vectors. $(\mathbf{x}, \mathbf{p})$ or, equivalently, by all vectors in a six-dimensional space whose coordinates are (x, y, z, p_x, p_y, p_z).

The first requirements on a state – that it should enable the results of any appropriate measurements to be predicted – can be met by defining 'appropriate' observables to be any functions $f(\mathbf{x}, \mathbf{p})$ of the position and momentum of the particle (e.g. angular momentum is $\mathbf{x} \wedge \mathbf{p}$) because the value of such an observable is uniquely given once the values of $\mathbf{x}$ and $\mathbf{p}$ are specified.

Since the states of a single particle moving in three spatial dimensions can be identified with the vectors in a six-dimensional vector space, the state of a pair of particles moving in three dimensions can be represented analogously by a single vector in a two × six = twelve-dimensional space. In general, the states of N particles are realised in a vector space of dimension $6N$. (NB. For motion in d spatial dimensions, the state space has dimension $2dN$.) Such spaces might not be the easiest thing to visualise, but it is fair to say that the relation between the mathematical model and the physical system is still fairly clear and 'intuitive'.

The discussion above has been for a set of particles, but general relativity is a field theory, and therefore it is important

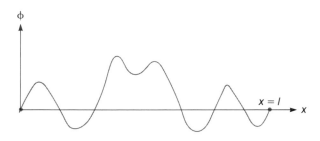

Figure 5.5. A one-dimensional string of length l is vibrating and ϕ denotes the amplitude of this vibration at a given time.

to analyse the concept of a state in a system of this type. Generally speaking, a 'field' is a function ϕ that associates a real number $\phi(x)$ with each point x in physical three-space, but, for ease of exposition, let us consider the simpler case where ϕ is just a function of a single real number x. An example is a vibrating string of length l in which x is the distance along the string and $\phi(x)$ is the amplitude of the displacement (figure 5.5).

The analogue of the particle equation (5.20) is a wave equation; for example,

$$\frac{\partial^2 \phi(x)}{\partial t^2}(t) = a \frac{\partial^2 \phi(x)}{\partial x^2}(t),$$

(5.23)

with the boundary conditions,

$$\phi(0)(t) = \phi(l)(t) = 0,$$

(5.24)

for all time t.

Equation (5.23) describes how the function $\phi(x)$ evolves in time, which explains the slightly cumbersome notation $\phi(x)(t)$. This association of a particular function of x with each time t can be viewed mathematically as describing a curve (whose points are labelled by t) in the space of all functions of the single variable x. However, it can also be viewed as a *single* function $\phi(x,t) = \phi(x)(t)$ of the *two* variables x and t. From the perspective of both special and general relativity, this is a particularly attractive way of writing a field theory because of the symmetric way in which the space and time variables appear as arguments of the field.

It is clear that the function $\phi(x)$ in the field theory plays a role analogous to the position x of the single particle, and this is strengthened by noting that the basic theorems dealing with a field equation like (5.23) guarantee the existence at each time t of a unique function of x, $\phi(x,t)$, given the values, at a fixed time t_0, of the functions of x, $\phi(x,t_0)$ and $\partial \phi(x,t_0)/\partial t$. Thus, just as the causality requirement on states was met in particle dynamics by transmuting equation (5.20) to the pair of equations (5.21) and (5.22), so it can be satisfied in the field theory by identifying

the states of the system with the set of all pairs of functions $(\phi(x), \pi(x))$ of the single variable x, and letting the dynamical evolution be the curve in this space satisfying the first-order (in time) equations,

$$\frac{\partial \pi(x,t)}{\partial t} = a \frac{\partial^2 \phi(x,t)}{\partial x^2},$$

(5.25)

$$\frac{\partial \phi(x,t)}{\partial t} = \pi(x,t).$$

(5.26)

A continuous stretched string has an infinite number of points, and, correspondingly, one might expect that the phase space of pairs of functions $\phi(x)$, $\pi(x)$ would have the mathematical structure of an *infinite*-dimensional vector space. This is indeed the case, although in exploring this idea it is necessary to take care in incorporating the observation that, precisely because the string *is* continuous, the positions of two 'neighbouring' points are closely related and cannot be viewed as being independent dynamical variables. The correct way of including this feature is to think of the different degrees of freedom of the stretched string as corresponding to the 'countably' infinite number of different possible modes of vibration of the string, rather than as the 'continuous' infinite number of points on the string. This idea can be described mathematically by using the boundary conditions in equation (5.24) to express each field as a Fourier series:

$$\phi(x) = \sum_{n=1}^{\infty} q_n \sin \frac{n\pi x}{l}; \quad \pi(x) = \sum_{n=1}^{\infty} p_n \sin \frac{n\pi x}{l}$$

(5.27)

and then note that specifying the infinite set of real numbers $q_1, q_2, \ldots, p_1, p_2, \ldots$ is entirely equivalent to specifying the pair of functions $\phi(x)$, $\pi(x)$. Thus we can think of the state space of this particular field theory as the infinite-dimensional vector space whose vectors have the coordinates $(q_1, q_2, \ldots, p_1, p_2, \ldots)$. The movement of the state vector in time is accommodated by allowing these coordinates to be functions of time, so that

$$\phi(x,t) = \sum_{n=1}^{\infty} q_n(t) \sin \frac{n\pi x}{l}$$

$$\pi(x,t) = \sum_{n=1}^{\infty} p_n(t) \sin \frac{n\pi x}{l}$$

(5.28)

and, indeed, in terms of these Fourier coefficients the original field equations (5.23) can be written as the infinite set of second-order *ordinary* differential equations

$$\frac{d^2 q_n(t)}{dt^2} + a \frac{n^2 \pi^2}{l^2} q_n(t) = 0; \quad n = 1, 2, \ldots,$$

(5.29)

while the first-order equations (5.25) and (5.26) can be written as (cf. equations 5.21 and 5.22)

$$\frac{dp_n(t)}{dt} = -a \frac{n^2 \pi^2}{l^2} q_n(t),$$

(5.30)

$$\frac{dq_n(t)}{dt} = p_n(t). \tag{5.31}$$

It is in this sense that a field theory can be regarded as an infinite-dimensional mechanical system.

General relativity can be recast into this first-order form, and the resulting structure has been extensively used in the study of quantum gravity (although whether or not one *ought* to violate the spacetime symmetry in this way has always been a question of lively controversy). The crucial step is to pick a particular time coordinate in the curved, four-dimensional spacetime manifold and then to rewrite the Einstein field equations so that they become first-order in this preferred time variable, just as we went from the second-order equation (5.23) to the first-order equations (5.25) and (5.26). The analogue of the field variable $\phi(x)$ turns out to be the metric tensor describing the generalised Pythagoras theorem of the curved three-dimensional physical space, and the analogue of $\pi(x)$ is a tensor field on this three-space that describes its curvature as viewed from the surrounding four-dimensional spacetime. Thus the basic canonical view of the dynamical structure of general relativity is of three-dimensional space 'curving and wriggling' its way through time.

In using infinite-dimensional vector spaces to describe the states of a field theory, we have introduced a modest degree of mathematical sophistication. But perhaps the reader will agree that there is still a fairly close intuitive connection (at least in principle) between the mathematical model and that which it models.

When we come to quantum physics the situation changes, and the lack of any intuitive relation between the mathematics and the physics is one of the major hurdles confronting a beginner. The abstract nature of the mathematics is grounded in *the* fundamental property of quantum theory: it does not predict the values of measurements made on the system, but only the *probabilities* of obtaining certain values. Of course, there is nothing striking in itself in the use of probabilistic language in physics; for example, the mechanistic explication of thermodynamics is firmly rooted in such ideas. But what is novel in quantum theory is the insistence that the probabilities are *intrinsic* in the sense that they cannot be improved by gathering any further information about the detailed functioning of the system. Thus they are deemed to be a fundamental property of the system itself (or of a statistical ensemble of such systems) rather than merely reflecting a contingent lack of knowledge on the part of the observer.

This has many deep philosophical implications (see Chapter 13 by Shimony) which can be focused in part on the profound difference between the concept of a 'state' in quantum theory and the analogous idea in classical mechanics. We shall return later to some of these philosophical considerations, but for the moment let us concentrate on the mathematical model that has been developed to accommodate this changed conception of a state. At this point, it should be emphasised that the idea of a mathematical 'space of states' is maintained in the quantum theory, and this is still required to carry whatever causal properties there may be. But the first requirement of a state *is* weakened, and a specification of a state now only enables the *probabilities* of the results of measurements to be calculated, not the results themselves; and it is these probabilities that evolve causally.

If one were starting *ab initio* to construct a theory that would yield probabilities, one might hunt through the mathematical literature looking for structures that yield real numbers lying between 0 and 1 (the possible ranges of a probability) and with the property that, if there are N different possible results for the measurement of some observable with probabilities $P_1, P_2, \ldots, P_N$, respectively, then

$$\sum_{n=1}^{N} P_n = 1. \tag{5.32}$$

This affirms that, with probability one, *some* result must be found.

One mathematical construction that suggests itself is to associate with each of the N possible results a subset X_n of some master space X, with P_n being defined as the (possibly) weighted ratio of the 'area' or 'volume' (or whatever is appropriate) of X_n to that of X itself. Taking ratios ensures that $0 \leqslant P_n \leqslant 1$ and condition (5.32) will be satisfied because of the additive properties of 'area', 'volume', etc. (assuming that the sets associated with different values of n are disjoint). This is essentially the mathematical idea that underlies conventional probability theory as applied to classical statistical physics. However, this is not the construction employed in quantum theory. Instead we exploit another 'obvious' (!) source of numbers lying between 0 and 1: the geometrical quantity $\cos \theta$ (or $\sin \theta$). More precisely, since $-1 \leqslant \cos \theta \leqslant 1$, it is $\cos^2 \theta$ that can be used as a probability. This suggests representing the states of a quantum system by vectors in a vector space, such that, to each possible result of a measurement of an observable, there corresponds another vector in the space, and with the probability of finding that particular result being the cosine squared of the angle between that vector and the original state vector. In general, this assignment of probabilities will not satisfy the crucial equation (5.32). However, it is useful to recall an elementary result from the cartesian geometry of three dimensions: if the vectors $\mathbf{a}$, $\mathbf{b}$, $\mathbf{c}$ in figure 5.6 are at right-angles to each other, and if $\mathbf{v}$ is any other vector, the cosines of the angles between $\mathbf{v}$ and these vectors satisfy

$$\cos^2 \theta_a + \cos^2 \theta_b + \cos^2 \theta_c = 1. \tag{5.33}$$

This result generalises to an arbitrary number of dimensions and is the key to understanding the basic mathematical structure of quantum theory, which is essentially as follows.

(i) The states of a quantum system are represented mathematically by vectors $\vec{\psi}$ in a vector space $\mathcal{H}$. The

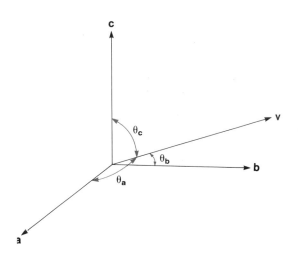

Figure 5.6. The cosines of the angles between the vector **v** and the three orthogonal vectors **a**, **b** and **c** satisfy the cosine rule, equation (5.33). This lies at the heart of the idea of using the cosine squared of an angle to represent probabilities in quantum theory.

dimension N of this space may be finite or infinite, and is determined by the system. It has nothing to do with the dimension of the system in the sense of classical mechanics. For example, the vector space describing the quantum motion of a single particle is infinite dimensional. (This is why even elementary quantum physics is so difficult!)

Any vector $\vec{\psi}$ can be expanded in terms of a set of basic vectors $\vec{e}_1, \vec{e}_2, \ldots, \vec{e}_N$ (the analogue of **i**, **j**, **k** in elementary vector space theory) in the form

$$\vec{\psi} = \sum_{n=1}^{N} \psi_n \vec{e}_n, \tag{5.34}$$

and the components $\psi_1, \psi_2, \ldots, \psi_N$ uniquely specify the vector.

(ii) The state vector evolves causally in time so as to preserve the intrinsic linear structure of the theory; i.e. if vectors $\vec{\psi}$ and $\vec{\phi}$ evolve in time t into $\vec{\psi}'$ and $\vec{\phi}'$, respectively, then the sum $\vec{\psi} + \vec{\phi}$ is required to evolve into $\vec{\psi}' + \vec{\phi}'$. These requirements of causality and linearity are met by postulating that the components of the state vector $\vec{\psi}$ satisfy a set of linear, cross-coupled, first-order differential equations

$$\frac{d\psi_n(t)}{dt} = \sum_{m=1}^{N} T_{nm} \psi_m(t) \tag{5.35}$$

in which the set of numbers T_{nm} plays a role analogous to that of the force in equation (5.21), i.e. it determines the dynamical evolution of the system.

(iii) If a measurement is made of a particular observable O, the result obtained will be one of a specific set $\lambda_1, \ldots, \lambda_N$. This assumes that the range of values for O is a discrete set. If it is a continuum (e.g. the position of a particle, or the value of a field), it can be reduced to the discrete case by dividing the continuum artificially into a finite or infinite set of nonoverlapping regions, and concentrating on measurements that determine in which region a result falls but do not distinguish between different real numbers lying in the same region. To each such possible result λ_i there corresponds a vector $\vec{u}_i$, and these vectors are all mutually orthogonal (i.e. at 'right-angles' to each other). If the state of the system is represented by a vector $\vec{\psi}$ and if the observable O is measured,

the probability of finding the value λ_i for O, given that the state is $\vec{\psi}$, is $\cos^2 \theta_{\psi, u_i}$, (5.36)

where θ_{ψ, u_i} is the angle between the vectors $\vec{\psi}$ and $\vec{u}_i$. And that more or less summarises the mathematical structure of quantum theory! The only significant 'cheat' in this exposition is that, in practice, the state-space must be a *complex* vector space in the sense that the expansion coefficients $\psi_1, \ldots, \psi_N$ in equation (5.34) have to be complex, not real, numbers. Euclidean geometry generalises fairly easily to the complex case, and the only real change is in the right hand side of equation (5.36). This must be replaced by the modulus $|\cos^2 \theta_{\psi, u_i}|$ since the 'cosine' is now a complex number, whereas probabilities must be real.

These axioms specify the type of mathematical structure that is to be used in quantum physics, but they do not say anything about how, for a specific system, one is to choose the correct space $\mathcal{H}$, the numbers T_{nm}, the special vectors $\vec{u}_1, \vec{u}_2, \ldots, \vec{u}_N$, and so on. This problem can be tackled in a variety of ways, and is an area in which the creative skills of the physicist are put to the test. Most work in quantum theory starts with a known *classical* system (e.g. electromagnetic field theory, or a particle moving in three dimensions) and then applies algorithms that have been developed over the years to construct the unknown $\mathcal{H}$, T_{nm} etc. in terms of elements of the given classical structure. That an approach of this type is viable is demonstrated by the striking success of modern atomic and molecular physics. There is not time to discuss here this classical → quantum path, but one or two points should be emphasised that apply irrespective of whether the quantum theory is built from some underlying classical system, or whether it is derived from an *a priori* quantum principle.

(i) If, at some time t, the state vector $\vec{\psi}_t$ points in the same direction as one (say $\vec{u}_j$) of the special set of vectors $\vec{u}_1, \ldots, \vec{u}_N$ of a particular observable O, then, from equation (5.36), a measurement of O is guaranteed to produce the associated result λ_j (since the probability of this happening is $\cos^2 \theta = 1$). In general, however, the vector $\vec{\psi}_t$ will not point along any of these directions, and a series of measurements on the system will yield a spread of results according to the probabilities in equation (5.36). (This effect is often referred to as the 'quantum fluctuations' in O.) Note that, even if the state vector *does* point along a special direction (so that the results of measuring O are determined exactly), this situation will typically change in time as the state vector moves around (figure 5.7).

(ii) The special vectors $\vec{u}_1, \ldots, \vec{u}_N$, $\vec{u}_1', \ldots, \vec{u}_N'$ associated with two different observables O and O' will in general constitute two different sets of orthogonal vectors (figure 5.8). Thus, even if a state vector is 'deterministic' for one of the observables, it will not have this property for the other one. For example, in figure 5.8 the state vector $\vec{\psi}$ points along the $\vec{u}_3$ direction, and so is guaranteed to yield the value λ_3 if O is measured; but there will be a statistical scatter among the values of O', should this be measured instead. This is the heart of the famous Heisenberg uncertainty principle which asserts the existence of 'complementary' pairs of observables such that, the more certain is the result of measuring one of them, the greater becomes the statistical scatter of measurements made on the other.

5.4 General remarks on quantum gravity

Having disposed of the preliminaries, we can begin to consider the problem of quantum gravity itself together with the difficulties that might arise in any attempt to cohere the very different structures of quantum theory and general relativity. The first observation is that the roles of space and time in quantum and classical physics are very similar. For example, the quantum mechanics of a single particle is concerned with the probability of finding the particle in some region of physical three-space at some given time. Similarly, the quantum theory of, say, electromagnetism yields the probabilities of obtaining various values of the electric or magnetic fields if these are measured in various regions of the same three-space.

In the case of a field theory, the preferential treatment of time can be avoided by the use of a slightly different (but mathematically equivalent) description in which the dynamical evolution of the system is carried by the observables, rather than the state vector. In the example considered in Section 5.3, this would correspond to using the second-order equation (5.23), rather than the first-order equations (5.25) and (5.26), as the basis for constructing the quantum theory from the given classical system. In the resulting 'covariant' quantum field theory, the space and time coordinates are handled on an equal footing and this is often claimed to be of particular importance in the quantisation of general relativity. For the moment, however, the relevant point is that, in either the canonical or the covariant schemes, spacetime appears purely passively as the fixed backcloth against which the drama of physics is played.

A passive view of spacetime accords with our understanding of those branches of physics concerned with special (rather

Figure 5.7. The state vector $\vec{\psi}$ moves around in time on the dashed trajectory. As it does so, its projections onto the special set of vectors change, and hence so do the probabilities of measuring the associated values of the corresponding observable.

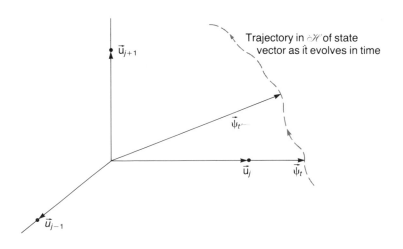

than general) relativity, and covariant quantum field theory has become one of the most potent tools with which to study the high-energy interactions of elementary particles. To understand how *particles* can be described in terms of a *field* theory, we must return for a moment to the discussion in Section 5.3 of the interpretation of quantum theory. The obvious observable in a field theory is the field itself, and the quantum theory will yield the probabilities of its values lying in arbitrary regions of three-space. But there are other observables of interest, of which perhaps the most important are the energy and the momentum of the field. These concepts are well-defined classically but, in

Figure 5.8. The state vector $\vec{\psi}$ points along one of the special set of vectors of the first observable and hence corresponds to a situation in which, with probability one, a definite result will be obtained if a measurement is made of this observable. On the other hand, if measurements are made on the second observable the results will be distributed statistically according to the cosines squared of the angles between $\vec{\psi}$ and the special vectors $\vec{u}'_1, \vec{u}'_2, \ldots$. Conversely, a vector pointing along $\vec{u}'_1$, say, will give a guaranteed value for the second observable, and it is the values of the first observable that will now be distributed statistically. State vectors which do not point along special vectors for either observable lead to an intermediate situation in which there is a statistical spread in both observables. This is the basis of the famous Heisenberg uncertainty principle.

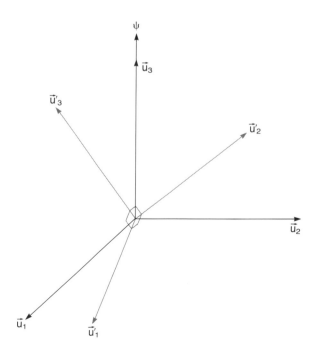

the quantum theory, they stand in a 'complementary' relation to the field variable in the sense that, as discussed in connection with figure 5.7, a state vector corresponding to a 'precise' value of the field will produce a statistical scatter for the values of energy–momentum, and vice versa.

Energy and momentum are ideas of the greatest importance in physics, and therefore considerable interest is attached to any state vectors that give precise values for these observables. In the special case when the underlying classical field equations are linear (e.g. in equation 5.23), the mathematical problem of constructing such vectors can be solved completely. The striking result is that the energy and momentum in such a state are related to each other in precisely the same way as are the energy and momentum of a *particle* in special relativity! Furthermore, there are linear superpositions of these states in which the energy and momentum are *localised* in regions in three-space, which suggests that these states correspond in some way to the quantum version of a particle (or collections of distinct particles). There are even vectors of this type that predict a definite value for the angular momentum of the system, a phenomenon that is interpreted by saying that the particles have a definite 'intrinsic spin'.

These particles are known as the 'quanta' of the field, and can in principle be observed. For example, the quanta of the electromagnetic field are identified with photons, and can be seen directly with the aid of an image intensifier.

This remarkable result suggests strongly that a quantum theory of gravity might involve the idea of the 'graviton' as the quantum of the gravitational field. However, Einstein's field equations are nonlinear while, as they stand, the remarks above apply only to a linear system of equations.

A direct consequence of this linearity is that the quanta in a state are 'transparent' to each other, i.e. they never collide, and their number does not change in time. This does not seem very promising from the viewpoint of explaining the many interactions among elementary particles, but the situation changes dramatically if the field equations are nonlinear. The ensuing quantum theory is very complex, and cannot be solved exactly (at least, not in a spacetime of dimension greater than two). However, the theory can be tackled using a type of approximation method in which the numerical results of interest are expanded in powers λ^n (where n is a positive integer) of some basic 'coupling constant' λ that sets the overall scale of the nonlinear parts of the equations. For example, we might consider a nonlinear version of the string equation (5.23) in which an additional term $\lambda(\phi(x)(t))^3$ appears on the right hand side. If λ is sufficiently small (i.e. if the nonlinear term is just a 'perturbation' of the linear theory) then the hope is that the terms proportional to λ^n will become rapidly negligible as n increases so that a respectable approximation to the complete answer can be obtained by just retaining the terms up to λ^N for some small value of N (the amount of work involved in performing the relevant calculations increases rapidly with N). The complete formulation of such a 'perturbation theory' is complex and subtle, and is discussed in more detail in the chapter by Taylor. For our purposes it suffices to remark that, within this framework, it can be shown that the quanta of the field *do* now interact with each other, and the total number need no longer be conserved. Thus, in using quantum field theory in elementary particle physics, one attempts to associate the various known elementary particles with fields, and then to construct nonlinear field equations that will, via quantisation, reproduce the observed particle interactions. Notice that, in this case, the field equations are constructed for the sole purpose of explaining the *particle* physics. There is usually no expectation that these equations will have any significance in the world of classical physics *per se*, and indeed the only fundamental field systems that are known to be relevant classically are those of electromagnetism and gravitation.

Contemplating these remarks in the context of general relativity, we can anticipate two different but complementary views of quantum gravity. One approach, favoured mainly by elementary particle physicists, is to regard gravity as 'just another field', and to treat it in the way that has been so successful with the electromagnetic interactions. The basic picture will be of gravitons interacting with each other and with the quanta of the other fundamental fields of matter (electrons, quarks, etc.), via the nonlinearities in the Einstein field equations which describe the coupling of the metric to itself and to the matter fields. Schemes of this type have been studied extensively, and the conclusions will be summarised in Section 5.5.

The alternative approach, preferred by most general relativists, is to regard the values of the *field* variables as the primary observables of interest. In this scheme the principal output of the quantum theory is a prediction of the probabilities of finding various values for the curvature of three-space as it 'curves and wriggles' through time. One of the attractions of this particular way of looking at quantum gravity is that it retains some of the seductive geometrical language of the classical theory, which tends to get lost in the particle oriented approach. However, this raises the general question of the extent to which classical geometrical concepts are compatible with quantum theory. And in answering this, we find ourselves in deep water.

The trouble is that there is more to the use of geometrical ideas in physics than meets the eye. For example, in the discussion of the curved space 'Pythagoras' theorem, it was implied that the spacetime curvature in equation (5.19) could be viewed as a departure from the flat, Minkowski, metric of equation (5.18). But there is a limit to the size that can be accepted for such a departure without violating some basic physical principles. For example, consider the metric tensor with the constant components

$$g_{xx} = g_{yy} = 1$$

$$g_{zz} = g_{tt} = -1 \qquad (5.37)$$

and all other components taking the value zero. Thus the analogue of equation (5.18) would be

$$(\delta s)^2 = (\delta x)^2 + (\delta y)^2 - (\delta z)^2 - (\delta t)^2, \qquad (5.38)$$

in which the occurrence of the minus signs means that both t and z must be viewed as 'timelike' directions! Physics in a spacetime of two spatial and two timelike dimensions would be very weird indeed, and the appearance in the theory of such a metric would normally be regarded as heralding its breakdown. This is not a problem in the classical theory of general relativity because such metrics are incompatible with the field equations. To be more precise, the number of spatial dimensions cannot change in time, and so as long as we start with the correct number (i.e. three), there is no danger of an anomalous value appearing suddenly as the system evolves.

However, in quantum theory, the phenomenon of quantum fluctuations means that things are quite different. Thus, if we attempt to probe the system at the scale of the Planck length, the statistical fluctuations might become large enough to cause a metric like equation (5.18) to change into one like equation (5.38). It is sometimes suggested that a transition of this type signals a change in the *topology* of the three-space as it evolves in time, rather as the two-dimensional sphere in figure 5.9 is changing into the torus.

Topology changes are not possible in classical relativity but they cannot be ruled out *a priori* in the quantum theory, and the idea that space has a 'foam-like' structure at the Planck length is one that has excited generations of theoretical physicists. If implemented, it would constitute a dramatic confirmation of the widely held expectation that space (or spacetime) should play a dynamical role in its own right, rather than being the passive container of classical physics. Unfortunately, it is unclear how to incorporate such a phenomenon into the mathematical structure of the quantum theory, and this remains a challenge for the future.

5.5   Quantum gravity from the perspective of particle physics

Researchers approach quantum gravity with a variety of *a priori* assumptions, and are therefore prone to disagree warmly about the whence, where, why and how of the subject. General relativists usually emphasise the role of geometry and the dynamical features of spacetime, whereas particle physicists are content to see spacetime as a fixed background in which gravitons and other quanta can interact and propagate. This divergence has far-reaching implications, and it is by no means clear that the votaries of these two schools are merely looking at the same subject in two different ways. On the contrary, at times it is hard to believe that they are talking about the same branch of physics!

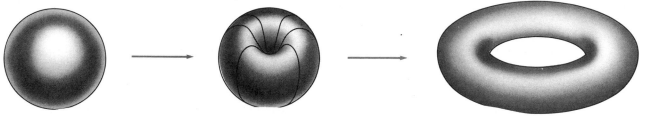

Figure 5.9. The two-sphere on the left hand side of the diagram is transformed into the two-torus on the right hand side by 'punching' through a hole as shown in the central figure. This is an example of a discontinuous transformation in topology.

We now attempt to summarise some of the more significant conclusions of both camps, beginning from the perspective of elementary particle physics. Not surprisingly, the starting point is the 'graviton' – an elementary particle whose relationship to the classical gravitational field is analogous to that of the photon to the electromagnetic field – and it is remarkable how much can be deduced about this exotic entity without mentioning general relativity at all. In a particle physics approach to the inverse square law of equation (5.1), the key observation is that static (i.e. time-independent) forces of this type can be obtained via the exchange of quanta of the associated classical field (figure 5.10).

If the exchanged quantum has mass μ, the force produced has the general form

$$F = \beta \frac{e^{-\mu r}}{r^2},\qquad (5.39)$$

where the constant β depends on various properties of the two particles and the way they couple to the exchanged quantum. Thus if the gravitational inverse square law is to be ascribed to the exchange of a graviton, the mass of this particle must be exactly zero. (The same argument applied to the electrostatic force in equation 5.2 shows that the photon is also massless.)

Like all elementary particles, the exchanged quantum has an intrinsic spin whose value (in units of $h/2\pi$) lies in the discrete series $0, \frac{1}{2}, 1, \frac{3}{2}, 2, \ldots$, and this has a significant effect on the force. Indeed, if the particle is a fermion (i.e. with spin in the series $\frac{1}{2}, \frac{3}{2}, \ldots$) there will be no static force at all ($\beta = 0$), which, for example, rules out neutrinos as the source of an inverse square law. The same can be shown to be true if the spin is greater than 2, and so the only possibility for the spin of the graviton is 0, 1 or 2. However, it can be shown that a spin-1 particle causes a *repulsive* force between two identical particles (which is why the photon has this spin), and so we can only reproduce the attractive gravitational force with a graviton whose spin is 0 or 2.

A detailed analysis shows that a spin-0 graviton is the quantum of a single field, whereas the extra angular momentum modes of a spin-2 graviton require it to be the quantum of a set of *ten* fields $h_{ab}(x)$ with $h_{ab}(x) = h_{ba}(x)$, and $a,b = 1, \ldots, 4$. But this is precisely the number of metric tensor components of a curved, four-dimensional spacetime! The conclusion is irresistible: the spin-2 graviton is related in some way to the quantisation of general relativity, while the spin-0 particle serves as the mediator of the gravitational force in a quantisation of the old Newtonian theory.

This unexpected connection between particle physics and general relativity can be strengthened as follows. First we rewrite the Pythagoras law for flat, Minkowski spacetime (equation 5.18) in the neater form

$$(\delta s)^2 = \sum_{a,b=1}^{4} \eta_{ab} \delta x^a \delta x^b,\qquad (5.40)$$

where η_{ab} is the Minkowski metric whose only nonvanishing components are

$$\eta_{11} = \eta_{22} = \eta_{33} = 1; \quad \eta_{44} = -1.\qquad (5.41)$$

Thus equation (5.40) is the 'flat-space' special case of the general equation (5.19) and, in so far as a curved metric describes a 'deviation' from flat-space (remembering our earlier cautionary remarks!), we can write

$$g_{ab}(x) = \eta_{ab} + h_{ab}(x),\qquad (5.42)$$

where $h_{ab}(x)$ is a measure of the strength of this deviation. Now substitute the decomposition (5.42) into the Einstein field equations, which become a set of very nonlinear partial differential equations for $h_{ab}(x)$. For sufficiently small values of $h_{ab}(x)$, all but the lowest-order terms can be neglected and we are left with a linear equation in $h_{ab}(x)$. Remarkably, this is *precisely* the one that we would have obtained had we started from the particle considerations above, and followed the well-established rules for constructing the field equations of a noninteracting, massless spin-2 particle!

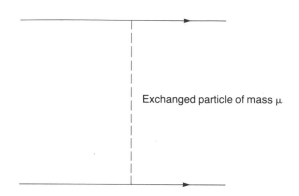

Figure 5.10. In modern elementary particle physics, static forces between elementary particles are viewed as arising from the exchange of a third particle. The range of the force is determined by the mass of this exchanged particle and is given in equation (5.39).

The important conclusion is that, by applying the 'classical → quantum' algorithms to the Einstein field equations, we obtain a quantum field theory in which

(i) the basic quantum is a massless spin-2 particle;

(ii) the exchange of this particle reproduces correctly the inverse square law of the static gravitational force;

(iii) the field which is subject to statistical quantum fluctuations is not the metric tensor itself, but rather the field $h_{ab}(x)$ that measures the *deviation* of the spacetime from flat Minkowski space;

(vi) the gravitons scatter and interact with each other according to the nonlinear terms in $h_{ab}(x)$ obtained when equation (5.42) is substituted into the Einstein field equations for the metric tensor $g_{ab}(x)$. Such nonlinear terms are an inevitable feature of general relativity and arise because *all* energy produces a gravitational field, and that includes the energy in the gravitational field itself!

From the perspective of general relativity, this reduction of quantum gravity to 'just another quantum field theory' is aesthetically displeasing and raises a number of critical technical questions. These hinge mainly on the status of the decomposition (5.42) and the implication that quantum gravity is the study of small quantum fluctuations around the flat, Minkowski spacetime, solution to the classical Einstein field equations. For example:

(i) As mentioned already, there is a danger that large fluctuations could change the number of 'time directions' and lead to an incomprehensible situation.

(ii) In using equation (5.42), we are committed to a spacetime whose global topology is the same as that of Minkowski space. But this rules out many of the solutions of Einstein's equations that are of interest in the study of cosmology.

(iii) In practice, the use of equation (5.42) in the quantum field theory means that calculations in quantum gravity are performed with the same type of weak field perturbation theory that is used in elementary particle physics. Such perturbative techniques are known to be problematical in the classical theory of general relativity (in discussions of gravitational radiation, black holes, cosmology, etc.), and it is hard to believe they will be any better when applied at the quantum level.

(iv) What has happened to the Planck length, $L_P = (Gh/c^3)^{\frac{1}{2}}$, which was expected to play such a key role in quantum gravity? A closer investigation shows that it *is* present in the theory, but not in the way we had hoped. In quantum field theory, a field whose quanta are bosons, such as $h_{ab}(x)$, has the dimensions of an inverse length, whereas the classical metric tensor η_{ab} is just a set of dimensionless numbers. Thus the right hand side of equation (5.42) is meaningless since we are trying to add terms with different dimensions. The correct expansion is in fact

$$g_{ab}(x) = \eta_{ab} + L_P h_{ab}(x), \qquad (5.43)$$

which *is* meaningful since, like η_{ab}, the product $L_P h_{ab}(x)$ is dimensionless. On substituting equation (5.43) into the Einstein equations, L_P appears in the ensuing nonlinear equations for $h_{ab}(x)$ as the coupling constant which determines the overall strength of the mutual interactions between gravitons, and in terms of which perturbative expansions are to be performed.

However, we had hoped that L_P would set the scale at which 'something odd' happened to the structure of spacetime, and there is no sign of that here. Indeed, the use of the background Minkowski metric in equation (5.5) guarantees that spacetime will play exactly the same passive role that it does in the rest of elementary particle physics.

It is clear that in reducing quantum gravity to 'nothing but' a traditional type of quantum field theory, we have failed to grapple with any of the deeper, *a priori*, worries about the subject, and this provides sufficient cause for many general relativists to reject approaches of this type. However, 'the proof of the pudding . . .' and if, in spite of all reservations, the perturbative theory worked, it would need to be taken seriously, if nothing more than as a springboard from which the more exotic features of quantum gravity could be explored.

Unfortunately, this is not the case, and the pudding that emerges from the mathematical machinations has a distinctly unpleasant flavour. As explained in the chapters by Taylor and

Georgi, quantum field theories have a regrettable tendency to produce infinite answers for almost everything, although, for special *renormalisable* systems, these unwanted pathologies can be tucked under the carpet without seriously damaging the predictive power of the edifice. Sad to say, this procedure fails for theories (like general relativity) in which the coupling constants have the dimensions of a positive power of length, and the entire structure collapses. Roughly speaking, these destructive infinities arise from integrals of the form

$$I = \int_0^{} \frac{1}{x^n} \mathrm{d}x; \quad n \geqslant 1, \tag{5.44}$$

in which the upper limit is not relevant to our present discussion. The zero lower limit is the crucial source of the infinities and it is important to know that the 'x' variable is a genuine coordinate in spacetime. Thus the ultimate origin of the infinities is the way in which conventional local quantum field theory probes right down to zero distances (or equivalently to infinite energies). This has lead to the frequent suggestion that in quantum gravity the infinities in *any* nonrenormalisable theory (including gravity itself) might be alleviated by introducing some sort of cutoff at the Planck length L_P so that, in effect, the lower limit of the integral in equation (5.44) would be L_P, rather than zero.

The general relativist might be tempted to respond 'serves you right' when he hears of the nonrenormalisability of gravity, which he will regard as an inevitable outcome of the use of equation (5.42) and the associated weak field perturbation theory. But the reaction of an elementary particle physicist is likely to be somewhat different. He will claim, with some justification, that in the praxis of physics it is always necessary to use *some* sort of perturbation theory, and it is no use dreaming of subtle, nonperturbative effects in quantum gravity until there is a well-defined perturbative structure on which to 'hang one's hat'. His thesis is that the fault lies, not with the perturbative methods, but with the Einstein field equations, which must therefore be modified until a system is found that *does* work.

This suggestion is guaranteed to cause fireworks, especially among those physicists who regard the existing theory of relativity as being more or less on a par with the Decalogue, and not to be subject to any form of exegesis. However, there is no real evidence justifying the use of the Einstein equations at subatomic scales, and it is entirely consistent with known experimental results to regard general relativity as an *effective* field theory (see Chapter 16 by Georgi) which should therefore be taken seriously only at low energies.

To resolve the nonrenormalisability by changing the Einstein equations (while keeping the low energy, classical predictions), there are a variety of possible options:

(i) add extra terms to the Einstein equations for the metric field alone;

(ii) add extra terms to the Einstein equations for specially chosen *matter* fields (i.e. fields whose quanta will be identified with elementary particles other than the graviton) as well as for the metric;

(iii) drop the Einstein equations all together, and invent a completely new set of field equations involving gravitons and other elementary particles. At low energies, this theory must reproduce the predictions of general relativity, but at high energies (and in particular at the Planck energy 10^{18} GeV) the results will (and must) be quite different. This means that the new field equations may be very different from the original Einstein equations.

Attempts have been made in all these directions, and have raised a number of intriguing questions and possibilities. A well-known example of the first approach is provided by the so-called 'higher derivative theories' of gravity in which the Einstein field equations are augmented by extra terms involving fourth-order derivatives of the metric tensor. If handled carefully, theories of this type can reproduce all the famous low energy predictions of general relativity (the perihelion shift of Mercury, the bending of light around the Sun, etc.) but with a high energy behaviour that is radically different. So different, in fact, that the quantum theory is now renormalisable, and therefore has the same theoretical credibility as quantum electrodynamics and the electro-weak theory. Unfortunately, this new theory has its own problems, and these are as unpalatable as the nonrenormalisability of the theory it is replacing. For example, it predicts the existence of graviton–graviton scattering processes in which the total probability for something to happen is less than one! It has been claimed that this difficulty can be surmounted by a careful nonperturbative analysis, but, for reasons that are perhaps as much psychological as scientific, this approach has not received the attention that it might deserve.

The second of the three options mentioned above – the addition to Einstein's equations of carefully selected matter fields – has been actively investigated over the last twelve years. Because of the equivalence principle, *every* piece of matter, be it microscopic or macroscopic, must interact with the gravitational field in a fairly precise way. But what is being suggested now is that this arises from the fundamental coupling of gravity to a few(?) special elementary particles (e.g. quarks) such that (i) the ensuing quantum field theory is renormalisable, and (ii) the effective coupling to objects built from these basic entities satisfies the equivalence principle.

One attraction of a scheme of this type is that the renormalisation requirement might be sufficiently restrictive that it can only be satisfied for a *unique* set of basic particles, which would then be *the* set from which all other particles (hadrons, leptons, etc.) are built. In this way one would obtain a field theory unifying the electromagnetic, weak, strong and gravitational forces.

This is an alluring possibility but, unfortunately, the needs of renormalisation are *so* restrictive that it is by no means clear that they can be met with *any* selection of matter particles. The difficulty is that the matter fields introduce extra infinities of their own, so that a theory of gravity plus matter is, if anything, even worse than that of gravity alone. What must be hoped is that, for a very special selection of matter particles, these infinities 'cancel' so that the final answer is finite (and correct!).

The idea of 'cancelling' infinities is not quite so outrageous as it might seem at first. As always, 'it all depends on what you mean by . . .'. Let us return for a moment to the divergent integrals in equation (5.44) and consider the particular case when $n=1$ (for convenience, we will set the upper limit on the integral equal to unity). Then

$$I = \int_0^1 \frac{1}{x}\,\mathrm{d}x = (\log x)\Big|_0^1 = \log 1 - \log 0 = 0 - (-\infty) = \infty \tag{5.45}$$

so that the answer is indeed infinite. Mathematically speaking, a better way of expressing this is to construct a new integral I_ε in which the lower limit is some small number ε with $0 < \varepsilon < 1$. This has the value

$$I_\varepsilon = \int_\varepsilon^1 \frac{1}{x}\,\mathrm{d}x = (\log x)\Big|_\varepsilon^1 = -\log \varepsilon \tag{5.46}$$

and, in saying that the value of the integral in equation (5.45) is infinite, we really mean that the limit of I_ε tends to infinity as ε tends to zero.

To fix our minds on a concrete situation, suppose that the integral is intended to give the probability of finding some specific configuration for the outgoing pair of gravitons in the graviton–graviton scattering process in figure 5.11. This Feynman diagram can be interpreted more or less literally with time flowing from left to right. Thus the incoming pair combine at A to form a third graviton. This travels to B where it turns back into a pair of gravitons that recombine at C to form a single particle. This in turn travels to D where it reverts to the pair of particles that finally emerges from the experiment and whose configuration we measure.

It is important to appreciate that one of the basic properties of quantum field theory is that the origin of the infinity in this particular contribution to the scattering process is the *loop* of gravitons in the centre of figure 5.11. If the loop is not present, (e.g. in figure 5.12) then the answer is finite (and, conversely, the more loops there are in a diagram, the more virulent is the infinity). Unfortunately, the physical scattering process is described by the sum of the individual processes in figures 5.11 and 5.12 (and of many others too), and so the infinite value predicted for the subprocess in figure 5.11 is a real problem.

Now suppose that there is a matter field in the system. This will couple to the gravitational field and generate a whole new series of subprocesses contributing to the graviton–graviton scattering. In particular, there will be a diagram (figure 5.13) in which the graviton loop is replaced by a loop of the matter particles.

Because of the loop this diagram will also be infinite and, purely for the purposes of illustration, let us suppose that it is given in terms of a particular pair of constants a and b, by the integral

$$I^{\text{matter}} = \int_\varepsilon^1 \frac{a}{x}(1 + bx^2)\,\mathrm{d}x = \left(a\log x + \frac{abx^2}{2} \right)\Big|_\varepsilon^1$$
$$= -a\log \varepsilon + \frac{ab}{2}(1 - \varepsilon^2), \tag{5.47}$$

so that, as expected,

$$I^{\text{matter}} = \underset{\varepsilon\to 0}{\text{Lim}}\ I_\varepsilon^{\text{matter}} = \infty.$$

Now we know that the actual scattering process will include the sum of the subprocesses in figures 5.11–5.13, but even the most optimistic theoretical physicist cannot make much sense of adding two genuinely infinite numbers. However, what we *can* do is to add together the integrals with $\varepsilon > 0$, and only take the limit $\varepsilon\to 0$ *after* performing the sum. Thus we have

$$I_\varepsilon + I_\varepsilon^{\text{matter}} = -(1 + a)\log \varepsilon + \frac{ab}{2}(1 - \varepsilon^2), \tag{5.48}$$

which in general will also become infinite in the limit $\varepsilon\to 0$. However, suppose that the quantum field theoretic calculation of the subprocess in figure 5.13 gives a value $a = -1$. Then a miracle occurs and we have

$$I + I^{\text{matter}} = \underset{\varepsilon\to 0}{\text{Lim}} -\frac{b}{2}(1 - \varepsilon^2) = -\frac{b}{2} \tag{5.49}$$

which is a finite number!

Sensitive readers are probably grinding their teeth by now at such brazen abuses to the mathematical formalism. However, the underlying physical motivation is clear: we do not really expect quantum field theory to work right down to *zero* distances; rather, we anticipate a series of modifications as the energy increases, culminating perhaps in a complete cutoff at the Planck length.

But even if we accept all this, how do we find sets of matter fields such that, in all processes of interest, a cancellation will occur in the coefficients of the infinite parts of the individual subprocesses? This is a very hard problem, and it is difficult to know even where to start. But one thing *is* clear: the desired miraculous cancellations will never occur if the signs of the infinite terms are all the same.

At this point, a fundamental feature of quantum field theory comes to our aid. It can be shown that the sign of the divergent parts of a loop process like figure 5.13 depends on whether the particle in the loop is a boson or a fermion, and this suggests trying to cancel the divergence in the graviton (a boson) loop

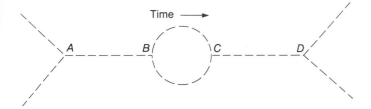

Figure 5.11. The two incoming gravitons combine at point *A* to give a third graviton that travels to *B* where it converts back into a pair that in turn recombine at *C*. This single graviton then travels to *D* where it transforms back into a real pair of outgoing gravitons. The presence of the loop in this particular contribution to the two-particle scattering process means that the values predicted are infinite.

Figure 5.12. This diagram also contributes to a two-particle scattering process but, unlike the analogous process in figure 5.11, its contribution is finite.

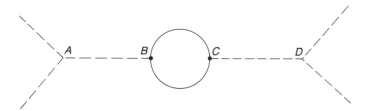

Figure 5.13. In this contribution to a two-graviton scattering process, there is a central loop of matter particles. This also is infinite.

with the divergence coming from the loop of some associated fermi particle. However, even with this knowledge, it is still a major task to find a set of matter fields which will enable *all* the divergences to be removed. For example, even if we managed to make the process in figure 5.11 well behaved, there is no *a priori* reason why that particular set of matter fields should help with the infinities of the other scattering processes.

To proceed further, a deeper insight is needed into ways in which the graviton can be 'associated' with other elementary particles. Drawing on earlier experience in particle physics, this might be expected to involve some sort of *internal symmetry* which would enable the graviton and its matter partner(s) to be regarded as different 'modes' of a single unified entity, just as, with the introduction of the $SU(2)$ iso-spin symmetry, the proton and neutron can be regarded as two states of a 'nucleon'. The obvious objection to this idea is that the graviton is a boson whereas we are expecting, at least some of, the matter fields to describe fermions. However, traditional internal symmetry schemes such as $SU(2)$, $SU(3)$, etc., associate bosons with bosons, or fermions with fermions, but never fermions with bosons.

For this reason, attempts to resolve the nonrenormalisability in this way would undoubtedly have ground to a halt had it not been for a remarkable discovery made in the early 1970s. By a fundamental revision of the idea of what constitutes a mathematical 'group', it is possible to construct symmetry schemes in which bosons and fermions *can* appear as different facets of an underlying object. The new structure is called 'supersymmetry' and has generated intense interest amongst both physicists and mathematicians.

When applied to the gravitational field, the ideas of supersymmetry show that the graviton should have a number of partners and, in particular, massless spin-$\frac{3}{2}$ fermions known as the *gravitino*. The imposition of supersymmetry on the Einstein field equations is very restrictive, and only a small number of such 'supergravity' theories exist. Indeed, one of the main attractions of supersymmetry is the lack of scope for 'adjusting' parameters to fit the theories to whichever piece of experimental information happens to be of interest at the time. Thus the theory is either right or wrong and there is no danger of getting trapped into the endless 'fiddling' that has marked not a few past approaches to elementary particle physics!

The most significant degree of freedom that *is* left in supergravity is the number N of spin-$\frac{3}{2}$ particles in the theory, which can lie anywhere between 1 and 8, inclusive. The general expectation is that nature has chosen the largest possible value: partly because this maximises the chances of removing the unwanted infinities, and partly because of the hopes that the additional partners to the graviton (whose number also increases with N) can be identified with the basic entities of electro-weak and strong interaction physics – and there are rather a lot of those!

Note that, because the gravitino is a fermion, it will not itself cause a static force, and so the classical predictions of general relativity are left intact. Indeed, it would be very difficult to construct a piece of equipment that would respond directly to the presence of gravitinos (it was hard enough to find the less exotic neutrinos), although the large scale presence of such particles could have important implications for cosmology and the 'missing matter' problem. However, the most crucial test of supergravity is whether it succeeds in removing the pathological infinities, and at first hopes ran high, particularly when it was discovered that the analogue of $N=8$ supergravity in gauge theories of the Yang–Mills type *was* finite. In fact, there were no infinities at all, not even the 'controllable' type that appear in quantum electrodynamics. But, alas, the proofs of finiteness are not transferable to supergravity and, after twelve years of intense study, it now seems to be generally agreed that even $N=8$ supergravity is not free of the infinities that plague ordinary quantum gravity. This has not yet been proved rigorously, but the prognosis is bad, and the number of people who view supergravity as *the* fundamental theory of physics is dwindling rapidly.

Of course, this is sad, but it by no means spells the end of supergravity as an ingredient in quantum gravity. Even if the $N=8$ theory is infinite, there is still a well-defined sense in which the pathological behaviour is not as bad as in the original pure Einstein theory. This has important implications for the third scheme mentioned above, in which the Einstein equations are dropped altogether and general relativity is required to appear only as the low energy limit of some, perhaps structurally quite different, theory. For why should it be just general relativity that is obtained in this way? Perhaps it will be *supergravity* that emerges as the effective field theory? This would do no violence to the classical tests of relativity, and there is a rapidly growing feeling that this is indeed what happens.

This third, iconoclastic possibility has a number of attractive features, not the least of which is that the nonrenormalisability of the old weak interaction theory was solved in precisely this way. The original 'four-fermi' model worked satisfactorily at low energies, but it had a bad high energy behaviour and produced the same type of uncontrollable infinities that were discussed above in the context of quantum gravity. The successful resolution involved, not the addition to the existing field equations of judicially chosen extra terms, but the construction of a completely new theory that reproduced the earlier low energy results but whose high energy predictions were quite different.

The key to the success of this theory is a set of new massive particles. However, the role of these extra fields is not to provide extra 'infinities' with which to cancel the existing ones. Rather, they introduce a fundamental change in the way in which the particles interact with each other. A good example is the well-known β-decay of the neutron into a proton + electron + neutrino. In the original picture, the main contribution was from a direct four-particle interaction in which all four particles 'met' at the same spacetime point (figure 5.14).

In the new theory, the four particles do not interact with each other directly at all. Instead there is an indirect effect mediated by the exchange of one of the new massive particles – the famous W-boson – which couples separately to the neutron–proton pair and to the electron–neutrino pair (figure 5.15).

If the difference between the energies of the incoming neutron and outgoing proton is much less than the rest mass of the W-meson (around 70 MeV), then the process depicted in figure 5.15 is not experimentally distinguishable from that in figure 5.14. However, at high energies the situation is quite different and the modified behaviour is sufficiently 'good' to ensure that (when loop graphs are included) the complete theory is now renormalisable. For good measure, it also unifies the weak and electromagnetic forces into a single 'electroweak' force.

It is very tempting to consider treating general relativity in an analogous way, with the hope that this would give, not only a renormalisable quantum theory, but also a unification of gravity with the other forces of nature: the extra massive particles required to make the theory well-behaved at short distances are perhaps precisely *all* the known elementary particles of physics! This is indeed a seductive idea. But how is it to be implemented? It is true, we know the theory (general relativity, or perhaps supergravity) whose low energy predictions must be reproduced, but this tells us little about the new field equations. Indeed, we do not even know how many new massive particles there should be, let alone how they should couple to the graviton and to each other. And the historical development of the successful electro-weak theory does not afford much comfort either. The discovery by Salam, Weinberg and Glashow required a number of brilliantly creative jumps, not just a simple deduction from the existing four-fermi theory.

Both supergravity and the electro-weak theory rely heavily on internal symmetry principles, and we might expect the same to apply here. However, the only really viable theory of this type involves a great deal more than just the application of group theoretic methods. Indeed, its structure is sufficiently novel and exciting to suggest that we may even be on the verge of a paradigmatic shift in our understanding of the fundamental constituents of matter.

I am speaking of the *supersymmetric string*, an idea which has generated an almost unparalleled level of enthusiasm and activity, and which, at the time of writing, seems to be occupying the thoughts of almost the entire theoretical physics community. In this new theory, the structureless 'point' particles of conventional quantum field theory are replaced by one-dimensional 'string-like' entities that can interact with each other and scatter according to a rather precise set of laws. Thus the familiar Feynman diagrams, (for example figure 5.12), are now replaced with string-scattering diagrams of the type shown in figure 5.16.

Unlike a pure 'pointlike' object, a string has internal degrees of freedom that can be examined with the aid of Fourier decomposition (cf. the discussion in Section 5.3 of a more conventional 'string'). A careful quantum mechanical analysis shows that a single quantised string is equivalent to an *infinite* set of 'normal' elementary particles whose masses and spins are related in a special way.

Infinite sets of particles of this type were widely used in the late 1960s and early 1970s as part of the S-matrix (i.e. nonfield

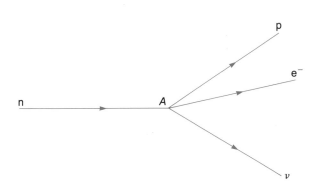

Figure 5.14. In the old picture of the β-decay of a neutron, the interaction between the incoming neutron and the outgoing proton, electron and neutrino, takes place at a single spacetime point *A*.

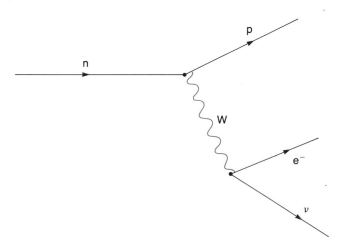

Figure 5.15. In the modern understanding of β-decay, the interaction between the four particles is 'spread out' in spacetime by means of the exchanged W-boson. At low energies, the two pictures in figures 5.14 and 5.15 give the same predictions but the high energy results are quite different. The infinities coming from loop diagrams can be removed in this new version.

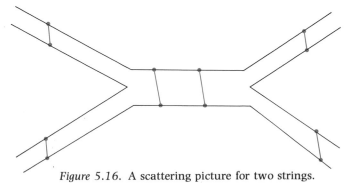

Figure 5.16. A scattering picture for two strings.

theoretic) approach to the study of the strong nuclear force. However, these particular string theories encountered a variety of difficulties and, with the development of quantum chromodynamics, the attention of the majority of high energy physicists turned back to mainstream quantum field theory. A minority persevered, however, and, about twelve years ago, discovered rather unexpectedly that many of the difficulties can be overcome if the string is interpreted as a theory of *gravity*, rather than of strong interactions, and in precisely the way we have been anticipating. Thus the graviton is now one of the string modes and, when the string-scattering processes are analysed in terms of their particle content, the *low energy* graviton scattering is the same as that computed from supergravity. However, the high energy results are changed radically by the infinite number of massive particles whose role is analogous to that of the W-meson in the electro-weak theory.

The significance of this result was not widely appreciated at the time, and the subject remained on the periphery of quantum gravity research for almost ten years. Then, almost overnight, the floodgates opened, and the hectic activity began. The catalyst was the discovery that these string theories may be genuinely *finite* if the internal symmetry group is chosen in a very special way and if the entire system is made supersymmetric. The final proof of this conjecture has yet to be given, but the prognosis is excellent, and it is clear that the supersymmetric string offers one of the most promising prospects yet for obtaining a viable quantum theory of gravity.

5.6 Quantum theory and the structure of space and time

Discussions of quantum gravity from the perspective of general relativity almost invariably focus at one stage or another on the role of the classical ideas of space and time. However, problems of this type are not just the prerogative of relativists, and one of the most intriguing, and as yet unanswered, questions in superstring theory concerns precisely the general status of spacetime concepts. In conventional quantum gravity, the graviton is the quantum of the metric tensor field (or, more precisely, of its deviation from flat space). But in string theories the graviton is just one of an infinite set of particles, and the question arises of whether the remaining, massive, members of this set also have a geometrical interpretation. This problem is complicated by a feature of these theories that I have so far neglected to mention: that to ensure internal consistency, the spacetime in which the supersymmetric string moves has to be ten dimensional!

The idea of 'many dimensions' has been a common feature of much recent theoretical speculation, but of course it does pose the intriguing question of why the extra dimensions are not 'seen' in the ebb and flow of daily life? This is usually answered by supposing that they are 'curled up on themselves' in a very small circle (presumably of Planck length size). The question of whether the global topological structure of the extra dimensions is exactly a set of circles, or whether it is something more complex, is currently a matter of some debate; but the general idea seems to work, provided that the extra dimensions are *spatial*: trying to have more than one time dimension is not very productive!

Motivated by such thoughts, we might turn our attention to the general problem of the implications of quantum gravity for our ideas of space and time. The central question here has always been whether a fundamental revision of spacetime concepts should *precede* attempts to quantise the gravitational field, or whether it should emerge 'after the event'. The big advantage of the latter approach is the existence of a great body of theoretical material against which new ideas can be developed and tested. As we have seen, particle physicists have had some success with this option and are now at the stage where really exciting questions can be asked about spacetime structure at the Planck length and beyond. On the other hand, there *are* genuine *a priori* objections to treating gravity as 'just another field', and therefore the former, more iconoclastic, option also has its attractions. But then we have to confront the very difficult question, 'Where should we start?'.

In practice, the answer is often determined by the preconceptions that people bring to bear on the subject, and by their expectations of what might be achieved by a fully successful quantum theory of gravity. There is no better example with which to illustrate this point than the famous 'Hawking radiation' from a black hole. 'Black holes' are, of course, thus named for a good reason, namely that the gravitational field is so strong that light rays attempting to emerge from the body are inevitably pulled back, giving rise to an asymmetric situation in which anything can fall into the hole, but nothing can escape. However, in 1974 Stephen Hawking made the astonishing discovery that if quantum theoretical considerations are taken into account, a 'black' hole is very far from being black; in fact, from a distance, it looks like a *hot* body, radiating particles with a temperature that is inversely proportional to the mass of the hole. There are a variety of quasi-pictorial ways of explaining this phenomenon, and perhaps the most illuminating is the following. One of the peculiar effects of special relativity combined with quantum field theory is that a purely static field (say the gravitational field of the Earth) can trigger the production of a particle and an anti-particle. Since the field is static, it has no energy that could be converted into 'real' particles but, because of the Heisenberg uncertainty principle it is nevertheless possible for a pair of 'virtual' particles to be produced, provided they recombine very shortly after their creation (figure 5.17).

Now consider the analogous process (figure 5.18) near the event horizon of a black hole (i.e. the 'one-way membrane' through which nothing can escape). Particle pairs will again be produced, and some of them will behave as before and recombine. But another possibility is that one of the pair will fall through the event horizon and become trapped inside the

black hole. It can therefore no longer recombine with its partner, which is free to wander off and become part of the outgoing radiation!

This is a remarkable process, but one might wonder where the energy has come from with which to produce the outgoing particles. The answer is related to the peculiar status enjoyed by energy in general relativity, and in particular to its dependence on the frame of reference. In the context of figure 5.18, it transpires that, if measured near the event horizon, the energy of the ingoing particle appears to be *negative* and is exactly balanced by the positive energy of the outgoing particle. But then the negative influx of energy *reduces* the mass of the black hole according to the usual $E = Mc^2$ law, and so the energy of the radiation observed at a distance comes ultimately from the rest mass of the black hole!

The Hawking effect has many implications for the study of the interaction of quantum theory and gravity, and it raises some intriguing questions about the very high energy behaviour of particles. For example, can we conceive a process in which two protons collide with sufficient force to produce a mini black hole, which then radiates with the Hawking spectrum? Such an effect would be very difficult to uncover within the weak field perturbative approach discussed in Section 5.5, and it gives some idea of the type of *a priori* structure that might be imposed in a more global approach to quantum gravity.

However, even if we accept the need for a fundamental revision of the role of spacetime concepts in quantum theory, we must still face the problem of where to start, and most work in this area has, of necessity, followed an indirect route. Typically, this involves using fairly conventional quantisation algorithms but in a way that retains the geometrical flavour of classical relativity as strongly as possible, and with the hope that this will eventually point towards the new ideas that are expected to arise in the unification of quantum theory and general relativity. In practice, this means that, of the two ways of interpreting quantum states discussed in Section 5.3, the preferred choice is the one in which the basic observables are

Figure 5.17. The static gravitational field of the Earth can induce a 'virtual' production process in which a pair of particles are produced at *A* and then annihilate back into 'nothing' at *B*.

Figure 5.18. If the source of the gravitational field is a black hole then, in addition to the 'virtual' process depicted in figure 5.17, there is a second possibility in which a pair of particles is produced at *C* and one of them falls into the black hole. Because of the 'one-way' nature of the event horizon, this particle can no longer recombine with its partner which is therefore free to drift away and be seen as a 'real' particle. This is the origin of the celebrated Hawking radiation.

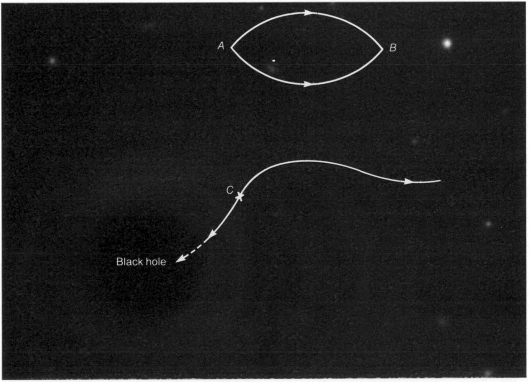

the values of the field; in our case this means studying the probabilities of measuring various ranges of values for the curvature of three-dimensional space as it evolves through time. Much thought has been given to this picture and to the problem of tackling the intensely difficult mathematical calculations. The single most important challenge is to develop approximation methods that can replace the weak field perturbation theory of the particle physics approach (with its fixed background spacetime structure) and at the same time maintain the geometrical flavour.

In any branch of physics, nonperturbative calculations are notoriously difficult, and it is not surprising that progress in this direction has been fairly modest. One of the more successful ideas has been to expand the components of the metric tensor in a type of Fourier series (cf. equations 5.27) and then to throw away all but, say, the first N modes. This has the effect of converting the infinite-dimensional field theory to a system with just a finite number (N) of degrees of freedom, and if N is sufficiently small the ensuing quantised field equations can be solved exactly. However, before this can be regarded as a fully legitimate way of tackling the problem, it is necessary to estimate the sizes of the effects that are neglected in studying the quantum fluctuations of only a finite set of modes; and this is very difficult. Indeed, the problem is in many ways analogous to the nonrenormalisability of the particle physics approach, and, drawing on the experience gained there, it seems unlikely that pure general relativity can be controlled in this way. Once again, the question of modifying Einstein's equations raises its head and, for example, it will be important to study supersymmetric string theories from a more nonperturbative, spacetime oriented point of view.

This remains a task for the future but, even for the unmodified Einstein theory, the mode truncation scheme has produced some intriguing results. For example, one can start to see how to tackle the important question of the effects of quantisation on the classical phenomenon of gravitational collapse. This would involve looking for vectors in the quantum state space that are 'special' vectors of the metric field observable with the property that the curvature, if measured in such a state, would be found to be infinite. One can then take a typical state vector and see if it is orthogonal to these 'singular configuration' special vectors. If so, the basic interpretative rule of quantum physics shows that the probability of finding such configurations is zero; presumably this is a signal that the quantisation has averted the classical singularity. The final outcome of investigations of this type is uncertain, but it is already apparent that some very peculiar things can happen in the neighbourhood of a singularity, including for example an extensive, quantum induced 'degeneracy' of the metric tensor in which the components are forced to vanish over large regions of space or spacetime. According to the 'Pythagoras theorem' interpretation of the metric, this would mean that any two points in such regions are at a zero distance from each other, even though they are distinct points!

One of the more speculative applications of these methods has been to the study of the very early universe (see Chapter 3 by Guth and Steinhardt), and indeed to the Creation itself! As remarked earlier, the 'big bang' is a striking example of a classical gravitational singularity, and an irresistible target for would-be quantisers of the gravitational field. It is even possible to contemplate giving some precise technical meaning to the question 'What is the probability of observing various configurations for the universe today, given that it originated in the past "out of nothing"?', i.e. the *Creatio Ex Nihilo* can perhaps be regarded as a type of quantum tunnelling from 'nothing' to 'something'.

The operative word here is 'technical', because, even setting aside theological considerations, it is clear that we must encounter some of the profound conceptual difficulties that accompany the present day formulation of quantum physics. Throughout the history of quantum theory, there has been a sustained, and often impassioned, debate as to what the mathematical structure 'really means', and in particular, how the probabilistic predictions are to be interpreted. The 'safe' approach is to treat them in much the same way as statistical laws are handled in the rest of physics, i.e. as statements of what will happen 'on the average' if an experiment is repeated a large number of times. But this means that, because of the insistence that quantum mechanical probabilities are *irreducible* (i.e. there are no hidden variables whose statistical fluctuations in the ordinary sense determine the observed results), one has to accept the fact that nothing can *ever* be said about the behaviour of a *single* system. Even in the context of ordinary atomic physics, this is surely one of the profoundest ontological pronouncements ever made by science. And in the context of the 'probabilities' of measuring various configurations for the entire universe (not an easy experiment to repeat), it becomes distinctly mind-boggling!

5.7 Where do we go from here?

Having touched the metaphysical heights, the time has come to descend to Earth and consider what has been learnt, and what the future might hold. From the perspective of elementary particle physics, it is clear that the supersymmetric string offers a most exciting and promising prospect for the construction of a quantum theory of gravity that is in harmony with our current understanding of the other fundamental forces of nature, and it is realistic to expect that the crucial conjecture on the finiteness of the theory will be confirmed, or denied, definitively in the near future.

From a more conceptual viewpoint, the move away from the 'structureless' point particle to a one-dimensional entity seems something to be welcomed; although whether it goes far enough is a moot point. A few preliminary investigations have been made into the possibility of extending this idea to include objects of two dimensions ('membranes') and above, but it is not yet clear whether theories of this type are really viable. The long term implications of the fact that in all cases the number of spacetime dimensions is greater than four, have not been fully assessed, both from the viewpoint of genuine high energy particle physics, and from the more exotic considerations of the history of the very early universe.

In general terms it is clear that the major interest in quantum gravity will always tend to focus on problems like the 'big bang' and the profound implications of the fundamentally different conceptual structures of quantum theory and general relativity. What is unclear is how much, if anything, will be left of these structures at the end of the day. The idea of a spacetime 'continuum' seems particularly vulnerable, and many attempts have been made to suggest ways of dispensing with this concept at a sub-microscopic level. Some of these (e.g. quantum topology) have been mentioned already, but there are many others. Perhaps spacetime is really a fractal (a mathematical space with a fractional number of dimensions), or a lattice, perhaps modelled on a finite number field; or perhaps, as in Penrose's twistor theory, there are no spacetime points as such, but rather systems of lines whose intersections are equivalent to points in the classical theory but which fail to intersect at all when subject to quantum fluctuations. And then there is the question of 'complexified' time. In conventional quantum theory the use of a complex number for the time variable is a well-established procedure, and is deeply connected with the phenomenon of 'quantum tunnelling'. But what does 'complex time' mean when space and time are curved? Could quantum topology transformations be described in this way?

We simply do not know. Neither do we know how much the basic ideas of quantum theory itself can be trusted in regimes so far removed from the world of atomic physics in which they originated. Indeed, it has been suggested more than once that the apparently fundamental linear structure of quantum theory is only an approximation to something else, and that it is precisely in the context of quantum gravity that this approximation will be seen to fail.

One of the most exciting things for a theoretical physicist is to be involved in the creation of a theory that, however modestly, actually works. But, in many respects, it is even more exciting to be present at the collapse of one of the great edifices of the subject, with the expectation that an Albert Einstein will rise one day to rewrite physics on the ruins of the old order. This is precisely how the inability to reconcile quantum theory with general relativity has presented itself for many years, but we still do not know if the disease is terminal. Perhaps the supersymmetric string will provide the cure; or perhaps it won't. In either event, it is going to be fun!

6 The new astrophysics

Malcolm Longair

6.1 The New Astrophysics

There is a well-known story about an Irishman who loses his key on his way home on a very dark night and is discovered searching for it underneath a lamp-post. 'Paddy,' his friend asks, 'did you lose your key here, then?' 'Sure, no,' Paddy replies, 'but it's the only bloody place I'd be able to find it, isn't it!' The story is almost an exact allegory for the impact of the New Astrophysics upon astronomy. Up till 1945, astronomers could only study the Universe at large in the optical waveband. The period since 1945 has seen the rapid expansion and exploitation of all the other wavebands available for astronomical study. It is the new disciplines of radio, millimetre, infrared, ultraviolet, X- and gamma-ray astronomies combined with optical astronomy which has led to the growth of the New Astrophysics. The astronomers and astrophysicists can now search for the key over a much wider area than ever before. It must remain a matter of debate as to whether or not they have found it, but in the process of searching they have discovered a huge range of new phenomena which were previously unknown and which may rightly be considered 'the New Astrophysics'.

What is the reason for this great upsurge in modern astrophysics? To put it at its very simplest, it is a question of the range of *temperatures* which are accessible for astronomical study. Figure 6.1(*a*) shows a plot of the temperature of a black-body against the frequency (or wavelength) at which most of the radiation is emitted. Figure 6.1(*b*) is a representation of the transparency of the atmosphere to radiation of different wavelengths and shows how high a telescope must be placed above the surface of the Earth so that the atmosphere is transparent to radiation of different wavelengths.

Until 1945, astronomy meant optical astronomy, and figure 6.1 shows that this corresponded to studying the Universe in the rather small wavelength interval 300 to 800 nm, and hence to black-body temperatures in the range 3000 to 10 000 K. Of course, a somewhat wider range of temperatures can be studied since bodies at temperatures outside this range emit some radiation in the optical waveband, but this range of temperatures is a fair representation of the temperatures of most of the objects observed at optical wavelengths. It is from these observations that most people derive their intuitive picture of the Universe, a picture dominated by stars, galaxies and hot gas, most of these components emitting at temperatures between 3000 and 10 000 K (see figure 6.2*d*).

The history of astronomy since 1945 has been dominated by the opening up of the whole of the electromagnetic spectrum for astronomical observation. In turn, each new waveband has leaned heavily upon the optical waveband because of the large information content of optical observations and because discoveries in the new wavebands can be related to the huge corpus of knowledge gained from optical studies.

What will become apparent as we describe these new vistas is the key contribution of technological advance in enabling the new astronomies to be undertaken. With new detector technologies and electronic computers, as well as the development of space science as a discipline in its own right, new challenges have been provided for instrument builders. In parallel with these endeavours, theoretical astronomy has capitalised upon these advances to explore totally new areas of parameter space.

Figure 6.1. (*a*) The relation between the temperature of a black-body and the frequency (or wavelength) at which most of the energy is emitted. The frequency (or wavelength) plotted is that corresponding to the maximum of the black-body curve at temperature T. Convenient expressions for this relation are:

$$\nu_{max} = 10^{11}(T/K)\,\text{Hz and } \lambda_{max}T = 3 \times 10^6\,\text{nm K}.$$

The range of wavelengths corresponding to the different wavebands – radio, millimetre, infrared, optical, ultraviolet, X- and gamma-ray – are shown. (*b*) The transparency of the atmosphere for radiation of different wavelengths. The solid line shows the height above sea-level at which the atmosphere becomes transparent for radiation of different wavelengths. (After Giacconi, R., Gursky, H. and van Speybroeck, L. P. (1968). *Ann. Rev. Astr. Astrophys.* **6**, 373.)

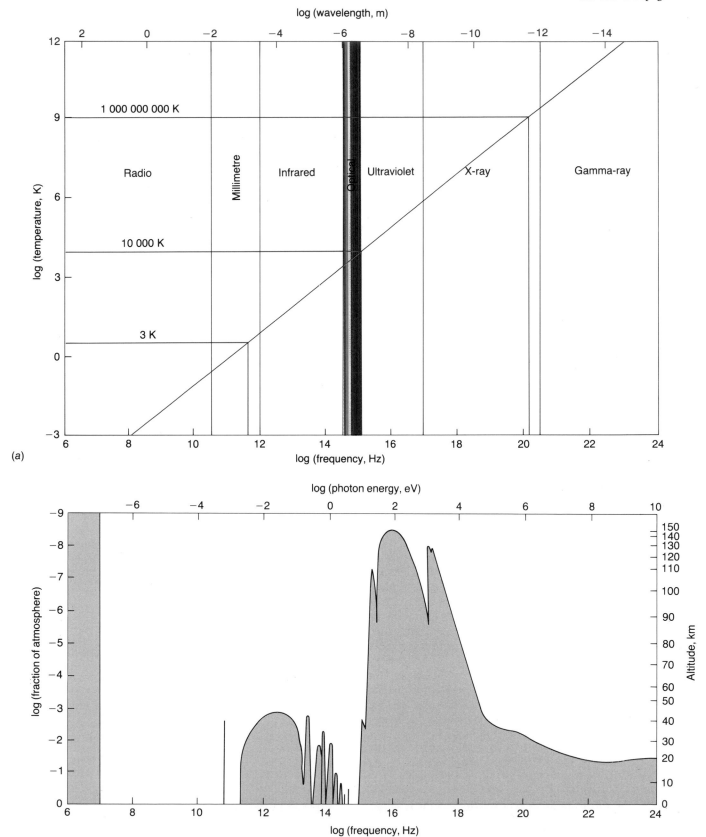

(a)

(b)

These advances are providing us with an understanding of the behaviour of matter under physical conditions which are inaccessible in the laboratory. Astrophysical and technological advances have gone hand-in-hand throughout the period since 1945 and it is a continuing process, which I confidently predict will extend far into the future.

To set the scene for the description of the new astrophysics, it is helpful to put the advances in their recent historical context. Readers interested simply in the new physics may advance to Section 6.2.

Radio astronomy

The first of the new astronomies was *radio astronomy*. Radio waves of extraterrestrial origin were discovered by Jansky in the early 1930s but this caused no great stir in the astronomical world. After the Second World War, radio astronomy developed very rapidly as great advances were made in electronics, radio techniques and digital computers. Radio emission was discovered from a wide range of different astronomical objects. Some of the radio emission processes could be related directly to phenomena observed at optical wavelengths – for example, the thermal radiation (or, more precisely, the free–free emission or bremsstrahlung) of hot electrons from regions of ionised hydrogen – but others were totally new. The radio emission was not generally of 'thermal' origin. It was soon established that, in most of the sources, the radio emission was synchrotron radiation, the emission of ultrarelativistic electrons spiralling in magnetic fields (figure 6.2a). Contrary to what might have been expected from figure 6.1, these radio observations provide information about the very hottest, indeed relativistic, plasmas in the Universe.

Observations made in the 1950s established that most of the discrete radio sources are extragalactic objects and two features were of particular significance. First, a number of the most massive galaxies known were found to be extremely powerful sources of radio waves. They are so powerful as radio sources that it is easy to detect them at cosmological distances, i.e. at distances such that the Universe was much younger than it is now when the radio waves were emitted. Simple calculations of the amount of energy necessary to power these radio sources showed that they must contain an energy in relativistic matter equivalent to the rest mass energy of about 100 million solar masses (i.e. about $10^8 M_\odot c^2 \simeq 2 \times 10^{55}$ J). (See Section 6.10 for a definition of $M_\odot$.) These galaxies must be able to convert mass of this order into relativistic particle energy. The second key fact is that the radio emission does not generally originate from the galaxy itself but comes from giant radio lobes which extend far beyond the confines of the parent galaxy. Through the 1960s and 1970s it was established that the sources of these vast energies were the active nuclei of these galaxies and the extended sources result from the expulsion of this energy from the nuclei in the form of jets of relativistic plasma.

Figure 6.2. The whole sky as observed in different astronomical wavebands. In each image, the plane of our Galaxy, the Milky Way, lies along the centre of the image and the direction of the centre of our Galaxy lies in the centre of each diagram. This form of projection is known as an Aitoff projection in which equal areas on the surface of the celestial sphere are preserved but the orthogonality of the coordinate system is distorted away from the Galactic equator. The coordinate system shown is known as *Galactic coordinates*. The north and south Galactic poles ($b = \pm 90^0$) are at the top and bottom of each diagram. The scale of Galactic longitude runs from 0^0 at the centre through $+180^0$ at the left of each image and then from $+180^0$ at the right of each image to 360^0 (or 0^0) at the centre.

Before 1945, only the optical waveband (*d*) was available for astronomical study. All the others have been opened up since that time as outlined in the text.

(*a*) Long radio wavelengths (408 MHz; 73 cm). This image is dominated by the radio emission of relativistic electrons gyrating in the interstellar magnetic field, a process known as *synchrotron radiation*. The radiation is most intense in the plane of the Galaxy but it can be seen that there are extensive 'loops' and filaments of radio emission extending far out of the plane. At high Galactic latitudes, there is a radio background component, most of it associated with radio emission of the Galactic disc and the halo of the Galaxy but some of it with an isotropic component of diffuse radiation. In addition, at high Galactic latitudes there are many discrete radio sources, some of which are visible on this image. The vast majority of these sources are of small angular diameter. If a survey of the discrete radio sources is made, their distribution is found to be isotropic (figure 6.73) and the integrated intensities of these sources can account for the isotropic background radiation at long radio wavelengths ($\lambda \geqslant 30$ cm). (Courtesy of Dr Glyn Haslam, Max-Planck-Institut für Radioastronomie, Bonn.)

(*b*) Millimetre wavelengths (38 GHz; 8 mm). This image of the sky was made by the RELIKT-1 spacecraft, which made a complete survey of the sky at this wavelength. The image is a 'differential' map in the sense that all the intensities are measured relative to the intensity of the Microwave Background Radiation at a fixed point in the sky which has brightness temperature 2.75 K. The colour coding of the intensity scale corresponds to a temperature difference between the maximum and minimum of $+8$ mK and -4.5 mK, respectively. The Galactic plane can be seen, the emission being associated with the free–free emission of extensive regions of ionised hydrogen. The image is, however, dominated by a 'dipole' component in the temperature distribution of the Microwave Background Radiation which is hottest in the direction $l \simeq 270^0$, $b \simeq 30^0$ and coolest in the direction $l \simeq 90^0$, $b \simeq -60^0$ (see also figure 6.74). The amplitude of this dipole component amounts to a temperature fluctuation of $\Delta T/T \simeq 10^{-3}$. This can be wholly attributed to the motion of the Earth through the frame of reference in which the Microwave Background Radiation is 100% isotropic at a velocity of about 350 km s^{-1}. (Courtesy of Dr I. A. Strugov and the Space Research Institute, Moscow.)

(*c*) Far-infrared wavelengths (60–100 μm). This image of the sky was generated from the IRAS all-sky survey and shows the distribution of radiation at 60 μm. The picture is dominated by emission from the Galactic plane, the radiation being the reradiated emission of heated dust grains. This map therefore delineates regions in which active star formation is proceeding. In addition, a broad band of radiation can be observed stretching across the map from top right to bottom left. This is the thermal radiation of zodiacal dust which is dust lying in the ecliptic plane of our own solar system and which is heated by the Sun. (Courtesy of NASA, the Jet Propulsion Laboratory and the Rutherford–Appleton Laboratory.)

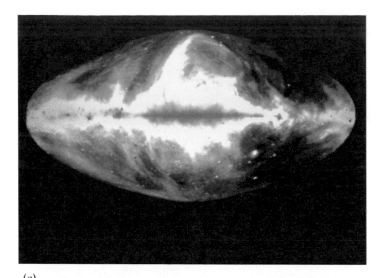

(a)

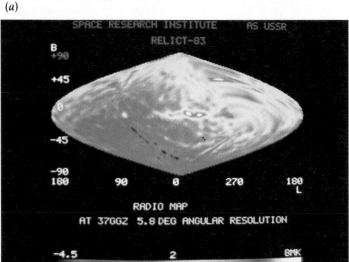

(b)

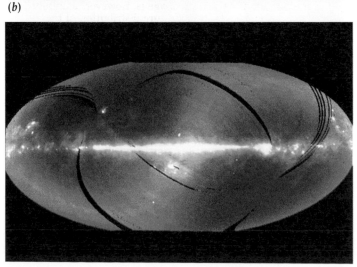

(c)

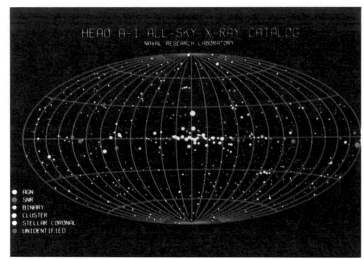

(d)

(e)

(d) Optical wavelengths. A painting of the whole sky as observed at optical wavelengths by M. and T. Keskula of the Lund Observatory. The painting reproduces accurately the optical images of nebulae and the observed brightness of stars down to about 10th magnitude. This image is convincing evidence that we live in a disc-shaped galàxy. The nearby dwarf companion galaxies to our own Galaxy, the Large and Small Magellanic Clouds, are seen in the Southern Galactic Hemisphere at about Galactic longitudes 290° and 310°, respectively. (Courtesy of the Lund Observatory.)

(e) X-ray wavelengths (1–5 keV). This distribution of bright X-ray sources was derived from the HEAO-1 survey of the X-ray sky. The image shows the distribution of the different types of X-ray source seen among the brightest objects. There is a concentration of the brightest sources towards the plane of the Galaxy and towards the Galactic Centre but at high Galactic latitudes the distribution of sources is isotropic within the limits of the available statistics. There is in addition an intense background component of diffuse X-ray emission. Its origin is as yet uncertain but discrete sources such as active galaxies must make up a significant fraction of the background. (Courtesy of NASA.)

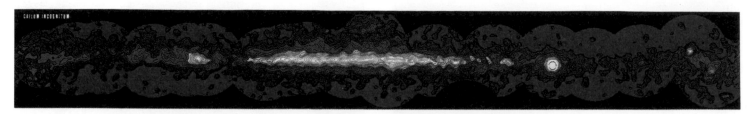

(*f*) Gamma-ray wavelengths (70 MeV–5 GeV). This map of the gamma-ray emission from the Galactic plane was made by the COS-B satellite. The emission consists of diffuse gamma-ray emission from the interstellar gas, most of it probably associated with gamma-rays produced in the decay of π^0 particles generated in collisions between cosmic ray protons and nuclei and the interstellar gas. In addition, twenty-five discrete sources of gamma-ray emission have been detected including the pulsars in the Crab and Vela supernova remnants and the quasar 3C 273. (Courtesy of Dr K. Bennett and the European Space Agency.)

These discoveries revealed a major new component of the Universe which had been previously unrecognised – *relativistic plasma*. It is present in all galaxies and in intergalactic space. These discoveries were the touchstone for the explosive growth of high energy and relativistic astrophysics over the last twenty-five years. The basic result of these new discoveries was that the energy demands and the time scales over which the energy had to be released were so extreme that conventional, astrophysical sources of energy, in particular nuclear energy, were inadequate, and relativistic phenomena associated with compact energy sources had to be investigated in detail.

The study of these radio sources led to further discoveries. Amongst the earliest of these was the fact that supernovae, or exploding stars, are very powerful sources of relativistic plasma. The study of the radio galaxies led to the discovery in the late 1950s of a class of galaxies known as *N-galaxies* which have very bright star-like nuclei in which there is a great deal of high energy activity as demonstrated by the observation of strong, broad emission lines in their spectra and their strongly varying optical continuum radiation. The culmination of these studies was the discovery of the *quasi-stellar radio sources*, or *quasars*, in the early 1960s, in which the starlight of the galaxy is completely overwhelmed by the intense nonthermal optical radiation from the nucleus. In some cases, the optical emission from the nuclear regions can be more than 1000 times greater than that of the parent galaxy (figure 6.3). These objects and their close relatives, the *BL–Lacertae*, or *BL–Lac*, *objects*, which were discovered in 1968, are the most powerful energy sources known in the Universe. Because of this, these objects can be observed at very great distances and provide important diagnostic tools for cosmology. The most distant quasars now known emitted their light and radio waves when the Universe was less than one-fifth of its present age.

Because they can be observed at such great distances, the quasars have given crucial information about the way in which high energy activity in galaxies has changed as the Universe grows old. They can also be used as probes of the gas lying along the line of sight from the quasar to the Earth through the observation of absorption lines in their spectra. Perhaps most

Figure 6.3. The quasar 3C 273. This deep image taken by Dr Halton Arp shows the quasar and its famous jet pointing towards the bottom right-hand corner of the image. The faint smudges to the south of the quasar are galaxies at the same distance as the quasar. (Arp, H. C. (1981). In *Optical Jets in Galaxies*, B. Battrick and J. Mort, eds., p. 55. ESA Publications SP-162.)

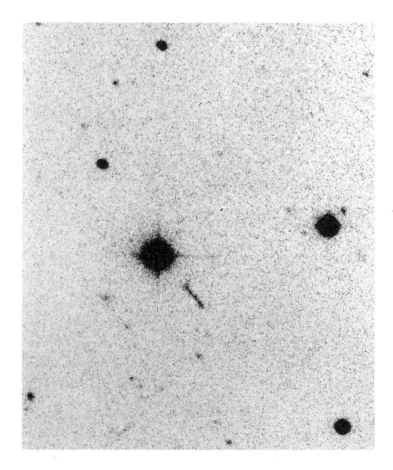

remarkable of all, convincing evidence for *gravitational lensing* has been discovered in the double quasar 0957 + 561. The two quasars are separated by only 8 arcsec and have identical spectra (figure 6.4). The two quasars are actually the images of a single background quasar but the light rays are deflected by a galaxy lying more or less along the line of sight to the quasar. This phenomenon, predicted by Einstein's general theory of relativity, has now been discovered in a few other distant quasars. The significance of these discoveries for theories of gravity is described in detail in Chapter 2 by Will.

As studies of active galactic nuclei were mushrooming in the 1960s, two other great discoveries were made in radio astronomy. The first was the discovery of the *Microwave Background Radiation* by Penzias and Wilson in 1965. Subsequent studies showed that this radiation is remarkably isotropic in the sense that it has the same intensity in all directions on the sky to a very high degree of precision (figure 6.2b) and has an almost perfect black-body spectrum. This radiation is the cool remnant of the equilibrium radiation spectrum formed early in the hot, dense early phases of the expanding Universe.

The second was the discovery of *pulsars*. In 1967 Bell and Hewish constructed a radio telescope to study very short time scale fluctuations imposed upon the intensities of compact radio sources by density fluctuations in the interplanetary plasma streaming out from the Sun, what is known as the solar wind. Early in these studies, sources consisting entirely of pulsed radio emission with very stable periods of about 1 s were discovered. They were soon identified conclusively as rotating, magnetised *neutron stars* and thus provided the first definite proof of the existence of these highly compact stars in which the central densities are as high as 10^{18} kg m^{-3}. A key point from the perspective of relativistic astrophysics was the fact that solar mass objects had been discovered with radii within a factor of three of the Schwarzschild radius of solar mass black holes. Thus, in these compact objects, general relativity is no longer simply a small correction to the equations of motion. The effects of general relativity are strong and these objects provide laboratories for the study of matter in strong gravitational fields.

One discovery of particular significance for general relativity, made by Hulse and Taylor in 1970, was that of the close *binary pulsar* PSR 1913 + 16, i.e. a pulsar in a close binary system. Accurate measurements of the arrival times of the pulses from this pulsar have shown that the system is losing energy at exactly the rate expected if the principal energy loss process is gravitational radiation from the rotating binary system. More recently a pulsar with period 1 ms has been discovered which has quite remarkable stability, the fractional period change $\Delta T/T$ amounting to less than one part in 10^{13} over a period of one year. These objects provide relativists with almost perfect clocks and are ideal laboratories for many subtle tests of general relativity (see Chapter 2 by Will).

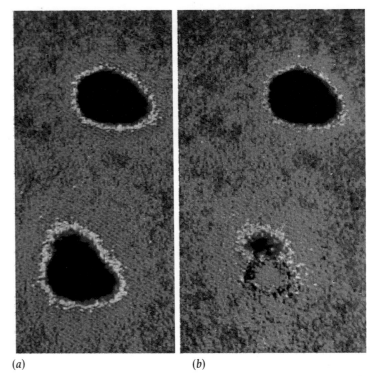

(a)   (b)

Figure 6.4. (a) The optical image of the double quasar 0957 + 561 A and B. The images are in fact point-like although they appear elliptical in this picture which has been enhanced by computer processing to show very faint features. It can be seen that there is a faint object just to the north of the southern component of the double quasar. (b) The optical image of the double quasar with the northern image subtracted from the southern image. This reveals that the 'smudge' to the north of the southern quasar is in fact a faint galaxy which is responsible for the gravitational lensing. (From CCD images taken by Dr A. Stockton and reproduced in Henbest and Marten (1983), p. 233.)

X-ray and gamma-ray astronomy

X-ray astronomy can only be carried out at very high altitudes because of photoelectric absorption of X-rays by the atoms and molecules of the Earth's atmosphere (see figure 6.1). Thus, it was only after rockets capable of lifting scientific payloads above the atmosphere became available that the exploration of the X-ray sky became possible. These rocket flights provided about five minutes of observation above the atmosphere but this was enough, even in the first rocket flights of 1962 and 1963, to show that the X-ray sky must be rich for astrophysical study. As in the case of the radio waveband, the sources which were first observed had not been predicted by astrophysicists. Amongst the earliest detections in the 1–10 keV waveband were the supernova remnant the Crab Nebula, the nearby radio galaxy M87, a number of stellar X-ray sources which seemed to be highly variable and the diffuse X-ray background radiation.

The full scope of X-ray astronomy became clearer in 1970 with the launch of the first dedicated X-ray satellite, the UHURU satellite observatory, which mapped the X-ray sky and provided systematic monitoring of variable X-ray sources (figure 6.2*e*). Some remarkable discoveries resulted from this mission. The variability of some of the Galactic X-ray sources was found to be due to the fact that the compact X-ray emitter is a member of an eclipsing binary star system. In a number of these cases, the X-ray binaries were found to contain 'pulsating' X-ray sources and these were soon identified with magnetised rotating neutron stars but, in the cases of the X-ray sources, the source of energy is the infall of matter transferred from the primary star, the process known as *accretion*. The X-ray emission is basically the thermal emission of very hot gas heated up as it falls onto the magnetic poles of the neutron star. The source of matter to fuel this accretion process is the primary star of the binary system. The traditional optical techniques of measuring the orbits of binary stars could then be used to estimate masses of the primary star and the X-ray emitter. In the case of the pulsating X-ray sources, the inferred masses are consistent with their being neutron stars but, intriguingly, there are a few X-ray sources in which the mass of the invisible secondary is inferred to be greater than the upper limit for stable neutron stars. These are all candidates for *black holes* in binary systems and as such are objects of the greatest astrophysical interest.

In extragalactic astronomy, the nuclei of active galaxies were found to be intense and often variable X-ray sources. It is likely that this emission is associated with the emission of ultra-relativistic gas generated close to the nucleus itself but the precise nature of the emission is not clear. One class of extragalactic source in which the emission processes are understood is the X-ray emission of diffuse hot gas in clusters of galaxies. The mass of the cluster forms a deep gravitational potential well in which the gas must be very hot if it is to form a stable extended atmosphere. This is what is observed in a number of great clusters. The X-ray emission is extended, filling the core of the cluster, and the gas responsible for the emission has temperature in the range 10^7–10^8 K. The emission process is thermal free–free emission (or bremsstrahlung) as is confirmed by the observation of very highly ionised iron lines from the cluster gas. In some of the richest clusters, there is as much mass in the form of hot intergalactic gas as there is in the visible parts of the galaxies in the clusters, i.e. about 10^{14} times the mass of the Sun. This offers opportunities for studying the physics of hot gas at temperatures $\sim 10^7$–10^8 K in bulk.

In 1978, the Einstein X-ray Observatory was launched. It provided the first high resolution images of many X-ray sources and made very deep surveys of small areas of sky. Many different classes of astronomical object were detected including regions of star formation and normal galaxies. Perhaps most significant of all was the fact that X-ray emission has now been detected from all types of star and not just from the binary X-

ray sources for which there are special reasons why there should be strong X-ray emission.

Gamma-ray emission from the plane of our Galaxy was first detected by the OSO III satellite in 1967. This was followed by the SAS-2 satellite which discovered the diffuse gamma-ray background and by the COS–B satellite which provided a detailed map of the Galactic gamma-ray emission and discovered about twenty-five discrete gamma-ray sources. These included the pulsars in the Crab and Vela supernova remnants and the quasar 3C 273. The gamma-ray image of the sky is dominated by the emission from the Galactic plane (figure 6.2*f*). At high photon energies, $\varepsilon \geqslant 100$ MeV, the principal emission mechanism is the decay of neutral pions generated in collisions between the nuclei of atoms and molecules of the interstellar gas and cosmic ray protons and nuclei. At lower energies, nonthermal processes, in particular the inverse Compton process and bremsstrahlung, can make important contributions to the background gamma-ray emission.

Gamma-ray line emission was discovered by the HEAO-C satellite, among the most remarkable discoveries being the electron–positron annihilation line at 511 keV and also the 1.809 MeV line of radioactive aluminium ^{26}Al, both of which have been detected from the centre of our own Galaxy. Another unexpected discovery was that of gamma-ray bursts which were detected by the US Vela and also by Soviet satellites. Over one-hundred of these events have now been observed over the last ten years but their nature is not yet established.

Ultra-high energy gamma-rays ($\varepsilon \simeq 10^{11-12}$ eV) have recently been detected by a remarkable ground-based technique. Gamma-rays of these energies initiate small electron–photon cascades in the upper atmosphere. The electrons are of such high energy that their velocities exceed the speed of light in air and consequently they emit optical Cerenkov radiation. The optical light emitted by these showers is detected at sea-level by simple telescope arrays. Several well-known sources, including Cygnus X-3, Hercules X-1 and the Crab pulsar, have been detected as gamma-ray emitters at about 10^{12} eV. An important result claimed by the Durham group is the discovery of a 12.6 ms pulsar at gamma-ray energies in the binary X-ray source Cygnus X-3. Finally, at even higher gamma-ray energies, $\varepsilon \simeq 10^{16}$ eV, the Kiel and Leeds University groups have claimed to detect gamma-rays from Cygnus X-3 using ground-based cosmic ray air-shower detector arrays. At these very high energies, the fluxes of gamma-rays are very low indeed, typically only about one gamma-ray per month being detected by the Haverah Park team.

Cosmic ray astronomy

Radio, X-ray and gamma-ray astronomy have resulted in many discoveries which can only be interpreted in terms of the presence of large fluxes of relativistic particles in galaxies. In parallel with these developments, cosmic ray studies opened up new areas of astrophysical importance through direct observation of high energy particles at the top of the atmosphere and in the environment of the Earth from satellites and, for the very highest energy cosmic rays, from the surface of the Earth by the large air-shower arrays.

Cosmic radiation (what we would now call cosmic rays) was discovered as long ago as 1912 by Hess, but the astrophysical understanding of the origin and propagation of these particles had to await the 1960s when cosmic ray particle detectors were flown in satellites. These observations established many crucial facts about the particles detected in the cosmic radiation. First of all, the energy spectra of the particles are almost exactly the same as the typical spectrum of high-energy particles inferred to be present in both Galactic and extragalactic nonthermal radio sources. In the region of the energy spectrum which is unaffected (or, to put it technically, unmodulated) by the propagation of the particles to the Earth through the solar wind ($E \geqslant 10^9$ eV), the energy spectra of the cosmic ray particles can be described by

$$N(E)dE = KE^{-\gamma}dE$$

with $\gamma \simeq 2.5$. This relation is found to be applicable for protons, electrons and nuclei with energies in the range 10^9–10^{11} eV. The direct relation with the relativistic gas inferred to be present in the interstellar gas is through two types of observation – the synchrotron radiation of cosmic ray electrons is detected in the radio waveband and the Galactic gamma-ray emission is attributed to the decay of neutral pions π^0 created in collisions between cosmic ray protons and nuclei and the nuclei of atoms, ions and molecules in the interstellar gas. The fact that these very different types of astronomy can be brought successfully to bear on these problems indicates that the cosmic ray particles observed at the top of the atmosphere are only part of a population of high-energy particles pervading the whole Galaxy.

Subsequent satellite observatories have determined the chemical composition and detailed energy spectra of cosmic ray nuclei. Remarkably, the chemical composition of the cosmic rays is similar to the abundances of the elements in the Sun, although there are some variations in the abundances at the higher energies. These observations provide evidence on the chemical composition of the cosmic rays as they left their sources and also about the modifications which could have taken place during propagation from their sources to the Earth. These observations are very important for high energy astrophysics because they are the only *particles* which we can detect which have traversed a considerable distance through the interstellar medium and which were accelerated in events such as supernovae and possibly pulsars in the relatively recent past, probably within the last 10^7 years.

At the very highest energies, cosmic rays are detected by large air-shower arrays on the surface of the Earth. The arrival rate of the most energetic particles is very low indeed, but particles with energies up to about 10^{20} eV have been detected. One important puzzle is the origin of these very high energy particles. Their arrival directions seem to be reasonably isotropic and, at these very high energies, these should not be significantly influenced by the magnetic field in our own Galaxy. It may be that these very high energy cosmic rays are of extragalactic origin. The acceleration mechanism for these particles is uncertain and poses a real problem for high energy astrophysicists.

Ultraviolet astronomy

As soon as observations from above the atmosphere became possible, one of the obvious developments was the extension of the classical techniques of optical astronomy to the ultraviolet spectral regions, 120–320 nm. Ultraviolet spectrographs were flown on rockets in the mid-1960s and were followed by the series of Orbiting Astrophysical Observatories culminating in the launch of the International Ultraviolet Explorer (IUE) in 1978.

As expected, a wide range of hot objects could be studied, but perhaps of most importance was the fact that a wide range of the common elements could be studied because their strong resonance transitions fall in the ultraviolet spectral region. Among the most important of these absorption lines have been those of deuterium which is of great cosmological significance – it is likely that deuterium was synthesised in the early stages of the Hot Big Bang. Active galaxies and quasars are particularly strong emitters in the ultraviolet waveband because the nonthermal radiation from the nucleus observed in the optical waveband extends to far ultraviolet wavelengths. This continuum radiation excites a wide range of ions and atoms which emit strong resonance lines in the ultraviolet waveband. These lines have proved to be particularly valuable diagnostic tools for the astrophysics of active galactic nuclei.

The waveband 120–300 nm has now been extensively studied spectroscopically by the IUE but the shorter ultraviolet wavelengths, $\lambda < 120$ nm, remain relatively unexplored. There are two reasons why this is a difficult waveband for astronomical observations. First, there is the problem of constructing an efficient telescope because most' materials are strongly absorbant for normal incidence optics at wavelengths shorter than about 120 nm. Second, the interstellar gas is expected to be opaque to radiation of wavelength less than 91.2 nm, corresponding to the wavelength at which the Lyman continuum begins, and this restricts the targets available for study. Fortunately, it appears that the interstellar medium is sufficiently clumpy for there to be 'holes' through which the more distant Universe can be observed.

Infrared astronomy

The development of infrared astronomy has been largely determined by the availability of detector materials for the wavelength range 1 μm to 1 mm. There are certain wavelength 'windows' through which observations can be successfully made from the surface of the Earth in the ranges 1–30 μm and 350 μm–1 mm (see figure 6.1b). In the range 30–350 μm, however, the Earth's atmosphere is opaque and observations can only be made from high-flying aircraft, balloons and satellites.

The distinctive problem to be overcome in infrared astronomy is the fact that the telescope and the Earth's atmosphere are strong thermal emitters of infrared radiation. For example, the radiation of a black-body at room temperature, say 300 K, peaks at a wavelength of 10 μm. Therefore, normally, the strength of the signal from an astronomical source is very much weaker than the background due to the telescope and the atmosphere.

The normal procedure is therefore to compare the signal from the source direction with a nearby empty piece of sky. Even if the telescope is placed above the Earth's atmosphere, there is still the background signal from the telescope and satellite itself and so it is necessary to cool the telescope and mirror to a low temperature. In the case of the Infrared Astronomical Satellite (IRAS), which was designed to explore the 10–100 μm waveband, the telescope was cooled to about 4 K to minimise the background signal.

Besides the ability to study cool objects in the temperature range 3000–2 K, there are two other important features of the infrared waveband. First, there is a great deal of dust in the interstellar gas in galaxies but the dust grains become transparent in the infrared waveband and hence it is possible to look deep inside regions which are obscured at optical wavelengths. For example, the dust grains between the centre of our own Galaxy and the Earth cause the signal to be attenuated by a factor of about 10^{12} in the optical waveband, whereas at an infrared wavelength of 3 μm, the intensity is reduced by a factor of less than ten. Second, at wavelengths longer than about 3 μm, dust grains become strong emitters rather than absorbers of radiation. They emit more or less like little black-bodies at the temperature to which they are heated by the radiation they absorb. They do not radiate at shorter wavelengths because, if the grains were heated to temperatures greater than about 1000 K, they would evaporate.

The development of sensitive semiconductor materials with narrow band-gaps, such as indium antimonide (InSb) and arsenic doped silicon (Si:As), have revolutionised studies in the infrared waveband. These developments are continuing apace and the first detector arrays are now available for astronomical use.

Even in the early 1970s, it was apparent that many objects emit most of their radiation in the infrared waveband. A discovery of special importance was the fact that, in regions where stars are forming, a vast amount of energy is emitted in the infrared waveband between about 5 and 100 μm. This emission is associated with the very youngest stars and their progenitors, the protostars, which form in cool dusty regions. The energy they release is absorbed and re-emitted by heated dust grains, resulting in intense far-infrared emission. Wherever stars are being formed, this characteristic signature of intense far-infrared emission is found. A particular example of this is the case of dusty irregular galaxies which can emit luminosities up to 10^{12} times the luminosity of the Sun in the far-infrared waveband, about one-hundred times greater than the optical luminosities of the galaxies.

The first complete survey of the far-infrared sky was carried out by the IRAS in 1983–4. It revealed intense emission from regions of star formation in our own Galaxy and nearby galaxies as well as a host of new detections of stars, galaxies, active galaxies and quasars (figure 6.2c).

Neutral hydrogen and molecular line astronomy

One of the great predictions of modern astronomy was made during the Second World War by van de Hulst who worked out which emission and absorption lines of atoms, ions and molecules might be detectable from astronomical sources in the radio waveband. The most significant prediction was that neutral hydrogen should emit line radiation at a wavelength of about 21 cm because of the minute change in energy when the relative spins of the proton and electron in a hydrogen atom change. Although this is a highly forbidden transition with a spontaneous transition probability for a given hydrogen atom of only once every twelve million years, there is so much neutral hydrogen present in the Galaxy that it should be detectable. In 1951, the 21 cm line of neutral hydrogen was discovered by Ewan and Purcell, and it has proved to be one of the most powerful tools for diagnosing not only the properties of the interstellar gas but also the dynamics of galaxies since the line is so narrow that it provides an excellent measure of the velocity fields inside galaxies.

Molecules had been known to exist in the interstellar medium from the absorption bands seen in the optical spectra of stars. The real significance of molecular line astronomy only became apparent, however, with the development of high precision radio telescopes and line-receivers working in the centimetre and millimetre wavebands. In 1967, the hydroxyl radical OH was first detected by radio techniques. In many ways, this was an unexpected detection because the signals were very strong indeed and variable in intensity. In fact, the inferred temperature of the source regions was greater than 10^9 K, indicating that some form of *maser* action must be pumping the energy levels of the molecule. As in cases of masers and lasers, the populations of the energy levels of the molecules must be far from equilibrium in such a way that enormous intensities in the lines are observed, far exceeding those expected from the thermodynamic temperature of the source region.

Soon, many more molecules were discovered through their molecular line emission in the centimetre, millimetre and submillimetre wavebands, including species such as ammonia, water vapour and even ethanol. Most of the molecular lines which have been discovered are associated with the rotational transitions of molecules, linear molecules with up to eleven carbon atoms having now been detected. The importance of these studies is that they have led to the development of the new discipline of *interstellar chemistry*. For the molecules to survive, it is essential that they should be shielded from the intense interstellar ultraviolet radiation. It is therefore not surprising that they are found in large abundances in dusty star formation regions into which ultraviolet radiation cannot penetrate. Different molecules enable different ranges of temperature and density to be investigated and, because of the very narrow line widths of the molecular lines in cool regions, the internal dynamics of these regions can be studied in detail from the frequency structure of the line emission. These observations are of special importance because, as will become apparent later, star formation is one of the key areas of contemporary astronomy which is most poorly understood.

The present generation of millimetre telescopes will soon be complemented by submillimetre telescopes which will enable the last of the wavebands accessible from the ground to be explored.

Optical astronomy

It is perhaps ironic that a survey of developments in modern observational astronomy should end rather than begin with the optical waveband. We should first note the great technical developments in instrumental technique which have taken place over the last twenty years. Until 1945, essentially all astronomy was undertaken using photographic techniques. The period 1960 to 1987 has seen the gradual replacement of photographic plates by electro-optical detectors. Photographic plates typically have a quantum efficiency of only a few per cent so that most of the light incident upon the plate is not detected. The most recent electronic detectors such as charge-coupled devices (CCDs) have quantum efficiencies of about 60–70% at the red end of the optical spectrum (500–1000 nm). Furthermore, detectors such as CCDs have a linear response to the intensity of the incident radiation whereas photographic plates are nonlinear in the sense that they saturate at high brightness levels.

The great advance represented by these new devices is that nowadays astronomers deal directly with digital data and can perform much more readily procedures such as sky subtraction and intensity and wavelength calibration of their data. This development would not have been possible without the continued development of electronic computers. Nowadays, it is routine to deal with arrays of, say, 1000×1000 picture

elements and each element can be separately processed by computer. This has resulted in a great increase in the quantity and quality of the data which the astronomer can study astrophysically. The only present limitation in the use of these electro-optical devices is that they have only a small field of view. Thus, the photographic plate still has advantages when very wide field images or very long spectra have to be obtained.

What is the present status of optical astronomy in the era of the new astrophysics? There is no question but that it continues to play a central role in all astrophysics. The fundamental reason for this is that a large fraction of the matter in the Universe is locked up in stars with masses within about a factor of ten of that of the Sun and these emit a large fraction of their energy in the optical waveband. Since they have long lifetimes, they are the most readily observable objects in the Universe. The stars are assembled into galaxies and these are the basic building blocks of the Universe. Furthermore, the evolution of stars is one of the most exact of the astrophysical sciences and so stars provide among the best probes of the evolutionary history of any stellar system.

Thus, whenever observations are made in the new wavebands, it is important to relate the objects detected to what is observed optically in the same region of space. One particularly important aspect of the procedure of associating optical objects with objects detected in other wavebands is that very often distances can only be estimated from optical observations, for example from the properties of the associated star or the redshift of a galaxy. The key point to be appreciated is the richness of the optical spectrum in emission and absorption features.

The new techniques have developed so rapidly and are producing so much new science that, ideally, one would want to look back at all the knowledge which has been obtained with photographic plates and reanalyse it using modern detectors and data reduction techniques. This will gradually come about and provide us with a securer base for the whole of astrophysics.

In summary, we can say that, although all the new wavebands have deepened our appreciation of the nature of our Universe, they are to a greater or lesser extent dependent upon optical observations to convert the observations into hard astrophysics.

Theoretical astronomy

Unquestionably the years since 1945 have been one of the most exciting periods in theoretical astronomy. There are three points which are worth noting. First, the theoretical developments have been largely stimulated by the great observational discoveries of the period. What I find particularly striking is the completely new range of astrophysical tools which have had to be developed by theoretical astronomers before the interpretation of the observations can be undertaken. All the discoveries of the new astronomies above have resulted in entirely new pieces of theoretical astrophysics and these in turn have deepened our understanding of the behaviour of matter in circumstances which are not found within terrestrial laboratories. Some obvious examples are the interiors and environments of neutron stars, the nuclei of galaxies, X-ray emission from accretion processes in X-ray binaries and so on. It is questionable whether or not advances in understanding general relativity would have been so rapid without the stimulus of active galaxies and the discoveries of modern astrophysical cosmology (see Chapter 2 by Will). These studies have also resulted in many unsolved problems which at this stage are being intensively studied. We will highlight these fundamental problems in the succeeding sections.

A second important feature has been the use of high speed computers in essentially all aspects of theoretical astrophysics. This has been of special importance in some of the most exact of the astrophysical sciences. An excellent example is the study of the internal structure and evolution of stars. The development of powerful computer codes for stellar models has resulted in precise predictions of the surface properties of most classes of star. Detailed comparison between theory and observation is now possible and is leading to more and more precise understanding of the processes of stellar evolution. This is, however, but one example and it is fair to say that, in essentially all branches of modern astrophysics where well-defined astrophysical problems have been formulated, theorists have provided the observers with good predictions which enable sensible astrophysical questions to be asked.

The third aspect is the impact of advances in theoretical physics upon theoretical astronomy. Essentially all the major advances in theoretical physics have had an important impact upon some aspect of astronomy. The importance of nuclear physics in understanding the processes going on in the centres of stars is obvious. But there are many perhaps less obvious examples – superconductivity and superfluidity in the centres of neutron stars (see Chapter 9 by Leggett), the role of the electro-weak theory of elementary particles in understanding how the energy of collapse may result in a supernova explosion (see Chapters 14, 17 and 18 by Close, Taylor and Salam, respectively), the role of plasma physics in understanding the dynamics of gases of relativistic particles. It is remarkable how many of these theories are playing a role in modern astrophysics and by extension how these new applications provide further tests of the theory. We will find numerous examples of this as the story unfolds.

In a spectacular example of the inverse process, Hawking's discovery of the evaporation of black holes by radiation is a fundamental piece of theoretical physics which developed from astronomical and cosmological studies (see Chapter 4 by Hawking). The most ambitious of these endeavours is the application of Grand Unified Theories of elementary particles (see Chapter 15 by Georgi) and quantum gravity (see Chapter 5 by Isham) to the very earliest phases of the Hot Big Bang. The relevant energies are so high that these theories can probably only be tested by using the very early Universe itself as a laboratory for ultra-high energy physics (see Chapter 3 by Guth and Steinhardt). The fact that these ideas are being taken seriously by the best workers in the field is some measure of the advance in understanding and ambitions of astrophysicists in the late 1980s.

6.2   The grand design

To begin this survey of the new astrophysics, we first describe the astrophysical context within which the new astrophysics can be appreciated. Obviously, this must be a very broad-brush description and should be complemented by the more learned texts listed at the end of this chapter. It is my intention that the contents of this section describe aspects of the astrophysical sciences which would be agreed by essentially all astronomers and astrophysicists. The subsequent sections will approach much closer to the frontiers of knowledge where we may be on less secure ground.

The large-scale distribution of matter and radiation in the Universe

The modern picture of the way in which matter is distributed in the Universe comes from a wide variety of different types of observation.

On the very largest scale, the best evidence comes from measurements of the isotropy of the Microwave Background Radiation. Very sensitive measurements of the distribution of this radiation have shown that it has the same intensity to better than one part in a thousand wherever one looks on the sky on all angular scales from about 1 arcmin to 360 degrees. Just below this sensitivity level, a global anisotropy of dipole form is observed so that the radiation intensity is slightly greater in a particular direction and slightly less in the opposite direction (figure 6.2*b*). This dipole anisotropy can be wholly attributed to the motion of the Earth at a velocity of about $350\,\mathrm{km\,s^{-1}}$ through a frame of reference in which the Microwave Background Radiation would be 100% isotropic. To virtually all astrophysicists, it is wholly convincing that the microwave background radiation is the cool remnant of the hot early phases of the Big Bang, but the precise origin of the radiation is not so important so far as the underlying theoretical framework of our cosmological models is concerned. The key point is that it provides us with strong evidence that, looked at on a large enough scale, one bit of the Universe must look very much like another.

When we look at the distribution of visible matter in the Universe we see a much less smooth picture. Figure 6.5 shows the distribution of visible matter derived from counts of galaxies made in the 1960s. The processed data provide a broad view of the distribution of galaxies in the Northern Galactic Hemisphere, i.e. the view looking up out of the plane of our own

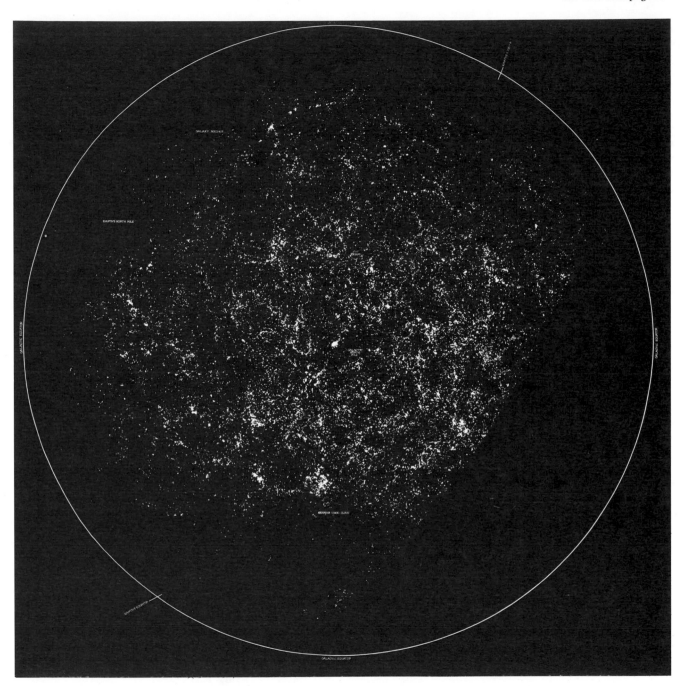

Figure 6.5. The distribution of galaxies in the Northern Galactic hemisphere derived from the counts of galaxies undertaken by Shane, Wirtanen and their colleagues at the Lick Observatory. Over one million galaxies were counted in their survey. The North Galactic pole is at the centre of the picture and the Galactic equator is represented by the solid circle bounding the diagram. The projection onto the sky is an equal area projection. This photographic representation of the galaxy counts was made by Dr P. J. E. Peebles and his colleagues. The large sector missing from the lower right-hand corner of the picture corresponds to an area of the southern hemisphere which was not surveyed by the Lick workers. The decreasing surface density of galaxies towards the Galactic equator is due to the obscuring effect of interstellar dust in the interstellar medium of our own Galaxy. The prominent cluster close to the centre of the picture is the Coma cluster of galaxies. (Seldner, M., Siebars, B., Groth, E. J. and Peebles, P. J. E. (1977). *Astron. J.* **82**, 249.)

Galaxy. Towards the centre of this picture, we obtain a reasonably unobscured picture of the distribution of galaxies. It can be seen that there are clearly irregularities in this distribution although, looked at on a large enough scale, one region looks very much like another. There is no question about the reality of a great deal of this structure. The dense knots correspond to regions of strong clustering of galaxies, for example the bright region towards the centre corresponding to the well-known Coma cluster of galaxies.

What has provoked a great deal of interest is the nature of the other structures on large angular scales apparent in this picture. To the eye, there appear to be holes and filaments in the distribution of visible matter. It has yet to be conclusively demonstrated from these data that these features are real. There may be some problems with patchy obscuration by interstellar dust which is now known to exist even at high galactic latitudes from the far-infrared survey of the whole sky made by the IRAS satellite. In addition, there are difficult problems in calibrating the counts of galaxies made in different fields.

One of the obvious problems in interpreting a picture like this is that it is a two-dimensional representation of the distribution of the galaxies and much more would be learned if their distances were known as well. Unfortunately, it is a very time-consuming job to measure the distances of galaxies, and this has only been undertaken for reasonably nearby samples of galaxies. An example of the results of such studies is shown in figure 6.6, which is taken from the Harvard-Smithsonian Center for Astrophysics survey of bright galaxies. These suggest that there are large holes or 'voids' in the distribution of

Figure 6.6. A sample of the data obtained as part of the Harvard-Smithsonian Center for Astrophysics survey of bright galaxies. The galaxies lie in different velocity (and hence distance) ranges: ∇ $3000 \leqslant v \leqslant 4000 \text{ km s}^{-1}$; $\triangle$ $4000 \leqslant v \leqslant 5000 \text{ km s}^{-1}$; $\bigcirc$ $5000 \leqslant v \leqslant 6000 \text{ km s}^{-1}$. It can be seen that the galaxies are not uniformly distributed but lie in 'chains' or 'filaments'. (From Davis, M. (1982). In *Astrophysical Cosmology*, H. A. Bruck, G. V. Coyne and M. S. Longair, eds., p. 117. Pontificia Academia Scientiarum.)

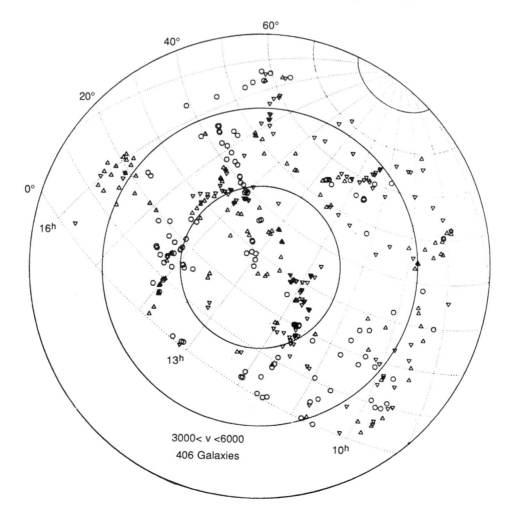

galaxies and that some of the long filamentary structures joining clusters of galaxies are real associations of galaxies. If this is indeed a correct picture, these are the largest known physical structures in the Universe, extending to scales of 100 Mpc (see Section 6.10) or greater. A number of studies have suggested that the distribution of galaxies possesses a sponge-like structure with the galaxies being located within the body of the sponge and there being large holes between the 'sheets' of galaxies.

On finer scales, the clustering of galaxies appears to occur on a very wide variety of scales from pairs and small groups of galaxies, like the Local Group of galaxies, to giant clusters of galaxies such as the Coma cluster, which can contain many thousands of members. The rich regular clusters are certainly self-gravitating bound systems. It is likely that most of the rich clusters of galaxies are bound systems but a number of them have an irregular, extended appearance and it is not so clear that these are self-gravitating, bound systems.

The term *supercluster* is used to define structures on scales larger than clusters. These may consist of associations of rich clusters or a rich cluster with associated groups and an extended distribution of galaxies. Some authors would classify the 'stringy' structures seen in figures 6.5 and 6.6 as superclusters or supercluster cells. From the physical point of view, the distinction between the clusters and the superclusters is whether or not they are gravitationally bound systems.

Notice that in this discussion of the large-scale structure of the Universe I have been careful to refer to the distribution of *visible* matter, and it is important to distinguish between this and the actual distribution of all the gravitating mass in the Universe. For a number of cogent reasons, there is good evidence that not all the gravitating mass in the Universe is visible. Therefore we cannot say with any certainty whether or not the actual distribution of mass is the same as that of the visible matter. Physically this is a very important question which has profound implications for the theory of the origin and stability of galaxies, clusters and other large-scale structures and quite possibly for fundamental physics. This theme of the *hidden* or *dark matter* will haunt much of this survey.

The galaxies

Galaxies are the basic building blocks of the Universe. Most of their visible mass is in the form of stars. It is the gravitational pull of the stars on one another which holds the galaxy together, although there may also be dark or hidden mass present in the outer regions of giant galaxies.

Many different types of galaxies have been identified, but the basic distinction which is obvious on a photograph is between *spiral* and *elliptical* galaxies. A smaller fraction of galaxies are known as *irregular* galaxies. The spiral galaxies (figure 6.7a) have a disc shape with a central bulge like our own Galaxy. The relative sizes of the disc and bulge vary from galaxy to galaxy. Their masses range from systems up to one-hundred times

more massive than our own Galaxy, which has mass about 100 billion times the mass of the Sun ($10^{11} M_\odot$) to dwarf systems which are only about ten million times the mass of the Sun ($10^7 M_\odot$). The discs of spiral galaxies rotate and this is what gives them their characteristic shapes. There are *normal* spirals and *barred* spiral galaxies in which the central bulge is elongated and the spiral arms trail from the ends of the 'bar' (figure 6.7b). The spiral arms are defined by the youngest, hottest, brightest types of star and also by the ingredients of regions of star formation, that is by gas clouds and dust. There is continuing star formation in the arms of spiral galaxies. The elliptical galaxies (figure 6.7c) have much smoother profiles with little evidence for dust, gas or spiral arms in general. They are spheroidal in shape and can have masses from about one-hundred times the mass of our Galaxy to only about ten million times the mass of the Sun. They are self-gravitating systems in which the random velocities of the stars prevent the galaxy collapsing.

There are galaxies which appear to be intermediate between the spirals and the ellipticals which are known as *SO* or *lenticular* galaxies (figure 6.7d). They look rather like spiral galaxies which have been stripped of their spiral structure; that is they possess stars in a disc and have a central bulge but have little evidence, or at best vestigal evidence, for spiral arms. Many of them are found in clusters of galaxies.

The *irregular* galaxies (figures 6.7e and f) are generally less massive than typical spiral and elliptical galaxies. They have an irregular structure and often possess large amounts of gas and dust. In a number of cases, in particular those referred to as 'peculiar' or 'interacting' galaxies, it is likely that the irregular appearance is due to the galaxy having had a recent strong gravitational interaction or 'collision' with a nearby galaxy.

There are, in addition, a number of special classes of galaxy, most of which are very much less common than the above types. Perhaps most interesting astrophysically are the galaxies with *active galactic nuclei*. Historically, the first class of galaxies with active nuclei to be discovered were the *Seyfert* galaxies. They appear to be spiral galaxies but possess star-like nuclei (figures 6.8a and b). When the spectra of these galaxies were studied, it was found that the emission lines are very broad and strong, unlike those found in any of the emission line regions in normal galaxies. Many of these galaxies have now been found and they are among the most important classes of active nuclei because they are relatively common.

The next class of active galaxy to be discovered were the *radio* galaxies. By the mid 1950s it was established that these galaxies must be sources of vast amounts of high energy particles and magnetic fields. A few of these galaxies had star-like nuclei and were called *N-galaxies*. They were similar to the Seyfert galaxies and also had strong broad emission lines in their spectra but the relation between these phenomena was not clear at that time.

In the early 1960s, the first *quasars* were discovered. Among the optical identifications of bright radio sources were a few

(a)

(b)

(d)

(e)

(c)

Figure 6.7. Examples of different types of galaxy. (*a*) The normal spiral galaxy M31 or the Andromeda Nebula. (Courtesy of the Mount Palomar Observatory.) (*b*) The barred spiral galaxy NGC 1365. (Courtesy of David Malin and the Anglo-Australian Observatory.) (*c*) The elliptical galaxies NGC 1399 and 1404 in the Fornax cluster of galaxies. These are the brightest members of the nearby Fornax cluster of galaxies. Both galaxies show a prominent distribution of globular clusters in their outer regions. (Courtesy of David Malin and the Anglo-Australian Observatory.) (*d*) The SO or lenticular galaxy NGC 3115. (Courtesy of David Malin and the Kitt Peak National Observatory.) (*e*) The irregular galaxy IC 1613. There is no regularity in the appearance of this galaxy. The irregular appearance is due to the large amounts of dust and gas present in the galaxy. (Courtesy of David Malin and the Anglo-Australian Observatory.) (*f*) The peculiar galaxy known as the Cartwheel. This is an example of a 'ring galaxy'. This peculiar structure is probably due to a 'collision' with one of the nearby galaxies. Simulations of galaxy collisions have shown that such a configuration could occur if one of the galaxies at the top of the picture had passed more or less through the centre of the galaxy. A circular wave-like disturbance would pass out through the disc and could give rise to a circular compression wave in which stars could be formed. (Courtesy of the Royal Observatory, Edinburgh.)

(f)

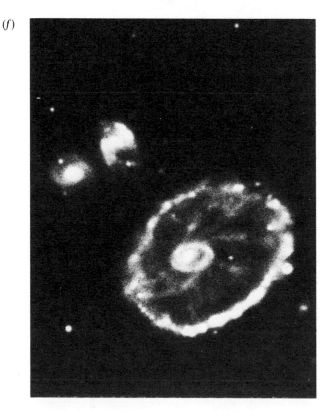

'stars' which had quite unintelligible spectra. The radio source 3C 273 (figure 6.3) was one of those which had been found in early radio surveys of the sky. 3C 273 looked exactly like a star on a photographic plate and was found to be variable in brightness. The remarkable discovery made by Maarten Schmidt in 1962 was that 3C 273 turned out to lie at a distance similar to that of the most distant galaxies whose distances were readily measurable at that time. The object was called a *quasi-stellar object* because it looked like a star but clearly could not be any normal sort of star at this very great distance. 3C 273 is more than 1000 times more luminous than a galaxy such as our own. Following this great discovery, many more quasars were found, all of them characterised by stellar appearance and very great distances. In addition to those which are strong radio sources, *radio-quiet* quasars were discovered in 1965 which are just as remarkable optically as the radio quasars but which were not powerful sources of radio emission. The quasars are among the most extreme examples of active galactic nuclei known. That they actually are the nuclei of galaxies has been convincingly demonstrated by the fact that the underlying galaxies in some of the more nearby examples have been detected.

Among the most extreme examples of active galactic nuclei are the objects known as *BL–Lacertae* or *BL–Lac objects*. These are similar to the quasars but they differ from them in two respects. First, they vary extremely rapidly in luminosity, variations being detected on time scales of days or less; they must therefore be very compact indeed. Second, their optical spectra are normally featureless and the continuum radiation is strongly polarised. It is plausible that, in the BL–Lac objects, we observe more or less directly the primary source of energy in active nuclei.

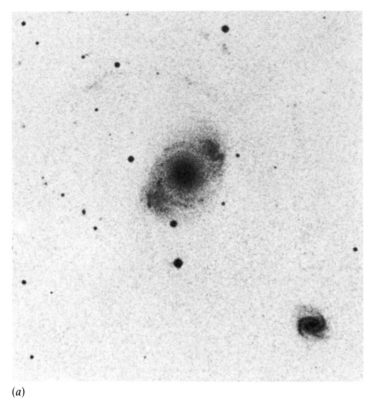

(a)

(b)

It should be emphasised that the most active of these systems are very rare indeed. It is because they are so luminous and have such distinctive properties that they can be discovered relatively easily. The astrophysical problem is to understand how they can produce such vast amounts of energy from very compact volumes. The current consensus of opinion is that these different types of active nuclei are all basically the same. There exists some ultracompact source of energy in these objects, and precisely what one observes depends upon the environment of the compact object and the way in which it obtains its fuel. Even rather quiescent nuclei like the centre of our own Galaxy possess small-scale versions of the phenomena we see in quasars and other active nuclei. It is likely that there is a continuity in activity which runs from systems such as the centre of our own Galaxy through the Seyfert galaxies and N-galaxies to the quasars and BL–Lac objects. The most convincing models for generating the energy needed to power these nuclei involve the presence of supermassive black holes with masses $\sim 10^6 - 10^{10} M_{\odot}$.

Stars and stellar evolution

The principal components of normal galaxies are stars, gas and dust. It is the stars which provide most of the mass of the galaxy and hence are responsible for the self-gravitating forces which bind the galaxies into stable associations of stars. The gas and dust play a key role in providing the material out of which new generations of stars form and also in providing a resting place for the debris expelled when stars die.

The study of stars and stellar evolution is central to all astronomy since stars are long-lived objects and a large fraction of the mass of the Universe is likely to be in this form. The study of the stars begins with the measurements of the total amount of light emitted by a star (its luminosity L) and also its surface temperature T. What makes the study of the structure and evolution of stars one of the most exact of the astrophysical sciences is the fact that not all possible combinations of temperature and luminosity are found among the stars. If we plot quantities which are equivalent to luminosity and temperature against one another, it is found that the stars occupy quite specific regions of this luminosity–temperature diagram (figure 6.9) which is also known as a *Hertzsprung–Russell* (or H–R) diagram or, equivalently, as a colour–magnitude diagram (see Section 6.10). The diagram shows where virtually all stars

Figure 6.8. Two images of the nearby Seyfert galaxy NGC 4151. This was one of the galaxies noted by Karl Seyfert in his first catalogue of galaxies with star-like nuclei and broad emission lines. (a) An image of NGC 4151 in a long exposure showing the inner structure of the disc of the galaxy. In an even longer exposure, spiral arms are clearly visible in its outer regions. (b) A short exposure image of NGC 4151 showing the star-like nucleus of the galaxy. (Courtesy of the Royal Greenwich Observatory.)

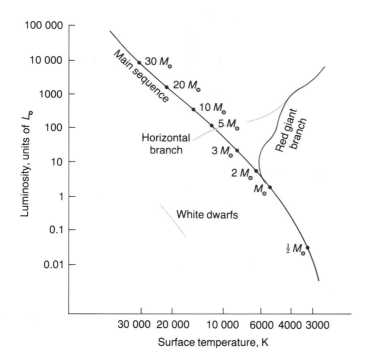

Figure 6.9. A schematic Hertzsprung–Russell diagram for stars. The diagram is also known as a colour–magnitude diagram or a luminosity–temperature diagram. Stars only occupy certain regions of this diagram as indicated. The names of the various sequences are shown.

lie. Most stars lie along the sequence which runs from bottom right to top left of this diagram which is known as the *main sequence*. What distinguishes stars along this sequence is their masses. The most massive stars are at the top left of the main sequence and the lowest mass stars at the bottom right. Our Sun lies about the middle of the sequence and is a very ordinary star.

Extending from about the location of the Sun on the luminosity–temperature diagram up towards the top right is what is known as the *giant branch*. The stars in this region of the diagram are large and cool. Extending across the diagram from the giant branch are stars of what is known as the *horizontal branch*. There are in addition stars which lie significantly below the main sequence which are very faint, blue, compact stars known as *white dwarfs*.

One of the main goals of the theory of stellar structure and evolution is to understand why it is that stars appear only in certain regions of this luminosity–temperature diagram and how they evolve from one part to another.

We consider first of all the main sequence. The source of energy for all main sequence stars is nuclear reactions occurring in their centres. The most common element in the Universe is hydrogen and the next most abundant helium-4

(^{4}He) with a cosmic abundance of about 25% by mass. All the heavier elements including species such as carbon, nitrogen, oxygen and iron amount to only about 1 to 2% by mass of all the elements. In the centres of main sequence stars, the temperature is so high that hydrogen is burned into helium, and the nuclear binding energy liberated in this process is in the end responsible for the light of main sequence stars. Theoretical models of stars using the best nuclear physics available from laboratory measurements can give a convincing quantitative explanation for the observed properties of main sequence stars. For most of its lifetime, a star remains at one point on the main sequence, a star like the Sun remaining there for about ten billion years. Along the main sequence, there is a one-to-one relation between luminosity (L) and mass (M) such that $L \propto M^\alpha$, where $\alpha \simeq 5$ for stars with mass roughly that of the Sun. This result is well understood theoretically.

When the star has converted about 12% of its hydrogen into helium, a limit known as the *Schonberg–Chandrasekhar limit*, it becomes unstable. The core contracts and its envelope expands to become a giant star. During this process, the star moves across the H–R diagram and then up the giant branch. This instability involves a number of changes in the internal structure of the star including the formation of deep convective zones within its envelope. All the evolutionary stages after the star moves off the main sequence take place very much more rapidly than its evolution on the main sequence. Since the available energy is just proportional to the mass of the star and radiation is the principal energy loss mechanism on the main sequence, this means that the lifetimes of stars are roughly proportional to $M^{-(\alpha-1)}$. Thus, stars with masses greater than $2M_\odot$ are expected to have lifetimes less than about 5×10^8 years.

For stars with masses roughly that of the Sun, the star undergoes other instabilities whilst on the giant branch during which it sheds mass from its outer layers. When this occurs, the star moves across to the horizontal branch and then evolves back towards the giant branch. During this evolution on the giant and horizontal branches, the helium in the star's core may be burned into heavier elements such as carbon, nitrogen and oxygen. After one or more excursions of this type, the star eventually ends up at the top right of the giant branch, an area occupied by long-period variables and unstable stars. When the star becomes unstable at this stage, the outer layers are blown off producing the characteristic *planetary nebula* phase of the star's evolution (figure 6.10), while the core of the star collapses to form a very hot white dwarf star which then cools and ends up to the lower left of the main sequence. This evolution is shown schematically in figure 6.11.

More massive stars evolve much more rapidly, and it is likely that stars which are more than about four times the mass of the Sun do not end up forming white dwarfs but explode as supernovae. It is far from certain that this simple picture is correct because the final outcome of the evolution of massive stars depends critically upon whether the stars can lose mass

Figure 6.10. The planetary nebula NGC 7293, also known as the Helix Nebula. The shell of hot gas is excited by the compact hot star in the centre. The star is so hot that it is a strong emitter of ultraviolet radiation which is responsible for heating, ionisation and excitation of the shell of gas. (Courtesy of David Malin and the Anglo-Australian Telescope Board.)

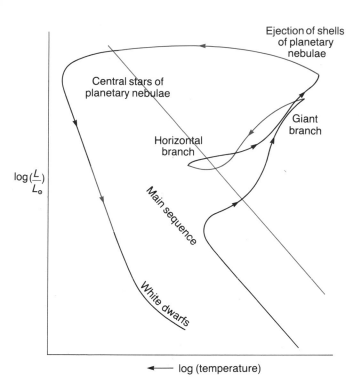

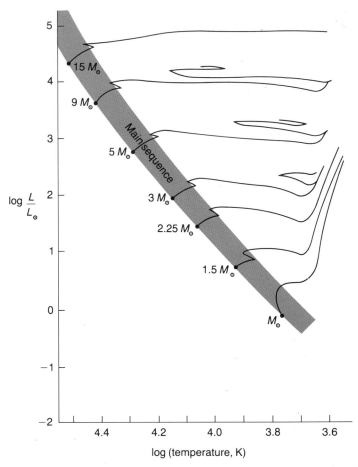

Figure 6.11. A schematic evolutionary track for a low-mass globular cluster star ($M \simeq M_\odot$). Regions of rapid evolution which may involve mass loss and in which the evolution is uncertain are indicated by red lines. Following evolution onto the giant branch and an excursion onto the horizontal branch, the star ends up at the tip of the giant branch. At this point strong mass loss takes place and a planetary nebula is formed. The central regions contract to become a hot helium star and then it evolves through to a white dwarf. The observed luminosities and colours of the central stars of planetary nebulae lie along the evolutionary sequence shown at the top left of the diagram. (After Michalas and Binney (1981), p. 152.)

Figure 6.12. Evolutionary tracks for stars of different initial masses on the Hertzsprung–Russell diagram. These models assume that there is no mass loss. Most of the lifetime of the star is spent on the main sequence and the subsequent phases of evolution on the giant branch take place over much shorter time-scales. (After Michalas and Binney (1981), p. 139.)

efficiently or not. Theoretical evolutionary tracks for stars of different main sequence masses are shown in figure 6.12.

We know rather precisely, however, the sorts of objects which can be produced at the end of a star's lifetime. There are three types of 'dead star'. In all three cases there is no longer any nuclear generation of energy in their centres. One form of dead star is a white dwarf in which internal pressure support is provided by electron degeneracy pressure; their masses are about the mass of the Sun or less. A second possible end point is as a *neutron star*. In this case internal pressure support is provided by neutron degeneracy pressure. These stars are very compact indeed, having masses about the mass of the Sun and radii about 10 km. They have been found in two ways. In the first, they are the parent bodies of radio pulsars which are rotating, magnetised, neutron stars. These are observed to emit very intense pulses of radio emission once per rotation period

which is about one second. In the second case, they are the compact 'invisible' secondary stars in binary X-ray sources in which the X-rays are produced by very high temperature gas which has been heated by matter from the primary star falling onto the compact neutron star. The third possibility is that the star collapses to a *black hole*. If the collapsing star has mass greater than about two and a half times the mass of the Sun, it is expected that a black hole will form unless there is some way in which it can lose mass so that a stable neutron star or white dwarf can be formed.

The interstellar medium

One of the areas of astrophysical research which has developed very rapidly over the last twenty years has been the study of the medium between the stars – the interstellar medium. Originally, it was thought that the interstellar medium was a rather simple, quiescent medium, but it is now clear from a wide range of different types of observation that there are many different phases and components. There are four main constituents of the medium – gas in all its phases (i.e. atomic, molecular and ionised gas), very high energy particles, dust and magnetic fields. The interstellar medium is not stationary but is constantly stirred up by winds blowing from stars, by stellar explosions and by large-scale perturbations such as the influence of spiral arms or interactions with companion galaxies. It is depleted by the formation of stars and replenished by various forms of mass loss. Let us briefly review the various components of the interstellar medium.

Gas   The coolest components observed are the *giant molecular clouds*, which are present throughout the interstellar medium of our Galaxy. These are observed by telescopes operating at millimetre and submillimetre wavelengths through their intense molecular line emission. Many molecular species have been detected in the interstellar gas in giant molecular clouds (see table 6.1), a number of these molecules having been detected only in the interstellar gas. Evidently, a great deal of organic and inorganic chemistry must be taking place in the interstellar gas. The typical giant molecular cloud has a mass of about one-million times the mass of the Sun and temperature in the range 10 to 100 K (figure 6.13). Most of the mass is in the form of molecular hydrogen which is shielded from the intense interstellar flux of ionising radiation by dust. Molecular line and infrared observations have shown that there exist within the large clouds much denser regions and it is within these that stars are formed. Hot molecular gas is observed close to the regions where stars have already formed, the gas being heated by the radiation of the protostars or newly formed stars as it pushes back the surrounding medium.

Hot ionised gas at temperatures of about 10 000 K is observed around young stars and the hot dying stars in the centres of planetary nebulae. The beautiful diffuse structures which are seen in young objects like the Orion Nebula or the Horsehead Nebula (figure 6.14a and b) and in old systems such as the shells of planetary nebulae (figure 6.10) are due to the excitation of strong emission lines of this ionised gas. In both cases, the hot stars are sources of intense ultraviolet radiation which ionises and heats the surrounding gas clouds and causes them to radiate intense line emission.

In addition to these hot components associated with the regions where stars form, there is very hot gas which is ejected in supernova explosions. These violent events, in which a whole star can be disrupted, liberate huge amounts of energy which is sufficient to heat the gas in a large volume around the explosion up to very high temperatures, $T \simeq 10^7$ K. The overlapping of the explosions of different generations of stars results in a considerable fraction of the interstellar gas being heated to high temperatures. This gas is detected through its soft X-ray emission at energies $\varepsilon \simeq 0.1$–1 keV and through the observation of absorption lines in the interstellar gas due to highly ionised species. Its temperature is typically about 10^6 K.

Table 6.1. *Interstellar molecules*

This list of known interstellar molecules is arranged in columns showing the numbers of atoms which make up the molecules. Different isotopic species have not been included, for example deuterated molecules. Those molecules indicated by an asterisk in brackets are ring molecules.

2	3	4	5	6	7	8	9	10	11	12	13
H_2	N_2H^+	NH_3	C_4H	CH_3OH	CH_3CCH	CH_3COOH	CH_3OCH_3	CH_3C_5N	HC_9N		$HC_{11}N$
CO	HCO^+	H_2CO	CH_2NH	CH_3SH	CH_3CHO	CH_3C_3N	CH_3CH_2OH				
CH	HCS^+	H_2CS	NH_2CN	CH_3CN	CH_3NH_2	$HCOOCH_3$	CH_3CH_2CN				
CN	HCN	HNCO	HCOOH	NH_2CHO	CH_2CHCN		CH_3C_4H				
CS	HNC	HNCS	CH_2CO	CH_3NC	HC_5N		HC_7N				
C_2	C_2H	C_3H	HC_3N		$CH_2(CN)_2$		C_2H_5OH				
CH^+	H_2O	C_3O	$C_3H_2(*)$				C_2H_5CN				
OH	SO_2	C_3N									
NO	H_2S	$HOCO^+$									
NS	HCO	C_2H_2									
SO	OCS	$HCNH^+$									
SiO	HNO	HSCC or									
SiS	HOC^+	HSiCC									
HCl	$SiC_2(*)$										
PN	$H_3^+(*)$										

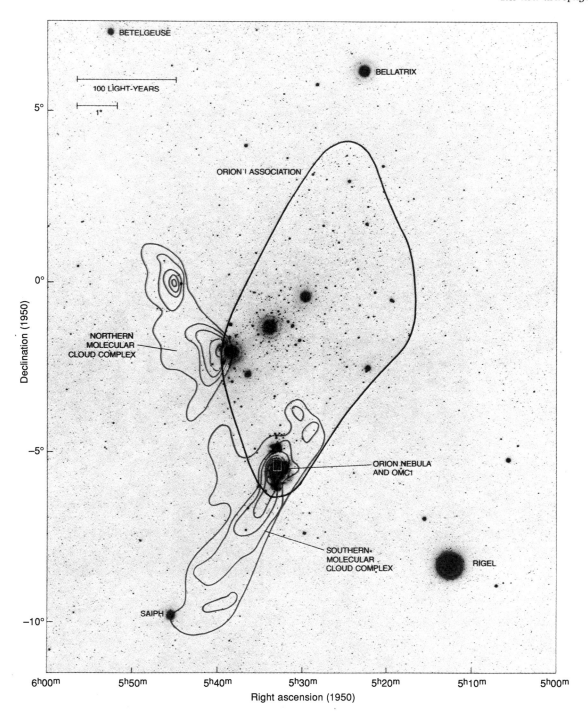

Figure 6.13. An example of the giant molecular clouds in the constellation of Orion. The enormous extent of the clouds can be appreciated from the photograph which shows the prominent stars in the constellation of Orion. The contours show the intensity distribution of carbon monoxide which is the most common molecular species after molecular hydrogen. Unfortunately, molecular hydrogen does not emit millimetre line radiation because the molecule has no dipole moment. Also shown on the diagram is a region known as the Orion 1 Association, which is known to contain many young stars. The Orion Nebula (figure 6.14*a*) can be seen half-way down Orion's sword. (From Wynn-Williams, C. G. (1981), *Sci. Am.* **245**, 31.)

(a)

Figure 6.14. (*a*) The Orion Nebula. In this optical photograph special photographic processing has been used to bring up faint features. This region of ionised hydrogen is excited by the four young bright stars close to the centre of the nebula, known as the Trapezium stars. The light of the Nebula is mostly due to the emission lines of hydrogen and oxygen. (Courtesy of David Malin and the Anglo-Australian Observatory.) (*b*) The Horsehead Nebula in the constellation of Orion. The dark patches are regions of dense interstellar dust in many of which protostars and young stars may be formed as revealed by molecular line and infrared observations. (Courtesy of the Royal Observatory, Edinburgh; photography by David Malin.)

Between the very hot regions and the giant molecular clouds, there is diffuse gas consisting of a mixture of ionised and unionised gas. These different components are not stationary but are shocked by supernova explosions and by strong stellar winds. These motions compress the gas and may assist in the formation of molecular clouds and in turn star formation.

Dust · The black areas seen in photographs such as those of the Orion and Horsehead Nebulae are due to obscuring dust. Dust is present throughout the interstellar gas and has the unfortunate effect of obscuring some of the most interesting regions we wish to study. Fortunately, dust becomes transparent in the infrared waveband and it is then possible to

(b)

observe deep inside the regions where stars are forming by making observations in the infrared waveband. In addition, the dust is heated by the radiation it absorbs and therefore it is a strong emitter at the temperatures to which it is heated, which are typically in the range from about 30 to 500 K. Indeed, in the far-infrared waveband, the dustiest regions become the most intense emitters rather than the most obscuring objects as occurs in the optical waveband. The dust radiates at temperatures less than about 1 000 K because at higher temperatures it would be evaporated and so would no longer exist. Dust plays a crucial role in the processes by which stars form – it protects the fragile molecules from being dissociated by the intense interstellar radiation field and, secondly, it acts as an efficient means

by which the protostar can lose energy. Indeed objects which may be protostars in the regions where stars are formed are characterised by very intense emission in the infrared waveband.

High energy particles In addition to gas and dust, there are also very *high energy particles* present throughout the interstellar medium. These are accelerated in supernova explosions and pulsars and are then dispersed throughout the medium. The electron component is detected by its radio emission which is due to the synchrotron radiation of these particles spiralling in the interstellar magnetic field. High energy protons are observed through the gamma-rays which they emit when they collide with ordinary matter in the interstellar gas. In these high energy collisions, pions of all types are produced and the neutral pions decay almost instantaneously into gamma-rays. These very high energy particles are very rare but they are so energetic that this relativistic gas has an important effect upon the dynamics of the interstellar gas.

The galactic magnetic field   Finally, there are a number of separate observations which indicate that there is a weak large scale magnetic field present in the interstellar gas. The evidence comes from observations of radio synchrotron emission of high energy electrons in our Galaxy and from the polarisation of the radio emission and of the light of nearby stars. Zeeman splitting of the 21 cm line of neutral hydrogen has also been measured in a number of hydrogen clouds. Typical values of the magnetic field strength are $B \simeq 10^{-9} - 10^{-10}$ T*. Although apparently rather weak, this magnetic field has an important influence upon the dynamics of the different components of the interstellar gas.

Star formation

The processes by which stars form is one of the most important unsolved problems of the astrophysics of stars and galaxies. There are many things which are uncertain about the processes of star formation from an observational point of view. We do not know how many massive and how many light stars are formed in any particular molecular cloud. We do not know how the rate at which stars form depends upon physical conditions within the cloud. We do not know how the molecular clouds themselves form. These are problems which can be addressed by observations in the new disciplines of infrared and millimetre astronomy.

Theoretically, the position is not very much better. The fundamental problem to be understood is the sequence of events which takes place between the formation of the molecular cloud with typical density of about 10^9 molecules m^{-3} and the formation of a new star which has density billions of billions of times greater. The denser regions of molecular clouds collapse under their own gravity but the problem is that, as they collapse, they heat up and the collapse slows down. There has to be some means by which the heat generated in the collapse can be removed from the cloud. This is almost certainly by radiation. The radiation is however trapped inside the collapsing cloud and so the interior heats up. The slow collapse and heating up continues until the protostar becomes so hot in the centre that nuclear burning of hydrogen into helium begins and the star begins its life as a main sequence star. Although

this sounds a very straightforward picture, there are many grave theoretical uncertainties. For example, it is not clear how one can get rid of any rotation or magnetic fields in the initial clouds which tend to prevent collapse.

These studies have implications far beyond their immediate relevance for the formation of stars. For example, in order to understand the evolution of galaxies, we have to know the rate at which new stars are forming under a wide range of different astrophysical conditions. This field of study is certainly one which is of the greatest importance for essentially all branches of contemporary astronomy and cosmology.

6.3   Stars and stellar evolution

The Sun as a test-bed for stellar structure and evolution

The theory of stellar evolution outlined in Section 6.2 forms the basis of our present understanding of the astrophysics of stars. Traditional astronomical techniques such as multicolour photometry and high resolution spectroscopy provide information only about the surface properties of stars such as their surface temperatures and surface gravities. It is the interpretation of these data in terms of the inferred internal structure of the stars which has proved to provide the strongest constraints on the theory of their internal structures. It would be of the greatest importance, however, if it were possible to test the theory more directly by studying the nuclear processes which take place in the centres of the stars and which are ultimately the source of their luminosities. There are now two methods by which these studies can be pursued – the study of *solar neutrinos* and the new discipline of *solar seismology* or *helioseismology*.

Solar neutrinos   Since the neutrinos produced in nuclear reactions in the centre of the Sun are very weakly interacting particles, they escape more or less unimpeded from its centre and therefore bring with them to the Earth information about reaction rates and physical conditions in the centre of the Sun. There are two problems. First of all, the same small cross-section which enables them to escape from the Sun also means that they are very difficult to detect when they arrive at the Earth. Second, the most readily available detector material which has been available in large quantities is sensitive to the rather high energy neutrinos ($\varepsilon = 14$ MeV) created in a rare side-chain of the main p–p reaction network which is the source of power in the Sun. The main p–p chain reactions (6.1), and the side-chain responsible for the high energy neutrinos, reactions (6.2), are as follows:

$$p + p \rightarrow {}^2H + e^- + \nu: {}^2H + p \rightarrow {}^3He + \gamma: {}^3He + {}^3He \rightarrow {}^4He + 2p$$

$$(\varepsilon_\nu = 0.420 \, \text{MeV max}), \tag{6.1}$$

---

*Astronomers usually work in a wide variety of units, very often c.g.s. units, and hence usually work in gauss. For consistency with the rest of this book, SI units are used throughout this chapter and hence magnetic fields are measured in tesla (T). For the benefit of astronomers, 10^4 gauss $= 1$ T.

and

$$^3\text{He} + {}^4\text{He} \rightarrow {}^7\text{Be} : {}^7\text{Be} + \text{p} \rightarrow {}^8\text{B} + \gamma : {}^8\text{B} \rightarrow {}^8\text{Be}^* + \text{e}^+ + \nu : {}^8\text{Be}^* \rightarrow 2{}^4\text{He}$$
$$(\varepsilon_\nu = 14.06\,\text{MeV}) \quad (6.2)$$

The 14 MeV neutrinos are detected on Earth through the nuclear transmutations which they induce in chlorine nuclei bound in a form of 'cleaning fluid', perchlorethylene C_2Cl_4:

$$^{37}\text{Cl} + \nu \rightarrow {}^{37}\text{Ar} + \text{e}^-. \quad (6.3)$$

The argon gas is swept out and the amount produced measures the rate of production of neutrinos in the Sun. These experiments have been carried out over the last twenty years by Dr Raymond Davis and his colleagues using a 100 000 gallon tank of C_2Cl_4 located at the bottom of the Homestake gold-mine in Colorado. Figure 6.15 shows how estimates of the detected neutrino flux and the theoretical predictions have varied over the last twenty years since these measurements were first attempted.

Perhaps the most important point is that a significant flux of solar neutrinos *is* observed but the number detected corresponds to about one-third of the predicted flux. The values quoted by Dr John Bahcall in a recent survey of this discrepancy, the famous solar neutrino problem, are

observed flux of neutrinos $= 2.1 \pm 0.3$ SNU
predicted flux of neutrinos $= 5.8 \pm 2.2$ SNU

($1\,\text{SNU} = 10^{-36}$ absorptions per second per target atom).

There appears to be a discrepancy outside the formal errors. This discrepancy has been the source of an enormous amount of speculation, and there is no agreed solution to the problem. Broadly speaking there are two aspects to the problem – either the nuclear physics could be in error or the astrophysics associated with the determination of physical conditions in the core of the Sun could be wrong. Whilst many novel ideas have been introduced to account for this discrepancy, it is likely that the best approach is to look for independent ways of testing observationally both the magnitude of the effect and the reliability of the solar models used to determine the central density and temperature of the Sun.

The most important check of the solar neutrino flux would be to study the much more numerous low energy neutrinos (0.420 MeV) produced in the main p–p chain of reactions (6.1). The predicted fluxes are much larger than those produced in the boron to beryllium decay but the problem is the availability of a suitable detector material. The best is gallium, which undergoes the following nuclear reaction with low energy neutrinos:

$$^{71}\text{Ga} + \nu \rightarrow {}^{71}\text{Ge} + \text{e}^-. \quad (6.4)$$

The detector would have to consist of about 30 tons of gallium, a very large quantity of this rare earth. An experiment of this type named GALLEX is currently being developed by a German, French, Israeli and Italian collaboration. In it, liquid gallium chloride $GaCl_3$ is used as a detector. The intention is to measure the solar neutrino flux to about 10% accuracy in four years, the

first solar observations being made in about 1990. This would provide a crucial independent test of the nuclear processes going on in the centre of the only normal star from which we can reasonably hope to detect significant fluxes of neutrinos. The resolution of the solar neutrino problem is crucial because much of our detailed understanding of stellar structure and evolution depends upon understanding our own Sun.

Solar seismology One of the most remarkable developments in the study of the Sun over the last ten years has been the discovery of solar oscillations. It is simplest to think of the Sun as a resonant sphere which, when perturbed, oscillates at frequencies corresponding to its normal modes of oscillation.

Figure 6.15. A diagram illustrating how the observed and predicted fluxes of solar neutrinos from ^8B decays in the centre of the Sun have varied over the last twenty years. The error bars are 1σ errors on both the theoretical and observed fluxes of neutrinos. × Observation; o theory. (Courtesy of Dr J. N. Bahcall.)

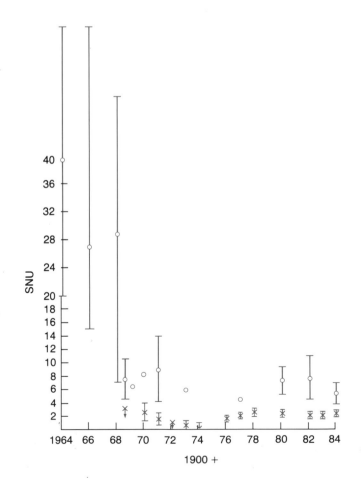

The predicted modes of oscillation of the Sun are worked out using the best available models of the solar interior. In these, the central regions are in radiative equilibrium but in the outer layers there are convective zones (figure 6.16). Figure 6.16 also shows schematically some of the different modes of oscillation of the Sun and indicates how they can be used to investigate different regions of density and pressure inside the Sun.

These oscillations have been observed by measuring very accurately the Doppler shifts of spectral lines in the atmosphere of the Sun over long periods and also by studying variations in the luminosity of the Sun. The periods of oscillation observed range from about five minutes to about ten hours. Fourier transforms of long runs of data are used to establish the resonant frequencies present in the data. Examples of the spectra of these oscillations are shown in figure 6.17 in which not only are the principal wave numbers seen, but also 'fine structure splitting' of these modes which are naturally attributed to coupling between the normal oscillatory modes of the static Sun and rotation within its interior.

The study of these oscillations has shown that the standard solar model provides a remarkably good fit to the data. This suggests that the solar models used in the estimation of the neutrino flux are a good first approximation to the real Sun. To obtain more stringent tests of the solar models, it is necessary to carry out these studies over very long time scales. To obtain the necessary continuity of observation, these observations must either be carried out using a world-wide network of observing stations or from space, the proposed SOHO space probe of the European Space Agency being the ideal type of mission for this purpose. These types of observation will enable the structure of the Sun to be determined deep into its interior.

Although our Sun is the prime target for these studies, the brightest nearby stars can also be investigated using similar techniques. In this case, the velocities are averaged over the whole surface of the star, a procedure similar to that carried out in some of the solar experiments. With the rapid advance of experimental technique for undertaking these studies, it is likely that in the future stellar seismology will become a key tool for studying the structure and evolution of nearby stars.

Figure 6.16. A schematic diagram showing the internal structure of the Sun and the zones in which heat transport is by radiation and convection. Also shown are examples of how different modes of oscillation of the Sun can be used to probe the conditions of density and pressure at different depths in the Sun. (Courtesy of the European Space Agency.)

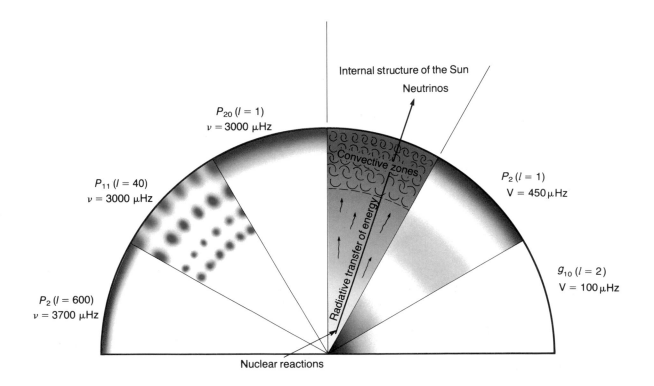

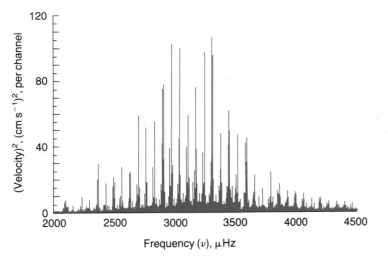

Figure 6.17. An example of the frequency spectrum of solar oscillations showing the fine-structure splitting associated with coupling between the rotation and normal modes of oscillation of the Sun. (Courtesy of the European Space Agency.)

Nucleosynthesis in stars

In a classical paper of 1957, Burbidge, Burbidge, Fowler and Hoyle described a wide range of nuclear processes which are important in synthesising the chemical elements during the course of stellar evolution. As might be expected, the physical conditions of density, temperature and chemical composition are crucial in determining which are the dominant processes.

During evolution on the main sequence, the prime energy source is the conversion of hydrogen into helium. For stars less massive than about $2M_\odot$, the proton–proton chain, reactions (6.1), provides the main route for the synthesis of helium. With increasing mass, however, the central temperature of the star increases and then the *carbon–nitrogen–oxygen cycle* (CNO cycle) becomes the dominant energy source because of the very steep dependence of the energy generation rate $\varepsilon(T)$ upon temperature, $\varepsilon(T) \propto T^{17}$, compared with $\varepsilon(T) \propto T^4$ for the p–p chain. The CNO cycle uses ^{12}C as a catalyst in the formation of helium through the following sequence of reactions:

$$^{12}C + p \rightarrow {}^{13}N + \gamma; \quad {}^{13}N \rightarrow {}^{13}C + e^+ + \nu; \quad {}^{13}C + p \rightarrow {}^{14}N + \gamma$$
$$^{14}N + p \rightarrow {}^{15}O + \gamma; \quad {}^{15}O \rightarrow {}^{15}N + e^+ + \nu; \quad {}^{15}N + p \rightarrow {}^{4}He + {}^{12}C. \tag{6.5}$$

The cycle proceeds through the successive addition of protons to the ^{12}C nucleus with two intermediate inverse β-decays which have the effect of converting two of the protons into neutrons.

The expected products of the p–p chain and the CNO cycle provide important tests of the role of these processes of nucleosynthesis in stars. In the case of the p–p chain, the main

effect is to enhance the ^{12}C abundance over its initial value. In the case of the CNO cycle, the abundances of ^{13}C, ^{14}N, ^{15}N and ^{15}O are enhanced relative to their initial abundances. Indeed, in the steady state CNO cycle, large $^{13}C/^{12}C$, $^{4}He/H$ and N/C abundances are expected. The problem is therefore to find stars in which it is possible to study the products of nucleosynthesis within the cores of stars rather than within their surface layers in which the relative abundances are most readily measurable.

The problem is a complex one because the observed abundances depend upon many factors. For example, what was the initial chemical composition from which the stars formed? Was it formed from a pure hydrogen–helium plasma or was it already enriched by the products of previous cycles of star formation? What is the role of convection and diffusion in dredging up material from the interior of the star to the surface layers? Has the star lost its outer layers, revealing the chemical elements in the interior? Despite these complications, there is good evidence for the types of abundance trends expected from the CNO cycle and the p–p chain in those classes of star where it is reasonable to expect that the products of nucleosynthesis are observable. Particularly interesting cases are those in which anomalously high values of $^{13}C/^{12}C$ and N/C abundance ratios are observed.

An outline of the evolution of massive stars was given in Section 6.2. Post-main sequence evolution proceeds by the process of successive burning of the elements to produce nuclei with higher and higher binding energies. For massive stars, the sequence of burning runs through helium burning to produce carbon, carbon and oxygen burning to produce silicon which can eventually be burned through to iron peak elements. These processes can be written

$$\left. \begin{aligned} ^{12}C + {}^{12}C &\rightarrow {}^{24}Mg + \gamma \\ &\rightarrow {}^{23}Na + p \\ &\rightarrow {}^{20}Ne + {}^{4}He \end{aligned} \right\} \quad T \geqslant 5 \times 10^8 \text{ K}$$

$$\left. \begin{aligned} ^{16}O + {}^{16}O &\rightarrow {}^{32}S + \gamma \\ &\rightarrow {}^{31}P + p \\ &\rightarrow {}^{31}S + n \\ &\rightarrow {}^{28}Si + {}^{4}He \end{aligned} \right\} \quad T \geqslant 10^9 \text{ K}.$$

In the case of silicon burning, which begins at a temperature of about 2×10^9 K, the reactions proceed slightly differently because the high energy gamma-rays remove protons and ^{4}He particles from the silicon nuclei and the heavier elements are synthesised by the addition of ^{4}He nuclei through reactions which can be schematically written as

$$^{28}Si + \gamma\text{'s} \rightarrow 7\,^{4}He$$
$$^{28}Si + 7\,^{4}He \rightarrow {}^{56}Ni.$$

It is therefore expected that, in the final stages of evolution of massive stars, the star will take up an 'onion-skin' structure with a central core of iron peak elements and successive surrounding shells of silicon, carbon and oxygen, helium and hydrogen. Calculations of the explosive burning of shells of carbon, oxygen and silicon have shown good agreement with the observed abundances of the heavy elements up to the iron peak. It is likely that these types of event occur in type II supernovae, and some of the evidence for this is described in Section 6.4. From the point of view of the present section, the important point is that these processes of nucleosynthesis lead to the production of elements up to the iron peak and these possess the greatest binding energy of the chemical elements.

To proceed beyond iron, two processes are important involving neutron reactions with iron peak elements. In these reactions, a neutron is absorbed and then the subsequent products depend upon whether or not the nucleus formed has time to decay before the addition of further neutrons takes place. The case in which the decay always occurs first is referred to as the *slow* or *s-process* and that in which several neutrons are added before β-decay terminates the sequence is known as the *rapid* or *r-process*. The latter is likely to be important in the extreme conditions during explosive nucleosynthesis where very high densities and temperatures are attained. This is believed to be the process which is responsible for the synthesis of neutron-rich species such as the heaviest elements of tin, ^{122}Sn and ^{124}Sn. The products of the s-process are normally estimated by calculations in which iron, by far the most abundant of the elements heavier than oxygen, is irradiated by neutrons. The products are sensitive to the irradiation time, but it has been shown that, if it is assumed that there is a range of irradiation times, the solar system abundances of the elements heavier than iron can be accounted for. Particular successes of this theory have been in accounting for the anomalously high abundances of heavy elements such as barium and zirconium and, in particular, for the unstable element technetium, Tc, the longest lived isotope of which has a lifetime of only 2.6×10^6 years.

The lowest mass stars and brown dwarfs

The theory of stellar evolution can account successfully for the variation of luminosity with mass along the main sequence. The more massive the star, the greater its central temperature and luminosity. For stars less massive than the Sun, the central temperature decreases with decreasing mass and there is a temperature below which the p–p reaction chain is unable to generate sufficient energy to provide pressure support for the star. The theory of stellar structure suggests that this lower mass limit for hydrogen-burning stars is about $0.08\,M_\odot$.

The search for intrinsically very faint stars is thus an important problem for the theory of stellar evolution. On the one hand, it is important to discover whether or not the prediction is correct; on the other hand, there is the important question of what happens to those 'stars' which have mass less than $0.08\,M_\odot$ if they exist at all. We know that planet-sized bodies exist within our own solar system, Jupiter having mass

$0.02\,M_\odot$. The big problem is that it is very difficult to observe intrinsically very faint objects. They have to be very nearby to be observable, and of course there is no guarantee that we will be located close to any of them.

One other important aspect of the discussion of the existence or otherwise of these very faint stars is whether or not they contribute significantly to the local density of matter in our own Galaxy. In the late 1920s Oort showed how it is possible to estimate the total amount of gravitating matter in the plane of our own Galaxy by measuring the velocity distribution of stars perpendicular to the Galactic plane. The value for the projected mass density in the vicinity of the Sun was found to be about $0.15\,M_\odot\,\mathrm{pc}^{-2}$, i.e. $0.15\,M_\odot$ of material in a column of cross-section $1\,\mathrm{pc}^2$ perpendicular to the Galactic plane. This value can be compared with the amount of mass present in stars and interstellar gas. Various estimates have suggested that these forms of matter contribute less than half the Oort value, although some authors believe the discrepancy may not be as large as this. This is the origin of the local hidden mass problem and one possibility is that this mass could be locked up in very low mass main sequence stars or stars which are of too low mass to become main sequence stars. These very low mass cool objects are collectively referred to as *brown dwarfs*, reflecting the fact that their colours are somewhat indeterminate.

Recent determinations of the luminosity function of faint stars, the function which describes the space densities of stars of different intrinsic luminosities, have been made by Gilmore and Reid and by Hawkins. It is found that the number of low luminosity stars decreases towards the 'limit' of $0.08\,M_\odot$ but in Hawkins' recent work there appears to be a population of stars present at very low luminosities (figure 6.18). These objects are very cool and have large proper motions on the sky, indicating that they are very nearby objects, exactly the properties expected of brown dwarfs. They seem to be of too low mass to be hydrogen-burning stars. There are few energy sources left for such stars. One possibility is that the stars simply radiate away their gravitational binding energy on a typical thermal time scale for the cooling of extremely low mass stars. There might be a contribution from deuterium burning since deuterium burns at a lower temperature than hydrogen. There is, however, very little deuterium, its cosmic abundance corresponding to only about 10^{-5} of that of hydrogen and so the deuterium would be rapidly consumed. Thus, these low mass stars are likely to be rather short lived, the cooling occurring in roughly the Kelvin–Helmholtz cooling time, this being the characteristic time it takes radiation to diffuse out of a star. Plainly, we need to know much more about these fascinating low mass stars. How much they could contribute to the local mass density is uncertain because of the uncertainties about their lifetimes. Simple estimates indicate that these very low mass stars could make a significant contribution to the local hidden mass problem – indeed, within the uncertainties, they could solve the problem.

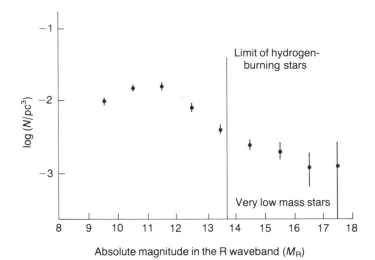

Figure 6.18. A recent determination of the low luminosity region of the luminosity function for stars. The points having $M_R < 14$ are due to Gilmore and Reid and the fainter points are due to Hawkins. The limit for hydrogen-burning stars is shown. (Courtesy of Dr M. R. S. Hawkins.)

Stellar evolution in globular clusters – the age of the Galaxy

One of the most difficult problems in observational astronomy is the determination of the distances of stars, and for this reason star clusters are of special importance in testing the theory of stellar evolution because all the stars in the cluster are at the same distance. It can be assumed that all the stars in a cluster formed at more or less the same time in the distant past from material of the same chemical composition. The ages of clusters range from very young systems, probably no more than about 10^6 years old, to very old clusters, the globular clusters, which are among the oldest stars in the Galaxy with ages about 10^{10} years. One of the goals of the study of clusters of different ages is to determine their luminosity–temperature diagrams and to compare these with the predictions of the theory of stellar evolution. The H–R diagrams for a number of clusters of different ages are shown in figure 6.19. It can be seen that these have a general shape which can be understood in terms of the theory outlined in Section 6.2. A key problem for observational astronomy is to provide accurate measurements of the colours and magnitudes for the stars in clusters so that the comparison of theory and observation can be made as accurate as possible. Examples of this comparison for the globular clusters NGC 6752 and 47 Tucanae (figure 6.20) indicate the types of problem which confront the astrophysicist. At faint magnitudes, the observational scatter increases and hence the location of the main sequence becomes more difficult to define. It is apparent that there is a significant scatter of the observed points about any mean line through the points. Is this scatter intrinsic to the cluster or is it due to variations in the chemical

composition of the stars or to different evolutionary histories for stars of similar masses? These are key questions for our understanding of stellar evolution.

The theoretical predictions are derived by working out the loci of stars of different masses on a theoretical luminosity–temperature diagram and then populating the different regions according to how long the star remains in that part of the diagram and the numbers of stars which began their lives at different points on the main sequence. These diagrams thus contain crucial information not only about stellar evolution but also about the initial distribution of masses of the stars in the cluster when it formed, i.e. about what is known as the initial mass function of stars.

The understanding of the details of the H–R diagrams of globular clusters proceeds by a process of symbiosis between the observation and the theory and some of these aspects will be treated in the next section. Of special importance for many aspects of astrophysics and cosmology is the determination of the ages of globular clusters from the location and shape of the distribution of stars near the main sequence termination point.

Figure 6.19. The Hertzsprung–Russell diagrams for star clusters of different ages. The differences in their H–R diagrams can be entirely attributed to their different ages. The youngest cluster is NGC 2362 and the oldest M67. (From Michalas and Binney (1981), p. 105.)

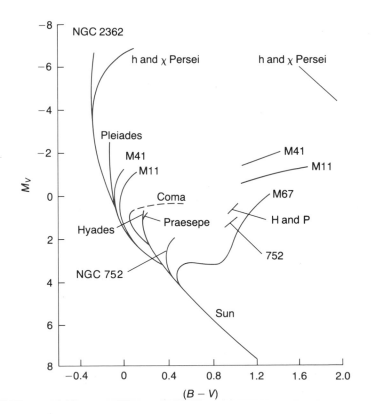

As described in Section 6.2, the main sequence termination point is a measure of the age of the globular cluster, and the oldest of these provide a measure of the age of our Galaxy and the Universe itself.

Typically, the ages of the oldest globular clusters are found to be between about fourteen and twenty billion years old. A prime objective of current research is to define the H–R diagrams with much greater precision for many more clusters using modern panoramic detectors such as CCDs on large telescopes. Observations with the Hubble Space Telescope will be of special importance because of its very high angular resolution which will enable very faint stars in dense and distant clusters to be distinguished.

Because they are so old, the globular clusters provide important information about some of the earliest generations of stars which formed in the Galaxy. In the oldest globular clusters, the heavy elements are much less abundant than they are in stars forming at the present day. This is consistent with a general picture in which the cosmic abundance of the elements has been built up over the last 10^{10} years through nucleosynthesis inside stars. It appears, however, that even in the oldest globular clusters, there are none which are totally devoid of heavy elements. Indeed it is very difficult to find stars which have chemical abundances less than about $1/100$ of the solar value. This is an important fact for studies of the formation of galaxies since it means that even in the oldest systems we can study directly, there must already have been some synthesis of the elements. Whether this can have taken place before or after the Galaxy formed is an intriguing and important question since

Figure 6.20. The Hertzsprung–Russell diagram for the globular clusters NGC 6752 (*a*) and 47 Tucanae (*b*). It can be seen that the scatter in the points increases at faint magnitudes largely because of the increase in the observational error associated with the photometry of faint stars. The solid lines show best-fits to the data using theoretical models of the evolution of stars from the main sequence onto the giant branch due to VandenBerg. (*a*) For NGC 6752, the locus (or isochrone) has an age of 1.6×10^{10} years and the abundance of metals is a factor of thirty smaller than the solar value. (From Penny, A. J. and Dickens, R. J. (1986). *Mon. Not. R. Astr. Soc.* **220**, 856.) (*b*) For the cluster 47 Tucanae, the best-fit isochrones have age $\simeq 1.2$–1.4×10^{10} years and the cluster is metal-rich relative to other globular clusters, the metal abundance amounting to about 20% of the solar value. (From Hesser, J. E., Harris, W. E., VandenBerg, D. A., Allwright, J. W. B., Shott, P. and Stetson, P. *Publ. Astron. Soc. Pacific*, **99**, 739.)

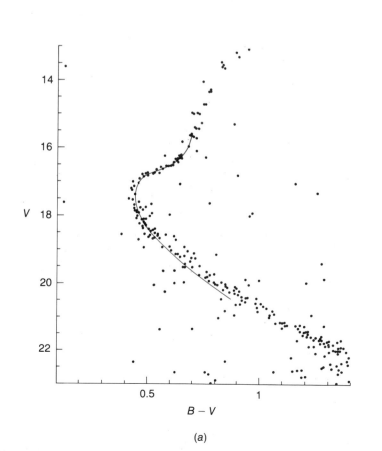

(a)

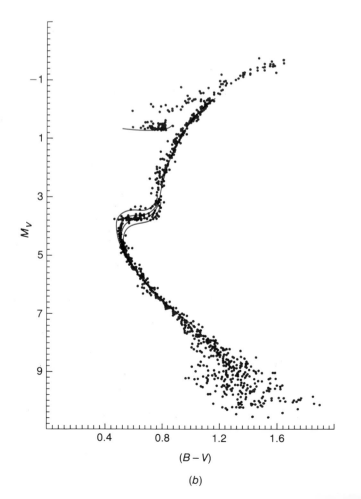

(b)

it bears upon the precise sequence of events which must have taken place when galaxies first formed. This hypothetical class of pre-galactic stars is often referred to as *population III stars*, population I stars being the young hot blue component and population II the old red population including the globular cluster stars. The nature of this theoretical population of stars is far from clear, and no member of this class has yet been observed. Their function would be to provide early enrichment of the primordial hydrogen–helium gas with some heavy elements which would make subsequent star formation much easier because the existence of heavy elements provides efficient routes for cooling of the gas.

Post main sequence evolution and mass loss

Stars lose mass from their surfaces throughout much of their lives. One of the most important discoveries of the Einstein X-ray Observatory was that essentially *all* classes of stars emit X-rays, the radiation generally originating from hot coronae or hot stellar winds. Thus, coronae similar to that observed about our own Sun are common to all classes of star. The immediate consequence is that there must be stellar winds and hence mass loss associated with essentially all classes of star. Stellar coronae are heated by waves originating in the convective layers close to the surface of the star, and this energy is dissipated above the photosphere leading to strong heating of the lower density gas in the immediate vicinity of the Sun. In the case of the Sun, the gas is heated to temperatures greater than 10^5 K so that it can no longer be bound to the Sun and a stellar wind, in our case the solar wind, is created. This may be termed *quiescent* mass loss since it occurs in all classes of normal star.

There are, however, much more violent forms of mass loss which are believed to be associated with the various evolutionary changes which stars undergo when they leave the main sequence. Some of the evidence is directly observational; other evidence is derived from theoretical arguments concerning the types of star which appear in different parts of the H–R diagram.

Direct observational evidence comes from a variety of sources. One of the most direct is the observation of the profiles of emission lines of 'P-Cygni' type (figure 6.21). In this type of profile, the emission line originates in the stellar atmosphere but the short wavelength side of the line is strongly modified by absorption by the same type of ions responsible for the emission line in the outflowing material which is moving along the line of sight towards the observer. The outflowing material absorbs not only the emission line radiation but also the underlying continuum of the star. Observations of this type have been made with particular success in the ultraviolet waveband by the International Ultraviolet Explorer (IUE) because the resonance lines of a number of the common elements fall in this waveband. As a result, mass loss rates have been determined for many classes of hot star.

Another piece of direct evidence for mass loss comes from the observation of expanding shells about highly evolved stars (figure 6.10). This is particularly the case for planetary nebulae in which roughly spherical shells of gas are observed moving outwards from the central star, the velocities being typical of the escape velocity from the surface of a star belonging to the giant branch.

Figure 6.21. Illustrating the P–Cygni profile of a strong emission line. The outflow of gas in the form of a wind causes absorption of both the emission line and continuum radiation to the short wavelength side of the line. This type of profile provides important information about the velocities, densities, and mass-loss rates of stellar winds.

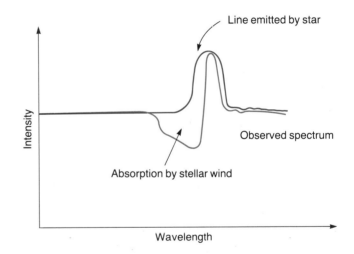

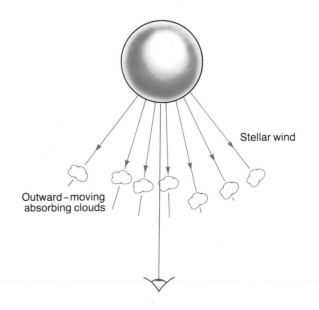

Another important technique is the observation of dust shells around giant stars. These are detected by their far-infrared emission, either in the wavebands accessible from the ground at 10 and 20 μm or from space infrared telescopes such as the Infrared Astronomy Satellite (IRAS). Dust particles condense in the cooling outflow from giant stars and these are then heated up by the stellar radiation from the giant star. The dust is heated to temperatures in the range 100–1000 K and this is readily detected as intense far-infrared radiation.

These observations are clear evidence for mass loss from stars on the main sequence and on the giant branch. There are, however, important theoretical reasons for believing that mass loss must be important. First of all, the sequence of nuclear burning within stars as they advance up the giant branch does not proceed continuously but as a discontinuous series of 'jumps'. For example, once the hydrogen in the core of a star is exhausted, hydrogen burning continues but now in a shell about an isothermal helium core. As evolution proceeds, the core contracts until the temperature is high enough for helium burning in the core to begin. This results in a major energy release within the star, and it moves to the left across the H–R diagram as the temperature increases. It is during these transitions from one form of burning to another that the star has to reorganise its internal structure, and in the process the outer layers of the star may be expelled.

One strong reason for expecting that this must have occurred during evolution on the giant branch comes from models of horizontal branch stars, which indicate that they are low mass stars with masses about $0.5\,M_\odot$ or less. Further evidence on the internal structure of these stars and their masses comes from certain special types of star which are found on the horizontal branch. These are the variable stars of *RR-Lyrae* type, which are only found within a certain 'instability strip' on the H–R diagram (figure 6.22). The models of these stars, which can account for their light curves, their luminosities and surface properties, are important clues to the nature of the horizontal branch stars and their masses. Since the main sequence termination points for even the oldest stars in the Galaxy have only just reached one solar mass, the horizontal branch stars must have lost mass from their outer layers. Thus, it is entirely reasonable to suppose that the horizontal branch stars orig-inate from stars on the giant branch which lose mass during the various convulsions undergone by the star as it evolves up the giant branch. The models of the horizontal branch stars indicate that they evolve back towards the giant branch and ultimately up towards its tip.

As the star moves towards the tip of the giant branch, it reaches the region occupied by long period variable and unstable stars. These are stars which are in the very final phases of evolution and there appears to be a continuity in properties between the various classes of objects found in this region of the diagram. The long-period variables and the OH/IR stars form a continuous sequence with increasingly long periods leading ultimately to a region of the H–R diagram populated by

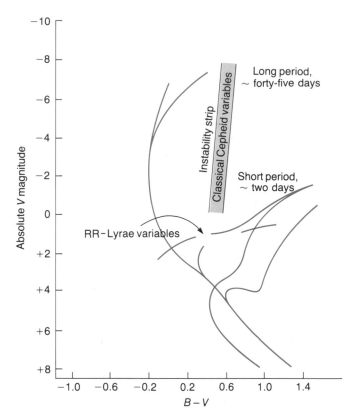

Figure 6.22. A Hertzsprung–Russell diagram illustrating the region occupied by Cepheid and RR-Lyrae variable stars. This region is known as the *instability strip*. The latter stars provide important information about the masses and properties of stars on the horizontal branch. (After Michalas and Binney (1981), pp. 125, 155; and Shu (1982), p. 169.)

unstable stars. In these stars, collapse of the core takes place with the expulsion of the envelope of the giant star, leading to the planetary nebula phase of evolution. Evidence that the cores of these stars collapse is the very high temperatures of the central stars in planetary nebulae and also the chemical abundances of the elements necessary to explain their surface properties. They appear to be essentially pure helium stars, the implication being that most of the outer layers of the stars have already been expelled. These very hot compact stars follow a sequence on the H–R diagram which indicates that they will end up as white dwarf stars (figure 6.11).

We have concentrated in the above discussion on the evolution of stars with mass roughly that of the Sun. Higher mass stars evolve much more rapidly and follow the evol-utionary paths shown in figure 6.12. The evolutionary tracks assume that mass loss is unimportant but it is now clear that for the most luminous stars known mass loss plays a major role in their evolution. There is an upper limit to the luminosities of stars which is set by a vibrational instability. It is believed that stars of about $60\,M_\odot$ are just stable but all stars of about this mass and greater exhibit enormous mass loss rates, values as large as $10^{-4}\,M_\odot$ year^{-1} being common and the extreme object η Carinae having a mass loss rate of about $10^{-2}\,M_\odot$ year^{-1}. These stars lose mass at such a high rate that they can lose their hydrogen envelopes during the process of what would normally be called main sequence evolution exposing the helium cores created in their centres. In some cases, they may evolve over towards the red giant region and then suffer further mass loss from their surfaces. Mass loss of this form or possibly associated with mass loss in close binary systems may be the origin of the class of star known as *Wolf-Rayet* stars which appear to be helium stars with high abundances of carbon or of nitrogen. Examples of these modified evolutionary tracks including mass loss are shown in figure 6.23. The problem of understanding the ultimate fate of high mass stars will be taken up in Section 6.4.

Throughout these processes of post main sequence evolution, it will be noted that two basic processes take place. The star attempts to achieve a state of higher and higher binding energy, the energy released in this process going towards providing pressure support for the star. The second aspect is that, in achieving this more highly bound state, the star loses mass. If we sum over all forms of mass loss from stars in our own Galaxy, it is likely that about $1-10\,M_\odot$ of material per year are returned to the interstellar medium. This is an important result because it means that the interstellar medium is constantly being replenished by stellar mass loss. Clearly, over a period of 10^{10} years, it is likely that a considerable fraction of the mass of the Galaxy will have been circulated through stellar interiors which provides a plausible explanation for the fact that the abundances of the elements in stars seem to have a fairly universal character. What we have not addressed in this section is how the observed chemical abundances of the elements are created. Obviously, many of the mass loss processes described above involve the expulsion of the outer layers of the stars and newly synthesised elements in the core of the star are not available for enriching the interstellar gas. It is likely that supernova explosions are responsible for much of the *chemical* enrichment (see Section 6.4) whilst the overall gaseous content of the interstellar gas is maintained by more quiescent mass loss.

Binary stars and stellar evolution

The above picture of stellar evolution refers to single stars. We know, however, that many of the stars in our Galaxy are in fact double and this can influence the evolution of the stars, particularly if they are members of close binary systems. The determination of the statistics of double stars is not a particularly straightforward task since they can only be detected, either by being a visual binary system (*visual binaries*) or if there is spectroscopic evidence showing periodic variations of the radial velocities of one or both stars (*spectroscopic binaries*). The

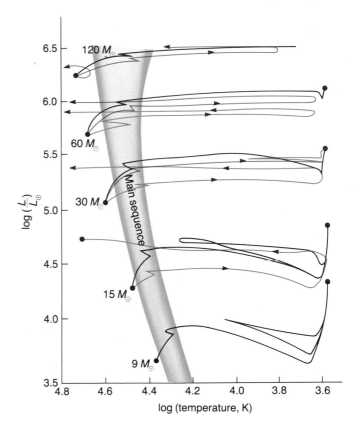

Figure 6.23. Examples of theoretical evolutionary tracks of massive stars once account is taken of the effects of mass loss during their evolution on the main sequence and in the red giant phase. The solid dot at the end of each track corresponds to the onset of carbon burning. The black lines indicate the evolution of the stars if there is no mass loss. The red line indicates the modification to the evolutionary tracks if there is moderate mass loss and the blue line the case of strong mass loss. The effect of mass loss is to make the surface temperatures of the stars rather greater than if no mass loss took place. As a result, the evolutionary tracks of the stars are shifted to the left across the H–R diagram and this is thought to be responsible for the absence of cool stars as luminous as the most luminous of the hot stars as would be predicted by the evolutionary tracks shown in figure 6.12. (After Maeder, A. (1981). *Astron. Astrophys.* **102**, 405–6.)

periods of the binary stars can range from a few hours in the case of close binary systems to hundreds of years, the upper limit almost certainly being set by the limited period over which precise observations have been carried out. As many as half of the stars in the Galaxy may be members of binary systems according to recent statistics. The great advantage of binary stars is that the determination of the orbits of the stars is one of the few precise ways of estimating stellar masses.

So far as stellar evolution is concerned, the greatest interest concerns the close binary systems in which the presence of the binary companion can strongly influence the evolution of the primary star. The separations of close binaries can range from a few times the radii of the stars to systems in which the stars share a common envelope which are known as *contact binaries*. Close binary systems are known containing a wide range of stellar types, from massive binaries containing O and B stars, through intermediate mass binaries to systems containing compact stars, either white dwarfs, neutron stars or black holes, to systems such as the binary pulsar PSR 1913+16 which consists of two neutron stars. We will deal with systems containing compact stars in the next section. Here, we consider how stars in binary systems can have very different evolutionary histories from isolated stars.

The most instructive way of understanding how binary stars can evolve is through consideration of the gravitational equipotential surfaces of the rotating system. In the rotating frame of reference, we add to the gravitational potential of each star the centrifugal potential associated with their binary motion. This results in the forms of equipotential surface shown in figure 6.24. There is a critical equipotential surface which encompasses both stars which is referred to as the *Roche lobe* of the binary system. The equipotential surfaces within the Roche lobe show that the shapes of the stars are distorted from spheres if they fill a significant fraction of their Roche lobe.

In the extreme case of contact binary systems, the common envelope of the binary lies outside the Roche lobe. These stars are recognised by their very short periods, which are less than about half a day, and their distinctive light curves which approximate much more closely to sine waves than to a typical eclipse light curve. This class of star is often referred to as W UMa-type binaries. The common envelope leads to a very different type of structure, for example the stars having a common convective envelope. This results in a number of important deviations from the properties of isolated stars. For example, the mass–luminosity relation becomes $L \propto M$, rather than $L \propto M^4$ as found for isolated main sequence stars.

For noncontact close binaries, the stars do not fill their Roche lobes and the stars evolve more or less as normal stars. The interesting phenomena occur as the more massive of the pair evolves off the main sequence. It becomes a red giant star at an earlier time than the less massive star and therefore expands to fill its Roche lobe. Matter always seeks the lowest gravitational potential and this is achieved if the matter passes through the Lagrangian point L_1 onto the secondary star. In this way, the mass of what was initially the less massive star increases whilst the mass of the primary decreases. In the case of massive binaries, this can lead to the secondary component becoming more massive than the primary. It is this sequence of events which can lead to apparently anomalous situations in which a low mass white dwarf is found as a companion to an intermediate or high mass star.

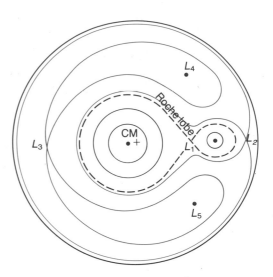

Figure 6.24. Illustrating the equipotential surfaces of a binary star system in a rotating frame of reference. The equipotential surfaces correspond to the sum of the equipotential surfaces of each star plus a centrifugal potential term to take account of the rotation of the stars about their common centre of mass (CM). In the above example, the mass ratio of the stars is 10:1. These surfaces define the shapes of the stars in the binary system. The equipotential surface which connects both stars is known as the *Roche lobe* of the system and the point at which the two lobes touch as the inner Lagrangian point L_1. Other turning points in the value of the potential are labelled L_2, L_3, etc. In close binary stars the common surface of the stars may well lie outside the Roche lobe. (After Shapiro and Teukolsky (1983), p. 400.)

There are many variations upon this theme of mass transfer in close binary systems. The secondary star itself can now evolve into a red giant and the reverse process of mass transfer back onto the original primary can occur. The end point could be the formation of a binary white dwarf or a white dwarf–neutron star pair. Another intriguing variation is if one of the stars undergoes a supernova explosion. These may be associated either with low mass or high mass stars as described in Section 6.4. In the process, a considerable amount of mass is ejected from the system and the binary may either remain bound or unbound. If the less massive star explodes due, for example, to mass transfer onto a white dwarf, the system can remain bound and this provides a means of creating a binary system containing a neutron star. If the more massive star explodes as a type II supernova (see Section 6.4), the system is likely to be disrupted and then the components will disperse with very high velocities, roughly the velocities of the stars in the binary orbit. This is a plausible explanation for the fact that isolated neutron stars are observed to have very high velocities.

Various classes of binary system have been associated with most of these permutations. For example, the *symbiotic stars* are believed to be associated with binaries in which mass transfer occurs between a red giant star and a compact companion which may be a dwarf main sequence star, a white dwarf or a neutron star. The *cataclysmic variables* probably consist of semi-detached binaries in which the 'cataclysmic' variability is associated with an accretion disc about a white dwarf. The *X-ray binaries* consist of a main sequence star and a neutron star.

It will be noted that, as mass transfer takes place, the inner regions of the star are exposed and material can be deposited onto the surface of the secondary component. These processes provide an explanation for some of the abundance anomalies found in the surface of some binary stars. This is also a process by which the massive luminous stars known as *Wolf–Rayet* stars can lose their hydrogen envelopes, exposing the products of advanced helium core burning. The theme of binaries containing compact stars will be taken up in more detail in the next section.

6.4 Dead stars

The formation of dead stars – supernovae

The stars described in Section 6.3 are held up by gas pressure, the source of energy to provide this pressure being derived from nuclear energy generation in their cores. As evolution proceeds off the main sequence, up the giant branch and towards the final phases when the outer layers of the star are ejected, the nuclear processes continue further and further along the route to using up all the available nuclear energy resources of the star – that is, once nucleosynthesis has proceeded as far as the production of iron, the most tightly bound of the chemical elements, there is no further nuclear energy source available which can be tapped to provide pressure support for the star. Therefore, the star collapses until some other form of pressure support enables a new equilibrium configuration to be attained.

We described in Section 6.2 the possible equilibrium configurations which can exist when the star collapses – these are white dwarfs, neutron stars and black holes. Whilst we have great confidence that these are the correct end-points of stellar evolution, we have much less understanding of the processes by which they are formed or of the progenitor stars from which they are formed. The problem can be understood by considering the ranges of masses which these forms of 'dead star' can take. In white dwarfs and neutron stars, the pressure which holds them up is the quantum mechanical pressure associated with the fact that the electrons, protons and neutrons are fermions, i.e. only one particle is allowed to occupy any one quantum mechanical state (see the next subsection). The white dwarfs are held up by electron degeneracy pressure and can have any mass less than about $1.5\,M_\odot$. Neutron stars, in which neutron degeneracy pressure is responsible for the pressure support, can

have masses up to about $1.5\,M_\odot$, possibly slightly higher if the neutron star is rapidly rotating. Thus, according to current understanding, dead stars more massive than about $2M_\odot$ must be black holes.

Whilst this knowledge is gratifying, it does not help us decide which sorts of star become white dwarfs, neutron stars or black holes. For example, low mass stars, $M < 2\,M_\odot$, may end up in any of the three forms. Even stars with masses much greater than $2\,M_\odot$ can form white dwarfs or neutron stars if they lose mass sufficiently rapidly. Theoretical calculations have shown that even $10\,M_\odot$ stars may lose mass very effectively towards the ends of their lifetimes and produce non-black hole remnants.

Some useful clues are provided by the statistics of stars of different types. Although in a number of cases the statistics are not very well known, the following general picture emerges. It should be recalled that the lifetime of a star of roughly the mass of the Sun is about 10^{10} years, roughly the age of the Galaxy. The numbers of white dwarfs observed locally are roughly consistent with the total number of stars of mass between about 1 and $4\,M_\odot$ which have completed their evolution since the Galaxy formed. The corresponding rate of formation of white dwarfs is also similar to the present rate of formation of planetary nebulae, again suggesting that they are important progenitors of white dwarfs. For neutron stars, the statistics are poorer. The observed number of neutron stars derived from the statistics of pulsars is roughly the same as the number of stars in the mass range 4–$10\,M_\odot$ which have died during the typical lifetime of a pulsar. This formation rate of pulsars is also similar to the rate of supernova events in our Galaxy.

These statistics are consistent with, but do not necessarily prove, a general picture in which low mass stars, 1–$4\,M_\odot$, evolve through the planetary nebula phase into white dwarfs. More massive stars, 4–$10\,M_\odot$, are the progenitors of neutron stars and many of these form in supernova explosions. It is normally assumed that stellar mass black holes are formed in supernova explosions associated with more massive stars, say $M \gtrsim 5$–$10\,M_\odot$, as must be the case if remnants of mass greater than $10\,M_\odot$ are formed.

The likely sequence of events which leads to the formation of white dwarfs was described in Section 6.3 and corresponds to what might be termed the 'peaceful' demise of stars. In contrast, it is evident that the formation of neutron stars and black holes must be associated with the rapid liberation of large amounts of energy, the binding energy of a $1\,M_\odot$ neutron star being about 10^{46} J. These events can be naturally associated with the violent events known as *supernovae* which give rise to *supernova remnants*. Supernovae are extremely violent and luminous stellar explosions in which the optical luminosity of the star at maximum light can be as great as that of a small galaxy. After the initial outburst, which lasts only a few days, the light decays exponentially with a half-life of about seventy-eight days. One of the more remarkable aspects of supernovae is that they appear to come in at least two varieties. In the case of type I supernovae, their properties are remarkably uniform with essentially identical light curves and intrinsic luminosities. On the other hand, type II supernovae exhibit a much more diverse range of properties.

The type I supernovae pose an intriguing problem since we have to devise a physical mechanism by which essentially identical explosions can occur. One appealing idea is that the explosion may result from the collapse of a white dwarf star which is accreting matter from a nearby companion. When the mass of accreted matter brings the total mass of the star above the critical mass for stability as a white dwarf, the star collapses, liberating about 10% of the rest mass energy of the star as a neutron star of about $1 M_\odot$ is formed. Since the progenitor stars are all of the same type and mass, this could account rather elegantly for the fact that their observed properties are so similar.

It is likely that type II supernovae are formed from more massive stars, probably with masses greater than about $4 M_\odot$. The processes which occur in the collapse and explosion of such a star are not fully understood but some remarkable physical processes must occur. In the centres of such collapsing stars, the temperatures and densities are very high. As the collapse proceeds, the densities and temperatures become sufficiently high for the inverse β-decay process to become important. In this, energetic electrons interact with protons to form neutrons through the reaction

$$p + e^- \rightarrow n + \nu. \qquad (6.6)$$

The result is that large fluxes of neutrons and neutrinos are produced. The densities are, however, so great that even the neutrinos cannot escape from the collapsing core, the dominant source of opacity involving the weak neutral currents which gives rise to scattering of neutrinos from nucleons (see Chapters 14, 17 and 18 by Close, Taylor and Salam). This is an important example of the impact of recent discoveries in particle physics upon astrophysics. Thus the neutrinos and the energy released in the collapse are trapped. How the energy is ultimately released in the form of a supernova outburst is not entirely clear, but one likely picture is that the very central regions collapse to form a neutron star which slightly overshoots and reverberates sending a shock wave out through the infalling material. The outer layers of the pre-supernova star are ejected at velocities exceeding its escape velocity.

One of the most exciting recent astronomical events has been the explosion of a supernova in one of the companion galaxies of our own Galaxy, the Large Magellanic Cloud, which was observed on 24 February 1987. The supernova, known as SN1987a, reached about third magnitude in May 1987 and is therefore the brightest supernova since Tycho's supernova of 1604 and the first of the bright supernovae to be studied in detail with modern instrumentation. It has provided a unique opportunity for studying the evolution of supernovae although,

ironically, it appears to be a peculiar type II supernova – examples of these anomalies are the facts that its light curve is unusual in that its luminosity remained roughly constant at magnitude 4 for about two months after the explosion, it exhibited a rapid decline in its surface temperature and was sub-luminous for the typical type II supernova. The supernova coincides exactly with the position of the bright blue supergiant star Sanduleak − 69 202 which has disappeared following the supernova as revealed by observations with the IUE, indicating that the progenitor of the supernova was an early-type B star.

One of the great pieces of good fortune was that, at the time of the explosion, neutrino detectors were operational at the Kamioka zinc mine in Japan and at the Irvine–Michigan–Brookhaven experiment located in an Ohio salt mine. Both experiments detected simultaneously a pulse of neutrinos, which lasted about 10 s, well above the background signal in the detectors. It is rather convincing that this neutrino pulse was indeed associated with the supernova explosion since the supernova was first observed optically some hours after the pulse. The neutrinos escape directly from the centre of the collapse of the supernova whereas the optical light has to diffuse out through the atmosphere of the supernova. These observations are uniquely important for the theory of stellar evolution. Perhaps most important of all, the inferred neutrino luminosity of the supernova corresponds more or less exactly to the flux expected when a neutron star is formed in the sense that it matches more or less exactly that expected when the binding energy of a neutron star is released. In addition, the energies of the neutrinos observed are in good agreement with the expected neutrino energy spectrum formed during neutron star formation. Preliminary results on the interpretation of the light curve suggest that the progenitor star must have been massive, $M \simeq 15 M_\odot$, consistent with the mass of the B-star Sanduleak −69 202. The best models may require the star to have a smaller envelope than is usual and a lower abundance of metals, consistent with the general trend of stars in the Magellanic Clouds. The subsequent evolution of the supernova and its atmosphere promise to provide astrophysicists with unique opportunities for understanding the evolution of expanding envelopes around exploding stars. Already, the neutrino results provide a convincing argument that the general overall picture of stellar evolution outlined above is certainly along the right lines.

Almost certainly the final collapse of the central regions of a collapsing star are complex. We have not taken account of any rotation present in the collapsing star nor the effects of any asymmetry in the collapse. The latter is of the greatest interest because any asymmetric collapse which results in a net quadrupole moment in the final collapse stages is a strong source of *gravitational waves*, indeed probably one of the most powerful of the more reasonable mechanisms for generating gravitational waves with intensities which might be detectable by the next generation of gravitational wave detectors (see Chapter 2 by Will). Unfortunately, there were no gravitational wave detectors operational at the time of the recent supernova outburst, although with present levels of sensitivity it is regarded as unlikely that gravitational waves would have been detected.

One of the more important aspects of supernova explosions is that they are probably the source of many of the heavy elements found in nature. When the core of a massive star collapses, the central temperatures become very high indeed and, in the ensuing explosion, *nonstationary* nucleosynthesis can take place. Whereas, in the interiors of stars, nucleosynthesis takes place over very long time scales, resulting in more or less equilibrium abundances of the elements, in nonstationary nucleosynthesis encountered in exploding stars, a much more diverse distribution of elements is found. This process, known as *explosive* nucleosynthesis, is particularly effective in producing many of the heavier elements in the periodic table. Examples of some of these calculations are shown in figure 6.25, indicating how this process can account in quantitative detail for the observed abundances of many of the heavy elements.

It would be important to find independent evidence for the process of explosive nucleosynthesis. A remarkable recent observation of the supernovae in M83 has suggested strongly that the above picture is indeed along the correct lines. One of the major puzzles has been why it is that the light curves of supernovae follow closely an exponential decay with a half-life of about seventy-eight days. One interpretation of this observation is that the exponential decline results from the radioactive decay of ^{56}Ni into ^{56}Fe, which has exactly the correct half-life to explain the observed exponential decay of the luminosity of supernovae. It is expected that large amounts of iron peak elements are synthesised in explosive nucleosynthesis and ^{56}Ni is one of the principal products. If this interpretation is correct, we might expect to observe strong iron emission in the spectra of supernovae. Such a feature has recently been observed in the supernova in M83 from observations made of the [FeII] line in the infrared waveband by Dr James Graham and his colleagues. This is convincing evidence for the formation, not only of heavy elements like iron in supernova explosions, but also of the exponential light curve of supernovae. Another important aspect of this work is that the supernova explosion provides us with a convincing mechanism of dispersing effectively the heavy elements through the interstellar medium.

Two other aspects of supernovae are worthy of special mention. The first is that the kinetic energy of the supernova

Figure 6.25. Examples of the products of explosive nucleosynthesis. In these computer models, shells of carbon, oxygen and silicon are rapidly heated to a very high temperature, as in a supernova explosion, and the nucleosynthesis of heavier elements takes place in an expanding, cooling shell. The peak temperatures reached were 2×10^9 K in the case of carbon burning (a), 3.6×10^9 K in the case of oxygen burning (b) and 4.7–5.5×10^9 K in the case of silicon burning (c). The circles represent the observed solar abundances of the elements and the crosses the products of explosive nucleosynthesis. It can be seen that there is encouraging agreement between the predicted and observed abundances. (From Arnett, W. D. and Clayton, D. D. (1970). *Nature* **227**, 780.)

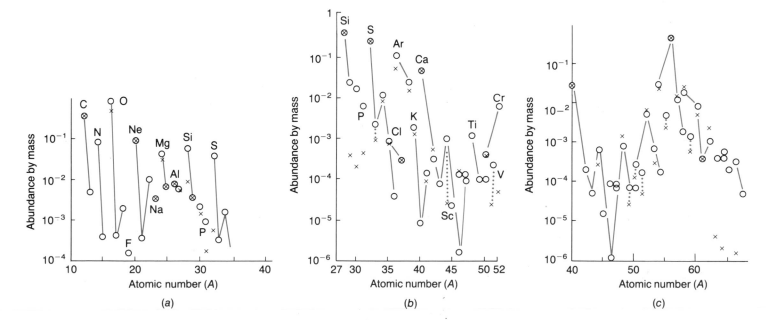

remnant is a powerful source of heating for the ambient interstellar gas. The shells of supernova remnants are observable until they are about 100 000 years old. At most stages they are observable as intense X-ray sources, in the early days through the emission of hot gas originating in the explosion itself and, at the later stages, through the ambient gas being heated to a high temperature by the shock wave which advances ahead of the expelled shell of gas (figure 6.26). In both cases the emission mechanism is the bremsstrahlung or free–free emission of hot ionised gas. Thus, the kinetic energy of the expanding supernova remnant is a powerful heating source for the interstellar gas, regions up to 50 pc or more about the explosion being heated to temperatures of 10^6 K or more. The overlapping of these expanding spheres of hot gas can result in a substantial fraction of the interstellar gas being heated to this temperature. This picture is consistent with the soft X-ray flux observed from the interstellar medium and the observation of highly excited ions such as five times ionised oxygen (O VI) and triply ionised carbon (C IV) in absorption in the interstellar gas.

Figure 6.26. An X-ray image of the supernova remnant Cassiopaeia A taken by the Einstein X-ray satellite observatory. Cassiopaeia A is a young supernova remnant, having exploded probably about 250 years ago. The X-ray emission is thermal free–free emission (or bremsstrahlung) of the hot gas expelled in the supernova explosion. (Courtesy of Dr F. Seward and the Harvard-Smithsonian Center for Astrophysics.)

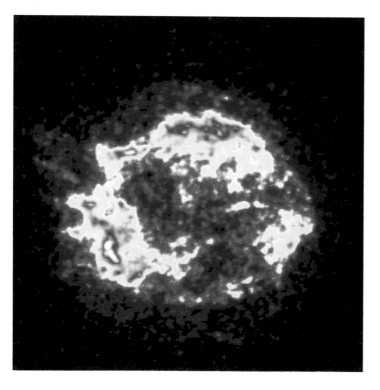

The second important aspect is that supernovae are sources of very high energy particles. The direct evidence for this is the observation of synchrotron radio emission from supernova remnants (figure 6.27). Synchrotron radiation is emitted when ultrarelativistic electrons spiral in magnetic fields and the inferred fluxes of particles are so high that they can only originate within the supernova remnant itself. In terms of the total energy requirement, it is likely that most of the high energy particles which are observed in the Galaxy originate in supernovae, although contributions from other sources such as pulsars and X-ray binaries might be important in explaining certain features of the cosmic radiation.

There are at least two candidate mechanisms by which these particles are accelerated to very high energies. One is acceleration in the vicinity of the neutron star which may form in the supernova explosion (see below). This is certainly the case in the Crab Nebula where there is a young pulsar present which supplies the relativistic particles needed to maintain the X-ray, optical and radio continuum radiation of the remnant. Alternatively, the particles may be accelerated in the shell of the supernova remnant. It has always been a problem to understand the origin of the power-law form of the spectrum of cosmic rays and of the radio emission from radio sources such as radio galaxies and supernova remnants. In a very wide range of different cosmic environments, the radiation spectrum of cosmic ray electrons corresponds to a power-law distribution of the form $N(E)dE \propto E^{-2.5}dE$. A similar law is found for the spectra of cosmic ray electrons, protons and nuclei at the top of the Earth's atmosphere.

One of the most important growth areas in these studies has been the acceleration of high energy particles in expanding supernova shells. By a remarkably elegant argument, Bell, and independently Blandford and Ostriker, showed that high energy particles can be statistically accelerated with a power-law spectrum similar to that observed in expanding supernova shells. The essence of the mechanism is that the particles receive a small increase in energy when they are swept up by the expanding supernova shock-wave. Some of them can then diffuse back upstream ahead of the shock and receive a further increase in energy when they are next overtaken by the shock front. Thus a small number of particles gain a large amount of energy whilst most of them receive only a small amount. There are many attractions in this mechanism. First of all, the process produces a power-law spectrum rather naturally, the exponent being remarkably independent of the local physical conditions and close to the 'universal' value of 2.5. Second, it is significant that the particles are accelerated in the diffuse conditions of interstellar space because, if they were accelerated close to the central object and no subsequent acceleration took place, they would suffer adiabatic losses which would convert their energy into the kinetic energy of expansion of the remnant.

Thus, supernovae and supernova remnants play a key role in many different aspects of the evolution of the interstellar gas.

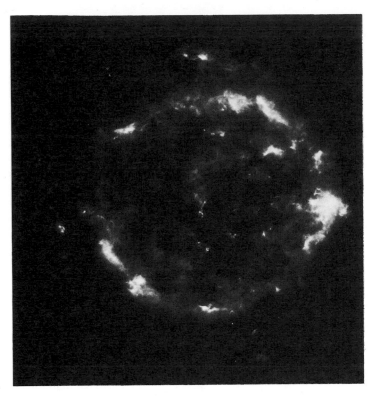

Figure 6.27. A radio image of the supernova remnant Cassiopaeia A. The radio emission is the synchrotron radiation of ultrarelativistic electrons gyrating in the magnetic field of the supernova remnant. There is a close similarity between the X-ray image of figure 6.26 and this image despite the fact that the radiation processes are quite different. The implication is that the hot gas is well mixed with the relativistic particles and magnetic fields. (Courtesy of Dr S. F. Gull and the Mullard Radio Astronomy Observatory.)

White dwarfs and neutron stars

In the cases of both white dwarfs and neutron stars, there are no internal heat sources – the stars are held up by *degeneracy pressure*. The significance of degeneracy pressure comes about very naturally because in the centres of stars at an advanced stage in their evolution the densities become large and the approximation of using the pressure formulae of a classical gas is inappropriate. The Heisenberg uncertainty principle ensures that at very high densities, when the interparticle spacing becomes small, the particles of the gas must possess large momenta according to the uncertainty relation $\Delta p \Delta x \simeq h/2\pi$. It is these large quantum mechanical momenta which provide the pressure of the degenerate gas.

In the case of white dwarfs, the densities are sufficiently high, $\simeq 10^9 \, \mathrm{kg \, m^{-3}}$, for the electrons to be squashed so close together that the uncertainty principle ensures that the quantum mechanical or degeneracy pressure of the electrons provides

pressure support for the star. Their structures are relatively easy to work out because the formula for the pressure of a degenerate gas is independent of temperature. The energy of the star is derived simply from the internal energy with which the star was endowed when it was formed. The cooling times for white dwarfs are about 10^9 to 10^{10} years, much longer than the thermal cooling time for a star like the Sun, largely because their surface areas are much smaller than those of main sequence stars. For the few white dwarfs which have been found in clusters, the ages of the clusters are of the same order as the cooling lifetimes of the white dwarfs.

With increasing density, the degenerate electron gas becomes relativistic, and ultimately at densities of about $10^{14} \, \mathrm{kg \, m^{-3}}$ the energies of the electrons are sufficiently high for the inverse beta decay reaction (6.6) to convert protons into neutrons. Up to these densities, the nuclei have formed a nondegenerate Coulomb lattice and the nuclei are the conventional stable elements such as carbon, oxygen and iron. As the density increases, however, the inverse beta decay reaction favours neutron-rich nuclei and equilibrium states are set up consisting of neutron-rich nuclei, a free neutron gas and a degenerate relativistic electron gas. In equilibrium, neutrons are exchanged between the neutron-rich nuclei and the free neutron gas, a process referred to as *neutron-drip*. These processes result in profound changes in the equation of state of the gas such that stable stars cannot form until much higher densities are attained, $\simeq 10^{17} \, \mathrm{kg \, m^{-3}}$, by which time the neutron-drip process has resulted in the conversion of most of the protons and electrons into neutrons. It is the degeneracy pressure of this relativistic neutron gas which prevents collapse under gravity. Exactly the same physics described above for the white dwarfs holds the star up, the only difference being that the neutrons are about 2000 times more massive than the electrons and consequently degeneracy sets in at a correspondingly higher energy. In both cases, the equilibrium configuration is determined by rough equality of the gravitational energy and the internal energy of the gas.

An important consideration for both types of star is why it should be that there is an upper limit to the masses of stable neutron stars and white dwarfs. The physics is the same in both cases and is associated with the form of the equation of state for a relativistic degenerate gas, both being determined by purely quantum mechanical forces. In the equilibrium configuration, there is rough equality of the internal (Fermi) energy of the gas and its gravitational potential energy. Because of the form of the equation of state of a relativistic degenerate gas, it turns out that the gravitational and internal energies depend upon the radius of the star in the same way. Therefore, if the gravitational potential energy exceeds the internal energy, the star cannot seek a stable state by collapsing to a higher density since the gravitational energy will still exceed the internal energy in the same ratio. As the mass M of the star is increased, its gravitational energy increases as M^2, whereas the internal thermal energy grows more slowly with increasing mass. Therefore, there must be an upper limit to the mass of stable degenerate stars. It is remarkable that the mass corresponding to this upper limit for stability, known as the *Chandrasekhar limit*, depends only upon fundamental constants

$$M \simeq m_B(hc/2\pi Gm_B)^{3/2} \simeq 2\,M_\odot,$$

where m_B is the average mass of the neutrons and protons in the star, i.e. the average baryonic mass. The determination of the internal structure of the white dwarfs and neutron stars depends upon detailed knowledge of the equation of state of the degenerate electron and neutron gases, and this has been the subject of much study. The case of white dwarfs is the more straightforward. The equation of state is well understood, the main uncertainty being the chemical composition of the star. For stars with masses about that of the Sun, it is unlikely that nuclear burning proceeds beyond carbon. On the other hand, it cannot be excluded that some of the massive stars which form iron in their cores are able to lose mass and angular momentum effectively and thus form white dwarfs which are composed of iron nuclei. As expected, when converted into an H–R diagram, the white dwarf sequence lies below the main sequence and can give a good account of the observed properties of white dwarfs (figure 6.28). In fact, the solid lines are no more than the cooling curves for black-bodies, the luminosity L being proportional to T^4.

Figure 6.28. Comparison of the theoretical H–R diagram for white dwarfs with their observed properties. The location of the cooling locus in the H–R diagram depends upon the mass of the white dwarf. (Shapiro and Teukolsky (1983), p. 70.)

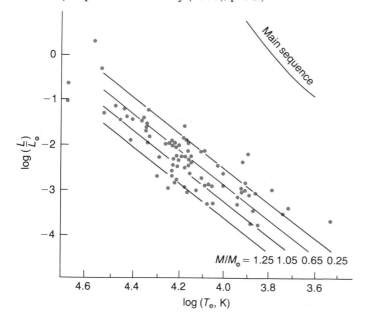

The internal structures of neutron stars are less well determined because of uncertainties in the equation of state of degenerate nuclear matter. The problems involved in determining the equation of state are elegantly presented by Shapiro and Teukolsky (1983), and figure 6.29 shows a representative example of the internal structure of a neutron star. Following Shapiro and Teukolsky, the various zones in the models may be described as follows:

(1) The *surface layers* are taken to be the regions with densities less than about $10^9\,kg\,m^{-3}$. At these large densities, the matter consists of atomic polymers of ^{56}Fe in the form of a close packed solid. In the presence of strong surface magnetic fields, the atoms become cylindrical. The matter behaves like a one-dimensional solid with high conductivity parallel to the magnetic field and with essentially zero conductivity across it.

(2) The *outer crust* is taken to be the region with density in the range $10^{10} \leqslant \rho \leqslant 4.3 \times 10^{14}\,kg\,m^{-3}$ and consists of a solid region composed of matter similar to that found in white dwarfs, i.e. heavy nuclei forming a Coulomb lattice embedded in a relativistic degenerate gas of electrons. When the energy of the electrons exceeds the difference in rest mass energies between the neutron and the proton $E > 1.29\,MeV$, inverse beta-decay increases the numbers of neutrons, reaction (6.6), and in the solid phase neutron-rich nuclei which would be unstable on Earth can form, for example, ^{62}Ni at $3 \times 10^{11}\,kg\,m^{-3}$, ^{80}Zn at $5 \times 10^{13}\,kg\,m^{-3}$ and ^{118}Kr at $4 \times 10^{14}\,kg\,m^{-3}$.

(3) The *inner crust* has densities between about 4.3×10^{14} and about $2 \times 10^{17}\,kg\,m^{-3}$ and consists of a lattice of neutron-rich nuclei together with a free degenerate neutron gas and a degenerate relativistic electron gas. As the density increases more and more of the nuclei begin to dissolve and the neutron gas provides most of the pressure.

(4) The *neutron liquid* phase occurs at densities greater than about $2 \times 10^{17}\,kg\,m^{-3}$ and consists chiefly of neutrons with a smaller concentration of protons and normal electrons.

(5) In the very centre of the neutron star, a *core region* of very high density ($\rho \geqslant 3 \times 10^{18}\,kg\,m^{-3}$) may or may not exist. The existence of this phase depends upon the behaviour of matter in bulk at very high energies and densities. For example, is there a phase transition to a neutron solid or to quark matter or to some other phase of matter quite distinct from the neutron liquid at extremely high densities? Many of the models of stable neutron stars do not possess this core region, but it is certainly not excluded that quite exotic forms of matter could exist in the centres of some neutron stars.

To add further spice to the picture, it has been shown that, in layers (3) and (4) above, the long-range attractive correlation forces between neutrons are sufficiently strong for the neutrons

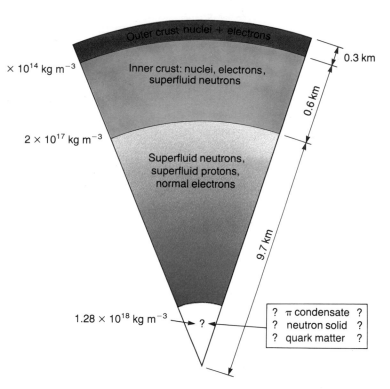

$\times 10^{14}$ kg m^{-3}

Outer crust: nuclei + electrons

Inner crust: nuclei, electrons, superfluid neutrons

0.3 km

0.6 km

2×10^{17} kg m^{-3}

Superfluid neutrons, superfluid protons, normal electrons

9.7 km

1.28×10^{18} kg m^{-3} ?

? π condensate ?
? neutron solid ?
? quark matter ?

Figure 6.29. A representative model showing the internal structure of a $1.4\,M_\odot$ neutron star. (Shapiro and Teukolsky (1983), p. 251.)

to form pairs with the result that the neutron gas is *superfluid*. By the same token, the protons in layer (4) are expected to be superfluid and superconducting (see **Chapter 9** by Leggett).

To this catalogue of exotic phenomena we should add a magnetic field. The early observation of polarised radio emission from radio pulsars, and, in particular, the observed rotation of the plane of polarisation of the radiation within pulses, was powerful evidence for the presence of a magnetic field in pulsars. Field strengths in the range 10^6 to 10^9 T were suggested from the observed rate of deceleration of pulsars. Further evidence for such strong magnetic fields has come from the observation of a cyclotron radiation feature in the X-ray spectra of the X-ray pulsar Hercules X-1 and others which indicate that field strengths of the order of 10^8 T are present in the source regions. Although these seem very strong fields indeed, there is no problem in accounting for them astrophysically. This is because, in the collapse of a star, the magnetic field is very strongly tied to the ionised plasma by the well known process of *magnetic flux freezing*. When a star collapses spherically, the magnetic field strength increases as $B \propto r^{-2}$, and so if a star like the Sun possesses a field of strength 10^{-2} T, there is no problem in accounting for a field strength of 10^8 T when the radius collapses to 10^{-5} times its initial radius. The neutron

stars must be threaded by intense magnetic fields which can influence the internal structure of the neutron star. It might be thought that the magnetic field would be expelled from the central regions of the neutron star because of the superconducting proton fluid inferred to be present. The presence of the normal relativistic degenerate electron gas, however, ensures that the magnetic field can exist within the central regions.

To complete the picture, we must add the rotation of the neutron star. It is the rotation of the neutron star which is responsible for the observation of the pulsed emission of pulsars at radio and X-ray wavelengths, the pulses being attributed to the swinging of a beam of radiation from the poles of the neutron star past the observer. It is interesting to compare the rotation periods of the neutron stars observed with the maximum rotational energy which they could possess. A rough estimate of this may be made by assuming that the neutron star would break up due to centrifugal forces when the rotational energy is roughly the same as the gravitational potential energy. For a $1\,M_\odot$ neutron star, the break-up rotational period is a fraction of a millisecond. This is much shorter than the observed rotation periods of all pulsars, although some pulsars with periods in the range 1–10 ms are now known.

Generally speaking, the *radio pulsars* are isolated neutron stars which are rotating and have strong magnetic fields. On the other hand, the *X-ray pulsars* are members of binary star systems and thus are subject to a wide variety of other phenomena which can influence the properties and evolution of the neutron star. In the latter case, other specific features to be added include the facts that the neutron star with its rotation and magnetic field is located in a frame of reference rotating about the centre of mass of the binary system and that the neutron star accretes matter from the primary star, the accreted matter bringing with it its own angular momentum and magnetic fields.

Thus, the discovery of pulsars and X-ray binaries has opened up vast new fields of study for physicists and astrophysicists – the challenge to the theorist and the observer is to devise ways in which, through careful study of the observed properties of these objects, further insight may be obtained into the behaviour of matter under extreme conditions which are inaccessible in terrestrial laboratories. These endeavours have been remarkably successful and some of the highlights are discussed below.

Isolated neutron stars – radio pulsars   Radio pulsars were more or less a complete surprise when they were discovered in 1967. They had been predicted as long ago as 1935 by Baade and Zwicky, soon after the discovery of the neutron, but the models of neutron stars indicated that, as compact stars, the only detectable emission should be the thermal radiation from their surfaces. In a prescient paper of 1967, before the announcement of the discovery of pulsars, Pacini in fact predicted that they might be observable at long radio wavelengths if they were magnetised and were oblique rotators.

On general grounds it is expected that neutron stars should be very hot when they form because they have to get rid of their gravitational binding energy and the cores of the collapsing stars are heated to extremely high temperatures during collapse. The neutron star cools by thermal radiation from its surface and by neutrino emission from its interior. Below temperatures of about 10^9 K, the neutron star is transparent to neutrinos and so they provide a very efficient means of getting rid of the thermal energy of the star. The neutrino emission processes are the dominant cooling mechanisms at high temperatures and involve a variety of neutrino emission processes including beta-decay modified by various neutrino interaction processes which occur in the neutron sea inside the neutron star. The predicted surface temperatures of the neutron stars are about 2×10^6 K after about 300 years and remain in the range about 0.5 to 1.5×10^6 K for at least 10^4 years. The youngest pulsars known, such as the pulsars in the Crab and Vela supernova remnants and the pulsar in the Magellanic Clouds, 0540–693, lie in this age range. Despite deep observations with the Einstein X-ray Observatory to detect this radiation, only upper limits to the temperatures of the surfaces of these neutron stars of about 2×10^6 K are available. There is thus no inconsistency with theory at the moment, but it will not require a large increase in sensitivity to provide a critical test of the theory of the cooling of neutron stars and consequently for the physics of condensed matter at nuclear densities.

It was not long after the discovery of pulsars that they were convincingly identified as isolated rotating magnetised neutron stars following proposals by Gold and Pacini. The key observations were the very stable, short periods of the pulses and the observation of polarised radio emission. The discovery of the pulsar in the Crab supernova remnant was of particular importance because it is one of the youngest supernova remnants known, the explosion of the star having been observed by Chinese astronomers in 1054. The age of the pulsar is therefore well known and it was also important that the Crab pulsar had the shortest period of all those known at that time. All radio pulsars are slowing down and the rate of decrease of their periods have been measured. For most pulsars, the characteristic time scale defined by the rate of change of their periods $\tau \sim P/\dot{P}$ is about 10^7 years. In the case of the Crab pulsar, the figure is about 2500 years, within order of magnitude of the age of the pulsar.

A particularly important parameter for pulsars is the rate at which this characteristic time scale is changing with time. It is most convenient to describe the slowing down by a *braking index n*, which is defined by $\Omega = -$ (constant) Ω^n, Ω being the angular frequency of rotation. The braking index gives information about the energy loss mechanism which is responsible for slowing down the neutron star. Among the most important of these is magnetic braking. In order to produce pulsed radiation from the magnetic poles of the neutron star, the magnetic dipole must be oriented at an angle with respect to the rotation axis and then the magnetic dipole displays a varying dipole moment at a large distance. This results in the radiation of electromagnetic energy from the dipole which is extracted from the rotational energy of the neutron star. For this form of *magnetic dipole radiation* the braking index is $n = 3$ – this appears to differ significantly from the value observed for the Crab pulsar, $n = 2.5 \pm 0.005$. It might be that part of the difference is associated with phenomena such as 'polar-wandering' of the magnetic dipole inside the neutron star but the situation is currently far from clear. It is interesting that, if the neutron star possesses a significant quadrupole moment in the early stages of formation and evolution, it radiates gravitational radiation for which the braking index is 5, plainly inconsistent with the observed value for the Crab pulsar.

If the magnetic braking mechanism is responsible for the slow-down of the neutron star, estimates can be made of the magnetic field strengths at the surface of the neutron star. These magnetic field strengths typically lie in the range from about 10^7 to 10^9 T for most pulsars, although for the millisecond pulsars much weaker field strengths are found. One particularly beautiful result for the Crab pulsar is that the rate at which it loses rotational energy, $dE/dt \sim 6.4 \times 10^{31}$ W, is similar to the energy requirements of the surrounding supernova remnant in nonthermal radiation and bulk kinetic energy of expansion, $dE/dt \sim 5 \times 10^{31}$ W. The origin of the continuous supply of high energy particles to the Nebula had been a major mystery prior to the discovery of the Crab pulsar because the radiation lifetimes of the particles emitting X-ray and optical synchrotron radiation in the nebula are much less than the age of the supernova remnant. The continuous injection of energy into the nebula from the pulsar solves this problem and in doing so provides one convincing mechanism by which particles can be accelerated in supernova events.

For a few pulsars, the slow-down rate is not continuous but exhibits a number of discontinuities known as *glitches* in which the pulsar period changes abruptly. The nature of the discontinuity is shown in figure 6.30 in which it can be seen that the pulsar eventually settles down to a steady slow-down following the abrupt glitch. These phenomena are attributed to changes in the moment of inertia of the neutron star as it slows down. An attractive picture for this process arises from the fact that there is only weak frictional coupling between the 'normal' component of the neutron star, the crust and the charged particles, and the 'abnormal' component, the superfluid sea of neutrons. The crust takes up an equilibrium configuration in which gravitational, centrifugal and the solid state forces in the crust are in balance. However, as the pulsar slows down, the centrifugal forces weaken and the crust attempts to establish a new shape with a lower moment of inertia. This phenomenon can take place in what is termed a *starquake*, in which the crust establishes its new shape by cracking the surface. Since the moment of inertia decreases, this results in a speed-up of the components attached to the crust, i.e. the crust itself, the charged particles and the magnetic field. However,

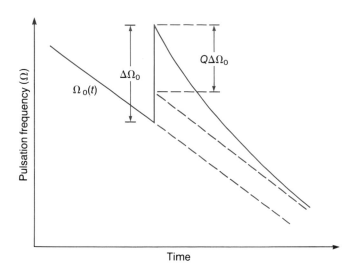

Figure 6.30. Illustrating the phenomenon of glitches observed in the Crab Nebula pulsar and a few other pulsars, including that in the Vela supernova remnant. The pulse period increases smoothly as the rotation rate of the neutron star decreases but there are sudden discontinuities in the pulse period after which the steady increase in period continues. The variation of the pulse period with time during these 'glitches' provides information about the internal structure of the neutron star. (Shapiro and Teukolsky (1983), p. 296.)

these components are weakly coupled to the neutron superfluid and so over a period of time the latter will be spun-up through the frictional force, resulting in a slow down of the crust and the associated magnetic field structure. It can be seen that this model can account qualitatively for the features of the glitches shown in figure 6.30. What is intriguing about this model is that these glitches give us physical information about the viscous interaction between the neutron superfluid and the normal component inside the neutron star.

This model seems to provide a good explanation of the glitches observed in the Crab pulsar, but it cannot account for those observed in the Vela pulsar because glitches are observed too frequently. In the latter case, it may be necessary to investigate other features of the coupling between the neutron superfluid and the crust. One of the remarkable features of the rotating neutron superfluid is that the rotation of each vortex within the neutron star is quantised. In the case of the Crab pulsar, the number of vortex lines per unit area is about $2 \times 10^9 \, \mathrm{m}^{-2}$, i.e. their spacing is about $10^{-4} \, \mathrm{m}$. The rotation of the neutron superfluid is the sum of the vorticity associated with all these vortices. The relevance of these vortices to the origin of pulsar glitches is through the question of how the vortices interact with the crustal material. In some models the vortices are 'pinned' to the nuclei of the crust, and in others they thread the spaces in between them. As the star slows

down, angular momentum is transferred outwards by the migration of the vortices. If the vortices are pinned, this process is jerky and may lead to small glitches.

The immediate environment of the pulsar is referred to as its *magnetosphere*, by analogy with the magnetic-dominated regions around the Earth. To an excellent approximation, the pulsar may be considered a nonaligned rotating magnet but the magnetic field strengths at the surface are very strong indeed. Much of the electrodynamics is best appreciated from the simpler case of an aligned rotating magnet. The physics is then exactly that of a *unipolar inductor*, but the induced electric fields at the surface of the neutron star are so strong that the force on an electron in the surface far exceeds the work function of the surface material and consequently there must be a plasma surrounding the neutron star so that electric currents can flow in the magnetosphere. The typical picture for the field distribution about a rotating magnetised neutron star is shown in figure 6.31. There is a certain radius, called the *light cylinder* or *corotation radius*, at which the velocity of rotation of material corotating with the neutron star is equal to the velocity of light. Within this cylinder the field lines are closed but those which extend beyond the light cylinder are open and particles dragged off the poles of the neutron star can escape beyond the light cylinder.

As is often the case in remarkable objects like pulsars, the most difficult part to understand is the signature which led to their discovery, i.e. the physical mechanism by which the radio pulses themselves are generated. The one clear requirement of all models of the radio emission mechanism is that the radiation cannot be incoherent radiation. An effective radiation temperature can be estimated from the known distances of the pulsars, the duration of the pulses and their observed intensities. Typically, brightness temperatures in the range 10^{23}–10^{26} K are found. This far exceeds the temperatures of thermal material which radiates in the radio waveband, and the way around this problem is to assume that the radiation is some form of coherent radiation. There are well established examples of coherent radiation mechanisms in other astronomical objects – for example, the hydroxyl OH line observed in regions of star formation and from around old stars and the processes occurring in solar flares.

The problem is to understand how coherent emission can occur in the vicinity of the magnetic poles of pulsars. One attractive scenario is associated with the enormous induced electric fields which are present at the surfaces of neutron stars. For example, the potential difference between the pole and equator of a magnetised neutron star with surface magnetic field 10^8 T rotating once per second is 10^{16} V. As a result, electrons are dragged off the surface of the neutron star and are rapidly accelerated to very high energies, more than 10^6 times their rest mass energies. As they stream out along the curved magnetic field lines, they radiate by the process known as *curvature radiation*, which is associated with the acceleration of the particles as they move in curved trajectories. In the strong

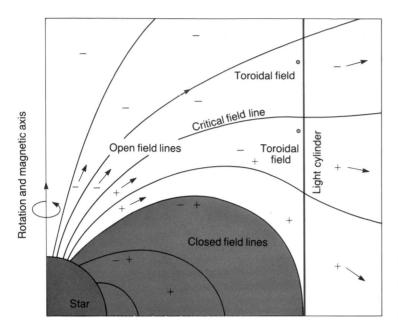

Figure 6.31. A diagram illustrating the magnetic field and charge distribution about a rotating magnetised neutron star according to the analysis of Goldreich and Julian (Goldreich, P. and Julian, W. H. (1969). *Astrophys. J.* **157**, 839). It is assumed that the magnetic axis is parallel to the rotation axis of the neutron star. The star behaves like a unipolar inductor and the charge distribution on the surface of the neutron star is shown. Electrons are removed from the surface of the star by the induced electric field in the surface layers so that there must be charge and current distributions within the magnetosphere of the neutron star. Particles attached to closed magnetic-field lines corotate with the star and form a corotating magnetosphere. The light cylinder is defined to be that radius at which the rotational velocity of the corotating particles is equal to the velocity of light. The magnetic field lines which pass through the light cylinder are open and are swept back to form a toroidal field component. Charged particles stream out along these field lines. The critical field line is at the same potential as the exterior interstellar medium and divides regions of positive and negative current flows from the star. The plus and minus signs indicate the sign of the electric charge in different regions about the neutron star. (From Shapiro and Teukolsky (1983), p. 285. After, Manchester, R. N. and Taylor, J. H. (1977). *Pulsars*, p. 179. W. H. Freeman and Co., San Francisco.)

magnetic fields close to the pulsar, these photons interact with the transverse component of the magnetic field producing electron–positron pairs. In turn, these produce photons by curvature radiation which generate electron–positron pairs and so on. The generation of these electron–photon cascades results in bunches of particles which may radiate coherently in sheets. This may well be the process responsible for producing the very high brightness temperatures observed in pulsars. An

attractive feature of this model is that it can explain why most pulsars have periods about 1 s and magnetic fields about 10^8 T – only if the magnetic field is strong enough and if the rotation period is sufficiently short will electron–photon cascades take place. According to this theory, there is a 'death-line' in the magnetic field–period relation for pulsars such that $BT_\mathrm{p}^{-2} \geqslant 10^7\,\mathrm{T\,s}^{-1}$, where T_p is the period of the pulsar in seconds.

The example of the Crab Nebula pulsar shows convincingly how magnetic torques exerted by the pulsar's magnetic field on its surroundings can account for its deceleration. It is likely that, as the pulsar grows older, its rotation period increases and the magnetic field decreases in strength, although it is not clear what mechanism is responsible for the latter process. One important result derived from the period distribution of pulsars is that there are very few short period pulsars such as that present in the Crab Nebula. Although more than 300 pulsars are now known and more than 150 supernova remnants have been observed in our Galaxy, there are only three good associations of supernova remnants with pulsars and these are all with the youngest known pulsars. Another surprise is how rare Crab Nebula-type supernova remnants are. This result suggests that only very rapidly rotating young pulsars form Crab Nebula-type remnants. Searches have been made for other young pulsars in our Galaxy but none have been found. Because of the absence of young pulsars, it is inferred that most pulsars must be created with periods about 1 s and not close to the break-up rotation speeds which correspond to periods $T_\mathrm{p} < 1$ ms. Evidently, the progenitors of pulsars are normally able to get rid of the angular momenta of their central cores in the process of collapse to form neutron stars.

The millisecond pulsars, of which three are now known, appear to be somewhat different objects from the standard pulsars. All of them have very stable periods from which it is inferred that they must have very weak magnetic fields. This makes good sense because if the magnetic field is weak, there is only weak coupling to the external medium and little deceleration. An attractive picture for their formation is that they are dead pulsars which were once members of binary systems. If mass transfer takes place from the companion star onto the dead pulsar, it will be spun up. In fact, a weak magnetic field is a great advantage for effective spin-up because the magnetic pressure determines the accretion radius about the star and, if this is weak, the angular momentum transfer can occur close to the surface of the neutron star resulting in a large spin-up. If the companion star then explodes, disruption of the system may occur, resulting in isolated millisecond pulsars. Notice that, in this picture, the dead pulsar becomes alive again because its period is spun up and it recrosses the 'death-line'. It will also be noted that this picture involves the dead pulsars having magnetic fields $\simeq 10^5$ T. These ideas have important implications for the role of neutron stars in the theory of stellar evolution, particularly the theory of the evolution of binary stars. Radhakrishan's invited discourse at the 1986 IAU

General Assembly provides an excellent survey of the full implications of the properties of pulsars for the evolution of a wide variety of binary stellar systems.

Neutron stars in binary systems – X-ray binaries   There is no reason to suppose that the neutron stars which are the optically invisible components of binary X-ray sources are different in nature from the isolated variety which are observed as radio pulsars. Thus, the distinctive phenomena which they exhibit are associated with the fact that they are members of binary systems. This is in fact an enormous bonus because it enables the properties of the neutron stars such as their masses, distances and physical environments to be determined with much greater certainty than for the radio pulsars.

The discovery of binary X-ray sources was made by the UHURU X-ray Observatory, and subsequent X-ray satellites such as SAS-3, Ariel V, HEAO-1, Hakucho, Tenma and EXOSAT have elucidated many of their properties. The distribution of bright X-ray sources as observed by HEAO-1 is shown in figure 6.2(*e*). It can be seen that they are concentrated towards the plane of the Galaxy with many of the brightest sources lying in the general direction of the centre of the Galaxy. Convincing evidence has been found for the binary nature of many of the Galactic X-ray sources and, because of selection effects, which can prevent sources being identified as binary systems, particularly if they are distant objects or if they are binaries with inclination planes at a large angle to the line of sight, it is likely that most of the point-like X-ray sources of the Galactic population are binary in nature.

The binary X-ray sources come in at least two varieties. In all cases the neutron star associated with the X-ray binary is invisible optically, and hence the optical observations refer only to the primary companion star. Many of the binary systems contain late O or early B type stars, these being among the most luminous and massive stars known with rather short main-sequence lifetimes, $\sim 10^6$–10^7 years. These are very rare classes of star and belong to the youngest of the stellar populations known. The second class is associated with cooler low-mass main-sequence stars, having masses, luminosities and temperatures similar to those of the Sun. Members of this class are referred to as *Galactic bulge sources* because they lie in the direction towards the Galactic centre and a number of them have been identified as belonging to globular clusters; the latter are members of the Galactic bulge population and are amongst the oldest stellar systems in our Galaxy. These low-mass X-ray binaries have the distinctive property that they exhibit the phenomenon of X-ray bursts.

The great advantage of X-ray sources being members of binary systems is the fact that their masses can be estimated with some degree of accuracy by the classical methods of dynamical astronomy. These systems are what are known as single-line spectroscopic binaries since high-resolution spectroscopy can provide the radial velocity of the primary star as a function of the phase of the star in its binary orbit. An added bonus comes if the X-ray source is an X-ray pulsar. The X-ray pulsars have periods which range from a fraction of a second to about fifteen minutes, the lower end of this range being similar to the periods found in the radio pulsars. Just as in the case of radio pulsars, the radial velocity of the X-ray source can be measured from the Doppler shift of the X-ray pulsar's period as a function of phase in the binary orbit. This means that the radial velocities of both components of the binary can be measured. There are other pieces of information which can be used. For example, it is very important to know the inclination of the binary orbit with respect to the line of sight from the observer. Compared with the size of the primary star, the X-ray source can often be considered a point object, and so the X-ray source may be occulted by the primary if the plane of the orbit lies close to the line of sight from the Earth. In a number of cases, such occultations of the X-ray sources are observed with periods equal to those of the binary orbits. In addition, the X-ray source itself may influence the surface properties of the primary star, either by distorting the figure of the surface into an ovoid shape, because of the gravitational influence of the neutron star, or possibly by heating up the face of the primary star closest to the X-ray source, thus causing that face of the primary star to be more optically luminous when pointing towards the Earth. There is convincing evidence for both of these phenomena among the binary X-ray sources.

All these pieces of information have been used to estimate the masses of neutron stars in X-ray binaries, and a recent compilation of mass estimates for X-ray pulsars in binary systems is shown in figure 6.32. Also included is the binary radio pulsar PSR $1913 + 16$, which is a close binary system consisting of two neutron stars. Most of the errors in the mass estimates are associated with uncertainties in the inclination of the binary orbits. It can be seen that the derived masses of the neutron stars are entirely consistent with the theoretical expectation that their masses should lie close to 1 $M_\odot$. I find this a remarkably beautiful example of the symbiosis between the traditional techniques of optical astronomy and the modern procedures of X-ray astronomy.

The case for associating the X-ray emission from these binary X-ray sources with *accretion* of matter from the primary onto the neutron star is wholly convincing as can be appreciated from the following simple arguments. First of all, if matter falls from the primary onto the secondary star, the kinetic energy which it gains is converted into thermal energy when the infalling matter hits the surface of the neutron star. The thermal energy released is equivalent to the gravitational binding energy of matter on the surface of the neutron star, which means that about 10% of the rest mass energy of the infalling matter is available for emission as radiation. This represents a very efficient source of energy, about ten times the efficiency available from nuclear energy sources.

The second simple calculation is to ask what the typical temperature of the radiating matter would have to be to

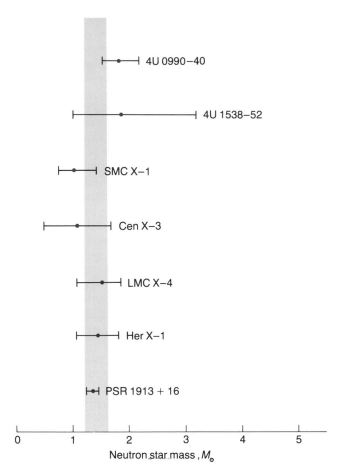

Figure 6.32. A diagram illustrating the mass estimates for the neutron stars in X-ray binary systems (and the binary radio pulsar PSR 1913 + 16) for which good mass determinations are available from their velocity curves and other information. The shaded area indicates likely masses for these neutron stars. (From Shapiro and Teukolsky (1983), p. 256, with additional information provided by Professor J. Trumper from EXOSAT observations.)

normally carried out assuming that the radiation pressure acting on the infalling matter is Thomson scattering of the emergent radiation by infalling electrons, on the basis that other sources of opacity will simply increase the radiation pressure and result in a lower value to the critical luminosity above which accretion is suppressed. The resulting critical luminosity is known as the *Eddington limit* and depends only upon the mass of the gravitating body:

$$L_{\text{Edd}} = 1.3 \times 10^{31} (M/M_\odot)\,\text{W}. \qquad (6.7)$$

It is remarkable that the luminosities of most of the binary X-ray sources in the Galaxy and the nearby Magellanic Clouds are more or less consistent with this upper limit. Their luminosities extend up to about 10^{31} W and the luminosity function cuts off sharply above this luminosity. The calculation which leads to the Eddington limit is a very general one and does not depend strongly upon the details of the accretion process or the means by which the X-ray emission is produced.

These three arguments show how naturally the accretion picture can account for the properties of X-ray sources. The process is efficient because up to 10% of the rest mass energy of the matter can be released, and the observed range of X-ray luminosities can be accounted for by accretion onto standard neutron stars. It is very natural that these ideas should be extended to accretion onto black holes, either those with masses of $10\,M_\odot$ or greater which may be present in some of the binary X-ray sources or to the supermassive black holes ($M \sim 10^8\,M_\odot$) which are likely to be present in the most active galactic nuclei.

The accretion process itself may occur in two different ways, both of which are illustrated in figure 6.33. We know that all stars emit stellar winds, 'quiescent' mass loss in the cases of stars like the Sun and much more powerful winds in the cases of luminous O and B stars in which the mass loss rates are observed to be as high as 10^{-5} to $10^{-7}\,M_\odot\,\text{year}^{-1}$. Therefore, in one model (figure 6.33*a*), the binary is embedded in the outflowing wind from the star. In this model, there is a bow shock wave in front of the neutron star and its associated magnetosphere and the accretion onto the neutron star takes place within this cavity. In the other model (figure 6.33*b*), the compact binary distorts the figure of the surface of the primary star. If the mass of the primary star is such that it fills its Roche lobe, the equipotential surface which joins the primary and secondary stars (figure 6.24), there will be matter transfer through the inner Lagrangian point L_1 from the primary star into the region in which the flow of matter is determined by the mass and rotation of the compact secondary. This process is known as *Roche lobe overflow* and the infalling matter forms an accretion disc about the compact secondary star.

The first model is applicable to the binary X-ray pulsars associated with luminous O and B stars, twenty-three of the twenty-six known X-ray pulsars being associated with these classes of high mass star. In order to produce pulsed emission,

account for the observed intensities of radiation from binary X-ray sources. If we take the typical luminosity of a strong X-ray binary to be 10^{30} W and assume that this is emitted as blackbody radiation from the surface of a neutron star, we find that, equating the flux from a black-body at temperature T to the above intensity, the lower limit to the temperature of the emitting region is about 10^7 K, i.e. it is entirely natural that the radiation should be emitted in the X-ray waveband.

The third simple argument concerns the steady-state X-ray luminosity of accreting compact objects. If the luminosity of the source were too great, the radiation pressure acting on the infalling gas would be sufficient to prevent the matter falling onto the surface of the compact object. The calculation is

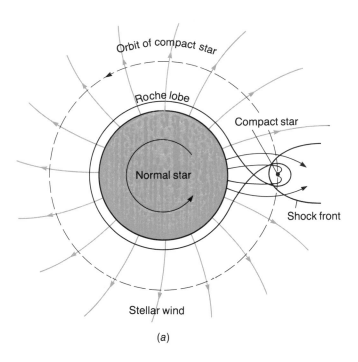

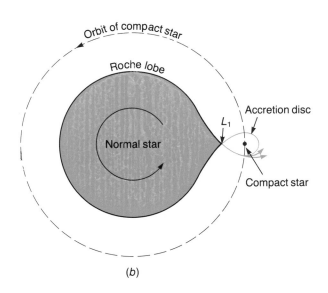

Figure 6.33. Illustrating two ways in which accretion onto stars in binary systems may take place. In (*a*), the primary star has a strong stellar wind and the neutron star is embedded in the strong outflow from the primary. In (*b*), the primary star expands to fill its Roche lobe and matter passes from the primary to the secondary star through the Lagrangian point L_1. An accretion disc is formed about the compact secondary star. (From Shapiro and Teukolsky (1983), p. 399.)

there must be a strong magnetic field present which is nonaligned with the rotation axis of the neutron star. A dipole magnetic field is the natural way of ensuring that matter can only funnel down onto the surface of the neutron star through the magnetic poles. This is a natural way of ensuring that the emission is concentrated in two 'hot-spots' on the surface of the rotating neutron star (figure 6.34). A second direct argument for the presence of a strong magnetic field derives from the observation of what is inferred to be an electron cyclotron radiation feature in the X-ray spectrum of Her X-1 and other X-ray binaries. The feature is observed at about 58 keV in the case of Her X-1, which would require a magnetic field in the source region of about 5×10^8 T. The details of radiation transfer in source regions with accretion, strong magnetic fields and rotation are not fully understood, but the consensus of opinion is that these ideas provide the correct framework for understanding the origin of the pulsed X-ray emission in X-ray binaries.

The flow of matter within the 'magnetospheric cavity' exerts a torque upon the magnetosphere of the neutron star and this leads naturally to another remarkable property of these systems which is that, in contrast to the case of radio pulsars which are observed to slow down as they radiate away their angular momentum, many of the high-mass X-ray binaries speed up and some of them display accelerations and decelerations. These phenomena are easily understood in the models described above because the matter accreted by the compact source possesses angular momentum. The exact amount of spin up depends upon the mode of coupling between the magnetic field of the neutron star, the infalling matter and the environment of the system. It also depends upon the steadiness of the accretion flow. Obviously, if there is no accretion of matter, there is no acceleration and the neutron star slows down. Some of the best observed pulsars show more or less steady acceleration but there are variations about the mean which can be attributed to fluctuations in the accretion rate. In some cases abrupt accelerations and decelerations are observed and these are probably associated with variations in the mass flow about the magnetosphere of the neutron star. Since variations in the mass-flow rate would then be related both to the X-ray luminosity of the pulsar and with variations in the pulsar period, the search for correlations between these observable quantities provides important probes of the behaviour of magnetised plasma in the vicinity of the neutron star. It will be noted that the construction of real models for the transfer of angular momentum requires an understanding of the effective viscosity of the magnetised plasma. The relevant physical processes are closely related to those which are found in plasma confinement experiments such as those carried out in machines such as the Joint European Torus (JET).

The second class of model, involving Roche lobe overflow, is applicable to the low-mass X-ray binaries referred to as Galactic bulge sources. The basic picture of accretion in an *accretion disc*

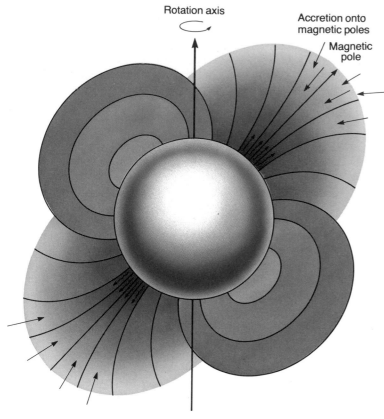

Rotation axis

Accretion onto
magnetic poles

Magnetic
pole

Figure 6.34. A schematic model illustrating how accretion onto a neutron star with a magnetic field nonaligned with the rotation axis can give rise to an accretion column which when observed at a distance gives rise to the observation of X-ray pulses at the rotation frequency of the neutron star. The accreting matter is channelled down the magnetic field lines to the magnetic poles of the neutron star where its binding energy is deposited, resulting in strong heating of the plasma. The problems of radiative transfer under these conditions are highly complex.

about the neutron star is illustrated in figure 6.35. In the steady state there is continuous flow of matter through the Lagrangian point, and the flow of matter into the accretion disc is believed to form a fairly narrow stream. The X-ray emission originates from the inner parts of the disc where most of the gravitational binding energy of the matter in the disc is dissipated by viscous forces within the disc. The hottest radiation originates from the regions closest to the neutron star, and indeed in some models the hard X-ray emission originates from very close to the surface of the neutron star itself. This is the basic model which can be vastly elaborated to account for the various remarkable properties exhibited by Galactic bulge sources. An example of this is the prediction of the X-ray light curve of the source as it is observed at different inclinations of the plane of the orbit to the line of sight to the observer. Occultations of the X-ray source occur when it is viewed in the plane of the binary orbit. Absorption features may be associated with the stream of cool gas flowing through the Lagrangian point to the accretion disc. This may well be the origin of the regular absorption features or 'dips' seen in some sources known as *dippers*.

Another remarkable example of modulation of the light curve of X-ray binaries is provided by the source Her X-1. In addition to the pulsar which has period 1.24 s and a binary period of 1.7 days, the source undergoes a 35 day cycle during which it is strong for 9 days and then relatively faint for the remaining 26 days. This behaviour is accompanied by changes in the pulse profiles. The likely explanation of this behaviour is the precession of the rotation axis of the neutron star in its binary orbit so that the X-ray beam from one of the poles is pointing directly towards the observer when the source is bright. As the precession changes the orientation of the magnetic poles with respect to the observer, the intensity observed from the bright pole decreases but radiation from the other pole is also observed. This model has important implications for the internal structure of the neutron star because the crust of the neutron star and the associated magnetic field must be decoupled (or unpinned) from the neutron superfluid since otherwise the neutron star would possess too much angular momentum and would not undergo appreciable precession. It is also necessary that the crust be sufficiently rigid to maintain a significant oblateness – otherwise there would be insufficient dipole moment for the precession torques to act continuously upon the neutron star.

The bright Galactic bulge sources do not contain X-ray pulsars but they have recently been shown to exhibit a remarkable unstable periodic behaviour which has been termed *quasi-periodic oscillation*. Sources such as Cyg X-2, Sco X-1 and GX5-1 show rapid fluctuations in their intensities, and power spectrum analyses of the source intensities as a function of time show, not a single sharp line spectrum indicative of a stable period, but rather a broad peak spanning a range of frequencies. The mean frequency of the oscillations moves to higher frequencies as the intensity of the source increases. For example, in the source GX5-1, the central frequency increases systematically from 20 to 36 Hz as the source intensity increases by almost 50%. This behaviour is observed in most of the sources exhibiting quasi-periodic oscillations, although there are some exceptions. Almost certainly these phenomena provide information about the flow of plasma from the accretion disc onto the neutron star. There must be inhomogeneities in the flow or oscillations would not be observed. Among the more promising ideas are those involving inhomogeneities in both the infalling matter and the magnetosphere about the neutron star. The width of the spectrum of the oscillations and the variations in the mean frequency trace out the dynamics of the accreting matter in the vicinity of the neutron star.

The X-ray burst sources, often referred to as *bursters*, appear to come in two varieties. The type I burst sources are the more common and occur at intervals of hours to days. The X-ray spectra steepen to low energies as the burst evolves, indicating that the temperature of the source region increases as a result of the burst and then cools. Type II bursters, of which only one example is known, exhibit much more rapid bursts with repetition periods of seconds to minutes. The energy of each burst is apparently proportional to the waiting time until the next burst, as if the burst had exhausted the supply of energy for the moment. Accretion onto a neutron star can account naturally for the short bursts of X-ray emission observed in bursters. There are two main classes of model which have been proposed to account for these phenomena. In one model, the X-ray burst is attributed to thermonuclear flashes occurring in the surface layers of the neutron star due to the accumulation of very hot gas in the accretion process. In the other, the burst phenomenon is attributed to instabilities in the accretion flows about the neutron star. The second model is directly related to the accretion picture described above but now generalised to the case of unsteady flow.

In the case of the model of surface nuclear explosions, the problem is to understand how the accreting gas can accumulate on the surface of the neutron star. A clue is provided

Figure 6.35. A schematic diagram illustrating the structure of a thin accretion disc about a neutron star or black hole. Thin discs are expected when the material of the disc is cool in the sense that the sound speed c_s in the gas is much smaller than the rotational velocity v. of the disc. In this approximation, the ratio of the thickness t to the radius D of the disc is approximately $t/D \simeq c_s/v_{.\theta}$. The material drifts slowly towards the centre at velocity v_r. The dissipation of energy associated with the outward transfer of angular momentum in the disc increases towards the central regions. In the simplest picture in which the radiation is thermalised, this results in a radiation spectrum $I(v) \propto v^{1/3}$. If the central object is a neutron star without a strong magnetic field, the disc can extend more or less down to the surface of the neutron star. In the case of a black hole, shown in the diagram, the last stable circular orbit about the black hole occurs at $r = 3r_g$, where r_g is the Schwarzschild radius of the black hole.

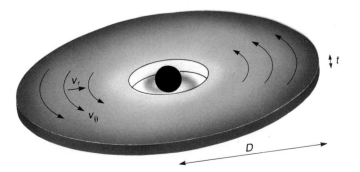

by the fact that no X-ray pulsars have been observed in the type I burst sources which suggests that the magnetic fields in the pulsars have decayed or, possibly, that the magnetic field is aligned with the rotation axis. It is normally assumed that the former is the case and that it is the continuous accretion of matter onto the surface of the star which raises its temperature and density until thermonuclear fusion reactions take place. It is important to note that, in contrast to the surface layers of isolated neutron stars which are composed of heavy elements, the infalling material is probably gas with the typical abundance of the elements found in stellar atmospheres, i.e. hydrogen and helium with about 2% heavy elements. The form of nuclear burning depends rather sensitively upon the accretion rate and is the subject of detailed nuclear calculations. Both hydrogen and helium burning can occur in the compressed layers of neutron stars. Once initiated, the nuclear reactions heat up the surrounding layers and this may initiate a run-away chain reaction through the outer layers of the star, and the temperature rise may be sufficient to initiate helium burning. One intriguing feature of this model is that it is possible that, in the course of the nuclear explosion, the X-ray luminosity of the neutron star may approach the Eddington limit. If the luminosity of an X-ray burster is observed to saturate at a particular value, this may be taken to correspond to the Eddington critical luminosity and then further information may be obtained about the mass and properties of the neutron star.

Perhaps the most extreme forms of high energy burst phenomena are those known as *gamma-ray bursts*. These were discovered by accident by the Vela series of surveillance satellites, the prime task of which was to monitor nuclear explosions which might violate the nuclear test-ban treaties. Gamma-ray bursts are rare events, no more than about ten to twenty being observed each year. The typical burst lasts for only a few seconds, but during that time the source becomes the strongest gamma-ray source in the sky. In addition, it has been shown from simultaneous observations made with X-ray satellites that essentially all of the energy is emitted at gamma-ray wavelengths. The gamma-ray bursts appear to be more or less uniformly distributed over the sky. They could therefore be of local origin if, for example, they were associated with the population of dead neutron stars. Because of the low angular resolution of the all-sky gamma-ray monitors, the positions of gamma-ray bursts are poorly known and none of them have been convincingly associated with known objects. There is some evidence for line emission from some of them, one of which has been tentatively identified with the gravitationally redshifted electron–positron annihilation line at 512 keV and another with an electron cyclotron line. These data, if correctly interpreted, suggest that the parent object may be a neutron star. This would be consistent with the rapid variability seen in some of the bursts. Some splendidly exotic proposals have been made to account for these bursts, including collisions of asteroids with neutron stars, thermonuclear explosions and major redistributions of the angular momentum within neutron stars.

It is important to note that the neutron stars discussed in this section are all basically *thermal* emitters, in the sense that it has not been necessary to call upon relativistic matter to account for any of their properties. There is, however, one remarkable exception and that is the binary X-ray source Cygnus X-3 (Cyg X-3). It is a close binary system in which the period of the binary is only 4.8 hours and the mass of the primary star is about $1 M_\odot$. It displays many of the characteristic properties of X-ray binaries but the unique feature is that it has been detected as a source of very high energy gamma-rays. The detection technique is to search for electron–photon air-showers initiated by high energy gamma-rays when they enter the atmosphere. Those gamma-rays with energies in the range 10^{11}–10^{12} eV initiate such cascades in the upper atmosphere and the Cerenkov radiation produced by the high energy electrons is detected on the surface of the Earth as showers of photons. In the Durham experiment, carried out in Utah in the USA, three independent telescopes were used and coincidence techniques between photon showers detected at the separate telescopes resulted in much higher signal-to-noise ratio than had been achieved in any of the previous experiments. A signal with the 4.8 hour periodicity from Cyg X-3 was detected and a **gamma-ray pulsar with period 12.6 ms was discovered.** Since the source lies in a very close binary system, presumably the pulsar has been spun up by the accretion of matter from the primary. Other sources including the Crab pulsar and Hercules X-1 (Her X-1) have also been detected as 10^{12} eV gamma-ray sources.

These results were surprising enough in their own right but even more surprising was the work by the Kiel and Leeds cosmic ray groups which showed that cosmic ray detectors at ground level also detected significant signals from the direction of Cyg X-3. The gamma-rays which initiate showers detectable at sea level are very energetic indeed, the typical energies of the photons being about 10^{15} eV. These are the highest energy photons to be detected from any astronomical source and they must originate from some ultra-high energy process occurring in the vicinity of the binary system. Whilst the 10^{12} eV photons can be produced by the standard nonthermal processes of high energy astrophysics, for example synchroton radiation in the strong magnetic fields around the neutron star, there are greater problems in understanding the origin of the 10^{15} eV photons. The reason for this is that the gamma-rays are of such high energy that they would result in electron–positron pair production in the strong magnetic field which would rapidly degrade the energy of the photons. A possible solution is that very high energy protons ($\sim 10^{17}$ eV) are accelerated in the strong magnetic field of the rapidly rotating pulsar and these produce cascades of unstable particles, similar to those found in accelerators at lower energies. Among these will be neutral pions which decay into high energy gamma-rays. This process avoids the degradation problem of the higher energy gamma-rays, provided the showers occur in regions of low magnetic field. From the point of view of the origin of cosmic rays, this discovery, if correct, is of first importance because, whatever the detailed model of the production process of the ultrahigh energy gamma-rays, it demonstrates that binary X-rays sources are important sources for the acceleration of very high energy cosmic rays. Until this discovery, this was no more than a reasonable hypothesis.

Black holes

The neutron stars are the last known forms of stable star. For more compact objects, the attractive gravitational force of gravity becomes so strong that no physical force can prevent collapse to a singularity. Black holes have been treated from the point of view of the general theory of relativity in Chapter 2 by Will. Here, I look at black holes simply from the point of view of the part they play in astrophysical problems. To summarise their properties very briefly, a spherically symmetric black hole of mass M possesses a characteristic radius known as the *Schwarzschild radius* $r_g = 2GM/c^2$, which represents the effective radius of the black hole. Putting in the values of the constants, we find that $r_g = 3(M/M_\odot)$ km so that solar mass black holes have $r_g = 3$ km. The key points about the Schwarzschild radius are as follows:

(i) Radiation cannot escape from within the radius r_g because the force of gravity is too strong. This is why the black hole is called 'black'.
(ii) Matter falling within the Schwarzschild radius cannot escape outside r_g. This is why it is called a 'hole'. Collapse to the central singularity takes place in a finite proper time for the infalling object, but to the external observer the collapse takes an infinite amount of time to reach r_g. The reason for this difference is that the gravitational redshift of radiation emitted from the radius r_g becomes infinite although there is no real physical singularity at this radius.

There are in addition rotating black holes, often known as *Kerr* black holes, after the discoverer of these solutions to Einstein's field equations of general relativity. They are somewhat more complicated than the spherically symmetric case because the surfaces of infinite redshift and the radius within which collapse to the singularity is inevitable do not coincide. In addition to its mass and angular momentum, the only other property possessed by an isolated black hole is its electric charge.

An important point astrophysically is that, if the black hole is rotating as fast as is allowed, the last stable circular orbit lies at a radius of only $0.54\, r_g$, whereas in the spherically symmetric case the last stable orbit lies at $3\, r_g$. This means that it is possible to tap much more of the rest mass energy of matter falling into a rotating black hole than onto the spherically symmetric variety. Detailed calculations show that, for spherically symmetric black holes, about 10% of the rest mass energy can be released in accretion onto the hole, whereas in the case of a maximally rotating black hole this value increases to about 43%. In addition, although the black hole cannot possess a dipole magnetic field, it can be coupled by a magnetic field to the surrounding medium and then it is possible to tap the rotational energy of the black hole to fuel astrophysical processes. Thus, black holes are potentially very powerful sources of energy for high energy astrophysical phenomena.

It may come as a surprise that astrophysicists write so confidently about objects which are in fact singularities in space-time. My own position on this question is very simple. There are two arguments which I find compelling about the existence of black holes. First of all, it has been proved that black hole type singularities are a very general property of all theories of gravity in which the force of gravity is attractive. It is the combination of the Einstein's mass–energy relation $E = mc^2$ and the attractive nature of gravity which are the root cause of the black hole phenomenon. The second argument is more intuitive. The fact that neutron stars have been discovered and that their masses match so well the values predicted by theory is for me a quite remarkable result. It will be recalled that the typical radius of a $1\,M_\odot$ neutron star is about 10 km, only three times the radius of the corresponding black hole. There must be other collapsing remnants of higher mass or perhaps remnants which collapsed more vigorously and so were unable to form stable neutron stars. There is also the possibility that accreting neutron stars in binary systems exceed the upper mass limit for stability, the neutron star analogue of the process believed to be responsible for the formation of type I supernovae through accretion onto white dwarfs.

An obvious way of searching for black holes with masses in the range, say, 1–$10\,M_\odot$, is to search for binary X-ray sources with invisible companions which are inferred to have masses greater than $3\,M_\odot$. A great deal of effort has been devoted to discovering whether or not any of the galactic X-ray binaries could contain black holes, and there are probably three strong candidates among the known X-ray binaries. These are the Galactic sources, Cyg X-1 and A0620-00 and the source LMC X-3 in the Large Magellanic Cloud. The signatures of these systems are that they are high-mass single-line spectroscopic binaries which do not possess X-ray pulsars. One possible clue is to search for 'flickering' in the X-ray intensity of the source on the basis that the typical time scale for variations in the X-ray intensity should be $\sim r_g/c \sim 10(M/M_\odot)$ microseconds. Evidence of flickering on the time scales of milliseconds has been claimed in a number of sources including Cyg X-1, but in a number of cases it has turned out that the flickering sources contain X-ray pulsars. This indicates that sources containing X-ray binaries can exhibit very short time scale fluctuations and that this is not a definite signature of black holes in X-ray binaries.

As a result, the best evidence comes from the classical astrometric and astrophysical techniques of analysis of single-lined spectroscopic binaries. Cyg X-1 has been the subject of particularly intensive study. The primary star is the OB supergiant star HDE 226868, and stars of this class normally have masses greater than $20\,M_\odot$. Adopting the most conservative assumptions about the distance of the system and about the mass of the primary star result in firm lower limits to the mass of the invisible companion $M \geqslant 3.3\,M_\odot$, significantly greater than the upper limit to the mass of neutron stars. Adopting the likely values for the distance and mass of the primary star results in a mass in the range 9–$15\,M_\odot$. Such a massive invisible companion could only be a black hole.

The models for disc accretion about stellar mass black holes are not dissimilar from those described earlier. The main distinction is that there is a last stable circular orbit at $r \simeq r_g$ and hence, once the matter of the accretion disc approaches this radius, it spirals inevitably inward towards the black hole. The construction of self-consistent models of accretion discs about black holes is highly nontrivial since the equations of magnetohydrodynamics and radiation transfer have to be solved in a strong gravitational field and the solution should be fully three-dimensional. Time dependent solutions would be highly desirable. Fortunately, the models of thin accretion discs are amenable to a considerable amount of analysis. The main uncertainty in the models of disc accretion is the value of the effective viscosity of the material of the disc. The physics of this process are schematic, probably involving turbulent viscosity associated with irregularities in the magnetic field. Fortunately, many of the most important results are insensitive to the exact value of the effective viscosity parameter α. Spectra for simple disc models are illustrated in figure 6.36 for the parameters described in the figure caption.

At low energies, the radiation is simply black-body radiation from the cooler outer regions of the disc. The radiation spectrum rising as $\nu^{1/3}$ is simply the superposition of black-bodies with the temperatures appropriate to dissipation within the disc. In the central and inner regions of the disc, the dominant source of opacity is Thomson scattering. The exponential cut-off at the highest energies originates in the highest temperature regions closest to the black hole. These models result in a remarkably good fit to the X-ray spectra observed from the candidate black holes found in X-ray binaries. There is much scope within the context of these models to account for other features of the X-ray properties of objects such as Cyg X-1.

6.5 Interstellar gas and star formation

The bright diffuse regions which lie close to the Galactic plane in an optical image of the sky (figure 6.2*d*) are regions of interstellar ionised hydrogen, but it has been the new astronomies which have led to a much fuller appreciation of the

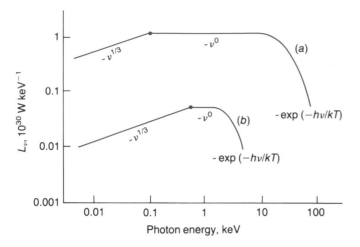

Figure 6.36. The spectrum of the radiation emitted by an accretion disc about a spherical (nonrotating) black hole. These models are due to Shakura and Sunyaev. In model (*a*), the mass of the black hole is assumed to be $1\,M_\odot$, the accretion rate is $10^{-8}\,M_\odot\,\text{year}^{-1}$ and the accretion takes place at the Eddington limiting luminosity $L \simeq 10^{31}$ W. In model (*b*), the mass accretion rate is greater, $10^{-6}\,M_\odot\,\text{year}^{-1}$, and the luminosity is 10^{29} W. The radiation generated in the outer cool regions of the disc results in a power law spectrum $L \propto \nu^{1/3}$. In the inner regions, electron scattering is the dominant source of opacity and the spectrum approximates $L \propto \nu^0$. The temperature of the exponential tail corresponds to the surface temperature of the innermost regions. (After Shapiro and Teukolsky (1983), p. 466.)

inhomogeneity of the interstellar medium. Professor Henk van de Hulst once remarked that, if you set out to detect an emission or absorption line from an atom, ion or molecule in astronomy, you are bound to discover it somewhere in the Universe. This statement is particularly true of the intersellar medium, as has already been alluded to in Section 6.2. Because it is far from equilibrium, a very wide range of density and temperature are present, those found by observation largely reflecting the characteristics of the observational tools used by the observer. The picture is complicated by three further components. First, a significant fraction of the interstellar medium, in particular the heavy elements, is tied up in the form of interstellar dust; second, the whole medium is permeated by the interstellar magnetic field which is tied to the ionised component of the interstellar gas; third, the medium is filled with cosmic rays, protons, nuclei and electrons, which form a relativistic gas which is also tied to the magnetic field.

It is no surprise, therefore, that there is a great deal of physics to be studied. Astrophysically, the understanding of the nature and properties of the interstellar gas is of the first importance since it is out of this medium that new stars are formed. It is continually replenished because of mass loss of various forms from stars and so the medium plays a key role in the birth-to-death cycle of stars.

A global view of the interstellar medium

The interstellar medium amounts to about 5% by mass of the visible mass of our Galaxy. In the plane of the Galaxy close to the Sun, its average space density amounts to about $10^6\,\text{m}^{-3}$ but there are very wide variations from place to place in the Galaxy. To understand the physics of the interstellar gas, let us consider the state of the gas from the point of view of the various physical processes which are likely to be important.

Large scale dynamics   Most of the gas in the Galaxy is confined to the Galactic plane and it moves in circular orbits about the centre of the Galaxy, the inward force of gravitational attraction being balanced by centrifugal forces. The gravitational potential in which the gas moves is defined by the mass distribution of the stars. The interstellar gas therefore acts as a probe of the gravitational potential field of the Galaxy and the distribution and radial velocities of interstellar neutral hydrogen and molecules provide the best information about the distribution of mass in the plane of the Galaxy. The neutral hydrogen distribution was determined as long ago as the 1950s and, more recently, carbon monoxide surveys have defined the distribution of the molecular gas. The picture which emerges is one in which the neutral and molecular gas is closely confined to the plane of the Galaxy, the typical half-widths being about 120 and 60 pc, respectively. However, they have very different distributions with distance from the Galactic centre. The neutral hydrogen extends from about 3 kpc from the centre to beyond 15 kpc, whereas the molecular component appears to form a thick ring between radii 3 and 8 kpc (figure 6.37).

The evidence of spiral arm tracers such as O and B stars and HII regions suggests that our Galaxy possesses a rather tightly wound spiral structure. Features possibly related to spiral arms have been observed in the distribution of neutral hydrogen (figure 6.38). There has been controversy as to whether or not the molecular clouds are associated with spiral arms but there is now agreement that they tend to be found in spiral arm regions.

The importance of these observations is that they indicate that the interstellar gas is influenced by large scale dynamical forces. Whilst the overall distribution of the gas is determined by the gravitational potential defined by the stars (and, possibly, the dark matter), some mechanism is needed to enhance the average gas density from about $10^6\,\text{m}^{-3}$ to values at least 100 or 1000 times greater in the regions of the spiral arms. One of the favoured mechanisms for achieving this is through a density wave which is set up in the distribution of stars in the Galactic disc. The density wave theory of spiral structure is the most successful attempt so far to account for the appearance of spiral arms in galaxies. The theory is based upon considerations of the stability of an axial perturbation within a disc galaxy and of whether or not a spiral perturbation in the distribution of stars is stable. The answer is far from trivial because the spiral wave in the stellar distribution tends to propagate either inwards or outwards, thus destroying the perturbation. It seems necessary that there must be some forcing mechanism which maintains the spiral pattern in the stellar distribution and this might be due either to the interaction with companion galaxies or possibly with perturbations associated with the ellipsoidal distribution of stars in the bulge component. Assuming that the density wave in the stellar disc is stable, we can then ask how the gas behaves under the influence of the density wave. A key point is that the sound speed in the neutral and cold gas is expected to be low. The gas therefore tends to collect in the potential minima of the density wave but the velocity it acquires under the influence of the density wave is such that the gas flow becomes supersonic and shock waves form along the trailing edge of the stellar density wave. A large increase in gas density behind the shock is expected because the compressed gas can cool. This picture provides an attractive explanation for the formation of clouds of neutral and molecular gas in the vicinity of spiral arms.

Whilst this picture has its attractions, much further study is needed to demonstrate that these are the processes which can form molecular clouds. There are certainly alternatives. For example, supernova explosions lead to strong shock waves propagating through the interstellar gas and, in the late stages of expansion, cooling of the compressed gas can lead to the formation of cool dense clouds. It has been pointed out by a number of authors that the greatest star formation rates appear to be found in the most irregular galaxies and not in those with the most beautifully developed spiral structures. It is likely that both processes are important for spiral galaxies.

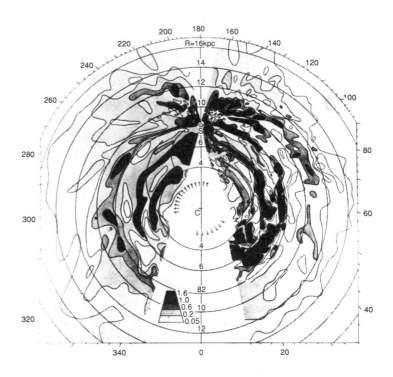

Figure 6.38. The distribution of atomic hydrogen in the Galactic plane. There are prominent features which may be related to the spiral structure of our Galaxy delineated by other spiral arm indicators such as young stars and regions of ionised hydrogen. (Oort, J. H., Kerr, F. J. and Westerhout, G. (1958). *Mon. Not. R. Astr. Soc.* **118**, 382.)

Figure 6.37. The radial distribution of atomic and molecular hydrogen as deduced from radio surveys of the Galaxy in the 21 cm line of atomic hydrogen (red histogram) and the molecular emission lines of carbon monoxide CO (blue histogram). (After Michalas and Binney (1981), pp. 535, 554.)

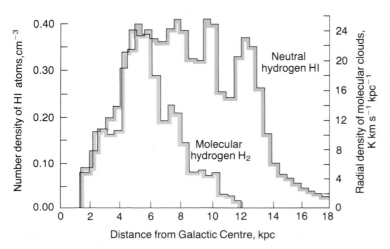

Heating mechanisms If left on its own, the interstellar gas would cool but this is in conflict with the observation of many different phases at a wide range of temperatures. In fact, there is an embarrassment of heating mechanisms for the gas.

The hottest gas is produced by supernova explosions. Supernova remnants are observed to be strong thermal X-ray sources, the gas temperature rising to 10^7 K or more. A shock wave runs ahead of the supersonically expanding shell of cooling gas and heats up the interstellar gas to high temperatures. In an elegant analysis, Cox and Smith first showed that heating by supernova explosions could lead to about 10% of the interstellar gas being heated to a high temperature. The collisions of the shells of supernova remnants can lead to reheating of the gas as the kinetic energy of expansion is converted into heat. Thus, they predicted that the hot component would form tunnels through the interstellar gas from the overlapping of supernova shells. It is entirely plausible that at least some part of the soft X-ray emission from the plane of our Galaxy is associated with this hot gas. It is also probable that the hot gas inferred to be present in the halo of our Galaxy from observations of highly ionised carbon CIV and oxygen OVI by the IUE represents strongly heated gas which has attained a dynamical equilibrium in the gravitational field of the disc and halo. It is natural that the hot gas should expand to form a hot halo since the scale height of hot gas is very much greater than that of the stars of the disc.

A second important heating mechanism is the ultraviolet radiation of young stars. The youngest of these are still embedded in the gas clouds from which they formed. The heated gas is easily recognised by the strong emission lines of hydrogen and oxygen. In the case of heating by ultraviolet radiation, the gas is raised to temperatures of about 10^4 K. Older blue stars, which are no longer embedded in ionised hydrogen regions, can ionise and heat the surrounding regions and this form of local heating is observed in the IUE spectra of certain O and B stars. Recently, the first detection was reported of a region of ionised gas about an X-ray binary source in which very high excitation species are observed, these being attributed to ionisation by the X-ray flux from the source.

Within the regions in which stars are currently forming, bipolar outflows may lead to local heating within these regions. A fourth mechanism is heating by the interstellar flux of cosmic rays. It is found that, even within the neutral and molecular gas, there is a significant density of free electrons. It is likely that the heating and ionisation of the material is associated with *ionisation losses* of the cosmic ray electrons as they permeate both the ionised and neutral phases of the interstellar gas. This source of free electrons in molecular clouds is important in understanding the chemical processes which take place in these regions.

There are other potential sources of heating. For example, the intergalactic flux of ultraviolet ionising radiation, mass loss from all types of star, including stellar winds, infall of matter from intergalactic space and so on. It will be recognised already that there are excellent reasons why the interstellar medium should be far from equilibrium.

Cooling mechanisms   In principle, the cooling mechanisms should be easier to understand because radiation is the principal means by which energy is lost and therefore simply by looking for line and continuum emission at frequencies close to the peak of the black-body spectrum appropriate to that phase of the interstellar gas, it should be possible to observe directly the cooling processes.

For very hot ionised gas, at temperatures in excess of 10^7 K, the principal cooling mechanism is the bremsstrahlung or free–free emission of the free electrons in the plasma. This process is observed in supernova remnants where, in addition, strong emission lines of twenty-four and twenty-five times ionised iron, FeXXV and FeXXVI, respectively, have been observed in the X-ray waveband at about 8 keV, confirming the high temperature of the gas. At lower temperatures, 10^4 to 10^7 K, the emission is due to bound–bound and bound–free transitions of hydrogen, helium and heavy elements. This temperature regime is much more difficult to study observationally because most of the radiation is emitted in the unobservable ultraviolet region of the spectrum. It seems likely, however, that at least part of the soft X-ray radiation detected in the plane of the Galaxy is associated with the radiation of gas at a temperature of about 10^6 K and the spectrum can be attributed to the bound–free emission of different elements, which when summed results in a smooth steep spectrum which extends to soft X-ray and far-ultraviolet wavelengths.

Much of the gas observed in bright regions of ionised hydrogen is found to be at a temperature of about 10^4 K. The reason for this is that the gas is excited by ultraviolet radiation from hot blue stars which have strong fluxes of radiation in the ultraviolet continuum. These hard photons ionise the gas and an equilibrium is set up between the ionisation due to the ionising flux of Lyman continuum radiation and the recombination of the ionised gas by collisions. This equilibrium is established at a temperature very close to 10^4 K. At this temperature the main cooling mechanism for the gas is line radiation, either in the resonance lines of hydrogen or the forbidden transitions of singly and doubly ionised oxygen ([OII] and [OIII], respectively). It is these lines which give ionised hydrogen clouds their characteristic red glow on colour photographs such as those shown in figure 6.14.

Below 10^4 K, the ionised gas recombines and therefore there are very few free electrons present. Between 10^3 amd 10^4 K, the principal radiation loss mechanism is the line emission from neutral or singly ionised carbon, nitrogen and oxygen. The emission lines are associated with forbidden transitions of low lying energy levels. Observations from high flying aircraft such as the Kuiper Airborne Observatory have shown that the lines of [OI] (63 and 145 μm), [CI] (609 and 370 μm) and [CII] (157.7 μm) are particularly strong and are likely to be among the most important coolants of the diffuse interstellar gas.

At temperatures below about 10^3 K, interstellar dust can survive, and it plays a key role in determining the state of the interstellar gas at low temperatures. At temperatures above 10^3 K, the dust grains are evaporated by collisions, but below this temperature they serve a number of different functions. First of all, dust absorbs ultraviolet and optical radiation and therefore, within dust clouds, molecules are protected from the interstellar flux of dissociating ultraviolet radiation. These dust clouds must be present throughout the Galactic disc because interstellar molecules are found in giant molecular clouds almost everywhere one looks in the Galactic plane. Within the dust clouds, there are two important cooling processes. The first is molecular line emission associated either with rotational transitions of asymmetric molecules such as carbon monoxide CO and water vapour H_2O or, in some cases, with the infrared forbidden rotational and rotational–vibrational transitions of molecular hydrogen H_2. In some regions these lines are so strong that they must be the dominant cooling mechanism.

The second process is the *reradiation* of the radiation absorbed by the dust grains. This is almost certainly the most efficient energy loss mechanism for stars which are in the process of formation or which have just formed. Stars form in the densest regions of giant molecular clouds and the ultraviolet radiation emitted by the stars is absorbed by the dust grains. These heat the dust grains to a temperature which is determined by the balance between the energy absorbed from the radiation field and the rate of radiation of the dust grain. The dust grains radiate more or less like little black-bodies, modified by the emissivity function of the material of the grain. A key point is that the grains radiate away the absorbed energy very rapidly at roughly the temperature to which they are heated which is typically about 50 to 100 K for the typical far-infrared sources found in dense molecular clouds. This means that the energy is reradiated at wavelengths of about 30 to $100\,\mu$m at which the dust is transparent and therefore allows the energy of the star to be very efficiently radiated away. This picture explains very convincingly why it is that intense far-infrared emission is the signature of the sites of star formation. In addition, many galaxies, particularly those in which it was previously inferred that there must be active star formation, the late type spiral and irregular galaxies, show extreme far-infrared luminosities which are naturally attributed to star formation.

The picture which emerges is one in which there are many different processes by which the interstellar gas can cool under different circumstances. Which process is most important depends very sensitively upon local physical conditions and the properties of the ambient radiation field.

Instabilities The stability of the interstellar gas is a key issue in understanding the processes by which regions of enhanced density begin the process of collapse to form stars. We have already discussed the global dynamical effects which can lead to the formation of giant molecular clouds, but we have still to consider how collapse within them can be initiated. There are several possibilities.

The most famous of the instabilities is the *Jeans' instability*, which is expected to occur in gas at any temperature and density. If a smooth medium is slightly perturbed, the self-gravitation of the perturbation causes the region to collapse but this is resisted by internal pressure gradients. The criterion for collapse is therefore that the gravitational force should exceed the internal pressure forces and collapse must occur on a large enough scale. There is a characteristic length scale known as the *Jeans' length* R_J, which is the largest stable scale before collapse under self-gravity becomes inevitable. The criterion depends only upon the speed of sound in the gas c_S and the density of the medium ρ:

$$R_J = c_S/(G\rho)^{\frac{1}{2}}. \qquad (6.8)$$

For the typical parameters of giant molecular clouds, $n \simeq 10^9\,\mathrm{m}^{-3}$ and $T \simeq 100\,\mathrm{K}$, the mass contained within the

Jeans' length is about $10^6\,M_\odot$, which is an interesting mass since it corresponds more or less to the mass of a galactic star cluster. In addition, it will be noted that the Jeans' mass decreases as the density of the cloud increases and therefore this instability leads naturally to continued fragmentation as the cloud collapses. These are encouraging results. Corresponding instabilities occur in gaseous discs which are in a state of differential rotation.

There are, however, other processes which can initiate the collapse of stars. For example, supernova remnants are a very effective means of creating regions of high density. The expanding shells of cooling gas are unstable to dynamical perturbations as the shell decelerates and this can form the sites of further star formation. The close association of supernova remnants with regions of star formation strongly suggests that this type of cooperative phenomenon may play an important role in initiating the collapse of stars.

So far, we have only considered dynamical instabilities but there are also thermal instabilities. If the interstellar gas finds itself in a temperature and density regime such that, as the density of the region increases, the region can lose internal energy efficiently by radiation, the region will become thermally unstable and collapse on a *thermal* rather than a *dynamical* time scale. This process is likely to be important in dense regions of the interstellar gas in which there are very efficient mechanisms of radiative energy loss such as cooling supernova shells and the dense regions in molecular clouds.

For completeness, we should note that there are also instabilities associated with the magnetic field and cosmic ray gas present in the interstellar gas. It turns out that the energy density of the magnetic field in the diffuse interstellar gas is of the same order as that of the cosmic ray gas. Parker first pointed out that this makes the interstellar gas subject to large scale dynamical instabilities. The magnetic field and cosmic ray gas are anchored to the Galactic plane by the thermal component of the interstellar gas. If the energy density of the cosmic ray gas exceeds the magnetic field energy density at some point, the pressure of the cosmic ray gas causes a magnetic loop to expand out of the plane. In this way, loops of magnetic field and high energy particles can be expelled into the galactic halo. This is a natural mechanism which explains at least two separate phenomena. First, it explains how cosmic rays can be effectively ejected from their source regions in the Galactic plane into the halo of the Galaxy. Second, it explains why there is rough equality between the energy density of the magnetic field and that of the cosmic ray gas. If the pressure of the cosmic rays increases at some point, a loop containing the excess pressure is expelled from the plane until equilibrium is restored.

In addition to these instabilities, there are streaming instabilities associated with the flux of cosmic ray protons, electrons and nuclei along the galactic magnetic field. These purely plasma phenomena have the effect of scattering the pitch angle distribution of the cosmic rays so that the bulk streaming velocity of the relativistic gas is restricted to roughly the sound speed in the magnetoplasma, which is the *Alfven velocity* $c_A = B/(4\pi\rho)^{\frac{1}{2}}$. This process helps to explain why the cosmic rays appear to arrive more or less isotropically at the Earth.

This summary emphasises the fact that the physics of the interstellar gas is complicated in the extreme. On reflection, it comes as no surprise that the interstellar medium presents such a complex appearance to the observer and that it is far from trivial to disentangle the important physical processes which lead to the formation of stars.

Interstellar chemistry

More than fifty species of interstellar molecule have been identified to date, mostly by millimetre observations, and new ones are being found all the time. Species which have been identified reliably are given in table 6.1, which includes organic molecules, inorganic molecules, free radicals and molecular ions; organic molecules are here taken as those containing carbon. Most of the entries in the table were found by searching at the frequencies of known transitions measured previously in the laboratory, but for a few of the exotic species such as HCO^+ and C_2H the procedure was the other way around, with laboratory confirmation following the astronomical discovery. There are hundreds of lines which have been detected in interstellar clouds but which cannot be identified at present for a variety of reasons; many of them may be transitions of species which have not yet been synthesised and characterised in the laboratory.

Table 6.1 contains a great diversity of types. There are entire molecules, which have paired electrons in all their filled orbitals, but there are also radicals, with one or more singly occupied orbitals. Radicals cannot be made in great abundance in the laboratory because they are reactive and therefore have short lifetimes, but in interstellar clouds the gas density and collision rates are so low that the radicals survive long enough to be detected. There is also a great range in the sizes of the known molecules; many consist of two atoms, but much larger ones are known and the record holder is $HC_{11}N$ with thirteen atoms.

Another feature is the existence of molecular ions. Most of the detected species are electrically neutral but a few singly charged positive ions are known. The discovery of these ions is crucial in understanding the chemical reactions which take place in the clouds and which build up the molecules from their constituent atoms.

Several clear patterns are discernible, which ultimately must be explained by a complete theory of interstellar chemistry. The Universe contains an overwhelming majority of hydrogen atoms and so the existence of many unsaturated species, i.e. species containing double or triple bonds, is remarkable. We might instead have expected these bonds to be saturated with

hydrogen atoms, giving $C_3H_7NH_2$, for example, rather than HC_3N. The most obvious pattern in table 6.1 is the sequence of acetylenic chain molecules, HCN, HC_3N, HC_5N, HC_7N, HC_9N, ..., suggesting that there is a simple mechanism operating which can readily lengthen an existing chain by adding or inserting two more carbon atoms. Another pattern is the frequent existence of thio-derivatives, molecules which differ from other known interstellar species by the substitution of a sulphur atom for an oxygen atom, such as the pair H_2O/H_2S. Sulphur and oxygen are known to have similar chemical properties and so this pattern is perhaps not so surprising.

Whilst the inventory of known interstellar molecules is in many ways rich and diverse, there are some conspicuous absences. Until recently, no ring molecule had been found. The *triangular* ring molecules silicon dicarbide SiC_2 and cyclopropenylidene C_3H_2 have been detected, however, rather than benzene-ring type molecules. There is still no benzene derivative on the list, no heterocyclic compound such as the six-membered ring pyridine C_5H_5N, nor any saturated ring compound such as cyclohexane derivatives. Many ring compounds are every bit as stable as the aliphatic compounds on the list and so the reason for their absence must simply be that they were not synthesised in the first place. Another conspicuous absence is that there are no distinctively biological molecules. This absence is not due to neglect: some species such as glycine, the simplest amino acid, have been sought by making sensitive observations at known transition frequencies, but without success.

In trying to understand the relative abundances of the chemical species found in the clouds, it is important to remember that the interstellar medium consists of slightly dirty hydrogen: the helium present can be ignored because it is virtually unreactive and the total number of atoms of heavier elements is about one-hundredth of the number of hydrogen atoms. It might be expected that hydrogen should dominate interstellar chemistry and this would be borne out if one were to perform a theoretical calculation of the thermodynamic equilibrium abundances of the various species at a representative temperature of $50\,\overline{K}$, for example. The only species predicted would be saturated hydrides such as CH_4, NH_3, H_2O, etc., in which every available chemical bond would be taken up with hydrogen and there would be no bonds between heavy elements. There would be no CO and none of the unsaturated multiple-bonded species such as HC_9N, and yet unsaturated species are present in the molecular clouds and CO is the most abundant interstellar molecule after H_2. Evidently the population of molecular species is not in thermodynamic equilibrium, which means that only a limited number of reactions are important and it is necessary to find out which they are.

It is easy enough to understand the general constraints on the gas-phase chemistry which can take place in a cold vacuum like the molecular clouds. Even in condensations, the gas density is lower than that in the best laboratory vacuum and is so low that only two-body collisions are important. The low kinetic temperature means that there is no energy available to drive reactions which require an input of energy; endothermic reactions (those which absorb energy overall) therefore do not take place, nor do those which give out energy but which require an activation energy input to get under way. These constraints are very restrictive and rule out almost all reaction mechanisms. A few reactions involving free radicals can still take place, on account of the reactivity of radicals, but the principal class of reaction appears to be between ions and neutral molecules. The electrostatic attraction between such species is the key point: it brings the molecules together with sufficient energy to overcome what would otherwise be a prohibitive barrier, and it bends the trajectories of the molecules towards each other, which substantially increases the probability of a collision and hence of a reaction. The reaction cross-sections are sufficiently large for ion–molecule chemistry to determine the abundance of many of the important species in the clouds, despite the low concentration of ions.

With these general principles in mind, it is useful to distinguish the different physical environments within which the chemical processes take place. The diffuse interstellar clouds are present throughout the Galactic plane and they are of low enough density that they do not obscure the stars which lie behind them. As a consequence, there is a significant flux of dissociating radiation within these clouds which means that only the simplest, most tightly bound molecules can exist. These include molecules such as CH, CH^+, CN, H_2, HD, CO, OH and C_2. In addition, atomic species such as C, C^+ and O have been observed through the submillimetre transitions described previously. The chemistry is therefore much more restricted than in the case of the dense clouds, but nonetheless the reaction networks for determining the abundances of the various species are complex. One of the major challenges has been to determine the deuterium abundance in the diffuse clouds since it is a key cosmological element. As an example, the reaction network for deuterium-bearing molecules is shown in figure 6.39. The reaction rate is determined by the ionisation rate which can be determined independently from the corresponding oxygen reaction network. The inferred deuterium abundance is $[D]/[H] \simeq 1.5 \times 10^{-5}$, in good agreement with the value derived from ultraviolet absorption lines.

Moving to denser regions, there are distinctive chemical processes which occur in the *envelopes* of giant molecular clouds. These envelopes are exposed to the interstellar radiation flux and therefore photodissociation of molecules such as CO are important. High abundances of C and C^+ are found close to warm clouds and this is believed to be due to photodissociation of molecules such as CO. One of the major puzzles is that although the C and C^+ lines are observed in the envelopes of the hot clouds, they are also observed *within* the warm clouds where it is expected that they would be combined into molecules. The reason for this is not yet established.

It is within the dense clouds that the full range of interstellar molecular chemistry is found. In these regions, the molecules

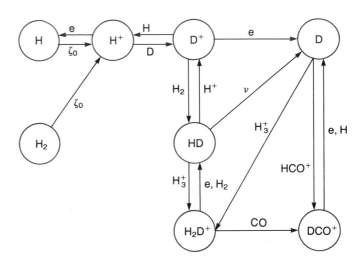

Figure 6.39. An example of the chemical reaction network for determining the abundance of deuterium-bearing molecules in diffuse interstellar clouds. The various atoms, ions and molecules involved in the reactions are indicated beside the arrows. The symbol ζ_0 represents the rate of ionisation of H and H_2 by cosmic rays. The symbol ν represents photodissociation by the interstellar flux of ultraviolet radiation. This type of network has to be studied in order to relate the observed abundances of species involving deuterium to its universal abundance. (From van Dishoeck, E. F. (1986). *Space-borne Submillimetre Astronomy*, N. Longdon, ed. p. 108. ESA Special Publication ESA SP-260.)

are fully protected from the interstellar ionising flux. The clouds are in fact almost neutral, with the abundance of ions being only about 10^{-8} of that of molecular hydrogen, the ionisation being due to the ionisation losses of cosmic rays. The scheme of ion–molecule reactions starts with the ionisation of the most abundant species, H_2, by a collision with a cosmic ray particle to give H_2^+, which is quickly followed by the reaction

$$H_2^+ + H_2 \rightarrow H_3^+ + H.$$

H_3^+ readily donates a proton to other species through reactions such as:

$$H_3^+ + CO \rightarrow HCO^+ + H_2$$
$$H_3^+ + N_2 \rightarrow HN_2^+ + H_2.$$

The detection of HCO^+ and HN_2^+ in about the expected abundances is clear confirmation that this class of reaction does take place, but what is not yet clear is just how many of the known species can be explained in this way.

Another major uncertainty is whether or not chemical reactions occurring on the surfaces of dust grains rather than in the gas phase are important. The principal uncertainty is that which reactions take place depends critically on the surface conditions on the grains because the surface can provide special sites which catalyse particular reactions. The surface conditions are, however, completely unknown; not even the identities of all the solid substances of which the grains are made are known. Another problem is that molecules which are made on the grains may stick there permanently as a layer of frost, owing to the very low temperature, and may never return to the gas phase at all. There is, in short, great uncertainty about this class of reaction: it may contribute a wealth of species or it may contribute nothing at all. One interesting result is that the abundances of many simple molecules such as CH, CN, HCN, HCO^+ and C_2H appear to have similar values in diffuse clouds, cold dark clouds and hot molecular clouds. This is somewhat surprising because it might be expected that these molecules would be depleted in dense clouds because they stick onto dust grains. Either they do not stick onto grains or they are rapidly deabsorbed.

The interest of these observations for chemistry is obvious. Their astrophysical importance lies in the fact that the chemistry leads to a determination of the physical conditions within the dense dust clouds where stars are forming. Because the dust clouds are transparent at millimetre wavelengths and the line widths of the radiation are very narrow, the dynamics of regions of different densities can be readily studied. The higher the frequency of the molecular line, the denser the region within which it was formed. These data provide key information about the physical processes which lead to the formation of stars.

Interstellar dust

One of the more perplexing aspects of the study of the interstellar gas is the fact that we are certain that there are large quantities of interstellar dust present in many classes of astronomical object but we have little definite knowledge of the exact composition and properties of the dust grains. It is known that they contain a significant fraction of the heavy elements of the interstellar gas and also that there must be a wide range of dust grain sizes present to account for the fact that the absorption coefficient extends reasonably smoothly from the ultraviolet and optical through to the infrared wavebands (figure 6.40). Superimposed upon this continuum absorption spectrum there are several prominent features. The most prominent is the strong absorption feature observed at about 220 nm which is present in the Galactic extinction curve. This feature corresponds rather well with a graphite resonance band, and it is commonly assumed that this is evidence for the presence of graphite grains. There are also dust absorption bands observed in the optical waveband, the most famous being the broad feature at 443 nm. This and about sixty other diffuse interstellar absorption features have remained unidentified despite an enormous amount of work by many authors. A major problem in identifying the nature of the interstellar grains is that the continuum absorption spectrum can be accounted for by a variety of different types of grains once the distribution of particle sizes is taken into account.

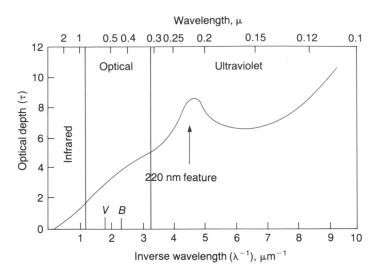

Figure 6.40. The observed absorption coefficient for interstellar dust grains as a function of wavelength. Superimposed upon the relatively smooth continuum absorption curve are a number of absorption bands, the strongest of these being the broad feature at about 220 nm. (After Greenberg, J. M. (1982). *Laboratory and Observational Spectra of Interstellar Dust*, R. D. Wolstencroft and J. M. Greenberg, eds. p. 2. Royal Observatory, Edinburgh Publications; and Spitzer, L. (1978). *Physical Processes in the Interstellar Medium*, p. 158. Wiley-Interscience Publications, NY.)

It is for this reason that the discovery of dust absorption and emission features in the infrared waveband are important because these are related to vibrational transitions associated with bonds in the solid material and enable identification of at least some of the constituents of the grains to be made (figure 6.41). For example, prominent emission and absorption features at 9.7 and 18 μm are observed in regions of ionised hydrogen and in diffuse and dense clouds – these have been convincingly associated with silicates. Water ice absorption features have been identified in molecular clouds at 3.07, 6.0, 12.5 and 42 μm. Ammonia molecules have been associated with grains through features observed at 2.95 μm and so on. It is apparent that interstellar dust grains can be composed of a wide variety of different materials. Among those which have been the subject of intense study are graphite and silicon grains, core–mantle grains in which a graphite or silicate core is surrounded by a water ice mantle, oxides or amorphous carbon grains and even large organic and biotic molecules.

Interstellar molecules may be formed on the surfaces of dust grains but their exact role in interstellar chemistry is unclear. One of the most intriguing recent results has been that there may exist very small dust grains in the interstellar medium. There are two lines of approach. Sellgren was the first to note that, when a very small grain absorbs an ultraviolet photon, the grain is rapidly raised to a high temperature and it then

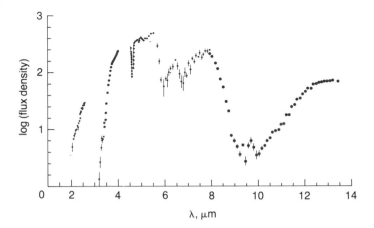

Figure 6.41. An example of the infrared spectrum of the young stellar object W33A, which is embedded in a dense dust cloud. The absorption features have been attributed to various constituent molecules which make up the grains. Examples of these attributions are as follows. The deep absorption features at 3.1 and 9.7 μm are associated with ice and silicon bands, respectively. The feature at 6.0 μm is also due to water ice, possibly with a contribution from ammonia, NH₃, in the long wavelength wing; that at 4.675 μm is associated with solid carbon monoxide, CO; and that at 4.62 μm may be associated with C≡N bond vibrations. The feature at 4.9 μm is attributed to sulphur-containing molecules, possibly to OCS. The feature at 6.8 μm is unidentified but may be associated with some form of hydrocarbon. (Prepared by Dr P. F. Roche from data from Capps, R. W., Gillett, F. C. and Knacke, R. F. (1978). *Astrophys. J.* **226**, 863 (2–4 μm); Geballe, T. R. (1985). *Astr. Astrophys.* **146**, L7 (4.55–5.0 μm); Soifer B. T., Puetter, R. C., Russell, R. W., Willner, S. P., Harvey, P. M. and Gillett, F. C. (1979). *Astrophys. J.* **232**, L53 (5–8 μm); Roche, P. F. and Aitken, D. K. (1984). *Mon. Not. R. Astr. Soc.* **209**, 338 (8–13.3 μm).)

various vibrational modes of the bonds in these molecules. For example, the 3.3 μm line may be associated with aromatic C–H stretch vibrations in the fundamental $v=1$ to $v=0$ mode, the 6.2 μm line with the aromatic C–C stretch vibrations, the 8.7 μm line with aromatic in-plane bending vibrations, the 11.3 μm line with aromatic C–H out-of-plane bending vibrations for nonadjacent, peripheral hydrogen atoms and so on. These vibrations are only excited if the molecules are small since, otherwise, most of the excitation energy is absorbed in the phonon lattice. These ideas tie up nicely with the idea that there are small grains in the interstellar gas. It is tempting to associate the small grains with the PAHs.

A number of general points are worth noting. First, solid carbon comes in a variety of different forms. In graphite, the solid consists of layers of carbon-ring planes held together by van der Waals forces (figure 6.43*a*). The lubricating properties of graphite are associated with the movement of these planes over one another. In amorphous carbon, the carbon-ring molecules are randomly oriented (figure 6.43*b*). The PAHs may be thought of as carbon-ring molecules chipped off either form of carbon. Thus, it is not at all implausible that PAHs should exist in the interstellar medium. It is not clear that the lines described above can only be produced in PAHs, amorphous carbon being another possibility. A second point concerns the dichotomy between the fact that no carbon-ring molecule has been identified in millimetre molecular line searches and yet there is growing evidence that the unidentified infrared lines are associated with the vibrations of aromatic compounds. A third point is that, although these ideas look promising and open up new areas of interstellar chemistry and solid state physics, the PAHs are unlikely to explain all the properties of interstellar grains.

emits radiation at a much higher temperature than the average thermal temperature of the grains. The small grain cools very rapidly until it next absorbs an optical or ultraviolet photon. Sellgren and her colleagues showed that this process could account for the extended continuum infrared emission seen about some nearby regions of ionised hydrogen.

The second route concerns the nature of the unidentified emission lines observed at infrared wavelengths in these same regions of ionised hydrogen. Prominent emission features have been observed at 3.3, 6.2, 8.6 and 11.3 μm (figure 6.42). One intriguing possibility is that these may be associated with the resonances of small associations of carbon-ring molecules known as polycyclic aromatic hydrocarbons (PAHs). These molecules consist of a small number of benzene rings bound into a single large molecule (figure 6.43*c*) and can contain from fifteen to fifty carbon atoms. They possess absorption and emission features at roughly the wavelengths of the unidentified lines and these have been tentatively identified with the

Figure 6.42. The emission spectrum of the planetary nebula HD 44179 in the near-infrared waveband showing the prominent unidentified emission lines at 3.3, 6.2, 7.7, 8.6 and 11.3 μm which may be attributed to various resonances of polycyclic aromatic hydrocarbons. (From Russell, R. W., Soifer, B. T. and Willner, S. P. (1978). *Astrophys. J.* **220**, 568.)

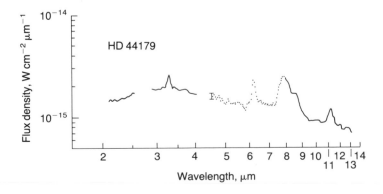

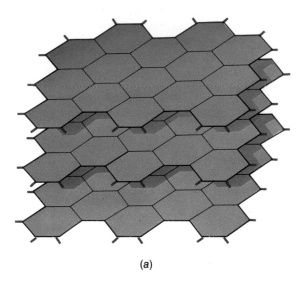

(a)

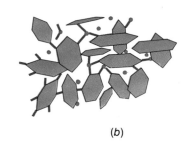

(b)

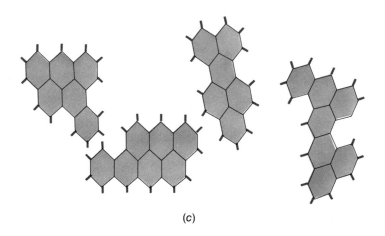

(c)

Figure 6.43. Illustrating the differences between three different forms of carbon: graphite grains, amorphous carbon and polycyclic aromatic hydrocarbons (PAHs). (*a*) In graphite, the benzene rings of carbon form large sheets which slide over one another and give graphite its lubricating properties. (*b*) In amorphous carbon, the benzene rings are randomly superimposed one upon the other but they are held together by carbon bonds often forming the typical tetrahedral structure found in diamond. (*c*) The PAHs are small associations of benzene rings.

Star formation

Astronomers have the bad habit of discoursing in great depth and subtlety about the problems that they can address but these might not be the most important to be solved. Of no problem is this truer than the understanding of the processes of star formation. This is perhaps the most important unsolved astrophysical problem because it pervades so much of the study of the origin and evolution of galaxies. Let us list some of the more important problem areas.

(i) What is the *initial mass function* of stars and how does it vary with location within a galaxy? The initial mass function is the function $\xi(M)$ which describes the birth rate of stars of different masses. It is a difficult function to determine observationally because we observe stars at very widely differing stages of evolution and corrections have to be made for the lifetimes of stars of different masses. A recent determination of the overall initial mass function for stars is shown in figure 6.44 in which it can be seen that the function is a monotonically decreasing function of increasing mass. In view of the uncertainties in its determination, some authors use the *Salpeter* initial mass function

$$\xi(M)dM \propto M^{-1.35}dM,$$

which is a reasonable approximation for stars with masses roughly that of the Sun. More recent determinations have suggested that the function can be described by a log-normal distribution function proposed by Miller and Scalo

$$\xi(\log M)dM \propto \exp[-C_1(\log M - C_2)^2]dM,$$

where C_1 and C_2 are constants. This form of distribution has the intriguing feature that it describes random multiplicative processes. What is uncertain is whether or not this function applies universally throughout the Galaxy. What are needed are better determinations of the function for regions of active star formation and this will become possible with the development of infrared array detectors which can map rapidly star formation regions at different locations within the Galaxy.

(ii) What is the dependence of the star formation rate upon physical conditions in the interstellar gas? An often quoted result is that of Schmidt who worked out the dependence of the star formation rate as a function of the interstellar gas density in our Galaxy. His result $d\xi/dt \propto n^\alpha$, where $\alpha = 2$ and n is the number density of the interstellar gas, may provide a rough global description of the dependence of the star formation rate in external galaxies as well, although significant variations in the index α are found. This form of relation describes the fact that we might expect star formation to be more rapid in denser regions. There is, however, some doubt as to whether or not this result applies in detail within the Galaxy, an exponent much smaller than two being found in more recent analyses. A further problem is that one can conceive of so many physical processes which could influence the star formation rate that the above law can only be taken as a global indicator of this dependence.

More uncertain is the dependence upon the chemical composition of the region. In many regions, the dominant energy loss mechanism is atomic, ionic or molecular line radiation or the continuum radiation of interstellar dust grains. It would be expected that, the greater the abundance of the heavy elements, the more effective the cooling and the greater the star formation rate. A particular example of this problem is the formation of the very first generation of stars which presumably have to form out of a pure hydrogen–helium gas without any interstellar dust grains or molecules to help radiate away the internal energy of the cloud.

(iii) What is the exact sequence of events which take place during the formation of stars? Ideally one would like to determine this empirically by observation. The basic problem is that the process of star formation takes place over a very short time scale, $\leqslant 10^5$–10^6 years, compared with the lifetime of typical main sequence stars. This means that statistically there will always be relatively few star forming regions which can be observed at different stages in their evolution and it may prove difficult to work out empirically an evolutionary sequence.

These are just three of the fundamental problems which have to be addressed before we can claim to have a firm observational basis for theories of star formation.

The contents of regions of star formation Figure 6.45 is a sketch of the contents of a region of star formation. The overall dimensions of the regions are defined by the giant molecular clouds, and it is within these that star formation takes place. The giant molecular clouds have vastly greater sizes than the prominent regions of ionised hydrogen which used to be thought of as the predominant regions in which stars form. The giant molecular clouds in the region of Orion are shown in figure 6.46, the Orion Nebula being the most prominent region of ionised hydrogen close to the densest region of the molecular cloud. These high angular resolution observations show that there is much fine structure within the cloud, these corresponding to regions of enhanced particle density, each of which is a

Figure 6.44. An estimate of the initial mass function of stars derived by Miller and Scalo showing their derived luminosity function and their best-fitting log-normal distribution. Also shown is the initial mass function of power-law form proposed by Salpeter. (After Miller, G. E. and Scalo, J. M. (1979). *Astrophys. J. Suppl.* **41**, 513–47.)

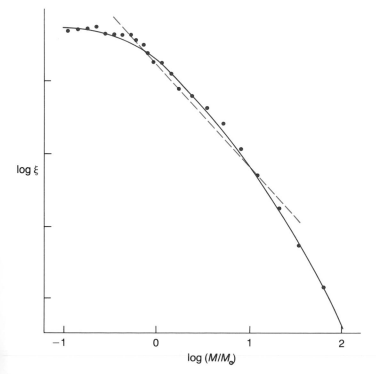

Figure 6.45. A schematic diagram illustrating the typical contents of a region of star formation. These include embedded stars, regions of ionised hydrogen (known as HII regions), protostars and bipolar outflow sources associated with protostars or very young stars. It is believed that the Orion Nebula is an example of an HII region which has been formed close to the nearside surface of the Orion giant molecular cloud. The sketch shows an example of this to the bottom right of the giant molecular cloud.

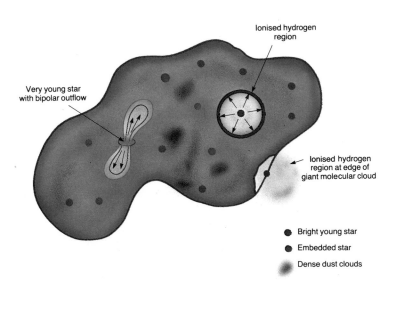

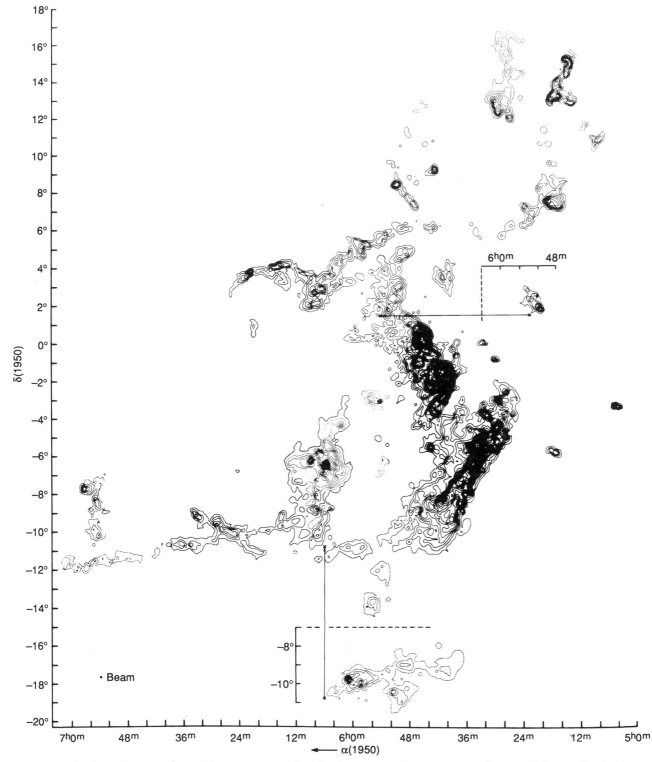

Figure 6.46. A high resolution image of the giant molecular cloud in the region of Orion. This contour map should be compared with that shown in figure 6.13. The higher resolution of the observations in this figure reveals that the giant molecular cloud contains many dense knots which may be the sites of the next generations of new stars. (Maddalena, R. J., Morris, M., Moscowitz, J. and Thaddeus, P. (1986). *Astrophys. J.* **303**, 375.)

potential site of star formation. There are large quantities of dust associated with the giant molecular clouds which protect the interstellar molecules from being photodissociated by the interstellar flux of ionising radiation. This means that we tend to see optically only those regions of ionised hydrogen which form close to the front surface of the clouds. The present picture of the Orion Nebula, for example, is that it is probably a 'blister' on the front surface of the Orion Giant Molecular Clouds. Thus, we expect there to be regions of ionised hydrogen associated with young stars which have formed recently, similar to the Orion Nebula, but embedded deep within the molecular clouds. In addition, there must be older stars within the clouds which formed at earlier times and which can be recognised by their spectral energy distributions at millimetre and centimetre wavelengths.

The youngest objects which have been recognised within the clouds are the hot far-infrared sources. Again the example of the Orion Molecular Cloud is instructive. Whilst the Orion Nebula is by far the most prominent feature optically, at far-infrared wavelengths most of the luminosity is associated with a region to the north-west of it, an object known as the Becklin–Neugebauer object (figure 6.47a). This compact far-infrared source has luminosity about 10^5 times that of the Sun. The spectrum is sharply peaked in the far-infrared region of the spectrum (figure 6.47b), typical of the emission spectrum of reradiated dust. The natural picture for such an object is that it contains a very young star or cluster of young stars which are still embedded in the dust shell out of which they formed. Unlike the case of the Orion Nebula, there is no region of ionised hydrogen surrounding the young stars, which suggests that the stars must be very newly formed. One of the intriguing questions is whether these objects are truly stars in the sense that hydrogen burning has begun in their cores or whether they are *protostars*. The distinction between the two cases is that in the case of a protostar the energy is derived from the dissipation of the gravitational binding energy of the star, i.e. its energy of collapse, whereas, in the case of a young star, the energy source is the nuclear burning of hydrogen into helium. The search for protostars has been one of the goals of infrared astronomy since it was realised that the intense far-infrared sources must be associated with what are among the youngest stellar systems known.

(a)

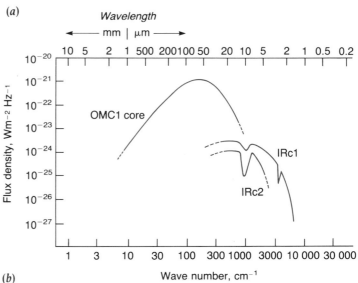

(b)

Figure 6.47. (a) An infrared image of the central region of the Orion molecular cloud (OMC) at $2.2\,\mu m$. The intense sources in the upper centre of the image are known as the Becklin–Neugebauer object and the Kleinmann–Low Nebula. These are by far the most prominent regions in Orion at far-infrared waveband, $\lambda \geqslant 10\,\mu m$. Towards the centre of the image is the intense emission associated with the four stars known as the Trapezium. These are recently formed stars which are responsible for the excitation of the intense optical emission from the Orion nebula. To the lower left of the image is the region known as Orion's bar which is an ionisation front associated with the ultraviolet emission of the trapezium stars. (Courtesy of Dr Ian McLean and the Royal Observatory, Edinburgh.)

(b) Infrared spectra of the central regions of the Orion Nebula. In all cases the emission is the reradiation of heated dust grains. In the case of the source labelled 'OMC1 Core', the dust is heated to typical temperatures of about $100\,K$. The discrete sources labelled IRc1 and IRc2 are hotter compact sources exhibiting strong absorption features at 3.1 and $9.7\,\mu m$. These absorption features are associated ice and silicate grains, respectively. These sources may be very young stars or protostars within the Orion complex. (From Wynn-Williams, C. G. (1981). *Sci. Am.* **245**, 35.)

These far-infrared sources have revealed a number of re-markable properties. Probably the most surprising is the fact that virtually all of them are associated with *bipolar outflows*. Observations at millimetre and infrared wavelengths have shown that there are molecular outflows from the infrared source which are remarkably well collimated in opposite directions (figure 6.48). Measurements of the velocity of outflow from the Doppler shifts of the molecular lines show that the velocities are highly supersonic, velocities of 50 to $100 \, \mathrm{km \, s^{-1}}$ being observed. The bipolar outflows came as a complete surprise, particularly since they were also observed in the forbidden molecular hydrogen lines observed in the infrared waveband. There must therefore be regions with temperatures $T \simeq 2000 \, \mathrm{K}$ which are assumed to lie at the interface between the outflowing material and the ambient molecular cloud. Recently, polarisation observations of the infrared molecular hydrogen emission in Orion have been made and these show, in addition to a molecular hydrogen reflection nebula, polaris-ation vectors parallel to the molecular outflow. This is inter-preted as evidence for a magnetic field in the outflow. A schematic representation of a typical bipolar outflow from a young stellar object is shown in figure 6.49.

Figure 6.48. The bipolar molecular outflow sources L 1551 and B 335 in which the molecular line emission originates from jets directed in opposite directions away from a central source which is assumed to be a very young star or protostar. In both cases, the redshifted and blueshifted lobes are indicated by red and blue shading, respectively. The strong infrared sources are indicated by black filled circles at the centres of the outflows. (*a*) In the case of L 1551, a molecular disc, shown in green, has been observed in the CS line about the infrared source. The CS molecules probe regions of higher density than the CO lines which define the global structure of the outflow. The knots with arrows are two Herbig–Haro objects moving outwards in the direction of the outflow as measured by their proper motions. These objects are some form of condensation which is observed optically and which are probably entrained in the outflow. (*b*) In the case of B 335, the bipolar outflow appears to increase in width with distance from the centre as expected in a freely expanding jet. (From Lada, C. J. (1985). *Ann. Rev. Astr. Astrophys.* **23**, 307, 311.)

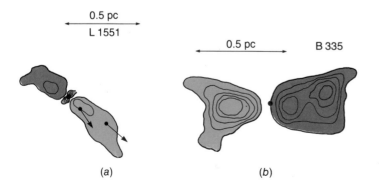

Another remarkable feature of these sources has been the detection of what may well be molecular rings about the axis of the bipolar outflow. This is also shown schematically in figure 6.49. There has been speculation that the disc of molecular gas may be the material out of which a planetary system about the young star may be formed. Further evidence for this type of phenomenon has come from observations made with the IRAS satellite which discovered what are believed to be dust discs about some of the brightest stars in the sky. In the case of Vega, the third brightest star in the sky, a far-infrared extended source has been observed with temperature about $80 \, \mathrm{K}$ which has been interpreted as a cool disc out of which a planetary system might form. Infrared speckle observations of young stars in a nearby low mass star forming region have revealed similar phenomena.

These infrared and molecular observations probably provide information about star formation about the time that hydrogen burning in the core of the stars begins. It is of the greatest interest to identify objects at an earlier stage in their evolution when the protostellar gas cloud is contracting. Recently, a candidate for such a system has been discovered in which it appears that molecular lines originating deep inside the cloud show evidence of collapse towards the central core of the cloud. This type of observation is crucial in understanding how much angular momentum the cloud has to lose in the process of collapse.

The problems of star formation There are three major problems which have to be solved in understanding how regions with densities about $10^9 \, \mathrm{m^{-3}}$ can collapse to stars with about 10^{30} times greater densities. First of all, there is an *energy* problem. To form a stable star, the protostar must get rid of its gravitational binding energy. Second, any cloud possesses some angular momentum and, because of conservation of angular momentum, the rotational energy increases during collapse. Unless there is some way of getting rid of angular momentum, the growth of rotational energy will halt the collapse in the equatorial plane. This is the *angular momentum* problem. Third, if there is a *magnetic field* present in the collapsing cloud, its field strength is amplified during collapse and this could become sufficiently strong to halt collapse in the equatorial plane.

Let us treat these problems separately. The best studied case is that of collapse of a spherical gas cloud. It is assumed that a density enhancement within a giant molecular cloud becomes unstable under gravity and then, according to the Jeans' instability criterion, it fragments into smaller mass regions. The Jeans' criterion, relation (6.8), indicates that, when the fragments have increased in density to $10^{-16} \, \mathrm{kg \, m^{-3}}$ at a temperature of $10 \, \mathrm{K}$, the Jeans' mass corresponds to $1 \, M_\odot$. According to small perturbation analyses, such a cloud simply collapses *en masse* and this may well be what happens if the mass of the collapsing cloud is much greater than the Jeans' mass. It should be noted, however, that the rate of growth of the

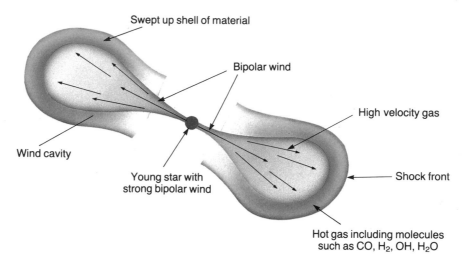

Figure 6.49. A schematic diagram showing the properties of a typical molecular outflow source. The outflow is supersonic and compresses the surrounding molecular gas. The heating of the molecular gas by the outflow and the cooling by molecular line emission results in a temperature of about 2000 K at which it can be observed through its infrared molecular line emission.

instability is strongly dependent upon the density and therefore, if a region of enhanced density is created, it tends to collapse more rapidly. In the case of masses which just exceed the Jeans' mass, computations of the collapse show that, even if the initial mass distribution is uniform, a density maximum rapidly forms in the very central regions. The origin of this behaviour can be understood in terms of the propagation of rarefaction waves into the cloud which enhance the central density. The net result is that the central regions collapse more rapidly forming a core region.

It is important to remember that the material of the collapsing gas cloud consists of molecular gas and dust. During the process of collapse, the material of the cloud is heated and the energy can be lost very efficiently by radiation so long as the cloud remains optically thin to radiation. The complications set in once the central collapsing core becomes optically thick to radiation because then the radiation is trapped and the core begins to heat up. We therefore end up with a situation in which the core begins to form a pressure supported structure whilst the matter in the envelope continues to collapse onto the core. The theoretical complication is that many of the processes now take place on roughly a collapse time scale and it is not at all clear that a steady state is ever attained until the star becomes a stable main sequence star.

The evolution of the central core of the protostar is more or less independent of the behaviour of the accreting envelope. Figure 6.50 shows what might be expected for the structure of the protostar. The core is indicated by the region labelled *hydrostatic core* and the regions outside this are all associated with the accreting flow. In the outer envelope, the matter and dust are optically thin and radiate away their energy very efficiently. The collapse in this region is therefore close to isothermal. Eventually the matter and dust densities increase to such values that the dust becomes optically thick, and the region with optical depth unity is referred to as the *dust photosphere*. Inside this, there is a dust envelope within which the temperature increases with decreasing radius until it becomes so hot that the dust evaporates, at $T \simeq 2300$ K for graphite grains, and then the radiative transfer is determined by the properties of the gas rather than the dust. The gas then falls in towards the hydrostatic core and, since the latter acts as a 'solid body', an accretion shock is set up which has the effect of dissipating the kinetic energy of the infalling gas and radiating away the binding energy of the gas as radiation. This radiation has to escape out through the accreting matter.

This picture makes it clear why the protostars are expected to be intense infrared sources. The dissipation of the binding energy of the protostar is carried away by radiation from the accretion shock and this energy is trapped and degraded in the dust envelope. The energy is radiated away at the temperature of the dust photosphere. The models of Adams and Shu (figure 6.51) show a broad maximum corresponding to the superposition of the emission from grains at different temperatures in the region of the photosphere with typical temperatures of about 100 K. These are similar to the spectra seen in a number of sources inferred to be protostellar objects.

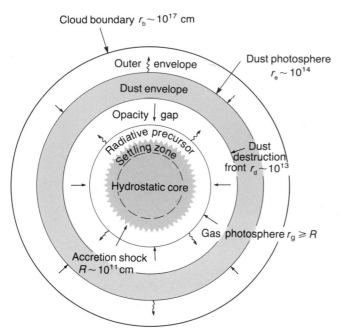

Cloud boundary $r_b \sim 10^{17}$ cm

Outer } envelope

Dust photosphere $r_e \sim 10^{14}$

Dust envelope

Opacity ↓ gap

Radiative precursor

Settling zone

Dust destruction front $r_d \sim 10^{13}$

Hydrostatic core

Gas photosphere $r_g \geqslant R$

Accretion shock $R \sim 10^{11}$ cm

Figure 6.50. A diagram illustrating the structure of an accreting protostar according to the analysis of Shu and his colleagues. The various regions are described in the text. (Stahler, S. W., Shu, F. H. and Taam, R. E. (1980). *Astrophys. J.* **241**, 641.)

Figure 6.51. An example of the predicted spectrum of a protostar from the models of Adams and Shu. In this model, the mass accretion rate is $10^{-5} M_\odot$ year^{-1} for a $1 M_\odot$ protostar. Similar spectra are found for a range of reasonable parameters for masses, 0.5–$3 M_\odot$, and accretion rates, 10^{-4} to $10^{-6} M_\odot$ year^{-1}. There is good agreement with the observed spectra of many of the far-infrared sources in which it is inferred that protostars or very young stars are present. (After Adams, F. C. and Shu, F. (1985). *Astrophys. J.* **296**, 655–69.)

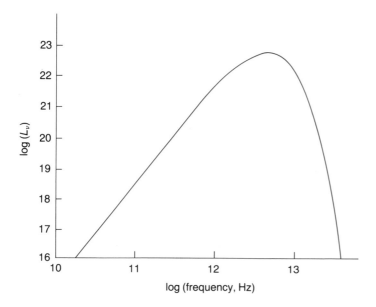

Thus, the energy problem is probably simply one of removing the kinetic energy of the infalling material through the thermal radiation of heated dust. The problem of removing the angular momentum and magnetic field associated with the collapsing cloud has yet to be solved. Some means has to be found of removing the angular momentum of the accreting material if it becomes so great that it would prevent collapse. In one picture, the formation of gaseous discs about protostars and young stars may be the means of removing much of the angular momentum of the accreting flow. The angular momentum of the infalling material can be transported outwards if it possesses sufficient viscosity. Another possibility is that, if the gas is strongly tied to the matter, the rotating gas cloud will wind up the magnetic field lines and then, when the energy density in the magnetic field is sufficiently high, the matter will be ejected out of the poles of the wound-up magnetic field structure (figure 6.52). This might be a means of solving the angular momentum and magnetic field problems simultaneously. It, or a process like it, may well be the origin of the bipolar outflows seen in young stellar objects.

There is much that remains uncertain in these pictures but it is encouraging that new observational capabilities are being developed at infrared and millimetre wavelengths for the study of these compact infrared sources which should help elucidate the relative importance of the different processes.

6.6   Galaxies and clusters of galaxies

An outline of the wide variety of different types of galaxy found in the Universe was given in Section 6.2. The diversity is not unexpected when one considers the many different components which can make up a galaxy. The numbers of stars can range from $\sim 10^6$ in the case of extreme dwarf galaxies to $\gtrsim 10^{13}$ in the most massive galaxies; varying amounts of interstellar gas can be present ranging from $\lesssim 0.01\%$ in elliptical galaxies to $\sim 30\%$ by mass in extreme irregular galaxies; the environments can range from essentially isolated galaxies, through pairs and small groups of galaxies to giant clusters such as the Coma cluster. Interactions between galaxies and with their environments can significantly influence their structure and evolution. The aim of the astrophysics of galaxies and clusters is to put some order into this diversity but, despite the enormous efforts of many astronomers, there are many simple questions to which no clear answers can yet be given. First we will quantify some of the properties of galaxies and then look at some specific questions.

The range of luminosities, and indirectly of masses, is best described by the *optical luminosity function* of galaxies. This function describes the number density of galaxies of different luminosities $\Phi(L)\mathrm{d}L$ and is shown in figure 6.53. It can be characterised by a function of the form

$$\Phi(L)\mathrm{d}L = AL^{-\alpha}\exp(-L/L^*),$$

outshine the next brightest members of the cluster. The Schechter function is intended to describe galaxies in general and not those special cases for which there are good astrophysical reasons why they are anomalously bright. Both spiral and elliptical galaxies span the whole range of luminosities but the irregular galaxies are mostly found at luminosities $L < L^*$.

The structures and internal dynamics of galaxies

In the simplest picture, spiral and elliptical galaxies can be decomposed into two stellar components: the *bulge* and the *disc* components. In the elliptical galaxies, the disc is normally absent, but in the spiral galaxies there is a wide range of ratios of the relative contributions which these make to the total light. It is normally supposed that the bulges of spiral galaxies are similar in nature to those of elliptical galaxies.

The dynamics of the disc components are much more straightforward than those of the bulges. The discs are in centrifugal equilibrium, the inward attraction of the gravitational field of the galaxy being balanced by the centrifugal forces acting on the stars. Indeed, the measurement of the rotation velocity as a function of distance from the centre of a galaxy is one of the best means of determining the mass distribution within galaxies. One note of caution is that, unfortunately, the mass distribution cannot be unambiguously determined from the rotation velocities, unless there is some independent evidence about how the total mass is distributed.

It might be thought that the internal dynamics of elliptical galaxies would be a relatively straightforward problem. These galaxies have an elliptical surface brightness distribution as their name implies, the ratio of the major to minor axes ranging from 1:1 to about 3:1. The mean velocity and velocity dispersion of the stars can be measured throughout the body of the galaxy. These measurements can be compared with the amounts of rotation and internal velocity dispersion which would be expected if the flattening of the elliptical galaxies were wholly attributed to the rotation of an axisymmetric distribution of stars. In the simplest models, it is assumed that locally the velocity distribution is isotropic at each point within the galaxy.

This type of analysis for elliptical galaxies and for the bulges of spiral galaxies results in the diagram shown in figure 6.54. The solid lines show the amount of rotation (v_m) necessary to produce the observed ellipticity of the system relative to the velocity dispersion σ of the stars. It can be seen that, for low luminosity elliptical galaxies and for the bulges of spiral galaxies, the ellipticity of the stellar distribution can be attributed to rotation. The most luminous ellipticals, $M_B < -20.5$, generally do not possess enough rotation to account for the observed flattening of the galaxies. This means that the assumptions of an axisymmetric distribution of stars and an isotropic velocity distribution of stars at all points in the galaxy must be wrong. As a consequence, these massive

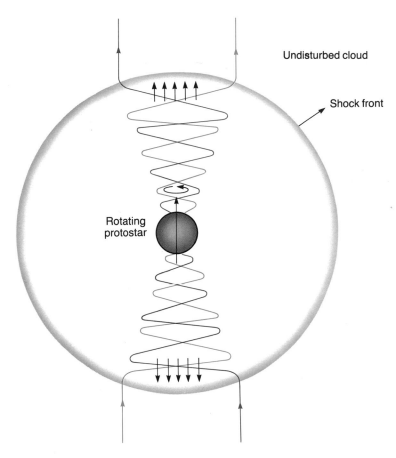

Figure 6.52. A sketch of a possible mechanism for the formation of bipolar outflows in protostellar objects. As the material collapses onto the protostellar object, the magnetic field is wound up and the magnetic field energy density increases. According to the analysis of Draine, the large magnetic pressure causes a shock wave to move out through the infalling plasma. The magnetic pressure increases, the winding-up continues, and eventually the pressure is released by the magnetic field and the plasma tied to it being ejected along the poles of the rotating object. (After Draine, B. T. (1983). *Astrophys. J.* **270**, 519.)

which was first introduced by Schechter. L is the luminosity of the galaxy, and the value L^* corresponds to the break in the luminosity function seen in figure 6.53 which has value $L^* \simeq 10^{10} L_\odot$. For values of $L < L^*$, the value of α is about 0.25 indicating that there is a long tail of faint galaxies which extends down to the dwarf galaxies. For reference, our own Galaxy has luminosity about a factor of two smaller than L^* so that it is typical of the spiral galaxies found in statistical samples but is by no means among the most luminous galaxies known.

The probability of finding the most luminous galaxies known is not well described by the Schechter function, the brightest galaxies in clusters being supergiant galaxies which often far

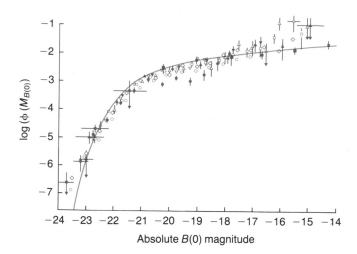

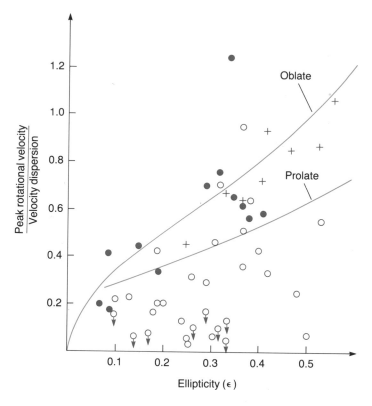

Figure 6.53. The luminosity function for galaxies. This function describes the space density of galaxies of different absolute magnitudes $M_{B(0)}$, meaning the total luminosity of the galaxy measured in the B waveband. The diagram shows a number of separate determinations (different symbols) of the form of the function which have been reviewed in depth by Felten. The solid line shows a best-fitting luminosity function of the type proposed by Schechter described in the text. (Based upon Felten, J. (1977). *Astron. J.* **82**, 869.)

Figure 6.54. A diagram showing the flattening of elliptical galaxies as a function of their rotational velocities. The open circles are luminous elliptical galaxies, the filled circles are lower luminosity ellipticals, and the crosses are the bulges of spiral galaxies. If the ellipticity were entirely due to rotation with an isotropic stellar velocity distribution at each point, the galaxies would be expected to lie along the solid lines. This diagram shows that, at least for massive elliptical galaxies, this simple picture of rotational flattening cannot be correct. (Adapted by Dr R. L. Davies (1986) from Davies, R. L., Efstathiou, G., Fall, S. M., Illingworth, G. and Schechter, P. L. (1983). *Astrophys. J.* **266**, 41.)

elliptical galaxies may well be *triaxial* systems, i.e. systems with three unequal axes, with anisotropic stellar velocity distributions. There is no problem in assuming that the velocity distribution is anisotropic because the time scale for the exchange of energy between stars through gravitational encounters is generally greater than the age of the galaxy. Therefore, if the velocity distribution began by being anisotropic, it would not have been isotropised by now. This situation contrasts dramatically with that of the atoms and molecules of a gas in which collisions very rapidly isotropise the velocity distribution.

Further evidence for the triaxial nature of massive elliptical galaxies comes from studies of the light distribution of the galaxies. In some galaxies, the ellipticity of the galaxy and the orientation of its major axis vary with distance from the centre. Another piece of evidence for the complexity of the shapes and velocity distributions within these galaxies comes from the observation of rotation along the minor as well as along the major axis in some ellipticals.

Fortunately, the theoretical position has been clarified by an elegant and original analysis by Schwarzschild. By applying linear programming techniques to the determination of orbits in general self-gravitating systems, he has shown that there exist stable triaxial configurations not dissimilar from those necessary to explain some of the internal dynamical properties of what appear on the surface to be simple ellipsoidal stellar distributions. His analysis shows that there exist stable orbits about the major and minor axes but not about the immediate axis of a triaxial figure. With this new understanding of the internal stellar motions of elliptical galaxies, galaxies can be characterised as oblate-axisymmetric, prolate-axisymmetric, oblate-triaxial, prolate-triaxial and so on but the determination of the precise figure requires accurate surface photometry and extensive long-slit spectroscopy. The fact that some systems are certainly triaxial opens up a very wide range of possible figures for galaxies so that nowadays the simplest position is to suppose that, unless proved otherwise, elliptical galaxies are triaxial systems.

Dark matter in galaxies and clusters of galaxies

One of the key issues of modern astronomy is whether or not all the mass in the Universe is readily detectable by the electromagnetic radiation it emits. The problem has been recognised for some time but it remains unresolved how much dark matter there might be and what its nature could be.

Let us start locally and proceed to larger and larger scales. In the vicinity of the solar neighbourhood, it has long been claimed that the local gravitational potential inferred from the distribution of stars perpendicular to the Galactic plane is greater than can be attributed to the mass of known types of star. The traditional position is that the local mass density projected onto the plane of the Galaxy should amount to about $0.15 M_\odot \text{pc}^{-2}$, whereas the contribution of known types of star and interstellar matter totals only about half this value. How severe this problem is is still a matter of debate. There remain significant uncertainties in the determination of the local luminosity function of stars and there are a number of new determinations of the local gravitational potential field in the vicinity of the Sun underway. As indicated in Section 6.3, ultra-low mass stars could make a significant contribution to the total mass. Most workers would agree that probably less than about half of the local mass density could be in the form of dark matter.

The position is clearer cut in the outer regions of spiral galaxies. The rotation curves describe the variation of rotational velocity with distance from the centre of the spiral galaxy. It is found that the rotation curves of many spiral galaxies, including our own, remain flat, i.e. $v_m \simeq$ constant, out to very large distances whilst the stellar luminosity decreases more or less exponentially from the centre. The significance of the flat rotation curve can be appreciated from Newton's laws of motion. Taking the galaxy to be spherical with the mass within radius r being $M(<r)$, the circular rotational velocity at distance r is found by balancing the inward gravitational acceleration $GM(<r)/r^2$ by the centrifugal force acting on a test particle of unit mass v^2/r so that we expect $v_m = [GM(<r)/r]^{\frac{1}{2}}$. For a point mass, $M(<r) = M_0$, we recover Kepler's third law of planetary motion, the orbital period $T = 2\pi r/v_m \propto r^{3/2}$. Thus, if we observe $v_m =$ constant, $M(<r) \propto r$ so that the mass within radius r increases linearly with distance from the centre. This contrasts strongly with the distribution of light in the disc and also in the galactic halo which decrease much more rapidly with increasing distance from the centre. It can be understood from these simple arguments that the ratio of mass to optical luminosity must increase dramatically in the outer regions of spiral galaxies. If averages are taken over the visible regions of galaxies, mass-to-luminosity ratios (M/L) less than about ten times that of the Sun ($M_\odot/L_\odot$) are found. This ratio must increase to much larger values at large r, values as high as 5000 being found at the outermost points at which some rotation curves have been determined. Similar results have been reported for the outer regions of elliptical galaxies from measurements of the velocity dispersions of the stars in these regions and from the distribution of extended thermal X-ray emission in the potential field of massive elliptical galaxies. The implication of these results is that there must be some form of dark matter in the outer regions of galaxies which contributes mass but not light.

Going up the scale of masses to clusters of galaxies, it has long been known that there must be more mass present than is contained in the visible parts of galaxies. Taking as an example the Coma cluster, the space density of galaxies increases smoothly towards the centre, resembling the spatial distribution expected of an isothermal gas sphere, i.e. a spherical self-gravitating gas distribution in hydrostatic equilibrium in which all the particles have the same temperature. The inference is that the cluster has relaxed to a bound equilibrium configuration and this is confirmed by comparing the *crossing time* of a typical galaxy in the cluster with the age of the Universe. The crossing time is defined by $t_{cr} = R/\langle v \rangle$, where R is the size of the cluster and $\langle v \rangle$ is the typical random velocity of a galaxy. For systems like the Coma cluster, the crossing time is less than about one-tenth the age of the Universe. This is clear evidence that the cluster must be a bound system or else the galaxies which form the cluster would have dispersed long ago.

It is therefore safe to apply the *virial theorem* to the cluster, which states that, for a self-gravitating system in statistical equilibrium, the gravitational potential energy must be twice the kinetic energy of the galaxies. Since the gravitational energy is roughly GM^2/R and the kinetic energy is roughly $\frac{1}{2}M\langle v^2 \rangle$, it follows that a value of M, the mass of the cluster, can be found. The masses of clusters such as the Coma cluster are found to exceed the mass which can be attributed to the visible parts of galaxies by factors of about twenty or more. This is most simply expressed in terms of the necessary values of the mass-to-luminosity ratio which would be required to bind the cluster. For the Coma cluster, the value of M/L would have to be about 300, compared with values of M/L for the visible parts of galaxies of at most about ten to twenty. To the mass of the galaxies, we should add the mass of the intergalactic gas and, in the case of the Coma cluster, the mass deduced from the X-ray thermal bremsstrahlung from the cluster amounts to about the same as the mass contained within the visible parts of galaxies. Thus, in the Coma cluster, the dark matter problem amounts to about a factor of ten, implying that most of the matter in the cluster is in the form of dark matter. This is typical of the values found in other rich clusters of galaxies.

Going to yet larger scales, the magnitude of the problem is more difficult to assess because, for superclusters, there is no guarantee that they are bound systems, and therefore the virial theorem must be applied with caution. Nonetheless, there are two approaches which give information about the large scale distribution of mass in the Universe. The first comes from the study of the velocities of infall of galaxies into large scale systems such as the local supercluster – this velocity is related to the mean density contrast between the average mass density of the Universe and the mass in the perturbation. The other is to apply the *cosmic virial theorem* to galaxies selected from the general distribution of galaxies in the Universe. For these, there exists a form of the virial theorem which relates their average kinetic energy with respect to the mean Hubble flow to the mean gravitational potential energy associated with the large scale distribution of mass. If the matter in the Universe is distributed like the visible matter, these procedures enable mass estimates of the mean matter density in the Universe to be made. Both methods lead to average values of the M/L ratio of about 200 to 600, far exceeding the typical values found for the visible parts of galaxies. It is evident from these arguments that, on the large scale, there must be more gravitating matter present than is revealed by the optical images of galaxies.

Finally, on the scale of the Universe itself, there is the question of the value of the density parameter Ω, which describes the mean density of matter in the Universe. As discussed in Section 6.8, if the arguments concerning primordial nucleosynthesis are accepted, the mean density of baryonic matter in the Universe corresponds to $\Omega \leqslant 0.2$ and hence to typical values of M/L of about 300–400. It is interesting that this is similar to the values found from studies of discrete systems such as clusters and from the cosmic virial theorem. On the other hand, some cosmologists take seriously the argument that Ω must be equal to unity because this is the value which comes out of the inflationary models of the early Universe and also because it is the only 'natural' value to come out of the standard world models (see Section 6.8 and Chapter 3 by Guth and Steinhardt).

There has been much speculation about the nature of the dark matter, possibilities ranging from 'conventional' forms such as ultra-low mass stars, planets and massive neutrinos to such 'exotic' forms as primaeval dark matter, possibly in the form of axions, cosmic strings and ultra-weakly interacting particles, or mini, intermediate and massive black holes. It is a real astrophysical challenge to devise observational techniques by which some of these possibilities can be tested. The most conservative picture is one in which dark matter is baryonic in nature. It could well be in the form of very low mass stars or planetary size objects which are very difficult to detect unless they are very nearby. Some aspects of this possibility were discussed in Section 6.3.

Next, it might be that the neutrino has a finite rest mass and, if the known types of neutrino had mass $\varepsilon \simeq 10-30\,\text{eV}$, this would provide sufficient mass to close the Universe, $\Omega = 1$. There are limits to how much mass could be present in this form in galaxies and clusters because of the phase space constraints upon the number of fermions which can be contained within a given volume. It would just be possible to explain the dark matter content of giant galaxies with such massive neutrinos. The spread in arrival times of neutrinos from the recent supernova in the Large Magellanic Cloud enables an upper limit to the neutrino mass to be determined and it lies close to the interesting range $\varepsilon \simeq 10-20\,\text{eV}$. A key test of this hypothesis is the laboratory measurement of the mass of the neutrino. Another variant of the massive neutrino hypothesis is that there might exist very massive unknown neutrino-like particles with masses $\varepsilon \simeq 1-10\,\text{GeV}$. These would have much smaller space densities than the known types of neutrinos but a case can be made that this number of massive neutrinos would survive from the early phases of the Hot Big Bang. It is intriguing that if these types of particles made up the dark halo of our Galaxy, the very massive neutrinos and other exotic particles with masses in the range $1-10\,\text{GeV}$ could be detected in ground-based cryogenic recoil experiments which have been recently become feasible. A positive result from these experiments would be of the greatest significance for particle physics and cosmology.

Constraints upon the contribution of black holes of different masses can be found from the statistics of gravitational lenses found in surveys of extragalactic radio sources. In a recent survey of thousands of radio sources designed specifically to discover gravitational lenses, Burke and his colleagues discovered only a few candidates. Hewitt, Burke and their colleagues have shown how these data constrain the contribution of galaxy-size and stellar-mass black holes to be less than roughly the critical value $\Omega = 1$. However, for many of the more exotic forms of dark matter, it is difficult to devise astrophysical tests which would set useful limits to their contributions to the mass density of the Universe.

The situation is plainly unsatisfactory because the dark matter may well play a key role in the origin and evolution of galaxies. I always feel that it is somewhat embarrassing for astronomers to have to admit that they do not know in what form most of the mass of the Universe really is. The astrophysical role of dark matter in cosmology, particularly in relation to the origin and evolution of galaxies, is discussed in more detail in Section 6.8.

The chemical evolution of galaxies

The chemical history of galaxies is an important tool in disentangling the sequence of events which must have taken place during galactic evolution. Evidence for different processes of nucleosynthesis was described in Section 6.3. The challenge in the case of galactic evolution is to synthesise these ideas into a coherent picture of the chemical enrichment of the interstellar gas and the stars which form from it in our own and other galaxies. The programme is not an easy one because it requires observations of very high quality, and the interpretation of the observations in terms of the relative importance of age, metallicity and excitation of the species observed is far from trivial. In addition, theoretical models of the chemical evolution of galaxies depend upon many factors which are at present poorly determined. For example, what is the dependence of the star formation rate upon metal abundance (or *metallicity*) and the average gas density? What is the *yield* of heavy elements as a function of stellar mass? How well mixed is the interstellar gas? Does the form of the initial mass function vary with location in a galaxy? Certain consistent trends are, however, now found for spiral and elliptical galaxies, and these seem to be consistent with observations of the distribution of the chemical elements in our own Galaxy.

In our own Galaxy, evidence for the chemical evolution comes from studies of the abundances of the heavy elements in different types of star and in regions of ionised hydrogen. Perhaps the most striking abundance gradients are observed in the globular cluster population in the bulge of the Galaxy. The globular clusters form a halo population about the Galaxy and are among the oldest of all stellar systems in the Galaxy. They are deficient in heavy elements relative to the solar abundances as would be expected if they formed very early in the evolution of the Galaxy. Abundance trends are observed such that those within the inner Galaxy have relatively higher abundances of heavy elements than those in the halo, which suggests that the former formed out of material which was already enriched in heavy elements. It is striking however, that, even among what are inferred to be the oldest stellar systems known, there are very few systems indeed in which the abundances of the heavy elements are less than about 1% of the solar values. This implies that even the very oldest systems we know of must have formed from material which was already somewhat enriched in heavy elements relative to the primordial hydrogen–helium gas. It is significant that, even in the most distant quasars known, the emission lines of the common elements appear to be as strong as those seen in nearby quasars, indicating that even when the Universe was about one-fifth its present age or less, the enrichment of the chemical elements was probably well underway.

In the case of stars formed within the disc of the Galaxy, it might be expected that they would show a steady increase in metallicity as the interstellar gas is enriched with heavy elements. Attempts have been made to derive the variation in metallicity with time from studies of intermediate age stars which are members of star clusters for which ages can be derived from fitting their colour–magnitude diagrams to models of stellar evolution in clusters. Although there are many complicating selection effects, these stars, the F-stars, show an increasing abundance of the heavy elements with cosmic epoch, indicative of chemical enrichment in the disc of the Galaxy. The study of the metallicity distribution of old stars in the disc is another useful tool for studying the enrichment history of the interstellar gas. There is found to be a remarkably narrow spread in metallicities of stars which have ages of the same order as the age of the Galaxy. This is somewhat surprising if the stars were formed at a uniform rate from material which is increasing in metal abundance with epoch. The lack of old disc stars with low metal abundances has been interpreted as evidence that there must have been some prompt initial enrichment of the chemical abundances in the disc of the Galaxy.

Galaxies show variations in colour with distance from their centres but it is important to distinguish variations due to the different populations of stars contributing to the light at different distances from the centre from those due to real variations in the chemical abundances of the elements associated with a particular class of object. Once these effects are taken into account, there is found to be evidence for abundance gradients in the discs of spiral galaxies and through the bulges of elliptical galaxies, the greater abundances being found in the inner regions. The exact cause of these abundance gradients is not established since a number of effects could be important. For example, is the yield of heavy elements the same for stars formed in the inner and outer regions of galaxies? What is the influence of the infall of unprocessed matter from intergalactic space?

The general concensus is that most of the abundance variations can be reasonably accounted for in terms of the processes of nucleosynthesis described in Section 6.3. This leads to the hope that it will eventually be possible to use these chemical tools to define in more detail the evolutionary history of the material from which stars form and also to set limits to the different types of stars formed in different galactic environments. In turn, it may be possible to derive from observation the chemical yields of different masses of star and the rates at which these stars must have formed.

Clusters of galaxies

Like the galaxies, groups and clusters of galaxies come in a variety of different forms. They range from the regular clusters, which have smooth galaxy density profiles and roughly circularly symmetric appearances, to the irregular clusters, which have a ragged appearance without any prominent central concentration of the galaxies. In terms of numbers they range from small groups with only a few members to giant clusters containing thousands of members. An important physical parameter which describes the state of evolution of the system is the *crossing time* defined earlier in this section. This can often be used as a means of deciding which associations of galaxies are physically bound groups, particularly for small systems. The regular symmetrical clusters are found to be those with the shortest crossing times so that there has been time for them to take up the configurations expected in statistical equilibrium. We will devote most attention to the rich clusters of galaxies since they exhibit the broadest range of features unique to clusters.

The masses of rich clusters range up to about $4 \times 10^{15} M_{\odot}$ for systems such as the Coma cluster. As discussed above, this is about twenty times the mass which can be ascribed to the visible parts of galaxies. As such, these are the most massive known bound systems in the Universe. The galaxies form a self-gravitating system, and it is significant that the radial distribution of galaxies can be modelled by that expected of an isothermal gas sphere. This is the simplest form of distribution which could be realised physically. It appears to be a good approximation in practice, the velocity dispersion of the galaxies being more or less independent of radius in the cluster, except possibly in the very central regions.

The most massive galaxies known are found in the centres of the giant rich clusters (figure 6.55). A number of analyses have shown that these galaxies are much brighter than would be expected from random sampling of the high luminosity end of the luminosity function and hence there must be some physical mechanism for enhancing the luminosity of the brightest member. An important clue is that, in the clusters with the most luminous galaxies, the second and third brightest members appear to be relatively fainter than the corresponding members in other clusters. This has suggested that the central massive galaxy grows at the expense of the next faintest members. These phenomena find a natural explanation in the theory of the dynamical evolution of clusters of galaxies.

The simplest picture of the evolution of a cluster of galaxies begins by assuming that the galaxies are point masses. When the collapse of the protocluster gets underway, large gravitational potential gradients are set up since the collapse is unlikely to be spherically symmetric and the system of galaxies relaxes under the gravitational influence of these large scale perturbations. This process, first described by Lynden-Bell, is known as *violent relaxation* and in it the galaxies settle down to an equilibrium configuration in which each galaxy has more or less the same average velocity. This is the theoretical justification for assuming in the simplest approximation that the cluster resembles an isothermal gas sphere. In this process, the cluster has to lose its gravitational binding energy. In practice, this means that the cluster has to lose about half of the kinetic energy generated in the collapse so that, in the stable configuration, the gravitational potential energy amounts to twice the kinetic energy of the galaxies in the cluster.

The galaxies form a bound system and move under the gravitational influence of the cluster as a whole. Every so often, however, they make close encounters with other galaxies and energy is exchanged between galaxies through their mutual gravitational attraction. Just as in a Maxwellian gas, the exchange of energy tends to lead to the equipartition of energy between the cluster members so that the less massive galaxies gain kinetic energy on average whilst the more massive galaxies lose energy. This process of energy transfer between galaxies (or stars in a star cluster) is known as *dynamical friction* and has the effect that the most massive galaxies which lose kinetic energy spiral in towards the centre of the cluster. This is the process by which the particles attempt to achieve a Maxwellian distribution of velocities. Another important feature of the process of dynamical friction is that it takes place most effectively between the most massive galaxies, and thus the dissipative effects are expected to be strongest for the most massive members of the cluster. The time scale for this process is very much longer than the crossing time of the cluster, and,

hence, since in most cases the regular clusters are only between five to twenty crossing times old, it is expected that this form of relaxation will only be important for the most massive members of the cluster, the other galaxies preserving the velocities which they acquired when the cluster came into dynamical equilibrium.

In real clusters, we have to take account of the finite extent of the galaxies. As a result, when the most massive galaxy drifts into the centre of the cluster followed closely by the next brightest members, there will be strong tidal interactions between them which will tend to disrupt the smaller galaxies and lead to coalescence of galaxies at the centre of the cluster. This process of the central galaxy consuming the next most massive members of the cluster is known as *galactic cannibalism* and provides an elegant explanation for the observed properties of the brightest galaxies in clusters. They are more luminous than normal bright galaxies because they consume less massive galaxies. Since they become more luminous at the expense of the next brightest members, this explains why the second and third brightest galaxies are fainter than in those clusters in which cannibalism has not taken place. The cannibal galaxies are more distended in appearance than normal giant elliptical galaxies because they have to absorb part of the relative kinetic energy of the galaxies when they coalesce. Multiple nuclei are often observed in the most massive galaxies in clusters and these are readily explained as fragments of galaxies consumed in the coalescence process (figure 6.56). This picture can also explain why the most massive galaxies in clusters have a small dispersion in luminosity. Their luminosities are boosted by consuming less massive galaxies. They become larger in physical size, and it was first pointed out by Hausmann and Ostriker that, when galaxies are observed through the same metric diameter, they have roughly the same luminosities, the increase due to cannibalism being offset by the fact that the galaxy increases in size so that less of the galaxy is contained within a fixed metric aperture.

One of the most important discoveries of X-ray astronomy has been that of the *hot intergalactic gas* in clusters of galaxies. The strong X-ray emission fills a considerable fraction of the volume of the cluster, and the X-ray spectrum is characteristic of the thermal bremsstrahlung of hot gas with temperature in the range 10^7–10^8 K (figure 6.57). The presence of this very hot gas is confirmed by the observation of X-ray emission lines of twenty-four and twenty-five times ionised iron, FeXXV and FeXXVI, at an X-ray energy of about 8 keV. It is remarkable how much hot intergalactic gas is present in rich clusters, in the case of the Coma cluster, the mass being comparable to that in the visible parts of galaxies. The origin of the gas is of considerable interest. In order to form a stable extended atmosphere within the deep gravitational potential well of a cluster, the gas has to be hot. This could either arise because of infall of gas from intercluster space or it could be expelled from the galaxies in the cluster. In either case, the gas is heated as it falls into the deep gravitational potential well of the cluster.

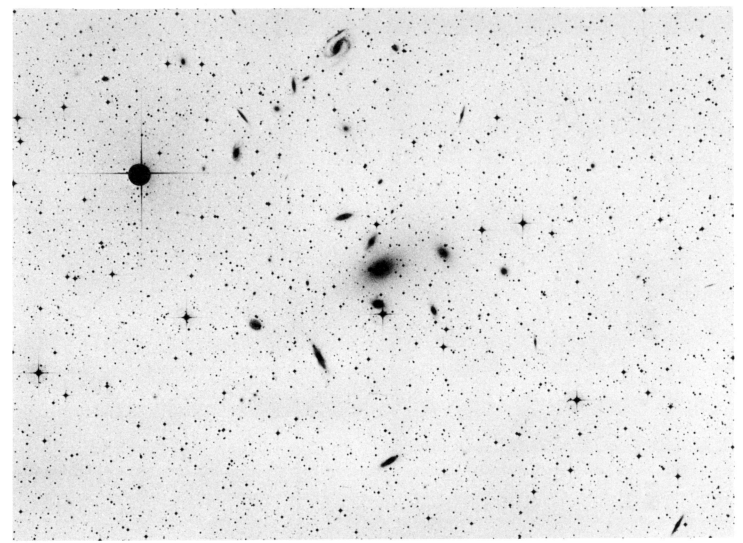

Figure 6.55. The rich cluster of galaxies in the constellation of Pavo in the southern hemisphere. The central galaxy is a supergiant or cD galaxy, which is very much brighter than all the other galaxies in the cluster. It is located at the dynamical centre of the cluster. (Courtesy of the Royal Observatory, Edinburgh.)

One important feature is the observation of highly ionised iron lines from the intracluster gas corresponding to abundances within about a factor of three of the local cosmic abundance of iron. This indicates that the intergalactic gas in the cluster has been enriched by the processes of nucleosynthesis which presumably took place within the galaxies. This suggests that significant amounts of mass loss of enriched material must have taken place from the galaxies during the lifetime of the cluster.

The hot gas has an important influence upon the galaxies in the cluster. One effect is that, as the galaxies move through the hot intergalactic gas, the *ram pressure* of the intergalactic gas

exerts a strong influence upon any interstellar gas present within spiral galaxies. In fact, this pressure is sufficiently large to remove the interstellar gas from spiral galaxies, an effect known as ram pressure *stripping* of the galaxies. This process results in disc galaxies which possess very little interstellar gas. These have been identified with the lenticular or S0 galaxies which are found in much greater numbers within rich clusters of galaxies than in the general field. The intracluster gas has the same ram pressure effect upon any double radio sources associated with the galaxies in the cluster. As the radio galaxies move through the cluster, the radio jets are swept back in the form of a trail behind the parent galaxy

Figure 6.56. The central galaxy in the distant cluster of galaxies ZwCl 0257+3542 observed by Dr James Gunn and his colleagues. This galaxy is very likely undergoing a process of coalescence into a supermassive cD galaxy. (From Gunn, J. E. (1977). *The Evolution of Galaxies and Stellar Populations*, B. M. Tinsley and R. B. Larson, eds., frontispiece, Yale University Observatory.)

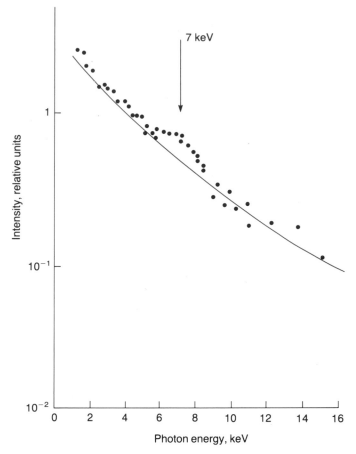

Figure 6.57. The X-ray spectrum of the Perseus cluster of galaxies showing the intense emission lines of the iron at about 7 keV. The underlying continuum spectrum is the bremsstrahlung (or free–free emission) of the highly ionised intergalactic gas within the cluster. (From Mitchell, R. J., Culhane, J. L., Davison, P. J. N. and Ives, J. C. (1976). *Mon. Not. R. Astr. Soc.* **175**, 30P.)

(figure 6.58). These observations enable limits to be set to the age of the radio source and provide information about the dynamics of relativistic particles and magnetic fields in the intracluster gas. An important point is that the intergalactic gas densities inferred from these observations are adequate to confine the outer radio lobes of even the most powerful radio sources (see Section 6.7).

A third effect associated with the hot intracluster gas is that it is expected that the gas density will increase towards the central regions and then, if it becomes large enough, rapid cooling of the gas by bound-free emission and bremsstrahlung can occur. The result is that *cooling flows* can be set up towards the central regions of the cluster and this can account for the observation of cool clouds of diffuse emitting gas close to the cluster centre. Fourth, the hot gas in the cluster has the effect of

Compton scattering any photons which pass through the intergalactic medium in the cluster. The most interesting example of this is Compton scattering of photons of the Microwave Background Radiation which, according to the **standard Hot Big Bang model of the Universe, were last** scattered at a redshift of about 1000 (see Section 6.8). The scattering of these photons of the microwave background has the effect of shifting the whole of the Planck distribution to a slightly higher temperature so that in the Rayleigh–Jeans region of the spectrum a small dip is expected whilst, in the Wien region, there should be a small increase in the intensity (figure 6.59). This effect was first predicted by Sunyaev and Zeldovich, who showed that the magnitude of the decrement in the Rayleigh–Jeans region of the spectrum is proportional to the Compton scattering optical depth, which in turn is just

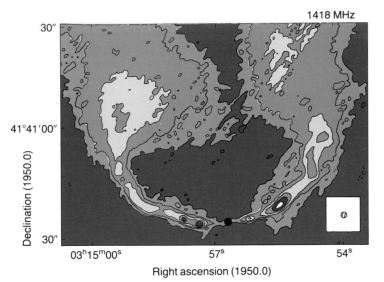

Figure 6.58. The radio source 3C 83.1B associated with the galaxy NGC 1265 in the Perseus cluster of galaxies. Sources with this morphology are known as radio trail sources, the parent galaxy being a double sided jet source. The radio jets are swept backwards because of the ram pressure of the intergalactic gas in the cluster. In the case of NGC 1265, the galaxy must be moving with a high velocity with respect to the intergalactic gas in the cluster since its projected velocity relative to the standard of rest in the cluster is about 2200 km s^{-1}. The ram pressure is associated with the rapid motion of the galaxy through the hot intergalactic gas in the cluster. (From O'Dea, C. P. and Owen, F. N. (1986). *Astrophys. J.* **301**, 845.)

Figure 6.59. Illustrating the effect of Compton scattering upon the spectrum of the Microwave Background Radiation as it passes through **a** region of hot intergalactic gas. Since the gas is much hotter than the radiation, the electrons scatter the photons of the Microwave Background Radiation to higher frequences. It should be noted that this process has the characteristic signature that it results in a decrease in the intensity of the radiation on the long wavelength side of the black-body maximum and an increase on the short wavelength side.

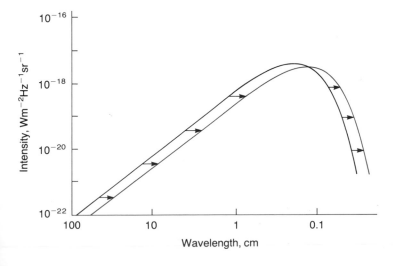

proportional to the pressure of the gas. The observation of the *Sunyaev–Zeldovich effect* is very difficult because the size of the effect is only about 1 mK, but it is claimed that a positive result has now been found in a few clusters which are known to be strong X-ray emitters. This provides yet another diagnostic tool for studying the properties of the intergalactic gas in the cluster.

Finally, it should be noted that hot intergalactic gas provides an excellent tracer of the gravitational potential distribution in gravitating systems. This has been used to good effect to measure the masses of galaxies and clusters in which the X-ray emitting gas fills a large fraction of the potential well of these systems.

6.7   Active galaxies and quasars

Basic considerations

The understanding of the extreme properties of the most active galactic nuclei almost certainly requires new physics in physical conditions quite different from those encountered elsewhere in the Universe. The key pieces of observational data which have forced astrophysicists to study new energy sources for active galaxies and quasars are their extreme luminosities and, even more important, the fact that this emission varies rapidly with time. The Seyfert galaxies, quasars, BL-Lac objects and other classes of active nuclei described in Section 6.2 all possess 'quasi-stellar' components in their nuclei which appear as point-like sources which vary in intensity on time scales of years, months or days. The most rapid known variations are found in some of the BL-Lac objects and in the X-ray variability of some Seyfert galaxies (figure 6.60*a* and *b*). Significant variations on the time scale of about one hour have been found in sources which lie at cosmological distances.

It is instructive to consider how these observations lead inevitably to consideration of supermassive versions of the black holes described in Section 6.4. We can combine two simple results. The first of these is the expression for the Eddington limiting luminosity, repeated here:

$$L_{\text{Edd}} = 1.3 \times 10^{31} (M/M_\odot) \,\text{W}. \qquad (6.7)$$

This sets a useful upper bound to the luminosity of a source of mass M since for greater luminosities the radiation pressure of the source itself will blow the object apart. Notice that the result is independent of the radius of the source.

The second result is the *causality* relation, which states that variations in the intensity of a source of size r cannot be observed to take place on time scales less than r/c since it takes this time for electromagnetic waves to travel from one side of the source to the other. The smallest size which a source of mass M can possibly have is roughly the Schwarzschild radius of an object of mass M and hence the lower limit to the time scale of variations from a source of mass M is the Schwarzschild radius divided by the velocity of light:

$$T \geqslant r_g/c \simeq 10^{-5} \, (M/M_\odot) \, \text{s}. \qquad (6.9)$$

Thus, according to relation (6.7), a $10^8 \, M_\odot$ black hole can have luminosity up to $\simeq 10^{39} \, \text{W}$ and, from relation (6.9), time variations in this luminosity could occur on time scales as short as roughly half an hour. To complete the picture we need only repeat what was stated in Section 6.4 about the efficiency of energy release in accretion onto black holes. In the case of spherical black holes, up to about 10% of the rest mass energy can be liberated whilst for maximally rotating black holes, the figure rises to about 43%. In addition, once a black hole has been spun up by the infalling matter of the accretion disc, the rotational energy of the black hole can be tapped subsequently if the rotation is coupled to the surrounding interstellar medium by a magnetic field. Thus, supermassive black holes have all the necessary properties for a successful theory of the most active galactic nuclei.

Direct observational evidence for masses of this order in the nuclei of Seyfert galaxies has come from long-term monitoring of their intensities and spectra. The most extreme types of Seyfert galaxies, the type I galaxies, have very broad, strong emission lines which are variable in intensity and also strong variable continuum radiation. The breadth of the lines is attributed to mass motions of gas clouds close to the nucleus itself. One remarkable result which has been obtained by a consortium of European astronomers using the IUE has been that there is a time delay between the variations observed in the continuum spectrum and the corresponding variations seen in the broad line spectrum in the type I Seyfert galaxy NGC 4151. This indicates convincingly that the broad line regions are photoionised by the continuum radiation from the nucleus and also enables a direct estimate of the distance of the clouds from the nucleus to be made since the ultraviolet photons travel from the nucleus to the clouds at the velocity of light. The combination of this distance with the velocities of the clouds enables the mass of the central region to be estimated and this turns out to be about $10^9 \, M_\odot$. Arguments of a similar type have been used by Wandel and Mushotzky to show that, for those Seyfert galaxies which are variable X-ray sources, masses determined from the causality relation (6.9) and dynamical masses derived from the line widths and inferred distances of the broad emission line regions from the nucleus, are in remarkable agreement over a considerable range of black hole masses, $10^6 \leqslant M/M_\odot \leqslant 10^{10}$ (figure 6.61a). Typically, the luminosities amount to about 1 to 10% of the Eddington limit (figure 6.61b).

As was indicated in Section 6.2, it is likely that many classes of galaxy possess 'mini-quasars' in their nuclei and even in the centre of our own Galaxy, there is some evidence for a compact massive object. Figure 6.62 shows a map of the distribution of molecular hydrogen in the vicinity of the Galactic nucleus made in the near-infrared waveband. Measurements of the

velocities of different parts of the molecular ring show that it is rotating, and the combination of these mass motions with the radius of the ring gives an estimate of the mass of the nuclear regions of about $10^6 \, M_\odot$. There is also evidence for a strong

Figure 6.60. Examples of the variability of the emission of active nuclei. (a) The optical variability of the radio quasar 3C 345 in the B waveband. (From Christiani, S. (1986). *Structure and Evolution of Active Galactic Nuclei*, G. Giuricin, F. Mardirossian, M. Mezzetti and M. Ramella, eds., p. 83. D. Reidel and Company.) (b) The X-ray variability of the Seyfert galaxy MCG-6-30-15. Strong and continuous variability can be seen in the X-ray variability down to time scales of several-hundred seconds. (Courtesy of K. Pounds and the European Space Agency (1987).)

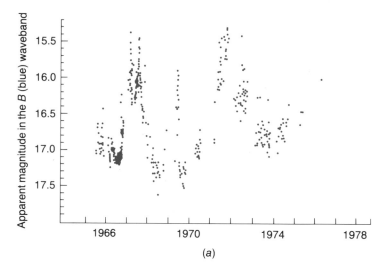

(a)

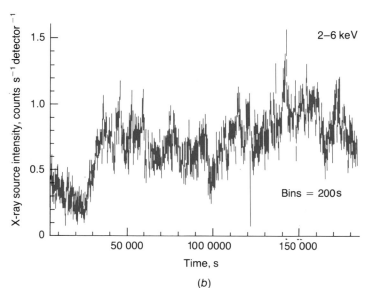

(b)

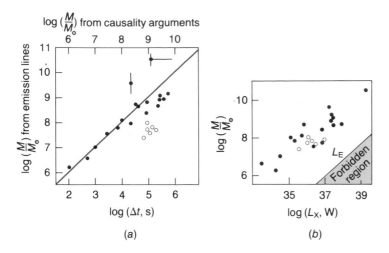

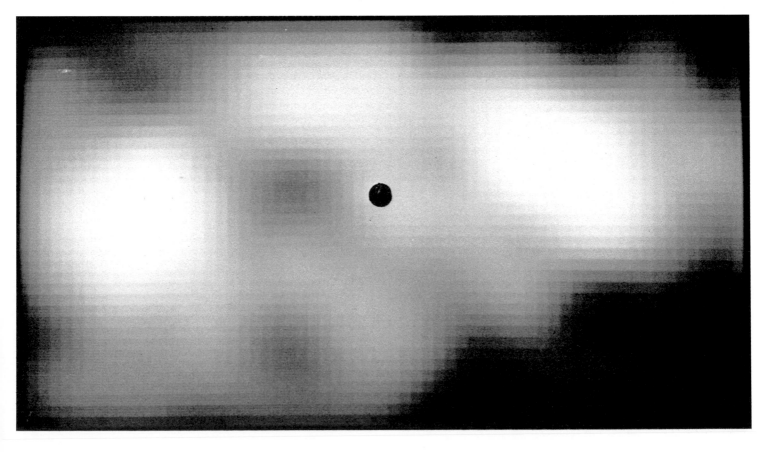

Figure 6.61. (*a*) Comparison of the mass estimates of active galactic nuclei from the variability of their X-ray emission and from dynamical estimates by Wandel and Mushotzky. The solid dots are quasars and type I Seyfert galaxies and the open circles are type II Seyfert galaxies. (*b*) Comparison of the inferred masses and luminosities with the Eddington limiting luminosity, $L_E = 1.3 \times 10^{31} (M/M_{\odot})$. It can be seen that all of these objects lie well below the Eddington limit. (Both diagrams after Wandel, A. and Mushotzky, R. F. (1986). *Astrophys. J.* **306**, L63–4.)

wind blowing from the nucleus itself, both from direct observation of broad velocity wings on the infrared HeII lines and from the fact that the region around the nucleus appears to have been swept clean of dust. In addition, there is a point-like radio source similar to those found in quasars but on a very much more modest scale and variable electron–positron annihilation line emission. The nucleus of our own Galaxy thus possesses a number of unique features which can be ascribed to some form of 'exotic object'. If it is a black hole, it is not yet clear how massive it might be. At one extreme it could have mass $\sim 10^6 M_{\odot}$, but it could equally well be that much of the mass is in the form of a compact star cluster and the black hole mass might be as low as $100 M_{\odot}$. It is obvious that the study of the nucleus of our own Galaxy provides a unique opportunity for studying the behaviour of matter in the vicinity of what is almost certainly the closest massive black hole to the Earth.

Figure 6.62. A map of the distribution of molecular hydrogen within 1 arcminute of the centre of our Galaxy from infrared molecular hydrogen line observations. Spectroscopic measurements show that the ring is rotating rapidly about the nucleus of our Galaxy. (From Gatley, I., Jones, T. J., Hyland, A. R., Wade, R., Geballe, T. R. and Krisciunis, K. (1986). *Mon. Not. R. Astr. Soc.* **222**, 299–306.)

Some words of caution are advisable, however, because it would, I believe, be wrong to suppose that accretion onto black holes or tapping the rotational energy of the black hole are the only ways of producing a great deal of activity within the central regions of galaxies. The history of active galactic nuclei is littered with theories which do not involve black holes and, whilst the most extreme examples of active nuclei almost certainly require the presence of accreting black holes, this is not necessarily the case for weaker nuclei. For example, among the early theories was the idea that the luminosity of active nuclei might be due to the formation of a large number of massive stars and their explosion as supernovae within a relatively short period. The problem with this picture was to explain the very luminous outbursts observed over very short time scales in objects such as BL-Lac objects. However, the idea that a great deal of star formation might take place in the nuclei of galaxies has been convincingly demonstrated to be the case from IRAS observations of the most luminous far-infrared galaxies detected in that survey. Indeed, for some of the most luminous IRAS galaxies, there is controversy as to whether they contain 'mini-quasars' or 'star-bursts'. This is but one example of non-black hole theories which may well be of importance for some classes of active nuclei, and it emphasises the point that one should not necessarily believe that all active nuclei have to be explained in the same way. Nonetheless, in this chapter, I will concentrate upon the black hole theories.

The prime ingredients of active galactic nuclei

It is convenient to divide the necessary ingredients of active galactic nuclei into two types – the *primary ingredients*, which probably have to be produced in the vicinity of the black hole itself, and *secondary phenomena*, which result from the interaction of these primary ingredients with the environment of the black hole. Figure 6.63 shows a schematic diagram of some of the components of any successful model. For the purposes of exposition, the primary ingredients will be classified as the *intense nonthermal continuum radiation* and *fluxes of high energy particles*. I will classify the secondary phenomena as being the interaction of these components with the surrounding medium, in particular gas clouds close to the nucleus and the ambient interstellar and intergalactic gas. The former gives rise to the strong emission line spectrum observed at optical, ultraviolet and infrared wavelengths whilst the interactions of beams of high energy particles with the interstellar and intergalactic gas give rise to beams and jets in extended radio sources.

The continuum spectrum   Figures 6.64(*a*)–(*d*) show typical spectra of types I and II Seyfert galaxies, a quasar and a BL-Lac object. In all cases most of the luminosity is in the continuum spectrum rather than in the lines; furthermore, the variability of the source is almost entirely associated with the continuum spectrum. The continuum spectrum is unlike any stellar or galaxy spectrum, the latter being the integrated light of many different types of star. Very often the continuum spectrum can be represented by a power-law within the optical and infrared wavebands, most objects having spectra which can be described by $I_\nu \propto \nu^{-\alpha}$ with the average value of α close to unity but with considerable dispersion about this value. For a number of bright objects, it has been possible to undertake more detailed decompositions of the continuum spectrum in the optical and ultraviolet wavebands and these show evidence for an excess at ultraviolet wavelengths as compared with what is expected from extrapolation of a power-law spectrum. The excess is probably similar in character to the 'blue bump' seen in the ultraviolet spectrum of 3C 273 (figure 6.65). A number of cases are known in which extremely steep spectra are found in the infrared waveband, optical to infrared spectral indices as steep as $\alpha \simeq 6$ being found. These are sometimes referred to as 'infrared quasars' since essentially all their energy is emitted in that waveband.

It is particularly important to know the continuum spectra of active nuclei in the ultraviolet waveband and beyond, but, whilst excellent observations in the wavelength range 120–300 nm have been made with the IUE, unfortunately no observations of active galaxies or quasars are available from about 120 nm to the X-ray waveband. Many quasars and active nuclei have been detected by X-ray satellites such as the Einstein X-ray Observatory and EXOSAT, and it appears that their X-ray spectra form a natural extension of the optical spectrum into the X-ray region. The optical-to-X-ray spectra of most quasars can be described by a power-law with spectral index $\alpha \simeq 1$, this form of spectrum indicating that roughly as much energy is radiated per decade of frequency in the optical as in the X-ray waveband.

The far-ultraviolet region of the spectrum is important for understanding the excitation of the gas clouds surrounding active galactic nuclei. For the most extreme quasars, a convincing case can be made that the strong line spectra result from photoionisation and excitation of the gas clouds surrounding the nucleus by the continuum radiation from the nucleus (figure 6.63). In this case, information can be obtained about the photoionising flux in the Lyman continuum and also about the far-ultraviolet behaviour of the continuum spectrum from the presence of high excitation lines. It is one of the attractions of the photoexcitation picture that a power-law continuum spectrum, containing photons with a very wide range of energies, can account for the observation of a very wide range of different ionisation states of elements in a single spectrum.

The continuum radiation is also known to be polarised. The strongest polarisation is found in the cases of the BL-Lac objects in which up to 30–40% linear polarisation has been observed. In a few important cases, polarisation changes have been followed through strong radio and infrared outbursts, and these prove important in understanding the magnetic field geometry of the source regions. It appears that the radiation associated

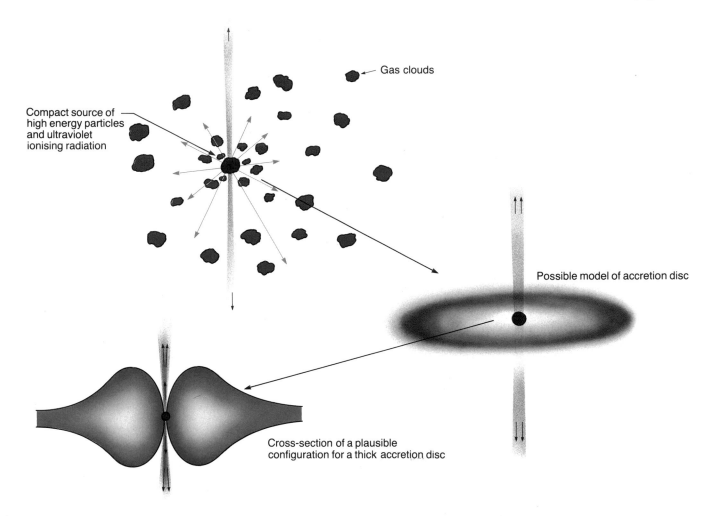

Gas clouds

Compact source of
high energy particles
and ultraviolet
ionising radiation

Possible model of accretion disc

Cross-section of a plausible
configuration for a thick accretion disc

Figure 6.63. A schematic diagram showing the necessary ingredients of a model of an active galactic nucleus. There must be a compact source of high energy particles and intense ultraviolet ionising radiation in the very centre. These may be produced close to the central engine, which is likely to be a supermassive black hole. The ultraviolet radiation may be the emission of the relativistic particles accelerated close to the central engine. The nucleus is surrounded by gas clouds, which are heated and excited by the ionising radiation from the nucleus. The clouds closest to the centre have high particle densities, values of about $10^{16}\,\mathrm{m}^{-3}$ being necessary to de-excite the forbidden line radiation from these regions. These are the origin of the broad-line emission seen in quasars, type I Seyferts and broad-line radio galaxies. Further out, the clouds are less dense, $\simeq 10^{10}$–$10^{12}\,\mathrm{m}^{-3}$, and these are responsible for the narrow-line regions observed in type II Seyfert galaxies and some of the narrow-line radio galaxies. It is inferred that there must be some mechanism responsible for the collimation of the beams of high energy particles. The inserts to the diagram show a possible structure for the accretion disc and a thick disc surrounding the black hole. These may be responsible for the collimation of the beams.

with the 'blue bump' is not polarised since the total percentage polarisation decreases when it contributes significantly to the total continuum intensity. This is consistent with the blue bump being mostly thermal radiation.

High energy particles It is likely that very high energy particles are accelerated close to the nucleus. Information about these particles comes from the nonthermal radiation itself. The characteristics of a power-law continuum spectrum, polarised emission and rapid variability are typical of those expected of the emission of ultrarelativistic electrons. The most direct evidence for the presence of ultrarelativistic electrons in the nuclei of active galaxies comes from very long baseline interferometric (VLBI) studies of radio quasars and BL-Lac objects at centimetre wavelengths. This radio astronomical technique enables very fine-scale angular structure in radio

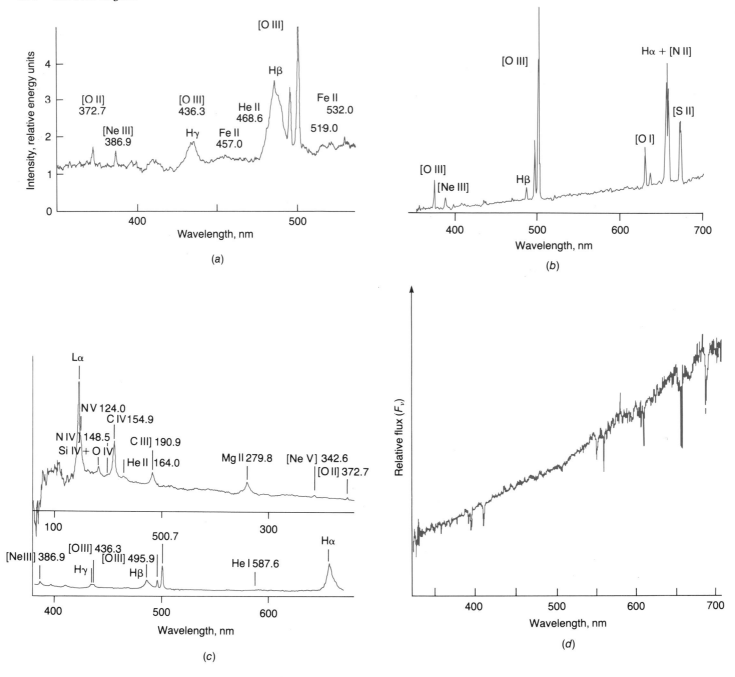

Figure 6.64. Typical spectra of different classes of active galactic nuclei. (*a*) The type I Seyfert galaxy III ZW 2 – the spectrum exhibits the characteristic broad emission lines of the resonance lines of hydrogen while the forbidden lines of oxygen and neon are narrow. (Osterbrock, D. E. (1978). *Physica Scripta* **17**, 140.) (*b*) The type II Seyfert galaxy Markarian 348 (Mk 348) – the lines are much narrower than in the case of type I Seyfert galaxies and forbidden lines of the same line width as the permitted lines are observed. (Osterbrock, D. E. (1978). *Physica Scripta* **17**, 138.) (*c*) A composite quasar spectrum extending from 100 to about 700 nm – the spectrum contains very broad emission lines with a strong nonthermal continuum. (From Baldwin, J. A. (1979). *Active Galactic Nuclei*, C. Hazard and S. Mitton, eds., p. 57. Cambridge University Press.) (*d*) The BL-Lac object PKS 0215 + 015 – there are no spectral lines observed in the spectrum which can be used to measure the redshift of the object. Narrow absorption line systems are observed at redshifts of 1.3449, 1.5494 and 1.6494. These are presumably foreground absorbers and hence the BL-Lac object must lie at a larger unknown redshift. The continuum spectrum can be represented by a power-law spectrum of the form $I_\nu \propto \nu^{-2.11}$. (From Gaskell, C. M. (1982). *Astrophys. J.* **252**, 447.)

sources to be measured, the longest baselines between radio telescopes now being limited by the diameter of the Earth. Angular resolutions of a fraction of a milliarcsecond have now been achieved. Combining the angular sizes of these ultra-compact radio sources with their flux densities, the *brightness temperature* of the source region can be determined, the highest values found being about 10^{11} K. It is interesting to compare this figure with the thermodynamic temperature of electrons which have kinetic energies equal to their rest mass energies, $kT \simeq m_e c^2$, i.e. $T \simeq 5 \times 10^9$ K. Assuming the emission is the radiation of high energy electrons, this proves that they must have ultrarelativistic energies, i.e. $E \gg m_e c^2$, since particles cannot radiate at brightness temperatures greater than their thermodynamic temperatures. It should be noted that the angular sizes which are currently accessible for distant quasars by VLBI techniques correspond to physical sizes which are much greater than the Schwarzschild radius of a $10^8 M_\odot$ black hole. The smallest physical scales which have been studied by VLBI correspond to about 1 pc, whereas the Schwarzschild radius of a $10^8 M_\odot$ black hole is only 10^{-5} pc. It seems entirely plausible, however, that the relativistic matter is accelerated close to the black hole because the continuum emission observed in the optical, infrared and ultraviolet wavebands also has all the characteristics of the emission of relativistic electrons.

The most popular mechanism for the emission of radiation by relativistic electrons in active galactic nuclei is the synchrotron radiation process in which they move in spiral trajectories through the magnetic field in the source. The radiation is associated with the acceleration of the electrons towards the axis of the spiral by the Lorentz force (the $\mathbf{v} \times \mathbf{B}$ force) acting on the electrons. This process has the characteristic feature that, because of the very strong beaming of the radiation associated with the relativistic motion of the electron, the radiation spectrum emitted by each particle is intrinsically broad band, i.e. $\Delta v / v \sim 1$. As a result, a power-law distribution of electron energies is guaranteed to produce a power-law emission spectrum. The radiation is linearly polarised in an aligned magnetic field with a maximum percentage linear polarisation of about 70%. The intrinsic polarisation angle of the radiation is perpendicular to the magnetic field direction for an optically thin source so that information about the magnetic field geometry can be obtained from these studies. It can be seen that these properties of synchrotron radiation match satisfactorily the observed properties of the optical, infrared and radio continuum emission.

It is important to note, however, that in the extreme conditions of active nuclei, it may be that other emission processes are important. For example, inverse Compton scattering, in which low energy photons are scattered to very high energies by relativistic electrons, is an attractive mechanism for producing X-ray emission and in the high photon densities of active nuclei is likely to be an important process. It is not excluded that other radiation mechanisms involving the

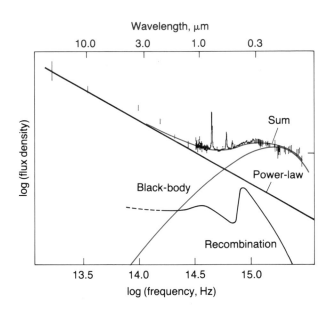

Figure 6.65. The optical–ultraviolet spectrum of the quasar 3C 273. The continuum has been decomposed into a 'power-law' component, a component associated with recombination radiation and a 'blue bump' component, which has been represented by a black-body curve. The prominent Balmer series in the optical waveband which led to the discovery of the high redshifts of quasars can be seen. (From Malkan, M. and Sargent, W. L. W. (1982). *Astrophys. J.* **254**, 33.)

emission of relativistic electrons could be important – for example, curvature radiation close to the black hole or possibly plasma emission mechanisms associated with the interaction of the relativistic particles with various forms of plasma wave. The basic point is that the broad-band continuum spectrum is primarily associated with the fact that ultrarelativistic electrons produce strongly beamed radiation. It is only necessary to perturb the relativistic particles to produce broad-band emission and the particles do not particularly care what the accelerating force is. The physical conditions close to active galactic nuclei are very different from those found in, say, the interstellar medium of our Galaxy or regions of intergalactic space in which the only reasonable emission mechanism is the synchrotron process. Thus, whilst it is certain that there are relativistic electrons in active nuclei and that synchrotron radiation is likely to be an important radiation mechanism, other radiation mechanisms may well be important.

One of the most important facts derived from radio VLBI observations is that the radio jets observed on a large scale in radio galaxies have their origin in their nuclei. The large-scale jets can be traced back more or less continuously to jet structures on a scale of about 10 pc or less, many of which show evidence that the radio components are moving away from the nucleus at velocities apparently in excess of the speed of light. We will return to this intriguing topic later, but for present purposes this observation gives us two further pieces of evidence. First, it is clear that *something* relativistic is escaping from the nuclear regions. Second, the observation of an axis of ejection of relativistic material strongly suggests that there is some axisymmetric structure such as an accretion disc or rotating black hole in the nucleus which provides a natural axis for the production of collimated beams of particles.

The black hole   The theory of accretion onto supermassive black holes is similar to the theory of accretion onto neutron stars and black holes described in Section 6.4, but there are important differences in that the black holes must be very massive and the radiation flux from the accretion disc is so intense that it strongly influences the structure of the disc. In fact, the problem of finding self-consistent solutions for accretion discs about supermassive black holes has not been solved. The problem lies in the fact that the pressure within the disc cannot be neglected and so thick disc solutions have to be found. A possible configuration which has many desirable features is the thick disc model shown in figure 6.63. The thick disc develops two oppositely directed funnels along the axis of the accretion disc and these may be important in collimating the beams of particles which are inferred to originate close to the nucleus. In pictures like this, it is not clear exactly what the origins of the various primary components are. The radiation from a simple thin disc may have a form similar to that shown in figure 6.36 for accretion discs about stellar mass black holes. Simple theory suggests that the temperature of the radiation scales as $M^{-1/4}$, and so if solar mass black holes produce X-ray emission, $10^8 M_\odot$ black holes may radiate in the ultraviolet region of the spectrum. One possibility is that the 'blue bump' seen in the spectrum of 3C 273 (figure 6.65) and other active nuclei may be the thermal emission of the accretion disc about a supermassive black hole.

Besides the emission of the disc, there may well be processes associated with the black hole itself which can give rise to the acceleration of charged particles. Although an isolated black hole cannot possess a magnetic field, a magnetic field can be associated with black holes if it is anchored to the surrounding ionised plasma. A massive rotating black hole thus bears some resemblance to magnetised rotating neutron stars which are certainly effective sources of relativistic electrons (figure 6.31). Lovelace, and Blandford and Znajec, have shown how particles can be accelerated in the vicinity of rotating black holes, and there is every possibility that the continuum radiation may originate in the radiation of these particles by synchrotron, inverse Compton or curvature radiation.

One of the ultimate aims of these studies is to learn more about the behaviour of matter and radiation in strong gravitational fields. So far this programme has not met with any conspicuous success but there remain real possibilities that some of the observable continuum components originate close

to the black hole. It would be of the greatest importance if, for example, emission lines associated with matter spiralling into the black hole could be observed. Unfortunately, it is very likely that any emission lines will be smeared out by the gravitational redshifts of the emitting particles in the strong gravitational field.

The emission line regions

One of the major areas of study in active galactic nuclei is the nature of their emission line spectra which originate close to the central regions. There is evidence for a wide range of physical conditions for different classes of active nuclei, the most important parameters being the intensity and spectrum of the ionising radiation and the particle densities within the gas clouds. The full apparatus of atomic and ionic spectroscopy can be employed to determine the physical conditions within the gas clouds. From these results, one may then proceed to develop models for the distribution of the gas clouds about the black hole which for these purposes may be simply taken to be a point source of ultraviolet ionising radiation.

The general picture which emerges is as follows (see figure 6.63). The emission line regions which originate closest to the black hole are responsible for the broad-line emission observed in the most active nuclei, the Seyfert I galaxies and the quasars. The most direct evidence for this is the fact that the emission profiles are found to be variable on a time scale of months and hence they must be very close to the nuclei. They must also be rather dense regions because there are no forbidden lines accompanying the strong broad emission lines. Forbidden lines, such as those of doubly ionised oxygen [OIII], are suppressed above a certain electron density at which collisional de-excitation depopulates the upper levels of the transitions more rapidly than the radiative transitions. For the broad-line regions, the electron densities are greater than about $10^{14} \, \mathrm{m}^{-3}$. Thus, in the nuclear regions within about 0.1 pc or less of the black hole, there are dense clouds with a large velocity dispersion. It has not been possible to distinguish what these motions are, whether they represent turbulent, rotational, expansion or contraction velocities, although careful study of the time evolution of the line profiles during an outburst from the nucleus can, in principle, distinguish between these.

Further out from the nucleus are the clouds responsible for the narrow emission-line spectrum. These emit strong permitted and forbidden lines with similar line widths and also exhibit a wide range of ionisation states. This is consistent with a picture in which the regions have particle densities $\sim 10^9$–$10^{11} \, \mathrm{m}^{-3}$ and are photoionised by the ultraviolet radiation from the nucleus. The dimensions of the regions can be roughly estimated from the strengths of the lines, the inferred particle densities and the fraction of the total volume occupied by clouds.

In many Seyfert nuclei, there is very often evidence for both types of gas cloud being present, the relative strengths of the broad and narrow line components varying from type I Seyfert galaxies which are pure broad line systems to pure type II Seyferts in which the narrow lines are dominant. Some of the radio galaxies which are very powerful radio sources have similar properties but there are important differences. Some of the radio galaxies have broad line spectra similar to the type I Seyfert galaxies. Others have spectra more akin to the type II Seyferts, but among the most distant and luminous radio galaxies there are several which possess very strong, narrow lines which do not originate in the nucleus. The emission line regions can extend to 50–100 kpc, greater than the size of the galaxy itself. It is these strong emission line radio galaxies which are the most distant galaxies yet observed, the largest redshifts obtained so far being about 3.4 (figure 6.66).

Finally, there are even weaker nuclei in which only a low ionisation spectrum is observed. In these 'low ionisation, narrow emission line region' galaxies (LINERs), the excitation mechanism for the clouds could be either collisions or photo-excitation by ultraviolet radiation.

It is important to note that the densities and temperatures which can be studied are largely limited by the lines available for study in the optical and ultraviolet spectrum. There may well be a wide range of densities and temperatures present in the gas clouds surrounding active nuclei but the diagnostic tools are only sensitive within well defined limits. There is no generally accepted picture of the role which these gas clouds play in the evolution of active nuclei. Are they associated with the material which may eventually form an accretion disc?

Figure 6.66. The composite optical spectra of the radio galaxies 3C 256 and 239 taken by Spinrad and his colleagues showing the ultraviolet spectrum redshifted into the visible spectral region. These are among the most distant galaxies for which redshifts have been measured, thanks to the strong narrow emission lines in their spectra. (From Spinrad, H. (1986). *Publ. Astr. Soc. Pacific* **98**, 269.)

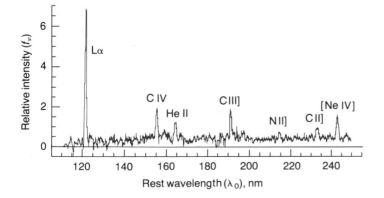

Radio emission from active galaxies and quasars

All galaxies are sources of radio emission for the same reason that our Galaxy emits radio waves – high energy electrons are accelerated in supernova explosions and these are dispersed throughout the interstellar medium where they radiate radio waves by the synchrotron process. However, these are very weak radio emitters indeed compared with what we mean by the term *radio galaxy* or *radio quasar* in which the radio luminosity can exceed that of our own Galaxy by factors of 10^8 or more. The production of the high energy particles is totally different in these powerful radio sources.

What is observed is that the nuclei of the radio galaxies and quasars are strong radio sources and jets or beams of relativistic material are ejected from them to form large-scale radio jets, double radio sources and so on. The example of the brightest extragalactic radio source in the northern sky, Cygnus A, observed by the Very Large Array is shown in figure 6.67. The important properties of the sources are, first, their very large radio luminosities, implying the production of enormous fluxes of relativistic electrons and magnetic fields and, second, the fact that the energy generated in the nucleus of a galaxy can lead to the expulsion of relativistic gas in collimated jets far beyond the confines of the parent galaxy and into the intergalactic medium. The largest radio structures known have physical diameters greater than 1 Mpc, the record being about 6 Mpc.

The facts that the radiation has a power-law spectrum, that the radiation is linearly polarised and that these huge radio structures are observed in the very diffuse conditions of interstellar and intergalactic space, make the identification of the radiation mechanism with synchrotron radiation wholly convincing. The theory of this process can be used to set lower limits to the amount of energy which must be present in the jets and radio lobes. The energy requirements are greatest for the largest and most diffuse sources and are $\geqslant 10^{54}$ J, corresponding to the conversion of at least $5 \times 10^5 M_\odot$ of matter into high energy particle and magnetic field energy. Thus, there must exist efficient mechanisms for converting rest mass energy into these forms and then injecting them into the extended radio lobes over a period of 10^7–10^8 years, the typical lifetime of extended radio sources.

The discovery of radio jets within the diffuse radio structure, for example in the VLA radio map of Cygnus A (figure 6.67), is a key observation for understanding how these radio structures come about. A schematic model for extragalactic radio sources is depicted in figure 6.68. A beam or jet of particles is emitted from the nucleus of the radio galaxy which 'burns' its way out through the interstellar and intergalactic gas much like a laser beam. At the interface between the beam and the static intergalactic gas a shocked region is set up, similar in shape to the magnetospheric cavity about the Earth's magnetic field. Particles are accelerated in the head of the beam and are left

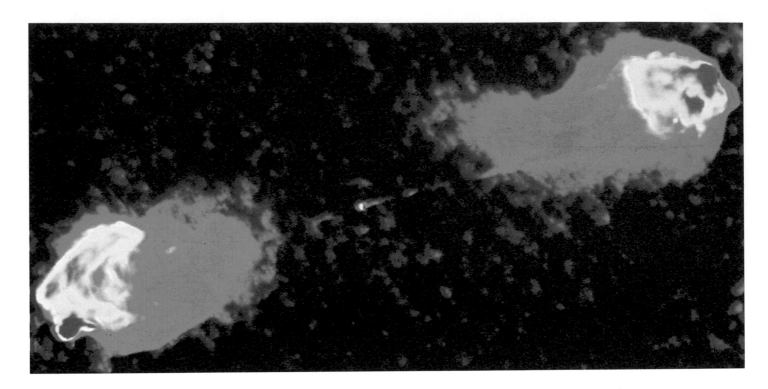

Figure 6.67. An example of the radio structure of the extended radio source Cygnus A as observed by the US Very Large Array (VLA). The radio source is associated with a giant elliptical galaxy which is the brightest member of a rich cluster. There is a compact radio source coincident with the nucleus of the galaxy, and a very narrow collimated jet can be seen extending from the nucleus to the extended radio lobe to the right of the image. VLBI observations show that even on the scale of 1 milliarcsec, the nucleus is elongated in the form of a jet in the direction of the long narrow jet seen in the figure. There are hot-spots within the lobes where particles are accelerated probably because of the interaction of the jet with the intergalactic medium. There is a great deal of structure in the lobes, probably associated with the escape of the particles from the hot-spots and the interaction of this relativistic gas with the surrounding medium. (From Perley, R. A., Dreher, J. W. and Cowan, J. J. (1984). *Astrophys. J.* **285**, L35.)

behind by the advancing jet. The relativistic gas left behind expands to form a large cavity of mixed relativistic plasma and magnetic fields which produces the large radio lobes observed in diffuse sources. When the beam switches off, the large diffuse lobes are left to dissipate their energy, either by synchrotron radiation or inverse Compton scattering of the Microwave Background Radiation or, most likely, simply by adiabatic expansion.

The schematic model shown in figure 6.68 has been the subject of detailed gas-dynamical studies, the aim being to find out how much can be understood when realistic parameters are adopted for the temperatures, densities and magnetic fields in the ambient plasma and for the relativistic gas which is ejected from the nuclear regions. The results of these calculations are of practical relevance to laboratory plasma physics in that the beams consist of relativistic material which has succeeded in remaining confined within a narrow jet despite the fact that relativistic plasmas are known to be notoriously unstable from laboratory experiments in plasma confinement. The most recent calculations have used three-dimensional computer codes to study the stability of the jets, and these have had some encouraging success in accounting for the detailed structures seen in the radio source components.

It is of great interest that small-scale counterparts of these sources may have been observed in the Galactic X-ray source Sco X-1 and the strange system SS 433. Both sources are associated with compact objects within our Galaxy and both exhibit a double radio structure similar in many ways to the extended extragalactic radio sources (figure 6.69*a, b*). The case of SS 433 is of particular interest because the radio jets also emit optical line radiation and so the velocities of ejection of the beams can be found directly from the Doppler shifts of the lines.

The basic problem is therefore to understand the nature of the jets which originate close to the nucleus itself. Figure 6.70 shows VLBI images of the central regions of the radio source 3C 273 taken at various epochs between 1977 and 1980. The radio structure consists of a jet of discrete radio blobs which move away from the nucleus in the direction of the much larger radio and optical component which lies at a distance of about 50.000 pc (see figure 6.3). This is convincing evidence that the large-scale radio lobes are powered by beams of particles ejected from the nucleus. The figure also illustrates one of the very surprising features of these compact radio sources. The source appears to have expanded in size by about 35 l.y. in only four years. This is an example of the phenomenon known as *superluminal* radio sources. The beam appears to be expanding away from the nucleus at a velocity about eight times the speed of light, which at first sight appears to contradict causality. This phenomenon has been observed in a number of the brightest compact radio sources and so it is not an uncommon phenomenon.

Naturally, these sources have excited the greatest interest among theoreticians who were not long in providing a large number of clever models to account for the observation of components which move at velocities apparently in excess of the velocity of light. In general terms, one may say that the observations demonstrate that *something* is moving at relativistic velocities from the nucleus, and this is encouraging from the point of view that ejection of relativistic material from active nuclei is a prerequisite of satisfactory models of extended radio sources. One model which has attracted particular attention is the *relativistic ballistic model* in which a jet of radio emitting material is ejected from the nucleus at an angle close to the line of sight. If the jet moves at a velocity close to that of light, the jet almost catches up with its own radiation, and it is observed to move away from the nucleus at a maximum velocity of γv, where $\gamma = (1 - v^2/c^2)^{-1/2}$ is the Lorentz factor and v is the velocity of the jet. The reason for this behaviour is illustrated in figure 6.71. Thus, if v is close enough to c, the jet can appear to move arbitrarily rapidly without any violation of causality. An attraction of this picture is that, in addition to producing superluminal expansions, the radio luminosity of the advancing component is strongly boosted by the large blueshift of the jet, and this can partly explain why many of the components appear to be one-sided. However, it is not at all clear that this model can explain everything. One problem is that there appear to be many such sources among the brightest radio sources and little room is left for the sources which are not pointing more or less towards us. The parent population from which we see only the small fraction of sources which are pointing more or less towards us is not at all obvious. Although this model has the advantage of simplicity and the attraction that it is directly related to the beams which must originate within the active nuclei of the radio sources, many other ingenious models are available. Among these are the *screen* models in which a beam of particles acts like the beam of a

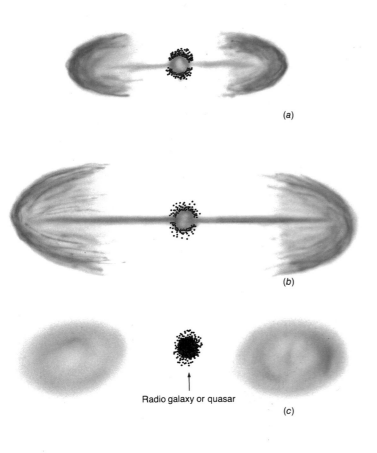

Figure 6.68. A schematic model illustrating the evolution of a double radio source. The jets observed in VLBI observations are assumed to be associated with the continuous supply of energy in the form of a relativistic beam of particles. (*a*) The beam burns its way through the intergalactic gas, and at the interface between the jet and the gas there is a contact discontinuity and a bow shock. Particles are accelerated in these regions which gives rise to the formation of 'hot-spots'. (*b*) The relativistic plasma expands producing an extensive region filled with relativistic particles and magnetic fields which form the extensive lobes seen in double radio sources. This configuration persists as long as the jet continues to supply energy to the outer lobes. (*c*) When the energy supply ceases, the diffuse lobes expand and decay in radio luminosity, principally because of adiabatic losses of the energy of the relativistic particles as the lobe expands.

These velocities are found to be about $0.21c$. In addition, it has been shown convincingly from the velocity curves of the system and directly from radio maps made at different epochs that the beam precesses about the mean axis of the jets (figure 6.69*b*). The study of these sources promises to be of the greatest importance in understanding the much more powerful double radio jets in active galaxies and quasars.

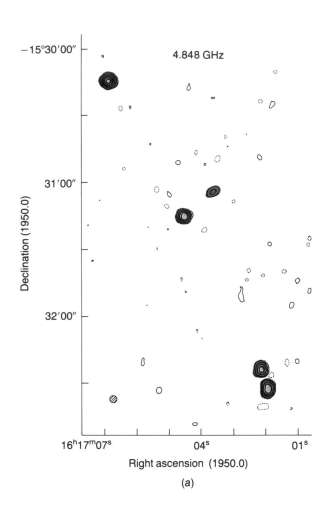

(a)

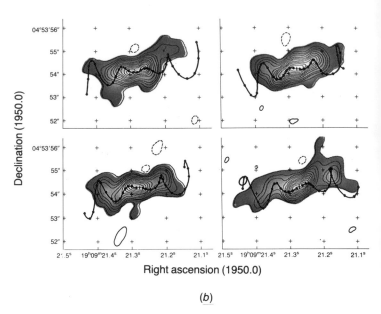

(b)

Figure 6.69. Radio maps of the Galactic radio sources (*a*) Sco X-1 and (*b*) SS 433. These sources may well be small-scale Galactic counterparts of the intense double radio sources found in extragalactic radio sources. (*a*) In the case of Sco X-1, the source is a double radio source with 'hot-spots' in the outer lobes and a compact source coincident with the X-ray source. (From Geldzehlar, B. J. and Fomalont, E. B. (1986). *Astrophys. J.* **311**, 144.) (*b*) In the case of SS 433, collimated outflow is observed, the direction of the jet precessing about the major axis of ejection as can be seen from the four radio maps which span a period of almost 200 days. The contour maps show the structure of the radio source at various phases in the cycle. The solid lines show a model of a precessing jet to account for the observed variations in structure. This phenomenon may also occur in extended extragalactic sources and account for the observation of multiple hot-spots in the extended radio lobes. (After Hjellming, R. M. and Johnston, K. J. (1981). *Astrophys. J.* **246**, L144.)

lighthouse and illuminates a distant screen so that the spot on the screen can move arbitrarily fast, just like the patch of light on clouds illuminated by a searchlight. In general terms, the proper interpretation of this phenomenon will give information about the kinematics of beams of particles ejected from active nuclei.

Although these observations give evidence for relativistic phenomena about 1 to 10 pc from the nucleus, this is still on a very much larger scale than the scale of the thick accretion discs and rotating black holes which may be the ultimate source of energy. Indirect evidence for relativistic streaming on the scale of light-days comes from analyses of the polarisation properties of the optical and infrared continuum of BL-Lac

objects. It is difficult to understand why the radiation is polarised at all because depolarisation of the radiation by the electrons responsible for the continuum emission itself by internal Faraday rotation is expected to be important in the extreme conditions of active galactic nuclei. The observation of polarisation and the rapid swings in the polarisation vectors of the continuum emission can be understood if the radiation is strongly beamed towards the observer by the relativistic bulk motion of the emitting electrons.

It is the combination of these separate pieces of evidence which strongly suggest that the beams of relativistic material originate very close to the accretion disc and the black hole in the active nucleus. It is inferred that this type of activity must

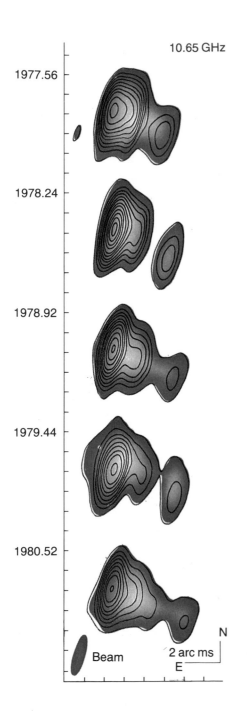

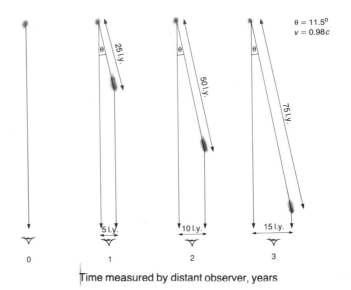

Time measured by distant observer, years

Figure 6.71. Illustrating how a jet ejected at a relativistic velocity at an angle close to the line of sight can result in motion which appears to be superluminal to a distant observer. If the jet is ejected at an angle close to the line of sight at a velocity close to that of light, the jet almost catches up with its own radiation. The illustration shows the geometry and parameters necessary for observing an apparent expansion velocity of five times the velocity of light.

continue over rather long periods because the extended radio components decay in radio luminosity by adiabatic expansion unless they are kept replenished by relativistic gas from the nucleus. The continued ejection of the relativistic gas in the same direction is guaranteed by the fact that the rotating black hole will maintain its axis in the same direction over very long periods. There has also been speculation that possibly some of the large-scale features associated with extended radio sources may be attributed to precession of the accretion disc, similar to the phenomena observed directly in SS 433.

We still have not answered the question of the physical origin of the relativistic jets in active galactic nuclei. The schematic model shown in figure 6.63 for a thick accretion disc with funnels along the axis of rotation may have something to do with the collimation mechanism, although it has yet to be established that these are stable structures. Two other possibilities are illustrated in figure 6.72. In one case, the production of the jet of particles is attributed to electromagnetic processes occurring close to the rotating black hole, the rotation axis providing a natural axis along which the beams of particles are ejected. In the second model, the jet could be created by gas dynamical processes occurring in the dense regions surrounding the nucleus. The continuous generation of energy leads to the formation of a region of very high energy density of high energy particles. This relativistic gas will burst through the

Figure 6.70. VLBI images of the nucleus of the radio quasar 3C 273 for the period 1977 to 1980. The radio component is observed to move a distance of about 35 l.y. in a period of four years implying a superluminal expansion velocity of about 8*c*. (From Pearson, T. J., Unwin, S. C., Cohen, M. H., Linfield, R. P., Readhead, A. C. S., Seielstad, G. A., Simon, R. S. and Walker, R. C. (1982). *Extragalactic Radio Sources*, D. S. Heeschen and C. M. Wade, eds., p. 356. D. Reidel and Company.)

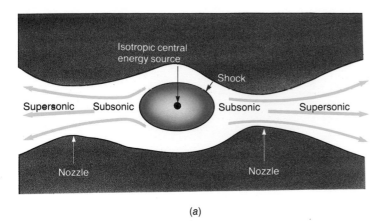

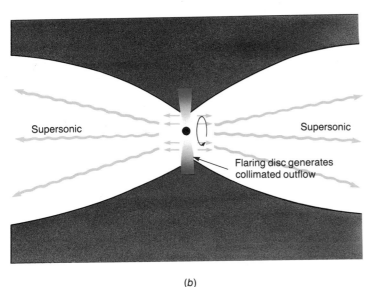

Figure 6.72. Two possible mechanisms for producing collimated jets of particles from the central regions of quasars and radio galaxies. (*a*) A gas-dynamical model in which a strong source of energy is located in a gaseous disc. The gas escapes from the region along the minor axis of the disc. In the process it may set up the type of 'nozzle' structure depicted in the figure. There is a shock surrounding the energy source. The flow of gas is subsonic until it passes through the nozzles beyond which the flow becomes supersonic as in a de Laval nozzle. (*b*) A 'flaring' disc model in which there is a powerful stream of relativistic gas ejected from the accretion disc about the black hole. Instabilities in the disc and the presence of a strong magnetic field lead to 'flares' on the surfaces of the disc which eject relativistic matter in the direction perpendicular to the disc. The resultant jet of relativistic gas escapes at the relativistic sound speed $c/\sqrt{3}$. (From Rees, M. J. (1976). *The Physics of Non-thermal Radio Sources*, G. Setti, ed., p. 107. D. Reidel Publishing Comapny.)

surrounding cloud in the direction of the poles of the cloud and form a collimated jet of relativistic gas. This model due to Blandford and Rees provides a possible link between the properties of the radio jets and the regions of high density gas inferred to be present in the broad line regions. It remains to be established how relevant these ideas are to the understanding of the origin of jets in active galactic nuclei.

The galaxies which contain active galactic nuclei

It is one of the remarkable aspects of active galactic nuclei that there are several well known correlations between their properties and the parent galaxies within which they are located, but there is no adequate theory to explain why this should occur. Perhaps the most distinctive of these is the difference between the radio galaxies and the Seyfert galaxies. The *Seyfert* galaxies appear to be almost exclusively spiral galaxies. In terms of optical luminosity, the most luminous of their active nuclei can be as luminous as the quasars, although it becomes very difficult to distinguish the underlying galaxy in the most luminous quasars. In general, they are very weak radio emitters. In contrast, the *radio* galaxies, which are strong radio emitters, appear to be exclusively associated with giant elliptical galaxies. In some of the radio galaxies, the nucleus has quasar-like properties, the N-galaxies, and the most luminous of these are so luminous that it is difficult to distinguish the underlying galaxy.

A second significant correlation is the fact that the quasars seem to be associated with the most luminous galaxies, whether or not they are radio sources. In a recent survey of the galaxies underlying low redshift quasars, Smith and his colleagues have shown that they are among the more luminous galaxies, all of them lying above the break in the optical luminosity function of galaxies (figure 6.53), and the most luminous being as bright as the brightest galaxies in clusters. Furthermore, for both the radio quiet and radio loud quasars, their luminosities appear to be independent of the luminosity of the host galaxy. They find that the radio quiet quasars are almost exclusively associated with disc galaxies and the radio quasars with elliptical galaxies which appear on average to be about a factor of two more luminous than the disc galaxies. The results are similar in many ways to the observation described above that only the most massive elliptical galaxies are the hosts of powerful double radio sources.

These must be important clues to the problem of understanding why different types of activity originate from the presence of massive black holes in galactic nuclei. One view is that the differences can be ascribed to the different ways in which the nuclear regions are supplied with fuel in the form of gas to power the accretion disc. Another view is that in some cases the energy source is the accretion process and in others the direct extraction of energy from the rotation of the black hole. These problems have yet to be resolved.

6.8   Astrophysical cosmology

The term *cosmology* can mean many different things to different readers. To the scholar of the humanities, it can mean the world picture which different cultures have developed to explain the origin of mankind and the human condition and is embodied in widely accepted mythologies. Even within the physical sciences, cosmology can be interpreted in different ways. Some authors interpret the term to mean studies referring to the global geometry and dynamical parameters of the Universe; others take it to mean the study of the origin and evolution of all the material contents of the Universe. I will adopt a very catholic meaning of the term as used in the physical sciences. I purposely entitled this section *astrophysical cosmology* since I wish to lay emphasis upon physical aspects of the origin and evolution of the Universe and its contents.

In many ways, a very good story can be told nowadays but some of the fundamental problems encountered in the generally accepted picture are just as difficult as they have always been. The big advance has been that, thanks to many observational discoveries and new observing capabilities, the questions can be asked more precisely than before and we have much better observational and theoretical tools for tackling them.

Basic observations

The standard model for the Universe begins by setting up the large-scale framework within which the questions of astrophysical cosmology can be addressed. There are two basic observations which are important in deriving this framework.

The isotropy and homogeneity of the Universe   The first concerns the *isotropy* of the Universe. Figure 6.5 shows the large-scale structure of the Universe as derived from observations of galaxies. There is structure in this distribution on the scale of clusters and superclusters of galaxies and greater, but it seems that on the very largest scales the distribution of galaxies is more or less uniform. Formally this can be tested by asking if there is evidence that the average surface density of galaxies N_g varies over the sky or, equivalently, if the angular two-point correlation function $w(\theta)$ for galaxies tends to zero on large angular scales. The latter function is defined in terms of the number of galaxies in solid angle $d\Omega$ at angular distance θ about any given galaxy

$$N(\theta)d\Omega = N_g[1 + w(\theta)]d\Omega. \qquad (6.10)$$

The function $w(\theta)$ describes the average enhancement in the surface density of galaxies at angular distance θ from any given galaxy and is a simple measure of the nonuniformity of the distribution of galaxies. N_g is the average surface density of galaxies once allowance is made for the clustering described by $w(\theta)$. The function $w(\theta)$ is found to be well approximated by

$$w(\theta) \propto \theta^{-0.77}. \qquad (6.11)$$

Although it is not an easy task to work out the isotropy of N_g, when averaged over large enough areas of sky, the galaxies appear to be uniformly distributed within the statistical uncertainties. Notice that $w(\theta)$ is a surface distribution of galaxies and can be related to the more physically meaningful function, the spatial correlation function, $\xi(r)$, which is defined by

$$n(r)dV = n_g[1 + \xi(r)]dV. \qquad (6.12)$$

For power-law correlation functions, $\xi(r) \propto w(\theta)r^{-1}$ so that for physical scales from about 1 to 10 Mpc, the overall distribution of galaxies can be described by

$$\xi(r) = (r/4h^{-1} \, \text{Mpc})^{-1.77}, \qquad (6.13)$$

where $h = (H_0/100 \, \text{km s}^{-1} \, \text{Mpc}^{-1})$. Thus, the average space density of galaxies is twice the mean density at a distance of $4h^{-1}$ Mpc from any galaxy. The exact range of validity of this power-law, especially at large physical scales, is a matter of debate.

The use of $w(\theta)$ is a rather crude measure of the distribution of galaxies and three-dimensional studies contain much more information about $\xi(r)$ and its variation at different locations in the Universe. Among the first results of the three-dimensional studies has been the discovery of large 'holes' or 'voids' in the distribution of galaxies, regions of the Universe which seem to be devoid of galaxies (figure 6.6). Some of these may extend to 50 or 100 Mpc in size. It may well be that the voids are simply the negative counterpart of superclusters. Indeed, there is strong evidence for significant clustering of very large-scale systems of galaxies on the scale of $\simeq 100$ Mpc. These are plainly major inhomogeneities in the distribution of visible matter.

The use of correlation functions formalises the observation that on the largest scales the distribution of galaxies tends to uniformity on the surface of the celestial sphere. The same result is found from studies of the distribution of the brightest extragalactic radio sources (figure 6.73). Away from the plane of our Galaxy, these are almost exclusively associated with distant galaxies and quasars, thus sampling a smaller but much more distant sample of the population of galaxies than simply by mapping all of them. The distribution shown in figure 6.73 shows no statistical departure from uniformity on any scale, the precision with which this statement can be made being simply limited by the number of objects which can be counted.

The most important information, however, comes from studies of the isotropy of the Microwave Background Radiation. This radiation brings to observers on Earth information about the large-scale distribution of matter and radiation in the Universe at the epoch when it was last scattered and, in most models, this occurs when the Universe was compressed by a factor of about 1000 relative to its present size. On the largest scale, $\theta = 360°$, a dipole component of the Microwave Background Radiation has been detected at the level $\Delta I/I \simeq 10^{-3}$ (figure 6.74). This is naturally explained in terms of the motion of the Earth at a velocity of 350 km s^{-1} relative to

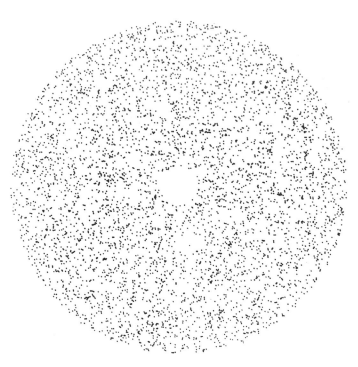

Figure 6.73. The distribution of radio sources in the fourth Cambridge catalogue in the northern celestial hemisphere. This part of the catalogue contains over 3000 sources of small angular diameter. In this equal area projection, the north celestial pole is in the centre of the diagram and the celestial equator around the solid circle. The area about the north celestial pole was not surveyed. The distribution does not display any significant departure from a random distribution. (Courtesy of Dr M. Seldner.)

an isotropic frame of reference. Notice that this is not the same as the motion of our Galaxy with respect to a frame of reference in which the Microwave Background Radiation is 100% isotropic. The Earth's motion is the vector sum of the Earth's velocity about the Sun, the Sun's motion in the Galaxy ($\simeq 220\,\mathrm{km\,s^{-1}}$), the Galaxy's motion in the Local Group of galaxies ($\sim 200\,\mathrm{km\,s^{-1}}$), the Local Group's motion with respect to the Local Supercluster of galaxies ($\sim 250\,\mathrm{km\,s^{-1}}$) and the Local Supercluster's motion with respect to the frame of reference in which the Microwave Background Radiation is 100% isotropic.

Recently measurements have been made of the kinematics of elliptical and spiral galaxies relative to galaxies at a distance of about 100 Mpc. A streaming velocity of about $600\,\mathrm{km\,s^{-1}}$ has been found for the elliptical galaxies relative to a frame of reference in which the Microwave Background Radiation would be isotropic but *not* in the direction of the maximum in the intensity of the Microwave Background Radiation. These

are recent results and the situation is as yet unclear. If such large-scale streaming motions of galaxies are real, they have profound implications for the large-scale distribution of matter in the Universe on very large scales. There is the interesting possibility that some of the voids seen in the distribution of galaxies are associated with these types of large-scale streaming motions. It is interesting to note that, if galaxies are selected at random from the general field, there is a scatter in their velocities relative to the mean Hubble flow amounting to a root mean square velocity of about 250–$300\,\mathrm{km\,s^{-1}}$. This is not so much smaller than the velocities derived for the mean streaming velocity of these samples of elliptical galaxies.

Besides this dipole term, no other significant anisotropies in the distribution of the Microwave Background Radiation have yet been detected. Fluctuations have been searched for on all scales from about 1 arcmin to $180°$ and only upper limits have been found so far. On most scales the present limits to the fluctuations in the Microwave Background Radiation are below the level $\Delta I/I \leqslant 10^{-4}$. As will become apparent later, these are key observations from the point of view of galaxy formation.

Thus, we can have considerable confidence that, looked at on a large enough scale, the Universe presents the same appearance in all directions with very high precision, i.e. it is *isotropic*. We should also check that the Universe is *homogeneous* but this requires us to have some knowledge of the distribution of matter with distance in the Universe. Perhaps the best evidence so far comes from the scaling law of the two-point correlation function (6.11) with distance. The correlation function can be compared at different apparent magnitude levels and the amplitude and scale of the functions compared. This comparison has shown that the two-point correlation functions scale as would be expected if the Universe were as inhomogeneous as described by equation (6.12), but that the same degree of inhomogeneity is found at different distances. So far, this comparison has only been possible out to distances of about 300 Mpc.

Hubble's law and the expansion of the Universe   In 1929, Hubble made his great discovery that the Universe of galaxies is not static but is in a state of uniform expansion. The basis for this result was the observation that all the galaxies are receding from our own Galaxy and that the further away a galaxy is from us, the greater its velocity of recession v, i.e.

$$v = H_0 r, \qquad (6.14)$$

where r is the distance of the galaxy and H_0 is a constant, appropriately known as *Hubble's constant*. A modern version of this relation is shown in figure 6.75, the galaxies plotted being the brightest galaxies in clusters which are found empirically to have more or less the same intrinsic luminosities. The velocity is measured from the shift of the spectral lines in the galaxy's spectrum to longer wavelengths, this shift being interpreted as

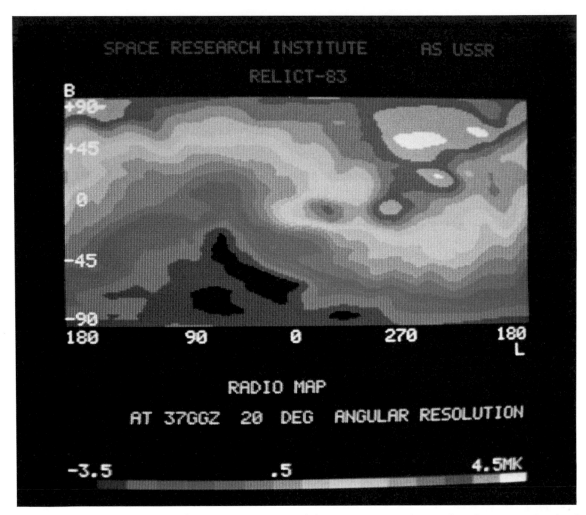

Figure 6.74. Illustrating the dipole component in the Microwave Background Radiation. This image of the whole sky derived from the RELIKT-83 experiment on board the PROGNOZ-9 spacecraft is derived from the same data shown in figure 6.2(*b*) but smoothed to a beam-width of 20°. The image is shown in Galactic coordinates with the plane of the Galaxy running along the centre of the diagram *B* = 0. The dipole component of the intensity distribution of the Microwave Background Radiation can be clearly seen. Once allowance is made for the presence of thermal sources in the Galactic plane, there is no evidence for any anisotropy in the distribution of the radiation other than the strong dipole term which corresponds to $\Delta T/T \simeq 10^{-3}$. (Courtesy of Dr I. A. Strugov and the Space Research Institute, Moscow.)

a Doppler shift. This shift is known as the *redshift* of a galaxy *z* and is defined by

$$1 + z = \lambda_0/\lambda_e, \qquad (6.15)$$

where λ_e is the emitted wavelength of a line and λ_0 is the wavelength at which it is observed. For small velocities, $v/c \ll 1$, $v = cz$. Thus, velocities of recession can be measured very accurately. The great problem in cosmology is to measure accurate distances, *r*. In figure 6.75, relative distances are found from the observed intensities, or apparent magnitudes, of the galaxies.

The expansion of the distribution of galaxies as a whole follows directly from Hubble's law and the isotropy of the distribution of galaxies. If the distribution of galaxies is to remain isotropic and the velocity–distance relation observed, it is necessary for the distribution of galaxies as a whole to expand uniformly, as may be confirmed by a simple thought experiment. This result is of profound cosmological importance. The Universe must have been very much more compact in the distant but finite past and this line of reasoning is part of the motivation for the Hot Big Bang model of the Universe.

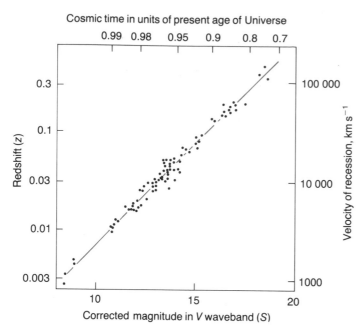

Figure 6.75. A modern version of the Hubble diagram for the brightest galaxies in clusters (after Sandage, A. R. (1968). *Observatory* **88**, 99). In this logarithmic plot, the observed flux density S from the brightest galaxies in clusters is plotted against redshift z. The straight line is what would be expected if $S \propto z^{-2}$. This indicates that the brightest galaxies in clusters have remarkably standard properties and that the distance of a galaxy is proportional to its redshift, which for small redshifts implies that velocity is proportional to distance.

For many problems, for example the estimation of the age of the Universe, it is important to know the value of Hubble's constant, H_0. This is a very difficult observation because of the problem of measuring extragalactic distances which are independent of redshift. It would take us too far afield to go into the huge amount of effort devoted to this problem and to explain the origins of the discrepancies between different estimates. In the end, the problem can be attributed to the lack of direct physical means of measuring distances. The best methods which have been employed to date involve assuming that certain properties of stars and galaxies are the same in nearby and distant systems so that ratios of distances can be measured and then absolute values estimated once good distance measurements are made for nearby objects. There are two schools of thought, one which finds H_0 to be somewhere in the range 45–60 km s^{-1} Mpc^{-1} and another which finds larger values, $H_0 \sim 75$–100 km s^{-1} Mpc^{-1}. Different approaches are taken by the authors favouring the different values, and there is certainly no agreement yet. My own view is that these discrepancies simply indicate the great problems of making a good measurement of H_0. Although the formal accuracy of the

estimates is good, the discrepancies are almost certainly dominated by systematic rather than random errors. Rowan-Robinson has given an excellent review of the origin of the discrepancies.

Hubble's constant is all-pervasive in astrophysical cosmology. The accurate determination of physical properties such as absolute sizes and luminosities of extragalactic objects depends upon a knowledge of H_0. Perhaps the most important aspect is the estimation of the age of the Universe. If the Universe expanded at its present rate since the beginning of the expansion, its age would just be $T_0 = 1/H_0$. This result is somewhat modified in the standard cosmological models because the expansion rate must have been greater in the past because of the decelerating effect of gravity; as a result, the inferred age of the Universe is less than $1/H_0$. Thus $T_0 = 1/H_0$ is a good upper limit to the age of the Universe in the standard model and has values 20×10^9 years if $H_0 = 50$ km s^{-1} Mpc^{-1} and 10×10^9 years if $H_0 = 100$ km s^{-1} Mpc^{-1}.

These values can be confronted with estimates of the ages of the oldest stars known in our own Galaxy. Most models of the evolution of stars in globular clusters suggest that their ages are between about 14 to 20×10^9 years, which would appear to pose problems for the higher values of H_0 (see Section 6.3). A literal interpretation of these results favours values of H_0 closer to 50 km s^{-1} Mpc^{-1}, but the argument is by no means conclusive and does depend upon an understanding of the evolution of stars in globular clusters.

The standard dust-filled world models of general relativity

The construction of the standard models of our Universe begins with the observed isotropy and expansion of the Universe. The assumption is then made that these properties are what would be observed by any typical observer located anywhere in the Universe at the present time. This assumption is known as the *cosmological principle* and gains considerable support from the isotropy of the expansion of the Universe and what is known about its homogeneity. In fact, these assumptions are sufficient to derive a general expression for possible forms of the metric of space-time for isotropic expanding universes, quite independent of the physics which causes the expansion. It is worthwhile writing down this metric, which is known as the *Robertson–Walker* metric, since it enables us to introduce a number of important cosmological parameters:

$$ds^2 = dt^2 - \frac{R^2(t)}{c^2}\left(dr^2 + R_c^2 \sin^2\frac{r}{R_c} d\theta^2\right), \qquad (6.16)$$

where ds is the interval or proper time between events in the sense of special relativity; t is cosmic time, i.e. proper time measured by an observer to whom the Universe appears isotropic; and r is a *radial comoving distance coordinate* which locates objects relative to other objects in a uniformly expanding Universe but does not change as the Universe expands. R_c is the radius of curvature of the global geometry of the Universe; the assumption of isotropy limits the possible large-scale geometries of the Universe to isotropic spaces in which R_c is real if the space is spherical and closed, imaginary if space is open and hyperbolic and infinite if space is flat. $R(t)$ it known as the *scale factor* of the Universe and defines the expansion of the Universe as a function of time, i.e. it defines the relative separation between points which move in such a way that the Universe appears to be isotropic to each of them. All the physics which determine the large-scale dynamics of the Universe is contained within the scale factor $R(t)$.

The metric determines the interval between events in the expanding Universe and hence enables relations between observables and the intrinsic properties of objects at cosmological distances to be determined. By cosmological distance, we mean distances such that the large-scale geometry of Universe cannot be neglected in our calculations which in practice means redshifts $z > 0.3$. One very useful result which comes directly from the Robertson–Walker metric is the relation between the scale factor and redshift

$$R(t) = \frac{1}{1+z} = \frac{\lambda_e}{\lambda_0}. \tag{6.17}$$

Thus, the redshift of a galaxy or quasar tells us directly the value of the scale factor of the Universe when the light was emitted relative to its present size. This means that, in principle, if it were possible to measure accurately the times when objects emitted their radiation, we could determine $R(t)$ directly. Unfortunately, present understanding of the astrophysics of galaxies and quasars is too primitive to make this procedure feasible, but it is not excluded that we may eventually be able to accomplish this if major advances are made in extragalactic astrophysics.

The next stage in the model building process is to prescribe the physics of the expanding Universe. The only known long-range force which can influence the dynamics of the Universe is gravity and hence the standard model adopts the best relativistic theory of gravity we possess, Einstein's general theory of relativity. As the chapter by Will demonstrates, general relativity has successfully passed all the tests which modern ingenuity and technology can devise. The standard model requires us to specify the components of the Universe which make up its inertial mass density. At the present time, the Universe is *matter-dominated* in the sense that the inertial mass density of ordinary matter exceeds that of all forms of radiation by at least a factor of 1000, the greatest contributor to the latter being the Microwave Background Radiation which has inertial mass density $\rho = \varepsilon/c^2 = aT^4/c^2 \simeq 6 \times 10^{-31}$ kg m^{-3}, where T is the temperature of the Microwave Background Radiation. As we will discuss later, the inertial mass density in the form of baryonic matter is at least 10^{-28} kg m^{-3}. Therefore, for recent epochs at least, we can use the approximation of a 'dust-filled' Universe, the term 'dust' meaning that the pressure of the matter can be neglected.

The dust-filled models come in three types which are parameterised by the present inertial mass density of the Universe ρ_0 (figure 6.76). There is a critical mass density $\rho_{crit} = 3H_0^2/8\pi G$, for which the Universe will just expand to infinity at which point its velocity of expansion tends to zero. If H_0 is taken to be 50 km s^{-1} Mpc^{-1}, ρ_{crit} is 5×10^{-27} kg m^{-3}; if H_0 is 100 km s^{-1} Mpc^{-1}, ρ_{crit} is four times greater. It is convenient to refer to the density of other models or different components of the Universe ρ to this critical value through a *density parameter* $\Omega = \rho/\rho_{crit}$. For densities greater than the critical value, $\Omega > 1$, the Universe will eventually stop expanding and collapse to a hot, dense phase. For values less than the critical value, $\Omega < 1$, the Universe expands forever and ends up with a finite velocity at infinity. For $\Omega = 1$, the Universe decelerates to zero velocity at infinity. A miracle of the standard models of general relativity is that there is a simple one-to-one relation between the curvature of space and the density parameter. If $\Omega > 1$, the geometry is closed and spherical and R_c is real; if $\Omega < 1$, the geometry is open and hyperbolic and R_c is imaginary, and, if $\Omega = 1$, the geometry is flat with R_c infinite. The other feature of these models is that, except for the empty model $\Omega = 0$, they are all decelerating models. The deceleration of the Universe is directly proportional to the inertial mass density and hence there is a critical deceleration associated with the model having $\Omega = 1$. If we define the *deceleration parameter* by $q_0 = -(\ddot{R}/\dot{R}^2)_{t_0}$, we find that $q_0 = \Omega/2$ and hence the critical deceleration parameter is $q_0 = \frac{1}{2}$. Notice that Hubble's constant is just the present value of $\dot{R}/R$, i.e. $H_0 = (\dot{R}/R)_{t_0}$ but that it is possible to define a Hubble's constant at any epoch by $H = \dot{R}/R$. In principle it is possible to measure the deceleration of the Universe if H_0 is measured at the present day and at an earlier epoch. It is important that q_0 and Ω can in principle be measured independently.

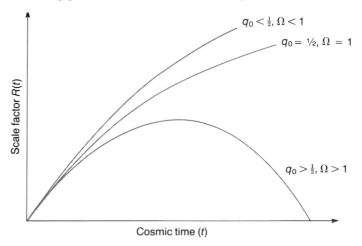

Figure 6.76. Illustrating the dynamics of the classical world models of general relativity. The models can either be parameterised by the density parameter Ω or the deceleration parameter q_0.

A particularly useful relation which comes out of these models is the relation between cosmic time as measured from the Big Bang and the redshift. This is shown on figure 6.77, which also indicates the redshifts to which various classes of extragalactic object can now be observed at the limits of capability of the present generation of telescopes. It can be seen that, for all models, a redshift of one corresponds to looking back to when the Universe was at most half its present age. The most distant galaxies studied have redshifts of one and greater but these are all amongst the most luminous of galaxies known. For the quasars, redshifts of four have now been measured, which means that the most distant quasars emitted their light when the Universe was a fifth its present age or less. Figure 6.77 makes the important point that, whilst the differences between the standard world models increase with increasing redshift, this also means that we are looking further and further into the past when the objects may not be the same as they are now.

Figure 6.77. The relation between redshift and cosmic time for world models having $\Omega = 0$ and $\Omega = 1$. The Big Bang occurred at $t = 0$ and the present epoch is taken to be $t = 1$. Also shown are the typical redshifts to which different types of galaxy and quasar have been observed.

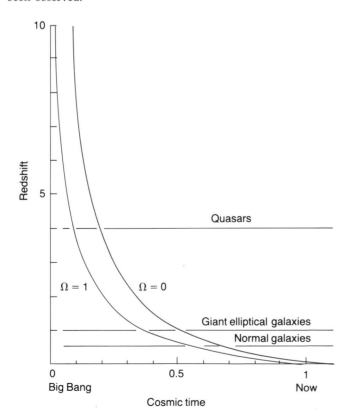

As expected, the standard models make different predictions about the observed properties of distant objects because the differences in dynamics and geometry of the standard models become important. Examples of these predictions, the redshift–flux density relation and the angular diameter–redshift relations, are shown in figure 6.78(a) and (b), where it can be seen that the differences are quite small until redshifts of 1 and greater are attained. For example, the difference in observed intensity from a standard quasar at a redshift of one amounts to a factor of two between the empty ($\Omega = 0$) and critical ($\Omega = 1$) models. The goal of the classical observational cosmologists was the determination of the parameters H_0, q_0 and Ω from astronomical observation and we will survey some recent results in the next section. Let us note some points about the above analysis.

The first point is that the above standard models are applicable during the matter-dominated era, but at early epochs the Universe was radiation dominated and different expanding solutions are applicable. We deal with this topic later.

Second, the Universe is, in fact, very far from uniform on small scales and we should investigate the influence of these irregularities upon observations in the standard models. It was shown by Zeldovich that the angular–diameter redshift relation is sensitive to the presence or absence of matter within the light cone subtended by the object. Suppose, for example, that all the mass of an $\Omega = 1$ model Universe is bound into discrete objects and none of them happens to fall within the light cone subtended by a distant object. Then, there is no minimum in the angular diameter–redshift relation and the relation mimics that of an $\Omega = 0$ model (figure 6.78b). The influence upon the intensities of distant objects is generally less important except in the case of *gravitational lensing*. The discovery of the double quasar $0957 + 561$ (figure 6.4), and the convincing demonstration that this is the double image of a single background quasar, stimulated considerable reinvestigation of the influence of gravitational lensing upon various cosmological relations. The effects are expected to be most important for point-like sources such as quasars and result in a statistical brightening of the quasar intensity. For this to occur with significant probability, the quasars must be distant. One effect is to increase the scatter in intrinsic luminosities in sensitive cosmological tests. Another, possibly significant, effect is that the gravitational lensing could influence the number counts of quasars if their luminosity function is steeper than $\Phi(L)\mathrm{d}L \propto L^{-3}\mathrm{d}L$. This brightening of the most distant and luminous quasars might partially account for the extreme evolution inferred for these objects.

A third point concerns the infamous *cosmological* or *cosmical* constant Λ, which Einstein introduced in 1917 in order to produce static solutions of the field equations of general relativity. The standard models are unstable because they either collapse to a singularity or expand to infinity. Einstein therefore made a simple modification to the gravitational field

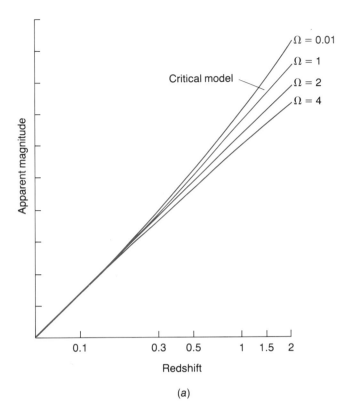

(a)

Figure 6.78. Examples of the differences between the observed properties of distant objects for different world models. (a) The redshift–flux density relation for sources with power-law spectra, $S \propto \nu^{-1}$. (b) The angular diameter–redshift relation for an object of physical size 10 kpc for different world models. It is assumed that Hubble's constant $H_0 = 50 \, \text{km s}^{-1} \, \text{Mpc}^{-1}$. In all cases except the empty world model ($\Omega = 0$), there is a minimum in the angular diameter–redshift relation for homogeneous world models. Also shown, in red, is the angular diameter–redshift relation for a highly inhomogeneous world model with $\Omega = 1$ in which there is no mass within the light cone subtended by the object.

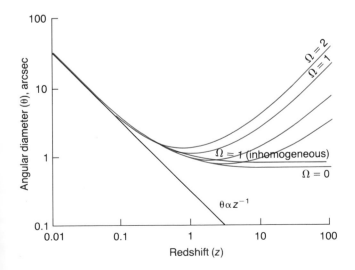

equations to incorporate a repulsive force which would counteract the attractive effect of gravity. In 1917, when the Λ-term was introduced, Einstein was in no position to know that the Universe is in fact unstable. The cosmological constant turns out not to perform the functions Einstein had hoped for, in particular the solutions were still unstable to perturbations and Mach's principle could not be incorporated into general relativity by this means (see Chapter 2 by Will). However, as Zeldovich remarked, 'Once the genie is out of the bottle, it is very difficult to put him back in again.' The cosmological constant has made regular reappearances in the cosmological literature, but there has so far been no absolutely compelling reason to believe that it should be present in the field equations. Nonetheless, the standard model makes the very strong prediction that $\Omega = 2q_0$ and this relation is testable. If this should turn out not to be the case, then we would have to modify Einstein's field equations of general relativity, and one possibility would be to introduce the Λ-term since in general $\Omega = 2q_0 + \frac{2}{3}\Lambda H_0^{-2}$. The Λ-term has gained renewed popularity recently since on its own it gives rise to exponentially expanding solutions for the scale factor $R(t)$. This has been termed by Zeldovich the *repulsive effect of a vacuum*. This is what is required in the inflationary models of the early Universe. However, this phase only lasts for a short period, and the Universe is then supposed to transform into the standard hot Big Bang model by means as yet unspecified.

A fourth aspect of the classical models is the question of how anisotropic the Universe could be in the early stages and yet end up as isotropic as we know it to be now. For example, could the Universe be so anisotropic in the beginning that the initial singularity could be avoided? These questions have been the subject of a great deal of detailed study by general relativists who have classified the various types of anisotropic models allowed by general relativity. It turns out that most of the anisotropic models end up producing large observable anisotropies in the Microwave Background Radiation and only a few of the anisotropic solutions would be compatible with present observations. One very general result has been proved by Hawking and Ellis, which is that under very general conditions, which our Universe is likely to satisfy, there must be an initial singularity from which the Universe expands. In fact, the result is much more general than simply applying within the context of general relativity – their results would be valid for essentially all theories in which gravity is an attractive force. It seems likely that the solution to the singularity problem lies in a proper quantum theory of gravity which as yet has not been formulated.

The classical cosmological tests

The determination of Hubble's constant H_0 has been discussed above. We now look at various ways in which attempts have been made to determine q_0 and Ω.

To make use of the various relations between intrinsic properties and observables shown in figure 6.78 in the classical approach to the determination of q_0, it is necessary to find standard properties of galaxies and other extragalactic objects which can be observed both nearby and at cosmological distances. Figure 6.77 indicates clearly the problem with this procedure. By observing objects at large redshifts, we are also looking at objects when the Universe was much younger than it is now and there is no guarantee that the 'standard' objects may not have evolved over this period. Therefore the problem of determining cosmological parameters by the standard procedure cannot be dissociated from an understanding of the astrophysics of the objects used in the tests.

The only objects which can be used to perform these tests at the moment are luminous objects such as the brightest galaxies in clusters, quasars and radio galaxies. It might seem that quasars would be ideal for these tests because they can be observed at very large redshifts and can be searched for in systematic ways. The big problem in using them is that there is a wide dispersion in their intrinsic luminosities which is greater than the difference between the world models. An example of this problem is shown in figure 6.79, which is the redshift–magnitude relation for a carefully selected sample of quasars which are strong radio emitters. In addition, there is definite evidence that the bulk properties of this population of objects has changed with cosmic epoch. The quasars could only be used in this type of test if some correlations were discovered which could identify objects of the same intrinsic luminosity and this has not yet been achieved. Another fundamental problem is that the astrophysics of quasars is poorly understood and one would much prefer to use objects for which their evolutionary behaviour was better understood.

The use of galaxies is, in principle and in practice, superior for the classical cosmological tests. The reason for this is empirical in that it is possible to select galaxies in such a way that they show relatively little scatter in their intrinsic luminosities. The procedures used most successfully to date involve selecting the brightest galaxies in clusters and those galaxies associated with strong radio sources. In both cases, the intrinsic dispersion in the optical luminosities of the galaxies has standard deviation of about 0.5 magnitudes so that, by accumulating sufficiently large statistical samples at large redshifts, it should be possible to discriminate between world models, if it can be established that the properties of these classes of galaxy have not changed with cosmic time, or at least, have changed in ways with cosmic time which can be distinguished observationally. The astrophysical reason why the galaxies are preferable to the quasars is that their light is starlight and it is possible to develop models for the evolution of their stellar populations based upon reasonable astrophysical hypotheses.

The problems in both cases lie in understanding the selection effects involved in the choice of galaxies used in the tests and the astrophysical evolution of the galaxies. In the case of the

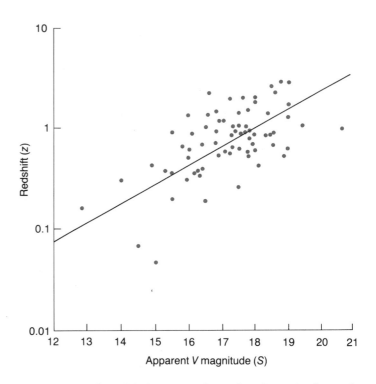

Figure 6.79. The redshift–magnitude (or flux density) relation for a complete sample of radio quasars. It can be seen that there is a positive correlation in the same sense as the redshift–flux density relation for the brightest galaxies in clusters but there is a very much greater dispersion. The solid line is arbitrarily drawn through the points with slope corresponding to $S \propto z^{-2}$. This relation for quasars is therefore not particularly useful for discriminating between world models. (After Wall, J. V. and Peacock, J. A. (1985). *Mon. Not. R. Astr. Soc.* **216**, 182.)

brightest galaxies in clusters, there are at least two evolutionary effects which can change their luminosities. The first is the stellar evolution of the stars which make up the optical luminosity of the galaxy and the second is the effect of the dynamical evolution of the galaxies which make up rich clusters of galaxies. The first of these turns out to be a reasonably straightforward correction, and can in principle be checked by making detailed studies of the spectra of distant galaxies. Since there were more young luminous stars in the past, this correction results in galaxies being intrinsically brighter in the past. The second evolutionary effect has the opposite effect in that, as a cluster evolves, the central massive galaxy captures more and more of the massive galaxies in the cluster which come within its gravitational sphere of influence. This process, known as galactic cannibalism, has the effect of making the central galaxy brighter at the expense of the next brightest members since it is the most massive galaxies which spiral most rapidly into the centre of the cluster as they lose energy by the process known as dynamical friction (see Section 6.6). In this process, the more massive galaxies lose kinetic energy by accelerating less massive galaxies and so gradually spiral into the centre. Again, in principle, this process can be quantified by observations of the galactic environment of the galaxies and their physical properties. In the case of the radio galaxies, which rarely lie in rich clusters, probably only the first evolutionary correction is important, although the question has to be asked whether or not the selection by a radio astronomical criterion raises new problems.

The above examples of the astrophysical problems which have to be addressed simply emphasise the difficulty of the classical approach to the determination of q_0 (or Ω). The samples of galaxies for which such tests can be carried out are limited, and one rarely has enough information to make strong statements about the different evolutionary effects which must be taken into account. The magnitude–redshift relation for the brightest galaxies in clusters extends to redshifts of 0.5 or slightly greater; for the radio galaxies, the relation extends to redshifts of about 1.5 with good statistics, and in figure 6.80(a) I show the relation derived by Simon Lilly and myself for a complete sample of radio galaxies. The photometry was carried out in the near-infrared waveband at $2.2\,\mu m$. There are two great advantages in making these observations in the $2.2\,\mu m$ (or K) waveband. First, as shown in figure 6.80(b), when galaxies are observed at redshifts greater than one, most of their radiation is emitted in the $1-2\,\mu m$ waveband rather than in the optical waveband. Second, the light observed in the $2\,\mu m$ waveband is associated with the oldest populations of stars in the galaxies. Therefore, the infrared luminosity of a galaxy is a much more stable measure of the average luminosity of the galaxy over cosmological time scales than observations in the optical waveband. The corrections for the evolution of the stellar population of the galaxies over cosmological time scales can be made in a quite straightforward manner. The most striking aspect of figure 6.80(a) is the fact that the dispersion in intrinsic infrared luminosities has more or less the same standard deviation of about 0.5 magnitudes at low ($z \lesssim 0.5$) and high ($z \gtrsim 1$) redshifts. This indicates that, despite the differences in the epochs at which these galaxies are observed, there is some systematic behaviour in their properties and this is encouraging. A literal interpretation of the observations suggests that the large redshift galaxies are more luminous than is predicted by the standard models with $q_0 = 0$ and $\frac{1}{2}$. The best-fitting relation shown in figure 6.80(a) assumes values of q_0 of 0 and $\frac{1}{2}$ and incorporates an evolutionary correction to account for the effects of the evolution of the stellar populations of the galaxies. We have interpreted this diagram as direct evidence for the evolution of the stellar populations of these radio galaxies.

None of the workers in this area claim good estimates of q_0. I believe it is fair to say that all the data are consistent with values of q_0 which lie in the range zero to one, which is probably a rather unhelpful conclusion from the point of view of determining q_0. My own view is somewhat different. I believe the achievement represented by this work is that it is now possible to study astrophysically objects at epochs significantly earlier than the present and to set real observational constraints upon their evolution. The challenge of the subject lies in disentangling the astrophysical from the geometrical effects. In my opinion, the astrophysical questions associated with the origin and evolution of galaxies are at least as important as the determination of q_0.

I have devoted most attention to the determination of q_0 through the redshift–apparent magnitude relation because it has produced the best results of all the classical tests so far as the determination of q_0 is concerned. Many others have been attempted, for example redshift–flux density relations at radio and X-ray wavelengths, angular diameter–redshift relations for galaxies, radio sources and clusters of galaxies, counts of sources at radio, optical and X-ray wavelengths. There has been no lack of effort devoted to these subjects but the results have been at least as inconclusive as the results from the redshift–apparent magnitude relation. Some important new areas have been opened up by these studies, in particular convincing evidence that there were many more active galaxies and quasars in the past. We will return to this topic later.

One final point about measuring q_0 should be made. Probably the most convincing way of measuring q_0 would be if physical sizes or luminosities could be measured at large redshifts which are independent of redshift. Thus, if the physics of some distant extragalactic object could be understood so well that its physical dimensions or luminosity could be estimated independent of redshift, measurement of the corresponding angular size or flux density would provide a distance which could be compared with the expectations of the standard world models. Let us give two examples of this approach. The physical size of the huge hot gas clouds in clusters of galaxies can be determined from the distribution and spectrum of its X-ray emission and from the decrement in the intensity of the Microwave Background Radiation in the direction of the cloud because of the Sunyaev–Zeldovich effect (see Section 6.6). A second example is the determination of the physical sizes of supernovae during their early expansion phases. A distance can be measured by observing the velocity of expansion of the supernova and comparing this with the luminosity changes as the supernova expands. In many ways, these physical techniques are more attractive than the use of the standard procedures and they are becoming practicable with the present and future generations of telescopes.

The related parameter which we can attempt to measure is the mean inertial mass density of the Universe Ω. This can be approached in a variety of different ways. First, there is the average mass density which is found from weighing galaxies and multiplying by their space density. Using the masses derived from the visible parts of galaxies, mass densities

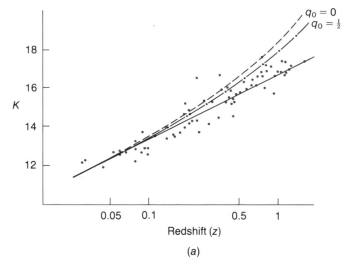

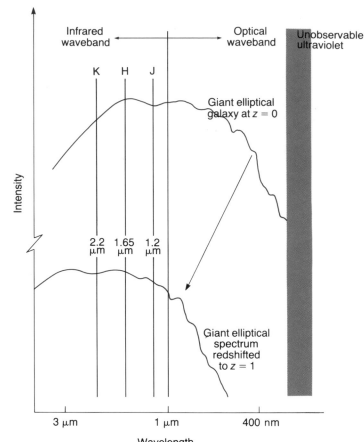

Figure 6.80. (*a*) The redshift–apparent magnitude relation for a complete sample of bright radio galaxies from the 3CR catalogue. The galaxies extend in redshift to about 1.5 and the apparent magnitudes were measured at an infrared wavelength of 2.2 μm. The dashed lines show the expected relation for the standard world models. The solid line is a best-fitting line for standard models with $\Omega = 0$ and 1 and includes the effects of stellar evolution on the old population of stars in the galaxies. (From Lilly, S. J. and Longair, M. S. (1984). *Mon. Not. R. Astr. Soc.* **211**, 833.) (*b*) Illustrating the effects of redshift upon the observed intensity of a giant elliptical galaxy. At $z = 0$, the spectrum peaks at about 1 μm. When such a galaxy is observed at a large redshift, its observed flux density decreases because of the inverse-square law and also because the energy spectrum is shifted to longer wavelengths by a factor $(1 + z)$. It can be seen that there is an advantage in making observations in the infrared wavebands such as H (1.65 μm) and K (2.2 μm).

$\Omega_g \simeq 0.01$–0.025 are found, well below the closure density. To this should be added the hidden mass present in galaxies and clusters of galaxies. How much hidden matter there is overall is not well known, but a factor of ten as compared with the visible mass of galaxies is typically found for clusters of galaxies, giving $\Omega_g(\text{tot}) \simeq 0.1$–$0.25$. This figure could still be a lower limit because there might be hidden mass in the space between rich clusters of galaxies. In principle, this could be detected by application of the cosmic virial theorem to galaxies selected from the general field (Section 6.6). The values found lie in the range 0.1–0.5. Finally, there could be some smooth underlying substratum, perhaps consisting of some form of ultraweakly interacting form of matter which is not clumped like the galaxies and which could contribute to the overall mass density of the Universe.

One important argument to be addressed later concerns the formation of the light elements in the early Universe. Remarkably, light elements such as helium, deuterium and helium-3 can be synthesised in roughly their observed proportions by primordial nucleosynthesis and this argument provides a firm upper limit to the mass density of the Universe in the form of baryonic matter, i.e. conventional matter such as protons and neutrons, of $\Omega_{\text{bar}}h^2 = \Omega_{\text{bar}} (H_0/100 \,\text{km s}^{-1} \,\text{Mpc}^{-1})^2 \lesssim 0.05$. Thus, the prediction is rather sensitive to the value of H_0. If $H_0 = 100 \,\text{km s}^{-1} \,\text{Mpc}^{-1}$ and the value of Ω from the cosmic virial theorem is correct, non-baryonic dark matter is required. If $H_0 = 50 \,\text{km s}^{-1} \,\text{Mpc}^{-1}$, $\Omega_{\text{bar}} \lesssim 0.2$ and there is not necessarily any need for nonbaryonic matter.

What is one to make of all this? It is important to emphasise that it is very difficult to make good estimates of Ω. I believe a concensus of opinion would be that Ω probably lies in the range 0.03 to about 1. From the point of view of our comparison of q_0 and Ω, it is not unreasonable to suppose that q_0 and $\Omega/2$ are the same within a factor of about ten. This is an important result in its own right because if other unknown forces were important in determining the large-scale dynamics of the Universe, there might be a much larger discrepancy than a factor of ten. Application of Occam's razor might tempt one to conclude that $q_0 = \Omega/2$ on the basis of present data but one would much prefer to have improved values of q_0 and Ω, difficult as they are to measure.

An alternative approach to measuring q_0 and Ω is to measure the age of the Universe from the oldest known objects. This procedure requires a knowledge of Hubble's constant H_0. For example, for the $q_0 = \Omega = 0$ model, the age of the Universe is just $T_0 = 1/H_0$; if $q_0 = \Omega/2 = 0.5$, the critical model, $T_0 = \frac{2}{3}H_0$. Thus, a literal interpretation of the age estimates of the oldest globular clusters, $T_0 \geqslant 14$–20×10^9 years, would pose problems for models with high values of both q_0 and H_0. For example, the value $H_0 = 100\ \mathrm{km\,s^{-1}\,Mpc^{-1}}$ poses a problem for all values of q_0 since $T_0 \leqslant 10^{10}$ years, which is smaller than the inferred ages of the oldest globular clusters. This argument limits values of Hubble's constant to $H_0 \leqslant 75\ \mathrm{km\,s^{-1}\,Mpc^{-1}}$.

Strong evolutionary effects with cosmic epoch

We have already described some of the evidence which has recently been found for the evolution of the stellar component of radio galaxies with cosmic epoch. This type of evolution could have been expected in the sense that younger galaxies are expected to have more luminous stars. Much stronger evolutionary effects have, however, been discovered among the population of active galaxies, particularly the radio galaxies and quasars.

A significant excess in the number of faint extragalactic radio sources was discovered in the 1950s and 1960s as compared with what would be expected for all the standard world models. This was interpreted as being due to the presence of many more radio sources at early epochs than there are at the present time. A similar phenomenon was found for the quasars, both those selected from radio surveys and from optical searches in the 1960s and 1970s. The present position is that all classes of quasar and strong radio source exhibit this phenomenon, and there is now sufficient data on these sources to interpret in some detail the way in which the populations of these objects have changed with cosmic time, although the physical reasons why these changes have taken place are far from being understood.

These evolutionary effects are very strong in the sense that the probability of powerful radio source and quasar activity must have decreased by a factor of about 100 or 1000 between a redshift of 2.5 and the present epoch, i.e. over the period when the Universe was about one-quarter of its present age to the present day. To put this another way, we are living at a rather dull epoch in the Universe compared to the amount of quasar and radio source activity which must have been going on when the Universe was only one-quarter of its present age. Some examples of the changes in the comoving space density of

quasars and radio galaxies with cosmic epoch inferred from radio source surveys for which much redshift and other statistical data are available are shown in figure 6.81(a). Similar results are found for radio quiet quasars, even more extreme evolution being inferred for the most optically luminous quasars (figure 6.81b). It appears that the evolution of the luminosity function of radio quiet quasars can be accounted for by assuming that the sources were on average much more luminous in the past. In the case of the radio galaxies and the quasars, the picture cannot be so simple because the *shape* of the luminosity function changes with increasing redshift (figure 6.81a). It is important to emphasise that these forms of evolution occur for very rare classes of extragalactic object and we know, for example, that the commoner weak radio galaxies do not exhibit these strong evolutionary trends.

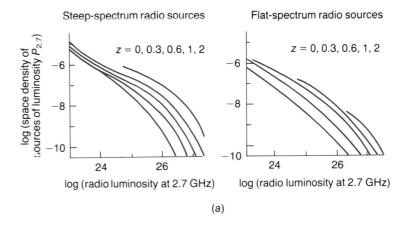

(a)

Figure 6.81. The evolution of the luminosity functions of quasars. (a) The evolution of the *radio* luminosity function of those quasars which are strong radio sources. (Courtesy of Dr J. A. Peacock.) (b) The evolution of the *optical* luminosity function for radio quiet quasars. (From Marshall, H. L. (1986). *The Structure and Evolution of Active Galactic Nuclei*, G. Giuricin, F. Mardirossian, M. Mezzetti and M. Ramella, eds., p. 627. D. Reidel and Co.)

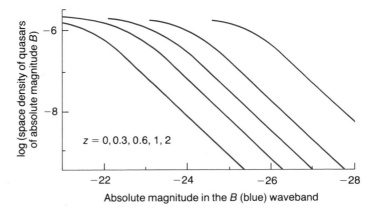

(b)

One of the most interesting questions is what happens at earlier epochs. The largest quasar redshift is 4.41, but it has proved very difficult to find larger redshift objects despite many careful systematic searches for them. Searches for these distant quasars have been carried out using radio and optical search techniques with limited success. In the process of carrying out these searches, the statistics on quasars at smaller redshifts have been greatly improved. What is clear is that the strong evolutionary increase in the number densities of quasars and radio galaxies with increasing redshift cannot increase at the same rate beyond redshifts of about 3. In fact, it is unlikely that there is an abrupt cut-off at a redshift of 4 but rather that the maximum probability of quasar and radio source activity occurs at redshift $z \sim 2.5$. An example of this type of evolutionary behaviour is shown in figure 6.82 for the radio galaxies and radio quasars, which shows that the decay of the population is very rapid to small redshifts and there appears to be a decrease by about a factor of five between redshifts of 2.5 and 4, i.e. at look-back times greater than about 70% of the age of the Universe. The absence of sources at very large redshifts probably reflects the combination of the decreasing comoving space density of quasars with increasing redshift beyond $z \sim 2.5$ and the small probability of finding them because at these large redshifts the volume elements increase very slowly and their observed intensities decrease rapidly with increasing redshift.

Figure 6.82. The evolution of the comoving space density ρ of radio sources which have steep and flat radio spectra. The shaded areas indicate the uncertainties in the estimates of the comoving space density which is for radio sources having radio luminosities at 2.7 GHz of $10^{26} \ \mathrm{W \ m^{-2} \ Hz^{-1}}$. A cosmological model having $\Omega = 1$, $q_0 = \frac{1}{2}$ is assumed. It can be seen that there is a maximum in the comoving space density of both classes of radio source when the Universe was about 25% of its present age and that at larger redshifts the comoving number density of sources decreases. (Courtesy of Drs J. A. Peacock and J. Dunlop.)

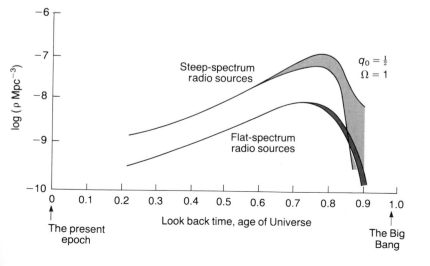

The causes of this behaviour are not understood, largely because the astrophysics of quasars and powerful radio galaxies are still at a relatively primitive state of development. The picture of supermassive black holes as the ultimate energy sources for the nuclei of quasars and radio galaxies is very appealing but their exact relation to the observed activity in galaxies is not clear. For example, it might be that the high rate of occurrence of quasars is due to the fact that there are plentiful supplies of gas available in galaxies when they are young and so the evolution of quasars might trace out the birth rate of galaxies. There remains, however, the question of how the supermassive black holes first formed in galaxies. My own view is that these phenomena are strongly related to the origin and early evolution of galaxies but there is certainly no agreement about this.

The canonical Hot Big Bang

Few astrophysicists would disagree with the statement that the discovery of the Microwave Background Radiation and the determination of its observational characteristics are by far the most significant discoveries in cosmology since Hubble's discovery of the expansion of the Universe. In my view, its singular importance lies in providing us with a framework within which we are able to discuss astrophysically the origin and evolution of galaxies and other large-scale structures in the Universe. Let us see how naturally this comes about.

The spectrum of the Microwave Background Radiation is essentially that of a pure black-body (figure 6.83), i.e. it possesses the characteristic signature of radiation which has been in thermodynamic equilibrium with matter. The radiation temperature is about $T_0 = 2.75$ K at the present day, and the isotropy of the radiation indicates that this radiation permeates the whole Universe. If we now contract the Universe back in time, the radiation is blueshifted according to relation (6.17), i.e. the energy of each photon changes with redshift as

$$\varepsilon(z) = h\nu(z) = h\nu_0(1 + z) = \varepsilon_0(1 + z). \qquad (6.18)$$

It is easy to show that, when a Planck spectrum is redshifted, the form of the spectrum is preserved, the radiation temperature varying as $T = (1 + z)T_0$ and the energy density (or inertial mass density) changing as $(1 + z)^4$. Since the inertial mass density in ordinary matter changes only as $R^{-3} \propto (1 + z)^3$, the inertial mass density in radiation increases relative to that of matter as $(1 + z)$. Therefore, at early enough times, not only does the radiation temperature become very high, but also the dynamics of the Universe become dominated by the inertial mass of the radiation. Thus, the early Universe is *radiation-dominated*. If $\Omega_{\mathrm{matter}} = 0.1$, the epoch of equality of the inertial mass densities of matter and radiation occurs about $z = 1000$; if $\Omega_{\mathrm{matter}} = 1$, it occurs about $z = 10^4$.

At a redshift $z = 1500$, the radiation temperature is about 4000 K and there are then sufficient high energy photons in the tail of the black-body spectrum with energy $h\nu \geqslant 13.6$ eV to

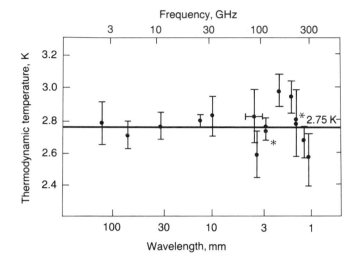

Figure 6.83. The observed spectrum of the Microwave Background Radiation. A selection of recent high precision measurements of the thermodynamic temperature of the Microwave Background Radiation is shown at wavelengths which span the peak of the spectrum. Those measurements indicated by asterisks are derived from measurements of the excitation temperature of interstellar CN molecules. The other measurements are either ground-based radiometer observations or bolometric observations made from high-flying balloons. All the observations are consistent with a black-body temperature of 2.75 K. (From Johnson, D. G. and Wilkinson, D. T. (1987). *Astrophys. J.* **313**, L1.)

ionise all the hydrogen in the Universe. Therefore, at earlier epochs, the matter of the Universe must consist of an ionised plasma bathed in a black-body radiation flux. Compton scattering of the electrons of the plasma by the radiation field ensures that the matter and the radiation remain at the same thermodynamic temperature despite the fact that they have different adiabatic indices. The strong coupling between the matter and radiation at $z \simeq 1500$ and earlier epochs means that there is a 'photon barrier' at $z \simeq 1500$. Scattering of photons at earlier epochs smooths out any structure that might be present in the spatial distribution of the radiation. Unless the intergalactic gas is reionised, the epoch $z \simeq 1500$ is the last scattering surface for photons of the Microwave Background Radiation which are observed at the present epoch.

There is no reason why we should not extrapolate backwards in time to the much earlier stages when our cosmological model predicts that the temperature becomes very high indeed. At a redshift of 10^9, the radiation temperature becomes about 3×10^9 K, and this is hot enough to cause the photodissociation of nuclei. Thus, no nuclei can have existed at epochs earlier than this. At $z = 3 \times 10^9$ and $T \simeq 10^{10}$ K, electron–positron pair production from photon collisions in the field of nuclei becomes possible so that a new thermodynamic equilibrium is set up between electrons, positrons and photons. This process of new

equilibria being set up as it becomes energetically feasible for particle–antiparticle pairs to be created from the thermal radiation field continues as far back as particle physics will allow us to extrapolate. It must be a matter of judgement how far one can extrapolate backwards in time with confidence, but there is one important point to note. Although the thermodynamic temperatures are very high, $T \geqslant 10^9$ K, the particle densities involved are not at all extreme. For example, if the Universe has a baryonic density corresponding to $\Omega_{bar} = 0.1$ today, the number density of protons at $z = 10^{10}$ would have been only about $10^{29} \, \mathrm{m}^{-3}$, a typical laboratory density, although it is at a very high temperature and in the presence of a very intense radiation field. Even at $z \simeq 10^{13}$, when the temperature is sufficiently high for baryon–antibaryon production to take place, the particle density is still subnuclear, i.e. less than the values typically found in the centre of neutron stars. For these reasons, nuclear astrophysicists have felt quite confident about following through the detailed nuclear physics of the Hot Big Bang model of the Universe from temperatures in excess of 10^{12} K to the present day. This standard thermal history of the Universe extending to scale factors as small as 10^{-16} is summarised in figure 6.84.

The following picture emerges for the evolution of the Hot Model of the Universe through the epochs when nuclear reactions are important. At every stage in the evolution of the model, all the types of particles which can come into thermodynamic equilibrium with the thermal radiation field do so. Generally, this means that the particles are as common as the photons of the Microwave Background Radiation so long as their typical energies exceed their rest mass energies. When the typical energy falls below the rest mass energy, the abundances of the species decrease exponentially. Furthermore, certain components 'freeze-out' if the time scales for the interactions which maintain the statistical equilibrium exceed the time scale of the Universe at that epoch.

If the calculations are started at $T \simeq 10^{12}$ K, the equilibrium is maintained between neutrons, protons, photons, electrons, positrons, muons and the various associated types of neutrino. The muons annihilate, and the next key event directly related to observable quantities is the freeze-out of the electron neutrinos which maintain the equilibrium between neutrons and protons. Although below the temperature corresponding to the mass difference between the proton and the neutron, the equilibrium is maintained by the weak interactions

$$\left. \begin{array}{l} p + e^- \rightleftarrows n + \nu \\ n + e^+ \rightleftarrows p + \bar{\nu} \end{array} \right\} \qquad (6.19)$$

The equilibrium abundances are simply given by Boltzmann's law

$$[\mathrm{n}]/[\mathrm{p}] = \exp(-\Delta m c^2 / kT), \qquad (6.20)$$

where Δm is the difference in mass between the neutron and proton. At a temperature of about 10^{10} K, the time scale for the weak interactions exceeds the age of the Universe (about 1 s)

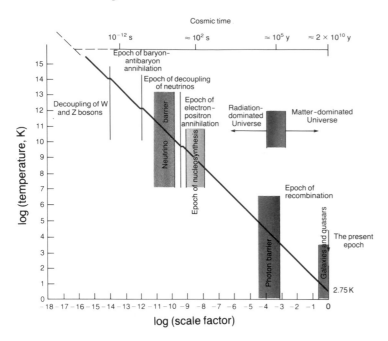

Figure 6.84. The thermal history of the radiation temperature of the Microwave Background Radiation according to the standard Hot Big Bang model. The radiation temperature varies as $T_r \propto R^{-1}$ except for abrupt jumps as different particle–antiparticle pairs annihilate at $kT \simeq mc^2$. The various epochs in the evolution of the Hot Big Bang model are indicated. Prior to the epoch of recombination, the matter and radiation are closely coupled by Compton scattering, and hence the matter temperature is maintained at the same value as that of the redshifted Microwave Background Radiation, despite the fact that they have different adiabatic indices. After the epoch of recombination, the matter and radiation decouple and cool independently, the matter cooling as $T \propto R^{-2}$ compared with $T \propto R^{-1}$ for radiation. Because the intergalactic abundance of neutral and molecular hydrogen is very low indeed at the present epoch, it is likely that the intergalactic gas is more or less fully ionised now. The gas must therefore have been heated up at some epoch prior to $z \simeq 4$. An approximate time scale is indicated along the top of the diagram. The neutrino and photon barriers are indicated. In the standard model, the universe is optically thick to neutrinos and photons, respectively, prior to these epochs.

and then the abundance of neutrons to protons is frozen at the value of [n]/[p] determined by the value of T at that time. Notice that the epoch of decoupling of neutrinos acts as an effective 'neutrino barrier' since prior to that epoch, the neutrinos are scattered and hence information is lost about any spatial structure in the neutrino distribution from earlier epochs.

According to expression (6.20), at 10^{10} K, there are about fifteen neutrons for every one-hundred protons. As the Universe continues to expand and cool, the temperature falls to low enough values that light nuclei can be formed from the protons and neutrons, most of the latter being combined into helium-4

nuclei, ^{4}He, through the proton–proton chain reaction (see Section 6.3). In the canonical Hot Big Bang model, the products of nucleosynthesis depend only upon the present average density of matter in the Universe, and the results are shown in figure 6.85. As can be seen, the helium abundance is about 25% for a wide range of densities. The reason for this constancy can be appreciated from the above analysis since the ultimate fate of most neutrons is to be bound into helium nuclei. In the above example, the predicted ^{4}He percentage by mass is $30/115 \times 100 \simeq 25\%$. Thus the helium abundance essentially acts as a thermometer which tells us at what temperature the decoupling of the neutrinos from the n–p reactions (6.19) takes place.

This is quite a remarkable result since the helium abundance has long been one of the great astrophysical problems. Wherever it is observed in the Universe its abundance appears to be about 23% or greater by mass. It is difficult to account for this abundance in terms of stellar nucleosynthesis because, as was described in Section 6.2, once stars have burned about 12% of their mass from hydrogen to helium, they become red giants and in the subsequent phases of nucleosynthesis helium is consumed rather than created. In addition, primordial nucleosynthesis accounts elegantly for the fact that there is a firm lower bound to the amount of helium observed everywhere in the Universe. If nucleosynthesis only took place in stars, there would surely be a much wider range of variation in helium abundance.

Only traces of other elements are produced because there is not time in the early phases of the Hot Big Bang for the triple-alpha process to jump the barrier caused by the lack of a stable ^{8}Be isotope. These trace elements are, however, very important. The amounts of deuterium, helium-3 and lithium synthesised in the early Universe depend critically upon the baryon density of the Universe. If the Universe is of high density, $\Omega \simeq 1$, the p–p reaction chain can consume essentially all the deuterium resulting in a predicted D abundance significantly less than 10^{-5} (figure 6.85). In a low density Universe, $\Omega \leqslant 0.1$, however, there is not time for the D and ^{3}He to be consumed, and significant abundances are expected as indicated in figure 6.85. This is another remarkable triumph for the canonical model. Deuterium and helium-3 are difficult elements to create by nucleosynthesis inside stars. They are such fragile elements that they are destroyed rather than created in stars, and this has posed a major problem for astrophysicists. A primordial origin can account for the observed cosmic deuterium abundance of about 1.5×10^{-5} by mass. An even stronger result, however, comes out of this calculation. If the baryon density of the Universe were significantly greater than $\Omega_{bar} \simeq 0.05 h^{-2} = 0.05\,(100\,\mathrm{km\,s^{-1}\,Mpc^{-1}}/H_0)^2$, less than the observed abundance of deuterium is synthesised. From figure 6.85, it can be seen from the dependence of the deuterium and ^{3}He abundances upon Ω_{bar} that $\Omega_{bar} = 1$ can be excluded. It should be noted that it may also be possible to synthesise the observed abundance of lithium-7 primordially. This element is found to have a constant abundance in old unevolved stars, suggesting a primordial origin as well.

These are quite astonishing results, and most astrophysicists believe that the combination of the three independent pieces of evidence – the expansion of the Universe, the Microwave Background Radiation and the abundances of the light elements – provide convincing evidence that the Universe went through a hot dense phase as described by the canonical Hot Big Bang model. Whilst recognising the elegance of this picture and adopting it in the subsequent discussion, it must be recognised that it is not without its problems. The strength of the model lies in the fact that by constraining the Hot Big Bang to reasonable values of the present day parameters of the Universe, a number of independent data can be explained. The concerns arise when one looks at some of the implicit constraints which have been placed upon the model.

Figure 6.85. The predicted abundances of the light elements synthesised in the early stages of the Hot Big Bang. The standard model assumes that the lepton number is zero and is evolved from an equilibrium state at a temperature of about 10^{12} K. The predicted and observed abundances of the elements are shown as a function of the density parameter Ω at the present epoch. (After Audouze, J. (1982). *Astrophysical Cosmology*, H. A. Bruck, G. V. Coyne and M. S. Longair, eds., p. 409–11. Pontifica Academia. Scientiarum Scripta Varia, Vatican City.)

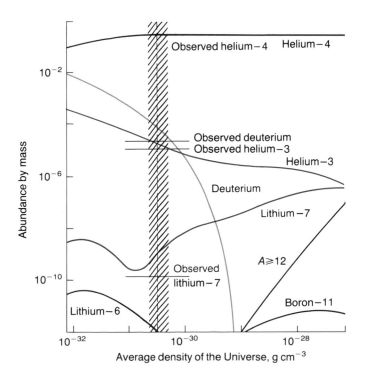

First of all, it is not accidental that the simulations start at 10^{12} K rather than 10^{13} K. If the model is extrapolated back to the epoch of baryon–antibaryon pair production from the thermal background, the Universe is flooded with baryon–antibaryon pairs (see figure 6.84). The reason for this can be understood from the fact that the ratio of the numbers of photons to baryons at the present epoch is $[\gamma]/[p] \simeq 10^8$. This ratio is more or less conserved back to the epoch of baryon–antibaryon production when for every photon there is one baryon or antibaryon. Thus, running the clocks forward, there must be a small initial asymmetry between matter and antimatter in the very early Universe such that for every 10^8 antibaryons, there are $10^8 + 1$ baryons. The 10^8 baryons annihilate with the 10^8 antibaryons producing photons which contribute to the thermal background radiation, and the one remaining baryon becomes the matter-dominated Universe we know today. This problem has been the subject of considerable discussion and, within the conventional Hot Big Bang model prior to the advent of Grand Unified Theories of elementary particles, there seems to be no alternative but to assume that this asymmetry between matter and antimatter was built into the initial conditions from which our Universe has evolved.

Another important point about the conventional model is that it only produces the excellent agreement between theory and observation if it is assumed that the lepton number is zero. Again, simplicity may dictate that we should only adopt a less 'natural' condition if we are forced to, but we might worry that the baryon number is certainly nonzero, although only at the level of one part in 10^8.

A third point is that the success of the canonical hot model can be used to constrain various aspects of particle physics. For example, the dynamics of the Universe at the epoch of nucleosynthesis could not have been much different from that of the canonical model if the observed abundances of the elements are to be produced. This constrains the number of types of unknown ultraweakly interacting particles which could be present in the Universe and also various classes of anisotropic world model. An interesting example of this procedure is in estimating the numbers of neutrino species which could have contributed to the total energy density of the Universe during primordial nucleosynthesis. If there are too many neutrino types, the expansion of the Universe becomes more rapid resulting in neutrino decoupling occurring at a higher temperature and hence *more* helium being synthesised than observed. Constraining the models to produce the observed primordial ^{4}He abundance restricts the number of neutrino species to three, the number now known to exist and matching nicely the number of quark flavours. The procedure can also be used to set constraints upon variations of the gravitational constant with cosmic epoch. The ultimate extrapolation of this process of using cosmological observations to constrain particle physics is to the very early Universe indeed, and this is the subject of Chapter 3 by Guth and Steinhardt.

Galaxy formation

The canonical Hot Big Bang model is thus remarkably successful in providing us with a convincing framework for cosmological studies. So far, however, it is a very dull Universe with no structure at all – no stars, galaxies or astronomers. The point of view taken by cosmologists is that the origin of the detailed structure of galaxies, stars, regions of ionised hydrogen and so on are purely astrophysical questions which may be addressed by the types of study described in the preceding sections. Once a large-scale perturbation has developed to such a density that its amplitude is large, i.e. the fractional fluctuations $\delta\rho/\rho \geqslant 1$, the cosmologist regards his job as largely done since it is then up to the astrophysicist to show how these nonlinear perturbations become the galaxies, stars and clusters we know.

The cosmologist's programme may appear modest, but it has turned out to be one of the most difficult of modern cosmology. The problems arise from two basic results. First, in 1902 Jeans showed that a stationary, uniform medium is unstable to gravitational perturbations on a sufficiently large scale. Collapse of a region of enhanced density under its own gravity is resisted by internal pressure gradients but, if one considers large enough scales, the attractive force of gravity ultimately becomes greater than the pressure gradient. For infinitesimal perturbations on a scale larger than a certain critical length, exponential growth of the perturbation ensues. This critical scale, known as the Jeans' wavelength R_J, depends only upon the speed of sound in the gas c_s and its density ρ

$$R_J = c_s/(G\rho)^{\frac{1}{2}} \simeq c_s t_{collapse}, \qquad (6.8)$$

where $t_{collapse}$ is the time it takes the perturbation to collapse under gravity. This is a very good thing from the point of view of star formation which may be considered to take place in a reasonably static medium as discussed in Section 6.5.

The problem is different in the case of an expanding medium. It was first shown by Lifshitz in 1946 that the density contrast $\delta\rho/\rho$ of the perturbed region increases with time but that the growth of the perturbation is slow because the background density against which it is trying to grow is continually decreasing. The net result is that the perturbation grows only *algebraically* rather than *exponentially* with time so that infinitesimal perturbations do not grow rapidly enough to finite amplitude in the age of the Universe. During the matter-dominated epochs the fractional density perturbation $\delta\rho/\rho$ grows only proportional to the scale factor, $\delta\rho/\rho \propto R \propto (1+z)^{-1}$ at redshifts such that $\Omega z > 1$. The complete analysis of the evolution of density perturbations involves applying rules similar to the above expressions for the Jeans' length and the rate of growth of instabilities for the radiation and matter-dominated phases and using the thermal history of the Universe as described by the standard Hot Model.

The following general picture emerges. Before the epoch of recombination and during the radiation dominated epochs, the Jeans' length R_J is just less than the horizon scale $r \simeq ct$ because

the sound speed is close to the velocity of light $c_s = c/\sqrt{3}$, the relativistic hydrodynamic velocity, and perturbations on the scales of galaxies and clusters of galaxies correspond to stable modes, i.e. to sound waves. This is a general result which arises because the matter is coupled to the radiation which provides virtually all of the inertial mass and pressure during the radiation-dominated epochs. To a good approximation, it turns out that the perturbations associated with these large-scale structures can only begin to grow after the epoch of recombinations, $z \simeq 1500$. Since we know that galaxies exist now, this means that by $z \simeq 1$, $\delta\rho/\rho \simeq 1$ and, hence, assuming $\Omega \simeq 1$, $\delta\rho/\rho \simeq 10^{-3}$ at the epoch of recombination. Notice that by adopting $\Omega = 1$ and $z(\delta\rho/\rho = 1) = 0$, we obtain the *smallest* fractional fluctuation at the epoch of recombination. Obviously, a complete analysis would produce more exact results, but the basic point is clear, which is that the existence of galaxies and clusters at the present day necessarily means that they must have evolved from perturbations which had amplitude at least $\sim 10^{-3}$ at the epoch of recombination. These should leave an imprint upon the spatial distribution of the Microwave Background Radiation, either because of the increase in temperature associated with adiabatic fluctuations or because of the Doppler shifts of the scattered radiation when the large-scale inhomogeneities begin to collapse. Detailed calculations of the expected amplitude of the fluctuations, independent of their nature, indicate that the present upper limits to intensity fluctuations in the Microwave Background Radiation can rule out this straightforward picture. It is evident that modifications to the simple scenario are required.

First of all, however, it is instructive to look at the nature of the fluctuations. Any random distribution of density, pressure and velocity can be decomposed into three types of perturbation which have quite different evolutionary behaviour with cosmic epoch. These are *adiabatic*, *isothermal* and *vortex* perturbations. The vortex fluctuations are unlikely to play a major role in the early evolution of the perturbation spectrum because it can be shown quite generally that they decay as the Universe expands and, to play an important role, they would have to be continuously regenerated from some large-scale energy source. During the radiation-dominated phase, the adiabatic perturbations are sound waves in which the pressure and density vary according to an adiabatic law. The isothermal perturbations are baryon inhomogeneities on an otherwise uniform background of photons. Most of the analyses of the origins of galaxies have concentrated upon the evolution of adiabatic and isothermal perturbations.

In the standard model, the two types of perturbation behave very differently during the epochs prior to recombination. The adiabatic perturbations are sound waves and therefore are susceptible to damping because the coupling between the matter and the radiation is strong but not quite perfect. The radiation, which contributes all the pressure in the perturbations prior to the epoch of recombination, can diffuse out of the perturbation and this results in damping. This process, which is known as *Silk damping*, has the effect of wiping out structure on all mass scales less than about $10^{13} M_\odot$ by the epoch of recombination. This means that, in the standard adiabatic model, it is large-scale structures such as clusters and superclusters which collapse first, smaller-scale structures such as galaxies forming once the collapse of the large-scale structure has become nonlinear (i.e. $\delta\rho/\rho \gg 1$). One attraction of this model is that it can account rather naturally for the large-scale distribution of galaxies in the Universe, including the 'voids' and elongated 'superclusters'. This comes about because the large-scale initial perturbations are unlikely to be perfectly spherical and therefore collapse first along their shortest axes. Thus, when the collapse becomes nonlinear, the adiabatic perturbations are expected to collapse into 'pancakes', and it is in these structures that galaxies form. Three-dimensional simulations show that a reasonable likeness to the large-scale structure of the Universe can be achieved in this model. It is a general feature of this class of model that galaxies and clusters form at late epochs. This is because large-scale systems form first and their average densities only become comparable with the average density of the Universe at redshifts less than about 10. The parameters can be arranged so that most galaxy formation occurs at redshifts of about 5 or less. Note, however, that this model would produce excessively large fluctuations in the Microwave Background Radiation if $\Omega \lesssim 0.1$, and therefore the simple model needs further modification. Another problem which this class of model has to address is the origin of the chemical abundances of the elements in the most distant systems known. It is remarkable that the abundances of the heavy elements in the gas clouds surrounding the most distant quasars appear to be more or less the same as those observed locally in our Galaxy. Since these elements are created by stellar nucleosynthesis, this suggests that there must have been considerable amounts of star formation and enrichment of the primordial gas even by a redshift of 4. Since the simple adiabatic model predicts late formation of galaxies and the stars within them, this observation requires careful arrangement of the sequence of events which form stars, galaxies and quasars.

The isothermal perturbations are stationary during the radiation-dominated phases and hence survive on all mass scales to the epoch of recombination. It is expected that the first objects to collapse are those with masses exceeding the Jeans' mass at the epoch of recombination which corresponds to $M \simeq 10^6 {}_\odot$. Thus, objects with masses similar to those of globular clusters form early in the Universe and then the large-scale structure of the Universe forms out of the clustering of these objects under gravity, bearing in mind that larger-scale masses are part of the fluctuation spectrum from which the $10^6 M_\odot$ objects form. Since the stars can form early in this model, there is no problem in accounting for the abundances of the elements in distant quasars. The difference between the isothermal and adiabatic models is that, whereas gas dynamics and dissipation of binding energy by radiation play a key role in

the formation of pancakes, i.e. of clusters and superclusters, the large-scale structure in the isothermal model has to form by clustering of objects which have already lost their binding energy. Thus, the process is appropriately described as hierarchical clustering and has been the subject of a number of computer simulation studies. The tendency is to form spherical systems by processes similar to those which take place in the evolution of clusters of galaxies and it is not so obvious how the long stringy structure and holes are formed.

Of the conflicts with the observational data, the problem of fluctuations in the Microwave Background Radiation is the most severe. One possible solution would be to suppose that the intergalactic gas was reionised subsequent to the epoch of recombination and Thomson scattering of the photons of the Microwave Background Radiation reduce the amplitude of the intensity fluctuations by a factor $e^{-\tau}$, where τ is the optical depth to Thomson scattering. The problem with this picture is that the heating would have to be very strong and occur soon after recombination in order to provide sufficient optical depth. This solution does not work in the case of the pure adiabatic pancake model since in it the clusters form late in the Universe and there are no suitable energy sources formed at early epochs. There is no reason in principle why a significant amount of reheating of the intergalactic gas could not occur in the isothermal model. One of the possibilities is heating of the intergalactic gas by an early generation of stars, often referred to as *population III* stars (see Section 6.3).

The most important development in seeking solutions to these problems has been the realisation that they should not be dissociated from the problem of the hidden or dark matter. The discussions of Section 6.6 and this section indicate that most of the matter in the Universe is probably in the form of dark matter and thus this matter is likely to have a profound impact upon the formation of galaxies. The problem is that the constraints which can be placed upon the nature of the dark matter are not particularly strong, and there is therefore a great deal of scope for creative astrophysics in finding solutions to the problem of galaxy formation. Indeed, I would caution that the whole story must be regarded as provisional because the basic physics is far from secure.

Let us classify the various forms which the dark matter could take. First, there is dark baryonic matter, for example ultralow mass stars, planets or rocks. This form of dark matter is principally constrained by the upper limits to the amount of baryonic matter which could be present in the Universe from the constraints imposed by the deuterium and helium-3 abundances, $\Omega_{bar} \lesssim 0.2$.

A second type of dark matter is massive neutrinos. There have been unconfirmed reports that the neutrino has a rest mass of about 30 eV. If this were the case, the local inertial mass density in known types of neutrino, the electron neutrinos, the muon neutrinos and the tau neutrinos which are all present in the standard Hot Model of the early Universe, would correspond roughly to the critical density $\Omega = 1$. A rest mass energy of

30 eV is also an intriguing value from the point of view that this means that the neutrinos would have been fully relativistic during the epoch of nucleosynthesis, and hence there would be no effect upon the predictions of primordial nucleosynthesis in the standard model. The key point is the fact that the neutrinos would become nonrelativistic at a redshift of about 10^5. Since the neutrinos would still have a large velocity dispersion when the perturbations begin to grow, this form of hidden matter is often referred to as *hot* dark matter.

Another variant on this theme would be if there existed unknown massive weakly interacting particles which decoupled from thermal equilibrium with the Hot Big Bang at a much earlier phase in the expansion. For each hypothetical class of particle, it is possible to work out how many of them would survive particle–antiparticle annihilation once the temperature fell below that corresponding to the rest mass energy of the species, $kT \simeq mc^2$. For certain masses of these particles, too much mass would survive, i.e. $\Omega \gg 1$. In one possible variant of this theory, a massive neutrino-like particle of mass 1–10 GeV could be present and result in a Universe with $\Omega \simeq 1$.

The most extreme form of dark matter is *cold* dark matter, which is characterised by the fact that it has essentially no velocity dispersion with respect to the mean Hubble flow. Examples of this include mini-black holes and the various exotic forms of matter predicted by Grand Unified Theories and supersymmetry theories of elementary particles. The point of principle which is important for cosmological studies is that these theories may permit the existence of many different types of ultraweakly interacting particles, among which are axions, photinos, etc. (see Chapters 14, 15, 17 and 18 by Close, Georgi, Taylor and Salam). No one knows whether or not these particles really exist, but at the moment particle physicists are optimistic about finding a successful theory in which they exist. An interesting example of the type of particle which might make up the dark matter is the lightest supersymmetric partner of the photon, the photino, which is predicted to be a stable particle. Its mass might lie in the range 1–10 GeV. It is interesting that unknown weakly interacting particles of mass $\simeq$ 1–10 GeV of cosmic origin could be detected in recoil experiments which have become practicable with the development of very low noise cryogenic detector systems. The cosmological studies described above have excited the interest of the particle physicists because at the moment the early Universe provides the only laboratory in which these ideas might be tested. For completeness, one should add other extreme forms of dark matter such as cosmic strings (see Chapter 3 by Guth and Steinhardt).

The way in which the dark matter helps is that, since it has no interaction with matter or radiation in the late Universe except through its gravitational influence, it can be supposed that it is the medium in which the perturbation spectrum develops prior to recombination. The baryonic matter and radiation are uncoupled to the dark matter and so they can possess a perturbation spectrum well below that which would cause problems in the conventional picture. After recombination, the matter and radiation are uncoupled and then the baryonic matter falls into the growing perturbations in the dark matter and rapidly attains the same level of density enhancement. In this picture the baryonic matter we see forms structures within the dark matter perturbations already present.

The alternatives of hot and cold dark matter again produce rather different predictions about the way in which the structure forms. In the case of the massive neutrino picture in which conventional neutrinos have mass $\simeq$ 30 eV, they remain relativistic so long as $kT > mc^2 \sim 30$ eV. This means that so long as $kT > 30$ eV, all small-scale perturbations within the horizon are damped out because the neutrinos escape freely at the velocity of light from any perturbation. The only scales which can grow are those on the scale of the horizon or greater. Scales smaller than the horizon can only survive when the neutrinos become nonrelativistic. The result is that in the Hot Model in which the neutrinos have rest mass 30 eV, the structures which survive at the epoch of recombination are very massive $M \simeq 10^{16} M_\odot$. These are the structures into which the baryonic matter falls, and the evolution is similar to the standard adiabatic model except that the large-scale structure is determined by the dark matter. The main problem which has been identified with this scenario is that numerical simulations suggest that it is too successful in producing narrow stringy structure in the Universe. By making simple assumptions about where galaxies form in this picture, the N-body simulations suggest that galaxies should be more strongly clustered than is observed but with a correlation function similar in shape to the relation (6.14). It is not clear how significant this problem is because astrophysical processes may well blur the predicted structure. The laboratory determination of the rest mass of the neutrino is of the greatest astrophysical and cosmological importance. The recent upper limits to the mass of the neutrino from the SN1987a supernova would still be consistent with this picture.

The cold dark matter picture successfully avoids the damping problem because the matter is cold to begin with and therefore the observed spectrum of the perturbations reflects the spectrum which came out of the early stages of the Big Bang. The picture is similar to the isothermal model in which the cold dark matter perturbations on a very wide range of scales develop to reasonable amplitude at the epoch of recombination and then the baryonic matter falls into these perturbations. The main problem which this theory encounters is the origin of the large scale holes and filaments in the Universe as was found in the isothermal picture involving hierarchical clustering. To overcome this problem, it has been supposed that galaxy formation may be biased towards those regions with the largest overdensity. This theory has achieved some success in explaining the small-scale structure but has problems in accounting for the large-scale structure and streaming motions already discussed.

The amount of theoretical analysis which has been undertaken of these problems is impressive, but it will be noted that we have not addressed the problem of the origin of the perturbations from which structure in the Universe develops. The types of simple analysis given at the beginning of the present section are sufficient to show that galaxies cannot develop from infinitesimal perturbations in the very early Universe. The amplitudes of the perturbations must be about $\delta\rho/\rho \simeq 10^{-4}$ when they enter the horizon or the large-scale structure of the Universe cannot be reproduced. The origin and mass spectrum of these finite-amplitude fluctuations have no explanation in the standard Hot Model. In all the analyses described above, it is assumed that these fluctuations are present in the beginning with a spectrum which will result in more or less the correct two-point correlation function at the present day and then the subsequent development of these perturbations follow. According to the standard model of the Universe, the initial fluctuation spectrum is something else which should be regarded as an arbitrary initial condition. Again, there are prospects that a proper understanding of the very early phases of the Hot Big Bang involving Grand Unified Theories, quantum fluctuations, phase transition and inflationary expansion may offer some explanation of this fluctuation spectrum (see Chapter 3 by Guth and Steinhardt).

It will be noted that the discussion is becoming more and more speculative and more bandages are being applied to the patient to prevent fatal confrontation with the observations. My aim has simply been to show why it is that astrophysicists have been driven to these considerations. It is not just that particularly exciting band-wagons happen to have turned up simultaneously in the areas of cosmology and particle physics – there are real basic problems which I regard as being among the most important of present day physics.

Basic problems

We have now uncovered several of the basic cosmological problems which have no solution within the context of the standard Hot Model of the Universe. These are:

(i) The origin of the baryon asymmetry in the Universe.
(ii) The origin of the spectrum of fluctuations from which galaxies and clusters form.

To these we should add

(iii) The origin of the isotropy of the Universe. Whilst this greatly simplifies the task of the cosmologist, there is a basic problem of *causality* in understanding why the Universe is so isotropic in the standard picture. This is because, as we go back in time, smaller and smaller regions of the Universe are causally connected so that it is not at all clear how regions of the Universe which are now well separated knew that they had to have the same average properties.

(iv) Why is the universe close to the critical model, $\Omega = 1$? *A priori*, there is no good reason why this should be so, and indeed it requires rather fine-tuning on the part of the initial conditions of the Universe to ensure that it will come out right.

In the standard picture, these four problems are solved by adopting suitable initial conditions which are fed into the model Universe in order to produce our Universe as it appears now. This is an unsatisfactory position, but there are at least two approaches to solving these problems. One is to adopt the *anthropic principle* according to which there will only be observers to look at the Universe in rather specially selected universes. For example, it is necessary for the existence of astronomers that there should have been enough time for biological evolution to have created astronomers. Arguments of this type can be used to some effect but most astronomers would regard them as an admission of defeat in the great challenge to find a physical origin for our Universe. The second approach is to use the remarkable new ideas of particle physics as embodied in the Grand Unified Theories and the attempts to quantise gravity. These topics are the subject of Chapters 3 and 5 by Guth and Steinhardt, and Isham, and I leave these cases to them. My own view is that there is much more than a glimmer of hope in these new ideas.

6.9 The astronomy of the future

Throughout this review, a dominant theme has been the impact of the new astronomies and their advancing technologies upon contemporary astrophysics. It is therefore appropriate to conclude with an outline of the types of observational facilities which may become available to astronomers through to the end of the present century. I fully expect theoretical advances to be strongly driven by ground- and space-based observations made with these facilities. I will take as my reference point a date ten years hence, i.e. the late 1990s.

Optical astronomy

In space, the NASA–ESA *Hubble Space Telescope* (HST) will be well into its mission, which is to provide astronomers with an optical and ultraviolet observatory with a fifteen year lifetime in space (figure 6.86). At the moment the launch date is uncertain, the only certainty being that the earliest launch date is in 1989. Because of the very high priority accorded to the HST programme, it is likely that it will be launched soon after that date, and it seems reasonable to expect that it will be in full operation in 1990. The telescope has a 2.4 m diameter primary mirror with full observatory-type facilities for imaging and spectroscopy with diffraction-limited optics in the waveband

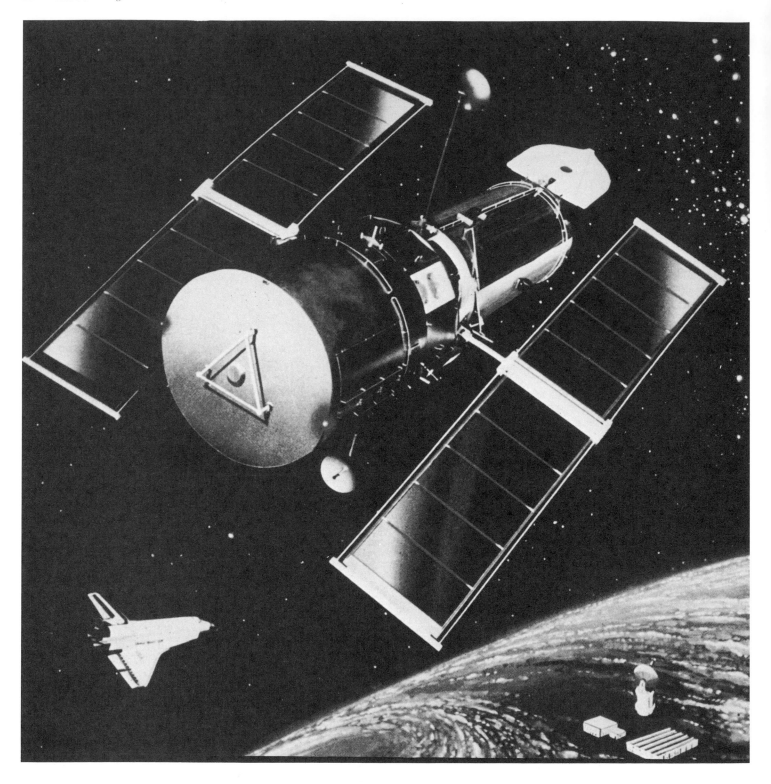

Figure 6.86. The NASA–ESA Hubble Space Telescope.

120 nm to 1 μm. It will open up many new areas of study which are likely to have a very strong impact upon all aspects of astronomy. It is likely that there will have been at least one instrument change by the late 1990s and possibly two.

The *Hipparcos* astrometric satellite of ESA will be launched in 1988 and will undertake its three-year mission of measuring very precisely the positions of the brightest 100 000 stars in the sky. The analysis of the data will take a further period of probably three years so that, by the mid to late 1990s, this very rich source of data of fundamental importance for astronomy will be available to all astronomers. It is expected to provide improved estimates of the cosmic distance scale and also stringent tests of stellar evolution upon which so much of our understanding is based.

In ground-based astronomy, the first of the *very large telescopes* (VLTs) will be in operation, i.e. optical–infrared telescopes with apertures of 8 m or larger. The 10 m *Keck Telescope*, which is currently under construction on the summit of Mauna Kea in Hawaii and which has adopted the segmented mirror concept with active control of the figure of the mirror, will be operational in about 1992. The *European Southern Observatory* (ESO) is planning a four-element array telescope, each of the elements being 8 m in diameter resulting in an equivalent aperture of 16 m. This project is not yet fully approved, but the plan is to complete the first of the 8 m telescopes in 1993 and to have the full array of four telescopes completed and in operation in 1998. There are numerous other proposals being discussed at the moment, a number of them involving the use of the large honeycomb 7.5 to 8 m blanks, the technology for which is being developed by Dr Roger Angel in Arizona. Among these is the US *National New Technology Telescope* (NNTT), which is planned to be of multi-mirror telescope design consisting of four elements each 8 m diameter. There are also several institutional initiatives within the USA. The *Japanese National Large Telescope* (JNLT) is a 7.5 m telescope to be located on the summit of Mauna Kea. I expect that, by the year 2000, a number of the VLTs will either be completed or at an advanced state of construction. The prime aim of the VLTs is faint-object imaging and spectroscopy, and they should enable the huge range of problems of contemporary astrophysics to be addressed. They will provide a particularly powerful tool for astrophysics in conjunction with HST observations.

Infrared astronomy

There will certainly be major advances in ground-based and space infrared astronomy throughout the period up to the year 2000. For both types of astronomy, one of the major reasons for these advances will be the impact of new detector technologies. New materials will become generally available with very low dark current performance which will result in at least an order of magnitude gain in sensitivity as compared with current indium antimonide (InSb) technology. These developments will

involve materials such as cadmium–mercury–telluride (CMT). A major incentive for the development of these materials for astronomical purposes will be the use of these detectors in the next generation of instruments for the Hubble Space Telescope. In addition to the higher sensitivities of the new materials, arrays of detectors will become available both for imaging and spectroscopy. In the case of spectroscopy, most of the devices now in operation on ground-based telescopes have of the order of ten elements, but we can confidently expect that this will increase to the order of 100 soon and, by the 1990s, to about 1000 elements and, in addition, two-dimensional spectroscopy will become possible. The combination of the new sensitive materials and the availability of large arrays will mean that infrared spectroscopy will more and more resemble optical spectroscopy in its information gathering capacity. For example, it will be possible to measure the infrared spectra of the faintest objects currently known in the near infrared. In the case of imaging in the infrared waveband, arrays of 64×64 pixels are already available. We can confidently expect that the arrays will continue to increase in size and also that new materials will be employed to extend the high sensitivity of the arrays into the far-infrared waveband. The latter will be crucial for capitalising upon space opportunities.

The second major advance in ground-based infrared astronomy will be the advent of the VLTs. These are all being designed to have excellent infrared performance. In the case of the short infrared wavelengths ($\lambda \leqslant 2\,\mu$m), where the observations are detector-noise limited, the improvement in signal-to-noise ratio is simply proportional to the area of the telescope whereas in the background-limited case the improvement is proportional to the diameter of the telescope. By the end of the 1990s, the VLTs will be equipped with large array detectors both for imaging and spectroscopy. These will be extraordinarily powerful facilities for astronomy.

The *Infrared Space Observatory* (ISO) of ESA will be launched about 1993. Whereas the HST will concentrate upon the short infrared wavebands ($1–3\,\mu$m), ISO will be dedicated to the longer wavelength bands, $3–200\,\mu$m, for which there is the greatest gain in sensitivity by going into space. The telescope will be cooled to about 4 K, and the four focal plane instruments will consist of an infrared camera ($3–20\,\mu$m), a short wavelength spectrometer ($3–30\,\mu$m), a long wavelength spectrometer ($20–200\,\mu$m) and a long wavelength photometer ($20–200\,\mu$m). The mission is a natural successor to the IRAS project and is designed to capitalise upon the wealth of data and new science produced by that mission. The projected lifetime of the mission is about twenty-one months, and so most of the observing programme will have to be very carefully preprogrammed for maximum efficiency.

The next major satellite observatory will be the *Space Infrared Telescope Facility* (SIRTF) of NASA. In specification, the telescope is not dissimilar from ISO but it has a somewhat larger primary mirror and it is planned to have a much longer lifetime in space with the capability of refilling the dewars with liquid helium in space. This is a natural successor to ISO but the project is not yet approved and a launch date has not been determined.

Ultraviolet astronomy

The Hubble Space Telescope will be the dominant ultraviolet telescope for observations in the 120–300 nm waveband. In many ways HST may be thought of as a 'super-IUE' with about thirty times the collecting area and with the capability for deep imaging with both the wide-field/planetary camera and the faint object camera. In the first generation of instruments there are also a high resolution spectrograph dedicated to the ultraviolet waveband 120–320 nm and a faint object spectrograph, each of them consisting of a linear array of 512 digicons. It is expected that, in the second generation of instruments for the HST, the spectrographs will have a two-dimensional capability, enabling much larger regions of the spectrum to be observed in an echelle mode and also enabling the spectra of extended objects to be obtained in a single exposure. It is very likely that these capabilities will be available through to the end of the HST mission since they are quite fundamental to science which cannot be undertaken in any other way.

There remains the problem of the far-ultraviolet waveband beyond 120 nm. This is a difficult waveband in which to work because there are few usable materials out of which spectrograph mirrors with normal incidence optics can be constructed. Possible materials include pure aluminium but, because of oxidisation, the mirrors would have to be coated with aluminium in space, a technology which is currently being developed. The alternative is to use grazing incidence optics, and this has been adopted as baseline for the ESA project *LYMAN*. The scientific reasons for such a mission are the presence of many of the resonance lines of the common elements in the far-ultraviolet waveband. In addition, there are important lines such as the Lyman lines of deuterium, which are of great cosmological interest, and also resonance lines of molecular hydrogen. This is a waveband which has yet to be properly exploited astrophysically. We have already had a foretaste of the type of science which can be undertaken from the results obtained with the Copernicus satellite, but it barely scratched the surface of the subject. With LYMAN it should be possible to observe the far-ultraviolet spectra of the brighter quasars whereas Copernicus was restricted in its capability to the brightest stars. LYMAN is planned to be an observatory mission with a long lifetime, the earliest launch date being the mid to late 1990s.

X- and gamma-ray astronomy

The next major project in X-ray astronomy will be the German–UK–USA *ROSAT* project. The dual goals of this project are, first, to provide a deep all-sky survey and, second, to provide an observatory-type facility for pointed observations. In the survey mode, the sources will be about 1000 times fainter than the faintest catalogued by the UHURU X-ray telescope. In addition, there is a far-ultraviolet camera for observations in the very far-ultraviolet waveband. The launch date is February 1990. The mission will last for about three years and so the results will be well understood by the late 1990s.

Other X-ray missions which are planned for the late 1980s and early 1990s include the US *Extreme Ultraviolet Explorer* (EUVE) and the *X-ray Timing Explorer* (XTE), and the Italian–Netherlands–ESA *SAX* project. In gamma-ray astronomy, the US *Gamma-ray Observatory* (GRO) should provide a significant advance over the pioneering work carried out by SAS-2 and COS-B.

The next generation of X-ray satellites will consist of large instruments whose primary purpose is to take pointed observations either in an imaging or spectroscopic mode. The *Advanced X-ray Astronomy Facility* (AXAF) of NASA will be the X-ray equivalent of the HST. It will undertake very deep imaging of all classes of object as well as making deep surveys of small areas of sky. To complement this programme there is an ESA 'cornerstone' mission known as XMM involving high sensitivity X-ray spectroscopic observations. The key-words of this mission are *high throughput* in the sense that the prime goal of the mission is spectroscopy in the 0.5–10 keV waveband with very large effective collecting area and high quantum efficiency. The current status of both projects is that they are seen as high priority programmes in both the USA and in Europe but neither has received final approval. AXAF is planned to be one of the first major projects to follow the successful launch of the HST. XMM is the second of ESA's planned cornerstone missions after the *SOHO-Cluster* mission for solar-terrestrial relations. This suggests that XMM could be launched in about 1998.

Radio astronomy

The large radio telescopes of the present generation are all designed to be long-lived observatory facilities. Instruments like the *Very Large Array* (VLA) will be operating with enhanced receivers and facilities throughout the period up to 2000, probably with an extension of the baselines available. For radio astronomical studies in the Southern Hemisphere the *Australia Telescope* (AT) will carry out many of the types of observation which are now carried out in the north by the VLA, and in addition it will have some long baselines which will provide it with angular resolution similar to that of the Jodrell Bank *MERLIN* array. There exists already the *European Very Long Baseline Interferometry Network*, and it will continue to develop through the next twenty years. The US *Very Long Baseline Array* (VLBA) is now being constructed and will provide a dedicated array of telescopes for VLBI. It is important that these facilities be guaranteed a long lifetime because in the VLBI area some of the most important studies concern time-varying phenomena such as superluminal velocities where the underlying physics can only be properly understood if systematic monitoring of a substantial number of sources is carried out over a period of a decade or more.

The natural extension of the VLBI programmes is to locate one of the elements of the array in space so that long baselines are obtained and, even more important, much more complete coverage of the aperture plane is obtained as the satellite moves in its orbit relative to the stationary ground-based telescopes. This is the goal of the ESA *QUASAT* mission, and there is a similar programme planned by the USSR, known as RADIOAS-TRON, with even longer baselines in space. By the year 2000, it is likely that VLBI incorporating at least one telescope in space will have been achieved.

Millimetre and submillimetre astronomy

There is currently a great flurry of telescope construction for the millimetre and submillimetre wavebands. These include two submillimetre telescopes for Mauna Kea, one for La Silla in Chile and one on the mainland USA. These telescopes will be fully occupied throughout the 1990s in opening up those regions of the submillimetre waveband which can be observed from the surface of the Earth. There are many important molecular lines to be observed as well as continuum sources such as dust clouds and active nuclei and also the study of the Microwave Background Radiation.

By the early 1990s, the prospects for ground-based milli-metre and submillimetre interferometry will also have been explored. The possibility of submillimetre synthesis telescope arrays and millimetre VLBI have already been considered. It is interesting that, even with a baseline of 100 m, angular resolution of 1 arcsec is obtained in the submillimetre waveband. What the longest practicable baselines are remains to be seen.

There remain, however, important submillimetre wave-bands which can only be studied from space. The precise determination of the spectrum and large-scale isotropy of the Microwave Background Radiation will be undertaken by the NASA *Cosmic Background Explorer* (COBE) which should fly in the late 1980s or early 1990s. Some of the main atomic and molecular coolants of the interstellar gas emit their radiation in the waveband $300-50\,\mu m$. These can be explored by high-flying aircraft, but the best solution is a submillimetre telescope in space. These have been studied in Europe, a project known as *FIRST* (Far-Infrared Space Telescope), and in the USA in a project known by the acronym *LDR*, meaning Large Deployable Reflector, referring to the technique by which a large sub-millimetre telescope might be put into space. These are both projects for the first decade of the twenty-first century.

Optical and infrared interferometry

The prospects for optical and infrared interferometry are now very good, but they have not been accorded the scientific priority which might reasonably have been expected by now. Part of the problem is that there is so much optical and infrared astronomy which needs to be done with existing 'single dishes' that there has not been the same urgency about interferometry as there was in the case of radio astronomy, where it was essential in order to achieve any angular resolution at all. In addition, interferometry is much more difficult in the optical and infrared wavebands, largely because of phase irregularities introduced by fluctuations in the refractive index of the atmosphere. It has now been demonstrated how these problems can be overcome. Therefore, the present and next generation of optical and infrared telescopes are being designed with inter-ferometry as a key objective. The ESO VLT will have an obvious interferometric capability, particularly if auxiliary telescopes of smaller diameter can be added on to the array. There are plans to construct modest infrared telescope arrays to exploit the capabilities for interferometry in the infrared waveband. These experiments should open up new regions of parameter space for ground-based observation. I would expect that by the year 2000 optical and infrared interferometry will be much more integral parts of astronomy than they are now.

More new astronomies

It is always difficult to predict where the next great discovery will come from which will expand further the parameter space available for astronomical observations. Three areas seem to be ripe for exploitation in the 1990s. There are already plans for very sensitive *gravitational wave telescopes*. The technology is now at the point where it seems feasible that gravitational waves from astronomical objects can be detected directly. Already the sensitivities are within a factor of about one-hundred of being able to detect supernovae in distant galaxies. These telescopes are complex and expensive but this is a wholly new dimension for astronomical study and it is only a question of time before the first direct detections of the waves are made. I would be surprised if they were not detected during the 1990s.

A second area is that of *ultrahigh energy gamma-rays*, which have just begun to be detected reliably by optical air shower techniques. The gamma-rays are of such high energy that they pose problems for conventional emission processes and give us information directly about the sites of the acceleration of very high energy particles.

A third area is the search for exotic forms of matter of cosmic origin using ground-based detectors. One of the few ways of detecting hypothetical particles such as axions and photinos may be through the measurement of cosmic fluxes of these particles. The technology now exists for such experiments and, although no one would deny that the ideas are to a large extent speculative, the importance of a positive result would be enormous for physics and astrophysics. One is forcefully reminded of the way in which cosmic ray studies led to great advances in particle and nuclear physics before the correspond-ing particles could be created in the laboratory. It might be that we are on the brink of a similar epoch if the searches for relic photinos, heavy neutrinos, axions, etc. were to prove successful.

Conclusion

We have come full circle in our review of the New Astrophysics, looking forward to the next surprises of the new astronomies. Whilst these provide the stimulus which often leads to fundamental new insights, there remains a huge programme of consolidation of our knowledge and understanding of the great achievements of the last forty years. I tell my students that my goal as an astrophysicist is to make the subject as rigorous as laboratory physics. In a number of areas this is now being approached but, in many of the most exciting new disciplines, the fundamental truths have not yet been uncovered. This process of consolidation is at least as important as the discovery of new phenomena which has characterised the New Astrophysics of the last forty years.

6.10   Appendix – Astronomical nomenclature

For historical reasons, astronomers have developed particular, convenient units for measuring distances, masses, luminosities and colours of celestial objects. A complete exposition is complex but the following brief notes provide the essential facts needed to comprehend this chapter.

Distances in astronomy

The unit of distance used in astronomy is the *parallax-second* or *parsec*. It is defined to be the distance at which the mean radius of the Earth's orbit about the Sun subtends an angle of one second of arc. In metres the parsec, abbreviated to pc, is 3.086×10^{16} m. For many purposes, it is sufficiently accurate to adopt $1 \text{ pc} = 3 \times 10^{16}$ m. The parsec is a recognised SI unit, and it is often convenient to work in kiloparsecs ($1 \text{ kpc} = 1000 \text{ pc} = 3 \times 10^{19}$ m) or megaparsecs ($1 \text{ Mpc} = 10^{6} \text{ pc} = 3 \times 10^{22}$ m).

Sometimes it is convenient to measure distances in light-years (l.y.), which is the distance light travels in one year: $1 \text{ l.y.} = 9.461 \times 10^{15}$ m. Thus, $1 \text{ pc} = 3.26$ l.y.

Another commonly used distance unit in astronomy is the *astronomical unit*, abbreviated to AU, which is the mean radius of the Earth's orbit about the Sun. $1 \text{ AU} = 1.49578 \times 10^{11}$ m. The very nearest stars to the Earth are at a distance of about 1 pc, and so they are about 2×10^{5} times as far away as the Earth is from the Sun.

Masses and luminosities

It is convenient to measure the masses of many celestial objects in terms of the mass of the Sun. One solar mass, written $1 M_\odot$, is $1.989 \times 10^{30} \text{ kg} \simeq 2 \times 10^{30}$ kg. Thus, Jupiter has a mass of about $0.02 M_\odot$, our own Galaxy has a mass of about $10^{11} M_\odot$, and the Coma cluster of galaxies has mass about $3 \times 10^{15} M_\odot$.

In the same way, it is convenient to measure the luminosities of celestial objects in terms of the *bolometric luminosity* of the Sun. The bolometric luminosity of the Sun, written as $L_\odot$, means the total luminosity of the Sun integrated over all wavelengths. Its value is $L_\odot = 3.90 \times 10^{26}$ W.

Magnitudes and colours

In general, we work in SI units throughout the text but it may prove necessary on occasion to use astronomical measures of flux density. To define flux densities, astronomers often work in terms of *magnitudes m* which are negative logarithmic measures of flux density defined by

$$m = \text{constant} - 2.5 \log_{10} S,$$

where S is the flux density of the object. Thus, a difference of five magnitudes corresponds to a difference in flux density of a factor of one-hundred. The magnitude system is normalised so that a standard star, chosen to be the bright star Vega (or α-Lyrae) in the constellation of Lyra, has zero magnitude at all wavelengths. In this way, magnitudes can be defined at all wavelengths. The very brightest stars in the sky have magnitude about 0. The faintest stars which can be seen with the naked eye have $m = 5$. The faintest stars which can now be observed on a four-metre telescope in a five minute observation using a CCD camera have $m \simeq 25$. To measure the intrinsic luminosities of celestial objects, astronomers often use *absolute magnitudes M*. These are defined to be the magnitude which the object would have if it were placed at a distance of 10 pc. Because of the inverse square law, an object of intrinsic luminosity L has flux density $S = L/4\pi r^2$ and therefore, for any object,

$$M = m - 5 \log_{10}(r/10),$$

where the distance r is measured in parsecs. For stars of different luminosities,

$$M = M(\text{Sun}) - 2.5 \log_{10}(L/L_\odot).$$

The absolute bolometric magnitude of the Sun is $M(\text{Sun}) = 4.75$ and hence in general we can write

$$M = 4.75 - 2.5 \log_{10}(L/L_\odot).$$

Astronomers define the *colours* of stars in terms of the difference in their magnitudes at different wavelengths. For example, a commonly used colour is the difference between the magnitudes of a star in the blue (B) waveband (centred on 440 nm) and the visual (V) waveband (centred at 550 nm). This colour index or colour $(B - V)$ is a measure of how blue or red the star is. As expected from figure 6.1, blue stars are hot and red stars are cool. This statement is quantified by working in terms of colours such as $B - V$.

Further reading

There is a huge amount of literature on astronomy and cosmology at the popular and the technical level. I give here a list of some texts which I have recommended to students and enthusiasts for further reading.

Selected books on the new astronomies

Field, G. B. and Chaisson, E. J. (1985). *The Invisible Universe.* Birkhauser Boston Inc.
  A descriptive book on the impact of the new astronomies upon contemporary astrophysics. It contains interesting anecdotes and personal reminiscences.

Henbest, N. and Marten, M. (1983). *The New Astronomy.* Cambridge University Press.
  An excellent introduction to the new types of astronomy which have become possible with the opening up of the whole of the electromagnetic spectrum for astronomical observations. Many excellent false colour photographs of objects observed in different astronomical wavebands are included.

Israel, F. P. (ed.) (1986). *Light on Dark Matter.* Proceedings of the 1st IRAS Conference. D. Reidel and Co.
  There is no simple up-to-date introduction to infrared astronomy. This symposium proceedings catches the flavour of the great advances represented by the IRAS satellite.

Kraus, J. D. (1986). *Radio Astronomy,* 2nd edn. Cygnus-Quasar Books, Powell, Ohio.
  An excellent textbook covering many of the basic concepts of radio astronomy techniques.

Learner, R. (1981). *Astronomy Through the Telescope.* Evans Brothers Limited, London.
  An interesting introduction to the history of the telescope from the point of view of telescope construction.

Ramana Murthy, P. V. and Wolfendale, A. W. (1986). *Gamma-Ray Astronomy.* Cambridge University Press.
  An up-to-date account of the state of gamma-ray astronomy and its perspectives for the future.

Tucker, W. and Giacconi, R. (1985). *The X-ray Universe.* Harvard University Press.
  A vivid and readable account of the development of X-ray astronomy and its future perspectives by one of its founders.

Walker, G. (1987). *Astronomical Observations – An Optical Perspective.* Cambridge University Press.
  A modern text-book on techniques of optical astronomy with interesting comparisons with other types of astronomy.

General references

Abell, G. O. (1982). *Exploration of the Universe,* 4th edn. Saunders College Publishing, Philadelphia.

Abell, G. O. (1984). *The Realm of the Universe,* 3rd edn. Saunders College Publishing, Philadelphia.
  These text books by Abell are among the best of the simple descriptive introductions to modern astronomy.

Audouze, J. and Israel, G. (eds.) (1985). *The Cambridge Atlas of Astronomy.* Cambridge University Press.
  A very beautiful atlas of many different classes of object in the Universe.

Kaufmann, III, W. J. (1985). *Universe.* W. H. Freeman and Co., NY.
  A well-illustrated introduction to astronomy with many excellent up-to-date illustrations.

Mitton, S. (ed.) (1977). *The Cambridge Encyclopaedia of Astronomy.* Jonathan Cape, London.
  Now somewhat out-of-date but still an excellent overview of modern astronomy written in nonspecialist language by a team of distinguished younger astronomers.

Shu, F. H. (1982). *The Physical Universe: An Introduction to Astronomy.* University Science Books, Mill Valley, CA.
  My own personal favourite elementary text-book for introducing students to the physics of astrophysics.

Stars and stellar evolution

Kippenhahn, R. (1983). *100 Billion Suns.* Weidenfeld and Nicholson, London.
  An excellent qualitative description of the basic physics of stars and stellar evolution by one of the great pioneers of the subject.

Shklovskii, I. S. (1978). *Stars: Their Birth, Life and Death.* W. H. Freeman and Co., NY.
  A nontechnical description of stars and stellar evolution by one of the great Soviet masters of astronomy and astrophysics.

Tayler, R. J. (1970). *Stars, Their Structure and Evolution.* Wykeham Publications Ltd., London.
  My own favourite introductory book on the theory of stellar structure and evolution. Professor Tayler provides a brilliant introduction to the mathematical theory at a level which is intended to be understood by advanced school-children. The quality of the writing and the science is such that I use it in my undergraduate teaching.

Interstellar medium

Spitzer, L., Jr (1982). *Searching Between the Stars.* Yale University Press, New Haven.
  A brilliant book which discusses carefully the basic physics of the interstellar medium. There is a wealth of material to ponder in this book.

Galaxies

All the general references above contain good descriptions of galaxies.

Michalas, D. and Binney, J. (1981). *Galactic Astronomy: Structure and Kinematics.* W. H. Freeman and Co., NY.
  A classic text-book on the properties of galaxies. The book is the first of two, the second concentrating upon the more mathematical aspects of the theory of the structure of galaxies.

Tayler, R. J. (1978). *Galaxies: Structure and Evolution.* Wykeham Publications Ltd., London.
  Another excellent introduction by Professor Tayler.

Active galaxies and quasars

There is no simple introductory text devoted to this subject. There are good chapters on these objects in all the general references above.

Longair, M. S. (1981). *High Energy Astrophysics.* Cambridge University Press.
  An introduction at the final-year undergraduate level to many of the physical tools and concepts needed in modern high energy astrophysics. A new edition is promised by the present author.

Rybicki, G. B. and Lightman, A. P. (1979). *Radiative Processes in Astrophysics.* Interscience, NY.
  This is an excellent text-book at the final-year undergraduate or first-year postgraduate level on radiative processes in astrophysics. The exposition of the fundamental physics of many of the radiative processes which are commonly used in the discussion of active galaxies and quasars is strongly recommended.

Shapiro, S. L. and Teukolsky, S. A. (1983). *Black Holes, White Dwarfs and Neutron Stars: The Physics of Compact Objects.* Wiley-Interscience, NY.
An outstanding but technical text-book on the detailed physics of white dwarfs, neutron stars and black holes. Interestingly, the authors concentrate mostly upon stellar mass black holes rather than supermassive black holes in active nuclei.

Weedman, D. W. (1986). *Quasi-stellar Objects.* Cambridge University Press.
An excellent survey of the observed properties of quasars. The book has the positive feature of avoiding excessive speculation as to their nature.

Cosmology

Barrow, J. D. and Silk, J. (1983). *The Left Hand of Creation: The Origin and Evolution of the Expanding Universe.* Basic Books, NY. (Heinemann, London, 1984)
An excellent account in nontechnical language of many of the important topics being actively discussed in contemporary cosmology. The book successfully describes how the modern theories of particle physics may help solve some of the great cosmological problems.

Harrison, E. R. (1981). *Cosmology: The Science of the Universe.* Cambridge University Press.
A nontechnical introduction to many of the basic ideas of modern cosmology.

Silk, J. (1980). *The Big Bang.* W. H. Freeman and Co., NY.
Another excellent account of the physics of the Hot Big Bang model with emphasis upon the physical problems encountered.

Weinberg, S. (1977). *The First Three Minutes.* Andre Deutsch, London.
An outstanding exposition of the case for the Hot Big Bang model of the Universe by one of the most distinguished of modern physicists. This book is already a classic of popular exposition.

Astronomical reference books

Allen, C. W. (1973). *Astrophysical Quantities.* The Athlone Press, London.
A useful reference for many pieces of astronomical information, particularly in relation to physical processes.

Lang, K. R. (1974). *Astrophysical Formulae.* Springer-Verlag, Berlin.
A useful compilation of astrophysical formulae presented without derivations but with ample references to the original sources.

The best source of up-to-date reviews of many aspects of astronomy and astrophysics is the excellent series of *Annual Reviews of Astronomy and Astrophysics.* These are technical reviews but are uniformly of very high standard. Volume 1 appeared in 1963 and the most recent volume, published in 1987, is volume 25.

7 Condensed matter physics in less than three dimensions

David Thouless

7.1 Introduction

General discussion

The space in which the real world resides has three dimensions, and it is not obvious why physicists should be at all concerned about physics in either less or more than three dimensions. In fact they have paid a lot of attention in recent years to how matter should behave in different numbers of dimensions, for a variety of reasons. It seems that the interest in four or more space dimensions is entirely theoretical, but the interest in two- or one-dimensional systems is practical as well as theoretical. The physics that has been developed is of importance to technology as well as to pure science; for example, many of the devices used in integrated electronic circuits involve two-dimensional systems. In this chapter a brief introduction is given to some of the theoretical ideas that lie behind recent work on low-dimensional systems, and some of the real systems to which these ideas can be applied are described.

Since many of the examples discussed in this chapter are taken from the theory of magnetism, there is a brief discussion of the background of this subject. A brief survey of the theory of phase transitions and critical phenomena is given also, although this overlaps to some extent with the chapter by Bruce and Wallace.

In Section 7.2 the general theoretical background of low-dimensional systems is described. One reason that impels physicists and applied mathematicians to look at systems in less than three dimensions is that very often they are simpler. There are two rather different reasons for this greater simplicity. The first is the rather obvious reason that space in lower dimensions is described by fewer coordinates, and so calculations are less laborious. The second is that the topology – the structure – of lower dimensional spaces is simpler, and this enables various tricks to be used that cannot be used in higher-dimensional spaces. This has been known and exploited for a long time. What has been realised in recent years is that many physical phenomena are controlled by a 'correlation length'. Examples

of this are given later in this chapter, and also in Chapter 8 by Bruce and Wallace. If the correlation length is greater than the thickness of a thin film, then the film will behave two dimensionally. If the film is thicker than this correlation length, then the film will behave more or less like the bulk (three-dimensional) material. This idea enables experimental physicists to study materials under conditions in which they should be behaving in a two- or one-dimensional manner, and to observe the crossover to three-dimensional behaviour. Another very important idea that has been recognised during the past ten or twenty years is the idea of 'universality', which implies that real systems have certain features in common with simple, idealised systems; again this concept is discussed in the chapter by Bruce and Wallace.

In Section 7.3 it is shown how these general ideas have been applied to the understanding of phase transitions in real systems. Much of the most interesting detailed work has been done on magnetic materials, because there is a wide variety of magnetic materials available and they are susceptible to very detailed measurement by a number of different techniques such as neutron scattering and magnetic resonance. There have also been a lot of measurements on the properties of gases adsorbed on to the surfaces of pure solids. Freely suspended films of 'liquid crystals', systems intermediate in their properties between liquids and solids, can also be studied, and these provide a rich and well controlled variety of phenomena. The superfluidity of helium films is a rather special problem, as it shows a new type of phase transition, unexpectedly susceptible to detailed theoretical treatment, which is controlled by the spontaneous appearance of 'vortices'. Superconductivity in thin films and in thin wires is also discussed in the same section.

In Section 7.4 various phenomena which are studied in two-dimensional electron systems are described. There are two different ways of producing what can be regarded as a two-dimensional metal. One method is to trap electrons above the surface of a liquid or a solid. Liquid helium provides the most intensively studied system, because the surface is so smooth, but other systems are also being studied. The other method is to use the electron 'inversion layers' that can be produced at the interface between two semiconductors or between a semiconductor and an insulator. These heterojunction and MOS (metal–oxide–semiconductor) systems form the basis of a lot of modern semiconductor technology, so their physics is of more than academic interest. Two aspects of these systems are described. It has been realised recently that two-dimensional metals are unlike ordinary three-dimensional metals in one important respect, because the electron wave functions are always 'localised' rather than extended through the system like the waves that describe an electron in an ordinary metal. The second aspect that is described is the 'quantum Hall effect', which was discovered experimentally a few years ago. In a strong magnetic field and at low temperatures the electric current flows in a direction perpendicular to the voltage – this is the Hall effect which has been known for more than a hundred years. What was very surprising was the discovery that the ratio of the current to voltage is a simple multiple of a quantity that depends only on fundamental atomic constants. This discovery is of great theoretical and practical interest. It even provides a readily reproducible and portable standard electrical resistance for calibration of measurements.

There is a final section in which an attempt is made to assess the significance of this subject.

Ferromagnetism and antiferromagnetism

Many of the examples studied in this chapter are taken from the theory of magnetism. One of the early partial successes of the quantum theory of solids in the late 1920s was the development of a theory of magnetic materials.

A ferromagnet is a material such as iron in which there may be a permanent magnetic moment. The situation is complicated by the fact that real magnetic materials are made up of small 'domains', each of which carries a relatively large magnetic moment. The lowest energy is obtained when these domains are lined up in such a way that their magnetic moments cancel, so that the whole sample has quite a small magnetic moment. A permanent magnet is one in which the domains have been forced to line up parallel to one another. It is however the magnetisation of the individual domains rather than the overall magnetisation which we shall be discussing. When the temperature is raised the magnetic moment (of a single domain of fixed size) decreases, and it goes smoothly to zero as the temperature approaches the 'critical temperature', which for iron is about 1050 K. Just above this critical temperature there are still signs of ferromagnetism. When a magnetic field is applied to any material a magnetic moment is induced, and the ratio of the magnetic moment to the applied field is known as the 'magnetic susceptibility'. Just above the critical temperature the magnetic susceptibility is very large. Indeed, it becomes infinitely large as the critical temperature is approached from the high temperature side.

The theory that I describe is not actually appropriate for ferromagnetic metals such as iron, cobalt and nickel, but it is more suitable for ferromagnetic insulators. The ferromagnetic metals do not actually play an important part in this chapter. In insulating magnetic materials there are atoms (usually the 'transition metal' atoms in the centre of the periodic table of elements) which carry a magnetic moment. They also have a corresponding angular momentum known as the 'spin', which is usually considerably larger than the basic spin of a single electron. The purely magnetic interaction between these magnetic moments is very weak, and is only important at temperatures as low as the boiling point of helium at 4 K. Quantum theory predicts that there should be a much stronger interaction between the magnetic moments mediated by the electrons. Under most circumstances this interaction favours opposite alignment of neighbouring spins, but in some cases it favours parallel alignment. This interaction is known by the name of the 'exchange' interaction, because of the way in which it is mediated by the electrons. When parallel alignment of neighbouring spins is favoured the configuration of lowest energy of a small domain will be the one with all the spins pointing in the same direction, and this will give rise to a very large total magnetic moment. This large magnetic moment gives rise to a large magnetic field which tends to demagnetise the material, which is why only small domains are completely magnetised.

The lowest energy configuration is the one which a domain will adopt at very low temperatures. As the temperature is increased there will be fluctuations away from this configuration of lowest energy, and so not all the spins will be pointing in the same direction. It was argued around the turn of the century by P. Curie and R. Weiss that the lower the proportion of spins pointing in the direction of magnetisation, the less would be the tendency of the spins to point in that direction. This argument led to predictions that the magnetisation should indeed go smoothly to zero at a definite temperature, and that the magnetic susceptibility should diverge as the temperature approaches the critical temperature from above. This theory is, however, based on a very crude approximation, and it was by no means certain that the conclusions would survive in a more careful theory.

It has been mentioned that exchange coupling more often favours opposite alignment of neighbouring spins than parallel alignment. Thus one might expect to find materials whose configuration at low temperatures has the form shown in figure 7.1, with all neighbouring spins opposed to one another. This is indeed the case, and they are known as 'antiferromagnets'. Insulating antiferromagnets are much more common than insulating ferromagnets. There is no overall magnetic moment, so their magnetic properties are much less spectacular, but also there is no demagnetising field produced, so the complications of domain structure do not arise. The antiferromagnetic order can be shown by the behaviour of the magnetic susceptibility, since the antiferromagnet is more readily magnetised by a magnetic field perpendicular to the direction along which the spins point than by a magnetic field parallel to that direction. At the critical temperature the magnetic susceptibilities in these two perpendicular directions become identical. The development of neutron scattering as a technique for studying the structure of solids allowed much more detailed information about antiferromagnets to be obtained, as neutrons are sensitive to the spin direction of the atoms which scatter them. There is now much more information about the behaviour of antiferromagnets than about ferromagnets.

As the theory has been presented so far, the spins of the atoms can be aligned in any direction in space. This is characteristic of the 'Heisenberg model' of magnetism. In fact the crystal structure of the magnetic solid produces additional forces on the magnetic moments of the atoms which may tend to produce alignment in a definite direction in space. If this tendency is strong one may get a situation where the spins are likely to be pointing in one of the two possible directions along a particular axis. This is the situation which is described by the 'Ising model', which is discussed later in more detail.

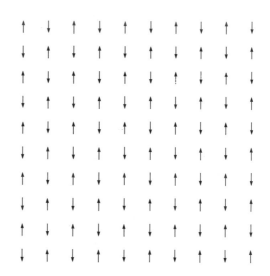

Figure 7.1. The state of lowest energy for a two-dimensional antiferromagnet. The spins point up on one sublattice and down on the other.

Phase transitions and critical phenomena

Changes of phase are very familiar from everyday life. The melting of a solid and the boiling of a liquid are both examples of phase transitions. When a solid is heated up to its melting point and begins to melt, heat must be supplied which is known as the latent heat of melting. While the latent heat is being supplied there will be solid mixed with liquid, both at the melting temperature. Much the same happens when a liquid is boiled. This kind of phase transition, characterised by a latent heat and coexistence of two distinct phases at the same temperature, is known as a 'first order' phase transition.

The disappearance of magnetism when a ferromagnet or antiferromagnet is heated proceeds in a quite different way. The two phases, ferromagnetic (or antiferromagnetic) and nonmagnetic are quite distinct, and so it is also a phase transition, but the magnetic moment goes continuously and smoothly to zero, and so it is known as a 'continuous' phase transition. There is no latent heat which has to be supplied, although the specific heat may be discontinuous at the transition. The two phases never coexist at the same temperature.

Continuous phase transitions have generated a great deal of interest in the past thirty years, as can be seen from the chapter by Bruce and Wallace. Much of this interest has centred on the question of 'critical exponents'. I have already mentioned that the magnetic susceptibility of a ferromagnet becomes infinite as the critical temperature T_c is approached. The susceptibility is proportional to some negative power of $T - T_c$, and the value of that power is the susceptibility exponent. The exponent is equal to unity in the Curie–Weiss approximation, but its actual value is somewhat larger. Similarly, the specific heat also behaves as a power of the same quantity, but the specific heat exponent is quite close to zero. A whole zoo of other exponents have been defined, but I shall not be discussing the relations between them in this chapter.

7.2 General theoretical considerations

Solvable models in lower dimensions – the importance of topology

In 1925 E. Ising published a paper which discussed a simple model of ferromagnetism, which ever since then has borne his name. It was an attempt to discuss the problem of a set of atoms in a solid, each with a magnetic moment and a corresponding spin. Each spin interacts with its neighbours in such a way that the energy is lowered if the spins are parallel. This is the formulation of the theory of ferromagnetism which was described briefly in the previous section, and for which Curie and Weiss obtained an approximate solution. In the Ising model each atomic spin can be in one of two states, either up or down. The energy of the system, in the absence of an external magnetic field, is taken to be proportional to the number of pairs of nearest neighbours in which the spins are in opposite states, minus the number of pairs in which the neighbours are in the same state. It is obvious that the state of the system which has the lowest energy is a magnetised state, in which all the spins point in the same direction. This is shown in figure 7.2(*a*). If there is no applied magnetic field the two possible states with all the spins up or all the spins down have the same energy, but if there is an applied field, even a very weak one, this will favour one spin direction rather than the other. Ising studied the one-dimensional model, in which the spins are all arranged in a row with only two neighbours to each spin, in detail. He showed that at any temperature above absolute zero there would be no spontaneous magnetisation. This one-dimensional model is the one illustrated in figure 7.2. Unfortunately he then went on to say that this result can be extended to the three-dimensional model. As far as I can tell, this paper, embodying the results of a PhD dissertation prepared under the direction of Prof. Lenz, was the only paper which Ising ever published in the physics literature, and so Ising purchased fame remarkably cheaply. In fact much effort was spent during the following fifty years in studying the magnetic phase transition, at nonzero temperature, of the Ising model in more than one dimension.

In 1936 R. E. Peierls produced an argument to show that in two dimensions there must be a spontaneous magnetisation at temperatures above zero. He did this by considering the boundaries between blocks of up spins and blocks of down spins. These are shown for spins on a square lattice in figure 7.3(*a*). The energy measured from the state with all spins aligned is just proportional to the total length of such boundaries in the system, since this gives the total number of oppositely aligned pairs. To calculate the thermodynamic properties of the system it is also necessary to know the entropy, which is the Boltzmann constant k_B times the logarithm of the number of different configurations of the same energy. The entropy cannot readily be calculated exactly in terms of the total length of the boundaries, but an upper limit for it can be found. On a square lattice there are four directions in which a segment of the boundary can go. Since the boundary cannot go back on its tracks there are actually only three directions a given segment of boundary can go, so the entropy per unit length is not more than $k_B \ln 3$ per unit length. This enabled Peierls to set a lower bound to the temperature at which magnetism in the two-dimensional Ising model is destroyed by raising the temperature.

Shortly after this H. A. Kramers and G. H. Wannier found a simple line of reasoning to determine the transition temperature at which the two-dimensional Ising model goes from the low temperature magnetic phase to the high temperature nonmagnetic phase. They showed that it is possible to set up a precise correspondence between the properties of the model at high temperatures and at low temperatures. At high temperatures, as is shown in figure 7.3(*b*), there is a small excess of favourably aligned over oppositely aligned pairs of neighbouring spins. At low temperatures, as is shown in figure 7.3(*a*), there is a small proportion of oppositely aligned pairs of neighbours. This correspondence, which is now known by the name of 'duality', gives a relation between the thermodynamic free energy in the low temperature magnetic state and the free energy at a corresponding temperature in the nonmagnetic state. Where the transition between magnetic and nonmagnetic states occurs this duality relation becomes an equation which can be solved to find the transition temperature. This was the first of a number of arguments which have exploited the idea of duality in two-dimensional systems.

In 1944 L. Onsager published a complete solution of the two-dimensional Ising model in the absence of a magnetic field. This was a difficult piece of work, and the argument, even in its more modern versions, involves a number of steps which are far from obvious. The solution showed a sharp transition at the expected temperature between the magnetised and unmagnetised states, with a logarithmic divergence of the specific heat at this critical temperature. It was also found that the magnetisation in the low temperature phase was proportional to the $\frac{1}{8}$ power of the difference between the temperature and its critical value. Thus the first firm information about the critical exponents which determine the relation between thermodynamic variables in the neighbourhood of phase transitions was obtained. Later it was found that the magnetic susceptibility diverged with an exponent of $\frac{7}{4}$, rather than the value of unity given by the Curie–Weiss theory. Slowly over the years a number of more complicated two-dimensional models gave exact answers, and added to our knowledge of the ways in which phase transitions occur.

Not surprisingly there is a much greater variety of one-dimensional models that can be solved. Whereas the two-dimensional problems mostly have a certain family resemblance to one another, there are several rather different classes of tractable one-dimensional systems. In fact the solution of the two-dimensional Ising model depends at one point in most versions of the argument on the solution of a one-dimensional problem which was found by H. A. Bethe in 1931.

The reason for the relative simplicity of one-dimensional systems is related to their simple topology – the well known fact that a line can be broken into two disconnected parts by making a break at one single point. For example, the absence of magnetisation in the one-dimensional Ising model can be explained in this way. At zero temperature all the spins will be aligned, as is shown in figure 7.2(*a*), but the creation of a

Figure 7.2. The one-dimensional Ising model. The state of lowest energy is shown in (*a*), and a domain wall between two oppositely magnetised domains is shown in (*b*).

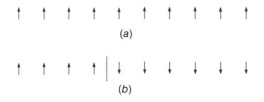

(*a*)

(*b*)

'domain wall', such as in figure 7.2(*b*), where two neighbouring spins are oppositely aligned costs only a finite amount of energy, and domain walls must therefore occur with nonzero concentration at any temperature above zero. To the right of a domain wall all information about what happens to the left of the domain wall is lost, and so there can be no preference for one spin direction rather than the other once there are any domain walls. In fact Ising's solution of the one-dimensional problem can be regarded as an application of the fact that successive domain walls are independent of one another. Other one-dimensional problems do not yield quite so readily, but most of them exploit this feature of one-dimensional space in one way or another.

Figure 7.3. Shown here are the domains in the two-dimensional ferromagnetic Ising model. The low temperature phase is shown in (*a*) and the high temperature phase in (*b*).

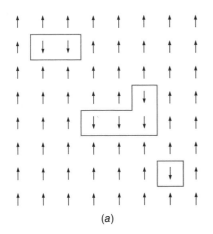

(*a*)

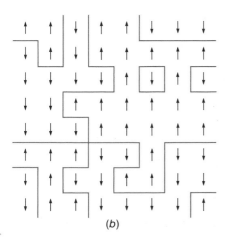

(*b*)

In two dimensions it requires a curve or a line to divide space into two regions. Peierls' argument for the existence of magnetisation at low temperatures in two dimensions was based on a consideration of the boundaries between regions of opposite spin, domain walls again, shown in figure 7.3(*a*). So was the argument of Kramers and Wannier for the value of the critical temperature. Onsager's solution involved considering lines of spins extending right across the system; for example, one of the horizontal rows of spins shown in figure 7.3(*a*) or (*b*). Again it can be argued that all information about what happens above the line is transmitted to the lower side through the line, so that a successive treatment of one line of spins after another is possible. The line is itself rather complicated, as it can be in very many different possible configurations. That is why the solution of a complicated one-dimensional problem is needed as an intermediate step in the solution of this relatively simple two-dimensional problem. This increasing complexity as the number of dimensions is increased also explains why, despite many claims of an exact solution, no one has solved the Ising model in three dimensions.

Correlation lengths

In the work that has been done over the past twenty years on the 'critical phenomena' that occur in the neighbourhood of a phase transition (see Chapter 8 by Bruce and Wallace), the correlation length plays a very important part. This length is the distance over which the effects of a disturbance spread, and it becomes very large in the neighbourhood of a critical point. For example, in a ferromagnetic system just above the critical temperature, if one spin is forced to point in a particular direction, then all the other spins within a distance equal to the correlation length will be aligned with it to a certain extent. Similarly there will be spontaneous fluctuations in which blocks of spins within a correlation length of one another will be coherently magnetised. At the critical temperature the correlation length becomes infinite, but below the critical temperature there is again a finite correlation length which gives the distance over which deviations from the uniform magnetisation occur. There is a corresponding correlation length in the gas–liquid system near its critical point, and this length gives the size of droplets of liquid which form spontaneously in the vapour phase, or of bubbles of gas which form spontaneously in the liquid phase. Because the correlation length close to the critical point becomes large enough to be comparable with the wavelength of visible light, the system takes on a cloudy appearance near the critical point, which is known as critical opalescence. There are many other problems in physics in which some sort of correlation length is important, and a few of these will be mentioned later in this chapter; it is a characteristic feature of critical phenomena that there is a region, the critical region, where the correlation length is very long.

We will consider the specific case of the Ising model of ferromagnetism, but the same ideas can readily be adapted to many other systems. In this model, as was pointed out earlier on, the spins are all lined up parallel at zero temperature. As the temperature is raised there is some reversal of the spins, but, except in one dimension, there remains an excess of spins in one direction and the system remains magnetised. There is a critical temperature at which the excess of spins in one direction disappears and the magnetisation vanishes. At the critical temperature the specific heat has some sort of sharp maximum, and the magnetic susceptibility, the rate at which the magnetisation changes when the magnetic field is changed, rises very rapidly, and becomes infinite at the critical temperature.

Now we consider how a large, but finite, system shaped like a cube behaves. It can be shown that except in the immediate neighbourhood of the critical point this system behaves very similarly to an ideal infinite three-dimensional system. As the temperature is lowered towards the critical temperature the correlation length increases, the specific heat rises slightly and the magnetic susceptibility rises rapidly. Once the correlation length is of the order of the length of the edge of the cube, however, it can increase no more, the specific heat reaches a smoothly rounded maximum, and the magnetic susceptibility increases much more slowly. Below the temperature at which this occurs the individual spins are locked together across the whole system, and the system behaves to a certain extent like a single large spin. Figure 7.4(*a*) shows how the magnetic susceptibility and correlation length behave for an infinite three-dimensional system and for a large cube.

Something very similar happens if we consider the spins to be arranged in an infinitely long bar of square cross-section instead of in a cube. As the temperature is lowered towards the critical temperature the system behaves very like the ideal three-dimensional system until the correlation length becomes comparable with the side of the square cross-section. At this point the spins are coherent across the cross-section. Therefore the magnetisation can be regarded as a quantity which is a function of the position along the bar, but which is constant across the bar. The system is therefore behaving one dimensionally below the temperature at which the correlation length becomes comparable with the diameter of the bar. Figure 7.4(*b*) shows the consequences of·this for the magnetic susceptibility and correlation length. Figure 7.5 shows the way the spins will be arranged in the low temperature region in which the correlation length is greater than the sides of the bar.

Similarly, if the spins are confined to the space between two parallel planes the system behaves three dimensionally when the temperature is high enough that the correlation length is less than the distance between the planes, but it behaves two dimensionally when the correlation length is greater than the distance between the planes. This situation is discussed in more detail in the following.

Layered materials and thin films

In practice there are two completely different ways of getting two-dimensional systems. In the first type of system, for example a layered magnetic material, there are planes of magnetic atoms in the three-dimensional material which are separated from one another by several planes of nonmagnetic atoms, so that magnetic moments in one plane are little influenced by the state of spins on the neighbouring planes. These are discussed in Section 7.3. Similarly there are two-dimensional metals in which electrons are free to move within a layer, which consists of a plane of metallic atoms, but can tunnel only slowly through intervening planes of nonmetallic atoms which separate the layers, so that the electrical conductivity in a direction parallel to the plane is high, but the conductivity in a transverse direction is orders of magnitude lower. In thin films, which form the second type of system, the layer is often, for practical reasons, more than a single atom thick, but it is really almost completely isolated from any similar layer. The electron inversion layers which are the basis of a number of important semiconductor devices are examples of two-dimensional metals of this sort, and are discussed in Section 7.4. They consist of a layer of electrons trapped on the surface of a semiconductor, very well insulated from any other electrically conducting layer.

The essential difference between these two types of systems can be seen if we consider what happens in magnetic systems when the transition temperature is approached from above. In the layered material the spins are correlated within each plane over a distance which increases as the temperature is lowered, while they are almost uncorrelated between different planes. This situation is shown in figure 7.6(*a*). However, the increasing coherence within a plane means that even the weak coupling between individual spins in different planes is increasingly effective. Eventually, before the two-dimensional critical point is reached, there is effective correlation between the planes, as shown in figure 7.6(*b*). In this region, very close to the critical point, the behaviour is three dimensional, although there is still a long correlation length for spins in the same plane, but a short correlation length in a direction perpendicular to the planes. The system crosses over from two-dimensional to three-dimensional behaviour when the correlation length in the perpendicular direction becomes comparable with the spacing between planes. As the temperature is lowered below the critical temperature the magnetisation remains coherent through the whole three-dimensional system. The fluctuations away from the average magnetisation are correlated only within the planes, and are uncorrelated between the planes once the temperature is well below the critical temperature.

For thin films the situation is quite different. Well above the critical temperature the correlation length may be shorter than the thickness of the film, so that the behaviour, as was explained in the previous subsection, is essentially three system as the phase transition is approached from the high temperature side. The effects of this one-dimensional behaviour are particularly striking, as the one-dimensional systems do not generally have phase transitions. Therefore the temperature may be very low before the correlations within chains are strong enough to allow effective coupling between the chains. Very low transition temperatures may therefore be produced in this way. In narrow channels, for example of square or circular cross-section, the situation is reversed, with bulk behaviour until the temperature is close to the three-dimensional transition temperature, and a crossover to one-dimensional behaviour, with no sharp transition or real long range order, when the correlation length becomes comparable with the diameter of the system. This is the situation which was shown in figure 7.5, which should be contrasted with figure 7.7.

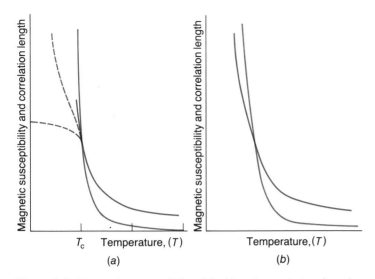

Figure 7.4. Magnetic susceptibility (black) and correlation lengths (red) for (*a*) a finite cube (dashed lines) compared with the infinite three-dimensional system (solid lines), and (*b*) a bar of square cross-section.

dimensional. As the temperature is lowered towards the critical temperature of the bulk material the correlation length increases, and there is a crossover from three-dimensional to two-dimensional behaviour when the correlation length is equal to the film thickness. Thus the crossover in thin films is from bulk behaviour at high temperatures to low-dimensional behaviour in the critical region, but in layered materials it is from low-dimensional behaviour at high temperatures to bulk behaviour in the critical region.

   For one-dimensional systems the situation is very similar. There are materials in which the atoms, magnetic or metallic, are arranged in chains separated from one another by inert material. These behave one dimensionally until the correlation length along the chains is so long that the weak coupling between the chains allows correlations in perpendicular directions to be established. When the temperature is lowered further a three-dimensional phase transition can occur. Figure 7.7 shows the way the spins are arranged in a ferromagnetic

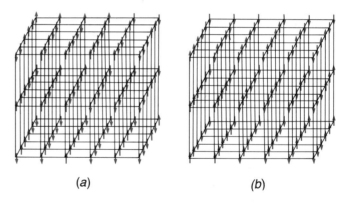

Figure 7.6. Spin arrangements in a layered material near the critical temperature. In (*a*) there are domains of parallel spins within each horizontal layer, but there is no correlation between the domains in neighbouring layers. In (*b*) the domains are slightly larger, and are correlated between the layers.

Figure 7.7. Spin arrangements in weakly coupled chains. Quite long domains within each chain are uncorrelated between chains until a lower temperature is reached.

Figure 7.5. Spins on a long bar at low temperatures. Two domains of up spins can be seen separated by a domain of down spins.

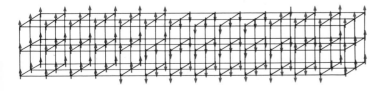

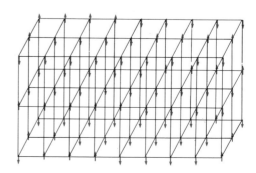

Universality

The idea of 'universal' behaviour of certain systems was developed by L. P. Kadanoff about twenty years ago. The original idea, which has since been worked out in much more detail, was that in the neighbourhood of a critical point, where some correlation length is much longer than any other length that affects the behaviour of the system, the details of what happens on these smaller length scales should be irrelevant, and only the appropriate description of the system on length scales of the order of the correlation length should be important. There is not supposed to be one single universal behaviour. There is a restricted number of universality classes which are determined primarily by the nature of the 'order parameter' which is used to describe the ordered system in the low temperature phase, and by the variables such as temperature and magnetic field which can control the approach to the critical point. The order parameter is the magnetisation in the examples we have considered so far; in other examples it is a more obscure quantity difficult to describe in simple terms. Systems in the same universality class have the same behaviour in the critical region, when an appropriate matching is made between the physical variables, and have the same critical exponents. The concept of a universality class is also discussed in Chapter 8 by Bruce and Wallace.

The Ising model provides a useful example of a universality class which has a number of interesting and diverse members. It has been recognised for a long time that the critical behaviour is not changed if the interaction of spins is not just with their immediate neighbours but is with a finite number of further neighbours. If the interaction between spins falls off with an inverse power of the distance the behaviour remains the same if the power is sufficiently high; but for a small power, so that the interaction falls off rather slowly, the behaviour is altered, and the system is no longer in the universality class of the Ising model. If the interaction between neighbouring spins favours opposite alignment instead of parallel the ordered state is antiferromagnetic instead of ferromagnetic; the lowest energy state of such a system is shown in figure 7.1. The order parameter is no longer the magnetisation, the difference between the number of spins pointing up and the number pointing down, but is a more complicated quantity. The order parameter for an antiferromagnet is defined by dividing the sites of the lattice up into two classes, which we call *A* and *B* as is shown in figure 7.8; the order parameter is then the excess of up spins on the sites labelled *A*, plus the excess of down spins on the sites labelled *B*. However, the critical behaviour is analogous and the system is in the same universality class. It is also not important that the spin variable of the Ising model can take on only two values, and there are equivalent models in which the variable can take on many different values or a continuous range of values. It is therefore not surprising that real ferromagnetic and antiferromagnetic systems, provided that they are sufficiently anisotropic to have a preferred axis of magnetisation, are observed to have the same critical behaviour as the ideal magnet of the Ising model.

There are also models in which the two states of the Ising spin are interpreted as the occupation of a lattice site of a solid by two different species of atoms, and these serve as models of binary alloys. The transition from the disordered high temperature phase of a 50–50 binary alloy to the ordered phase in which neighbours are preferentially of the opposite type is very like the phase transition in an Ising antiferromagnet. For example, the phase transition in beta-brass is indeed observed to behave in the expected way. Figure 7.9 shows the arrangement of copper and zinc atoms in this alloy. In the low temperature, ordered, phase (figure 7.9*a*) there is an excess of copper atoms on the *A* sites and of zinc atoms on the *B* sites (or the other way around), and the order parameter is defined as the degree of this excess. At high temperatures both types of sites are equally populated by the two types of atoms (figure 7.9*b*). Also it can be argued that this can serve as a model of binary liquid mixtures, so that these should belong to the same universality class, despite the absence of an underlying lattice structure in this case. Similarly it has been argued that instead of two different species of atoms, one could consider one sort of atom and 'vacancies' (sites with no atom present), and so make a model of the gas–liquid critical point that should be in the same universality class as the real transition from gas to liquid. All these parallels between the behaviour of different systems in the universality class of the three-dimensional Ising model have been confirmed by experiments and numerical calculations.

The behaviour of the two-dimensional Ising model is quite different from that of the three-dimensional Ising model, so it is a member of a different universality class, and the behaviour of the one-dimensional model is again different. In the Ising model the spins can only point up or down along a fixed direction, but in the Heisenberg model, which was discussed in Section 7.1, the spins are free to point in any direction in space. The order parameter for the Heisenberg model is therefore a vector, a quantity which can point in any direction. The critical properties of the Heisenberg model are quite different from those of the Ising model, particularly in lower-dimensional systems, and so the Heisenberg model provides a new set of universality classes (different classes for different numbers of dimensions of space). It also turns out that a compromise between the Ising and the Heisenberg models in which the spins are constrained to lie in a plane but can point in any direction in the plane – the planar spin model – is of importance for many systems. The planar spin model is used to describe the order which occurs in superfluid helium and in superconducting metals (see Chapter 9 by Leggett). The planar spin model gives new universality classes which are different from the universality classes of both the Ising model and of the Heisenberg model. This does not by any means exhaust the possibilities that have been explored, but it covers many of the examples which are discussed in this chapter.

Figure 7.8. The two sublattices *A* (shown in red) and *B* (shown in black) of an antiferromagnetic system in two dimensions.

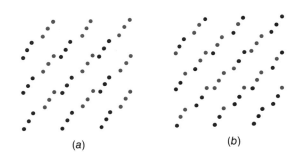

Figure 7.9. (a) The ordered and (b) the disordered states of beta-brass. In this three-dimensional lattice each of the copper atoms in (a) has eight zinc atoms as its neighbours, and each of the zinc atoms has eight copper atoms as neighbours.

This idea of universality classes is obviously of importance for the analysis of systems of low dimensions. In layered magnetic materials the detailed assumptions of the Ising model do not give a good description of the system. However, in the temperature range in which the correlation length in the plane is considerably longer than the range of interaction between the spins, but the correlation length between planes is less than the spacing, the behaviour is expected to be, and is observed to be, that of the Onsager solution of the two-dimensional Ising model. Similarly, in thin films of helium close to the transition to the superfluid phase the behaviour is that of an ideal two-dimensional superfluid transition, provided the correlation length is long compared with the thickness of the film. It is for this reason that we can expect the theoretical study of idealised systems to give us detailed information about real systems of reduced dimensionality.

7.3 Phase transitions in real systems

Magnetic systems

Magnetic systems provide a lot of examples of phase transitions in reduced dimensions. There are many examples of layered and chain materials that have been studied in detail, and rather fewer examples of magnetic thin films. There are a number of techniques which are readily available for these bulk materials with well separated planes or lines of magnetic atoms, but which are much harder to apply to thin films because there is usually much less material in the thin films. Specific heat measurements, for example, are much easier to make on a bulk material than on a film whose thermal properties may be masked by the substrate on which it resides.

Elastic neutron scattering is a powerful technique for studying the properties of a magnetic system. Neutrons, unlike X-rays, are sensitive to the spin direction of the atom which scatters them, and so the alternating arrangement of spins in an antiferromagnet, such as is shown in figure 7.1, can be observed directly by neutron scattering. This enables the sublattice magnetisation of an antiferromagnet, as defined in Section 7.2, to be observed, and the correlation length is capable of more or less direct measurement. This is, however, a technique which can only be used if there are fairly large samples available, so layered materials are much more suitable for study than thin films.

Many different types of materials have been studied. One example is Rb_2MnF_4, in which the planes of magnetic manganese atoms are not only well separated from one another, but are also arranged in such a way that there is a cancellation between the interactions of atoms in different planes. The atomic moments tend to line up perpendicular to the planes, so this provides a good example of a magnetic system that is in the universality class of the Ising model. The arrangement of the spins is sketched in figure 7.10. Another case involves planes of $CuCl_4$ separated by layers of organic material whose thickness can be varied within a rather wide range by varying the number of CH_2 groups in the organic material. This system is in the universality class of the Heisenberg model, in which the spin is free to point in any direction in space. It can be made to behave two dimensionally down to lower and lower temperatures by varying the separation between the planes. The spin arrangements in such a material are shown in figure 7.11.

Figure 7.12 shows the sublattice magnetisation measured for Rb_2CoF_4 by neutron scattering in the neighbourhood of the transition temperature. These experimental results fit very closely with the form calculated for the two-dimensional Ising model from Onsager's exact solution. In particular the slope of the plot in the logarithmic scale which has been chosen to display the results agrees closely with the theoretical exponent of $\frac{1}{8}$ which gives the relation between the magnetisation and the displacement of the temperature from the critical temperature. At temperatures so close to the critical temperature that the deviations from perfect antiferromagnetic alignment in different planes are correlated, this slope should increase to the much larger value (by a factor of about 2.5) that it has in the three-dimensional Ising model. However, this crossover to three-dimensional behaviour occurs so close to the critical temperature in this system that it cannot be unambiguously seen in the results of this set of experiments. Materials of this sort also seem to have magnetic susceptibilities which are close to the theoretical two-dimensional forms over a surprisingly wide temperature range.

Figure 7.13 shows the magnetic susceptibility measured for a number of materials which consist of layers of copper atoms, with a coupling between spins that favours parallel spins within each layer, separated by inert molecules which consist of chains of carbon atoms whose length can be chosen. In this way it is almost possible to study the behaviour of an ideal two-dimensional magnetic system for which the spins are free to point in any direction, and which should therefore fall into the universality class of the Heisenberg model. It is generally accepted by theorists that the transition in the two-dimensional Heisenberg model only occurs at zero temperature, so that the susceptibility should simply increase rapidly as the temperature is lowered without ever reaching a maximum. In fact, even the very weak coupling between the different planes which occurs when the planes are separated by many carbon atoms is strong enough to produce a three-dimensional phase transition, so a quite careful analysis of the results is necessary to show that they do conform to the expected theoretical pattern.

In the magnetic chain compounds, even when the environment of the magnetic spin is sufficiently anisotropic that they behave like an Ising system rather than like a Heisenberg system, there is no one-dimensional phase transition to be approached. The correlation length increases as the temperature is lowered until the coupling between different chains takes over. This is the situation illustrated in figure 7.7. There are a number of systems known which behave like one-dimensional Ising systems at reasonably high temperatures, but then they have three-dimensional phase transitions, for the reasons given in Section 7.2. There are rather more systems known to behave like an ideal one-dimensional Heisenberg system down to quite low temperatures before the coupling between chains forces a three-dimensional phase transition. The exact solution to the one-dimensional Heisenberg model is not known in such detail as that of the Ising model, but very precise calculations can be made, so comparison with experimental results is not difficult. Figure 7.14 shows how the specific heat data for copper sulphate ($CuSO_4$) and for copper ammonium sulphate ($Cu(NH_3)_4SO_4$) can both, by a simple choice of the temperature scale, be fitted on to the same calculated curve. At very low temperatures the results for $Cu(NH_3)_4SO_4$ show the sharp peak which is due to the three-dimensional phase transition.

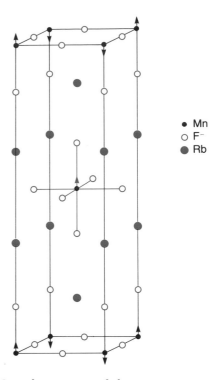

Figure 7.10. Crystal structure and the spin arrangement of lowest energy in rubidium manganese fluoride (Rb_2MnF_4). The layers of magnetic manganese atoms are well separated, and the spins are perpendicular to the layers and arranged antiferromagnetically within the layers. (After Lines, M. E. (1969). *J. Appl. Phys.* **40**, 1352, figure 6.)

Adsorbed gases

Gases adsorbed on to solid substrates also provide a wide variety of two-dimensional systems that can be studied. Much of the interesting information comes from very thin layers of gas, one atom thick, or with fewer gas atoms than are needed to form a single complete layer. This has two important consequences. The first is that the interaction of the substrate with the adsorbed atoms is rather strong, and so the crystalline structure of the solid surface is of great importance in determining the behaviour of the adsorbate. The second is that the amount of adsorbed material is rather small compared with the amount of substrate material, so that observations of specific heat, neutron scattering or X-ray scattering have to be made in the presence of a very strong background term due to the substrate. Techniques such as low energy electron diffraction do not suffer from this disadvantage, as the electrons do not penetrate very deep into the material but are scattered close to the surface.

Various tricks may be used to increase the surface to volume ratio of the substrate so that more material can be adsorbed.

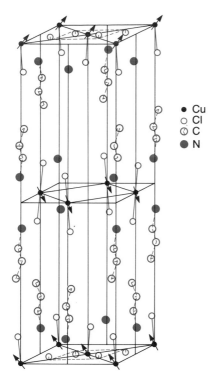

Figure 7.11. Crystal structure and low temperature spin arrangements in a material containing layers of magnetic copper ions separated by organic molecules of variable length. The spins are coupled ferromagnetically within the layers, but are weakly coupled antiferromagnetically between layers. The spins are free to point in any direction, so the magnetic properties correspond to those of the Heisenberg model. (After de Jongh, L. J., Botterman, A. C., de Boer, F. R. and Miedema, A. R. (1969). *J. Appl. Phys.* **40**, 1363–5, figure 1.)

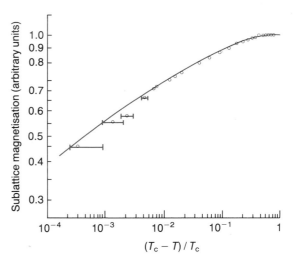

Figure 7.12. Sublattice magnetisation for rubidium cobalt fluoride (Rb_2CoF_4), measured by neutron scattering (open circles), compared with the theoretical result (solid line) for the two-dimensional Ising model. The horizontal axis represents the difference between the temperature and the critical temperature, plotted on a logarithmic scale, while the sublattice magnetisation is also plotted logarithmically on the vertical axis. (After Samuelsen, E. J. (1974). *J. Phys. Chem. Sol.* **35**, 785–93, figure 1.)

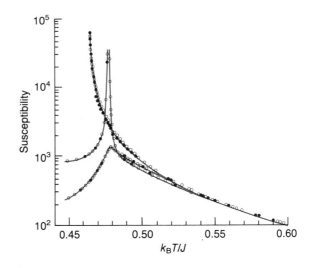

Figure 7.13. Magnetic susceptibility of materials with layered copper ions, with two different separations of the layers. For the smaller separation the susceptibility has a maximum at the antiferromagnetic transition temperature, and below that temperature has a value that depends on whether the field is parallel to or transverse to the sublattice magnetisation. For the larger separation the susceptibility continues to increase as the temperature is lowered. J is the strength of the coupling between neighbouring spins. The open and closed symbols refer to measurements on different samples. (After de Jongh, L. J. (1976). *Physica B* **82** 247–61, figure 4.)

One commonly used substrate is exfoliated graphite, in which some of the layers of carbon atoms have been forced apart from one another by first forcing in organic materials and then baking them out. This results in a free surface after every hundred or so layers of carbon atoms, so that a cubic centimetre can have a surface area of $50\,m^2$ or so. This is obviously very helpful for increasing the scattering of neutrons or X-rays by the adsorbate, but it has the unfortunate result that the extent of unbroken regular surface is not very great. The very best material made in this way has surfaces which extend for only a fraction of a micrometre. In contrast, good single crystals of graphite are available which have surfaces many millimetres across.

The influence of the periodic potential due to the interaction of the substrate with the adsorbed atoms is obviously an unwelcome complication when one wants to study the transitions analogous to melting and to the gas–liquid transition in a two-dimensional system. It does, however, enable physicists to study 'lattice gas' systems in the real world. In the lattice gas atoms can only be adsorbed at the sites of an already existing regular lattice. It has already been mentioned that the Ising model was interpreted in this way as a simplified model for transitions such as occur in brass. Atoms adsorbed onto the surface of graphite will be held on the lattice sites formed by the centres of the hexagons of carbon atoms on the graphite surface, at least if the density of adsorbed atoms is fairly low. These centres form a triangular lattice (the centres of the hexagons are arranged at the corners of a set of equilateral triangles) in the manner shown in figure 7.15. If a second atom comes into a hexagon already occupied by an adsorbed atom it cannot be at all tightly bound, and so the possibility of two atoms on one site can safely be ignored. Most adsorbed atoms are sufficiently large that an atom on a hexagonal site immediately neighbouring an occupied site is also somewhat repelled. Therefore an appropriate model of the system involves sites which can be either occupied or not, with a positive energy when two neighbouring sites are occupied and a negative energy when two next-to-nearest neighbour sites are occupied. This is obviously a variant of the two-dimensional Ising model on a triangular lattice, but it turns out that it behaves in a quite different way, and belongs to a different universality class, from the familiar Ising model with nearest neighbour interactions.

This has led to the consideration of a class of generalisations of the Ising model known as Potts models. These models cannot be directly interpreted in terms of the lattice gas in the same way that the Ising model can be, but an analysis of their behaviour near the critical point indicates that their critical behaviour should correspond to that of the lattice gas, and hence to that of the atoms adsorbed on a crystalline surface. In the n-state Potts model each site of a lattice can be in any one of n equivalent states. The energy is simply proportional to that number of pairs of neighbouring sites which are in the same state, with a negative constant of proportionality, so that the configuration with the lowest energy has all sites in the same state; it does not matter which state. For $n = 2$ the Ising model is recovered, since the two states are just the two different spin directions. For $n = 3$ or 4 there is a critical point, with a critical behaviour that is very different from that of the Ising model. Until recently the critical behaviour of the Potts models was only known from numerical estimates, some of which were somewhat misleading, but it is now known precisely. In particular the specific heat, which for the Ising model diverges as the logarithm of $T - T_c$, the temperature difference from the critical temperature, diverges like a negative power of this temperature difference for the three- and four-state Potts models. This power is $-\frac{1}{3}$ for the three-state model and is $-\frac{2}{3}$ for the four-state Potts model.

For $n > 4$ there is no critical point, but there is a 'first order phase transition', such as occurs with the melting of a solid or the boiling of a liquid. In a first order phase transition there is a sudden change from one phase to the other; in the case of the Potts model from the ordered phase, in which more sites are in

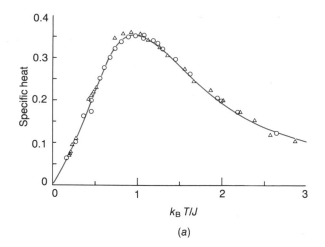

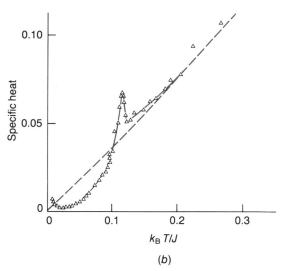

Figure 7.14. (*a*) Specific heat of two different materials consisting of chains of magnetic copper ions, compared with the calculated specific heat for a one-dimensional Heisenberg model (solid curve). (*b*) The specific heat of one of these materials at very low temperatures, showing the effect of the coupling between chains in producing a departure from the one-dimensional behaviour (represented by the dashed straight line) to give a three-dimensional phase transition. (After de Jongh and Miedema (1974), figure 20.)

one state than in the other $n - 1$ states, to the disordered phase, in which all n states are equally populated. As was pointed out in Section 7.1, there is no very long correlation length associated with a first order phase transition, but there is a latent heat, and the two different phases can coexist at the same temperature.

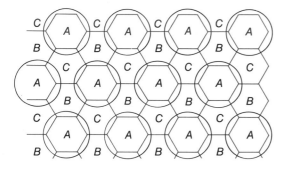

Figure 7.15. Three sublattices on a graphite surface. A krypton atom can be adsorbed on to one of the hexagonal spaces between the carbon atoms, but it is sufficiently large that another atom cannot then be adsorbed on to one of the adjacent spaces. An energetically favourable configuration in which one-third of the spaces are filled is shown in this figure. Either of the other two sublattices could equally well be filled.

For systems such as helium or krypton adsorbed on graphite the ground state involves the occupation of every third hexagonal site between the carbon atoms. There are clearly three equivalent states with equal energy, according to whether the occupied sites are those marked *A*, *B* or *C* in figure 7.15. At nonzero temperatures there will be some atoms on the wrong sites but there will be more on the right sites than on either of the wrong set of sites. Figure 7.16 shows how the atoms might be arranged on the surface at higher temperatures. Above the critical temperature the three sets of sites will be equally occupied and there will be no preferential occupation of one set. The 'sites' of the Potts model are not the sites provided by the hexagons of the graphite substrate, but they are groups of three of these sites, an *A*, a *B* and a *C* taken together. Which of the three sites of the graphite lattice is occupied by a gas atom determines which state the 'site' of the Potts model is in. If the *A* site is occupied in one particular triplet of graphite sites, then the corresponding 'site' of the Potts model is in the state *A*, and so on. Figure 7.17 shows the specific heat of helium adsorbed on graphite at one-third coverage as measured by M. Bretz. The large peak in the specific heat can be fitted as a power law divergence, and the exponent found was very close to the theoretical value of $-\frac{1}{3}$; significantly, at the time the experimental data were analysed an exponent much closer to zero was favoured by many theorists.

If krypton is adsorbed onto the graphite at one-third coverage, and then the adsorption of helium on to this surface is studied, only two of the three possible types of sites are available for the helium atoms; all the *A* sites are already occupied by krypton atoms, and so only the *B* and *C* sites are available for the helium atoms. Analysis shows that the two-state Potts model, or in other words the Ising model, is appropriate as a

description of the subsequent helium adsorption. Measurements of the specific heat of helium on graphite which already has krypton on one-third of the sites have shown that the specific heat divergence is indeed logarithmic. There are also systems which should have the transition appropriate to the four-state Potts model, because the sites on the substrate can be divided into four classes. However, none of the theoretical possibilities are actually found to have a critical point, as a first order phase transition intervenes before it is reached.

At higher densities a variety of things happen. Because the spacing between the adsorbed atoms no longer fits neatly with the available minima of the potential due to the substrate there is a phase transition to a phase which is incommensurate with the substrate, that is it is a solid phase with a lattice parameter that is not related to the lattice spacing of the substrate surface. This solid can also melt and go into a fluid phase. When the adsorbed gas is not monatomic but consists of molecules, like nitrogen or oxygen with their own structure, there are more complicated possibilities involving the alignment of the molecule, either with the substrate or with the crystal structure of the adsorbate. A number of these systems have been studied experimentally, and theoretical analysis has been carried out with some success. There are a large number of theoretical possibilities in this situation, and a detailed discussion cannot be given here. One important question, which is considered further later in this chapter, is whether the incommensurate solid in two dimensions can melt at a critical point, or if it always melts at a first order phase transition like a three-dimensional solid. The situation is still somewhat uncertain, but it can be said that there is no strong evidence for a critical point.

It is also possible to study magnetic properties of adsorbed gases, and in 1988 H. Godfrin, R. R. Ruel and D. D. Osheroff showed that a thin film of ^{3}He behaves like a two-dimensional Heisenberg ferromagnet.

Figure 7.16. Higher temperature configurations for an adsorbed gas on graphite. A few of the atoms have shifted on to a different sublattice.

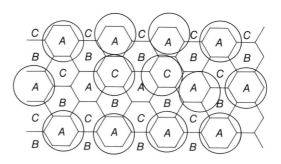

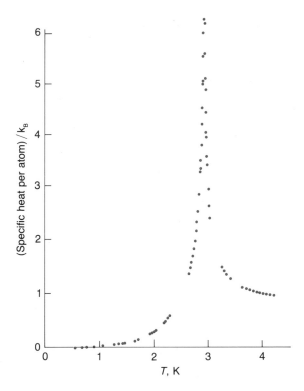

Figure 7.17. Specific heat of helium on graphite. The sharp rise of the specific heat is characteristic of the transition for the three-state Potts model. (After Bretz, M. and Dash, J. G. (1971). *Phys. Rev. Lett.* **27**, 647–50, figure 2.)

Liquid crystal films

It has been known for a long time that there are substances intermediate in their properties between liquids and crystals, and these are known as liquid crystals. There is considerable variety in the type of structure that they can have, but in general they have the anisotropy of a complex crystalline solid but no crystalline long range order, and they can flow like a liquid. Generally they are composed of quite long molecules, which for the purposes of this discussion will just be regarded as rigid rods. One of the simplest forms of liquid crystal is known as the 'nematic' phase; in this phase the molecules are partially aligned, that is they tend to point in a common direction, but their position in space is more or less random. This is shown in figure 7.18(*a*). A more complicated set of phases is known as 'smectic'; in these the molecules are arranged in layers, which may be able to slide more or less freely over one another. Examples of various types of smectic phases are shown in figure 7.18(*b*), (*c*) and (*d*). The smectic phases are subdivided according to whether the average orientation of the molecules is perpendicular to the layers or has a component parallel to the layers, and according to the degree of spatial ordering of the

molecules within the layers or between layers. Liquid crystals are generally three-dimensional materials, but it has proved possible to make freely suspended thin films of them, like soap films supported on their rims by a loop of wire, which are very uniform in their thickness, and which can be as little as two molecules in thickness. These thin films, for which there is no problem about the effect of a substrate, can be studied to find out if they possess two-dimensional phases which are essentially different from their three-dimensional phases.

One example of a phase transition in liquid crystals is the transition between the smectic *A* phase, shown in figure 7.18(*b*), in which the molecules are oriented with their axes perpendicular to the layers, and with no positional order of the molecules within layers, and the smectic *C* phase, shown in figure 7.18(*c*), in which the molecules have an average component of orientation parallel to the layers, but there is still no positional order within the layers. Even in a bulk liquid crystal, the coupling between the orientation in different layers is often rather weak, so that as the temperature is lowered in the high temperature (smectic *A*) phase each layer develops an orientation of the molecules parallel to the plane of the layer almost independently of what happens in the other layers. The transition may be regarded as a two-dimensional transition except in a narrow region around the critical point; the analogy with magnetic layer compounds discussed in Sections 7.2 and 7.3 is quite close. In the case of very thin films a few molecules thick the transition is truly two dimensional even close to the critical point. This is an example of a transition which is in the universality class of the planar spin model, intermediate between the universality classes of the Ising model and the Heisenberg model. The order parameter is represented by a vector which is confined to a plane. This was mentioned in Section 7.2, and it is discussed in more detail in the next subsection. It is indeed true that its properties in two dimensions are intermediate between those of the Ising model, which has a transition of the usual sort at a nonzero temperature, and the Heisenberg model, for which the transition temperature is thought to be zero. For the planar spin model the transition temperature is nonzero, but the nature of the transition is unusual.

Various types of solid order are also observed within the layers. In those cases in which the solid order also extends between the layers we would regard the material as a true solid rather than as a liquid crystal, but there are many materials for which it can be seen that although the different layers have some influence on one another, there is not complete solid ordering. This type of ordering is illustrated in figure 7.18(*d*). The structure of these materials can be studied by means of X-ray scattering, even for the thin films, since radiation from a synchrotron provides a very satisfactory source of X-rays, and the free suspension of the films means that there is no background of scattering from a substrate. The transition between the liquid-like phases, with no positional order in the layers, and the solid-like phases, which show sharp features in

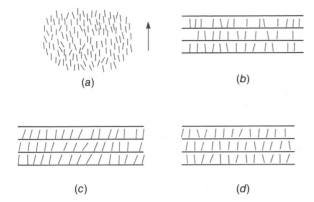

(a)

(b)

(c)

(d)

Figure 7.18. (*a*) The arrangement of molecules in the nematic phase of a liquid crystal; they are aligned parallel to an axis shown by the arrow, but are not otherwise ordered. (*b*) In the smectic *A* phase the molecules are arranged in planes perpendicular to the direction of alignment. (*c*) In the smectic *C* phase the axis of alignment is tilted with respect to the planes in which the molecules are arranged. (*d*) In the smectic *B* phase the molecules are regularly spaced within the planes. ((*a*)–(*c*) after de Gennes, P. G. (1974). *The Physics of Liquid Crystals*. Oxford University Press, figures 1.2, 1.9, 1.10.)

rest of the system, in the way shown in figures 7.2(*b*) and 7.3(*a*), there is a positive contribution to the energy of the system proportional to the number of spins on the boundary. In one dimension this leads to an energy independent of the size of the domain of reversed spins, since there are just two pairs of wrongly aligned spins, one at each end of the domain. In two dimensions this energy is proportional to the length of the boundary, and in three dimensions it is proportional to the area of the boundary. If the spins are free to point in any direction, or in any direction in a plane, the energy of such a boundary can be greatly reduced by a gradual turning over of the spins in the manner shown in figure 7.20. The energy of the boundary is reduced by a factor equal to the thickness of the boundary.

Figure 7.19. A model made with ball-bearings to illustrate the arrangement of molecules within a plane of a 'hexatic' phase. In such a phase there are clear axes visible making angles of 60° with one another, but there is no positional order. (Photograph courtesy of Prof. D.R. Nelson.)

the X-ray scattering patterns, is generally first order. The system goes abruptly from one type of phase to the other, and there is no critical point characterised by a very long correlation length. In some cases there is an intermediate phase with well defined crystal axis directions in the plane of the layers, but without the positional order of a solid; this is illustrated for a single plane of molecules in figure 7.19. This hexatic phase, which has been observed in systems with as few as three layers of molecules, was predicted for two-dimensional systems by B. I. Halperin and D. R. Nelson. It was also predicted that there should be a continuous transition (critical temperature) both between the solid and hexatic phases, and between the hexatic and liquid phases, but this has not been observed; in most cases the transition seems to be of first order.

Long range order and vortices

Fifty years ago or so it was observed, by F. Bloch, R. E. Peierls and L. D. Landau, that there was a very important difference, particularly important in two dimensions, between systems such as the Ising model, in which the magnetisation can point in one of two directions, and models such as the Heisenberg model or the planar spin model, in which the magnetisation is free to point in different directions in space, so that it can be rotated through a small angle. In the Ising model at low temperatures there are two possible states of lowest energy, but if the spins in one region are reversed relative to the spins in the

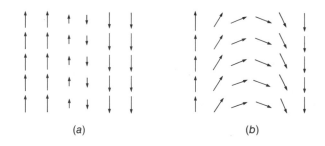

(a)   (b)

Figure 7.20. (a) The way in which the magnetisation varies across a wall between two oppositely magnetised domains in the two-dimensional Ising model. (b) The domain wall for the planar spin model (or for the Heisenberg model).

In two dimensions the boundary energy proportional to the length of the boundary for the Ising model makes it unfavourable for a high concentration of boundaries to occur at low temperatures. It is only when the temperature T is high enough that $E - TS$ is negative, where S is the entropy associated with a unit length of boundary and E is the energy, that the system becomes unstable against the formation of a lot of boundaries and the magnetisation is destroyed; this was essentially the argument given by Peierls for the existence of a nonzero transition temperature for the Ising model which was mentioned in Section 7.2. On the other hand, for the Heisenberg model the spin direction can vary slowly in space at the cost of rather little energy, and this was shown to lead to the destruction of the spontaneous magnetisation at any temperature above zero. The result, and related results for other two-dimensional systems, were proved rigorously by D. Mermin, H. Wagner and P. Hohenberg about thirty years later. In 1968 it was shown by F. J. Wegner that these slow fluctuations of the direction of the magnetisation do not lead to a finite correlation length, but lead to correlations between the spins which fall off like some temperature-dependent power of the distance just as they do at a critical point.

In 1971 it was shown by V. L. Berezinskii, and a year later in independent work by J. M. Kosterlitz and myself, that for the planar spin model the infinite correlation length should be destroyed at sufficiently high temperatures by a new process, the spontaneous formation of 'vortices'. A vortex is a pattern of spins such as those shown in figure 7.21 in which the spin direction rotates by 360° along any path which surrounds the centre of the vortex. The vortex is positive or negative according to the sense of the rotation, so that figure 7.21(a) and (b) represent positive vortices, while figure 7.21(c) represents a negative vortex. The energy of a free vortex is proportional to the logarithm of the area of the system, and the entropy of a vortex is also proportional to the logarithm of the size of the system. As a result the free energy $E - TS$ of a vortex is large and positive at low temperatures, and so no free vortices occur in an infinitely large system. When this quantity changes sign the vortices can occur spontaneously, and a phase transition

occurs in which the correlation length becomes finite. We also showed that pairs of vortices of opposite sign play a very important role in determining the behaviour of this sort of system.

For the Heisenberg model, as was shown by A. M. Polyakov, the situation is quite different from the planar spin model, because local spin reversals can occur by a mechanism which uses the third dimension for the spin direction to make the energy quite low. Figure 7.22 shows a pattern of spins with such a reversal in the centre. These spin reversals will be spontaneously excited at any nonzero temperature, and lead to a finite correlation length at all temperatures, so there is thought to be no phase transition in the two-dimensional Heisenberg model.

Similar arguments can be applied to solids in two dimensions. The thermal excitation of sound waves leads to a destruction of the sharp Bragg peak of X-ray scattering which is characteristic of the long range order found in three-dimensional solids. There does, however, remain a more subtle type of order which leads to a power-law rise in the X-ray scattering amplitude as the Bragg peak is approached. This behaviour has been confirmed in recent experiments using high intensity X-rays from synchrotron sources on two-dimensional targets. The long range order should eventually be destroyed by the spontaneous formation of 'dislocations', whose energy and entropy also depend logarithmically on the size of the system. However, as was stated in an earlier discussion, the expected continuous phase transitions do not seem to have been observed, but first order phase transitions always seem to intervene.

Figure 7.21. Vortices in the planar spin model. The two arrangements shown in (a) and (b) are equivalent, and the two arrangements shown in (c) are equivalent. Vortices of positive sign are represented in (a) and (b), and vortices of negative sign are represented in (c). (After Nelson (1983), figure 4.)

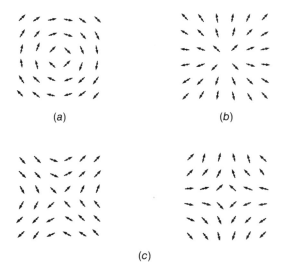

(a)   (b)

(c)

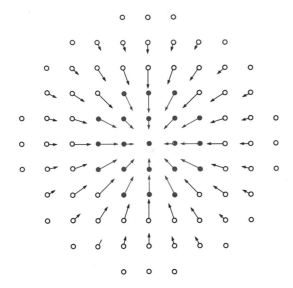

Figure 7.22. This shows how the spin direction can be smoothly reversed in a small region in the two-dimensional Heisenberg model in a way that is not possible for the planar spin model. The open black circles represent spins pointing up from the plane of the diagram, while the red solid circles represent spins pointing down from the plane of the diagram. The length of the arrows represent the projection of the spin into the plane.

Superfluid helium films

The theory described in the previous subsection was developed as a theory of helium films, and although it has potentially a much wider range of application, that has been the system for which it has been most successful. This is an example of the use of the principle of universality discussed in Section 7.2; the same theory can be used for magnetic systems and for superfluid helium. Over the past ten years a large number of experiments have been done on helium films under very different conditions, on different substrates, with different thicknesses of films, with different degrees of purity, and with different methods of measurement. Many of the peculiar features of this type of transition have been seen.

The nature of the superfluid state of liquid helium is discussed in Chapter 9 by Leggett. Because the helium atom is so light, it behaves quite unlike other materials at low temperatures and remains a liquid instead of solidifying. Below a temperature of about 2 K it becomes a 'superfluid'. This superfluid liquid can flow extraordinarily readily, and has a number of other properties which show it to be quite distinct from a normal liquid. There is in fact a sharp, and much studied, phase transition between the high temperature normal liquid phase of helium, and the low temperature superfluid phase.

A theoretical explanation of superfluidity was given by F. London. At very low temperatures each of the helium atoms can be regarded as being in the same quantum state described by a 'condensate wave function'. This condensate wave function is a function of position in three-dimensional space. Like any wave function in quantum theory, it is a complex number, which is a quantity that can be represented by a vector lying in a plane. The length of the vector is known as its 'modulus', and the angle it makes with an arbitrary direction is known as its 'phase'. The rate of flow of the superfluid helium is proportional to the rate of change in space of this phase. As the temperature is raised more and more atoms get excited from the condensate wave function, and so the condensate is represented by a shorter and shorter vector. At the transition from the superfluid state to the normal state of liquid helium the vector has shrunk to zero length. This description is very similar to the description that would be given of the magnetic phase transition of a planar spin model, and so the superfluid–normal transition in liquid helium is in the same universality class as the planar spin model, which was defined in Section 7.2.

It was observed a long time ago that this description of superfluid helium would lead to the conclusion that the superfluid should flow with the type of flow known as 'potential flow' in hydrodynamics. This conclusion does not agree with the observed properties of liquid helium on a large scale, although it is true in sufficiently small regions of the fluid. In 1947 L. Onsager suggested that there should be 'quantum vortices', which are lines (or curves) in space around which the phase of the condensate wave function makes a complete rotation. The observed large scale flow patterns of liquid helium could then be reproduced by superimposing some density of quantum vortices on the potential flow patterns. Figure 7.23 can be regarded as giving a representation of the phase and magnitude of the condensate wave function in the neighbourhood of a vortex. Many years passed before direct confirmation of the existence of quantum vortices could be obtained, but their properties are now well understood. In particular it has been shown that the flow of superfluid round a ring is stable, and will persist for a time which is practically infinite, unless quantum vortices can move across the system to destroy the flow.

The freedom of superfluid helium to flow in very thin films was one of the earliest observed manifestations of its unusual properties. The films have to be more than a single atom thick, since the first layer of atoms is so tightly bound to the substrate that it is more like a solid layer than a liquid, but any subsequent layers of helium are remarkably free to move. In a normal liquid, as in liquid helium in its normal (high temperature) phase, viscosity prevents appreciable flow in very thin films on a solid substrate. The superfluid flows very easily, and special surface waves, known as third sound waves, can be propagated in it. The propagation of third sound provides one of the ways in which the proportion of superfluid, the superfluid density, can be measured as a function of temperature. Another important method of measuring the superfluid density depends on the use of the Andronikashvili oscillator. This consists of a solid cylinder or other shape suspended from a torsion fibre with

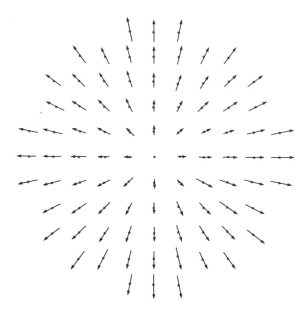

Figure 7.23. This is a diagram representing the magnitude and phase of the wave function near a quantum vortex. Each arrow can be rotated through an arbitrary angle (compare the equivalent arrangements shown in figure 7.21a and b).

a helium film adsorbed on it. When the solid substrate oscillates the normal fluid moves with it, but the superfluid slips over the surface rather than moving with the surface. This has two effects. Firstly the measured moment of inertia of the system is the sum of contributions from the solid substrate and the normal fluid film, but does not include a contribution from the superfluid component. Secondly, there is a contribution to the damping of the system due to the movement of the superfluid relative to the normal fluid. Both these methods have been used to get detailed information about the transition between the normal and superfluid states of helium films.

The vortex unbinding transition which was discussed in the previous subsection should have important consequences for the superfluid helium film. At low temperatures there are no free vortices, because their energy is too high for them to occur in thermal fluctuations. According to the basic principles of statistical mechanics, the probability of occurrence of a fluctuation depends on the ratio of its energy to the absolute temperature. As there are no vortices present there is no mechanism for the spontaneous decay of superfluid flow in low temperature helium films, provided the flow velocity is sufficiently low; at higher flow velocities there will be appreciable generation of free vortices by thermal fluctuations. There is a drop in the superfluid density as the temperature is raised, but there is still no spontaneous formation of vortices, until a certain ratio of the superfluid density to the temperature is

reached. At this point there is a sudden disappearance of superfluidity. Therefore there is a definite relation, with no adjustable constants, between the superfluid density and the temperature. This is completely different from the situation in bulk superfluid helium, where the superfluid density goes to zero like a power of the temperature difference. When there are free vortices spontaneously generated by thermal fluctuations, or, what amounts to the same thing, when the bound vortex pairs can dissociate, any superfluid flow in a thin film will be quickly destroyed by the movement of these free vortices, and so the film should behave as a normal film.

Figure 7.24 shows the results of measurements by D. J. Bishop and J. D. Reppy using an Andronikashvili oscillator. Rather than using a solid cylinder as a substrate, they used a long strip of mylar wrapped up in a tight roll, so that a reasonable ratio of substrate mass to helium mass could be obtained. The moment of inertia drops rather sharply as the temperature is lowered, and there is a fairly sharp peak in the damping in the same temperature range. A theoretical analysis by V. Ambegaokar, B. I. Halperin, D. R. Nelson and E. Siggia shows that the rounding of the drop in the moment of inertia and the nonzero width of the peak can be explained quantitatively if account is taken of the finite oscillation period of the measuring apparatus. Figure 7.25 shows the results of these experiments, combined with other experiments based on the observation of third sound. A good fit is obtained to the predicted linear relation between the superfluid density just below the transition and the transition temperature.

Superconductivity

Bulk superconductivity gives an example of a system with a continuous phase transition with one rather special feature. The basic entities in superconductivity theory are pairs of electrons rather weakly bound (see the chapter by Leggett), so that the pair spreads out over a region many hundreds of atomic spacings across. As a result, there is only an unobservably narrow region in temperature around the transition temperature in which critical behaviour is expected. Therefore the observed properties behave smoothly right up to the transition temperature, so that, for example, the specific heat just has an abrupt discontinuity at the transition temperature, and there are none of the usual symptoms of critical behaviour.

The effect of reduced dimensionality, such as occurs in superconducting thin wires and thin films, or in chain or layer compounds, is to broaden this critical region; it is still narrow, but not unobservably narrow. Because of the large size of the electron pairs the wire or film does not have to be very thin for reduced dimensionality to be observed; as a rough approximation the correlation length is equal to the size of the Cooper pair divided by the square root of the ratio of the shift of the temperature from the critical temperature to the value of the critical temperature. Supercurrents are destroyed by quantum flux lines moving across the system, just as superfluid flow in

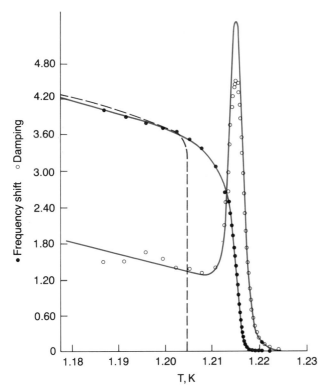

Figure 7.24. Comparison between the frequency shift and damping of an Andronikashvili oscillator measured by Bishop and Reppy and the predicted results for the vortex unbinding transition. The frequency shift, proportional to the superfluid density, goes to zero in a rather narrow temperature interval, and there is a sharp peak in the damping over the same temperature interval. These both occur at a temperature slightly higher than would be obtained from a static measurement, for which the theoretical superfluid density is shown by the dashed curve. (After Bishop, D. J. and Reppy, J. D. (1980). *Phys. Rev. B* **22**, 5171–85, figure 12.)

superconducting state while the polymers are behaving one dimensionally. Eventually a low enough temperature is reached that there is sufficient coupling between electron pairs in different chains that a bulk phase transition to the super-conducting state takes place. Particularly in the chain compounds, but also in thin wires, there is a wide range of temperatures above the transition temperature (or transition region for the wires, since the wires have no sharp transition temperature) where local fluctuations to the superconducting state can be observed.

Superconducting thin films have a lot of similarities to helium films, with a few important differences. One important difference is that because of the large size of the electron pairs, and consequently long correlation length, a superconducting thin film can be two orders of magnitude thicker than a helium film and still display the same degree of two-dimensional behaviour. A second difference is that, while a quantised vortex in liquid helium has an energy which gets larger and larger as the area of the film increases, the energy of a quantum flux line in a superconducting thin film does not increase indefinitely. In principle this leads to a nonzero probability for the existence of free flux lines, and so there should be no sharp phase transition. However, the energy of a flux line in a thin film is very large, and so this is not really an important consideration. For practical purposes the flux lines can be thought of as equivalent to the vortices in a helium film, and the transition from the superconducting state to the normal state takes place by the same mechanism as the transition in helium films.

Superconducting layered materials also exist. One class of examples of these consists of layers of niobium or tantalum

Figure 7.25. The relation between superfluid density and temperature obtained from a number of different experiments. These results come close to the theoretical (solid) line. Different symbols relate to different experiments. (After Bishop, D. J. and Reppy, J. D. (1980). *Phys. Rev. B* **22**, 5171–85, figure 11.)

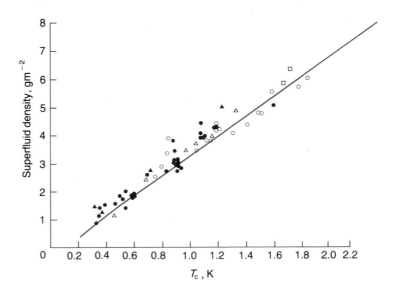

helium is destroyed by the motion of quantised vortices. Quantum flux lines are very like the quantised vortices of liquid helium described in the previous subsection, except that they carry a definite amount of magnetic field with them when they move. In a thin wire there is no true phase transition, but the transition from normal behaviour to superconducting behaviour is spread out over a temperature range. Even at the lowest temperatures there is a calculable probability of flux lines moving across the system to destroy supercurrent flowing in a narrow ring, but this probability diminishes rapidly in the region in which the transition occurs, and it is often unobservably small at low temperatures. Similarly, there are superconducting polymers which are made up of long molecules with the atoms arranged in chains along which the electrons are free to move. The electrons move across from one chain to another with much more difficulty, so these superconducting polymers behave similarly to the magnetic chains mentioned in Section 7.2. There is no transition to the true atoms separated by double layers of sulphur or selenium atoms. These are materials which in the normal state have a reasonable metallic electrical resistivity in directions parallel to the layers, but considerably higher resistivity in the perpendicular direction. This difference can be enhanced by the insertion of layers of organic material between the layers of nonmetallic atoms. These materials display a two-dimensional superconducting behaviour in the temperature region above the eventual three-dimensional transition temperature.

In 1986 J. G. Bednorz and K. A. Müller, working at the IBM Research Laboratories in Zurich, found a new class of materials which are superconducting at much higher temperatures than any that had been found previously. This discovery led to an explosion of activity, and materials have been found which are clearly superconducting around 100 K, which is four times the highest temperature for superconductivity of any materials known before 1986, and several groups have reported signs of superconductivity at temperatures close to room temperature. The reasons for this behaviour are not yet clear, but it is agreed that these materials are layered materials of the sort described in the previous paragraph. They are based on the compound La_2CuO_4 or closely related compounds. The pure material is an insulator, but when certain impurities are added it becomes a conductor, with the crystal structure of Rb_2MnF_4 shown in figure 7.10. The copper atoms are arranged in a square array on well separated planes, just as the manganese atoms shown in the figure are arranged. Electrical conduction takes place primarily along the planes of these copper atoms. Although the reason for the observed high temperature superconductivity is not generally agreed, there is not much doubt that the two-dimensional nature of the system must be important in any detailed explanation. It is also thought to be significant that these materials sometimes behave like layered anti-ferromagnets.

7.4 Electrons in two dimensions

Metals, semiconductors and insulators

The quantum theory of solids was developed around 1930, shortly after the development of quantum mechanics. It enabled physicists to understand in some detail, among many other things, the electrical properties of solids. It had been known for many years that the high electrical conductivity of a metal could be explained in terms of the motion of electrons, but many of the details were paradoxical, and the reasons for the difference between metals and insulators were not known. One of the most important discoveries was W. Pauli's 'exclusion principle', which said that there could not be more than one electron in each quantum state. In this respect electrons are quite unlike helium atoms, which, as was said in Section 7.3,

may all collapse into the same quantum state and so produce superfluidity. The quantum states of an electron in a metal are waves which travel all over the metal. The wavelength of the wave and its direction of propagation can be varied to get the different quantum states. Waves with a long wavelength have the lowest kinetic energy, and it is these states of low kinetic energy which are preferentially occupied by the electrons. If there is a high density of electrons, as there is in a metal, some of the electrons must go into states of rather short wavelength and high kinetic energy. At room temperature the energy k_BT of typical thermal excitations is a factor of thirty or so smaller than the maximum kinetic energy of the electrons. The electrons more or less fill all states up to this maximum energy, and we refer to such a system as a 'degenerate electron gas'. Such a system carries an electric current very easily, as an electric field will give all the electrons a little extra velocity.

The other important ingredient of the quantum theory of solids is a consequence of the fact that the solid is made up of a regular array of atoms. The energy of the waves which form the quantum states of the electrons does not vary continuously as the wavelength is reduced. Instead there are 'energy bands', with 'energy gaps' between the bands. An energy gap is a range of energies for which there are no quantum states of the electrons. If the lowest energy bands are completely filled with electrons up to a certain gap, and there are no electrons above the gap, then the electrons are prevented from adjusting to an electric field by the exclusion principle, and so we have an insulator. The bands for an insulator are shown in figure 7.26(*a*). The highest filled energy band is known as the 'valence band', and the lowest empty band is known as the 'conduction band'. In a metal, as shown in figure 7.26(*b*), one of the bands is partially filled with electrons; those bands with lower energy are completely filled, and those with higher energy than the partially filled band are completely empty.

Semiconductors, such as silicon, selenium, or gallium arsenide, are intermediate between metals and insulators. A pure semiconductor is like an insulator with a rather small gap between the valence band and the conduction band. In most applications semiconductors are 'doped' by the introduction of impurity atoms which are specially chosen to affect the electrical properties. In a 'p-type' semiconductor the impurity atoms have empty electron levels just above the top of the valence band of the pure semiconductor, as is shown in figure 7.26(*c*). Thermal excitations cause electrons to be excited from the top of the valence band to the empty levels associated with the impurity atoms. This leaves holes in the valence band which act as positive carriers of electricity, which is why they are called p-type. In 'n-type' semiconductors the levels associated with the impurity atoms are just below the bottom of the conduction band, and each of these impurity levels contains an electron. These electrons from the impurity atoms can be excited into the conduction band, where they act as negative carriers of electricity.

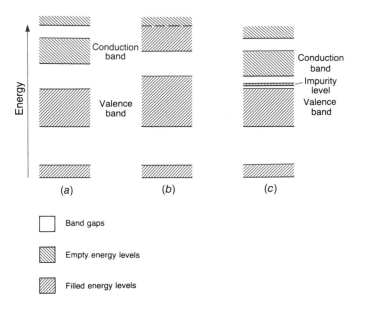

Band gaps

Empty energy levels

Filled energy levels

Figure 7.26. Energy bands and band gaps for (*a*) an insulator, (*b*) a metal, and (*c*) a p-type semiconductor.

Trapped electrons

Electrons can be confined to the neighbourhood of a surface by various means, and they then form a two-dimensional system. There are two sorts of system that have been extensively studied. A typical example of the first sort is made by the trapping of electrons above the surface of liquid helium at low temperatures. Electrons are attracted to the surface of liquid helium by the small polarisation they induce in the liquid, but they cannot penetrate into the liquid because it is a good insulator. They can be bound more tightly by the application of an additional electric field normal to the surface. It is also necessary to provide a guard ring to produce a tangential field component so that the electrons do not drift out to the edge of the helium surface and so on to the container. A typical arrangement is shown in figure 7.27. Other insulating surfaces or interfaces can be used to produce the trapping, but liquid helium has the advantage of a very smooth surface, and the low density of helium atoms in the vapour above the surface makes the mean free path of the electrons very long. They are free to move parallel to the surface, and a typical density of trapped electrons is of the order of 10^{12} per square metre.

The second method used involves the trapping of electrons at a semiconductor surface. In MOS (metal–oxide–semiconductor) devices, which are widely used in integrated circuits, a p-type semiconductor, usually silicon, has an insulating layer, such as silicon dioxide, grown on top of it. A metal electrode, known as the gate, is grown on top of the oxide layer. A positive voltage is applied to the gate so that the electric field lowers the energy of the conduction band close to the surface. If the field is strong enough electrons are trapped in a narrow layer on the semiconductor side of the interface between the semiconductor and the oxide. These trapped electrons are then free to move parallel to the surface, and a current can be made to flow along the surface between two electrodes, which are known as the source and the drain. This type of device is shown in figure 7.28. It has the great advantage that the number of electrons trapped in the surface can be controlled by adjusting the voltage applied to the gate. This is how a MOS device is used as a switch or an amplifier, but the ability to control the number of electrons at the surface is also very useful for studies of two-dimensional physics.

A different type of device is the heterojunction, which is the junction between two different semiconductors, such as gallium arsenide and gallium aluminium arsenide. In this case the number of electrons cannot be readily controlled, but it can be arranged so that the electrons trapped in the interface are in a region of very high purity, and so are much less scattered than they would be in a MOS device. In these semiconductor devices, MOS devices or heterojunctions, a much higher density of electrons can be obtained than is possible above a helium surface, of the order of 10^{15} or 10^{16} per square metre. The dielectric constant is also much more than it is in helium vapour, so the electrons interact less strongly with one another than they would at the same density above the helium surface.

At the typical densities for electrons trapped at a liquid helium surface the potential energy of interaction between electrons is much larger than the kinetic energy due to the exclusion principle. At the higher densities at semiconductor surfaces the opposite is the case. As a result the electrons at the helium surface can be regarded as forming a classical two-dimensional Coulomb system, for which the quantum effects are unimportant. In contrast, the electrons on a semiconductor surface can be regarded as a degenerate two-dimensional electron gas, provided the temperature is sufficiently low; temperatures below the boiling point of liquid helium (about 4 K) are needed for most of the studies mentioned in this section. For both liquid helium surfaces and for semiconductor surfaces the electrons are held tightly enough to the surface that the region in which they are confined is narrow in comparison with the average spacing of the electrons. Motion in the third dimension can be neglected except at high temperatures or at high excitation energies, and the systems are really two dimensional.

It has been known for a long time that the state of lowest energy of a classical Coulomb system is a solid, with a triangular lattice structure in two dimensions. Increasing temperature or increasing importance of the kinetic energy due to the exclusion principle (the kinetic energy is more important at high electron densities than at low density) leads to a melting of this lattice, but theoretical calculations of the conditions needed for melting have differed from one another by a factor of fifty or so. In 1978 C. C. Grimes and G. Adams studied the vibrational motion of electrons trapped on the helium surface

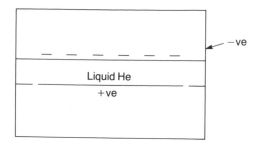

Figure 7.27. The trapping of electrons above a liquid helium surface. The positive electrode below the surface of the helium is necessary to stop the electrons from drifting out along the helium surface to the sides of the container. It can also be used to bind the electrons more tightly to the helium surface.

as the temperature was varied. As the temperature was lowered they found the sudden appearance of modes of vibration which are characteristic of a solid, so they could identify the temperature at which this occurred as the freezing point of the electron solid. The dependence of the freezing temperature on electron density showed clearly that it was due to the Coulomb interactions in the classical limit. It also appears that the melting temperature is in good agreement with the temperature which is predicted by the dislocation unbinding theory of melting which was mentioned in Section 7.3. It is not yet unambiguously established whether the melting transition is a continuous transition at a critical point, as would be predicted by that theory, or if it is the more usual first order phase transition which seems to occur in other examples of two-dimensional melting.

It would be very interesting to observe the melting of the two-dimensional electron solid at very low temperatures as the density is increased and the kinetic energy due to the exclusion principle becomes more important. At present there is a gap in the densities between the low density regime observed on the helium surface and the high density regime found on semiconductor surfaces, and so this transition has not been observed. Various experiments have been planned to fill this gap and observe the transition.

Disorder and localisation

No solid has the perfectly regular structure of an ideal solid. Departures from regularity arise both from the presence of impurity atoms, and from the thermal motion of the atoms which causes them to deviate from their ideal positions. This 'disorder' is an essential ingredient of the quantum theory of solids. In a metal it was argued that the disorder would determine the electrical conductivity. The less disorder there was the more electrons would be accelerated by an electric field, and so the lower would be the electrical resistance of the metal.

Since the thermal motion of the atoms gets less as the temperature is lowered, the resistance of a metal should get less, until, at very low temperatures, there is left a residual resistance which is entirely due to the presence of impurities in the metal. The real situation is complicated by the fact that most real metals are either magnetic (ferromagnetic or antiferromagnetic) or superconducting at very low temperatures. Nevertheless this basic picture was unchallenged for thirty years.

In 1958 it was shown by P. W. Anderson that the effect of strong disorder (which might be due to a very high concentration of impurities) in a system such as a metal would be to localise all the electron wave functions. That is, each quantum state, which is a solution of the wave equation for the electron, would be confined to a certain region of space and fall off exponentially with distance outside that region. It had been universally expected that the quantum state would be a wave spread over the entire volume of the metal. An apparently unrelated argument was put forward a short time later by R. Landauer and by N. F. Mott and W. D. Twose. They argued that in a one-dimensional system with a random potential all stationary solutions of the wave equation are exponentially localised however weak the disorder. Both of these results have now been put on a firm mathematical basis.

Figure 7.28. A metal–oxide–semiconductor field effect transistor (MOSFET). The inversion layer is produced at the surface between the silicon and the oxide layer by means of the electric field produced by the voltage applied to the metal gate. Current is passed through the inversion layer between the source and the drain. To the left of this figure is shown how the electric field bends down the conduction band near the surface, so that the electrons are trapped there. (After Kawaji, S. (1983). Localization in two-dimensional systems: silicon MOS inversion layers in weak and strong magnetic fields. In *Symposium on Recent Topics in Semiconductor Physics*, Kamimura, H. and Toyozawa, Y., eds., pp. 105–44. World Scientific, Singapore, figure 1*a*.)

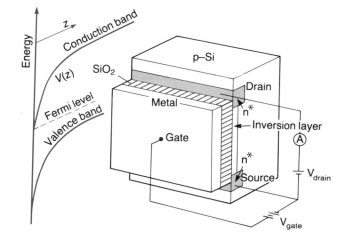

From the second of these results I drew the paradoxical conclusion that the zero temperature electrical resistance of any thin wire would only be proportional to its length for lengths which gave a resistance up to about twenty thousand ohms, and beyond that length the resistance should increase as the exponential of the length. This is shown schematically in figure 7.29. In practice it is not easy to work at sufficiently low temperatures for these effects to be directly observable, but they can be observed indirectly from the increase in resistance as the temperature is lowered. The main effect of reducing the temperature at very low temperatures is to increase the period of time for which this zero temperature theory is applicable, and the maximum length for which the theory is applicable is the distance which an electron travels in that time. Beyond that length the resistance is proportional to length just as it is at high temperatures. This characteristic time generally increases like an inverse power of the temperature, and the result of combining this temperature dependence of the characteristic time with the exponential dependence of the resistance on length for large lengths is a resistance that increases rather slowly as the temperature is lowered until the temperature is low enough that electrons can travel the localisation length. When the temperature is lowered beyond this point the resistance increases like an inverse power of the temperature. This general pattern of behaviour is observed in very narrow systems, either made from microscopically fine metallic wires, or from confined channels at semiconductor surfaces.

A link between the weak localisation found even for weak disorder in one dimension and the localisation found even in three dimensions for strong disorder was made in a theory proposed by E. Abrahams, P. W. Anderson, D. C. Licciardello and T. V. Ramakrishnan in 1979. They developed a scaling theory of the electrical resistance of electrons in disordered materials in analogy with the scaling theories of critical phenomena (see the chapter by Bruce and Wallace). The correlation length, which plays such an important part in the theory of critical phenomena (see Section 7.2), is replaced by the 'localisation length' – the distance over which the wave function falls off. In this scaling theory the resistance always increases exponentially with the size of the system if the resistance is large enough. When the resistance is small it depends on size in a manner that is characteristic of the number of dimensions, and which is only slightly different from the size dependence familiar from elementary physics. In three dimensions the resistance of a cube decreases as the system gets larger, since the cross-sectional area through which the current may pass increases more rapidly than the distance across the cube. For a one-dimensional system the resistance is proportional to length. Because of this length dependence the resistance of the one-dimensional system always gets large enough to cross over into the exponentially increasing regime if the length is sufficiently large, as is shown in figure 7.29. On the contrary, in three dimensions if the resistance is low for small cubes it will be even lower for larger cubes, and the region of

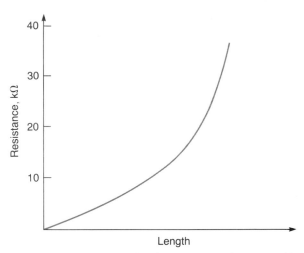

Figure 7.29. Resistance of a thin wire as a function of length at zero temperature. The familiar linear increase of resistance with length only occurs when the resistance is low, and an exponential increase occurs when the resistance is higher.

exponential growth is never reached unless the resistance on a small scale, say in cubes whose size is of the order of the distance an electron goes between collisions with impurities, is sufficiently large. There should be an intermediate regime, analogous to the critical region in a phase transition, in which either there is exponential localisation (exponential increase of the resistance with size) which is only perceptible on very long length scales, or, on the other side of the critical point, the resistance decreases very slowly with size initially, so that it is only on very large length scales that the system behaves like a metal. These different kinds of behaviour are shown in figure 7.30. As in the one-dimensional case the physical length scale is not usually set by the size of the sample, but by the distance an electron travels before thermal effects change its energy – the 'inelastic scattering distance'.

The two-dimensional case is particularly interesting. In the ordinary theory of electrical conduction the resistance of a square film of fixed thickness is independent of the size of the square, because the resistance is proportional to length and inversely proportional to width. In the scaling theory of localisation it is argued that the resistance should increase slowly with size even if the resistance of a small square is rather small. The larger the resistance gets the more rapidly it increases, until eventually it gets large enough to reach the exponentially growing regime. This kind of behaviour is shown in figure 7.31. It is therefore concluded that all states in a two-dimensional system are exponentially localised, although the localisation length may well be so large that it cannot be directly seen. It can be seen indirectly by measuring the resistance as a function of temperature: because the inelastic scattering length increases as the temperature is lowered, the resistance increases like the logarithm of the temperature. This is indeed seen experimentally, both for thin metallic films and for electrons confined to the surface of a semiconductor.

There are various additional complications that have been taken into account in recent work. The interactions between the electrons have an important effect that was not taken into account in the original scaling theory. Spin-dependent interactions of the electrons have an important effect, and can even cause the resistance of a two-dimensional system to become smaller as the length scale increases if they are strong enough. Finally, a magnetic field drastically reduces the localisation effects.

Figure 7.30. Dependence of the resistance of a cube on its size. Below the critical resistance the behaviour is metallic, and localisation, with an exponential increase of resistance, occurs above the critical resistance.

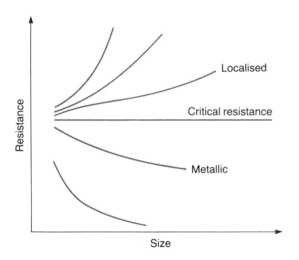

Figure 7.31. Size dependence of the resistance of a square. For low resistance there is a slow increase of resistance with length, but eventually the increase becomes exponential.

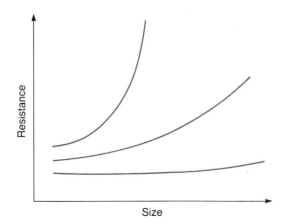

Quantum Hall effect

It is a well known result of classical electrodynamics that a uniform magnetic field B causes particles of mass m and charge e to move in circular orbits with an angular frequency eB/m, known as the 'cyclotron frequency'. If an additional electric field E is applied in a direction perpendicular to the magnetic field there is a drift of these circular orbits with velocity E/B in the direction perpendicular to both fields. The resultant motion is shown in figure 7.32. In most real metals, even at very low temperatures, the electrons are scattered in a time which is short compared with the period of this orbit. Therefore the resultant current when crossed electric and magnetic fields are applied is largely in the direction of the electric field, but there is a small component, known as the Hall current, in the perpendicular direction. These were the arguments that led E. H. Hall to look for this effect more than a hundred years ago, although the detailed interpretation of the results had to await the development of the quantum theory of solids sixty years later.

In a two-dimensional system, such as a MOS device or a heterojunction, with the magnetic field applied perpendicular to the interface, there are several important modifications of this description which are introduced by quantum theory. In the first place, for the ideal free two-dimensional electrons the magnetic field makes the electron energy levels have discrete values, which are all integer multiples of Planck's constant times the cyclotron frequency. Each of these levels, known as 'Landau levels', is highly degenerate, and can be filled by eB/h electrons per unit area, where h is Planck's constant. When this degeneracy factor is combined with the drift velocity E/B, which is the same in quantum theory and in classical theory, it is concluded that each filled Landau level should contribute a current density equal to e^2E/h in a direction perpendicular to the electric field. This means the ratio of the Hall current to the electric potential that induces the current is an integer multiple of e^2/h if all Landau levels are either completely filled or completely empty, as they may be at sufficiently low temperatures. Additional quantum effects are introduced by the electron spin, whose two possible directions cause each Landau level to be doubled; in semiconductors the splitting of the different spin components results in a considerably smaller energy difference than the splitting between the main Landau levels proportional to the cyclotron frequency. In silicon crystals, but not in gallium arsenide, there is a further doubling of the Landau levels due to the energy band structure; this splitting is known as valley splitting.

Until recently there was no suggestion that these simple results for an idealised two-dimensional system would hold for a real system. In 1980 it was found by K. von Klitzing, G. Dorda and M. Pepper that, for a silicon MOS device in a strong magnetic field perpendicular to the interface, variation of the gate voltage gave regions in which the current was accurately perpendicular to the electric field, and the ratio of current to field constant. Furthermore, in these regions the constant value of this ratio was very close to the integer multiples of e^2/h predicted by the simple theory. Even the first published results give this to an accuracy of one part in 10^5, and more recent work has confirmed its value to better than one part in 10^7. Figure 7.33 shows the results obtained in this experiment. Simultaneous voltage measurements were made parallel to the current, denoted by V_P, and transverse to the current, denoted by V_H. For certain values of the gate voltage, which presumably correspond to completely filled Landau levels, the value of V_P is very small, and the corresponding values of the Hall voltage V_H are flat, and accurately equal to some fixed value divided by an integer n. The highest of these flat regions in the Hall voltage corresponds to $n=2$, with the lowest two states with one spin direction filled. The next one has $n=3$, and has one of the two valleys for the opposite spin state filled. For $n=4$ both spin states and both valleys for the lowest Landau level are filled. Beyond that much of the structure due to the two spin states and two valleys gets lost, but the flat regions for $n=8$ and $n=12$ are clearly visible. For this work von Klitzing was awarded the Nobel Prize in 1985.

It would be a gross over-simplification to say that the simple picture obtained on the basis of ideal free electrons in a uniform magnetic field gives a correct explanation of these quantised Hall currents. We know, for example, that impurities in the semiconductor lead to a disorder which should produce localised electron states. Nevertheless it can be shown without much difficulty that these quantised values of the Hall current predicted on the basis of an over-simplified theory are not destroyed or changed by the effects of disorder, the existence of localised electron levels, or by the interactions between electrons, provided that none of these effects is too strong. The effect gives the possibility of determining fundamental constants, such as the fine structure constant, with an accuracy comparable with the best available methods. Already the effect provides a readily reproducible standard of resistance whose accuracy is considerably better than the best available outside standards laboratories.

In heterojunctions it is possible to arrange for the electrons to have even higher mobilities, that is to travel even further before being scattered, than they do in MOS devices. D. C. Tsui, H. Stormer and A. C. Gossard found in 1982 that with very high mobility devices they got flat regions in the Hall voltage which corresponded to rational fractions instead of integers. This fractional quantum Hall effect was first observed with the value $\frac{1}{3}$, and that is known to be correct with an accuracy of one part in 10^4, but many other fractional values have now been observed. The denominators are mostly odd, and values of the denominator up to nine have been identified. This is much less well understood than the integer Hall effect, but a more or less satisfactory theory has been proposed by R. B. Laughlin. It is generally agreed that the fractional quantum Hall effect is due to the interactions between the electrons.

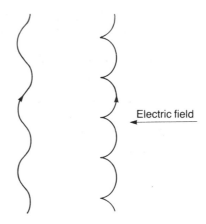

Figure 7.32. Classical motions of electrons in crossed electric and magnetic fields. The magnetic field is perpendicular to the plane of the diagram, and the electric field is horizontal. The pattern on the right is what will occur if the electron starts from rest.

Figure 7.33. The original measurements of the quantum Hall effect. The Hall voltage is shown by the steadily decreasing black curve with flat regions, while the voltage in the direction of the current is shown by the grey curve. The regions of quantised Hall current (flat regions) correspond to the regions of zero voltage in the direction of the current. (After von Klitzing, K., Dorda, G. and Pepper, M. (1980). *Phys. Rev. Lett.* **45**, 494–7, figure 1.)

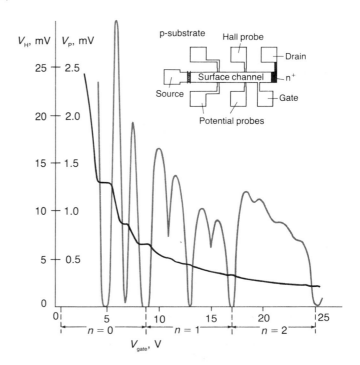

7.5 Why are low-dimensional systems studied?

There are many different levels at which one can answer the question of why a particular area of science has received a lot of attention. One answer, equivalent to the statement that Everest must be climbed because it is there, is that some of the problems that area presents can be solved, and that the solution of the problems is sufficiently difficult that it presents a challenge to good physicists. That is certainly part of the answer in this case, and one of the earliest reasons for looking at low-dimensional systems was that solutions could be found to problems that could not be answered at that time in three dimensions. It is clear that the availability of solutions was never a complete answer, because it was also hoped that the insight gained from these 'unrealistic' problems would shed light on important problems about the nature of ferromagnetism, phase transitions, and so on, in real three-dimensional systems. To a large extent these hopes have been fulfilled. Few people would dispute that Onsager's solution of the two-dimensional Ising model prepared the way for important advances in the understanding of real systems.

As in many other areas of physics the interplay between theory and experiment has greatly enriched the subject. In the early days it was an almost entirely theoretical subject, and important developments have been made without any reference to experiment. The study of experimental systems that appeared to behave two dimensionally or one dimensionally challenged many of the theorists' preconceptions, and forced, in some cases, completely new ways of thinking about the problems.

It is also true that this is not just an isolated area of physics, but there are many connections, both theoretical and experimental, to other areas of physics. The obvious connections to the corresponding problems in the physics of three-dimensional systems have already been mentioned. It looks as if reduced dimensionality must play an important role in the behaviour of high-temperature superconductors. There are some much less obvious connections, even to problems involved in the theory of the early universe (see Chapter 3 by Guth and Steinhardt) and to elementary particle theory (see Chapter 17 by Taylor), but these more subtle connections cannot be exposed here. On the experimental side it is clear that a study of adsorbed films involves problems concerned with the nature of the surface on which the adsorption takes place, and study of magnetism in layered materials involves problems about how the layers are arranged in a real crystal. In this way connections with other areas of physics and chemistry are inevitably built up. The study of low-dimensional systems is not just a quiet corner of science, but is closely connected to a lot of other vigorous activity.

It is also fair to ask if this is a useful field of study. At present I can only point to one useful development that comes directly from work on low-dimensional systems. In Section 7.4 it was mentioned that the quantum Hall effect provides a high precision standard of electrical resistance which is more accurate and more readily reproducible than any which existed before. This is not the complete answer, as there is a close relation between experimental work in this area and high technology. One might call the relationship parasitic, but I prefer to call it symbiotic. There is no doubt that many of the most interesting experiments exploit expensive techniques which were developed for other reasons – synchrotron radiation, high flux neutron beams, very high magnetic fields, electron beam lithography, molecular beam epitaxy, and so on. It is also true that many of the systems studied, such as magnetic thin films, liquid crystals, superconducting films, electrons on semiconductor surfaces, thin metal films, and so on, have actual or potential uses in recently developed devices of many sorts. The conditions under which they are studied by physicists who are interested in the sort of problems which have been discussed in this chapter are very different from the conditions under which they are used in devices. However, the work shows up unexpected characteristics of the materials which may be of importance in the development of devices. It should also be remembered that work on fundamental problems with such systems encourages some people, who might otherwise not get involved in such matters, to take a serious interest in the physics of devices. There is little doubt that the pressure to make electronic devices more and more compact will make the lessons learned in this area of physics increasingly relevant to modern technology.

Further reading

This list is divided into two parts. The first consists of surveys intended for the general physicist or scientist. The second consists of review articles intended for the more specialised reader.

General surveys

Bechgaard, K. and Jerome, D. (1982). *Sci. Am.* **247**, (1), 52–61.
  A survey of the recently discovered superconducting polymers, which are essentially one-dimensional superconductors mentioned on p. 227.
Birgenau, R. J. and Horn, P. M. (1986). *Science* **232**, 329–36.
  This is concerned with rare gas solid monolayers, and takes further some of the points raised in pp. 218–22 and 224.
Birgenau, R. J. and Shirane, G. (1978). *Phys. Today* **31**, (12), 32–43.
  This relates to pp. 217–8 and gives a survey of experimental work on magnetic chains.
Brinkman, W. F., Fisher, D. S. and Moncton, D. E. (1982). *Science* **217**, 693–700.
  This is concerned with the melting of adsorbed monolayers, liquid crystals, and trapped electrons, and so relates to pp. 218–24.

Conwell, E. M. (1985). *Phys. Today* **38**, (6), 46–53.
   A survey of polymeric semiconductors, a form of one-dimensional
   conductor which could have been discussed in Section 7.4.

Dash, J. G. (1985). *Phys. Today* **38**, (12), 26–32.
   This relates to pp. 214–5, and discusses how adsorbed films become
   more three dimensional as they become thicker.

Narayanamurti, V. (1984). *Phys. Today* **37**, (10), 24–32.
   An introduction to some of the many useful things that can be done by
   forming layered semiconductors.

Pepper, M. (1985). *Contemp. Phys.* **26**, 257–93.
   A survey of localisation and quantisation in silicon inversion layers,
   which relates to pp. 230–3.

Pindak, R. and Moncton, D. (1982). *Phys. Today* **35**, (5), 56–62.
   This is a survey of work on thin liquid crystal films, and relates to pp.
   222–3.

von Klitzing. K. (1986). *Rev. Mod. Phys*, **58**, 519–31.
   von Klitzing's Nobel Prize lecture on the quantum Hall effect (pp.
   232–3).

Specialised reviews

Ando, T., Fowler, A. B. and Stern, F. (1984). *Rev. Mod. Phys.* **54**,
   437–672.
   A very detailed survey of inversion layers and trapped electrons. The
   bibliography is worth admiring for its length.

Barber, M. N. (1980). *Phys. Repts.* **59**, 375–409.
   This is a theoretical review of two-dimensional phase transitions
   covering the ideas mentioned in Section 7.2.

Berlinsky, A. J. (1979). *Repts. Progr. Phys.* **42**, 1243–83.
   A review of one-dimensional metals and charge density waves in these
   materials. Another topic that could have been covered in Section 7.4.

de Jongh, L. J. and Miedema, A. R. (1974). *Adv. Phys.* **23**, 1–260.
   A comprehensive review of magnetic phase transitions with special
   emphasis on layered and chain materials, so it relates to pp. 214–5 and
   217–8.

Nelson, D. R. (1983). Defect mediated phase transitions. In *Theory
   of Phase Transitions and Critical Phenomena*, Domb, C. and
   Lebowitz, J. L., eds., Vol. 7. pp. 1–99, Academic Press, NY.
   A detailed survey of defect mediated phase transitions, discussed in pp.
   223–7.

8 Critical point phenomena: universal physics at large length scales

Alastair Bruce and David Wallace

8.1 Introduction: the new physics of critical point phenomena

At atmospheric pressure and temperature H_2O exists as a liquid: water. The liquid state, or *phase*, remains the favoured one at atmospheric pressure for temperatures less than 100°C. At 100°C water boils. For temperatures in excess of 100°C H_2O exists in its vapour phase: steam. The change of state which occurs at $T_b = 100$°C is known as a *phase transition*. It is accompanied by an abrupt change in the fluid density: the volume of a given mass of the vapour phase is some 1600 times greater than that of the same mass of liquid, and its density is correspondingly smaller. The boiling temperature T_b depends upon the pressure p. At pressures in excess of that of the atmosphere the boiling temperature is larger than 100°C; moreover the jump in the fluid density which occurs on boiling is less than it is when the phase change takes place at 100°C. This trend continues as the pressure is increased: the difference in densities of the liquid and vapour phases on either side of the boiling temperature $T_b(p)$ decreases smoothly until, at a pressure p_c of 218 atmospheres, for which $T_b = 374$°C, this difference vanishes altogether. The parameters p_c and $T_c = T_b(p_c)$ together locate the *critical point* of the fluid (figure 8.1*a*).

At room temperature iron exists as a ferromagnet. The distinctive feature of a ferromagnetic phase is its magnetic moment, manifesting itself in the capacity of the material to attract other ferrous materials and in its disposition (apparent in the behaviour of compass needles) to orient itself so that its magnetic moment is aligned with an external magnetic field. As the temperature is increased the magnetic moment of iron decreases smoothly until at a temperature $T_c = 770$°C it vanishes altogether. The parameter T_c locates the critical point of the magnet (figure 8.1*b*).

This article is concerned with the behaviour of a physical system at, or close to, a critical point. It is not immediately obvious that the study of such behaviour can reasonably be claimed to constitute 'new' physics: the study of critical point phenomena in both fluids and magnets has a history extending back well into the nineteenth century. Nor do our cursory opening remarks offer any hint as to why the physics of the critical point, old or new, should in any case be of particular interest and significance.

In fact the claim which the study of critical phenomena makes for inclusion in this volume can be very easily if superficially warranted. Testimony to the existence of ample 'new' physics in this area is to be found in the enormous research literature it has spawned over the last decade, stimulated in a large measure through articles published in 1971 by Kenneth Wilson of Cornell University. As immediate testimony to the interest and significance of these developments we may simply cite the award to Wilson of the 1982 Nobel Prize for physics. In the course of this article we shall describe both the conceptual motivation (the 'why') and the technical apparatus (the 'how') underlying this remarkable upsurge in interest and activity in the field of critical phenomena. Their key ingredients may, however, be usefully identified at this point.

The 'why' question finds its answer largely in the notion of *universality*, in two rather different senses of that word. Firstly, the significance attached to the critical point has grown with the recognition that the problem it poses is a 'universal' one, that is, one generic to a whole class of problems in the physical sciences. Thus, phenomena as disparate as the turbulent motion of fluids and the interaction of quarks confront the theoretical physicist with technical difficulties which have the same essential flavour as those encountered in the critical region. Secondly the intrinsic appeal of the critical point has been greatly enhanced by a growing appreciation that the associated phenomena are 'universal' in a rather more specialised but equally appealing sense. Specifically, it is now well established that the behaviour of many physically very different systems, near their respective critical points, exhibits remarkable similarities. Thus, for example, the critical behaviour of a magnet and that of a fluid are (with certain caveats) similar to a degree which far transcends the superficial parallels brought out by our introductory remarks.

The 'how' question finds its answer in the emergence of a powerful new theoretical technique, due very largely to Wilson, and known as the *renormalisation group*. The development of this technique undoubtedly represents the single most significant advance in the theory of critical phenomena and one of the most significant in theoretical physics generally over the last fifteen years. It is a method specifically designed to tackle the technical difficulties posed by the class of problem of which the critical point is the paradigm; in it one finds a beautiful resolution of the conceptual problem posed by the existence of universality in critical point behaviour.

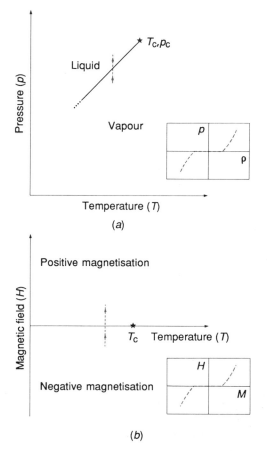

Figure 8.1. (*a*) The critical point of a fluid. Along the boiling curve $T_b(p)$ the liquid and vapour phases of a fluid coexist. A plot (inset) of density as the pressure is varied at constant temperature (the grey path) shows a jump at the line of phase coexistence. The coexistence curve ends in a critical point where the difference in the densities of the two phases vanishes and the two phases merge into a single gas phase.

Figure 8.1. (*b*) The critical point of a magnet. Below a critical temperature T_c a magnet possesses a magnetic moment. The direction of the magnetic moment can be switched by a magnetic field, H. Along the line of zero magnetic field different phases, distinguished by the direction of the magnetic moment, coexist. As the magnetic field is varied at constant temperature (the grey path) the orientation of the magnetisation, M, is reversed at the line of phase coexistence (inset). The line of coexistence ends at T_c where the magnetic moment vanishes, and the different phases merge into a single paramagnetic phase.

These remarks define the landmarks in the territory to be explored in the following sections. In Section 8.2 we look in greater detail at the phenomena displayed in the critical region. The essential points are illustrated by reference to results of computer simulation studies of a simple model system with a critical point (the *Ising* model). We introduce the two key quantities, the *order parameter* and the *correlation length*, instrumental in describing critical point behaviour; and we identify more explicitly the technical and conceptual problems which a theory of the critical region must confront. In Section 8.3 we explore in some depth the *configurations* (patterns) formed by the microscopic degrees of freedom near a critical point, drawing extensively on the results of computer simulation studies. We introduce, and express in configurational terms, the essential concepts of the renormalisation group: *coarse-graining, system flow, fixed point* and *scale-invariance*; we see how, within this framework, one may address both the technical and conceptual problems of the critical region. In Section 8.4 we show how these concepts knit together to form the renormalisation group method itself; the presentation emphasises the essential strategy of the method rather than the technical details, a selection of which the reader (interested in such matters!) will find in Section 8.5. Finally, in Section 8.6 we attempt to give some impression of the remarkable range of problems, in the field of critical phenomena and beyond, which have already yielded, or are potentially susceptible, to the renormalisation group method.

8.2   Background

Concepts

In theoretical physics the assault on a problem of interest traditionally begins (and sometimes ends) with an attempt to identify and understand the simplest model exhibiting the same essential features as the physical problem in question. In the field of critical point phenomena this is a strategy which (as a corollary of universality) brings distinctively rich rewards; we adopt it here. The simplest model system with a critical point was introduced by Lenz, although it is now invariably known as the Ising model. Though devised as a simple model of a ferromagnet (cf. the chapter by Thouless in this volume), in keeping with its much wider significance we shall describe it in general (rather than specifically magnetic) terms. The model envisages a regular array of points in space (a lattice); the number of points, N, is large. The space dimension of the array is a significant matter of choice; the essential physics is apparent in a two-dimensional array (occupying a *plane*), and we will use this case for illustrative purposes. The precise nature of the pattern formed by the lattice sites (the *symmetry* of the lattice) is more a matter of taste and convenience; we will consider a square lattice. We label the lattice sites by an index i $(=1,2,\ldots,N)$ and associate with each site i a variable s_i. Each variable may adopt one or other of two values, $+1$ or -1. Neighbouring variables exert an influence on one another. This influence is expressed through the association with each pair of neighbouring sites i and j an energy $-Js_is_j$, where J is some positive constant. The interaction energy is thus lower ($-J$) or higher ($+J$) according to whether neighbouring variables have the same value or opposite values. This completes the definition of the model.

The model has a critical point, discernible for large enough N; it occurs at a temperature $T_c = 2.269 J/k$, where k is Boltzmann's constant. For the moment it is less important to explain the basis of this claim than it is to explain what it means. What, then, *is* a critical point? The following definition is formally adequate if conceptually a little opaque: a critical point is the terminus of a line which exists in a space of properties (notably the temperature) controlling the state of the system, and which separates two (or more) distinct phases. That this line of phase coexistence 'terminates' in a critical point signals the fact that, at this point, the distinction between the two phases vanishes. This state of affairs is most clearly evident in the case of the fluid (figure 8.1*a*). Along the coexistence curve $T_b(p)$ there are *two* clearly different phases, liquid and vapour, distinguished primarily by their different densities; at the critical point (T_c, p_c) this density difference disappears, and the two phases coalesce into a *single* phase. In the case of the ferromagnet (figure 8.1*b*) the various phases which coexist for temperatures less than T_c are distinguished by the various possible orientations of the magnetic moment; the distinction between the phases vanishes as the magnetisation (the net magnetic moment per unit volume) itself disappears with the approach to T_c. In the case of the Ising model two phases coexist below the specified critical temperature: in one phase a majority of variables have a value $+1$; in the other a majority have value -1. With the approach to criticality the size difference of the two populations shrinks and, above T_c, the populations are essentially equal.

It is useful to have a quantitative measure of the difference between the phases coalescing at the critical point: this is the role of the *order parameter*, Q. In the case of the fluid the order parameter is taken as the difference between the densities of the liquid and vapour phases. In the ferromagnet it is taken as the magnetisation. In the Ising model it is defined by the fractional excess of $+1$ variables over that of -1 variables. Clearly, thus defined, the order parameter vanishes as the critical point is approached along the line of phase coexistence.

As its name suggests the order parameter may be thought of as a measure of the kind of orderliness that sets in when a system is cooled below a critical temperature. Our next task is to give some feeling for the principles which underlie this ordering process. Here we must appeal to the fundamental result of statistical (thermal) physics. The probability p_a that a physical system at temperature T will have a particular microscopic arrangement, labelled a, of energy E_a, is

$$p_a = Z^{-1} e^{-E_a/kT}. \tag{8.1}$$

The prefactor Z is the *partition function*; since the systems must always have *some* specific arrangement the sum of the probabilities p_a must be unity implying that

$$Z = \sum_a e^{-E_a/kT}, \tag{8.2}$$

where the sum extends over all possible microscopic arrangements. In utilising these equations it is generally (though, as we shall see, not invariably) correct to suppose that the physical system evolves rapidly (on the timescale of typical observations) amongst all its allowed arrangements, sampling them with the probabilities prescribed by equation (8.1); the 'observed' value of any physical property will thus be given by averaging the property over all arrangements a, weighting each contribution by the appropriate probability p_a.

Consider, specifically, the Ising model order parameter. An 'arrangement' a is defined by some particular set of values $s_1^{(a)}$, $s_2^{(a)}, \ldots$ of the variables; the 'configurational' energy E_a is then given by the total interaction energy

$$E_a = -J \sum_{\langle ij \rangle} s_i^{(a)} s_j^{(a)}, \tag{8.3}$$

where the sum extends over all pairs of adjacent sites. The probability p_a (equation 8.1) is the probability of finding the N variables to have the specified values; the sum on a (equation 8.2) extends over the complete set of possible assignments of all the N variables. The order parameter is then given by

$$Q = \sum_a Q_a p_a, \tag{8.4a}$$

where

$$Q_a \equiv \frac{1}{N} \sum_i s_i^{(a)} \tag{8.4b}$$

gives the fractional difference between the populations of $+1$ and -1 variables in the arrangement a. Sums like equation (8.4*a*) are not easily evaluated: if they were we would not be writing this article. Nevertheless some important insights follow painlessly. In the limit in which the temperature T is low (on the scale set by J/k) it is clear from equation (8.1) that the system will be overwhelmingly likely to be found in its minimum energy arrangements ('ground states'). There are two such arrangements: one in which all the s variables are $+1$ and another in which all the variables are -1. Although these fully ordered arrangements have the same energy, and are thus equally likely, a system found in one ground state at some instant is extremely unlikely to find its way to the other ground state (an important exception to the general rule that a system evolves rapidly amongst its arrangements) since to do so it must pass through arrangements of much higher energy, which have a probability which is very small (indeed utterly negligible in the limit of low T and large N). Thus, at $T = 0$, the system will certainly be found in, and will remain in, one fully ordered arrangement. (*Which* one is a matter of the system's history.) The order parameter Q (equations 8.4a,b) then necessarily has magnitude 1. Now consider the high temperature limit. The enhanced weight which the fully ordered arrangement carries in the sum (8.4a), by virtue of its low energy, is now no longer sufficient to offset the fact that arrangements in which the fractional population difference Q_a has some intermediate value, though each carrying a smaller weight, are vastly greater in number. A little thought shows that the arrangements which have essentially equal populations of $+1$ and -1 are by far the most numerous. At high temperatures these (essentially) fully disordered arrangements dominate the sum in equation (8.4a) and the order parameter is zero.

The competition between 'energy-of-arrangements' weighting (or simply 'energy') and 'number-of-arrangements' weighting (or 'entropy') is then the key principle at work here and, indeed, in thermal physics generally. The distinctive feature of a system with a critical point is that, in the course of this competition, the system is forced to choose amongst a number of macroscopically different sets of microscopic arrangements.

To illuminate the manner in which this ordering process occurs it is helpful to appeal to computer simulation studies of the Ising model. Computer simulation is widely used in condensed matter physics. A detailed discussion of the method lies beyond the scope of this chapter. Briefly, one uses an algorithm (a set of instructions) which, presented with some initial microscopic arrangement for the system concerned, generates a different microscopic arrangement. The earliest arrangements in the sequence generated by repeated application of this algorithm reflect the specific choice of the initial arrangement. However, the algorithm is so designed that, irrespective of the initial arrangement, the arrangements appearing later in the sequence do so with the frequency (probability) defined by equation (8.1). Desired properties may then be determined by computer-measuring the average value of the property in question over a suitably large number of arrangements. The availability of this technique does not itself solve the critical point problem: if it did, this chapter, for one, would be redundant. Nevertheless, much has been and can be learned from it, particularly through its application to the Ising model, which lends itself very naturally to this form of analysis. In this instance the basic ingredients of the algorithm take the form of a set of *probabilities* that an *s*-variable in a given state and with its immediate neighbours (those with which it interacts) in a given arrangement will, in unit time, jump to its other state.

In figure 8.2 we show three arrangements ('configurations') of the *s*-variables generated in a computer simulation study of a square lattice Ising model. Each of the three arrangements is representative of the spectrum of arrangements appropriate to a given temperature. Regions of black (white) signify *s*-

variables with value $+1$ (-1). With these conventions a fully disordered arrangement (which, we have argued, is typical at very high temperatures) would appear as a random patchwork of black and white, while a fully ordered arrangement (to be expected at very low temperatures) would appear completely black or completely white. Our three figures reveal the subtlety of the patterns favoured at temperatures in between these two simple extremes.

Figure 8.2(a) shows an arrangement typical of a temperature a little *above* the critical temperature T_c, defined earlier. As the committed reader may check there are essentially equal numbers of $+1$ and -1 variables: the areas of black and white are essentially the same, indicative of a phase with vanishing order parameter Q. However, the pattern is not what would be appropriate at very high temperatures. In this limit the likely colours (signs) of the variables surrounding some chosen site will be independent of the colour (sign) of the variable at that site, since the probability of an arrangement is effectively independent of its energy in this regime. At temperatures in the vicinity of T_c, however, it is apparent (cf. figure 8.2a) that a site of one colour tends to have (a majority of) neighbours of the same colour. This effect manifests itself in the islands, or clusters, of the same colour discernible against an otherwise random black and white background.

Now consider figure 8.2(b), which shows an arrangement typical of the critical temperature T_c itself. The essential equality of the black and white populations is still plausible, if less convincingly so*. A dramatic change in the pattern is nevertheless obvious: the relatively small islands of black and white apparent above T_c now have a spectrum of sizes amongst which the largest are very large, extending indeed throughout the system.

Finally consider figure 8.2(c), which shows an arrangement appropriate to a temperature a little below T_c. The duel between black and white, whose climax is the critical point, is now resolved: there is a clear-cut preponderance of black over white, signalling an ordered phase characterised by a nonzero value of Q. The order is not complete, however, as (we argued) it would be at very low temperatures: the remaining white islands, smaller in size and number than those present at the critical point, imply an order parameter less than unity.

---

*Here we have sacrificed truth on the altar of clarity. At the critical point the system is dominated by two sets of configurations, the one set having a significant preponderance of black and the other a significant preponderance of white. Occasionally, as the *s*-variables flip from state to state and the black and white islands grow or shrink, the system will find its way from one of these sets of configurations to the other. The equality of the populations of black and white is not, now, in general apparent in any one configuration, but emerges when the appropriate average is taken over very many configurations. We have endeavoured to convey the true average behaviour in a single figure. Thus we have (here, and in figure 8.5 below) carefully selected our 'typical' configuration so as to catch the system in the act of changing between its black-enriched and its white-enriched configuration set.

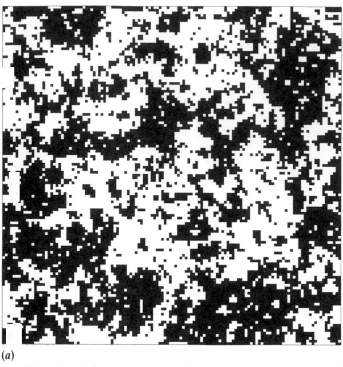

(a)

(b)   (c)

The behaviour revealed in figure 8.2 is typical of all systems with a critical point. Quite generally the approach to the critical point is signalled at a configurational level by a growth in the spatial extent (and, incidentally, temporal persistence) of regions which, in some measure, have the character of one of the ordered phases that coalesce at the critical point. In the fluid these regions take the form of bubbles of the liquid or vapour phase; in the magnet they consist of magnetised micro-domains, differentiated by the direction of their magnetic moments. Depending on one's perspective this unbounded growth in microstructure may be seen either as responsible for the smooth creation of an ordered phase (cf. the approach to the Ising critical point from above) or as responsible for its smooth destruction (the approach to T_c from below).

It is useful to have a quantitative measure of the extent of this spatial structure: this is the role of the *correlation length*, ξ, so called because it expresses the typical distance over which the behaviour of one microscopic variable is correlated with (influenced by) the behaviour of another. Settling for an intuitively rather than formally satisfactory definition, one may think of the correlation length as a measure of the typical linear dimension of the largest piece of correlated spatial

Figure 8.2. Configurations of the Ising model. The patterns depict typical arrangements of the *s*-variables generated in a computer simulation of an Ising model on a square lattice of $N = 512^2$ sites, at temperatures of (a) $T = 1.2\,T_c$, (b) $T = T_c$ and (c) $T = 0.95\,T_c$. In each case only a portion of the system containing 128^2 sites is shown. The square surrounding each lattice site is white or black according to whether the *s*-variable at that site is -1 or $+1$, respectively. The typical island size is a measure of the *correlation length*; the excess of black over white (below T_c) is a measure of the *order parameter*.

structure (e.g. the largest black or white island in figure 8.2a, the largest white island in figure 8.2c). The divergence of this length as the critical point is approached is the key to both the problems and the beauty of the critical region.

Objectives

The essential objectives of a theory of the critical region may be identified by a closer scrutiny of the behaviour of the order parameter and the correlation length.

The order parameter, Q, (the excess of black over white in our pictorial representation of Ising model configurations), vanishes as the critical temperature T_c is approached from below (strictly, along the line of phase coexistence). It is a matter of experimental fact that, very generally, the order parameter approaches zero as a *power* of the *reduced temperature* $t \equiv (T - T_c)/T_c$, which measures the deviation of the temperature from its critical point value. Specifically one finds that (figure 8.3)

$$Q \simeq Q_- |t|^\beta \quad (t < 0). \tag{8.5}$$

The wavy-line equality signifies that this power-law behaviour sets in only sufficiently close to the critical point; the minus sign subscript is a reminder that this relationship holds for negative t. We shall discuss the significance of the *critical index β* and the associated *power-law amplitude Q_-* in a moment.

Consider now the correlation length ξ, which grows arbitrarily large as the critical temperature is approached from above or from below. Again experiments establish power-law behaviour of the form (figure 8.3)

$$\xi \simeq \xi_- |t|^{-\nu} \quad (t < 0) \tag{8.6a}$$
$$\xi \simeq \xi_+ |t|^{-\nu} \quad (t > 0). \tag{8.6b}$$

These relationships define the second key critical index ν and the amplitudes ξ_- and ξ_+ associated, respectively, with the limiting behaviour below and above the critical point.

The critical point amplitudes and indices appearing in these equations have an altogether different status. The amplitudes vary from one physical system to another. This is wholly reasonable: the amplitude Q_- sets the basic scale of the variable which exhibits the ordering process; the amplitudes ξ_+ and ξ_- set the maximum scale upon which the incipient order is evident for a given reduced temperature. It is clear that these scales must reflect the nitty-gritty details of the system concerned. The study of the factors which set these scales, for any given physical system, is not uninteresting. However, like the determination of the factors locating the critical point itself, it poses difficulties of an altogether different kind from those of concern here, which emerge in the nature of the critical indices.

The indices display critical point universality: their values are remarkably insensitive to system-specific details. Thus, experiments have established that a whole range of fluids and magnets, and indeed many other systems with critical points, display behaviour which may be described by a single pair of

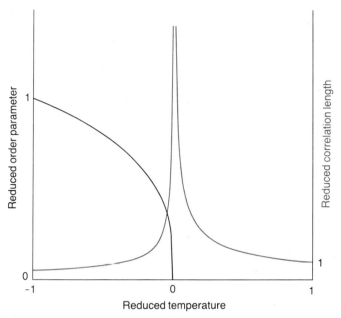

Figure 8.3. Critical behaviour of the order parameter and the correlation length. The order parameter vanishes with the power β of the reduced temperature t as the critical point is approached along the line of phase coexistence. The correlation length diverges with the power ν of the reduced temperature. The behaviour shown is qualitatively typical of all critical points. The specific values of the indices ($\beta = 0.33$ and $\nu = 0.63$) and the amplitude ratio ($\xi_-/\xi_+ = 0.51$) implicit in the figure are appropriate for the three-dimensional Ising model. These numbers uniquely define the reduced quantities plotted: the reduced order parameter is the order parameter Q divided by the amplitude Q_- and is thus normalised to unity at $t = -1$; the reduced correlation length is the correlation length ξ divided by the amplitude ξ_+ and is thus unity at $t = +1$. Data for a wide variety of (sufficiently near critical) fluids, magnets and other systems are found to collapse onto these universal curves.

index values, $\beta \simeq 0.33$ and $\nu \simeq 0.63$. However, the indices are not insensitive to *all* microscopic details. As discussed by Thouless in this volume, magnets with a *layered* (effectively planar, two-dimensional) microstructure exhibit markedly different indices, $\beta \simeq 0.12$ and $\nu \simeq 1$.

These remarks draw on but a very small fraction of the extensive experimental data which have been painstakingly accumulated in studies of the critical region over the past decade, as an essential complement to the theoretical developments with which we are primarily concerned here. Nevertheless the tasks with which the theory is confronted are clear.

Firstly, the theory has to furnish a computational scheme for the critical indices which describe critical point behaviour. Secondly the theory is faced with potentially deeper questions relating to the universality apparent in the indices: Why are the microscopic details of a physical system so irrelevant to features of its critical behaviour that systems as disparate as the fluid and the magnet are characterised by the same magic numbers? Why, in contrast, should a system's space dimension have such a crucial influence on its critical index values? What other features of a physical system *are* relevant to the universal critical behaviour it displays? Which measurable quantities share with critical indices the remarkable property of universality?

Difficulties

The tasks we have identified pose distinctive difficulties. To determine the level at which these difficulties arise it is useful to refer back to the fundamental equations (8.1) and (8.2). These equations are at the heart of the theory of phase transitions, and indeed of any problem in thermal physics. In implementing them (in the fashion prescribed, for example, in equation 8.4a) one may encounter difficulties at two levels.

Firstly there is the problem of specifying the manner in which the configurational energy E_a of the system depends upon the arrangement a of its microscopic constituents. This is not the problem at issue here. The manner in which the energy of a fluid depends upon the positions of its constituent atoms, or the way in which the energy of a magnet reflects the orientations of its magnetic ions can, in fact, be modelled with some confidence, at least in the case of the simpler fluids and magnets. In any case the phenomenon of universality makes it plain that such details are largely *irrelevant* to critical point behaviour. Thus we may set the tasks in hand in the context of simple model systems, with the confident expectation that the answers which emerge will have not merely qualitative but also quantitative relevance to nature's own systems. In this spirit (partially presupposing universality, while seeking to understand it) we proceed here, and in much of what follows, in the specific context provided by the Ising model whose configurational energy has the particularly simple form given in equation (8.3). The problem is then well posed: it is a matter of evaluating sums like that in equation (8.4a). It is at this level that the difficulty arises.

The difficulty is a matter of length scales. The problem has three characteristic length scales. The first, and most important, is defined by the correlation length ξ. The second length scale is implicit in the extent of the critical region, namely the range of reduced temperatures t in which universal power-law phenomena (typified by equations 8.5 and 8.6) are apparent. The condition, defining this regime, that $|t|$ should be 'sufficiently small' is equivalent to the requirement that ξ should be 'sufficiently large'. The second length, L_{min}, sets the scale for this statement: the critical region is the regime in which ξ is large compared to L_{min}. Physically L_{min} is the largest length (excluding ξ) identifiable in the microphysics of the problem, typically a lattice spacing or range of interaction. (In our Ising model these are the same.) There is a third length scale, peripheral to the essential physics, but requiring identifi-

cation at this point. This third length, L_{max}, is the size of the system (fluid, magnet, Ising model, . . .) under study. Clearly L_{max} sets an upper limit for the correlation length. More precisely one can expect the correlation length to exhibit the divergent growth described by equations (8.6) (and, indeed, more generally one can expect power-law forms like equations 8.5 and 8.6 to be valid) *only* if ξ is small compared to L_{max}. The authentic critical region of pure power-law behaviour is, then, defined by the window in which $L_{min} \ll \xi \ll L_{max}$.

Consider first the difficulties which this state of affairs presents to a direct assault by computer simulation along the lines sketched earlier. In real laboratory experiments it is quite possible to satisfy the window condition: careful temperature control allows the realisation of correlation lengths 10^2 or 10^3 times the relevant microscopic length L_{min}, but still very small compared to the typical macroscopic size L_{max}. By contrast, in a computer experiment, it is virtually impossible to fulfil the window condition convincingly. To realise a regime where, say, $L_{max}/\xi \sim \xi/L_{min} \sim 10^2$ requires a simulation handling 10^8 variables for a two-dimensional problem (or 10^{12} variables in three dimensions!). Such requirements make unrealistic demands upon the technology of even this supercomputer era. If, then, one is to proceed along this avenue one must interpret the window condition much less stringently. There are then two possibilities. One may simply accept the distortions imposed by the finiteness of L_{max} ('finite-size effects') and settle for results of qualitative, or at best semi-quantitative, significance. Alternatively one must appeal to a theoretical framework going beyond the bare bones of statistical mechanics underlying the simulation procedure, and allowing one to transcend these, and other, limitations of computer-generated data. The renormalisation group provides just such a framework. Indeed, the synthesis of computer simulation and renormalisation group methods (to be discussed briefly in Section 8.5) offers arguably the most powerful and versatile approach to the critical point and related problems.

The obvious alternative to a simulation-based approach to the problem is an analytic one in which the task set by sums such as equation (8.4a) is tackled by essentially algebraic (rather than numerical) operations. Again, though less obviously, the problem encountered is one of length scales. As it stands equation (8.4a) requires one to perform a multiple sum over the variables $s_1, \ldots, s_N$, each of which must be allowed to take on the values $+1$ and -1. A full-frontal attack on this task soon runs into trouble: the sum on s_1 may be performed with ease, but the result is a function (of the remaining variables $s_2, \ldots, s_N$) whose form is considerably less pleasant to contemplate than the argument of the original sum. This trend continues: each variable eliminated (each sum performed) leaves as a legacy a nastier function of the variables which remain. No discernible limiting behaviour is apparent in the terms of the sequence accessible to even the most energetic investigation.

This unimaginative approach founders on two rocks. The first is that the sum in equation (8.4a) extends over many variables. The second is that these variables interact with one another. These rocks cannot easily be avoided: they are ingredients of the essential physics. To see this, consider figure 8.2(b), in which much of the essential physics of the critical point is apparent. The crucial feature we must now note is the existence of correlated microstructure (islands of black or white) not simply on the scale of the correlation length ξ, but *on all scales* intermediate between ξ and the lattice spacing L_{min}. This many-length-scale structure can be characterised only by equally many variables. Moreover it is clear that the configurational likelihood of structure of one prescribed scale (say a white droplet of some prescribed size) depends upon the likelihood of there being structure of other scales (a larger black droplet). Thus variables characterising structure on different length scales interact with one another in an essential way. These remarks identify the key difficulty posed by the critical point and the class of problems which it typifies: these problems are characterised by microstructure which has many different length scales and which can be described only with the aid of correspondingly many interacting variables.

The renormalisation group technique is specifically designed to tackle this type of problem. In essence it does so by reorganising sums such as that in equation (8.4a) so as to deal successively with the contributions made by configurational features of larger and larger scales. We shall develop the conceptual and technical framework in succeeding sections. At this juncture, however, we must forestall a possible misapprehension. The renormalisation group is by no means the source of all knowledge of the critical region (even if, arguably, it accounts for most of the wisdom). Two other sources must be mentioned.

First of all it is possible to reorganise the problem posed by sums such as that in equation (8.4a) so as to deal successively with contributions ordered according to the temperature at which they become important. In the course of nearly three decades, such high or low temperature series expansions (perturbation expansions about the fully disordered high temperature limit or fully ordered low temperature limit) have provided a solid bedrock of information on the values of critical indices and related critical point parameters. The insights which they offer are, however, very limited by comparison with those emerging from the renormalisation group, and even their computational power has only been realised fully with the motivation and guidance furnished by the renormalisation group framework.

Secondly there is a very limited number of instances in which sums such as that in equation (8.4a) have proved amenable to exact evaluation. The most celebrated instance is, in fact, the two-dimensional Ising model, which we have discussed here, a number of whose properties (including the order parameter) have yielded to the inspired efforts of a few brilliant theorists,

spearheaded over forty years ago by Lars Onsager. Thus, in particular, the critical temperature is known to have the form quoted earlier while the values of the critical indices introduced in equations (8.5) and (8.6) are known to be $\beta = \frac{1}{8}$ and $\nu = 1$ for this model. Such successes rest heavily upon the particular simple features of the model in question; they do not provide a general framework for understanding the critical region and the universality it displays. (The correspondence which the reader may have noted between the cited $d = 2$ Ising index values and the values identified earlier as appropriate for planar magnets substantiates but does not explain the universality phenomenon.) Nevertheless, like series expansion studies, they have proved invaluable in the assessment of the more general if less rigorously grounded framework to which we now turn.

8.3 The renormalisation group: a configurational view

The behaviour of any macroscopic physical system reflects the configurations (arrangements) which its microscopic constituents tend to adopt. This is the central thesis of statistical physics and its key equations (8.1) and (8.2). There is an obvious corollary: the tricks and truths of statistical physics have (or frequently have) an illuminating configurational significance. In this spirit we proceed now to introduce the key concepts and insights of the renormalisation group in a configurational context where pictures rather than equations form the natural language.

Coarse-graining

The solution to our problem, like the problem itself, is a matter of length scales. We have already met three important scales; we must now introduce a fourth. In contrast to the three other lengths, which characterise the system itself, the fourth length, L, characterises the *description* of the system. It may be thought of as typifying the size of the smallest resolvable detail in a description of the system's microstructure.

Consider the Ising model arrangements displayed in figure 8.2(a)–(c). These pictures contain *all* the details of each configuration shown: the resolution length L in this case has its smallest conceivable value, coinciding with the lattice spacing L_{min} (which, hereafter, we shall take to define the unit of length). In the present context the most detailed description is not the most useful: the essential signals with which we are concerned are hidden in a noise of irrelevant detail. A clue to the nature of the irrelevant noise, and how to eliminate it, is to be found in the nature of the properties (equations 8.5 and 8.6) whose behaviour we seek to compute and understand. Both the order parameter and the correlation length characterise *large-length-scale* configurational trends (the overall excess of black over white; the size of the largest droplets). The *explicit form* of the small scale microstructure is irrelevant to the behaviour of these quantities. The small scale microstructure is the noise. To eliminate it we simply select a larger value for the resolution length (or 'coarse-graining length') L.

There are many ways of implementing this *coarse-graining* procedure. (There are correspondingly many forms of the renormalisation group.) We choose a variant of a scheme originally due to Leo Kadanoff, in whose 1966 paper are to be found the seeds of the renormalisation group idea applied to phase transitions. We divide our sample into blocks of side L, each of which contain L^d sites. (We shall use the symbol d for the space dimension to preserve generality; our illustrations will always invoke the planar, $d = 2$, case.) The centres of the blocks define a lattice of points indexed by $I = 1, 2, \ldots, N/L^d$. We associate with each block lattice point centre, I, a coarse-grained (or block) variable $S_I(L)$ defined as the spatial average of the local variables it contains:

$$S_I(L) = L^{-d} \sum_i^I s_i, \qquad (8.7)$$

where the sum extends over the L^d sites in the block I. The set of coarse-grained coordinates $\{S(L)\}$ are the basic ingredients of a picture of the system having spatial resolution of order L.

The coarse-graining procedure is easily implemented on a computer. In so doing we are immediately faced with the fact that, whereas the local variables s_i have only two possible values, the block variables $S_I(L)$ have many more ($L^d + 1$, to be precise). It is not actually essential to record the precise value of each coarse-grained variable: for many purposes a knowledge of the *signs* of the block variables is sufficient and a two state representation is adequate. (This is the philosophy underlying the so-called majority rule form of renormalisation group transformation.) For our purposes, however, it is helpful to preserve some more of the richness of the coarse-grained configurations. Accordingly in displaying the consequences of the procedure we need a more elaborate colour convention than we used in figure 8.2. We will associate with each block variable a shade of grey drawn from a spectrum ranging from black to white; darker (lighter) shades signify more positive (more negative) variables.

The results of coarse-graining configurations typical of three different temperatures are shown in figures 8.4 and 8.5. Two auxiliary operations are implicit in these results. The first operation is a *length scaling*: the lattice spacing on each blocked lattice has been scaled (shrunk) to the same size as that of the original lattice, making possible the display of correspondingly larger portions of the physical system. The second operation is a *variable scaling*: loosely (we shall return to this point in subsequent sections) we have adjusted the scale of the block variables so as to match the spectrum of block variable values to the spectrum of shades at our disposal.

Consider first a system marginally above its critical point at a temperature T chosen so that the correlation length ξ is approximately six lattice spacing units. A typical arrangement

(without coarse-graining) is shown in figure 8.4(ai). The succeeding figures, 8.4($a$ii) and 8.4($a$iii), show the results of coarse-graining with block sizes $L = 4$ and $L = 8$, respectively. A clear trend is apparent. The coarse-graining *amplifies* the consequences of the small deviation of T from T_c. As the coarse-graining length is increased, the ratio of the size of the largest configurational feature ($\sim \xi$) to the size of the smallest ($\sim L$) is reduced. The ratio ξ/L provides a natural measure of how 'critical' is a configuration. Thus the coarse-graining operation generates a *representation* of our physical system that is effectively *less* critical the *larger* the coarse-graining length. The limit point of this trend is the effectively fully disordered arrangement shown in figure 8.4($a$iii), and, in an alternative form, in figure 8.4($a$iv), which shows the limiting distribution of the coarse-grained variables: the distribution is a Gaussian which is narrow (ever more so the larger the L-value) and centred on zero. This limit is easily understood: when the system is 'viewed' on a scale, L, larger than ξ the correlated microstructure in which the incipient order is expressed is no longer explicitly apparent (although it remains implicit in the *scale* of the coarse-grained field). Each coarse-grained variable is essentially independent of the others. Moreover since each coarse-grained variable is a sum of many local variables, each of which is correlated only with a minority fraction (of the order of $(\xi/L)^d$) of the others, its distribution is necessarily Gaussian.

A similar trend is apparent below the critical point. Figure 8.4(bi) shows a typical arrangement at a temperature $T < T_c$ such that ξ is again approximately six lattice spacings. Coarse-graining with $L = 4$ and $L = 8$ again generates representations which are effectively less critical (figure 8.4$b$ii and iii). This time the coarse-graining smooths out the microstructure which makes the order incomplete, again pushing the effects of the eliminated microstructure into the scale of the block variables. The limit point of this procedure is a homogeneously ordered

Figure 8.4. Coarse-graining above and below the critical point. The sets of patterns, (a) and (b), depict configurations generated in a computer simulation of a 512^2 square lattice (two-state) Ising model (a) just above and (b) just below the critical point. The first figure, (i), in each sequence shows a typical configuration of the *local* variables, s; only a portion of the lattice, containing 64^2 sites, is shown. The subsequent figures in each sequence represent configurations of the coarse-grained variables, $S(L)$, formed by space-averaging the local variables over blocks of side $L = 4$ ($a$ii, $b$ii) and $L = 8$ ($a$iii, $b$iii) lattice spacings, and compressing the representation so that the block lattice size L coincides with the original lattice spacing; the $L = 4$ and $L = 8$ configurations thus reflect, respectively, the states of 256^2 and all 512^2 of the original local variables. Above the critical point the configuration flow under coarse-graining is directed to a fully disordered fixed point near which the block variables have a normal distribution centred on zero ($a$iv). Below the critical point the configuration flow approaches a homogeneously ordered fixed point, near which the block variables are normally distributed about the order parameter ($b$iv). The widths of both distributions tend to zero in the limit of large L.

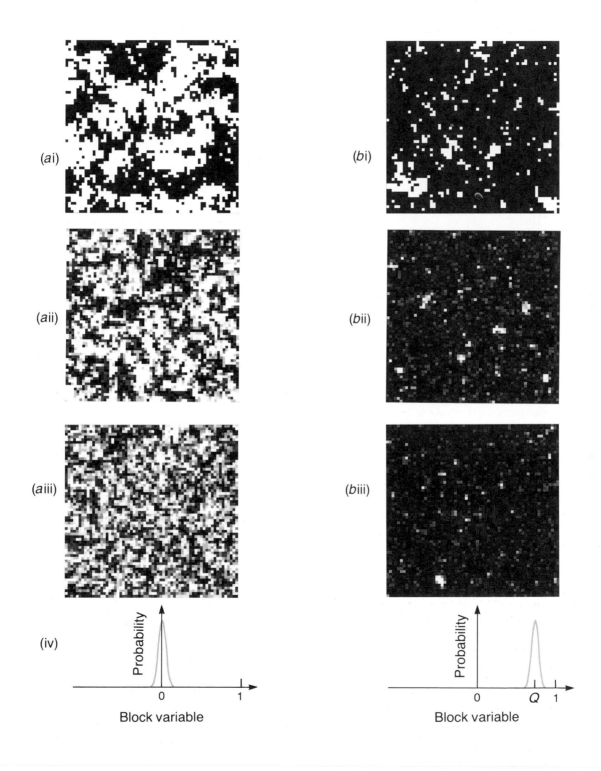

arrangement in which the block variables have a random (Gaussian) distribution centred on the order parameter (figure 8.4*b*iv).

Consider now the situation *at the critical point.* Figure 8.5(*a*i) shows a typical arrangement; figures 8.5(*a*ii) and (*a*iii) show the results of coarse-graining with $L = 4$ and $L = 8$, respectively. Since the correlation length is as large as the system size itself the coarse-graining does not produce less critical representations of the physical system: each of the figures displays structure over *all* length scales between the lower limit set by L and the upper limit set by the size of the display itself. A limiting trend is, nevertheless, apparent. Although the $L = 4$ pattern is qualitatively quite different from the pattern of the local variables (the latter is a two-colour picture!) the $L = 4$ and $L = 8$ patterns display qualitatively similar features. These similarities are more profound than is immediately apparent. A statistical analysis of the *spectrum* of $L = 4$ configurations (generated as the local variables evolve in time) shows that it is (almost) identical to that of the $L = 8$ configurations (given the block variable scaling, of which more anon). This state of affairs is expressed in figure 8.5(iv), which shows the near coincidence of the distributions of block variables (grey-levels) for the two different coarse-graining lengths. The implication of this limiting behaviour is clear: the patterns formed by the ordering variable at criticality look the same (in a statistical sense) when viewed on all sufficiently large length scales.

Let us summarise. Under the coarse-graining operation there is an evolution or *flow* of the system's configuration spectrum. The flow tends to a limit, or *fixed point*, such that the pattern spectrum does not change under further coarse-graining. These *scale-invariant* limits have a trivial character for $T > T_c$ (a perfectly disordered arrangement) and $T < T_c$ (a perfectly ordered arrangement). The hallmark of the critical point is the existence of a scale-invariant limit which is neither fully ordered nor fully disordered but which possesses structure on all length scales. The answers we seek are to be found in the nature and, indeed, the very existence of this scale-invariant limit.

Universality and scaling

Armed with the coarse-graining technique we now address the problems posed by the universality phenomenon. We seek to understand how it is that systems as different, microscopically, as the fluid and the magnet can nevertheless display critical point behaviour which (in certain respects) is quantitatively identical. We will not examine, explicitly, the behaviour of a fluid or a magnet. The essential points can be more readily demonstrated by a comparative study of two different models. To that end we introduce now a variant of the Ising model. In our original model the local variables assume just two values (1 and -1). In the new model the local variables take on three values (1, 0 and -1). In other respects the models are the same.

To distinguish between them we will (unconventionally) refer to them as the two-state and three-state Ising models. Like the fluid–magnet pair the two models have properties which are clearly (if less dramatically) different: for example, the critical temperatures of the three-state model is some 30% lower than that of the two-state model (for the same coupling J). However, again as with the fluid–magnet pair, there is abundant evidence that the two models have the same universal properties. Let us explore what is different and what is the same in the configurations of the two models.

The configurations of the local variables s_i are clearly qualitatively different for the two models: at this level the two-state character of the one model and the three-state character of the other are quite apparent (figures 8.5*a*i, *b*i). Now, however, consider the coarse-grained configurations. Figures 8.5(*b*ii) and (*b*iii) show coarse-grained configurations (with $L = 4$ and $L = 8$, respectively) for the three-state model at its critical point. We have already seen that the coarse-graining bears the configuration spectrum of the critical two-state Ising model to a nontrivial scale-invariant limit. It is scarcely surprising that the same is true for the three-state model. What is remarkable is that the two limits are the same! The coarse-graining erases the physical differences apparent in configurations where the local behaviour is resolvable, and exposes a profound configurational similarity.

Figure 8.5 Coarse-graining at the critical point. (*a*i) shows a typical configuration of the local variables in a 64^2 section of a (two-state) Ising model at the critical temperature. (*a*ii) and (*a*iii) show coarse-grained configurations of this model, characterised by coarse-graining lengths $L = 4$ and $L = 8$, respectively. In addition to the scaling of the block lattice size, the block variables are scaled so that the width of their distribution remains constant. Given these auxiliary scaling operations, under coarse-graining the configuration spectrum flows to a fixed point with structure on all length scales. The limiting behaviour is manifested in the similarity of the $L = 4$ and $L = 8$ configurations, and in a more quantitatively explicit form in (iv), which shows the distributions of the $L = 8$ block variables ($\square$) and that of the $L = 16$ variables (—). The accompanying chart indicates the correspondence between the scaled block coordinate values in the probability distribution, and the grey levels in the coarse-grained patterns. (*b*i), (*b*ii) and (*b*iii) are constructed in a similar fashion from a study of a *three-state* Ising model, at its critical temperature. The *local* configuration spectrum (*b*i) is qualitatively quite different from its two-state counterpart (*a*i). Under coarse-graining the configuration spectrum approaches the *same* limit as does the two-state model, as one sees in the qualitative similarity of (*b*iii) and (*a*iii) and in the close agreement of the distribution of the $L = 8$ block variables for the three-state model ($\bigcirc$ in (iv)) with the limiting distribution for the block variables of the two-state model. The minor differences between the distributions reflect residual memory of local system-specific detail; such 'corrections to scaling' die off as a power of the coarse-graining length.

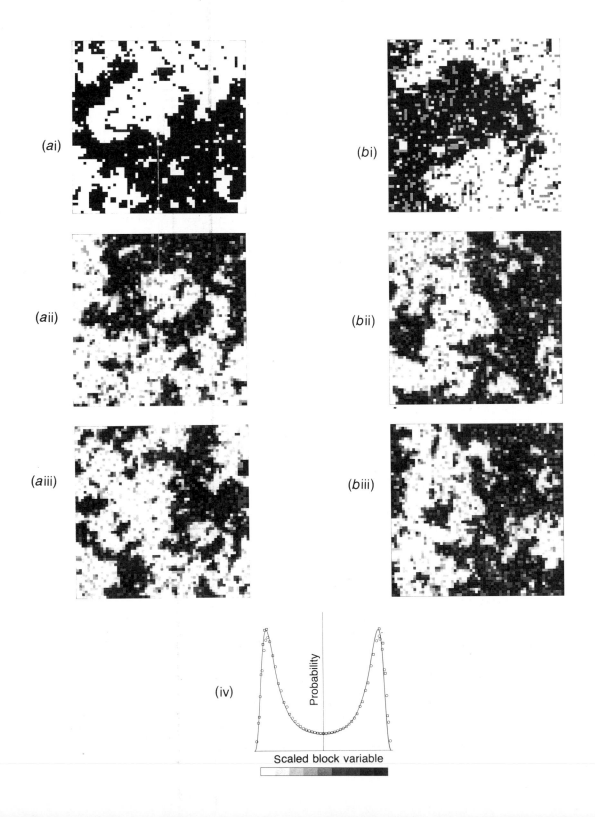

(ai)

(bi)

(aii)

(bii)

(aiii)

(biii)

(iv)

Probability

Scaled block variable

Some qualification is necessary here. There is one configurational difference – a difference between the overall *scales* of the block variables of the two problems – which the coarse-graining operation itself does not erase. This difference is significant: the variable scale defines one of the two key system-specific (nonuniversal) scales in the problem, manifesting itself in the nonuniversality of the amplitude Q_- (equation 8.5). The difference is, nevertheless, superficial. Its superficiality is apparent from figure 8.5, from which it is clear (most explicitly in figure 8.5iv) that its effects can be suppressed (and the near-perfect statistical similarity of the configurations exposed) merely by means of the *variable scaling* we introduced as an auxiliary part of the coarse-graining procedure.

The similarity of the configurations of the different systems is not restricted to the critical temperature itself. Suppose we have a two-state model and a three-state model each somewhat above their critical points, at some reduced temperature *t*. The two systems will have somewhat different correlation lengths, ξ_1 and ξ_2 say, manifesting the nonuniversality of the amplitude ξ_+ (equation 8.6*a*), which may be taken to define the second key system-specific scale of the problem. (The amplitude ξ_- turns out to be a universal multiple of ξ_+.) This difference is less than meets the eye. Although it survives the coarse-graining itself, it can be suppressed by the *combined* effects of the coarse-graining and the auxiliary *length scaling* operation. Specifically we choose coarse-graining lengths L_1 and L_2 for the two models such that $\xi_1/L_1 = \xi_2/L_2$. We adjust the scales of the block variables (or our grey-level control!) so that the typical (root-mean-square) variable value is the same for the two systems. We adjust the length scale of the systems (stretch or shrink our snapshots!) so that the sizes of the minimum-length-scale structure (set by L_1 and L_2) looks the same for each system; the sizes of the maximum-length-scale structure in the two systems will then necessarily also look the same. One can show that these operations take care of all the differences: the configurations of the two systems again look (statistically) identical. Precisely what they look like depends upon our choice of the ratio of correlation length to coarse-graining length (cf. figures 8.4*a*ii and *a*iii): what you see depends upon how you look!

We have set this discussion in the context of computer experiments on model systems. Now let us formulate the implications for laboratory experiments on nature's systems. The similarities in the critical behaviour of fluids and magnets can, it now appears, be traced to an underlying similarity in their coarse-grained configurations. In a magnet the relevant configurations are those formed by the coarse-grained magnetisation (the magnetic moment averaged over a block of side *L*). In a fluid the relevant configurations are those of the coarse-grained density (the mass averaged over a block of side *L*) or, more precisely, its fluctuation from its macroscopic average. The patterns in the latter (bubbles of liquid and vapour) may be matched to the patterns in the former (microdomains of the magnetisation), given appropriate scaling operations to camouflage the differences between the length scales and the differences between the variable scales (indeed, the variable *units*: density and magnetisation) in the two problems.

This claim is informed surmise, not experimental fact. Real experiments do not offer the complete configurational information accessible in computer experiments and required to warrant the claim fully. Nevertheless the circumstantial evidence is strong: the picture we have developed implies the universality (modulo two scales) which is, indeed, observed in properties that reflect (and can, in principle, be computed from) coarse-grained configurations. In particular it implies the universality of critical point indices, to which we shall shortly turn.

There is, however, one final issue we must touch on here (some elaboration follows in Section 8.6). It should be quite clear that the coarse-grained patterns of the fluid and magnet will not match those of the model we have studied explicitly: the latter are two dimensional (planar), the former are, generally, three dimensional. The space dimension *d* is, we should now recognise, one hallmark of a physical system which is not erased by coarse-graining and which therefore can (and does) reflect itself in the values of 'universal' quantities, such as critical indices. The space dimension is in fact one of a small set of features of a physical system which are sufficiently deep-seated to survive coarse-graining and which together serve to define the system's universal critical behaviour, or *universality class*. The constituents of this set are not all identifiable *a priori*. The claim of the other principal member is, however, immediately plausible: it is the number of components, *n*, of the order parameter. In the majority of instances cited in this chapter the order parameter is a *scalar* (for the fluid a density difference; for the Ising model a population difference) for which $n = 1$. In some ferromagnets the symmetry of the underlying lattice constrains the magnetisation to lie parallel or antiparallel to a particular axis; for such *uniaxial* magnets the order parameter (the magnetisation) is, then, again a scalar. However, in ferromagnets with a higher symmetry the magnetisation may have components along two axes, or three axes, implying a *vector* order parameter, with $n = 2$ and $n = 3$, respectively. It is clear *a priori* that the order-parameter *n*-value will be reflected in the nature of the coarse-grained configurations, and thus in the universal observables they imply.

These remarks imply what is in fact the case: the fluid, the uniaxial magnet and the three-dimensional Ising model fall into the same universality class. Some of the evidence substantiating this claim is to be found in the data, presented in figure 8.11, at the conclusion of Section 8.5.

Critical indices

We now turn to show how the prototype universal quantities, the critical indices, may be computed from the properties of the coarse-grained configuration spectrum. To do so we must express the contentions of the preceding section in a more mathematically explicit form.

Consider, then, coarse-grained variables of our Ising model system, or, indeed, of any system belonging to the same universality class. In what follows we will need to refer only to the behaviour of a single typical coarse-grained variable which we shall denote by $S(L)$ (dropping the redundant suffix I appearing in equation 8.7). We suppose that the system is sufficiently close to criticality (at sufficiently small reduced temperature t) that the important observables are dominated by large-scale microstructure: this is the critical-region condition, $\xi \gg L_{min}$. We restrict our attention to variables sufficiently coarse-grained to wipe out local system-specific details: thus we require $L \gg L_{min}$. The configurational universality developed above may then be expressed in the claim that, for *any* system of the stipulated class, and for *any* L and t satisfying the prescribed conditions, scale factors $A(L)$, $B(L)$ and $C(L)$ may be found such that the coarse-grained variable distribution $p(S(L))$ can be written in the form

$$p(S(L)) \simeq C(L)\tilde{p}(A(L)t, B(L)S(L)), \qquad (8.8)$$

where $\tilde{p}$ is a function unique to the universality class. The role of the two factors A and B (C is something of a technical triviality) is to absorb the two basic nonuniversal scales identified in the preceding section. The information we seek is implicit in the manner in which these scale factors depend upon the coarse-graining length, L. For the more technically minded the relevant arguments are developed in box 8.1. The key results are

$$A(L) = A_0 L^{1/\nu} \qquad (8.9a)$$

$$B(L) = B_0 L^{\beta/\nu} \qquad (8.9b)$$

and $C(L) = B(L)$. The amplitudes A_0 and B_0 are system-specific ('nonuniversal') but L-independent constants which together absorb the nonuniversality of the correlation length and order-parameter amplitudes appearing in equations (8.6) and (8.5).

These results reveal that the basic critical indices (in the forms $1/\nu$ and β/ν) serve to characterise the ways in which the configuration spectrum evolves under coarse-graining.

Consider, first, the index β/ν. Precisely at the critical point there is only one way in which the coarse-grained configurations change with L (assumed sufficiently large): the overall scale of the coarse-grained variable (the black–white contrast in our grey-scale representation of the configurations spectrum) is eroded with increasing L. Thus the configurations of coarse-graining length L_1 match those of a larger coarse-graining length L_2 only if the variable scale in the latter configurations is amplified. The required amplification follows

---

Box 8.1. **The scale factors $A(L)$, $B(L)$ and $C(L)$**

We sketch here the arguments substantiating the forms (8.9a,b) claimed for the scale factors appearing in equation (8.8).

Consider first the scale factor $A(L)$. The function of this factor is to absorb the system-specific dependence of the coarse-grained configuration spectrum upon the reduced temperature. The configuration spectrum depends upon the reduced temperature because the configurations look different for different correlation lengths, and the correlation length varies with the reduced temperature. More specifically we have argued that the configuration spectrum should depend upon the *ratio* of the correlation length to the coarse-graining length, $z = \xi/L$: two systems with *different* correlation lengths will look statistically similar if viewed with correspondingly different coarse-graining lengths. This claim can be true only if $A(L = \xi)t$ is a temperature-independent and system-independent constant. Recalling the way in which the correlation length depends upon the reduced temperature (equations 8.6a,b) one finds that, for the combination $A(L = \xi)t$ to be independent of t (and thus ξ) and independent of the system-specific details implicit in the amplitude ξ_+ (or, equivalently, ξ_-), one must have $A(L) = A_0 L^{1/\nu}$, where A_0 is proportional to $\xi_+^{-1/\nu}$ and is thus itself a system-specific amplitude.

Next let us deal briefly with the factor $C(L)$. This factor is needed if (as is convenient) we require that both p and $\tilde{p}$ are authentic probability density functions, correctly normalised so that the integral of p over all $S(L)$ and the integral of $\tilde{p}$ over all $B(L)S(L)$ both be unity. Implementing these conditions, and invoking equation (8.8) one finds that $C(L) = B(L)$. In effect the function $\tilde{p}$ is obtained from the function p by stretching the latter by a factor of $B(L)$ along the abscissa and shrinking it by a factor

of $B(L)$ along the ordinate, thus maintaining normalisation.

Now consider the scale factor $B(L)$. The function of this factor is to absorb the system-specific dependence of the configuration spectrum upon the scale of the ordering variables. To establish the form of this factor we consider the block-variable average value $\overline{S(L)}$. Recalling the defining relation (8.7), and noting that the average value of any one site variable must coincide with the average value of any other, one sees that $\overline{S(L)} = \overline{s_i}$, where s_i is some (any) site variable. Similar considerations show that the order parameter too (equations 8.4a,b) has the simple representation $Q = \overline{s_i}$. The block variable average is then nothing but the order parameter itself. Now the block variable average is defined by the first moment of the probability distribution p (i.e. the product of $p(S(L))$ with $S(L)$, integrated over all $S(L)$). Making use of the representation in equation (8.8) we then find that $Q = \overline{S(L)} = B^{-1}(L)f(A_0 L^{1/\nu}t)$, where f is a universal function (defined by the first moment of $\tilde{p}$), and we have invoked our results for $A(L)$ and $C(L)$. The apparent L-dependence of the right hand side of this relation is clearly illusory: the order parameter Q does not depend upon our choice of coarse-graining length. However the t-dependence of our expression is real enough: we expect that the order parameter varies with the power β of the reduced temperature (equation 8.5). The latter observation implies that the function f must have the property $f(y) \sim y^\beta$. The former observation then, in turn, implies that $B(L)$ must vary as $L^{\beta/\nu}$ in order to cancel the L-dependence of $f(A_0 L^{1/\nu}t)$. More precisely we must make the assignment $B(L) = B_0 L^{\beta/\nu}$, where B_0 is a system-specific amplitude which may be related to the nonuniversal order-parameter amplitude Q_- (equation 8.5).

from equations (8.8) and (8.9b); it is $B(L_2)/B(L_1) = (L_2/L_1)^{\beta/\nu}$. The index β/ν thus controls the rate at which the scale of the ordering variable decays with increasing coarse-graining length.

Consider now the index $1/\nu$. For small but nonzero reduced temperature (large but finite ξ) there is a second way in which the configuration spectrum evolves with L. As we noted in Section 8.3, coarse-graining reduces the ratio of correlation length to coarse-graining length and thus results in configurations with a less critical appearance. More precisely we see from equation (8.8) that (setting aside the associated variable-scale erosion) increasing the coarse-graining length from L_1 to L_2 while keeping the reduced temperature constant has the same effect on the configuration spectrum as keeping the coarse-graining length constant while amplifying the reduced temperature t by a factor $A(L_2)/A(L_1) = (L_2/L_1)^{1/\nu}$. One may think of the combination $A(L)t$ as a measure of the effective reduced temperature of the physical system viewed with resolution length L. The index $1/\nu$ controls the rate at which this effective reduced temperature grows with increasing coarse-graining length.

These conclusions are illustrated in figure 8.6. Here we show how the coarse-grained configuration spectrum of a $d = 2$ Ising model viewed with one coarse-graining length can be matched to the configuration spectrum of the same model viewed with a larger coarse-graining length, when the reduced temperature and the coordinate length scale are amplified in the fashion prescribed above. The indices which secure the degree of matching shown are $\beta = \frac{1}{8}$ and $\nu = 1$, which, the reader will recall, are the exactly established values for the $d = 2$ Ising model.

At this point let us pause to examine what we have and what we have not achieved in the course of this third section. We have established that the solution to the central conceptual problem of the critical region (the origins of universality) and the prototype technical problem (the determination of the indices) is, in principle, to be found in the evolution of the configuration spectrum under coarse-graining. This is the central truth of the renormalisation group. In the process we have encountered the key concepts (configuration flow, fixed points, scale-invariance, . . .) in terms of which the solutions may be framed. These are the key concepts of the renormalisation group. However, although we have established *where* to look and *what* to look for we have contrived to dodge the question of *how* one is to expose ('see') the relevant coarse-grained behaviour. Of course the coarse-grained physics is easily exposed if, as supposed in our illustrations, we have access to the full computer-generated configuration spectrum. But it is neither aesthetically satisfactory nor (in some instances) practically feasible to appeal to the computer. Accordingly we require apparatus that will expose the coarse-grained physics given not the configuration spectrum itself, but simply the configurational energy E which controls the configuration

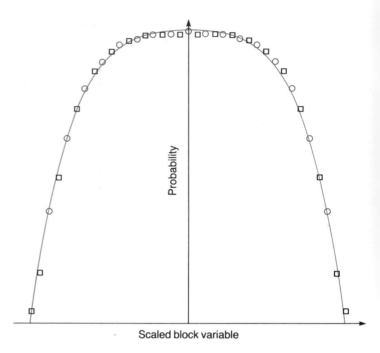

Figure 8.6. Configurational matching and critical indices. The two data sets represent the probability distributions of the block variables of two different Ising systems. The red set reflects the behaviour of the coarse-grained variables associated with blocks of side $L_2 = 16$ at a reduced temperature $t_2 = 0.0471$. The black set describes the behaviour of coarse-grained variables associated with blocks of side $L_1 = 8$ at a reduced temperature $t_1 = A(L_2)t_2/A(L_1)$ $= 2t_2$, with the implicit identification $\nu = 1$; the scale of the $L_2 = 16$ block variables has been stretched with respect to that of the $L_2 = 8$ block variables by a factor $B(L_2)/B(L_1) = 1.09$, with the implicit identification $\beta/\nu = \frac{1}{8}$. The residual discrepancy between the two data sets reflects the existence of corrections to the asymptotic (large L) scaling form.

spectrum through the fundamental relationships in equations (8.1) and (8.2). This is precisely the function of the renormalisation group technique.

8.4   The renormalisation group: the effective coupling view

In this section we wish to expose the key features of the beautiful and simple mathematics which encapsulates the ideas developed in the preceding section, and which forms now the standard theoretical research tool known as the renormalisation group. We begin by introducing the idea of the effective energy function for coarse-grained configurations; the renormalisation group transformation describes how the coupling strengths in this effective energy function evolve, or 'flow', as we increase the coarse-graining length scale. We then continue with a concrete example, which involves only very simple algebra and illustrates the roles of a fixed point of the renormalisation group transformation in determining the critical temperature and exponents (specifically, the correlation length exponent v). We then discuss in general terms how the flow of coupling strengths (near an appropriate fixed point) mirrors the universal aspects of the configurational flow observed in Section 8.3.

Effective couplings for coarse-grained variables

Let us begin by returning to our fundamental equation (8.1). It will prove helpful to make a cosmetic change: we absorb the factor $1/kT$ into the energy by introducing a 'reduced' energy function $\mathcal{H} = E/kT$ so that the probability p for an arrangement becomes $p = Z^{-1}e^{-\mathcal{H}}$, where we have suppressed the arrangement label, a. The first step is then to imagine that we generate, by a computer simulation procedure for example, a sequence of configurations with relative probability $\exp(-\mathcal{H})$. We next adopt some coarse-graining procedure which produces from these original configurations a set of coarse-grained configurations. We then ask the question: what is the energy function $\mathcal{H}'$ of the coarse-grained variables which would produce these coarse-grained configurations with the correct relative probability $\exp(-\mathcal{H}')$? Clearly the form of $\mathcal{H}'$ depends on the form of $\mathcal{H}$; thus we can write symbolically

$$\mathcal{H}' = R(\mathcal{H}). \qquad (8.10)$$

The operation R, which defines the coarse-grained configurational energy $\mathcal{H}'$ in terms of the microscopic configurational energy function $\mathcal{H}$, is known as a renormalisation group transformation.

The rationale for the name is technical. The word 're-normalisation' signifies that the parameters in the energy function $\mathcal{H}'$ reflect the influence of degrees of freedom (describing microscopic details eliminated in the coarse-graining procedure) not explicit in $\mathcal{H}'$. The word 'group' signifies that, for example, a double application of the operation with coarse-graining length L lattice spacings could be realised, in principle, by a single application with coarse-graining length L^2 lattice spacings.

What the operation is called is less important than what it does. What it does is to replace a hard problem by a less hard problem. Specifically, suppose that our system is near a critical point and that we wish to calculate its large-distance properties. If we address this task by utilising the configurational energy $\mathcal{H}$, and appealing to the basic machinery of statistical mechanics set out in equations (8.1) and (8.2), the problem is hard. It is hard because the system has fluctuations on all the (many) length scales intermediate between the correlation length ξ and the minimum length scale $L_{\min}$ (the lattice spacing, say). However, the task may instead be addressed by tackling the statistical mechanics of the coarse-grained system described by the energy $\mathcal{H}'$. For the large-distance properties of this coarse-grained system are the *same* as the large-distance properties of the physical system, since the coarse-graining operation preserves large-scale configurational structure. In this representation the problem is a little easier. For, while the correlation length associated with $\mathcal{H}'$ is the same as the correlation length associated with $\mathcal{H}$, the minimum length scale of $\mathcal{H}'$ is bigger than that of $\mathcal{H}$, by virtue of the coarse-graining operation. Thus the statistical mechanics of $\mathcal{H}'$ poses a not-quite-so-many-length-scale problem, a problem which is effectively a little less critical and is thus a little easier than that posed by the statistical mechanics of $\mathcal{H}$. The benefits accruing from this procedure may be amplified by repeating it. Repeated application of the operation R will eventually result in a coarse-grained energy function describing configurations in which the correlation length is no bigger than the minimum length scale. The associated coarse-grained system is far from criticality. Its properties may be reliably computed by any of a wide variety of approximation schemes available for dealing with noncritical systems. These properties *are* the desired large-distance properties of the physical system. In effect, repeated application of the renormalisation group operation captures successively the effects of larger and larger scale fluctuations upon the largest-scale fluctuations of concern; as *explicit* reference to fluctuations of a given scale is eliminated by coarse-graining, their effects are carried forward *implicitly* in the parameters of the coarse-grained energy.

In order to put some flesh on this formalism, and to provide a framework for a simple illustrative calculation, let us return to the lattice Ising model explored in the previous sections. In the model considered there the energy function depended on the product of nearest neighbour spins. The coefficient of this product in the energy is the exchange coupling, J. In principle, other kinds of interactions are also allowed; for example, we may have a product of second neighbour spins with strength J_2 or, perhaps, a product of four spins (at sites forming a square whose side is the lattice spacing), with strength J_3. Such interactions in a real magnet have their origin in the quantum mechanics of the atoms and electrons and clearly depend upon the details of the system. For generality therefore we will allow a family of exchange couplings $J_1, J_2, J_3, \ldots$, or J_a, $a = 1, 2, \ldots$. In the reduced energy function the equivalent coupling strengths are $K_a \equiv J_a/kT$. Their values determine uniquely the energy for any given configuration. Now consider the coarse-graining procedure. Let us suppose that this procedure takes the form of a 'majority rule' operation (cf. Section 8.3) in which the new spins are assigned values ± 1 according to the signs of the magnetic moments of the blocks with which they are associated. After this coarse-graining procedure, the new energy function $\mathcal{H}'$ will be expressible in terms of some new coupling strengths K_a' describing the interactions amongst the new spin variables (and thus, in effect, the interactions between blocks of the original spin variables). The renormalisation group transformation simply states that these new couplings depend on the old couplings: K_1' is some function f_1 of all of the original couplings, and generally

$$K_a' = f_a(K_1, K_2, \ldots) \equiv f_a(\mathbf{K}), \quad a = 1, 2, \ldots, \quad (8.11)$$

where $\mathbf{K}$ is shorthand for the set $K_1, K_2, \ldots$. We note that it is not only useful to allow for arbitrary kinds of interactions; if we wish to repeat the transformation several (indeed many) times, it is also necessary because even if we start with only the nearest neighbour coupling in $\mathcal{H}$ the transformation will in general produce others in $\mathcal{H}'$.

One further point is worth stressing before we illustrate this method by a simple example. In practice, the functions f_a in equation (8.11) cannot be calculated exactly; one has to resort to some suitable approximation scheme. However, in contrast to thermodynamic quantities like magnetisation or susceptibility which are *singular* in the critical region, the functions f_a are expected to be *regular*. (It is frequently possible to justify this expectation *a posteriori*, or by physical arguments.) The standard approximation schemes which are inadequate for calculating thermodynamic quantities are therefore potentially usable here. This idea is a very important part of the renormalisation group method.

A simple example

This example illustrates how one can perform the renormalisation group transformation (8.11) directly, without recourse to a 'sequence of typical configurations'. The calculation involves a very crude approximation which has the advantage that it simplifies the subsequent analysis.

Consider an Ising lattice model in two dimensions, with only nearest neighbour interactions as shown in figure 8.7. We have divided the spins into two sets: the spins $\{s'\}$ form a square lattice of spacing 2, the others being denoted by $\{\tilde{s}\}$. One then *defines* an effective energy function $\mathcal{H}'$ for the s' spins by performing an average over all the possible arrangements of the $\tilde{s}$ spins

$$\exp(-\mathcal{H}') = \sum_{\{\tilde{s}\}} \exp(-\mathcal{H}). \quad (8.12)$$

This particular coarse-graining scheme is called 'decimation' because a certain fraction (not necessarily one-tenth!) of spins on the lattice is eliminated. This formulation of a new energy function realises two basic aims of the renormalisation group method: the long-distance physics of the 'original' system, described by $\mathcal{H}$, is contained in that of the 'new' system, described by $\mathcal{H}'$ (indeed the partition functions are the same) and the new system is further from criticality because the ratio of correlation length to lattice spacing ('minimum length scale') has been reduced by a factor of $\frac{1}{2}$ (the ratio of the lattice spacings of the two systems). We must now face the question of

how to perform the configuration sum in equation (8.12). This cannot in general be done exactly, so we must resort to some approximation scheme.

The particular approximation which we invoke is the high temperature series expansion which was touched upon towards the end of Section 8.2. In its simplest mathematical form, since $\mathcal{H}$ contains a factor $1/kT$, it involves the expansion of $\exp(-\mathcal{H})$ as a power series:

$$\exp(-\mathcal{H}) = 1 - \mathcal{H} + \mathcal{H}^2/2 - \mathcal{H}^3/6 + \ldots.$$

We substitute this expansion into the right hand side of equation (8.12) and proceed to look for terms which depend on the s' spins after the sum over the possible arrangements of the $\tilde{s}$ spins is performed. This sum extends over all the possible (± 1) values of all the $\tilde{s}$ spins. The first term (the 1) in the expansion of the exponential is clearly independent of the values of the s' spins. The second term $(\mathcal{H})$ *is* a function of the s' spins, but gives zero when the sum over the $\tilde{s}$ spins is performed because only a *single* factor of any $\tilde{s}$ ever appears, and $+1 - 1 = 0$. The third term $(\mathcal{H}^2/2)$ does contribute. If one writes out explicitly the form of $\mathcal{H}^2$ one finds terms of the form $K^2 s_1' \tilde{s} \tilde{s} s_2' = K^2 s_1' s_2'$, where s_1' and s_2' denote two spins at nearest neighbour sites on the lattice of s' spins, and $\tilde{s}$ is the spin (in the other set) which lies between them. Now, in the corresponding expansion of the left hand side of equation (8.12), we find terms of the form $K' s_1' s_2'$, where K' is the nearest neighbour coupling for the s' spins. We conclude (with a little more thought than we detail here) that

$$K' = K^2. \quad (8.13)$$

Of course many other terms and couplings are generated by the higher orders of the high temperature expansion. If our aim is to produce reliable values for the critical temperature and exponents it is necessary to include these terms. However, our aim here is to use this simple calculation to illustrate the renormalisation group method. Let us therefore close our eyes, forget about the higher order terms and show how the renormalisation group transformation (8.13) can be used to obtain information on the phase transition.

The first point to note is that equation (8.13) has the *fixed point* $K^* = 1$; if $K = 1$ then the new effective coupling K' has the *same* value 1. Further, if K is just larger than 1, then K' is larger than K, i.e. further away from 1. Similarly, if K is less than 1, K' is less than K. We say that the fixed point is *unstable*: the flow of couplings under repeated iteration of equation (8.13) is *away* from the fixed point, as illustrated in figure 8.8.

To expose the physical significance of these results let us suppose that the original system is at its critical point so that the ratio of correlation length to lattice spacing is infinite. After one application of the decimation transformation, the effective lattice spacing has increased by a factor of two, but this ratio remains infinite; the new system is therefore also at its critical point. Within the approximations inherent in equation (8.13)

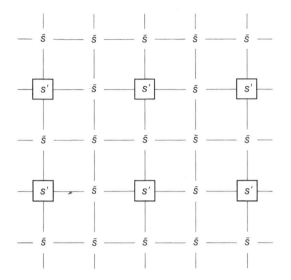

Figure 8.7 Coarse-graining by decimation. The spins on the original lattice are divided into two sets, $\{s'\}$ and $\{\tilde{s}\}$. The $\{s'\}$ spins occupy a lattice whose spacing is twice that of the original. The effective interaction between the $\{s'\}$ spins is obtained by performing the configurational average over the $\{\tilde{s}\}$.

Figure 8.8. Coupling flow under the transformation (8.13). The fixed point $K^* = 1$ represents the critical point coupling within the approximations leading to equation (8.13). Systems with couplings less than (greater than) the fixed point have temperature above (below) this approximation to the critical temperature. Under iteration of equation (8.13), the effecting couplings flow to the high and low temperature fixed points, $K = 0$ and ∞, respectively.

the original system is an Ising model with nearest neighbour coupling K and the new system is an Ising model with nearest neighbour coupling K'. If these two systems are going to be at a common criticality, we must identify $K' = K$. The fixed point $K^* = 1$ is therefore a candidate for the critical point K_c, where the phase transition occurs. This interpretation is reinforced by considering the case where the original system is close to, but not at, criticality. Then the correlation length is finite and the new system is further from criticality because the ratio of correlation length to lattice spacing is reduced by a factor of two. This *instability* of a fixed point to deviations of K from K^* is a further necessary condition for its interpretation as a critical point of the system. In summary then we make the prediction

$$K_c = J/kT_c = 1. \qquad (8.14)$$

We can obtain further information about the behaviour of the system close to its critical point. In order to do so, we rewrite the transformation (8.13) in terms of the deviation of the coupling from its fixed point value. Simple algebra shows that

$$K' - K^* = 2(K - K^*) + (K - K^*)^2. \qquad (8.15)$$

For a system sufficiently close to its critical temperature the final term can be neglected. The deviation of the coupling from its fixed point (critical) value is thus bigger for the new system than it is for the old by a factor of two. This means that the reduced temperature is also bigger by a factor of two:

$$t' = 2t. \qquad (8.16a)$$

But the correlation length (in units of the appropriate lattice spacing) is smaller by a factor of $\frac{1}{2}$:

$$\xi' = \frac{1}{2}\xi. \qquad (8.16b)$$

Thus, when we double t, we halve ξ, implying that

$$\xi \propto t^{-1} \qquad (8.17)$$

for T close to T_c. Thus we see that the renormalisation transformation predicts scaling behaviour with calculable critical exponents. In this simple calculation we estimate the critical exponent $v = 1$ for the square lattice Ising model.

This prediction is actually in agreement with the exactly established value (cf. Section 8.2). The agreement is fortuitous. A more sober measure of the crudity of the calculation is to be found in the fact that the prediction in equation (8.14) for K_c is larger than the exactly established value by a factor of more than two. In order to obtain reliable estimates more sophisticated and systematic methods must be used; some of these will be indicated in Section 8.5.

Universality and scaling

The crude approximation in the calculation above produced a transformation, equation (8.13), involving only the nearest neighbour coupling, with the subsequent advantages of simple algebra. We pay a penalty for this simplicity in two ways: the results obtained for critical properties are in rather poor agreement with accepted values, and we gain no insight into the origin of universality. In order to expose how universality can arise, we should from the start allow for several different kinds of coupling J_a, and show how the systems with *different* J_a can have the *same* critical behaviour. The understanding which results is encapsulated in figure 8.9; it is the purpose of the rest of this section to explain what the figure is, how it emerges from equation (8.12), and why it enables us to understand universality.

Figure 8.9 is a representation of the space of all coupling strengths K_a, in the energy function $\mathcal{H} = E/kT$. This is of course actually a space of infinite dimension, but representing three of these, as we have done, enables us to illustrate all the important aspects. First let us be clear what the points in this space

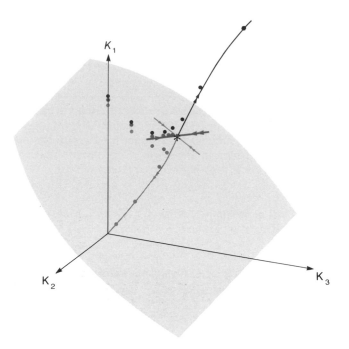

K_1

K_2

K_3

Figure 8.9. General flow in coupling space. The space of all couplings is represented here by three axes K_1, K_2 and K_3 corresponding to nearest, second and third neighbour couplings, say. Points on the K_1 axis close to (far from) the origin correspond to the nearest neighbour Ising model at high (low) temperatures. The shaded sheet is a schematic representation of the critical surface – the set of points where the correlation length is infinite; it intersects the K_1 axis at the critical coupling of the nearest neighbour Ising model. The critical fixed point of the renormalisation group transformation is shown, with its three 'eigendirections', along which the flow is a simple contraction or expansion; the directions along which the renormalisation group flow contracts lie in the critical surface. The set of red points represents schematically the successive couplings obtained if one starts at the nearest neighbour critical point; they iterate into the critical fixed point (eventually along the least irrelevant direction, shown as the heavy red line). If we start from a nearest neighbour coupling which is just above or just below the critical temperature, we initially move towards this fixed point, but ultimately move away to the high or low temperature fixed point (along the black eigendirection) under the influence of the relevant perturbation. The large-length-scale behaviour of all systems which lie close enough to the critical surface is described by effective couplings lying on the one-dimensional line emanating from the critical fixed point along the relevant eigendirection.

represent. Suppose we have some magnetic material which is described by a *given* set of exchange constants $J_1, J_2, J_3, \ldots$. As the temperature T varies, the coupling strengths $K_a = J_a/kT$ trace out a straight line, or ray, from the origin of the space in the direction $(J_1, J_2, \ldots)$. Points on this ray close to the origin represent this magnet at high temperatures, and conversely points far from the origin represent the magnet at low temperatures. The critical point of the magnet is represented by a specific point on this ray, $K_a = J_a/kT_c$, $a = 1, 2, \ldots$. The set of critical points on all of the possible rays forms a surface, the critical surface. Formally, it is defined by the set of all possible models (of the Ising type) which have infinite correlation length. It is shown schematically as the shaded surface in figure 8.9. (In the figure it is a two-dimensional surface; more generally it has one dimension less than the full coupling constant space, dividing all models into high and low temperature phases.)

Our immediate goal then is to understand how the renormalisation group can explain why different physical systems near this critical surface have the same behaviour. Before doing so a caveat is important. If some of the coupling strengths J_a are strongly negative then the competition among the different kinds of couplings can produce ordered states which are much more complicated than the ferromagnet. Correspondingly one must expect that the form of the critical surface and the behaviour near it may be much richer than the 'simple' magnetic transition we aim to understand here. More precisely therefore we want to understand why universality might hold

in some region of the critical surface, for example not too far from the K_1 axis which corresponds to the Ising model with nearest neighbour coupling.

Let us turn now to the schematic representation of the renormalisation group flow in figure 8.9. Suppose we start with a physical system, with coupling strengths K_a, $a = 1, 2, \ldots$. What the renormalisation group transformation (8.12) does is generate a new point in the figure, at the coupling strengths $K_a^{(1)} = f_a(\mathbf{K})$; these are the couplings appearing in the effective energy function describing the coarse-grained system. If we repeat the transformation, the new energy function has coupling strengths $K_a^{(2)} = f_a(\mathbf{K}^{(1)})$. Thus repeated application of the transformation generates a flow of points in the figure: $\mathbf{K} \to \mathbf{K}^{(1)} \to \ldots \to \mathbf{K}^{(n)} \ldots$, where the superscript (n) labels the effective couplings after n coarse-graining steps. If the change in coarse-graining scale is b (> 1) at each step, the total change in coarse-graining scale is b^n after n steps. In the process, therefore, the ratio of correlation length to coarse-graining scale is reduced by a factor of b^{-n}.

The dots in figure 8.9 identify three lines of renormalisation group flow starting from three systems differing only in their temperature. (The flow lines are schematic but display the essential features revealed in detailed calculations.) Consider first the red dots which start from the nearest neighbour Ising model at its critical point. The ratio of correlation length to coarse-graining scale is reduced by a factor b at each step, but, since it starts infinite, it remains infinite after any finite number of steps. In this case we can in principle generate an unbounded number of dots, $\mathbf{K}^{(1)}, \mathbf{K}^{(2)}, \ldots \mathbf{K}^{(n)}, \ldots$, all of which lie in the critical surface. The simplest behaviour of such a sequence as n increases is to tend to a limit, $\mathbf{K}^*$, say. In such a case in the transformation $\mathbf{K}_a^{(n+1)} = f_a(\mathbf{K}^{(n)})$, as n increases, both $\mathbf{K}^{(n)}$ and $\mathbf{K}^{(n+1)}$ must tend to this limit, which must therefore obey

$$K_a^* = f_a(\mathbf{K}^*) \qquad a = 1, 2, \ldots. \qquad (8.18)$$

This point $\mathbf{K}^* \equiv K_1^*, K_2^*, \ldots$ is therefore a *fixed point* which lies in the critical surface; we begin to make contact with the simple example in the preceding subsection. The number of equations in (8.18) is the same as the number of unknowns, so we would usually expect a solution to be an isolated point, and this is what we have illustrated in the figure. Lines of fixed points do arise in some interesting physical models such as the two-dimensional XY model (which has an order parameter with $n = 2$ components); they illustrate the richness in possible renormalisation group behaviour, and give a qualitatively different flow picture, but do not undermine our main conclusions.

By contrast, consider the same magnet as before, now at temperature T just greater than T_c. Its couplings K_a will be close to the first red dot (in fact they will be slightly smaller) and so will the effective couplings $K_a^{(1)}, K_a^{(2)}, \ldots$ of the corresponding coarse-grained systems. The new flow will therefore appear initially to follow the red dots towards the same fixed point. However, the flow *must* eventually move away from the fixed point because each coarse-graining now produces a model further from criticality. The resulting flow is represented schematically by one set of black dots. The other set of black dots shows the expected flow starting from the same magnet slightly below its critical temperature.

We are now in a position to understand both universality and scaling within this framework. We will suppose (some subsequent qualification will be necessary) that there exists a *single* fixed point in the critical surface which sucks in *all* flows starting from a point in that surface. Then *any* system at its critical point will exhibit large-length scale physics (large-block spin behaviour) described by the *single* set of fixed point coupling constants. The uniqueness of this limiting set of coupling constants is the essence of critical point universality. It is, of course, the algebraic counterpart of the unique limiting spectrum of coarse-grained configurations, discussed in Section 8.3. Similarly the scale-invariance of the critical point configuration spectrum (viewed on large enough length scales) is expressed in the invariance of the couplings under iteration of the transformation (after a number of iterations large enough to secure convergence to the fixed point).

To understand the behaviour of systems near but not precisely at criticality we must make a further assumption (again widely justified by explicit studies). The flow line stemming from any such system will, we have argued, be borne towards the fixed point before ultimately deviating from it after a number of iterations large enough to expose the system's noncritical character. We assume that (as indicated schematically in the streams of red and blue lines in figure 8.9) the deviations lie along a single *line* through the fixed point, the *direction* followed along this line differing according to the *sign* of the temperature deviation $T - T_c$. Since any two sets of coupling constants on the line (on the same side of the fixed point) are related by a suitable coarse-graining operation, this picture implies that the large-length-scale physics of all near-critical systems differs only in the matter of a length scale. This is the essence of near-critical point universality. (The reader may wonder what has become of the ordering variable scaling which, we saw in Section 8.3, is generally necessary to map one near-critical system onto another. The answer is that this scaling is built into the coarse-graining transformations envisaged here which impose the constraint that the coarse-grained spins have unit magnitude.)

The mathematics which justifies these intuitive expectations is as follows. First, define a coarse-graining scheme and calculate the transformation functions f_a in equation (8.11) (presumably approximately, with controlled errors). Second, solve equation (8.18) for the fixed points. Third, for each critical fixed point (there may be more than one, see below), find the special directions at the fixed point, along which the renormalisation group flow is simply expanding or contracting, as indicated schematically in figure 8.9. For the mathematically minded, these are the eigenvectors of the transformation (8.11), linearised about the fixed point, as in equation (8.15), i.e. the eigenvectors of the matrix $\partial K_a'/\partial K_b$ evaluated at $\mathbf{K}^*$. The special combinations of coupling strengths which expand away from the fixed point are called *relevant* perturbations; those which contract in are called *irrelevant*. (The *marginal* case must and can also be dealt with.) A fixed point which has only one relevant perturbation, corresponding to a deviation of temperature from criticality, controls the universal behaviour of a standard critical point: this is the situation envisaged in figure 8.9. The *rate* at which this relevant perturbation grows under the transformation (prescribed by the associated eigenvalue of the matrix $\partial K_a'/\partial K_b$) determines the critical exponent ν (cf. equations 8.15, 8.16 and 8.17 for the simple example in the preceding subsection). The other special directions, which all lie in the critical surface, contract in towards such a fixed point; the rate at which they contract in determines how unimportant they are, the associated eigenvalues giving the *correction to scaling* exponents. These exponents control the rate at which the configurations of the two- and three-state Ising models converge to the same universal spectrum under repeated coarse-graining, as described in Section 8.3 (cf., in particular, figure 8.5).

Four further remarks may clarify some questions in the reader's mind. First, the fixed point controlling standard critical behaviour actually has *two* relevant perturbations. The reason is that, like a deviation of T from T_c, an applied magnetic field also drives a magnet away from criticality (cf. figure 8.1b). Accordingly this variable (or its analogue in nonmagnetic systems) must also be tuned to bring about criticality; it is also, therefore, a relevant perturbation. It turns out that the eigenvalue associated with this perturbation determines the other principal critical index, through the ratio β/ν.

Second, not every fixed point of a renormalisation group transformation need be a critical point. For example, the simple equation $K' = K^2$ has two fixed points in addition to $K^* = 1$: $K = 0$ (∞) is the high (low) temperature fixed point. Note that these two fixed points are stable in this case, reflecting the evolution of all systems above (below) T_c to disordered (totally ordered) configurations under coarse-graining. In fact, in some models, the high and low temperature fixed points may be the only ones which exist. This is the case for the one-dimensional Ising model, which remains disordered at all nonzero temperatures. (In the language of Section 8.2 the 'number of arrangements' weighting wins over the 'energy of arrangements' weighting at any nonzero temperature.) In more than one dimension the Ising model has a phase transition; we say that the *lower critical dimension* of the nearest neighbour Ising model is one.

Third, we must now record the qualification that the critical surface will in general contain not only the normal critical fixed point discussed above, but also fixed points having more than two relevant perturbations. These fixed points describe *multi-critical* points, where special critical behaviour can be seen by tuning more than two variables. Such fixed points usually lie on the boundary of the basin of attraction of a more stable fixed point. A classic example is the transition from normal fluid to superfluid behaviour in He^3–He^4 mixtures; a tricritical point is found for a special value of temperature, pressure and He^4 concentration.

Finally, we note that we have emphasised fixed points in our discussion of the sequence of effective couplings because this leads to a deep understanding of scaling and universality. However, one should always be alert to other behaviour such as limit cycles or chaotic motion, in other many-length-scale problems with different and potentially exciting physics.

8.5   Practical methods for renormalisation group calculations

In the previous section, we illustrated the renormalisation group method in a simple calculation based on the decimation scheme. Although this calculation is useful for exposing the importance of fixed points and their stability under the coarse-graining procedure, the results obtained are not quantitatively reliable. In this section we wish to indicate the kinds of calculations which can yield reliable results, in the sense that the errors due to any approximations can be controlled. Our main aim is to give some impression of the breadth of possible approaches, but the topics covered are far from exhaustive. We shall continue to suppress mathematical details as far as

possible. Nevertheless this section is more technically demanding than its predecessor. The readers who, thus warned, wish to turn directly to the more general issues addressed in the final section can do so without loss of continuity.

The practical renormalisation group methods fall broadly into two categories – 'real-space' and 'Fourier-space', with wide variations in each category, and indeed some useful overlap. In real-space methods the configurations of the system are described by a set of variables associated with the sites of a lattice (e.g. the spin variables in the Ising model). The coarse-graining procedure then involves some form of spatial averaging over groups of neighbouring variables. The decimation scheme discussed in Section 8.4 offers the most primitive illustration of this strategy; more refined variants are described in the following subsections. In Fourier-space methods the configurations are described in terms of the Fourier-components (plane waves of magnetisation, or of density in the case of a fluid) from which they can be built. The coarse-graining procedure involves an averaging out of the effects of the shorter wavelength components. The techniques involved are outlined in the final subsection.

A preliminary remark on the choice of techniques is in order. In the context of a specific physical system one approach may seem more natural than the other. In the case of the magnet, for example, where nature supplies a definite lattice, the real-space formulation may appear the more appropriate; in the fluid, where there is no intrinsic lattice, the Fourier-space approach seems the natural one. However, if one's aim is to understand the critical behaviour of the system, one should realise that the microscopic authenticity of the approach is not the most vital consideration. For it is the essence of universality that the *same* large-length-scale physics will emerge irrespective of nitty-gritty microscopic details. Freed of the obligation to work with a microscopically realistic model one may thus choose one's approach so as to make the computation of the universal properties of interest as simple and effective as possible. In practice one finds that the universal physics tends to be more amenable to the real-space formulation in 'lower' space dimensions, and to Fourier-space methods in 'higher' space dimensions.

'Pencil and paper' real-space methods

Some of the practical difficulties faced in analytic calculations become immediately apparent if one attempts to extend the decimation calculation beyond the lowest order, which produces an effective nearest neighbour coupling K^2 according to equation (8.13): at fourth order in the expansion, one can generate *new* couplings between second and third neighbour spins. Since the renormalisation transformation is generally to be iterated, one must then *start* with a Hamiltonian with several couplings. This only exacerbates the problem of the proliferation of coupling constants. One should however appreciate that this problem is one of algebraic complexity, rather than a barrier in principle. The reason is that the effective energy function generated by the coarse-graining should be *short-range* (in fact of the order of the correlation length of the *partial* configuration sum involved). The problem is therefore a practical one of numerically reliable truncation, i.e. identifying the most important couplings in the fixed point and the flow generally, and estimating the error due to the neglect of the others. The last in particular is generally very difficult to do analytically in real-space methods, and in most instances is not achieved.

Notwithstanding this limitation, imaginative work has been done in this area with some very impressive results. For example, the decimation scheme has been studied extensively by K. G. Wilson; using L. P. Kadanoff's generalisation of the transformation, he looked systematically at the effect of including up to 217 types of interactions for the Ising model on a square lattice. One of the earliest methods which met with partial success consisted of performing an exact blocking on a finite lattice. The range of couplings generated by this procedure is then automatically truncated to those supported by the finite lattice itself; for example a two-dimensional square of four spins can support only four even interactions – nearest and second neighbour and the product of the four spins. One of the most elegant approaches is a variational method due to Kadanoff which can be applied in principle in any dimension and truncates to all interactions possible on a hypercube (a square in two dimensions, a cube in three, etc.). Finally there are a very few exact results, the most notable of which probably is an exact blocking for the two-dimensional Ising model on a triangular lattice due to H. J. Hilhorst, M. Schick and J. M. J. van Leeuwen. This particular approach involves an infinitesimal change of scale: a lattice of (large) side L is mapped onto one of side $L-1$ so that the scale factor $b = L/(L-1) = 1 + 1/(L-1)$. Exact results for the critical temperature and the critical index v follow from the mapping. (Of course, the exact direct solution of this model was already known.)

Clearly these various approaches exploit a wide range of potential coarse-graining procedures and approximations. Here we limit ourselves to emphasising that the particular coarse-graining procedure chosen must ensure that critical fluctuations on increasing length scales are captured by successive transformations. An amusing example may illuminate this point. Suppose we take an Ising model as previously, but one with *negative* nearest neighbour coupling favouring *antiparallel* (antiferromagnetic) alignment of neighbouring spins. The antiferromagnetic character of the low energy configurations would be lost if one defined a coarse-grained spin as a sum of the neighbours in a 2×2 cell (as in Section 8.3). One solution explored in the literature considers a cell formed from five second neighbours as in figure 8.10. All spins in such a cell are correlated ferromagnetically relative to

one another, and the resultant cell spins are correlated antiferromagnetically with their nearest neighbour cell spins. The coarse-grained lattice is also a square, with lattice constant $\sqrt{5}$ times that of the original.

Computational real-space methods

We described the methods above as analytic or 'pencil and paper', although their serious implementation almost inevitably involves some computational work to deal with the complexity of the algebra. Here we are concerned with intrinsically computational approaches in the sense that the partial or weighted configuration sum which defines the coarse-graining procedure is estimated numerically.

Because it follows directly from the discussion in Section 8.3, it is natural to begin with the method of finite size scaling. This approach starts by recognising that in any numerical simulation one is necessarily limited to a finite number of degrees of freedom in a box of limited spatial extent. In such a finite system, the true critical singularities – for example, the divergence of the correlation length or of thermodynamic quantities such as susceptibility or specific heat – are inevitably rounded off; the location of the maximum is also typically shifted from the exact bulk critical point. This shift and the neighbourhood where rounding takes place become smaller as the size of the system is increased. The brute force approach of fitting the expected bulk singularities to numerical data from larger and larger simulations is too crude and expensive in computer time to form a viable method; the increase in the number of variables is compounded with the increasingly slow evolution of the large distance correlations accumulating from the short-range (local) Monte Carlo updating procedure – the phenomenon of 'critical slowing down'.

The finite size scaling method incorporates explicitly the finite size of the system, in the same spirit as in Section 8.3: the dependence of all quantities on the size of the system, L, is recognised but this extra dependence appears only in the ratio $\xi/L(= z$ in Section 8.3). The formalism is very similar to that in equations (8.8) and (8.9), but with L interpreted as the size of the system; the L dependence is not illusory here! Finite size scaling can therefore be thought of as a single coarse-graining step to the system size.

In order to illustrate the formalism, recall from equations (8.4) and (8.5) that the bulk order parameter, Q, vanishes as $Q_- |t|^\beta$, as $T \rightarrow T_c$ from below. The finite size scaling form is (some care with boundary conditions is required here)

$$Q(t,L) = |t|^\beta \tilde{Q}(|t|^{-v}/L), \qquad (8.19)$$

where the bulk correlation length has been written in terms of its t-dependence. Equation (8.19) reproduces the bulk behaviour provided that $\tilde{Q}(z) \rightarrow Q_-$ as $z = |t|^{-v}/L \rightarrow 0$, the infinite volume ('thermodynamic') limit. In the finite system, the limit $t \rightarrow 0^-$ must exist. The asymptotic form of $\tilde{Q}(z)$ for z large must

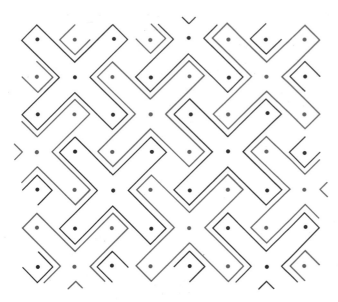

Figure 8.10. Possible coarse-graining scheme for an antiferromagnet. In an antiferromagnet the nearest neighbour coupling, J_1, is negative, so that the spins at the black and red sites of this square lattice tend to be *anti*parallel. The coarse-graining procedure shown replaces local clusters of five predominantly parallel spins by new block spins which occupy a (tilted) square lattice whose spacing is $\sqrt{5}$ times that of the original lattice. The nearest neighbour block spins tend to be antiparallel – the antiferromagnetic nature of the system is preserved under the transformation.

therefore cancel the $|t|^\beta$ prefactor. This requires $\tilde{Q}(z) \sim z^{\beta/\nu}$ implying

$$Q(0,L) = L^{-\beta/\nu} g_0, \qquad (8.20)$$

where g_0 is another constant. Thus simulations on lattices of different size L at the bulk critical temperature enable an estimate to be obtained for the exponent β/ν from the calculated order parameter. Of course, the value of T_c may not be known *a priori*, but it also can be estimated numerically by exploiting equation (8.19). The present state of the art in this approach in the three-dimensional Ising model (which cannot be solved exactly) involves lattices of up to 128^3 sites, and very high statistics (more than 10^7 sweeps through the entire lattice, on parallel computers which perform in excess of 200 million Monte Carlo update steps per second). Exponents obtained are accurate to within a few parts per thousand. Many variants of this approach have been applied to a wide range of problems.

The finite size scaling method can be proved within the framework of the renormalisation group; the Monte Carlo renormalisation group (MCRG) is a very powerful technique which explicitly implements a renormalisation group analysis on configurations generated by a computer simulation ('Monte Carlo') method. The idea was first proposed by S. K. Ma and was developed into an effective tool by Wilson and R. H. Swendsen. We summarise its key features.

Recall that to calculate critical exponents one needs the stability matrix $\partial K'_a / \partial K_b$ evaluated at the fixed point. Ostensibly, in a numerical simulation this appears to require a sequence of simulations searching a necessarily high-dimensional space for the critical fixed point $\mathbf{K}^*$, and mapping out the flow of effective couplings near this fixed point; although we have not indicated how in principle one might actually implement these steps, it should be clear that any direct approach would involve a prohibitively large number of simulations and hence of computing resource. Fortunately, both of these problems can be elegantly bypassed by the following procedure.

The first point is that we don't have to *search* for $\mathbf{K}^*$. Provided we *start* at a critical point (say the critical point of the nearest neighbour Ising model), the effective couplings of the coarse-grained configurations will *automatically* flow towards $\mathbf{K}^*$, according to figure 8.9. Therefore the derivatives $\partial K_a^{(n+1)} / \partial K_b^{(n)}$ approach the required limit $\partial K'_a / \partial K_b$ (evaluated at $\mathbf{K}^*$) as the blocking level n is increased, without any further tuning.

The second point is that in order to calculate the matrix of derivatives $\partial K_a^{(n+1)} / \partial K_b^{(n)}$, it is not necessary to know the couplings themselves. Instead we can focus on the expectation values of the various interactions. Specifically, consider the energy function at the nth blocking level. We write

$$\mathcal{H}^{(n)} = -\sum_a K_a^{(n)} S_a^{(n)},$$

where, for example, $S_1^{(n)}$ is the sum of the nearest neighbour pairs of spins on the lattice after n blocking steps, $S_2^{(n)}$ is the sum of second neighbour pairs etc. Now consider the derivative $\partial \langle S_c^{(n+1)} \rangle / \partial K_b^{(n)}$ and use the chain rule:

$$\frac{\partial \langle S_c^{(n+1)} \rangle}{\partial K_b^{(n)}} = \sum_a \frac{\partial K_a^{(n+1)}}{\partial K_b^{(n)}} \frac{\partial \langle S_c^{(n+1)} \rangle}{\partial K_a^{(n+1)}}. \qquad (8.22)$$

This is a matrix equation which can be solved for the required stability matrix, provided we can calculate the derivatives of the expectation values on both sides of this equation. This we can do by exploiting the definition of the configurational average

$$\langle S_a^{(n)} \rangle = \frac{\sum e^{-\mathcal{H}^{(n)}} S_a^{(n)}}{\sum e^{-\mathcal{H}^{(n)}}}$$

to show that

$$\frac{\partial \langle S_c^{(n+1)} \rangle}{\partial K_b^{(n)}} = \langle S_c^{(n+1)} S_b^{(n)} \rangle - \langle S_c^{(n+1)} \rangle \langle S_b^{(n)} \rangle.$$

The required derivatives are thus expressed in terms of exotic, but measurable, correlation functions involving block spins at the nth and $(n+1)$th level.

This discussion has assembled the basic MCRG formalism. The approach is a powerful one permitting systematic control of all approximations in the calculation. It has also many practical advantages. For example, one can allow for an increasing number of effective couplings in a single simulation simply by increasing the number of expectation values which are measured. Moreover the critical point does not have to be known *a priori*: it can be obtained directly with high precision using the same basic formalism. Further refinements include optimising the coarse-graining procedure to minimise the effects of the transient flow into the fixed point. The field is currently an active one, and is potentially applicable to a wide range of problems outwith critical phenomena.

Fourier-space methods

Fourier-space methods are most naturally implemented in models where degrees of freedom are described by a 'density' defined at all points in space – magnetisation density or fluid density, for example. The general term for such a quantity is a 'field' and the Fourier method is also known as 'field theory' because of its origins in the quantum field theory used to describe the interaction of subnuclear particles. (We shall return to this connection in Section 8.6.) We will write our field as $\phi(\mathbf{x})$, where $\mathbf{x}$ denotes position in space; the computations actually focus on the Fourier-components of the field, which we write as $\hat{\phi}(\mathbf{q})$, where the wave-vector $\mathbf{q}$ is an inverse measure of the wavelength of the Fourier-component. To do calculations, we must identify an appropriate energy function $\mathcal{H}\{\phi\}$ for a given density configuration and define the average over all density configurations which yields quantities of physical interest.

The prototype energy function in this case is identified with Landau, Ginzburg and Wilson (LGW) and has the form

$$\mathcal{H} = \int d^d x \{ \tfrac{1}{2}\lambda(\nabla\phi)^2 + \tfrac{1}{2}r\phi^2 + \tfrac{1}{4}u\phi^4 - H\phi \}, \qquad (8.23)$$

where $(\nabla\phi)^2 = (\partial\phi/\partial x_1)^2 + \ldots + (\partial\phi/\partial x_d)^2$ and d is the spatial dimension. Although this looks completely different from the Ising energy function, it is based on the same principles. Think of ϕ as a magnetisation density (technically, for a uniaxial magnet). Apart from the last term, which models the energy in an applied magnetic field H, the terms are symmetric under $\phi \to -\phi$. If λ is positive, the first term implies that it costs energy for ϕ to vary in space, i.e. homogeneous ferromagnetic ordering is favoured; conventionally, units are chosen so that $\lambda = 1$. For zero field H it is clear that the minimum energy configuration is $\phi(\mathbf{x}) = 0$ or $\phi(\mathbf{x}) = (-r/u)^{\frac{1}{2}}$, a nonzero constant, according to whether $r > 0$ or $r < 0$. Although the minimum energy configuration does not itself determine the equilibrium properties (recall the crucial role of entropy discussed in Section 8.2) we may thus anticipate that a phase transition will occur close to a point at which the coefficient r changes sign. Accordingly we may expect that the strong temperature dependence of the

model will reside in the parameter r. More specifically we assume that r is a smooth function of T with a power series expansion in t, for small t. This assumption is, in fact, explicitly justified by arguments deriving the continuum model energy equation (8.23) from microscopically more realistic models (for magnets or fluids). Such arguments, which involve an initial coarse-graining over the shortest wavelength fluctuations, also show that higher powers of ϕ than those displayed in equation (8.23) are present in principle. However, the presupposition of universality gives us the licence to discard them: the theory allows us to check this supposition explicitly by showing that such interactions are indeed irrelevant. Finally, the fact that $\mathcal{H}$ is an integral over space of a local function of ϕ at the point x is intended to capture the intrinsically short-range nature of the ordering forces.

The average over all field configurations we denote by $\int D\phi$. This is a subtle object; it means 'add up the contributions for all possible functions ϕ'. It is clearly inappropriate here to provide a more mathematical definition; we comment only that it is not misleading to think of $\int D\phi$ as a multi- (in fact, infinite-) dimensional integral and that, with it, the formalism for physical averages parallels that for the Ising model, e.g.

$$\langle Q \rangle = Z^{-1} \int D\phi \, Q\{\phi\} \exp(-\mathcal{H}\{\phi\}), \qquad (8.24)$$

where

$$Z = \int D\phi \exp(-\mathcal{H}\{\phi\}) \qquad (8.25)$$

is the partition function.

One further remark is needed to provide a platform for an illuminating discussion later. The model defined by equation (8.23) apparently makes no reference to the coarse-graining length scale at which the continuous field description becomes appropriate. Now, by definition all fluctuations on a length scale smaller than this have already been performed. Such fluctuations should therefore be *excluded* from the remaining configuration sum $\int D\phi$. The same effect can be achieved by giving *extra* energy to configurations which vary on shorter length scales, so that they are damped out by the weight factor $\exp(-\mathcal{H}\{\phi\})$. A possible solution is to add to $\mathcal{H}$ a term

$$\mathcal{H}_\Lambda = \tfrac{1}{2}\Lambda^{-2} \int d^d x (\nabla^2\phi)^2, \qquad (8.26)$$

where $\nabla^2\phi = \partial^2\phi/\partial x_1^2 + \ldots + \partial^2\phi/\partial x_d^2$: if ϕ varies on length scales less than $1/\Lambda$, i.e. $|\nabla^2\phi/\Lambda| > |\nabla\phi|$, this extra term dominates. The parameter Λ, which has dimensions of an inverse length, or wave-vector, is called the cut-off. It is the fundamental dimensional quantity setting the scale for *all* the parameters in $\mathcal{H}$. An elementary exercise in dimensional analysis of the terms in the dimensionless quantity $\mathcal{H}$ readily yields

$$[\phi] = [\Lambda]^{\frac{1}{2}(d-2)}; \ [r] = [\Lambda]^2; \ [u] = [\Lambda]^{4-d}. \qquad (8.27)$$

Let us turn now to the question of performing the averages in equations (8.24) and (8.25). It should come as no surprise to the reader that in general these configuration sums cannot be done exactly. We must resort to some approximation scheme. The following scheme may suggest itself. Since $\mathcal{H}$ is an integral over space, it clearly increases as the volume of the system. Thus, in the large volume limit, the largest weight in the configuration average is associated with the configuration of lowest energy. Hence, as lowest approximation, we could imagine that $\int D\phi$ is completely dominated by the configuration of lowest energy. Since $(\nabla\phi)^2$ always costs energy because it is positive, the minimum energy configuration has $\phi = M$, a constant which satisfies

$$0 = \frac{d}{d\phi}\{\tfrac{1}{2}r\phi^2 + \tfrac{1}{4}u\phi^4 - H\phi\}|_{\phi=M} = rM + uM^3 - H. \quad (8.28)$$

With the identification that the critical point divides $M = 0$ from $M \neq 0$ (for $H = 0$), we must have $r = (T - T_c)$, up to an overall constant and neglecting terms of order $(T - T_c)^2$. Hence equation (8.28) becomes

$$H = (T - T_c)M + uM^3. \quad (8.29)$$

This is the Landau–Ginzburg or *mean field* equation of state relating magnetisation density M to T and H. It can be written in the scaling form

$$\frac{H}{M^\delta} = f\left(\frac{T - T_c}{M^{1/\beta}}\right) \quad (8.30a)$$

with the identifications

$$\delta = 3; \quad \beta = \tfrac{1}{2}; \quad f(x) = x + 1; \quad (8.30b)$$

(u can be set to 1 with a suitable choice of units).

This mean field theory may give a qualitatively correct description of a phase transition, but it is in general simply wrong quantitatively for one, two or three dimensions (recall that in $d = 1$ we expect no phase transition at any temperature). Clearly there must be corrections to the approximate result expressed in equation (8.28), and indeed there are: the variations about a given configuration are *also* exponentially extensive in the size of the system, so that the naive expectation about minimum energy dominating must be modified. These variations can be calculated perturbatively, as a power series in the parameter u, using what is called the Feynman graph expansion, which was developed in the late 1940s and early 1950s. Clearly there would be no case for including this chapter in a book on the 'New Physics' if this Feynman graph expansion directly solved the problem, and it does not. Indeed, how could it? – universal numbers can't be given in terms of an expansion in u whose value varies from system to system!

A more illuminating discussion can be made by recalling the dimensional analysis of equation (8.27). Since all terms in an expansion in u must have the same overall dimension, the factor $[\Lambda]^{4-d}$ from every power of u must be cancelled out by another factor. There are only two parameters available, Λ itself from the cut-off term (equation 8.26), or $r \sim T - T_c$. It is the latter which gives the problem, both potentially and in

practice, because it means that every power of u can be accompanied by a factor $(T - T_c)^{-\frac{1}{2}(4-d)}$, according to the dimension of r in equation (8.27). Thus the *effective* expansion parameter is $(\Lambda^2/(T - T_c))^{\frac{1}{2}(4-d)}$, which becomes unboundedly large as $T \to T_c$, for any spatial dimension $d < 4$; the perturbation expansion breaks down catastrophically! Conversely, in a world of more than four spatial dimensions, each term in perturbation theory behaves in the same way as the leading term. In particular for $d > 4$, the form (8.30a,b) for the equation of state remains unchanged (in the asymptotic limit $t \to 0$, $H \to 0$), by the perturbation corrections. The delimiting case $d = 4$ is called the *upper critical dimension*; above it, mean field theory is exact for the leading critical behaviour. Along with the lower critical dimension (at and below which there is *no* phase transition), it also provides a useful characteristic of a universality class.

Returning to physical dimensions, we are faced with a classic strong coupling problem, in which the effective expansion parameter can become apparently arbitrarily large in the critical region of interest. It is this problem which the renormalisation group tackles.

Just as in real-space methods, there are many ways of implementing the renormalisation group using the Fourier methods natural for a continuous field. The simplest conceptually is to start from the model defined by equation (8.23), in which we suppose (given the initial coarse-graining implicit in the model's derivation) that the field has Fourier-components $\hat{\phi}(\vec{q})$ limited to a ball with wave-vectors $|\mathbf{q}| < \Lambda$. A blocking transformation can then be defined by performing a partial configuration sum over all fields whose Fourier-components lie in a shell $\Lambda/b < |\mathbf{q}| < \Lambda$. This partial average generates an effective energy function of a field whose Fourier-components lie in a ball Λ/b. A second stage is necessary but trivial; rescale x (or q) so that the ball radius (cut-off) $\Lambda/b \to \Lambda$ again, and the field ϕ so that the $(\nabla\phi)^2$ term in the effective Hamiltonian again has coefficient $\tfrac{1}{2}$. These rescalings are the analogues of the space and variable rescalings discussed in Section 8.3.

Such a transformation generates effective parameters $r'(r,u)$ and $u'(r,u)$, which can be calculated as a power series in u, as well as possibly new kinds of interactions. The qualitative features of this renormalisation group formalism are the following. First, the functions r' and u' are *not* singular at the critical point. Second, u is an irrelevant coupling for $d > 4$; in fact $u' = b^{4-d}u + O(u^2)$. This is the renormalisation group statement that mean field theory is valid in high dimensions. Third, for $d < 4$, the perturbation of u away from the fixed point $u^* = 0$ is relevant; the system crosses over from mean field theory to new behaviour.

In order to find out quantitatively what this new behaviour is, we must find a nontrivial fixed point to which u flows under a renormalisation group transformation. One approach is simply to calculate the functions r' and u', in three dimensions, to as high an order in u as possible, and to find the fixed point and behaviour near it (i.e. critical exponents) by numerical analysis of the power series in u. Remarkable calculations to the sixth nontrivial order by B. G. Nickel have made this a powerful technique. The second approach due to M. E. Fisher and K. G. Wilson is to recognise that something must simplify near the upper critical dimension $d=4$; this is one motivation for thinking of d as a continuous variable and asking what happens if $\varepsilon \equiv 4-d$ is small (and positive). One then discovers that the new stable fixed point value of u is of order ε also. All quantities such as exponents can then be calculated as power series in ε, starting with their mean field values. For example, the first two terms (four are now known) for the exponent v for the model defined by equation (8.23) are

$$v = \tfrac{1}{2} + \tfrac{1}{12}\varepsilon + \tfrac{7}{162}\varepsilon^2 + O(\varepsilon^3). \qquad (8.31)$$

This is a particularly neat way to obtain a qualitative understanding of physical behaviour in three dimensions; for simple systems enough terms can be calculated to provide also quantitative results when ε is set to 1 ($d=3$). For example, equation (8.31) gives $v=0.626$, which is actually remarkably close to the best currently available estimates for the Ising universality class assembled in figure 8.11. The accord between the results obtained by theoretical studies of different model systems (formulated on a lattice in the case of the series calculations, and in the continuum for the field theory calculations) and experimental studies of different real systems (fluids, uniaxial magnets) offers firm evidence in support of universality. The precision of the claims made is some testimony to the sophistication of the theoretical techniques we have discussed (and the experimental techniques we have not).

8.6 Many-length-scale phenomena: the different flavours

In this article we have chosen to illustrate the problems posed by many-length-scale physics, and the techniques necessary to handle these problems, in the context of the Ising model and the universality class of which it is the simplest representative. Although this class of systems is a large one (and the fact that it is now largely understood is correspondingly impressive), it does not by any means exhaust the range of forms of many-length-scale phenomena found in nature, and susceptible to renormalisation group methods. In this concluding section we attempt to give some impression of the diversity of such phenomena. We begin by developing observations made in Section 8.3, identifying some of the different universality classes of many-length-scale phenomena occurring within the phase transition context. We then look briefly at many-length-scale phenomena outwith the context of phase transitions, but still in the framework of condensed matter physics. Finally, we leave the territory of condensed matter, and turn to relativistic quantum theories of the fundamental forces. As discussed elsewhere in this book (for example in the chapter by Taylor), there is now good evidence that the strong and weak nuclear forces and the more familiar electromagnetic force are all

described by so-called gauge theories. How can such disparate physical phenomena be accommodated within a single unifying framework? As we shall see, the answer lies in the deep connection between relativistic quantum theories and phase transitions in statistical systems; the different properties of the fundamental forces are simply characteristic of different possible phases of gauge theories.

Figure 8.11. The critical index v for members of the $d=3$ Ising universality class. The table gathers together the results of comparatively recent theoretical and experimental studies of systems believed to belong to the three-dimensional Ising universality class. The series expansion studies of the lattice Ising model are by Fisher and Chen. The field theoretical renormalisation group calculations are by Le Guillou and collaborators. The Monte Carlo renormalisation group calculations are due to Pawley and collaborators. The result for the liquid–vapour critical point in SF$_6$ is derived from light scattering studies by Cannell. The study of the phase separation transition of the binary fluid is due to Chang and collaborators. Finally, the result for the uniaxial (antiferro) magnet FeF$_2$ is due to neutron scattering measurements by Belanger and Yoshizawa. The horizontal line is the best (least-squares) representation of the six data points.

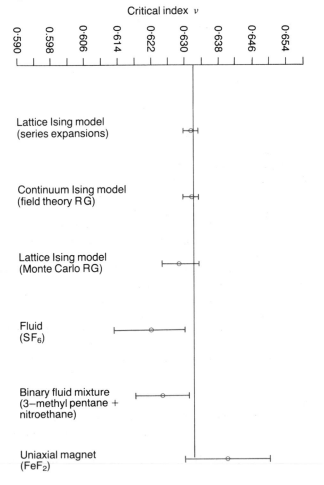

The universality classes for phase transitions

Physical systems undergoing phase transitions do not all fall into the Ising universality class. The configurational physics viewed on large enough length scales and (equivalently) the limiting flow in the space of coupling constants are insensitive to much microscopic detail; but (as observed in Section 8.3) there are some essential features which are conserved by the configurational and coupling constant flow, and are consequently reflected in the limiting critical behaviour. In this subsection we identify some of these key features, and give examples of the variety of universality classes which they serve to delineate.

The first (and practically dominant) feature is the *dimension d of the space* over which the interacting degrees of freedom are spread. Thus the 'Ising universality class' actually incorporates a number of subclasses distinguished by the dimension d. Physical realisations of the $d = 3$ Ising universality class are identified in figure 8.11. Physical realisations of the $d = 2$ class (of which the square lattice Ising model is the prototype member) are to be found in the ranks of planar magnets discussed in the chapter by Thouless in this volume. Magnets consisting of essentially isolated chains of Ising spins also exist, offering realisations of the $d = 1$ Ising universality class, which however exhibits no finite temperature phase transition, since $d = 1$, it will be recalled, is the lower critical dimension for Ising systems. Finally we note that physical realisations of the Ising universality class at its upper critical space dimension $d = 4$ also exist (!) in a sense to be clarified below.

The second key feature, also briefly identified in Section 8.3, is *the number, n, of components of the order parameter*. The Ising model order parameter is a scalar, with a single component. Ferromagnets such as europium oxide (EuO) have order parameters formed from ($n = 3$)-component spins. The spins are strictly quantum mechanical operators, but this turns out to be an 'irrelevant' microscopic detail. Accordingly (modulo refinements registered below) such systems can be described by (i.e. fall into the same universality class as) the generalisation of the Ising model associated with the name of Heisenberg, in which the Ising spins are replaced by three-component classical vectors of unit length. The structural phase transition displayed by K_2SeO_4 offers an example of the case of an ($n = 2$)-component order parameter formed from the amplitude and the phase of the profile of the distortion which occurs as the temperature is lowered through the critical point. The superfluid phase transition occurring in liquid helium (He^4) is also described by an order parameter with two components, in this case the amplitude and phase of the wave-function describing the condensate. There is experimental evidence (substantial in the latter instance) that these systems fall into one universality class, of which the simplest representative is the *XY* model, in which the microscopic degrees of freedom are represented by ($n = 2$)-component classical spin vectors.

The Heisenberg and *XY* model systems have an important symmetry property: their configurational energy is invariant under an operation in which all the spins are rotated through the same angle. The configurational energies relevant to the structural instability in K_2SeO_4 and the superfluid phase transition in He^4 have analogous symmetries (reflecting the energetic irrelevance of changes in the *phase* of the respective order parameters). In general, however, systems with n-component order parameters do not possess this rotational invariance. Thus, for example, the energy of a real ($n = 3$)-component magnet depends, to a degree, upon the orientation of the magnetisation, as is reflected in the fact that the magnetisation tends to be directed along one of a number of well-defined ('easy') axes. The *symmetry* of the configurational energy constitutes a third feature, potentially important to the nature of the large-length-scale physics. Any interactions which break the rotational symmetry may be preserved, in fact amplified, by the coarse-graining procedure so that the universal critical behaviour is described by a fixed point configurational energy with this reduced symmetry. Alternatively the symmetry-breaking perturbations may be irrelevant, so that they are erased by the coarse-graining procedure which bears the system to a fixed point whose configurational energy has full rotational invariance. (There is actually a third possibility: the symmetry-breaking perturbation may drive the coupling constant flow into a region where the critical behaviour is truncated abruptly by the occurrence of a 'discontinuous' phase transition, characterised by a jump in the order parameter.) Which scenario is realised in any given situation depends upon the other key features (such as d and n) in a nontrivial way which can only be established by explicit renormalisation group calculations. Here we must confine ourselves to the observation that, in the important class of ferromagnets with cubic symmetry (typified by EuO) such symmetry-breaking interactions either are formally irrelevant or lead to a fixed point which lies close to that characterising rotationally invariant systems.

The fourth potentially important feature which we identify is the *range of the interactions* amongst the microscopic degrees of freedom. As long as these interactions decay sufficiently rapidly with the separation of the interacting variables, the universality class remains that of the appropriate short-range interaction model: thus, for example, the addition of next neighbour interactions to the Ising model does not in general change the system's universality class. However, if the interactions do not decay sufficiently rapidly with separation the large-length-scale physics is materially altered. Indeed, such 'long-range' interactions do exist in real magnets. Although the dominant spin–spin interaction (the quantum-mechanical exchange force) is short-ranged, there exists also the essentially classical interaction appropriate to pairs of magnetic dipoles. This interaction is long-ranged and is important to the critical behaviour. Its consequences differ dramatically according to the value of the number n of order parameter components. For ($n = 3$)-component ferromagnets, the appropriate universality class, though nominally different from that of a system with only short-range interactions, is characterised by numbers (indices) which differ insignificantly from their short-range (Heisenberg) counterparts. For the ($n = 1$)-component ('uniaxial') magnet, on the other hand, the consequences of the magnetic dipole interactions are stark: within the renormalisation group framework one can show that their effect is to increase by one the effective space dimension of the magnet. Thus a three-dimensional uniaxial magnet with dipole–dipole interactions falls into the universality class of a *four*-dimensional strictly short-range (Ising) model, whose behaviour is classical mean-field (to within subtle correction terms which reflect the nonclassical behaviour that sets in fully below four space dimensions). These predictions have been corroborated by studies of the uniaxial ferromagnet LiTbF$_4$.

We have already stressed the potential significance of the rotational symmetries of the configurational energy. Its *translational symmetry* is also of key importance. The systems discussed thus far all have such a symmetry: thus, for example, we have tacitly supposed that the interactions of any one spin with its neighbours are the same (i.e. are characterised by the same coupling constants) as the interactions between any other spin and its neighbours. Many systems exhibiting phase transitions do not have this basic symmetry. In particular, in some magnetic alloys (for example a dilute solution of magnetic manganese in nonmagnetic copper) the magnetic ions interact with coupling constants that are effectively randomly distributed (over positive and negative values). The behaviour of such ('spin glass') systems is radically different from their homogeneous counterparts, so different in fact that, despite a decade of committed theoretical activity, many quite basic questions regarding the nature of the phase transition, and the nature of the ordered phase itself, remain to be resolved.

The five key features we have identified arguably serve to define the basic boundaries between universality classes. It is not appropriate here to refine these boundaries further. There is, however, one more general issue which we can and should address briefly at this point.

Consider a system described by a configurational energy of the form $\mathscr{H} = \mathscr{H}_0 + \mathscr{H}'$, where the universality class associated with the energy $\mathscr{H}_0$ is different from that associated with $\mathscr{H}$. (In the language of Section 8.4 $\mathscr{H}'$ represents a *relevant* perturbation and the fixed point of $\mathscr{H}_0$ is a *multicritical* fixed point in the space of the configurational energy $\mathscr{H}$.) As a concrete example we may think of the ferromagnet, associating $\mathscr{H}_0$ with the configurational energy due to the short-range (exchange) forces, and $\mathscr{H}'$ with the long-range (dipolar) forces. Now suppose that $\mathscr{H}'$ is small compared to $\mathscr{H}_0$ (the dipolar forces are weak compared to the exchange forces – the situation typical of most real ferromagnets). It is natural to expect that, notwithstanding the claims of universality, the behaviour of

the system will, in *some* sense, be dominated by the configurational energy $\mathscr{H}_0$. This is true. The smaller is the perturbation $\mathscr{H}'$ the greater is the length scale on which one has to view the system before the effects of $\mathscr{H}'$ begin to manifest themselves. In more practical terms, the smaller is $\mathscr{H}'$, the smaller is the range of reduced temperature in which its effects are apparent. Thus, for example, as the critical point of a weakly dipolar magnet (such as EuO) is approached there is an initial critical region in which the behaviour is that of a system with only short-range interactions (described by the fixed point of the Heisenberg model); there is also a second critical region accessed at much smaller reduced temperatures where the effects of the dipolar interactions are fully developed (and the behaviour is controlled by the appropriate dipolar fixed point). In between these two regimes there is a *crossover* region, typically several decades of reduced temperature in width. Renormalisation group methods allow one to handle, quantitatively, such crossover behaviour, which itself displays universal characteristics.

Other many-scale phenomena in condensed matter

There are numerous other areas in condensed matter physics, besides phase transitions, which involve many different scales and exhibit analogous universal scaling behaviour. We discuss briefly five examples in this section.

Percolation phenomena are important in many processes, from the spread of communicable diseases to the recovery of oil from porous rocks. The key ideas can be illustrated most simply in the context of a lattice model. Consider a two-dimensional square lattice. Suppose we take counters and place them at random on the sites of the lattice. When a counter is placed on an 'unoccupied' site, that site becomes 'occupied'. We define a *cluster* of occupied sites as the set of counters which can be reached by nearest neighbour steps from one counter to the next. If p, the fraction of all sites which are occupied, is small, then the clusters will be small and contain only a small number of counters. If more counters are placed on the lattice, the typical maximum cluster size increases. The percolation concentration p_c is the value of the coverage at which this maximum cluster size diverges: when the fraction of occupied sites is greater than or equal to p_c an infinite cluster exists on which one can step from counter to counter across an arbitrarily large lattice. The key point for our purposes is that there are clusters of *all* sizes, up to the typical maximum size which depends on the coverage p. (Technically, for $p > p_c$, this maximum size is defined by excluding the infinite cluster.) It is the existence of clusters of many length scales which is responsible for universal scaling properties in percolation. Although percolation is essentially a geometric phenomenon making no reference to thermal effects, there is a strong analogy with phase transitions, with the coverage p playing the role of inverse temperature. For example, the typical maximum cluster size is analogous to the correlation length, and diverges (as p approaches p_c) with an exponent v_p whose exact value is approximately 1.35 in two dimensions. The analogue of the order parameter is the fraction of sites in the infinite cluster. It is zero for $p < p_c$, and vanishes as p decreases to p_c with an exponent β_p. Thus at p_c there is (just) an infinite cluster, but it covers a vanishingly small fraction of the lattice sites.

Polymers are chemicals which form long flexible chains built of identical units, 'monomers'. They have long been an important sector of the chemical industry. The most familiar is probably polyethylene $(CH_2)_N$; the quantity N in the chemical formula just represents the end-to-end length measured *along* the chain. A deep understanding of many properties of polymers has emerged from a renormalisation group approach which recognises and deals with the many-length-scale phenomena they exhibit. To be specific, let us consider such a molecule in a solution and let us suppose that the monomer junctions are completely flexible, i.e. the relative orientation of adjacent units is unconstrained. Then the simplest description of the molecule at any instant is in terms of a 'random walk' – the position of the molecule is described by a drunken stagger of N steps. This random walk problem is also realised in the Brownian motion of a single particle suspended in or floating on a fluid. The problem can be solved exactly; for example, the *straight* end-to-end distance, which measures the linear size of the molecule, increases with the number of steps N as $N^{\frac{1}{2}}$. However, this random walk description of a polymer fails to take into account that the polymer in solution may be *self-avoiding*, i.e. the monomers in the polymer chain may have a greater affinity for the solution in which the chain is suspended than for monomers at other points in the chain. The net effect is that segments on the chain which could cross in a random walk are repelled from one another, so that the chain becomes more distended than a random walk. This problem clearly involves two very different length scales, the monomer size which sets a minimum distance scale, and the physical polymer size. That the problem involves intrinsically many length scales is best seen by considering distance along the chain: in the random walk, crossing may occur between segments separated by *any* distance along the chain and the real polymer size is influenced by repulsive forces between potentially any pair of segments on it. In the polymer problem, the number of monomer units, N, is analogous to the inverse of the reduced temperature in a phase transition problem, and the typical straight end-to-end polymer size corresponds to the correlation length, so that the universal exponent relating them is naturally also denoted by v. In three dimensions, v is approximately 0.588. In fact, P. G. de Gennes has shown that such polymers belong to the universality class of n-component Heisenberg (or Landau Ginzburg Wilson) models, in the limit $n \to 0$. In this context it is interesting to note that the mean field value $v = \frac{1}{2}$, corresponding to simple random walk behaviour, is valid above the upper critical dimension four

because polymer crossings are there of sufficiently low density to be technically irrelevant.

The onset of *chaotic behaviour* is another topic where the renormalisation group has yielded important insight. This is a complex problem; the particular scenario on which we focus is the 'period doubling route to chaos', which was studied in pioneering computational work by M. Feigenbaum. A wide variety of dynamical systems including lasers and convective currents have subsequently been shown to exhibit this behaviour. A general characteristic of such systems is the existence of an oscillatory motion with period τ. In the case of convective flow, this motion is a transverse oscillation of the 'convective rolls', which sets in when the temperature gradient driving the convection is larger than some threshold. As the temperature gradient is further increased, a second threshold is reached at which this oscillatory motion becomes unstable and a new component appears with a frequency which is half of the original one: thus the period of the motion doubles at this second threshold. As the temperature gradient is increased further, the amplitude of this new component increases until a third threshold is reached at which a new component appears with a frequency one-quarter of the original one. In mathematical models of this process, there is an infinite sequence of thresholds at each of which the period doubles as a new frequency component enters, this sequence accumulating at a point where chaotic motion sets in. This is clearly a problem of many scales, the frequency components now corresponding to different *time*scales. The scaling phenomena in this problem were discovered by Feigenbaum in numerical studies of simple mathematical models. For example, if θ_n is the temperature gradient at which the nth period doubling appears and if θ_c is the threshold value for chaotic motion, then $\theta_c - \theta_n \propto \delta^{-n}$, where δ is a universal number: $\delta = 4.6692016091 \ldots$. The renormalisation group again provides a beautiful description of this universal scaling behaviour.

Fully developed *turbulent flow* also involves many length scales in an intrinsic way, although here the applicability of the renormalisation group remains a matter of some controversy. Turbulent flow is created in a fluid when it is forced to flow past a body with high enough speed. For example, if fluid flows through a pipe at low speeds, the resistance to flow is determined by the intrinsic fluid viscosity. But, at high enough speeds, unstable vortices are generated; these interact strongly to produce turbulent eddies in a cascade process, in which eddies are created on successively smaller length scales. This process is limited by some dissipation length scale, where the energy is converted into random molecular motion. Under these circumstances, the resistance to flow and hence the *effective* viscosity is determined by the collective interaction of the eddies on many length scales. The fundamental problem is to describe this process in the framework of the Navier Stokes equation for viscous fluid flow. A key development was made by Kolmogorov in 1941: he obtained a scaling solution of the

Navier Stokes equations, in which the energy density associated with motion of wave-vector $\mathbf{q}$ (i.e. of length scale $1/q$) is proportional to $q^{-5/3}$. This power law appears to describe rather well the experimental results in the inertial range (between the large-scale eddies produced by the 'stirring' forces and the microscopic dissipation scale) although it remains unclear whether it is exact. The value $\frac{5}{3}$ has also been obtained in some (approximate) renormalisation group studies. This is a technologically important area where advances in fundamental understanding could have significant impact on computer-aided design.

Finally it is appropriate to mention the phenomenon known as the *Kondo effect*, after the Japanese physicist Jun Kondo. The electrical resistance of a metal depends upon its temperature. As the temperature is lowered the thermal vibrations of the atoms become weaker and their contribution to the electrical resistance falls. In a nonmagnetic metal containing a small concentration of magnetic impurities (for example copper containing some iron) it is found that the resistance actually reaches a minimum value and then increases on further cooling. Qualitatively it is clear that this effect should be attributed to the contribution to the electrical resistance made by processes in which electrons are scattered by the magnetic impurities. However, to account quantitatively for the effect one must again confront a many-scale phenomenon. The difficulty here is most naturally thought of in terms of energies rather than lengths. The electrons responsible for the current flow typically have kinetic energies of the order of a few electron volts; the *change* in their energies, resulting from scattering processes in which the orientation of the spin of a magnetic impurity is flipped, is the spin-flip energy, of the order of 10^{-4} eV. The calculation of the electrical resistance (and other properties) thus requires one to handle all energy scales between these extremes. This challenge was first met, in 1974, by K. G. Wilson with the renormalisation group methods he pioneered in the phase transition context.

The fundamental forces: phases of gauge theories

There are two related concepts which we wish to describe in the remaining paragraphs. The first is that the relativistic quantum theories of the interactions of elementary particles are in fact examples of critical phenomena. The second is that the different fundamental forces, with their very disparate properties, can be understood as different possible phases of models with the *same* underlying structure – the gauge theories discussed in Chapter 17 by Taylor.

As a starting point, let us observe that a full description of the interactions of elementary particles requires both *relativity*, because the collisions and decays can involve very energetic particles whose velocities are close to that of light, and *quantum mechanics*, because these processes also occur at subatomic distances. The formalism which incorporates both of these requirements is called relativistic quantum field theory. The prototype for such theory is quantum electrodynamics, which describes the interactions of electrons with the electromagnetic field. In this theory, Maxwell's equations for the electric and magnetic fields are quantised; in the conventional formalism, at the heart of the quantisation are creation and annihilation operators which can create and destroy the fundamental quanta of the field, photons. The electrons are treated on a similar footing: they (and their antiparticles, positrons) are the fundamental quanta of the electron field. It is this creation operator formalism which expresses the equivalence of energy and mass implicit in Einstein's equation $E = mc^2$, manifested for example in the creation of real particles in a collision process at sufficiently high energy.

Such theories typically cannot be solved exactly if the fundamental field quanta interact with one another, which is the only case of real interest. In seminal papers in the late 1940s and early 1950s, Feynman, Schwinger, Tomonaga and Dyson showed how one could calculate the effects of the interactions as a power series in the coupling strength. In order to make sense of this perturbation theory, the then mysterious process of 'renormalisation' had to be introduced because each order of perturbation expansion was found to be infinite. These divergences could be swept away by introducing 'renormalised' couplings and particle masses. In order to place this recipe on a sound mathematical footing it was necessary to introduce a cut-off Λ on the allowed momentum; at the end of the calculation, Λ could be removed by letting it tend to infinity. In spite of the very formal-looking prescription, the procedure was found to work exceptionally well for quantum electrodynamics, where the expansion parameter is the fine structure constant $\alpha = e^2/2\varepsilon_0 hc$ (e is the electron charge, ε_0 is the electrical permittivity of free space, h is Planck's constant and c is the velocity of light) whose value, roughly $1/137$, is small, so that the perturbation expansion converges rapidly in practice. A 'renormalisation group' formalism was also introduced at this time, by Stückelberg and Petermann and by Gell-Mann and Low, in an attempt to understand high momentum scattering processes. Their approach was based on the observation that there is actually a family of ways to remove these divergences, all ways describing in the end the same physics.

The seeds of our present deep understanding of the renormalisation process were also sown by Feynman at that time, following earlier work of Dirac. Feynman showed that there is another way to formulate the same relativistic quantum field theory, which makes no reference to creation operators. His way is based on the idea that, in quantum theory, almost anything can happen; the key is to get the correct form for the likelihood of the process. Consider for example the passage of an electron through a pair of closely spaced narrow slits. In the Dirac–Feynman picture, an interference pattern is formed on a screen because of the superposition of the possible paths of the electron through the slits. A path here is described by the function $q(t)$, which gives the position q at time t. Similarly, in this formalism, a relativistic quantum field theory is obtained by summing over all histories of the field, which we denote generically by $\phi(\mathbf{x},t)$, where ϕ represents the field value at the point $\mathbf{x}$ at time t. This sum over histories is in fact extremely close to the configuration sum for the field formulation of statistical mechanics which we discussed in Section 8.5. In particular, it is also a functional integral.

It is illuminating to compare the structure of the functional integral in the two contexts. In equilibrium statistical mechanics there is no time and the natural measure of distance is the length l defined by the 'invariant' $l^2 = x_1{}^2 + x_2{}^2 + \ldots + x_d^2$; the corresponding quantity in relativity turns out to be $x_1{}^2 + x_2{}^2 + \ldots + x_d^2 - c^2 t^2$, where d denotes the spatial dimension in both cases. Suppose now we take units in which the velocity of light c is 1 and replace t by $x_{d+1} \equiv it$, where $i = \sqrt{(-1)}$. Then $x_1{}^2 + x_2{}^2 + \ldots + x_d^2 - c^2 t^2$ becomes $x_1{}^2 + x_2{}^2 + \ldots + x_d^2 + x_{d+1}^2$, just like length (squared) in $d+1$ spatial dimensions. Remarkably, it turns out that this 'imaginary time' relativistic quantum field theory is indeed mathematically equivalent to statistical mechanics in one higher dimension; quantum fluctuations mandated by Heisenberg's uncertainty principle are the analogues in the former of thermal fluctuations in the latter.

We are now in a position to appreciate the 'renormalisation' procedure and its relationships to critical phenomena. In statistical mechanics, nature presents us with a microscopic length scale, such as a crystal lattice spacing. Cooperative phenomena near a critical point create a correlation length. In the limit of the critical point, the ratio of these two lengths tends to infinity. Similarly, for all physical relativistic quantum theories to date, we have noted that in the first instance it is necessary to introduce an artificial upper limit Λ on the allowed momentum equivalent to an ultra-microscopic length scale, $h/(2\pi\Lambda c)$. The philosophy is that real physics is recovered in the limit in which this artificial scale is small compared to the natural physical scale of the problem defined by the Compton wavelength of the particles in the theory, $h/(2\pi mc)$, where m is the particle mass. Accordingly, we must tune the ratio of these two length scales, Λ/m, towards infinity. It is in this sense that all relativistic quantum field theories have to describe critical points, with associated quantum fluctuations on arbitrarily many length scales.

Having exposed this relationship between quantum field theory and critical phenomena, let us turn now to the specific problem of gaining a unified picture of the fundamental forces as gauge theories.

Quantum electrodynamics, the theory of the electric and magnetic interactions between electrons and photons, is the prototype gauge theory. At the heart of what a gauge theory actually is are deep concepts of symmetry and geometry which it is inappropriate for us to try to review here. For our purposes it is adequate to say that the basic equations of all gauge theories appear to demand the existence of at least one 'gauge boson' which

(i)   is massless and thus moves with the velocity of light,
(ii) has unit intrinsic spin,
(iii) acts as the mediator of the force between the 'charged' particles of the theory, producing a Coulomb-like force decaying as the inverse square of particle separation.

In quantum electrodynamics, the gauge boson is the photon, and its exchange is responsible for the familiar Coulomb force between charges.

Ostensibly, gauge theories would seem to be ruled out as candidates both for the strong nuclear force which binds quarks inseparably together to form the nuclear particles – proton, neutron, pion, etc., and for the weak force, which is responsible for decay of radioactive nuclei: neither has associated massless gauge bosons and the forces are both short-range, roughly the nuclear size 10^{-15} m for the strong force between protons, and even shorter range for the weak force.

The key idea which unlocks this dilemma is the realisation that gauge theories can exhibit different phases depending on the value of parameters, just as magnetic materials can exhibit two different phases with distinct properties, according to the temperature. Quantum electrodynamics is in one possible phase, which we call now the Coulomb phase. The integration of the electromagnetic and weak forces in the Glashow, Salam, Weinberg model is achieved by the weak sector of the theory existing in a different phase, now called the Higgs phase, because the only way we can realise this phase theoretically necessitates the existence also of a spinless particle, the Higgs boson. In this phase the associated gauge bosons appear as the massive W and Z particles. The strong nuclear force is realised in yet another phase of a gauge theory (quantum chromodynamics), called the confining phase. In this phase, the matter particles (in this case, quarks) are permanently bound in overall 'neutral' (in this case, 'colour neutral') combinations, the observed nuclear particles; the gauge bosons (the eight gluons in this case) are also permanently bound into colour neutral states by their mutual forces.

In order to make these claims more plausible, we show in figure 8.12 the phase diagram obtained for a model of charged spinless particles interacting with photons ('scalar electrodynamics'). This diagram is obtained by the same kind of Monte Carlo methods as in statistical mechanics, by approximating space-time by a finite box of discrete points. This lattice is of course a mathematical artefact; the previous discussion implies that we must eliminate it by studying the 'lattice gauge theory' in the neighbourhood of its critical points. These phase diagrams reveal that this particular gauge theory can exist in all three of the above phases.

In many respects the confining phase appears to be the most mysterious, because in it none of the fundamental quantities, the quarks and gluons, can exist as isolated particles. In fact we now understand that this is in reality the most natural phase of

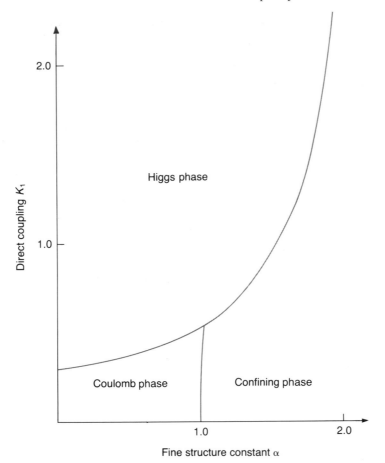

Figure 8.12. Phase diagram for scalar electrodynamics. Scalar electrodynamics is the theory describing the electric and magnetic interactions of charged scalar particles. The phase diagram shown is obtained by numerical simulation on a four-dimensional lattice approximating space-time. The horizontal axis represents the 'fine structure constant' $\alpha = e^2/(2\varepsilon_0 hc)$, where e is the electric charge of the scalar particle; the physical value of α is $1/137$. At strong charge-coupling (i.e. at large values of α) the theory is in the confining phase in which charged particles are permanently bound in 'molecules' which have zero net charge. In this phase it is impossible to isolate charged particles. The vertical axis is the direct coupling K_1 between particle fields on nearest neighbour sites, and is a measure of the scalar particle mass. At low values of α, this parameter determines whether the theory is in the Coulomb phase or the Higgs phase.

gauge theories. Indeed, for the theory of the gluon interactions responsible for the strong nuclear force, there is now good evidence that it is the *only* possible phase of the theory; four-dimensional space-time is its lower critical dimension, in which no phase transition can occur, at any coupling (cf. the one-dimensional Ising model). Fortunately quantum electrodynamics, like scalar electrodynamics, can exist also in a Coulomb phase; this is what makes possible the wonderful world of chemistry and biology in which we can live.

Further reading

A transcript of the lecture delivered by K. G. Wilson on the occasion of the presentation of the 1982 Nobel Prize in physics is to be found in *Reviews of Modern Physics* (1983), vol. 55, p. 583. This article sets out the basic ideas of the renormalisation group method, sketches their historical context, and gives an extensive list of relevant references covering core research work, review articles and books.

9 Low temperature physics, superconductivity and superfluidity

Anthony Leggett

9.1 Low temperatures: order and disorder

At first sight, low temperature physics is a narrow and parochial branch of science which deals only with a very small corner of the physical universe. Suppose for example that we were to plot temperatures vertically on this page according to the usual absolute (Kelvin) temperature scale, so that absolute zero – the ultimate limit of 'coldness', beyond which matter cannot be cooled even in principle – falls at the bottom of the page, the freezing point of water (273 degrees absolute, 273 K) almost three-quarters of the way up and its boiling point (373 K) at the top. Then the subject-matter of low temperature physics, by most accepted definitions, would correspond to perhaps the last two lines of the page, and many of the phenomena I shall be talking about in this article would occur at temperatures so close to the bottom of the page that it would need a magnifying glass to distinguish them from zero. By contrast, on this scale even the temperature of an ordinary domestic oven would lie a foot or two off the top of the page, and the highest temperature achieved on Earth (about 2×10^8 K, in a nuclear explosion) would be something like fifty miles away! So one might think that the tiny fraction of the temperature scale occupied by 'low temperature' physics would be only of extremely specialist interest.

Needless to say, this way of looking at things is totally misleading. It is much more sensible to plot temperatures on a *logarithmic* scale, on which each factor of ten one goes down (or up) is indicated by a constant interval on the scale. Such a scale is shown, for temperatures around and below room temperature, in figure 9.1, with some of the phenomena I shall be talking about marked. I have also marked the approximate dates at which the points marked by arrows on the left were attained in bulk matter; it is apparent that, on a logarithmic scale, more progress has been made in the last twenty-five years than in the whole of the previous history of mankind. Probably no other frontier area of physics can claim a comparable rate of progress in extending its area of coverage. (If the comparison is made in terms of 'decibels per dollar' the contrast is even more striking!)

Please note particularly one feature marked on the scale: the temperature, roughly 3 K, of the cosmic black-body radiation. This is believed to be the present temperature (in so far as one can be defined) of outer space, and according to almost all cosmologies in current fashion the universe as a whole was never cooler than this (see Chapters 3 and 6 by Guth and Steinhardt and by Longair). If this is correct, then when we cool bulk matter to temperatures below the point marked we are in effect creating new physics which Nature herself has never explored: indeed, if we exclude the possibility that on some planet of a distant star alien life-forms are doing their own cryogenics, then the new phases of matter which we create in the laboratory at low temperatures *have never previously existed in the history of the cosmos.*

Why do we expect new phases of matter to occur at low temperatures? The answer lies in the competition of order and disorder (or, if you will, of energy and entropy). Consider for example the atoms in a piece of iron, which can be thought of for the purposes of the present discussion as tiny bar magnets. The forces between the magnets will tend to align them all in the same direction; on the other hand, if the iron is in equilibrium with its environment at some (absolute) temperature T, it will be continually receiving random inputs of energy which are proportional to T, and this will tend to disturb the alignment. To put it in slightly more technical terms, it turns out that at any temperature T the system will behave in such a way as to minimise the quantity (the so-called 'free energy')

$$F = E - TS, \tag{9.1}$$

where E is the mean energy associated with the state of the system and S is its entropy, which is a measure of the 'disorder' associated with it; neither E nor S depends directly on temperature. We can see that at sufficiently high temperatures the second term is more important and the system will therefore choose its state so that its entropy (disorder) is a maximum; since it is clear, intuitively speaking, that the state of maximum disorder is one in which the atomic magnets are oriented at random, it is this state which occurs (see figure 9.2) and the piece of iron has no net magnetisation. On the other hand, if we cool the system to a sufficiently low temperature, the second term will be negligible compared to the first and the system will choose its state so as to minimise its energy; for material like iron, this is achieved when all the atomic bar magnets are oriented parallel to one another (figure 9.2). Thus, the piece of iron as a whole behaves like a single bar magnet. At somewhat higher (but still low) temperatures there will be some thermal disorder, so that the atomic magnets will not all point exactly parallel, but there will still be a tendency to align and the iron will still possess an overall magnetisation. It is only when we reach a definite temperature (the so-called Curie temperature) that the overall magnetisation finally vanishes and the iron becomes totally disordered*. The details of how this happens are discussed in Chapter 7 by Thouless and in Chapter 8 by Bruce and Wallace.

*This discussion is rather oversimplified. Actually, a bar of iron will often not show a net magnetisation even below the Curie temperature. This is because, in the absence of a strong external magnetic field, it tends to form 'domains', that is, regions which each individually have a net magnetisation, but in different directions so that the total magnetisation of the bar averages to zero.

The transition from disorder at high temperatures to order at low temperatures can be exemplified in many other phenomena. For example, if we put a large number of water molecules together in a bucket at room temperature, we get of course a liquid, in which the molecules are arranged more or less at random relative to one another, because such a state maximises the disorder. But if we cool the system to 273 K (0° C) it freezes: molecules arrange themselves in a regular crystalline array, because by doing so they minimise their energy of interaction. Another well known example is that of a 'binary alloy', that is a mixture of equal numbers of atoms of two different kinds, such as copper and zinc, where the interatomic forces are such that each atom prefers to sit near to an atom of the opposite species rather than one of its own (i.e. it has a lower energy when it does so). In such a case, at high temperatures, the atoms will be totally disordered, with each copper atom being as likely to be found next to another copper atom as to a zinc atom, while at lower temperatures we will find an ordered array, the two species alternating in a regular manner. All these examples are illustrated in figure 9.2.

Figure 9.1. The logarithmic temperature scale. Each of the equal intervals marked represents a factor of ten in the absolute temperature (indicated in kelvin). The dates on the left are the approximate dates at which the corresponding temperatures were first obtained.

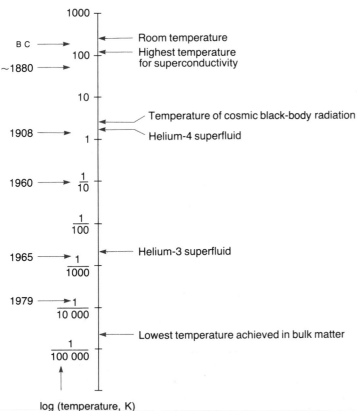

Figure 9.2. Order, disorder and temperature. Each of the pairs of diagrams in the same row schematically represents the state of the same physical system at high temperatures (on the left) and at low temperatures (on the right). In each case, the low temperature phase is much more 'ordered'.

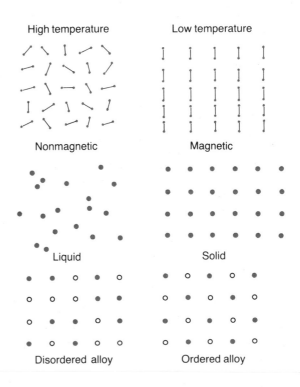

Although we have seen that ordered phases tend to occur 'at low temperatures', the point at which the transition from the disordered to the ordered phase occurs need by no means fall within the region conventionally denoted by 'low temperature physics'. In fact, none of the above examples does so: as we all know, water freezes at a temperature not so different from room temperature, the Curie temperature of iron is about 1000 K absolute (about three and a half times room temperature in absolute units) and the temperature of the 'order–disorder' transition in brass (copper–zinc alloy) about three times room temperature. So 'low temperatures' in the conventional sense are not a necessary precondition to see ordered phases. What then do we gain by going, say, to 20 K and below? Well, of course, in the first place we would expect to see many more kinds of ordered phase. Crudely speaking, the (absolute) temperature at which the transition to order takes place is usually proportional to the energy of interaction, or force, which is responsible for favouring the type of order in question. For example, certain dilute magnetic salts contain, like iron, atomic magnets which interact with one another, but (because they are far separated from one another) the forces of interaction are only about one-millionth of the strength of those in iron: consequently the transition from disorder to order occurs at about one-millionth of the Curie temperature of iron, that is at about one-thousandth of a degree absolute. There are many other examples in which the forces of interaction are so weak that the ordered phase only occurs at very low temperatures.

However, this feature as such would not make the low temperature region *qualitatively* different from the rest of the subject matter of physics; most of the ordered phases which occur below 20 K show qualitative features quite similar to those which occur at room temperature or above. What really makes the low temperature region unique is the occurrence of a special class of ordered phases which (at least until very recently) have never been seen at all elsewhere, the so-called 'quantum liquids', or more precisely the subclass of these which goes under the generic name of 'superfluids' (which in this context includes superconductors). Quantum liquids are the class of systems which shows a peculiarly quantum-mechanical kind of order, and superfluids are the subclass of these which shows it in a peculiarly striking way: in particular, they show the effects of quantum mechanics on a macroscopic scale. A very substantial part of all research in the low temperature area is devoted to these systems, and from now I shall concentrate exclusively on them.

9.2   What is a quantum liquid?

To understand the concept of a 'quantum liquid' we must first recall one of the basic ideas of elementary quantum mechanics, namely that a microscopic particle, such as an electron, proton or neutron, is in some sense represented by a wave. This feature was originally inferred from a complex of experiments which showed that electrons could undergo diffraction from periodic crystalline lattices in a way almost identical to that of X-rays; nowadays, of course, there is an overwhelming mass of evidence for the hypothesis, the most direct and spectacular of which are probably experiments done with neutrons using a device called a neutron interferometer (see figure 9.3). In this device a neutron beam is treated almost exactly as a light beam is treated in the familiar Young's slits experiment; that is, it is split into two beams which are allowed (in effect) to pass through well separated slits and are then recombined at a screen. If the number of neutrons arriving at the screen is plotted as a function of transverse position, the resulting pattern of distribution looks almost identical to the intensity distribution (interference pattern) of light in the Young's slits experiment. Moreover, just as in the latter experiment, if the beam from either of the two slits is blocked off, the interference pattern disappears and we get a more or less uniform distribution. The most natural interpretation is that, just like light, the neutron beam is associated with a wave field which is transmitted through *both* slits, and that the two waves can undergo constructive or destructive interference when they recombine. We can obtain a quantitative as well as qualitative understanding of this and similar experiments if we postulate the *de Broglie relation* between the mass m and velocity v of the particle in question and wavelength λ of the associated wave; if we neglect relativistic effects (an approximation which is almost always valid in a low temperature context) this reads

$$\lambda = h/mv, \qquad (9.2)$$

where h is Planck's constant (about 6.6×10^{-34} Js). This relationship is believed to be valid not just for electrons, protons and neutrons but for atoms, molecules and any heavier objects built out of them, provided that the quantity m is interpreted as the total mass of the object in question.

Figure 9.3. Schematic picture of a neutron interferometer. The observed behaviour of the intensity (number of neutrons received) at the counter shows clearly the effects of interference between the two beams (indicated by broken lines), indicating that each neutron is behaving like a wave.

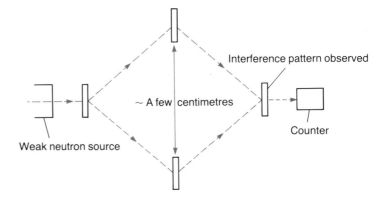

An obvious question now arises: why do we not see the effects of this 'wave–particle duality' routinely in everyday life? To answer this, let's recall an elementary feature of the theory of wave motion which is well illustrated by the everyday observation that we can hear round corners (e.g. through an open door) but cannot similarly see round them. The reason for this (more accurately, part of the reason) is that when a wave approaches an aperture in an otherwise opaque screen (e.g. the open doorway) its behaviour depends critically on whether its wavelength λ is much shorter than the width a of the aperture, or is comparable to it (or greater). In the first case, the wave crests simply travel through the aperture undeflected and continue to propagate in their original direction of travel; see figure 9.4(a). In other words, the wave behaves almost exactly as would a point particle incident on the same slit (this is called 'the limit of geometrical optics'). When λ is comparable to (or greater than) a, on the other hand, the wave suffers diffraction, that is it is dispersed in all directions (figure 9.4b). For a door of ordinary dimensions (say one metre in width) a light wave, with a wavelength $\sim 10^{-5}$ cm, behaves in the first way, but a sound wave (wavelength ~ 1 m) behaves in the second way. (Of course in practice there are further complications associated with the difference in absorption and reflection mechanisms for the two kinds of wave, but they are not relevant here.)

The upshot of this is that, even if a particular kind of entity is 'really' a wave, it will appear to behave much like a classical particle unless its passage is affected by apertures, or obstructions, whose dimension a is comparable to, or smaller than, the appropriate wavelength λ. (This is why Newton was able to hold a particle theory of light – in his day, it would have been extremely difficult to do experiments with apertures anywhere near 10^{-5} cm.) Now, in the case of an entity with finite mass, such as an electron or atom, we saw that λ was related to the velocity in equation (9.2). If the system is in contact with an environment in thermal equilibrium at temperature T, then the typical velocity is given by equating the kinetic energy, $m\mathrm{v}^2/2$, to the mean thermal energy $3k_{\mathrm{B}}T/2$ (where k_{B} is Boltzmann's constant, about 1.4×10^{-23} J K^{-1}). Putting all these conditions together, we find that the condition to see the wave-like behaviour of 'particles' is roughly

$$T \lesssim h^2/3mk_{\mathrm{B}}a^2 \equiv T_0. \qquad (9.3)$$

How should we estimate the quantity a in this expression? At an atomic level, what the electrons and atoms are diffracted by is other electrons and atoms, so we should presumably take a to be of the order of the mean spacing between them. This is typically of order of 2 or 3 Å (1 Å $= 10^{-10}$ m) for liquids or solids and much larger for gases. The temperature T_0 so obtained is called the 'degeneracy temperature' of the system in question, and below it the system is said to be 'degenerate'.

Thus, putting in numerical values of Planck's constant h and Boltzmann's constant k_{B} in equation (9.3), we see that for electrons (mass $\sim 10^{-30}$ kg) the condition (9.3) is fulfilled, in liquids or solids, for all temperatures at which those phases

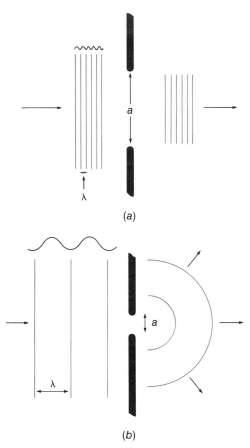

(a)

(b)

Figure 9.4. A wave propagating through an aperture. (a) When the wavelength λ is much smaller than the aperture width a, the wave simply travels onward in a straight line, just as would a particle passing through the same aperture. (b) In the opposite case (λ of order a or longer) we get the phenomenon of diffraction.

exist. (In a gas, electrons are closely tied to individual atoms, and move as part of them, except at temperatures too high for equation 9.3 to be fulfilled.) For atoms the condition is much more stringent: noting that the mass of an atom is roughly A times the proton mass ($\sim 1.6 \times 10^{-27}$ kg), where A is the atomic number, we find that condition (9.3) reduces roughly to

$$T \lesssim 50/A \qquad (9.4)$$

for a solid or liquid, where T is the temperature in kelvin. For atoms in the gas phase condition (9.3) can almost never be fulfilled, because as we lower the temperature the 'saturated vapour pressure' of the gas, or equivalently the mean density of gas in equilibrium with the liquid or solid phase, rapidly decreases and hence the right hand side of the inequality (9.3) decreases (because of the factor a^{-2}) much more rapidly than the left hand side. There is one striking exception to this rule, which has been discovered only in the last few years: a gas of hydrogen *atoms* (not molecules!) appears to be relatively stable against recombination at low temperatures in a high magnetic field, and in this case there is no liquid or solid phase, so the density can be kept constant as we lower the temperature. However, although there are high hopes that condition (9.3) can eventually be reached in this system, this has not been achieved to date. I will return to this intriguing system later.

Condition (9.3) is, essentially, the condition for the entities we are talking about to show the effects of quantum mechanics in a marked way; as we have just seen, it is always satisfied, in liquids and solids, for the electrons, and is satisfied for atoms provided the temperature is low enough that condition (9.4) is met. The fact that the electrons/atoms behave quantum mechanically has many experimental consequences: for example, the specific heat associated with the electrons in a liquid or solid is very much less than the classically expected value at all temperatures, and that associated with the vibrations of the atoms in solids, though roughly equal to the classical value at high temperatures, falls far below it at low temperatures.

To obtain a 'quantum liquid', however, we need one further ingredient: quantum indistinguishability. It is a fundamental principle of quantum mechanics that the basic ingredients of matter cannot be 'tagged' or labelled, and therefore that the state of the system in which (say) electron 1 is at point 1 and electron 2 at point 2 is completely indistinguishable from the state in which 1 is at point 2 and 2 at point 1. (This can be made plausible – though the argument is not rigorous – by the observation that, whereas it would make sense to keep track of the identity of two particles colliding with one another, if two wave groups run through one another, there is no meaning to the question 'which is which?'.) It turns out (here one needs to invoke some rather subtle considerations from elementary particle theory) that this factor profoundly affects the behaviour of an assembly of particles of the same kind, and the effect depends critically on the kind of particle involved, and in particular on its intrinsic spin (angular momentum). In fact, all elementary particles known to us have an intrinsic spin which, measured in units of $h/2\pi$, has either an integral or a half-integral value: in the former case the particle is called a Bose particle (or 'boson'), in the latter a Fermi particle (or 'fermion'). The electron, the proton and the neutron all have spin $\frac{1}{2}$ and are therefore fermions: the best known elementary particles which are bosons are the photon (light quantum), which has spin 1, and the pi-meson, with spin zero (see Chapter 14 by Close). However, it also turns out that complex systems made up of elementary particles must also have either integral or half-integral spin and can therefore be similarly classified: in fact, a system containing an odd number of fermions has half-integral spin and is therefore a fermion (no matter how many bosons it also contains), while one containing an even number has integral spin and is a boson. (This result would follow at once if we assumed that the spins of the component particles are always parallel or antiparallel to one another, since then,

for example, it would follow immediately that an odd number of particles of half-integral spin must itself have half-integral spin. In fact, of course, the spins must be added like vectors, but the correct quantum-mechanical treatment produces the same conclusion.) Thus, for example, an atom of ordinary hydrogen (one proton and one electron) is a boson, while an atom of deuterium (one proton, one neutron and one electron) is a fermion.

A large assembly of identical fermions in thermal equilibrium will obey *Fermi statistics*, and a large assembly of bosons will similarly obey *Bose statistics*: the effects are often spectacularly different in the two cases. For example, let us consider the simplest possible case, in which a large number N of particles move freely and independently in a box of large volume V. Suppose, first, that we described these by classical mechanics. Then each particle could have any (vector) velocity v, and correspondingly any energy ($mv^2/2$). Suppose we assume the system is in thermal equilibrium at some temperature T, and ask for the number of particles which have a velocity vector in a certain small range which corresponds to an energy E. The answer, as given by standard statistical mechanics, turns out to be

$$n(E) = \text{const.} \times e^{-E/k_B T} (= \text{const.} \times e^{-mv^2/2k_B T}). \quad (9.5)$$

This is plotted, as a function of energy E, in figure 9.5(*a*). (The standard 'Maxwell distribution' usually quoted in textbooks has an extra factor of v^2 in front. This is because it is the answer to a slightly different question than the one we have just asked.)

Next, suppose that we try to describe the particles by quantum *mechanics* but imagine that they are somehow still distinguishable ('classical statistics'). The principal difference, now, is that the velocity and hence the energy cannot take any value but are 'quantised' that is can only take discrete values: one easy way to see this is to observe that the 'de Broglie wave' which represents the particle in quantum mechanics must 'fit into' the box, i.e. the wavelength λ must be such that the length L of the box is an integral number of half-wavelengths. Then from equation (9.2) it follows at once that the velocity v is restricted accordingly, and hence so is the energy $mv^2/2$. So there are only certain allowed states i, with particular energies E_i (note, though, that under certain circumstances two different states may have the same energy). We may then ask the question (which is the quantum analogue of the question asked about the classical system): If the system is in thermal equilibrium at temperature T, what is the average number n_i of particles in a particular quantum state i which has energy E_i? Again, statistical mechanics gives a straightforward answer:

$$n_i = \text{const.} \times e^{-E_i/k_B T}, \quad (9.6)$$

i.e. exactly the same as in equation (9.5) except that the energy E can only take the discrete values E_i. Since for a large volume the spacing between the allowed values of E_i is extremely tiny, it is convenient for purposes of plotting the graphs (only!) to regard the energy E as still a continuous variable; however, it should be clearly borne in mind that we are plotting the number *per quantum state*, not the number in unit energy range. (The two are different because the number of allowed quantum states per unit energy range is in general itself a function of energy. This is the quantum version of the point about the classical Maxwell distribution mentioned above – see the sentence in parentheses below equation 9.5.) Thus, the result (9.6) can also be represented by figure 9.5(*a*). The average energy per particle is rigorously $3k_BT/2$ for the classical distribution (9.5), and is negligibly different from this for the 'distinguishable' quantum distribution (9.6) except at temperatures which for any reasonably macroscopic volume are so low ($< 10^{-11}$ K) as to be of no practical interest.

For N indistinguishable particles the situation is radically different, and depends critically on whether they are bosons or fermions (obey Bose or Fermi statistics). In the case of fermions it turns out that no more than one particle can occupy a particular quantum state (the so-called 'Pauli exclusion principle'). Since the number of states in any given range of energy, though very large, is finite, this means that at zero temperature (when the system would like to have the lowest possible energy, cf. equation 9.1) all the lowest N states are filled: that is, if the energy of the Nth state is E_f (the so-called 'Fermi energy') then all states i with energies E_i below E_f are filled with one particle each, while all those with energy above E_f are empty. At finite temperature T such that k_BT is much less than E_f the distribution is of the general form shown in figure 9.5(*b*): some states close to, but below, the Fermi energy are now partly empty, while some close above it are partly filled. As the temperature is gradually raised this 'smearing' of the zero temperature distribution gradually increases, until at very high temperatures ($k_BT \gg E_f$) we recover the form given in figure 9.5(*a*). (The temperature above which the effects of indistinguishability are lost is, in order of magnitude, the same as the one appearing on the right hand side of equation 9.3, and indeed for the simple model considered the above argument is really just a more rigorous version of the earlier one which led to equation 9.3. But in the more general case the two arguments are different, cf. below.)

Whereas Fermi statistics cannot tolerate more than a single particle per quantum state, Bose statistics actually prefers a large number in the same state. The average number of particles per state n_i is shown as a function of energy E_i for Bose particles in figure 9.5(*c*): note that at low energies n_i is *greater* than in the 'distinguishable' case, figure 9.5(*a*). Figure 9.5(*c*) actually applies for intermediate temperatures. At very high temperatures it goes over into the curve of figure 9.5(*a*) just as for fermions. At *low* temperatures, if the number of particles in question is fixed, something much more interesting happens, namely the phenomenon of 'Bose condensation' which I shall discuss in the next section. Note that in all of figures 9.5(*a*)–(*c*), the number of particles in a single quantum state is 'of order one' (not 'of order N').

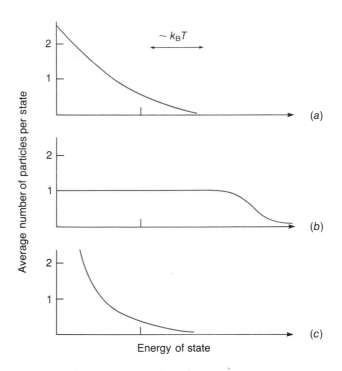

Figure 9.5. The average number of particles per state at a given temperature for particles obeying (*a*) classical, (*b*) Fermi and (*c*) Bose statistics. Note that in the Fermi case there is at most one particle per state, and that the number of particles in the lower energy states is greater in the Bose case than in the classical one.

In the light of the above considerations we would expect, at first sight, that the behaviour of matter in bulk at low temperatures should depend critically on whether the basic units composing it are bosons or fermions. But in most cases this does not seem to be true: for example, the behaviour of a piece of diamond at low temperatures appears to be much the same whether the atoms composing it are of isotopic number 12 (six protons, six neutrons and six electrons, hence a boson) or of mass 13 (six protons, seven neutrons and six electrons, hence a fermion), and what differences there are are easily explained in terms of the different atomic mass in the two cases. So, what has gone wrong? The crucial point is that the effect of the different 'statistics' (Fermi or Bose) can be dramatically seen *only to the extent that the particles can easily change places* (as they could in the simple situation envisaged above, where they all move freely and independently in a box). If they never change places, then they as it were never find out that they are supposed to be indistinguishable and modify their behaviour accordingly; alternatively, one could say that to the extent that they never change places, it is possible to 'tag' a particle by its location just as in the classical case.

Consider first the case of atoms. In most *solids*, the atoms are closely tied to particular sites (this is true even for non-crystalline (glassy) solids, where the sites do not form a regular crystalline lattice), and the probability of two atoms changing places is very small indeed. Only for the very light elements (hydrogen and helium) is there any appreciable likelihood of this happening, and even then the effects of indistinguishability in the solid state, though they can be seen, are not spectacular. In *gases*, by contrast, the atoms change places very easily, but for the reason given earlier the density is almost always so low that no quantum effects can be seen anyway. Thus the only serious candidates among the gases to see the effects of indistinguishability are atomic hydrogen (a Bose system, see above) and atomic deuterium (a Fermi system). What one really needs is a *liquid*, i.e. a system which is sufficiently dense that quantum effects can be seen in the first place but also sufficiently mobile that atoms change places fairly readily: a system satisfying both conditions (at temperatures satisfying condition 9.3), is called a *quantum liquid*. Unfortunately, all the known elements (and compounds) except one freeze at temperatures either well above or comparable to the temperature T_0 in equation (9.3) above which quantum effects are negligible and are therefore not candidates. The one exception is the element helium, which because of its low atomic mass and weak interatomic forces does not freeze under its own vapour pressure (to freeze it, we need to raise the externally imposed pressure to about thirty atmospheres); thus, as far as we know, it remains liquid down to absolute zero temperature. It is our great good fortune that Nature has provided us with two different stable isotopes of helium: the commonly occurring one, helium-4, contains two protons, two neutrons and two electrons and is therefore a boson, while the rare light isotope, helium-3 (largely produced by the beta-decay of tritium from nuclear reactors) contains two protons, one neutron and two electrons and is therefore a fermion. (A third isotope, helium-6, does exist but decays radioactively with a half-life of the order of one second, so that it is very difficult to do experiments with it.) We should expect that a liquid made up of helium-4 atoms should behave in a spectacularly different way from one composed of helium-3 atoms, and we shall see below that this is indeed the case; thus, the helium isotopes are our paradigm of a quantum liquid and are in effect a perfect laboratory for examining the effects of quantum (Fermi/Bose) statistics.

Turning from atoms to electrons, we note that in an *insulator* the electrons are tied, crudely speaking, to particular atoms and hence can change places readily only within the atom. This, of course, gives rise to many interesting effects on the structure of the atom and its energy levels, but it is not directly relevant to the effects we are interested in here, which are a consequence of the indistinguishability and exchange of a *macroscopic* number of particles. Thus the systems of interest to us are the electrons in solid or liquid *metals*, where they can migrate more or less freely (in gases, as we observed earlier, electrons are closely tied to particular atoms and have to be regarded as part of them). As we saw above, these electrons are 'degenerate' at all temperatures up to the boiling temperature.

If we are willing to leave the terrestrial laboratory, there is at least one more system which deserves to be called a quantum liquid, namely the neutrons which are believed to compose the core of the so-called neutron stars (see the chapter by Longair). Because of the very high density, the degeneracy temperature in this case is enormous ($\sim 10^{10}$ K). Despite this, it is believed that the behaviour of the neutrons in the degenerate regions is qualitatively very similar to that of laboratory quantum liquids, and in particular that, being fermions, they may show phenomena similar to the superfluidity of liquid helium-3, which is discussed below. Unfortunately, for obvious reasons controlled experiments on this system are in practice out of the question, so I shall not discuss it further here.

In table 9.1 is a list of the laboratory quantum liquids, with the type of quantum statistics they obey and the degeneracy temperature in degrees absolute (to an order of magnitude). Although the general properties of quantum liquids as such are well worth study, the most spectacular (and unique) phenomenon they can display is the complex of effects which goes under the name of *superconductivity* for a charged system (electrons) and *superfluidity* for an uncharged one (atoms). Not all quantum liquids display these effects: the electrons in many solid metals (and all liquid ones) do not enter the superconducting state, and although atomic hydrogen is confidently predicted to undergo a superfluid transition at high enough densities and low enough temperatures, the necessary combination is still some way from being achieved. (In atomic deuterium, if superfluidity occurs at all, it is still further out of reach.) Moreover, even in those quantum liquids where the phenomenon is observed, it may occur only at temperatures very low compared to the degeneracy temperature, particularly in Fermi systems: as we saw, electrons in metals are degenerate up to temperatures of the order of tens of thousands of kelvin, but until two years ago no metal was known in which superconductivity occurs much above 25 K. In the last two years the situation has changed very dramatically in this respect (see section 9.4 below), but the highest temperature at which superconductivity is reliably established is still much smaller than the degeneracy temperature. In columns 4 and 5 of table 9.1 I indicate the occurrence or not of superconductivity or superfluidity and temperature of onset, if any.

In the next three sections, I discuss the properties of the three known terrestrial 'superfluids', in increasing order of complexity: liquid helium-4, the electrons in superconductors and liquid helium-3.

Table 9.1. *Laboratory quantum liquids*

System	Statistics	Degeneracy temperature (K)	Super-conducting/ superfluid?	Onset temperature (K)
Electrons in metals	Fermi	$\sim 10^4$	sometimes	< 125
Liquid helium-4	Bose	~ 3	yes	2.17
Liquid helium-3	Fermi	~ 5	yes	2.6×10^{-3}
Atomic hydrogen	Bose	$\sim 5 \times 10^{-2}$	?	?
Atomic deuterium	Fermi	$\sim 5 \times 10^{-2}$	?	?

9.3 Liquid helium-4: Bose condensation

If we put a large number of helium atoms, of isotopic mass 4, in a box, then at temperatures above about 4 K they will form a gas. As we cool below this temperature they form a liquid (with, of course, a few atoms left as a gas above the liquid), but this liquid does not have any particularly spectacular properties; in this phase it is called helium-I (He-I). However, if we cool it further, below a specific temperature close to 2 K, conventionally called T_λ (because the graph of specific heat near there, when plotted against temperature, has a shape resembling the Greek letter λ), the liquid suddenly starts to display quite abnormal and spectacular properties: it flows through tiny capillaries without apparent friction, climbs, in the form of a film, over the edge of vessels containing it ('film creep'), spouts in a spectacular way when heated under certain conditions ('fountain effect', see figure 9.6) and displays a host of other abnormal properties, some of which I shall discuss in detail below. This complex of effects is generally lumped together under the name of superfluidity, and the liquid in its superfluid phase is known as helium-II (He-II). The transition between the two phases is called the λ-transition. As we shall see, it is generally believed that the phenomenon of superfluidity is directly connected with the fact that the atoms of helium-4 obey Bose statistics, and that the λ-transition is due to the onset of the peculiar phenomenon called Bose condensation.

Before embarking on a discussion of this, let me describe one aspect of superfluidity which, though at first sight much less spectacular than some of the phenomena mentioned above, is actually far more profoundly puzzling and a more direct clue to what is going on. Suppose we pour some liquid into a cup and set the cup on the axis of a turntable, such as that of a record-player; having done this, we set the turntable into rotation. If the liquid is an ordinary one such as water, it will initially appear to remain at rest with respect to the room, but within a few minutes it will come into equilibrium with the rotating cup and rotate at the same speed. If, however, the same experiment is done with superfluid helium at low temperatures ($T \ll T_\lambda$) in a suitably sized and suitably rotating container, then the liquid will never come into rotation however long we wait; it appears to stay forever at rest with respect to the outside room. Now at first sight this is not specially surprising; we already saw that He-II appears to suffer no friction when flowing through thin capillaries, so we could argue that this is just another example of the same phenomenon; since the walls of the container exert no frictional forces on the liquid, they cannot bring it into rotation. That this is not the correct explanation is demonstrated by reversing the order of the experiment, that is by setting the container into rotation while the helium is above T_λ and subsequently cooling through T_λ while still rotating the container at constant speed. If this is done, then the liquid, while above T_λ, behaves perfectly normally and comes fully into rotation with the container. But as soon as the temperature is dropped below T_λ, it starts to slow down, and at very low temperatures ($T \ll T_\lambda$) will appear stationary with respect to the outside room – even though the container continues to rotate! Since the liquid has got *out* of step with the container, it is clear that the explanation cannot be simply zero friction; in fact, this peculiar behaviour cannot have anything to do with friction, viscosity or any other properties which refer to nonequilibrium situations, but must be a manifestation of the *equilibrium* properties of the system.

Let me now try to indicate how the Bose statistics obeyed by the helium atoms can be used to account for this phenomenon. Consider, as in the preceding section, a large number N of bosons moving freely and independently in a large volume V. As we saw, in this case the number of particles per quantum state in thermal equilibrium is given by a graph of the form in figure 9.5(c): note, again, that the low energy states are more populated than they would be for the corresponding system of distinguishable particles. As the temperature is further lowered, the concentration in the low energy states is further increased, while that in the high energy states is decreased, in such a way that the total number of particles remains constant, as of course it must. However, it turns out that, for any given values of N and V, there comes a point at which this process has to stop: the number of particles in any given energy level is fixed as a function of temperature, and as a result the total number of particles occupying the levels cannot be greater than some number $N_{max}(T)$ which decreases along with T. At some temperature T_0 the quantity N_{max} becomes equal to the total number of particles in the system, N, and below T_0 we have $N_{max} < N$: if we continue to use the distribution formula illustrated in figure 9.5(c) at such temperatures, there are simply not enough quantum states available to accommodate all the particles. Clearly, we have a problem!

Figure 9.6. The fountain effect: liquid helium spouting out of a vessel through small pores in the plug when heat is applied to the inside. (Photograph courtesy of Prof. J. F. Allen, FRS.)

photons, which can be created or destroyed in collisions, the total number N simply adjusts so that it is never greater than $N_{max}(T)$. In fact, as far as we know the phenomenon of Bose condensation in the strict sense is unique to liquid helium-4. (A phenomenon somewhat similar to Bose condensation can, however, occur in the photons in a laser under suitable (nonequilibrium) conditions. See Chapter 10 by Knight.)

What determines the Bose condensation temperature T_0? It turns out that it is proportional to $(N/V)^{2/3}$, i.e. to a^{-2}, where a is the distance between particles, and to the inverse of the mass m of the particles involved. In fact, apart from numerical factors of order one it is just the degeneracy temperature T_0 given by equation (9.3) (which is why I have used the same symbol for it). If we put in the numerical values of m and a for liquid helium-4 at its vapour pressure, we find that the resulting value of T_0 is about 3 K. Since the experimentally observed value of the temperature T_λ of the transition to the superfluid phase is about 2.17 K, it is extremely plausible that the onset of superfluidity is in fact associated with the onset of Bose condensation in the liquid; the fact that the numbers do not come out exactly right is not surprising, since the calculation of T_0 was for a gas of noninteracting particles whereas in fact helium atoms interact strongly.

The phenomenon of Bose condensation, and the closely related phenomena which occur in some Fermi systems (see the next two sections), are absolutely crucial to our modern understanding of the anomalous phenomena which occur in superfluids and superconductors, and particularly of the way in which they display the effects of quantum mechanics on a macroscopic scale. Before embarking on specific topics, it may be useful to indicate in a general way why this is so. Imagine that you are on a mountain-top looking down at a distant city square on market day. The crowd is milling around at random, and each individual is doing something different; from that distance it is very difficult to make out precisely what. Now suppose, however, that it is not market day but the day of a military parade, and the crowd is replaced by a battalion of well drilled soldiers. Now every soldier is doing the same thing at the same time, and it is very much easier to see (or hear) from a distance what that is. The physics analogy is that a normal system is like the market day crowd – every atom is doing something different – whereas in a Bose condensed system the atoms (or, more accurately, the fraction of them which is condensed at the temperature in question) are all forced to be in the same quantum state, and therefore resemble the well drilled soldiers: *every atom must do exactly the same thing at the same time!* As we shall see, this means, among other things, that effects which are far too small to be detectable at the level of single atoms may be quite easily observable in a Bose condensed (or similar) system; this is one feature which makes such systems so unique and exciting.

Let us now try to apply this idea to a specific phenomenon, namely the 'refusal to rotate' described above. For technical reasons it is convenient to consider, rather than a simple cup, a

The resolution of the problem is remarkably simple: below T_0 the system adjusts by taking all the particles which cannot be accommodated by the distribution formula and putting them in the single quantum state which has the lowest energy (the 'groundstate'). Since these surplus particles are a finite fraction of the whole (in fact, at zero temperature all of them), we reach the remarkable result that a *macroscopic* number of particles (of order N, which typically is of order say 10^{23}) occupy a *single* quantum state. This phenomenon is known as *Bose condensation*; although I have led up to it by considering a thermal equilibrium situation, it can also occur under conditions where the system is somewhat out of equilibrium, provided that the overall temperature is below the condensation temperature T_0. It should be remembered that Bose condensation can occur only for particles (such as atoms) whose total number cannot be changed by collision processes; for Bose particles such as container of annular shape (figure 9.7) with an annular width d small compared to its radius R (this saves us from having to worry in detail about the motion of the particles in the radial direction). Consider a system of N atoms in such a container. As we saw above, quantum mechanics requires that each atom be represented by a de Broglie wave, whose wavelength is related to the atom's velocity by equation (9.2). On the other hand, if the wave is to 'fit into' the annular region properly it is clear that as we go round the ring once the wave must turn back into itself, i.e. there must be an integral number n of wavelengths in the circumference (see figure 9.7): $n\lambda = 2\pi R$. But the frequency f of rotation (i.e. the number of times per second the particle passes a given point) is clearly just $v/2\pi R$: putting the above relations together with equation (9.2), we obtain the result

$$f = nh/4\pi^2 mR^2 \equiv nf_0, \tag{9.7}$$

where h is Planck's constant, m is the mass of the particle and R is the radius of the container. It is important to note that n may be zero as well as $1,2,3 \ldots$. Thus, the frequency of rotation is *quantised* – only certain special values can occur. The energy of rotation, $mv^2/2$, is also quantised: the energy E_n of the state with $f = nf_0$ is given, according to the above relations, by

$$E_n = \frac{mv^2}{2} = \frac{m(h/m\lambda)^2}{2} = \frac{m(nh/2\pi R)^2}{2} = \frac{n^2 h^2}{8\pi^2 mR^2} \equiv n^2 E_0. \tag{9.8}$$

Some readers will recognise the above argument as similar to one which is often used, concerning an electron in an atom, to show that the energy levels are quantised. In that case we would put m equal to the electron mass and R of the order of the atomic radius, about 10^{-10} m; the resultant value of f is about 10^{16} rev s^{-1}, and the value of E_0 is about 10^{-18} J, much greater than the mean (classical) thermal energy $k_B T$ at room temperature. As a result, under normal circumstances the electron in (say) a hydrogen atom is always in its groundstate ($n = 0$, no rotation at all). (The 'old' (Bohr) quantum theory of the atom excluded the value $n = 0$, so in that theory the groundstate corresponds to $n = 1$. However, modern quantum mechanics admits also $n = 0$.)

Now consider our actual situation, namely a ring of a reasonable radius R (let us say 1 mm for definiteness) containing atoms of mass, let us say, about 10^{-26} kg (the helium-4 atom has mass $\sim 6 \times 10^{-27}$ kg). If we put these numbers into formulae (9.7) and (9.8), we find that the 'quantisation unit' of frequency, f_0, is about 10^{-3} rev s^{-1}, and the corresponding unit of energy, E_0, is about 10^{-32} J. This is very tiny compared to the classical thermal energy, $k_B T$, at room temperature, or even at the lowest temperatures achievable to date (at a temperature of 10^{-4} K, the thermal energy is still about 10^{-27} J). So, in a 'normal' liquid such as water at room temperature (or for that matter He-I, i.e. liquid helium-4 above T_λ), the vast majority of the atoms or molecules are in states which correspond to very large values of n: if the ring is at rest, the average number rotating clockwise and counterclockwise with the same frequency will be equal. For a degenerate Bose system, as we shall see, the situation is quite different.

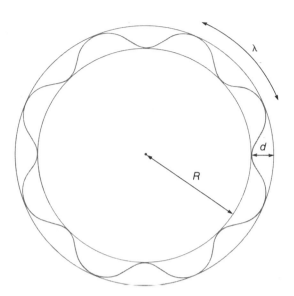

Figure 9.7. A de Broglie wave representing a particle moving in an annular container. For the wave to 'join up with itself' smoothly, we need $n\lambda = 2\pi R$, where n is an integer.

Now, what happens when we rotate the ring slowly, let us say for definiteness clockwise? It turns out that the state of the system which is thermodynamically most stable is that state in which the *mean* velocity of rotating of the liquid corresponds as closely as possible to the velocity of rotation of the container (i.e. the liquid will come into equilibrium with the rotating container as far as the states available to it allow it to do so). In a normal liquid, this is achieved very easily, independently of the speed of rotation, by simply biasing the distribution slightly: crudely speaking, we take a few of the counterclockwise-rotating atoms and put them into clockwise-rotating states. Whatever the speed of rotation, it is possible to do this in such a way that the liquid, when viewed from the walls of the rotating container, will look as if it is at rest (except for a small curvature of the surface due to the centrifugal force). This is, of course, exactly the observed behaviour of ordinary liquids like water, once they have come to equilibrium with the rotating container. Thus, the fact that the rotational motion is actually quantised has no observable effect, at any rate at any temperature attainable to date.

A Bose condensed liquid like He-II, however, will behave quite differently. Suppose for simplicity that the temperature is far below T_0, so that all the atoms are forced to be in the same quantum state. When the ring is at rest, they will all be in the state of zero rotation, corresponding to $n = 0$ in equations (9.7) and (9.8). Now suppose that the ring rotates with a frequency between zero and f_0. Now *all* the atoms are forced to choose, in unison, either all to stay in the state $n = 0$ or all to switch (say) into the state $n = 1$. They will choose the option which brings their mean rotational speed (or frequency) closest to that of the container: that is, for f less than $f_0/2$ they will stay in the state $n = 0$, and the liquid will appear to be at rest with respect to the laboratory, while for f greater than $f_0/2$ all the atoms will, if possible, switch into the state $n = 1$, so that (if f is less than f_0) the liquid will actually appear to be rotating *faster* than the container! Although for a container of reasonable size the value of f_0 is very small (of the order of $10^{-3}\,\mathrm{rev\,s}^{-1}$, as we saw) experiments which confirm the main features of this analysis have in fact been done. There appears little doubt, therefore, that in this and similar experiments liquid He-II is showing (in one sense of the phrase, see Section 9.6) 'quantum-mechanical effects on a macroscopic scale': the liquid as a whole is behaving just as would an electron in a single atom.

One point about the above argument might legitimately puzzle the reader. I repeatedly referred to an atom in the $n = 0$ state as being at rest, or not rotating, with respect to the laboratory. But an ordinary (terrestrial) laboratory is not an inertial frame in the sense of relativity theory, since it is rotating once per (sidereal) day with respect to the 'fixed stars'. Is it therefore legitimate to apply the concepts of quantum mechanics in this frame at all? Should we not, rather, first work out the problem in the frame of the fixed stars and then transform the results to the laboratory frame? If we do this, then it follows that when the container is at rest (in the laboratory) the Bose condensed liquid will be at rest with respect to the fixed stars and therefore rotating once per day in the frame of the laboratory! (For simplicity, I consider here a laboratory at the North Pole – elsewhere, there will be complications similar to the ones occurring in the analysis of the Foucault pendulum problem.) In fact, this conclusion is almost certainly correct, although at present the experimental evidence is not completely conclusive. Thus, among its many other surprising properties, liquid He-II can in principle serve as a detector of absolute rotation. (Because of the curvature of the surface mentioned above, even a classical liquid can detect absolute rotation: however, helium is the only liquid which can detect it by a *quantum-mechanical* mechanism.)

We still have not explained what is experimentally the most striking and obvious property of He-II, and the one which is primarily responsible for its description as 'superfluid', namely the ability to flow without apparent friction in a variety of situations (e.g. in small capillaries or, as a film, on rough surfaces) where an ordinary liquid would be so slowed by friction that it does not flow at all. The simplest example of this property, from a conceptual point of view, is the following. Suppose we take the annular container of liquid helium-4 described above and start rotating it at a frequency f close to nf_0, where n is not zero, at a temperature above the λ-transition. The liquid, being normal at this temperature, will rotate with the container, i.e. at frequency f. If we now cool through T_λ, continuing to rotate the container, and continue cooling to a low temperature, then according to the above argument all the helium atoms will choose that quantum state, namely the one

with frequency nf_0, which is closest to the frequency of rotation of the container; thus, the liquid will rotate at frequency nf_0. Now suppose that at this stage we stop the rotation of the container. We know for sure that the rotating state of the liquid cannot now be its true thermodynamic equilibrium state (the latter is, of course, the state $n = 0$, as earlier verified); nevertheless, it is an experimental fact that the liquid will continue to rotate, with frequency nf_0, for as long as we care to watch it. Since a normal liquid under these conditions would be rapidly brought to rest by the friction of the walls, this phenomenon is, just as much as capillary flow and film creep, a manifestation of the ineffectiveness of friction in the superfluid phase.

The explanation of this phenomenon in terms of the Bose condensation of the helium atoms is quite subtle, and some of the details are still somewhat controversial; however, the following simplified argument is probably not qualitatively misleading. First, it is virtually certain that in the model we have been using so far, in which the helium atoms are treated as moving independently and without interaction, subject only to Bose statistics, the rotating state would *not* persist but would rapidly decay to the thermodynamic equilibrium (nonrotating) state (as would an electron in an atom in similar circumstances). To obtain the phenomenon of superfluidity it is essential to invoke the interactions of the atoms with one another, and in particular to invoke the fact that any process in which the density of the liquid is changed, locally, from its equilibrium value is bound to cost an appreciable amount of energy. Now, the most obvious way to go from the rotating to the nonrotating state of the whole liquid (though actually not the only one) is to keep all the atoms in the same quantum state (as Bose statistics prefers) but to deform this state continuously from the original state of nonzero n (let us say for definiteness $n = 1$) to the $n = 0$ state. It turns out that any such process of deformation must inevitably take us, on the way, through states of much higher energy, and therefore such processes are forbidden (or, to be precise, astronomically improbable except very close to the λ-transition).

The reason why it is impossible to deform the state continuously from $n = 1$ to $n = 0$ without a huge cost in energy is essentially a *topological* one. To appreciate it, it is necessary to be a little more explicit than we have been so far about the nature of the de Broglie wave which in quantum mechanics represents a particle. The point which is essential in the present context is that the amplitude Ψ of the wave is a *complex* quantity, so that at any point we write

$$\Psi = A\mathrm{e}^{\mathrm{i}\Phi} \tag{9.9}$$

(A, Φ real quantities). The quantities Ψ, A and Φ are in general functions of position (and possibly also of time); the phase Φ is well defined everywhere, *except* at points where $A = 0$, where it is clearly meaningless. The quantity $A^2(\mathbf{r})$ has the physical meaning of the probability of finding a particle at $\mathbf{r}$, which in the case of a Bose condensed liquid is proportional to the mean density of particles there: thus, the only points at which Φ is not defined are those, if any, where the density is zero. Now, when we say that the nth quantum state corresponds to n wavelengths of the de Broglie wave fitting into the circumference of the ring, what we mean more precisely is that in such a state, if θ indicates the angle around the ring, we have

$$A = \text{const}, \quad \Phi = n\theta \qquad (9.10)$$

so that the phase Φ increases by $2n\pi$ as we go once around the ring. It may be convenient to visualise the ring as a hula-hoop and to imagine that the de Broglie wave is represented by a string which we wind around the hoop, gluing it to the surface as we go (see figure 9.8)*; then $\Phi(\theta)$ corresponds to the angle made by the string in the hoop cross-section (figure 9.8), and the nth quantum state corresponds to the string twisting n times around the hoop as we go around it once (in figure 9.8, $n = 2$). Now, it is intuitively obvious that once we have wound the string (say) once around the hoop and tied the ends together (state $n = 1$), it is impossible to deform it into the state $n = 0$ (where it is not wound around at all) without either cutting the string or breaking the hoop – which in the case of the de Broglie wave would correspond to making A zero. If we neglect this

Figure 9.8. Schematic representation of the behaviour of the phase Φ of a de Broglie wave as a function of the angle θ travelled around the annulus (hoop) of figure 9.6 (see text for explanation). It is clear that we cannot change the number of 'turns' without cutting the string or breaking the hoop.

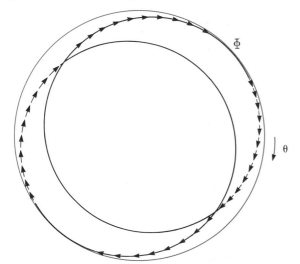

*It should be strongly emphasised that this is only an analogy: while the angle θ does represent the real (geometrical) angle around the loop, the angle Φ is only the phase of the abstract quantum-mechanical wave function represented by equation (9.9), and in reality has no geometrical significance. Representing it as the 'winding angle' around the cross-section of the ring, as in the figure, is only a guide to one's intuition and has no real significance.

possibility for a moment, the number of 'turns' made by the phase of the de Broglie wave (the 'winding number') is a quantity which for topological reasons cannot be changed – a so-called *topological invariant*.

Thus, the only way to deform the state from $n = 1$ to $n = 0$ is to pass through, on the way, states in which the density is zero at at least one point on the ring. If we were dealing with a single particle such as an electron in an atom, such states would pose no special problem and would require no extra energy; thus, an electron in a state $n = 1$ can quite easily make the transition to an $n = 0$ state. However, for a system of N interacting atoms, as in liquid helium-4, to create a 'hole' where the density is zero requires us to pile up the particles, i.e. to compress the liquid, elsewhere in the ring, and this now costs us a large amount of energy, very much greater than the typical thermal energy which can be supplied by the environment. Thus, the possibility of such states occurring is quite negligible, and the topological conservation law ensures that the atoms remain forever in the $n = 1$ state.

9.4 Superconductivity

The behaviour of most metals at temperatures of the order of room temperature and below is accounted for by a rather simple picture. We imagine that each of the atoms which compose the metal gives up one or more of its electrons, which can then migrate freely throughout the whole system, and keeps the rest closely tied to itself, thus forming an 'ionic core' which is positively charged. These cores may form a regular lattice (in which case the metal is crystalline – most of the well known metals are of this type), or they may be arranged irregularly, as happens for certain alloys. In any event, the 'free' electrons which have been given up by the atoms behave, to a first approximation, very much as if they were moving independently in free space. As we saw in Section 9.2, under such conditions their distribution between the quantum states available to them will be given by the Fermi distribution shown in figure 9.5(b). Since all the states which are below the Fermi energy, E_f, by much more than the thermal energy k_BT are automatically full and all those above it by a similar amount automatically empty, all the action involves only the small fraction (of the order of k_BT/E_f) of the electrons within $\sim k_BT$ of the Fermi energy, i.e. only these are free to respond to changes in the conditions of the system. For example, the specific heat attributable to the electrons in a metal is only about a fraction (k_BT/E_f) of the classically expected value.

This simple picture gets a lot of things qualitatively right, but to get any kind of quantitative agreement with experiment it is necessary to supplement it in at least two respects. First, the electrons cannot really ignore the presence of the ionic cores: what in fact happens (in a crystal) is that the de Broglie wave representing the electron deforms itself near the cores so as to take the diffraction by the atoms into account and, having done so, manages to propagate much as if it represented a truly free electron. However, the 'free' electron now behaves as if it had an 'effective mass' which is different from its real mass. Typical values of the effective mass range from about 0.1 to 10 times the real mass, depending on the metal (though see below). The second essential modification is to note that the electrons do not in fact move completely freely, even after this correction, but are subject to random scattering; this is due both to the thermal vibrations of the ionic cores and to imperfections in the crystalline lattice (either chemical impurities, or defects in the structure). In the case of an 'amorphous' (noncrystalline) solid there really is no crystalline lattice there to start with, and the electron can travel only a short distance (a few interatomic distances) without being scattered, even at low temperatures. In a typical crystalline metal, the electron 'mean free path' (the distance it can travel without being scattered) is only of the order of a few atomic spacings at room temperature, because of the scattering by the thermal vibrations of the lattice; however, at low temperatures the latter are frozen out, and in a very pure crystalline metal the electron mean free path can be as long as a centimetre. The above picture can of course be made quantitative, and explains most of the general features of the behaviour of metals.

However, there is a class of metals which at low temperatures displays the complex of effects known as *superconductivity*. Like superfluidity in helium-4, superconductivity occurs only below a certain definite 'transition temperature' (usually denoted T_c) which depends on the metal in question; above the superconducting transition temperature the metal behaves exactly like any other metal (just as He-I is qualitatively like any other liquid) and it is in practice very difficult to predict ahead of time whether any given metal will become superconducting, still less at what temperature. In fact, over the 75 years from 1911 to 1986 the highest transition temperature known crept up gradually from 4 to 25 K. Over the last two years, however, the situation has been totally revolutionised with the discovery of a class of copper oxide (and, very recently, other) materials which show superconductivity up to at least 125 K (40% of room temperature!). Whether the explanation of the phenomenon given below applies equally to these new materials, or whether the mechanism of superconductivity in them is of a radically new kind, remains at the time of writing an open question.

The most spectacular property of the superconducting state, and the one which gives it its name, is the fact that the electrical resistance appears to be exactly zero. In fact, if a ring of superconducting material is cooled below its transition temperature and a current set up in it (e.g. by varying the magnetic flux through the ring), it will continue to circulate for as long as one cares to observe it. Similarly, if a piece of superconducting metal is incorporated in an otherwise normal (i.e. non-superconducting) circuit, then no voltage is observable across the superconductor even when a quite large current is flowing through it. In a normal metal the electrical resistance is, of course, finite: the electrons are accelerated by the electric field

(voltage drop) but are scattered by impurities and lattice vibrations, and the balance of these two effects determines the resistance,. which is, roughly speaking, proportional to the efficacy of scattering. Thus it seems that in the superconducting state the usual scattering mechanisms are somehow rendered ineffective – just as in He-II the frictional effect of the walls was rendered inoperative.

An effect with almost equally spectacular consequences is the so-called 'Meissner effect', that is the total exclusion of the magnetic flux from a superconducting body. If a piece of ordinary (nonmagnetic) metal is placed in a magnetic field, the field lines to a good approximation go straight through it as if it were not there (figure 9.9*a*), and any forces on the body due to the field are very small. However, if we try to carry a simply shaped specimen of superconducting material into a field, or alternatively switch on the field while the specimen is nearby, the magnetic field lines distort so as totally to avoid the specimen! (See figure 9.9*b*.) Now, as I have described it, this effect could be just a consequence of the zero resistance of the superconducting state; if we try to change the magnetic flux through a piece of conducting material, we set up eddy currents which tend to screen the field out, and there will be a finite time-lag before the field penetrates the interior of the body. This time-lag is inversely proportional to the resistance of the material, and while it is quite short (at most of the order of seconds) for most metallic bodies in laboratory situations, it can be thousands of years for some astrophysical objects. We could, therefore, argue that since the resistance in the superconducting state is zero, the relevant time-lag should be infinite. However, just as in the case of the analogous 'refusal to rotate' phenomenon in superfluid helium, this explanation can be excluded by performing the experiment in the reverse order, that is by applying the magnetic flux in the normal state (i.e. above the superconducting transition temperature T_c) and then cooling through T_c. If this is done, then above T_c the magnetic field penetrates the sample as it would any normal metal (figure 9.9*a*), but the moment the temperature falls even slightly below T_c the field is completely expelled, giving the situation of figure 9.9(*b*). Thus, the Meissner effect, like the absence of rotation in He-II, is an equilibrium phenomenon and not simply equivalent to the absence of resistance. Because it takes work to expel the field, this effect can be made to have quite spectacular consequences: for example, if a small bar magnet is gradually lowered into a superconducting bowl, there comes a point at which the work necessary to expel the field of the magnet from the bowl becomes so large that it can no longer be supplied by the gravitational potential energy gained by lowering the magnet further, and at this point the magnet floats above the bowl (see figure 9.10)! The above description of the Meissner effect applies only to the simplest case, in which (*inter alia*) the sample is 'simply connected' (i.e. has no holes through it).The behaviour of a superconducting ring under these conditions is even more fascinating: although the field is still excluded from the ring itself, it can pass through the hole

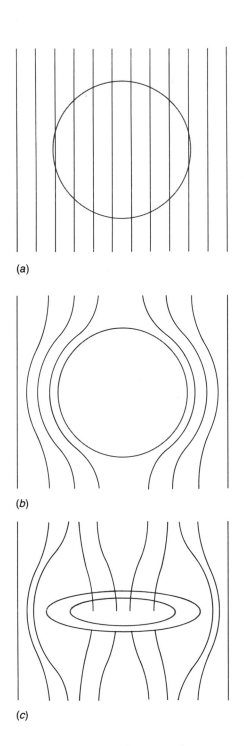

(a)

(b)

(c)

Figure 9.9. Behaviour of a piece of superconducting metal in a magnetic field. (*a*) Above the transition temperature T_c, when the metal is normal, the sample is completely penetrated by the field (whatever its shape). (*b*) Below T_c, when it is superconducting, a spherical sample completely expels the field. (*c*) For a ring-shaped specimen in the superconducting state, some of the field lines can go through the hole.

(figure 9.9*c*), but only in such a way that the total magnetic flux threading the hole is an integral multiple of the quantity $h/2e$ (the so-called 'flux quantum', about 2×10^{-15} Wb). This striking phenomenon, which was actually predicted and observed only after a satisfactory theory of superconductivity had been achieved, is called flux quantisation: crudely speaking, the system will generally choose that value of the quantised flux which is nearest to the flux originally passing through the hole in the normal phase.

It is clear that many aspects of these phenomena are reminiscent of the behaviour of helium-4 in its superfluid phase. In fact, one can see that the phenomenon of zero resistance is the direct analogue of zero friction in the flow of He-II. It is less obvious, but nevertheless true, that flux quantisation is the exact analogue of the 'refusal to rotate' of He-II. Although a detailed understanding of this statement requires more background than I am assuming here, a crude and intuitive argument can be made as follows. As we know from atomic physics (the Larmor precession) the effect of imposing a magnetic field on a charged particle in a circular orbit of radius r is much the same as if we looked at the system from a rotating system of reference but without the magnetic field. The quantitative equivalence is that the magnetic flux Φ through the orbit is related to the speed of rotation f of the rotating frame by

$$f = e\Phi/4\pi^2 mr^2, \qquad (9.11)$$

where e is the charge on an electron.

Thus, plausibly, the effect of imposing a flux Φ through a superconducting ring of radius R is similar to that of filling an annular container of the same radius with He-II atoms and rotating it at the speed given in equation (9.11). Now we saw in Section 9.3 that in this case the helium could achieve its 'optimum' behaviour, i.e. come exactly into rotation with the container, if and only if f was an integer multiple of the quantity $f_0 \equiv h/4\pi^2 mR^2$ (equation 9.7). Correspondingly, we might guess that 'optimum' behaviour could be achieved in the superconducting case only if the flux threading the loop satisfies the condition

$$\Phi = n(h/e). \qquad (9.12)$$

This conclusion is, in fact, correct (or nearly correct: see below). However, contrary to what one might perhaps think, the 'optimum' behaviour corresponds to zero circulating current not in the rotating frame but in the *laboratory* frame; thus, when condition (9.12) holds, no electric currents flow in the superconducting ring and so no extra flux is generated. Now, what happens when equation (9.12) is *not* fulfilled, i.e. when we have created (e.g. by bringing up a bar magnet) an external magnetic field such that the flux produced by it does not satisfy the quantisation condition (9.12)? It turns out that what happens is that electric currents then flow on the inner surface of the superconducting ring and produce an extra magnetic flux which adds to the externally imposed one – in just such a way that the total flux which threads the ring (including the surface region where the currents flow) *does* exactly equal equation (9.12) (or nearly so, see below). For a simply connected sample such as a sphere, there is no inner hole for the flux to go through, so the only possibility is to have $n = 0$, i.e. the magnetic field is totally excluded from the sample – the Meissner effect.

There is, alas, one thing wrong with this argument – it leads to quantisation of flux in integral multiples of h/e, not $h/2e$ as is observed experimentally! This is a strong indication that the entities in a superconducting metal which correspond to single atoms in the helium case cannot be single electrons (charge e) but must rather be *pairs* of electrons (charge $2e$). To see how this can be, it is necessary to say something about our modern microscopic understanding of the superconducting state.

We have seen that all the signs indicate that this state has much in common with the superfluid state of helium-4, whose peculiarities we attributed to the onset of Bose condensation. But the electrons in metals obey Fermi, not Bose, statistics, and if they move freely and independently their energy distribution is the Fermi distribution shown in figure 9.5(b), which is quite unlike the Bose distribution (for $T < T_0$) of figure 9.5(c). So how can be idea of Bose condensation possibly be relevant here?

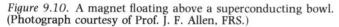

Figure 9.10. A magnet floating above a superconducting bowl. (Photograph courtesy of Prof. J. F. Allen, FRS.)

The key to the mystery lies in the realisation in 1957 by Bardeen, Cooper and Schrieffer (usually abbreviated BCS) that if the forces which act between the electrons in the 'active' region, that is in states near the Fermi energy, are attractive (as they can be in a metal, despite the Coulomb repulsion, because of a rather subtle effect connected with the polarisation of the ionic lattice), then these electrons will tend to bind together to form something like diatomic molecules. Whether or not in any particular metal the forces are in fact attractive is a delicate matter, and depends not just on the behaviour of the electrons themselves but also on that of the ionic cores, which by their vibrations can transmit forces between the electrons; however, let us suppose that there is indeed a net attraction. Then what the electrons actually do is to form, not strictly 'dielectronic' molecules but rather *Cooper pairs*: this means that each electron tends to find itself preferentially near another electron, but not, as it were, any particular one: we have a *collective* bound state. We can try to form some kind of intuitive picture of the Cooper pairs in the superconducting state by visualising each electron as forming, indeed, a 'dielectronic molecule' with another electron, but with such a huge radius that between the partners there are, literally, millions of other electrons, each forming their own pairs. Since it is impossible to 'tag' the electrons, it is intuitively clear that in this situation the question 'with which other electron is this one paired?' is quite meaningless. Apart from this feature, which as we shall see has profound consequences, the picture of Cooper pairs as like giant diatomic molecules is quite helpful to one's intuition. In particular, it makes sense to ask the question 'what does the internal structure of the pair look like?'. It turns out in all known superconducting metals (or more accurately all these known until very recently – see below) the pairs form in such a way that the two electrons involved have oppositely oriented spins and zero rotation relative to one another (zero relative orbital angular momentum). A consequence of this is that the internal state of the pair is completely isotropic – it looks the same whatever direction it is viewed from. In fact, in any given metal at a given temperature there is only one possible form of internal structure; anything else would cost far too much energy to be stable. Once we have specified the metal and the temperature, we need say no more to specify the pair structure.

One might think, however, that just as in an ordinary gas of (say) molecular hydrogen the various molecules can have the same internal structure and behaviour, yet move as wholes with quite different velocities, etc., so in a superconducting metal the Cooper pairs, however well defined and unvaried their internal structure, could have quite different velocities as a whole. The BCS theory of the superconducting state, however, tells us that this is not so: *the Cooper pairs must all behave in exactly the same way*, not only as regards their internal structure and motion but as regards their motion as wholes (centre-of-mass motion). If they did not do this there would be no energy advantage in forming the pairs in the first place. This situation is clearly very reminiscent of the behaviour of the atoms in helium-4 in the Bose condensed phase. Indeed, one way of thinking about the superconducting state is to take the idea of the Cooper pairs as diatomic molecules to its logical conclusion and observe that a molecule made up of two particles of spin $\frac{1}{2}$ (such as electrons) must, as we saw in Section 9.2, have integral spin and therefore behave like a boson. Viewed in this way, superconductivity is simply due to Bose condensation of the pairs, just as superfluidity in liquid helium-4 is due to Bose condensation of the helium atoms. However, it should be emphasised that the analogy is only partial: there is no analogue of the He-I phase, where the bosons exist but are not condensed – in a superconducting metal, the Cooper pairs either do not exist at all or are automatically condensed.

Once this is realised, the qualitative explanation of the main properties of the superconducting state follows the same lines as that given in the preceding section for superfluid helium-4. In particular, flux quantisation (and the Meissner effect, which is in a sense a special case of it) is a result of the fact that the motion of the pair as a whole is represented by a de Broglie wave which must 'fit in' to the circumference of the ring and that all pairs must, since they are Bose condensed, choose the *same* state. Similarly, the stability of the current-carrying state is, just as in helium-4, basically of topological origin: to depress the density of Cooper pairs to zero requires much more energy than is usually available, and so long as the pair density is everywhere finite the 'winding number' which represents the number of twists made by the phase of the wave as we go around the ring is conserved for topological reasons.

What, then, is new about the superconducting state? One property which distinguishes the (charged) Cooper pairs from (neutral) helium atoms is, of course, that they interact strongly with external electric and magnetic fields, and we have already seen that this can lead to fairly spectacular effects such as superconducting 'levitation'. There are other effects, however, which although perhaps less obviously dramatic are actually even more intriguing in the light they shed on the basic concepts of quantum mechanics. Consider, for example, the experimental setup shown in figure 9.11, in which a super-conducting metal, interrupted by two special elements known as Josephson junctions*, encloses a hole through which a magnetic field is applied. The experiment consists in measuring the current which flows between the points marked A and B as a function of the magnetic field through the hole, when everything else is held constant. The point of the arrangement shown is that the current is transported by the electrons in the metal, which of course never get into the region of the hole,

*A Josephson junction may for present purposes be thought of as a sort of barrier through which electrons can pass, but with much more difficulty than through the bulk superconducting metal. In practice it is often a thin layer of oxide sandwiched between the two pieces of bulk metal.

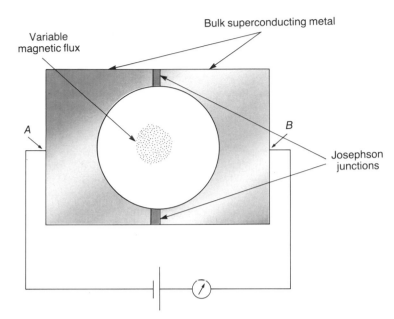

Figure 9.11. An experiment which shows that the behaviour of electrons is not completely determined only by the electric and magnetic field which they directly experience. The magnetic field is confined entirely to the hole, which the electrons never enter.

while the magnetic field is entirely confined to the hole and is zero in the metal itself; thus, the electrons are never in a position to feel the magnetic field, and one would think, therefore, that the current which flows through the circuit should be quite unaffected by the field. But the experimental data spectacularly refute this conclusion – the current shows a very well defined and reproducible dependence on the field! This phenomenon is actually a macroscopic version of a quantum-mechanical effect, known as the Aharonov–Bohm effect, which has also been demonstrated for single electrons, and which apparently forces us to drop at least one of two very deeply ingrained assumptions about the theory of electromagnetism, namely (*a*) that all electromagnetic effects are in principle completely described if we know the value of the electric and magnetic fields as a function of position and time, and (*b*) that in order to be affected by a field, a particle such as an electron must enter the region where the field exists – it cannot 'know' about fields which exist in a region which it goes around but never actually enters. The experiment just described shows that the failure of these assumptions, which is a purely quantum-mechanical phenomenon, can even have macroscopic consequences. It is possible that superconductors can be used for even more spectacular tests of the foundations of physics, and I will return to this question in Section 9.6.

One postscript should be added to our discussion of the theory of superconductivity. I have given the impression that in all metals in which it is known to occur, the phenomenon of superconductivity is well explained by the formation of Cooper pairs, and moreover that these pairs always form in an isotropic state, with spins paired off and zero relative rotation. Until about seven years ago this would have been a fairly safe statement. However, in the last seven years a whole new class of superconducting metals – the so-called 'heavy-fermion superconductors' – has been discovered. These metals (along with various others which do not go superconducting) are quite anomalous even in the normal (nonsuperconducting) state – in particular, the conventional interpretation of data on their specific heat would give them effective masses which can be as much as a thousand times the real electron mass (hence the name 'heavy fermions'). Moreover, many of the properties of the superconducting state of these metals are quantitatively very different from those of previously known superconductors. At the time of writing, therefore, there are at least three possible hypotheses about the mechanism of superconductivity in these systems, none of which can be said to have been conclusively proved or refuted by experiment: (*a*) it is Cooper pairing and the pairs are isotropic, just as in previously known superconductors; (*b*) the mechanism is Cooper pairing, but the pairs are formed in an anisotropic state (as we shall see below is the case for liquid helium-3), or perhaps in some even more sophisticated arrangement; (*c*) the mechanism is not Cooper pairing at all but some previously unknown possibility. These remarks apply with even greater force to the new class of 'high-temperature' superconductors discovered in the past two years.

9.5   Liquid helium-3: an anisotropic superfluid

For many years, the liquid phase of helium-3, the light isotope of helium, was the Cinderella of low temperature physics. The helium-3 atom, which contains two protons, one neutron and two electrons, has a spin of $\frac{1}{2}$ and is therefore a fermion, so a collection of helium-3 atoms should obey Fermi statistics, and provided the system remains a gas or liquid, the distribution of atoms over the available quantum states should be given by the Fermi distribution shown in figure 9.5(*b*), just as for the electrons in a metal. In fact, if we ignore for the moment the fact that electrons are charged whereas the helium-3 atoms are neutral, the two systems are very similar indeed and we should expect them to show similar properties at low temperatures. This, indeed, turns out to be the case, and over the whole temperature range from about 3×10^{-3} K up to about 0.1 K the behaviour of liquid helium-3 is qualitatively very similar to that predicted for a degenerate gas of fermions moving freely and independently in the volume available to them. (Above 0.1 K the approach to nondegeneracy, which is complete by about 5 K, already begins to affect the quantitative properties.)

Thus, even at temperatures far below those corresponding to the spectacular phenomena discussed in the preceding two sections (125 K for superconductivity, 2 K for superfluidity in helium-4) liquid helium-3 persists in behaving in an obstinately 'normal' way.

This picture changed dramatically in 1972, when it was discovered that below the very low temperature of 3×10^{-3} K liquid helium-3 possesses not one but (at least) *three* new phases, with properties radically different from those of the 'normal' phase. Although the phenomenon of superfluidity has not, at the time of writing, been established quite as directly and spectacularly as in the case of helium-4, there is very little doubt that it does occur (with some reservations indicated below) in all three of the new phases, and they have therefore been referred to collectively since their original discovery as 'superfluid helium-3'.

The occurrence of a superfluid phase in liquid helium-3 at a sufficiently low temperature had in fact been predicted with some confidence on theoretical grounds. We saw that in a metal, if the forces of interaction between the electrons near the Fermi energy are attractive, then these electrons will tend to form Cooper pairs, which promptly undergo Bose condensation, and the metal will be superconducting. Using this analogy, it was observed that in helium-3, since the forces of interaction between atoms near the Fermi energy are likely to be at least partly attractive, Cooper pairs should form in this system too at a sufficiently low temperature and should undergo Bose condensation just as in a superconducting metal. Since the atoms are neutral, the result should be the analogue of superconductivity for a neutral system, that is superfluidity; qualitatively speaking, the liquid should show the same kinds of anomalous behaviour as occur in helium-4 below the λ-transition. All the experimental evidence available to date seems consistent with this analysis, i.e. with the assumption that the superfluid phases of helium-3 are indeed Cooper-paired phases. The fact that the transition to the Cooper-paired state does not occur until a very low temperature is not particularly surprising: the degeneracy temperature of liquid helium-3 is only about 10^{-4} of what it is in metals, so it is not surprising that the transition temperature scales in approximately the same way (3×10^{-3} K versus a few kelvin for most superconducting metals).

However, there are very important differences between the Cooper pairs which form in a superconducting metal and those which are believed to form in the superfluid phases of helium-3. As we saw in the preceding section, in all metallic superconductors (with the possible exception of the 'heavy-fermion and/or high-temperature ones) the pairs form in an isotropic state, with spins opposed and zero relative rotation, so that this relative state is unique and is not characterised by any internal 'degree of freedom'. By contrast, there is by now overwhelming evidence that the pairs which form in each of the superfluid phases of liquid helium-3 are anisotropic: the (nuclear) spins of the two atoms involved are at least partly parallel, so that the 'molecule' has a net spin of one (rather than zero as is the case in superconductors), and moreover there is a relative rotational motion (relative angular momentum). Thus the 'molecule' certainly does *not* look the same when viewed from different directions. Crudely speaking, we should expect each pair to be characterised by a particular axis (unit vector) which tells us how the total nuclear spin of the pair is oriented, and also by a second unit vector which specifies the axis around which the two atoms are rotating relative to one another ('orbital axis'). (As a matter of fact, it turns out that for one of the new phases it is meaningful only to specify the *relative* orientation of the nuclear spins and the rotational motion, but that is a technical complication which need not concern us.) Now, the crucial point is that the pairs are automatically Bose condensed, and hence must all behave in the same way, not only as regards their motion as a whole (centre-of-mass motion) *but also as regards their internal structure and orientation* (see figure 9.12). Thus, each of the two characteristic axes mentioned above must be the *same* for all the pairs (or more precisely all the pairs in the region of space we are considering). This means that the liquid as a whole acquires two characteristic vectors, one of which governs its spin properties and the other its other ('orbital') properties. In general, the behaviour in (say) a magnetic field applied parallel to the characteristic spin axis will be quite different from that in one applied in a perpendicular direction: similarly, a heat flow parallel to the orbital axis will produce effects quite different from one perpendicular to this. The properties of the liquid in any given region of space are thus anisotropic, in a way determined by the orientation of the characteristic vectors in this region. That a liquid should show anisotropic properties is, of course, not particularly surprising or new: the so-called liquid crystals also have this property, which is now routinely exploited in, for example, digital display devices. However, the new phases of helium-3 are unique (among terrestrial substances, at least) in combining such properties with the phenomenon of superfluidity; in fact, the specifically 'superfluid' properties are themselves markedly anistropic. Consequently, the liquid in these phases is said to be an *anisotropic superfluid*.

The fact that the internal structure and orientation must be identical for all the pairs has many remarkable consequences. One is that the anisotropic superfluid phases are extremely sensitive to very tiny effects which would be quite negligible in any ordinary system. Consider for example the magnetic interaction between the spins of the nuclei. If we think about an ordinary diatomic molecule, let us say hydrogen, then this interaction would tend to favour a state in which the spins lay parallel to one another and perpendicular to the axis around which the molecule is rotating; that is, such a state would have slightly lower energy than the alternative possibility (spins parallel to rotational axis). However, the energy difference (call it ΔE) is extremely tiny – in practice it is at least four orders of magnitude smaller than the thermal energy $k_B T$ at the lowest temperatures currently attainable. Thus, it is totally unable to compete with the thermal disorder, and in a gas of molecular hydrogen one would find that the nuclear spins are oriented quite at random relative to the rotation axis, as in figure 9.12(a). Now consider superfluid helium-3. The magnetic interaction energy of the nuclear spins is still very tiny; however, the crucial difference, now, is that we cannot choose the orientation of the Cooper pairs independently for each pair. In fact, if pair 1 has its nuclear spins oriented parallel to the rotation axis, then so must pair 2 and . . . so must *all* the pairs ($\sim 10^{23}$ of them) in the system (figure 12b)! Thus, there are only two possibilities – all spins parallel to the rotational axis, or all perpendicular to it – and the energy difference between these alternatives is not ΔE, but rather ΔE times something like 10^{23}, that is a very large energy, certainly much greater than the thermal energy at 3×10^{-3} K where the superfluid phases occur. Thus, the magnetic interaction effect certainly can compete with the effects of thermal fluctuation, and indeed it is this 'negligible' effect which actually determines the orientation, and hence many of the properties, of the liquid in many cases. As an example, the nuclear magnetic resonance behaviour of superfluid helium-3 is qualitatively quite different from that of a normal liquid.

Another amusing manifestation of the same kind of effect is the question of the 'orbital magnetism' of superfluid helium-3. It is known that the rotation of an ordinary homonuclear diatomic molecule (such as the hydrogen molecule) gives rise to a magnetic moment along the axis of rotation, since the rotation of the positively charged nuclei is not exactly cancelled by that of the negatively charged electrons. This effect is very small, but its existence can actually be confirmed by magnetic resonance techniques. Now, in an ordinary gas such as molecular hydrogen, the axis of rotation is oriented at random, so that the resulting magnetic moments cancel on average and the gas as a whole has no magnetic moment. By contrast, in superfluid helium-3 the axis of rotation must be the same for all Cooper pairs; consequently, the tiny magnetic moments associated with the rotation should add up and the liquid is predicted to be in effect a permanent magnet (albeit a very weak one). Existing experiments are consistent with this prediction, although they cannot be said to have proved it beyond reasonable doubt.

As a final example of the unexpected and amusing consequences of the nontrivial internal structure ('internal degrees of freedom') of the Cooper pairs in superfluid helium-3, let us consider the question of the stability of supercurrent flow in an annular container, as we did at the end of Section 9.3. There, we saw that so long as the density of particles was not depressed to zero at any point, the state in which the atoms were condensed into a quantum state with nonzero rotation was stable for topological reasons. At first sight, one would think (and in fact it was thought for the first two or three years of existence of the new phases) that the same conclusion ought to apply to superfluid helium-3. However, it turns out that there is a crucial difference between helium-4 and superconductors,

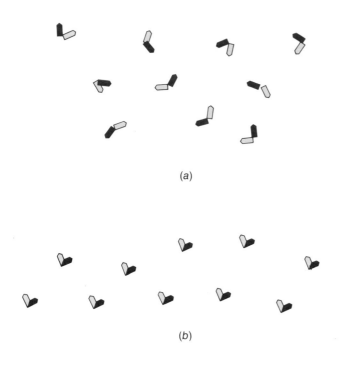

(a)

(b)

Figure 9.12. (a) In a gas of ordinary molecules, the orientations of the rotational axis and the spin axis are random, both absolutely and relatively to one another. (b) In a system of Cooper pairs, which are Bose condensed, the absolute, and hence the relative, orientation must be identical for all the pairs.

where the internal structure of the condensed objects is trivial, on the one hand and helium-3 on the other (more precisely, two of the three phases: the third is actually very similar to superconductors in this respect). Namely, whereas the 'phase' of the de Broglie wave which we introduced in Section 9.3 is in the former case an abstract quantity without direct physical significance (the device of representing it as a 'twist' in real geometrical space was of course merely an aid to one's intuition), in the case of the anisotropic pairs in superfluid helium-3 the phase does have a real geometrical significance: it is equivalent in some sense to a rotation of the state of the pair around the orbital axis. For example, if for helium-3 in an annular container the orbital axis is held fixed in a particular direction which is constant in space, then a quantum state in which the phase goes through $2n\pi$ as we go once around the ring (i.e. a state 'with winding number n') is one in which the state of the pair is in effect rotated n times around the orbital axis as we go once around the ring. Suppose that we start with a state in which, say, n is equal to two. Can we deform the state of

the system continuously, without forcing the pair density to go to zero at any point, so as to reach the state with n equal to zero (no current)? So long as the orbital axis is held fixed, the situation is exactly the same as in helium-4 (we can, as there, model the phase by a string twisted around a hula-hoop) and the same discussion applies word for word: the answer is no. However, it turns out that if we are prepared, in the course of the deformation, to allow the orbital axis itself to twist in space, the answer is yes! (This can be demonstrated, given sufficient patience, by using a series of suitably marked tops. Actually, n can change only by an *even* number, so the states $n = \pm 1$ are in fact stable as well as $n = 0$.) Thus, in principle, 'superfluid helium-3' should not be superfluid at all – the current carrying states are generally not topologically stable. Unfortunately, for various reasons an explicit experimental demonstration of this feature is very difficult, and the actual situation with regard to the stability of superflow is complicated and at the time of writing not too well understood.

The above examples only give a taste of the many exotic properties of our first anisotropic superfluid, liquid helium-3. It is certain that it will remain a playground for both theorists and experimentalists for many years to come!

9.6   Applications and implications

In this section I shall try to outline briefly some of the ways in which progress in understanding the phenomena of superconductivity and superfluidity may help us in other areas of science and technology.

In the first place, the phenomenon of superconductivity itself, and some of the more sophisticated effects associated with it, has a number of direct technological applications. For example, it allows one to pass very high electric currents with no dissipation of energy; this is already routinely used in making large electromagnets, and may well be used in the future in the context of the transmission and storage of electrical power. It is also entirely conceivable that the phenomenon of 'superconducting levitation' in Section 9.4 will find large-scale applications; until recently the main obstacle to such applications was the need to cool large masses of metal down to well below 25 K, the highest temperature then known for superconductivity to occur. The discovery of superconductivity in the new copper-oxide materials up to at least 125 K may ease this particular problem considerably, since these temperatures are well within the range of liquid-nitrogen cooling. On the other hand, at the time of writing it is unclear whether other problems, such as the need to make robust and durable large-scale elements, can be successfully overcome using the new materials. However that may be, the advances of the last two years make the quest for the superconducting technologist's ultimate goal – room-temperature superconductivity – look considerably less of a pipe-dream.

A different kind of application arises from the extreme sensitivity of the superconducting phase to small forces; as was mentioned in the context of superfluid helium-3 in Section 9.5, this feature is in essence a consequence of the fact that all the atoms (in helium-4), or in the case of a superconductor or helium-3 Cooper pairs, have to be doing the same thing at the same time. A particularly useful example of this is the extreme sensitivity of superconducting circuits incorporating Josephson junctions (like the circuit shown in figure 9.11) to small changes in the magnetic flux through the circuit; by a suitable experimental arrangement, changes in flux of the order of a small fraction of the flux quantum (about 2×10^{-15} Wb) can be easily measured. Since the area of such a circuit is typically of the order of a few square millimetres, this means that changes in the average magnetic field as small as 10^{-13} T (10^{-9} G, or about one part in a billion of the Earth's magnetic field) can be quite routinely measured. This has numerous applications in cardiography, the neurophysiology of the brain, geophysical prospecting and elsewhere; indeed, the use of superconducting magnetometers has become standard whenever extreme sensitivity is required.

The direct technological applications of the phenomenon of superfluidity, in either of the helium isotopes, are at present less numerous, the main traditional one being the exploitation of the excellent heat transfer properties of the superfluid phase in cooling other systems (e.g. superconducting magnets) to temperatures of the order of 1 K. A more recently developed application exploits the anomalous flow properties to effect isotopic purification of helium-4 to a degree several orders of magnitude better than is possible for any other element: this ultrapure helium-4 can then be used as a 'bottle' for neutrons, an application now being envisaged in particle-physics experiments. In addition, it seems by no means inconceivable that the macroscopic quantum properties of the superfluid phases may in the future be put to practical use; in particular, the ability of superfluid He-II to detect absolute rotation by a quantum mechanism could perhaps in principle be of use in space-flight navigation.

Turning now to the applications and implications of research in this area for other fields of physics and of science, we find a striking example in the close ties which have developed in the last three decades with research in particle physics. In fact, one of the most important ideas in contemporary particle physics, the 'Higgs mechanism' by which the so-called gauge boson acquires a finite mass (see Chapter 17 by Taylor) is closely associated, both historically and conceptually, with the Meissner effect in superconductors (which, from this point of view, consists in the photon effectively acquiring a mass). A number of other ideas in the theory of superfluidity, such as the concept of topological conservation, also have close parallels in a particle-physics context, and there has been a constant and fruitful exchange of ideas between these two apparently disparate fields over the last thirty years or so.

This kind of cross-fertilisation rests on the similarity of the formal aspects of the problems in the two areas. A rather different kind of application of low temperature physics to the study of high energy processes may arise from the extreme sensitivity of superfluid systems to ultraweak forces. For example, there is strong evidence that the 'weak interaction' which exists between elementary particles, unlike the electrical and gravitational forces with which we are familiar in ordinary life, in some sense 'knows the difference' between a right and left handed screw and between the forward and backward directions of time (cf. Close's chapter). While this feature can be inferred from the study of collision and decay processes involving only a few particles, in any normal macroscopic system any effects of the weak interaction are totally swamped by the much stronger electromagnetic forces. However, in principle a superfluid system should be able to amplify these effects up to a level where they might be detectable. Moreover, if there are other such effects which are so weak that they have so far escaped detection at the single-particle level, it is by no means inconceivable that they could show up at the macroscopic level in a superfluid. There have been several theoretical proposals of effects along those lines, and while none of them can be said at the time of writing to have firm experimental confirmation, one or two at least seem sufficiently close to the edge of detectability that it is probably only a matter of time and energy before they are in fact detected.

An even more exciting potential application of low temperature systems, in particular superconductors, lies in the area of the foundations of quantum mechanics. In representing a microscopic system such as an electron by a de-Broglie wave, quantum mechanics in effect allows it not to 'have' a particular property until this property is actually measured (see Chapter 13 by Shimony). (For example, in the well known 'Young's slits' thought experiment an electron will, if its trajectory is measured, be found to pass through one slit or the other, yet the interference pattern built up by the arrival of electrons on the screen indicates that the de Broglie wave representing the electron passed through *both* slits. See also the discussion of the neutron interferometer in Section 9.2.) If we believe that quantum mechanics is a universal theory, then this property should in principle not be confined to the level of electrons and atoms: in fact, even obviously macroscopic objects such as tables and chairs should, under appropriate circumstances, not 'be' (or should not be described as being) in definite states! Now, it turns out that in practice, although the quantum formalism does indeed lead to this bizarre result, under all normal circumstances it has no experimental consequences – that is, it is impossible to construct an experiment on macroscopic objects which would be the analogue of the Young's slits experiment for an electron. For this reason most physicists shrug off the paradox as 'philosophical'. However, it has become clear in recent years that there is a very real possibility that by an appropriate arrangement of superconducting circuits one might be able to arrange an experimental test of this bizarre quantum-mechanical prediction – that is, one could establish (or not) that such a circuit 'was not' in a definite macroscopic state until it was measured to be so. If such an experiment could be done, and came out in favour of the quantum-mechanical prediction, it would considerably sharpen the paradox; if it came out *against* quantum mechanics (after all the many theoretical and experimental loopholes had been plugged!) this would be even more exciting, since it would indicate that quantum mechanics is not a universal theory of the world but that, at some level of scale and/or complexity, new effects, at present completely unknown to us, begin to play a role. At the time of writing, some 'circumstantial' experiments have been done in this area (in particular, it has been confirmed experimentally that the characteristically quantum-mechanical phenomenon of the tunnelling of a system through a classically impassable barrier can occur also in the case of (reasonably) 'macroscopic' objects), and it appears very likely that the decisive (in principle!) experiment will at least be attempted in the next two or three years.

Finally, let me indicate briefly some of the possible extensions of the theory of superconductivity and superfluidity to other systems. One such system was already mentioned, namely the new class of 'heavy-fermion' superconductors; these may be forming Cooper pairs with nonzero angular momentum, like the atoms in superfluid helium-3, and if so we should have our first genuine 'anisotropic superconductors', which might then be expected to combine many of the spectacular properties of ordinary superconductors and of superfluid helium-3. Another 'exotic' system is the neutrons in neutron stars, which for a certain range of density are predicted to form an anisotropic superfluid state. Yet other systems have been considered as possible candidates for Cooper-paired phases, some of them with the hope (cf. above) that the pairing would take place even at room temperature. However, one may hope that the theory of superconductivity and superfluidity may give us guidance in a less restricted sense. It is probably fair to say that the systems considered in this chapter, in particular superfluid helium-3, represent the most 'sophisticated' phases of matter known to us for which we can begin to claim a detailed quantitative understanding. There are, of course, very many other forms of matter, ranging from relatively simple enzymes to the human brain, of whose behaviour we have only a very rudimentary and qualitative understanding, if any at all. In the study of low temperature systems, we have acquired some idea of *some* of the ways in which the quantum mechanics of individual particles and the collective interactions between them can combine to produce qualitatively novel behaviour even at the macroscopic level. It may not be too arrogant to hope that this understanding may be useful, in ways which at present we cannot even envisage, in the study of these even more fascinating biophysical systems.

10 Quantum optics

Peter Knight

10.1 The nature of quantum optics

Optics before the laser was concerned entirely with the production, manipulation and detection of noise. The light our eyes receive from natural processes is wildly fluctuating in amplitude and phase, reflecting the chaotic, random environment which gave birth to these photons. Light is emitted by atoms which were excited either by collisions which transfer kinetic energy to the internal state of the atoms, or by photon absorption. Each different atom in the light source is excited in quite unrelated and independent events. Once excited, the atom emits radiation in two ways, either spontaneously or by stimulated emission induced by the surrounding radiation. In the natural world outside the laboratory, spontaneous emission is the dominant process, with each photon spontaneously emitted being completely uncorrelated in phase with any of its neighbours. The total field radiated by many atoms reflects this lack of correlation through extensive intensity and phase fluctuations as the individual fields from each atom irregularly interfere. When the number of excited atoms is increased, however, the atoms begin to communicate through their common radiation fields and their radiative dynamics become tightly correlated as stimulated emission takes over from spontaneous emission. A kind of phase transition occurs in which a common radiative field is established and ordered coherent light is produced. Such nonequilibrium processes are responsible for the laser threshold, with a dramatic transition from disordered random emissions to correlated coherent emission as the number of excited atoms is increased. Quantum optics is concerned with the nature of such optical correlations, the description of coherence and the properties of photons and their interactions with atoms. I will describe in this chapter the processes involved in the generation of photons, the properties of laser light and how its extreme coherence and brightness may be exploited.

10.2 Photon emission and lasers

The classical electrodynamical theory of light due to Maxwell treats light as a wave motion at a speed $300\,000\,\mathrm{km\,s^{-1}}$, with the speed c related to the wavelength λ and the frequency v by $c = \lambda v$. The electromagnetic wave is a combined periodically varying electric field E and magnetic field B (figure 10.1). Visible light has a frequency $\omega \sim 3 \times 10^{15}\,\mathrm{s^{-1}}(5 \times 10^{14}\,\mathrm{Hz})$. If we confine the electromagnetic radiation in a cavity formed by two mirrors separated by a distance L, then only standing waves exist within the cavity, provided an integer number n of half-wavelengths fit into the length $L = n(\lambda/2)$. This is true only for a set of discrete frequencies that describe the modes which are the natural states of the radiation able to exist in the cavity. The density of modes depends on the square of the frequency, so that it is much more straightforward to isolate a single radiation field mode in a cavity for microwaves than for light.

Figure 10.1. Electromagnetic field propagation: the electromagnetic wave is made up of an electric field E and a magnetic field B. The electric and magnetic fields oscillate in orthogonal directions.

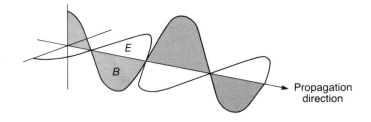

289

The quantum theory of light takes the mode description of radiation as its starting point. Each mode of frequency ω is quantised with its energy in discrete quanta $\hbar\omega$ (about 3×10^{-19} J for optical frequency photons), where $\hbar$ is Planck's constant. The single excitation of a mode represents the presence of a photon in that mode. In quantum theory an oscillator even in its lowest energy state has a residual zero point energy which prevents precise localisation of position and momentum. The optical oscillation of an electromagnetic field mode has just such a zero point energy of $(\frac{1}{2}\hbar\omega)$ associated with the random fluctuations of electric and magnetic fields even in the lowest, vacuum state. The vacuum in quantum electrodynamics is pictured as a ferment of activity with large uncertainties in the strength of the electromagnetic field, and these uncertainties are the source of quantum noise in quantum optical systems such as lasers. The association of photons with excitations of normal modes of the electromagnetic field is essential for a proper understanding of interference: interference occurs whenever there are at least two different modes which support the photon energy. Each has a probability amplitude, and the total probability amplitude is made up of the sum of amplitudes so that the probability contains interference cross-terms.

The source of optical photons is emission from atoms in excited states. Atoms can be excited by a number of different mechanisms. For example, in a gas discharge lamp electrons in the electrical discharge collide with atoms and leave them in excited states. Photons from an outside source can also excite atoms if their energy matches the transition energy required. The microscopic processes involved in the optical excitation and de-excitation of atoms were first studied by Einstein. Atoms in their lowest energy ground states can make transitions to excited states by absorption of incident radiation. The rate of absorption R_a depends on the photon flux through the density of radiation ρ at the required resonance energy: $R_a = B\rho$, where B is the Einstein 'B-coefficient', which depends on atomic parameters governing the ease with which transitions can occur. Once excited, the atom can lose its extra energy by stimulated emission, again at rate $B\rho$. But even in the absence of incident radiation an excited atom can lose its extra energy by spontaneous emission at rate A quite independent of photon energy density ρ. The spontaneous decay transitions can be heuristically regarded as caused by the excited atom being perturbed by the residual vacuum fluctuations of the quantised ground state vacuum of the electromagnetic field.

Spontaneous emission is completely isotropic: there is no preferred direction of emission when all the equally probable polarisation states of the radiated light are taken into account. Stimulated processes, on the other hand, have a built-in preference for emission into the direction of the incident flux of photons. A collection of many atoms can be excited so that the emitted light contains contributions from both stimulated and spontaneous transitions (see figure 10.2). We concentrate on

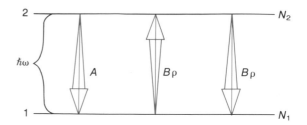

Figure 10.2. Stimulated and spontaneous transitions between atomic states. The spontaneous rate A is independent of incident photon flux ρ, the stimulated rates upwards and downwards are given by $B\rho$. The A and B coefficients for these transitions were first given by Einstein.

the two levels participating in the transition dynamics: we imagine there are N_1 atoms in the lower energy state 1 and N_2 in the excited state 2. In thermodynamic equilibrium at temperature T there are far more ground state atoms than excited ones: the relative numbers are given by the Boltzmann factor $(N_2/N_1) = \exp(-\hbar\omega/kT)$, where k is the Boltzmann constant. Under these circumstances, if the incident photon flux is low, spontaneous emission dominates the de-excitation and photons are emitted isotropically. A thermal light source, with $N_2 \ll N_1$, is a source of independent random spontaneous emissions, each uncorrelated with the other, and represents just optical noise.

Lasers work by creating large departures from thermodynamic equilibrium. No matter how hard a two-level system is excited, the Boltzmann factor prevents the number of excited atoms exceeding the number of ground state atoms in thermodynamic equilibrium. Theoretically even at infinite temperatures the populations are only equalised. We get round this equilibrium restriction by employing more than two atomic states in the dynamics (see figure 10.3). The atoms are excited (either optically, by collisions with electrons in an electrical discharge, or other means), from the ground state 0 to a highly excited state 2. The Boltzmann factor demands $N_2 \ll N_0$. Nevertheless, the number in the highly excited state greatly exceeds that in state 1. We have found a way of producing population inversion in states 2 and 1. The excited atoms spontaneously decay from state 2 to state 1, radiating photons in all directions. The laser geometry is chosen to be a long thin rod. By chance, any photons spontaneously emitted might encounter further excited atoms and stimulate them to decay, emitting more photons but in the same direction. Photons spontaneously emitted into the side directions are just lost. In this way an avalanche of stimulated emission will produce an amplified beam of light down the axis of the laser rod (figure 10.4). The light produced by stimulated emission is in phase with the incident stimulating light. In a laser, the initial phase is spontaneously generated by the first emission, and subsequent emissions lock on to that arbitrarily chosen phase, with small fluctuations caused by rare spontaneous emissions into the on-axis laser modes. To encourage the stimulated emission, mirrors are often used to reflect the light beam back into the active atomic medium to induce further emissions in repeated transits of the light. This encourages the formation of highly directional beams of stimulated photons characteristic of lasers. Any emission in directions other than that of the mirror reflection direction will not be redirected back into the atomic gain medium and will not be further amplified.

A 1 W laser with an output beam 1 mm in diameter (typical of some continuous wave gas lasers in which photons are generated continuously, the gain being replenished by a constant pump mechanism) has a very small beam divergence: at a distance of ten metres the beam diameter has spread through diffraction to a diameter of only 5 mm. A much brighter thermal light source such as a conventional tungsten filament light bulb could emit 100 W, but this light is radiated in all directions, so that at ten metres distance the power through a 5 mm section is less than a millionth of a watt. This is why even low power laser beams can be seen from enormous distances, and are employed in surveying theodolites and even in lunar ranging where the earth–moon distance is measured with an accuracy of a few centimetres.

Laser light is not only highly directional, but is made up of an extremely narrow range of frequencies. Normal atomic spontaneous emission takes place in a time $\Delta t \sim 10^{-9}$ s, and the spontaneously radiated electromagnetic field has a frequency spectrum centred on the atomic transition frequency, with a width $\sim 1/\Delta t$ of about 100 MHz. There are additional sources of frequency broadening of the emitted light spectrum. In a solid, there are crystal strains and inhomogeneities which distort the energy levels of the emitting ions. Ions in different parts of the crystal differ in their transition frequencies, so that the light is emitted in a band of frequencies reflecting the frequency width of these inhomogeneities in the crystal. In a gas, atoms move with a distribution of velocities v, and radiate a field which is shifted in frequency by the Doppler effect by a fractional shift $(\Delta \omega / \omega) = v/c$. Atomic velocities are approximately 10^3 m s^{-1} so $\Delta \omega / \omega$ is about 10^{-5} and the Doppler shift $\Delta \omega \sim 10^{10}$ Hz. Of course, there is a distribution of velocities and a range of emitted frequencies with a width approximately equal to $\Delta \omega$. The light from such atoms therefore has an extremely broad bandwidth. In a laser, stimulated emission could take place throughout this range, with gain maximised at the centre of the Doppler distribution of transition frequencies. If there are no laser mirrors, the linewidth of the laser radiation will narrow as stimulated emission is more likely at the centre of the frequency spectrum than out in the wings. This is known as gain narrowing. If the laser operates with reflecting mirrors at each end there are further restrictions on the possible frequencies allowed in the laser cavity. Only radiation whose wave-

length is just right to allow an integer number of half-wavelengths to fit between the mirrors ($n\lambda/2 = L$) is supported as a cavity mode. Many such modes may be possible within the gain frequency width $\Delta \omega$ (see figure 10.5). The mode spacing is $c/2L$. Modes whose frequencies fall inside the range where the optical gain exceeds the losses caused by absorption or diffraction can be amplified and the laser light will consist of the interfering superposition of light from the available modes. If the laser cavity length L is small, only one mode will see gain and will be amplified by stimulated emission. Such single-mode laser action can result in extremely monochromatic light output.

Figure 10.3. Three-level atomic pumping to achieve population inversion. The atoms are initially concentrated in the ground state 0, from which they are pumped either by collisions or radiation to an excited state 2. Stimulated and spontaneous transitions to state 1 form the laser action, and atoms are removed from 1 back to the ground state by a relaxation.

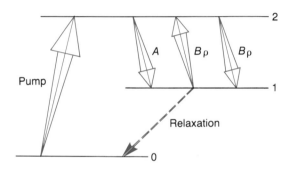

Figure 10.4. Schematic description of light amplification by stimulated emission and losses through absorption and spontaneous emission. ○ excited atom; ● ground state atom, absorbing photons. (Adapted from Heavens, 1970.)

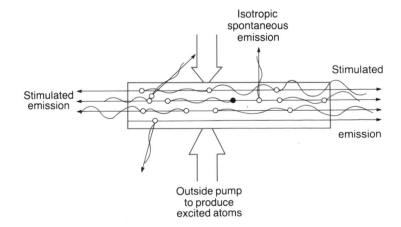

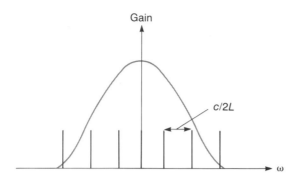

Figure 10.5. Frequency dependence of optical gain and available mode frequencies which are separated by $c/2L$. The gain is centred at the natural atomic transition frequency and has a broad width. Whether or not several modes can operate at frequencies for which there is substantial gain depends on the cavity length L.

Laser action has been reported in a huge variety of atomic media, with frequencies from the far infrared to the soft X-ray regions of the spectrum. Shorter wavelengths are much more difficult to generate: it's just so much harder to generate population inversion when the atomic transition frequencies are so large. Lasers can be miniaturised – for example, the low power semiconductor laser diode used in fibre-optic telecommunications is much less than a millimetre in any dimension. High power lasers can be quite enormous: the largest is the neodymium-doped glass laser NOVA constructed at the Lawrence Livermore National Laboratory in California. NOVA generates 100 kJ pulses of infrared light (with 1.05 μm wavelength) in 3 ns pulse lengths, so the peak power can approach 100 terawatts. The light is amplified in chains of glass rod and disc optical amplifiers in ten separate beams of up to 74 cm diameter which can be focused on to a microscopic target for laser-fusion studies. In laser fusion, the target is heated to thermonuclear temperatures and pressures by the huge flux of incident photons. This enormous laser cost $176 million to build and has a pulse repetition rate of less than a shot a day!

10.3   Quantum properties of light and radiative processes

Photons

The modern quantum theory of light has unified the old classical notions of 'wave' and 'particle' in the concept of probability amplitudes. In any optical experiment the quantised electromagnetic field is represented in terms of the normal modes which represent the solution to the wave equation for light; the occupation of these normal modes is in discrete amounts, and the photon is the quantum of energy of these modes. In some circumstances the quantum of energy can be in one of a number of alternative modes. For example, a wavefront from a source can be used to illuminate two closely spaced slits, see figure 10.6. Very crudely we can imagine the light from each slit to a point on the screen as defining two possible paths or modes. Each of these has a certain probability amplitude a_i, $i=1,2$, whose absolute square represents the probability that that particular mode contains the quantum of excitation. We add probability amplitudes $(a_1 + a_2)$ to find the total probability $|a_1 + a_2|^2$ of finding the quantum at the screen. The intensity at the screen is the consolidation of many repeated experiments, each photon taken singly producing quantum interference. The interference is *not* between photons: each photon interferes only with itself through the basic quantum uncertainty of which path through the optical apparatus is taken (see Shimony's Chapter 13, where fundamental problems in quantum mechanics are illustrated by this experiment).

It is extremely difficult to test the photon concept using interference experiments. A very early two-slit interference experiment was carried out by G. I. Taylor in 1909 with an extremely feeble source of light to see whether the interference fringes altered when only one quantum at a time is present in the apparatus. His light source was so weak that his exposure time of the photographic plate was up to 2000 hours (about three months). He described his light source in the following way: 'A simple calculation will shew that the amount of energy falling on the plate during the longest exposure was the same as that due to a standard candle burning at a distance slightly exceeding a mile.' In no case did Taylor find any diminution in the sharpness of the interference pattern, supporting our contention that interference is the aggregated sum of many single-photon interference effects. Many later experiments have confirmed this pioneer work.

The photon, being a single occupation of one of the radiation field states, is indivisible. In 1974 John Clauser performed an ingenious experiment in which he tried to split a photon in two, with no success. In his experiment, photons were produced by an atomic emission from a very highly excited state which decayed by emitting two different frequency photons in sequence. The arrival of the first photon effectively signalled the incipient arrival of the second. Interference experiments can then be performed with this second photon, with counting equipment being turned on by the detection of the first photon. In this way we can guarantee that only a single photon is being studied. This single photon is incident on a glass slide which acts as a beam splitter. Classically a proportion of the light will go in each direction and a coincidence recorded at detector photomultipliers PM1 and PM2 placed behind the beam splitter (figure 10.7). Quantum mechanically *no* such coincidence is possible: there is a finite probability of the photon being reflected to PM1 *or* transmitted to PM2. Experimentally no coincidences were seen by Clauser.

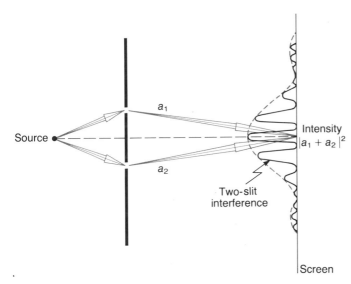

Figure 10.6. Two-slit optical interference. Light from a source illuminates two slits in an obstructing barrier and diffracts through to form two interfering beams on a screen. The amplitudes of the light through the screens are a_1 and a_2 and the intensity is the absolute square of the sum of these amplitudes. Interference generates light and dark fringes when the amplitudes add constructively and destructively.

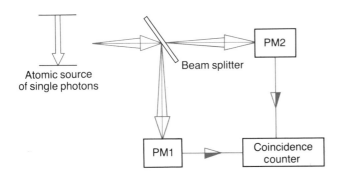

Figure 10.7. An experiment to demonstrate the indivisibility of photons. No simultaneous counts are registered in this experiment to beam split single photons generated from an atomic source and detected by two photomultipliers PM1 and PM2.

Photons and atomic transitions

More recently, experimentalists have been able to measure quantum jumps in single atoms. Atoms excited by lasers tuned to atomic transition frequencies can emit many photons by resonance fluorescence: the atom can spontaneously emit a photon and return to the ground state from which it can be re-excited and fluoresce again, and so on. At low laser excitation

intensities, the process of resonance fluorescence is well understood. If the excitation field is monochromatic, the atom absorbs a photon at the excitation frequency, and energy conservation demands that the emitted photon has the same frequency. The emitted fluorescence has the same spectral profile, and in particular the same narrow width as the excitation, and can be much narrower in width than the 'natural' width given by the reciprocal of the excited state lifetime.

When the exciting laser intensity is increased, the fluorescence completely alters. In a strong resonant laser field nonlinear scattering occurs in which several photons rather than single photons are involved. Multiphoton processes become as important as single-photon processes. Energy conservation holds overall for the multiphoton process, but the fluorescence spectrum is no longer nearly monochromatic. The atom can coherently interact many times with the laser radiation, absorbing and emitting by stimulated transitions before spontaneously radiating a photon.

An atom of principal transition frequency ω_0 interacting with a coherent laser field of frequency ω is stimulated to emit and reabsorb photons. The rate at which transitions are coherently induced between the two atomic levels is known as the Rabi frequency and is proportional to the square root of the laser intensity. One can think of the atom heuristically as a damped driven quantum oscillator, the damping being supplied by spontaneous emission and the driving force provided by the interaction with the laser field. At low intensities, spontaneous emission is more important than stimulated processes, and the atom behaves as an over-damped oscillator. A classical oscillator will prefer to vibrate at its own natural frequency if left to its own devices. But, when it is driven by an external force which has its own frequency, the classical oscillator settles down to an imposed harmonic motion at the driving frequency. A charged oscillator will then radiate an electromagnetic field whose frequency equals this driving frequency. The electron charge distribution in the atom oscillates just like this classical oscillator, radiating a field centred upon the driving frequency. At higher intensities the stimulated Rabi frequency can exceed the spontaneous emission rate. The atom is excited by photon absorption, but the photon flux is so high that other photons stimulate emission before spontaneous decay can occur, and the atom behaves just like an under-damped oscillator. The Rabi oscillations show up not only in the population of the atomic levels (the inversion) but also as a modulation of the quantum dipole moment. When the atomic dipole moment oscillates at the driving laser frequency it radiates a field at that frequency. If the atomic dipole has an additional time-dependence, this is reflected in extra frequencies in the radiated field. The simple sinusoidal modulation created by the atomic Rabi oscillations leads to new satellite frequencies above and below the driving frequency called sidebands. This so-called 'dynamic Stark splitting' has attracted a great deal of interest.

Ordinary Stark splitting in atoms is caused by external electric fields which severely distort atomic charge distributions. This distortion shifts the atomic energy by an amount called the Stark shift. We have seen above that even oscillating electric fields in laser radiation can induce such effects in fluorescent emission of photons. Theorists find the problem of intense field fluorescence challenging as an example of quantum fluctuations influencing strong-coupling physics typified by the laser excitation.

The emission spectrum is calculated directly from the time development of the quantum atomic dipole moment excited by the laser field. Contributions to the fluorescence arising from the modulated dipole moment are found at frequencies displaced above and below the laser frequency by the Rabi oscillatory frequency. A second contribution derives not from the dipole but from the population in the excited atomic state which radiates directly at the driving frequency. These two contributions lead to a strong-field fluorescence spectrum with a central spontaneous emission line and two subsidiary sidebands separated from the central component by the Rabi frequency. Theory demonstrates just how this happens in detail and predicts a spectrum made up of three Lorentzian-shaped lines: a strong central component at the laser frequency and two sidebands with peak intensities $\frac{1}{3}$ that of the central line, see figure 10.8. The factor of $\frac{1}{3}$ derives from the relative importance of the dipole moment evolution and the excited state population. Populations decay twice as fast as dipole moments in laser excitation. The central component derives from both dipole and population terms in a quite complicated and unexpected way, whereas the sidebands derive from the dipole only.

Figure 10.8. Laser excited atomic fluorescence spectrum. A beam of atoms is illuminated by a laser whose frequency is chosen to match resonantly the atomic transition frequency. The atoms radiate a fluorescent field which is examined as a function of frequency ω.

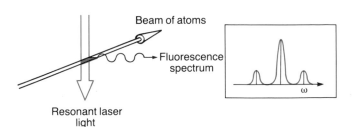

Experimentally atoms in an atomic beam are used to study this dynamic Stark splitting (to avoid Doppler widths and collisions), and are excited by a laser field resonant with an atomic transition frequency. The spectrum of the fluorescent light is examined as a function of emitted photon frequency ω. The central component is found to have the natural width γ. The sidebands have a width $3\gamma/2$ and are displaced from the central peak by the Rabi frequency Ω, which is proportional to the amplitude of the laser electric field strength. Experimental work using dye-laser excited atomic beam methods broadly confirm the original theoretical predictions. The triplet splitting as the intensity increases has been clearly observed. The coherence of the incident laser is important in these basically multiphoton processes. Intensity fluctuations will obviously be important as a 'smearing out' effect: the a.c. Stark splitting is proportional to the square root of the instantaneous intensity. Similarly, phase fluctuations will create jitter in atomic dipole moments and correlation functions.

Antibunching and photon correlations

It is surprising to find that very few measurable optical effects require field quantisation and the idea of a photon for their explanation. One notable example of this rare class is photon antibunching in intensity correlation of resonance fluorescence. Whereas most optical coherence and correlation experiments have an adequate semi-classical interpretation, antibunching requires quantisation of the emitted fluorescence and the idea of a 'quantum jump'.

To see photon antibunching, the same experimental apparatus used to study resonance fluorescence is employed. The radiation emitted from the resonantly excited atoms is collected, split in two, and imaged onto two photomultiplier tubes whose output is electronically correlated. The intensity correlation of interest is described by the second-order correlation function $g^{(2)}(\tau)$. The $g^{(2)}(\tau)$ correlation function is a measure of the joint probability of detecting a photon at time t and another at time $t+\tau$ later, and must therefore vanish as $\tau \to 0$ for light from this single-atom source. This is because an atom can only emit one photon at a time and requires a delay τ to enable it to be re-excited to the excited fluorescent state. Formally, the correlation function is defined as the average of the product of the light intensities at the two times t and $t+\tau$, divided by the square of the average intensity to give a normalised dimensionless form

$$g^{(2)}(\tau) = \frac{\langle I(t+\tau)I(t)\rangle}{\langle I(t)\rangle^2}.$$

There are quantum states of the electromagnetic field which do not have a classical description, but the experimental realisation of such states has proved difficult. States exhibiting negative photoelectric correlation ($g^{(2)}(0) - 1 < 0$, where $g^{(2)}(0)$ is the correlation measured with zero delay time τ) or 'antibunching' are much discussed examples of an effect requiring field quantisation: it is hard to visualise a function $I(t)$ representing the classical intensity whose mean square is less than the square of its mean. The resonance fluorescence radiated by a single two-level system driven by a coherent field possesses the required statistical properties to exhibit antibunching. This can be demonstrated easily: the atom emits a photon at time t and is unable to radiate again immediately after having made a quantum jump down to the lower state.

The measurement of $g^{(2)}(\tau)$ consists of two parts: detection of the first photon (ensuring the de-excitation of the source atom) and a delay while the atom is re-excited by the resonant driving field for the subsequent re-emission of a second photon. So

$$g^{(2)}(\tau) \propto p(t,\tau),$$

where $p(t,\tau)$ is the probability of finding an atom, initially in its lower state at t, re-excited at $t+\tau$. The antibunching effect in resonance fluorescence is a striking example of wavepacket reduction in action. When the first photon is detected at time t, it is certain that at that time the atom is not in a superposition state but is instead in the ground state. The probability of the next photon to be detected at $t+\tau$ is proportional to the conditional probability of being excited at time τ after being *not* excited at time t. We expect first a lack of coincidence counts at short times $t+\tau$, and further expect the Rabi oscillation of $p(t,\tau)$ to be reflected in $g^{(2)}(\tau)$. At longer times $t+\tau$, the atom can be restored to its excited state by the resonant laser field and so be quite likely to emit a second photon. At such times, the light would be 'bunched' or strongly correlated. In general, $g^{(2)}(\tau)$ should exhibit periodic bunching and antibunching as the driven atom undergoes Rabi oscillations. For a sufficiently intense resonant driving field for which the Rabi frequency Ω is much greater than the spontaneous decay rate γ,

$$g^{(2)}(\tau) = 1 - [\exp(-3\gamma\tau/2)]\cos\Omega\tau.$$

If there is more than one atom in the observation region the interpretation is a little harder, and since fluorescence could then be produced by two or more atoms this would increase $g^{(2)}(0)$ and obscure the antibunching.

More recently it has become possible to study the photons emitted by a single ion trapped in space and excited by a laser field. By carefully designing electromagnetic fields in a confining geometry of electrodes, we can make a confining potential well in which cold ions in an ultrahigh vacuum can be trapped. If the observer is carefully dark-adapted, and looks through a well designed microscope objective lens, he or she will see the fluorescence emitted by the ion excited by a laser as a minute twinkling atomic 'lighthouse'. H. Dehmelt proposed an ingenious experiment which exploits this atomic fluorescence to construct an atomic amplifier to measure exceedingly weak transitions, and in doing so made individual quantum jumps visible to the naked eye. The eye of course is a superb detector of weak radiation and sees the stream of photons from the single ion, when given time to adapt to the darkened laboratory. In Dehmelt's atomic amplifier, two transitions are possible, driven

by two different lasers, one to a strongly fluorescing state which decays so quickly that it can scatter up to 10^8 photons per second, the other to a metastable state (figure 10.9) which can live for several seconds. The strongly fluorescent state dominates the radiative dynamics. But whenever the atom makes a transition (a quantum jump) to the metastable state, this strong fluorescence is terminated and a period of darkness ensues. This period of darkness is ended when the atom spontaneously decays from the metastable state 2 back to the ground state 0, from which the transitions to the strongly fluorescent state are resumed. The spontaneous decay is a random process, and so the resumption of strong fluorescence takes place at random times. As soon as the laser field resonant with the transition to the metastable state 2 is turned on, our atomic 'lighthouse' begins to flash on and off at random times. The transition to the metastable state removes millions of strong fluorescence photons, and in this way the *lack* of millions of photons constitutes a visible, macroscopic signature of a weak transition.

Rydberg atoms as sensors of transitions

Quantum optics and laser spectroscopy have made it possible to study radiative transitions in very highly excited atoms which form an ideal test-bed for quantum effects in the interaction of photons with atoms. Highly excited atoms with one electron orbiting at large distances from an unexcited core are termed Rydberg atoms and have remarkable properties. Their size gives them a large polarisability, and they are easy to ionise in weak electric fields. If they possess a large angular momentum they do not return rapidly to their ground state by radiative decay simply because decay photons cannot easily carry away all that excess angular momentum. Rydberg atoms provide a unique vehicle for studying radiative processes over long times. All atomic systems are immersed in a bath of black-body radiation, a background thermal field characterised by the ambient temperature of the body. For the most part physicists interested in radiative transitions of excited atoms just ignore the existence of this field, principally because the number of black-body photons at room temperature with a frequency close to the atomic transition under study is so small, so that stimulated processes due to background thermal fields are negligible. Nevertheless, there will always be *some* ambient photons at the transition frequency, and if the transition frequency is small enough the distribution of thermal photon numbers ensures that stimulated emission and absorption will begin to take over from purely spontaneous radiative decay. Recent experiments have indicated that transitions between very highly excited Rydberg states are sensitive to background thermal fields. At the transition frequencies that interest them, the number of photons at each frequency in the apparatus (typically only a few centimetres in size) at room temperature can be as high as ten. The stimulated absorption and emission of this background radiation rapidly redistributes the population initially prepared by laser excitation in a single atomic level among its neighbours and considerably shortens the lifetime of some states. Consequently it would seem quite difficult to observe the decay of a single isolated Rydberg state in these circumstances: after a very short time a distribution of states rather than a single state exists.

It is interesting to the theorist that the atom immersed in a radiation field is quite different from an isolated atom. The radiation bathing the atom also dresses it, so that what started as an isolated state is scrambled into a distribution of nearby states with transition frequencies determined at least in part by the local environment. Hot atoms are not just moving faster than cold atoms (and therefore the ensemble emitting a wider Doppler-broadened line); each hot atom is distinctly different from a cold atom in its radiative properties.

Transitions between atomic states are induced by incident electromagnetic fields. Even in the absence of an incident field, the vacuum fluctuations and zero point energy of the quantised field interact with atoms and are responsible for spontaneous emission and for small radiative corrections to atomic energy levels. Ordinary stimulated transitions can be altered by changing the incident field frequency or intensity, but spontaneous decay would appear to be unalterable unless the field vacuum state can be changed. D. Kleppner of MIT proposed a method for doing just this, by placing the excited atom in a small high-Q cavity which 'squeezes out' some of the radiation modes and changes the mode density. (The Q-factor determines how long a cavity can store radiation. The radiation decay rate is ω/Q, where ω is the frequency.) The cavity can either frustrate or enhance the decay by a factor of the order of Q of the cavity, depending on the cavity detuning from atomic resonance. The effect of mode confinement on decay lifetimes provides new and direct information on the nature of spontaneous decay.

The idea of altering the radiation field vacuum state is an old one, going back at least to work in the 1940s on quantum

Casimir forces between conducting plates. An enclosing cavity alters the boundary conditions of the problem and changes the appropriate mode expansion for the field. The Casimir force derives from the change in the density of modes between two plates. As the plates are brought together, low frequency modes with wavelengths longer than one-half the separation between the plates are excluded. Their zero point fluctuation energy (equal to $\frac{1}{2}\hbar\omega$ for each mode) is lost. The suppression of such mode energy leads to a small but measurable force between the plates. An elementary example of mode suppression is a one-dimensional cavity supporting only standing waves: an excited atom in such a cavity will have a strongly position-dependent lifetime, which is enhanced or diminished depending in part on whether the atom is sited at a node or antinode of the field mode.

Kleppner's suggestion was to 'freeze out' vacuum modes by using a waveguide cavity with characteristic dimensions small compared with the decay radiation wavelength. This seemed to be unattainable with optical frequencies involved in normal decay.

In 1981, Kleppner and his research group used laser-excited Rydberg atoms to observe these mode-confinement effects. They report a sudden increase in the induced lifetime of a Rydberg atomic transition between two conducting plates when the resonant wavelength is increased beyond the waveguide cut-off (twice the plate separation d). The conducting plates are unable to support field modes of wavelength less than $2d$ in the direction normal to the plates. In addition, those modes parallel to the plates have their density altered but not turned off. Thermal radiation can only be supported on these modes so that transitions induced by black-body radiation inside the plates are sensitive to the confining geometry.

The technique used to observe these waveguide effects by Kleppner and his group is particularly ingenious and is outlined schematically in figure 10.10. The sodium 29d Rydberg state is populated by stepwise absorption from two pulsed lasers of wavelengths λ_1 and λ_2 which drive the transitions 3s to 3p to 29d in an atomic beam travelling between two conducting plates separated by $\frac{1}{3}$ cm. The whole apparatus is placed in a shielded box at 180 K so that the photon occupation number (the number of radiation field quanta in the apparatus) at wavelength $\lambda \sim 2d$ is approximately 86.

All radiation polarisations in the atomic transition are possible. The transition 29d to 30p at a wavelength of $\frac{2}{3}$ cm ($\sim 2d$) is chosen for study and is excited by the thermal field at a rate of 300 per second. (Note that the atomic labels are explained in box 10.1.) Transitions to other levels at $\lambda \neq 2d$ are of course also induced and contribute to a background which does not depend on the plate separation. The plate separation is kept fixed but the transition wavelength is varied by exploiting the large polarisability of Rydberg states to Stark shift energy levels in a small d.c. voltage across the plates. The cut-off frequency $v_c = c/2d$ is $1.48 \, \text{cm}^{-1}$ (where c is the velocity of light).

Figure 10.9. Atomic configuration employed to demonstrate quantum jumps. The ground state 0 is excited by a laser to a strongly fluorescent state 1; the emitted photons are readily detected. When a second long-lived metastable state 2 is excited, the strong fluorescence is terminated. The period of darkness in this fluorescence is the signature of quantum jumps to state 2.

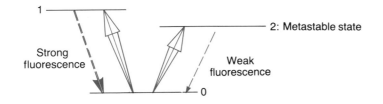

Box 10.1 Atomic energy level classifications

Atomic states have discrete energies

$$E = -Rn^{-2}, \quad n = 1,2,3 \ldots.$$

where the Rydberg constant $R = e^4 m/2\hbar^2$ in terms of the electron mass m and charge e. The integer n is the principal quantum number. Atoms are also described by the orbital angular momentum L of the electron and labelled by a letter S (for $L = 0$), P ($L = 1$), D ($L = 2$) and so on. Finally a subscript indicates the total angular momentum including electron spin. So, for example, $25P_{3/2}$ means a highly excited atom with $n = 25$, orbital angular momentum $L = 1$ and total angular momentum $\frac{3}{2}$.

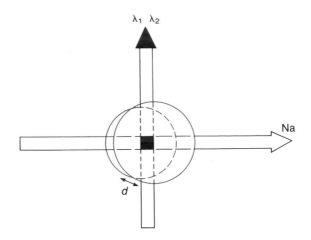

Figure 10.10. Schematic outline of an experiment in which a sodium atomic beam is excited by lasers of wavelength λ_1, λ_2 to the 29d state between two conducting plates separated by a distance d. The 29d is a highly excited Rydberg state with very long wavelength transitions to nearby Rydberg levels. The long wavelength transitions can be suppressed by the confining plates if these are so closely spaced that the transition wavelength is unable to fit inside.

As the electric field across the plates is varied from 0.7 to $5.7\,\mathrm{V\,cm^{-1}}$, the transition frequency varies from $0.97\nu_c$ to $1.11\nu_c$. The transition rate for the production of 30p atoms is measured by selectively ionising only those 30p atoms making the transition, and counting the ions. As the transition frequency ν approaches the waveguide cut-off frequency ν_c, the rate of transition suddenly changes (figure 10.11): on the right-hand side, black-body photons can fit between the plates, and on the left-hand side they cannot, decreasing the 30p production rate. As ν increases beyond ν_c, the modes normal to the plate surface are switched on.

The Einstein relationships between stimulated and spontaneous transition rates imply that spontaneous decay was similarly affected in this experiment. The experiment of Kleppner *et al.* is a beautiful example of how Rydberg atoms are sensitive to their immediate environment, making them the perfect test-bed for what would otherwise be minute and unmeasurable effects.

Figure 10.11. Radiative transition signals for sodium 29d to 30p excitation induced by thermal radiation between parallel conducting plates as the transition frequency ν is tuned through the cut-off frequency $\nu_c = c/2d$. Note the drop in transition rate below the cut-off frequency. Below this point transition wavelengths in the direction perpendicular to the plates cannot be supported and the total transition rate is diminished.

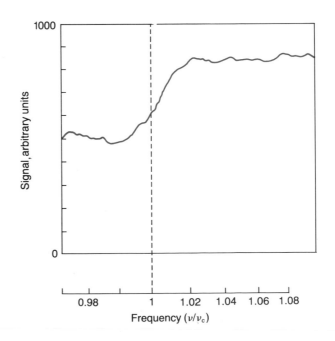

10.4 Nonlinear optics

The subject of nonlinear optics is concerned with the effects arising from the nonlinear response of a medium to electric and magnetic fields, at least one of which is in the optical frequency range. Any nonlinear system requires a stimulus of a certain magnitude before its nonlinearity becomes significant. Atoms and molecules reveal their nonlinear properties when subjected to fields which are not entirely negligible in comparison with their own internal fields ($E_{\mathrm{atomic}} \sim 10^{11}\,\mathrm{V\,m^{-1}}$). Because of the high sensitivity of detection equipment, applied fields of the order of $10^6\,\mathrm{V\,m^{-1}}$ are sufficient in many cases. A few nonlinear optical phenomena which involve intense d.c. fields were observed in the nineteenth century but the power density of optical frequency fields of the required strength is in the region of $100\,\mathrm{kW\,cm^{-2}}$, a figure which is unobtainable from conventional thermal light sources. Modern nonlinear optics has developed as a direct consequence of the invention of the laser: even the earliest laser systems were easily capable of creating power densities of the necessary magnitude.

Nonlinearities are well known in electrical amplifiers, where they are responsible for a lack of fidelity in signal amplification at high gain levels. Imagine an electrical circuit with a slightly nonlinear response as shown in figure 10.12. A purely linear response to an input signal x would produce an output signal $y = a_1 x$, where a_1 is a constant. For example, an input signal of the form $V \cos \omega t$ would produce a linear output at the same frequency, $a_1 V \cos \omega t$, merely scaled in magnitude. A small cubic nonlinearity in the response of the form

$$y = a_1 x - a_3 x^3 \qquad (10.1)$$

produces a response to a sinusoidal input of the form

$$y = a_1 V \cos \omega t - a_3 V^3 \cos^3 \omega t, \qquad (10.2)$$

and the last term can be written as $(\frac{1}{4} \cos 3\omega t + \frac{3}{4} \cos \omega t)$. A small $(a_3 \ll a_1)$ cubic nonlinearity therefore generates a response not only at the fundamental frequency ω but also at the third-harmonic frequency 3ω.

An intense electric field can similarly produce a nonlinear response. We can regard the atomic or molecular charge clouds in an optical medium as distorting, rather like electrons on springs. The polarised distortion of the charge clouds cannot be entirely linear because the medium distorts in an anharmonic fashion and saturates just like an electrical circuit. In a linear dielectric we can write the optical polarisation P in terms of the incident electric field E as

$$P = \chi^{(1)} E, \qquad (10.3)$$

where $\chi^{(1)}$ is the linear susceptibility. A nonlinear dielectric has a polarisation P given by some complicated function of the incident electric field E. We find it convenient to write the polarisation P of a nonlinear medium as a power series expansion in the applied electric field E of the form

$$P = \chi^{(1)} E + \chi^{(2)} E^2 + \chi^{(3)} E^3 + \dots . \qquad (10.4)$$

In figure 10.13 we show how a linear and a nonlinear optical polarisation is induced by an oscillatory optical field. We can write the induced polarisation in figure 10.13(b) as a Fourier frequency series

$$P = P_0 + P_1 \cos \omega t + P_2 \cos 2\omega t + \dots , \qquad (10.5)$$

as shown in figure 10.14.

An example of an optical medium with a quadratic non-linearity is quartz. Peter Franken and his colleagues at the University of Michigan in 1961 used quartz to generate the second harmonic at 347.1 nm. They observed predominantly the input signal coming out of the crystal, but in addition detected a very small intensity at twice the input frequency. This is the second harmonic beam. Their experiment had a conversion efficiency of $1:10^{14}$; modern second-harmonic generation

efficiencies are of the order of 50%. A large conversion efficiency requires large input electric fields, and thus nonlinear optics had to await the development of the laser. For comparison, the Coulomb electric field in an atom responsible for binding is 5×10^{11} V m^{-1}, whereas a 1 MW cm^{-2} laser intensity represents a peak electric field of order 3×10^6 V m^{-1}.

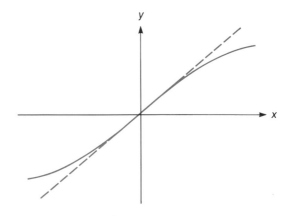

Figure 10.12. Nonlinear response y to input x. Over a small range of input values near the origin, the output y is linearly related to the input. But at the higher values of the absolute value of the input x, the output increases less rapidly and 'saturates'.

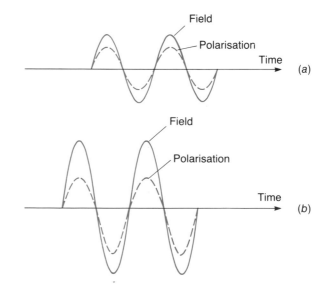

Figure 10.13. Relationship between electric field and induced polarisation for (a) a linear medium and (b) a nonlinear optical medium. For the linear medium above, the polarisation faithfully mimics the applied field and has exactly the same frequency but is merely scaled down in amplitude. The nonlinear response is entirely anharmonic, with more rounded peaks and sharper troughs.

If we imagine an input applied electric field made up of a d.c. and an optical component

$$E = E^0 + E^\omega \cos\omega t \qquad (10.6)$$

what is the response polarisation? Using equation (10.6) as the input in the polarisation expansion, equation (10.5) we find a variety of harmonic and cross-term effects conveniently shown in table 10.1.

Table 10.1. *Induced nonlinear polarisation*

Terms	Frequency	Effect	Date of discovery
$\chi^{(1)}E^0$	0	d.c. polarisability	?
$\chi^{(1)}E^\omega\cos\omega t$	ω	optical polarisability (refractive index)	?
$\chi^{(2)}(E^0)^2$	0	d.c. hyperpolarisability	?
$\chi^{(2)}E^0E^\omega\cos\omega t$	ω	linear electro-optic effect Pockels' effect	1893
$\chi^{(2)}(E^\omega)^2$	0	optical rectification	1964
$\chi^{(2)}(E^\omega)^2\cos2\omega t$	2ω	second-harmonic generation	1961
$\chi^{(3)}(E^0)^2E^\omega\cos\omega t$	ω	d.c. Kerr effect: change in refractive index proportional to (d.c. field)^2	1877
$\chi^{(3)}(E^0)(E^\omega)^2\cos2\omega t$	2ω	d.c.-induced second-harmonic generation	1968
$\chi^{(3)}(E^\omega)^3\cos3\omega t$	3ω	third-harmonic generation	1962
$\chi^{(3)}(E^\omega)^3\cos\omega t$	ω	optical frequency Kerr effect (light field changes medium through which it travels to give self-focusing)	?

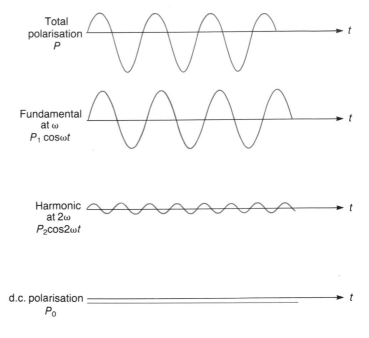

Figure 10.14. Decomposition of nonlinear polarisation into its harmonic components. The total polarisation is an anharmonic waveform which is made up of a set of basic frequencies. The most important is the fundamental but mixed in is a set of less intense harmonics and finally a static d.c. polarisation induced in the nonlinear medium.

10.5 Multiphoton processes

When weak radiation interacts with atoms or molecules, at best only one or two photons are absorbed or emitted. Laser radiation fields are so intense that many photons can be simultaneously involved in a transition. A dramatic example may be seen when a short pulse from a high power laser with a duration of a few nanoseconds and a pulse energy of 1 J is focused in air. A bright blue flash and a loud bang are produced and ultraviolet radiation and X-rays are emitted as the air breaks down to form an ionised plasma. The energy of each photon in the laser pulse is typically 1 eV, yet the ionisation potential of the oxygen and nitrogen molecules in the air is many times this amount. To produce at least the initial stages of the breakdown, the laser radiation ionises atoms and molecules by the concerted action of many photons absorbed simultaneously. The description of how an atom can absorb so many photons forms a major part of quantum optics.

The probability that an electron in an atom may be excited from its ground state to a higher energy state by a light field depends on how intense the light is, and on the mismatch, or detuning, of the energy of each photon from the transition energy required to reach the excited atomic state. Provided that the photon energy matches this transition energy, the probability of being excited increases with time until the excitation 'flow' induced by the light balances the inherent decay mechanisms by which the excited atom can lose energy. If the light is sufficiently intense, a second (or more) photon may be absorbed before the excitation is lost in decay as the atom is excited to even more energetic states, and these are two-photon or multiphoton transitions.

Multiphoton transitions involving the net absorption or emission of more than one photon were first predicted by Maria Göppert-Mayer in 1931 in her Göttingen thesis. One of her predictions was the spontaneous two-photon decay of the metastable excited state (the $2S_{\frac{1}{2}}$ state) of hydrogen with a decay rate which, although minute, is larger than that of any other decay mechanism in this transition. The normal mechanism for spontaneous emission involves single-photon decay in a dipole transition. Here this is forbidden because the initial and final states have the same symmetry and angular momentum and will not permit the emission of a single photon which must carry off angular momentum. Only the two-photon decay is permitted between these two states, with each photon carrying off a balancing amount of angular momentum. Metastable states are important in astrophysics where isolated atoms can be excited, yet are too far away from other atoms to collide and lose their energy. It is thought that the blue continuum emission observed in the spectra of gaseous planetary nebulae with wavelengths from 360 nm to 480 nm is produced by Göppert-Mayer's two-photon decay mechanism. No discrete line spectra are produced in these decays because only the sum of the energies of the two photons has to match the atomic transition energy. Such decays from these enormously long-lived hydrogenic metastable states have now been observed in the laboratory.

Once two-photon processes were observed to occur spontaneously, it was natural to look for them in induced transitions forced by an intense applied radiation field. In fact, they were first observed by accident in 1950 by Hughes and Grabner, who were exciting molecules of rubidium fluoride by oscillating radiofrequency radiation. They found unexpected transitions at one-half the expected photon energy. Rabi suggested to them that these were two-photon transitions and this was confirmed by later experiments, both in molecules and in alkali atoms. When the two-photon absorption was studied in atoms where intermediate excited states existed between the ground state and the two-photon excited state, it was found that a suitably resonant intermediate state hugely increased the two-photon absorption. The near-resonant intermediate state makes it easier for the atom to be excited by two photons and complete the transition in a near-sequential absorption. All of these early multiphoton processes were induced by the only intense coherent radiation available at that time: radiofrequency fields. As soon as laser radiation became available, intense optical multiphoton absorption became possible. Kaiser and Garrett in 1961 used a ruby laser to excite transitions in a $CaF_2 : Eu^{++}$ crystal transparent at ruby laser wavelengths but strongly absorbing at twice the optical energy. Extremely large numbers of photons can be involved in the multiphoton processes in which atoms are ionised by intense laser pulses, and offer a real challenge to the laser physicist. In the initial stages of multiphoton ionisation, the Coulombic forces which bind the electron to the ionic core are much stronger than the laser radiative interaction. As the excitation proceeds, the electron is promoted to higher-lying energy levels in which the radiative interaction is greater than the Coulombic interaction. The role of unperturbed system and perturbation are reversed during the multiphoton transition.

The current workhorses for high intensity multiphoton physics in the optical range are the Nd : glass laser, the ruby laser, and more recently pulsed dye lasers. The Nd : glass system produces 1.17 eV photons at very high intensities (10^9 to 10^{16} W cm^{-2}) and has a small wavelength tuning range between 1052 nm and 1065 nm and has been used widely, especially by the leading multiphoton group at Saclay in France, to study rare-gas multiphoton ionisation involving up to twenty-two photons. At the upper intensity range, laser electric fields are produced which exceed the Coulomb electric field which binds the electron (about 10^9 V cm^{-1}) so that the electron 'tunnels' out in the combined fields very rapidly.

The basic description of multiphoton absorption involves quite complicated atom-field interaction physics but can be illustrated using simple Feynman diagrams. For two-photon absorption, for example, figure 10.15 shows the excitation from the initial state labelled 0 to an intermediate state 1 and thence to a final state 2 by absorbing two photons of frequency ω. In general there will be many intermediate states 1 accessible through one-photon absorption. Each of these different intermediate states contributes a different route between the same initial state 0 and the same final state 2. Different routes can interfere constructively or destructively in the total transition probability (the modulus squared of the total summed transition amplitudes for each route). The multiphoton absorption rate as a function of laser frequency will scan these coherently excited intermediate states, with maxima when the frequency matches the intermediate state resonance frequency, and deep interference minima when several routes are excited and interfere destructively. In figure 10.16 the two-photon ionisation rate (actually a generalised cross-section) is plotted versus photon energy for the caesium atom. Here two photons of optical frequency radiation are sufficient to ionise caesium, so the final state 2 corresponds to the electron removed in two-photon ionisation.

For this simple two-photon ionisation case, the rate of ionisation R_2 will depend on the square of the laser intensity (each absorption is proportional to intensity),

$$R_2 = \sigma_2 I^2,$$

where the proportionality constant σ_2 is a generalised cross-section which contains all of the atomic factors including the detuning of the laser from intermediate state resonance frequencies. Clearly it is easiest to climb the 'ladder' of excitation from 0 to 2 if state 1 can be reached en route by a resonant absorption, and this (inverse) dependence on the intermediate state detuning is responsible for the resonance peaks in figure 10.16. The rate of a general N-photon absorption is given by

$$R_N = \sigma_N I^N.$$

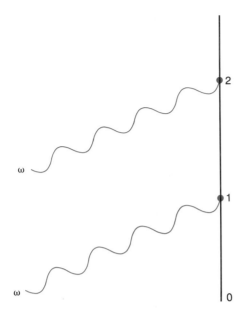

Figure 10.15. Feynman diagram for two-photon absorption. The atom starts in state 0 and is represented by the vertical line. Photon absorptions are indicated by the wavy lines. The first absorption promotes the atom to the excited state 1, the next absorption to state 2.

It is worth noting the size of the generalised cross-section σ_2 for caesium. It varies from $10^{-47}\,\text{cm}^4\,\text{s}$ near a resonance to $10^{-51}\,\text{cm}^4\,\text{s}$ at an interference minimum. To observe a significant ion signal given such small cross-sections requires rather high photon fluxes and correspondingly high peak power pulsed lasers.

An important experimental parameter in multiphoton physics is the 'order of nonlinearity' K of the multiphoton absorption process, defined as the slope of a log–log plot of the final state yield N_f ((proportional to R_N) versus intensity:

$$K = \ln N_f / \ln I.$$

The simplest (perturbative) model of N-photon absorption gives

$$R_N = \sigma_N I^N,$$

so $\ln R_N = \ln \sigma_N + N \ln I$ and so K should equal N.

This is verified for many multiphoton processes provided the laser frequency (or multiples of it) is chosen to avoid near-resonances, but, as explained later, breaks down near resonance because of intensity-dependent modifications (dressing) of the atomic target states.

So far we have discussed only the excitation of atoms by perfectly coherent radiation. In fact, high intensity lasers often operate on many different modes with differing frequencies, so laser pulses can be quite noisy. Nonresonant multiphoton

The generalised cross-section σ_N has units of $\text{cm}^{2N}\,\text{s}^{N-1}$, where the photon flux I is measured in photons $\text{cm}^{-2}\,\text{s}^{-1}$. It is extremely difficult to calculate σ_N reliably for large N as the calculation involves $(N-1)$ infinite sums of atomic intermediate states and detunings of multiples of laser frequencies from atomic resonance frequencies. Sophisticated atomic models are required in such calculations. But the general features remain those visible in figure 10.16: resonance peaks and interference minima.

The experimental observation of the resonance maxima in atomic multiphoton ionisation is quite straightforward. It is harder to see the minima because even a small amount of molecular contamination (for example by dimers, which are molecules made up of pairs of the atoms under scrutiny) can lead to the production of ions through processes enhanced by molecular resonances, and to the infilling of minima. The deep minimum in figure 10.16 at a photon energy of 2.6 eV in the two-photon ionisation of caesium arises from cancellation of the 6P and 7P excited intermediate states. This was observed by a group of experimentalists at Saclay in France using a dye laser of intensity $10^{10}\,\text{W cm}^{-2}$ and laser wavelengths from 460 nm to 540 nm. They were forced to use special sources of atomic caesium in a beam to reduce dimer molecule background, and when this was done they found results in good agreement with the theoretical results.

Figure 10.16. Generalised cross-section for two-photon ionisation of caesium as a function of laser frequency. The generalised cross-section is a measure of the size of the two-photon absorption. When the photon energy matches transition energies in the atom, the absorption is hugely facilitated and we observe large resonance peaks. Between these resonances destructive quantum interference strongly suppresses the absorption.

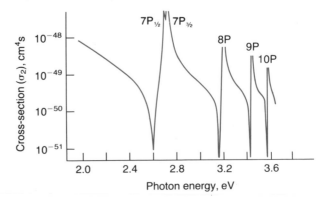

absorption is extraordinarily sensitive to radiation field fluctuations. The rate of an N-photon transition depends on the instantaneous field intensity I, not the average intensity $\langle I \rangle$, so that the average final state yield depends not on $\langle I \rangle^N$ but $\langle I^N \rangle$. If the light fluctuates in a chaotic way with Gaussian statistics, the probability of having a particular laser electric field is Gaussian:

$$p(E^2) = \text{const} \times \exp(-E^2/\langle E^2 \rangle)$$

or in terms of intensity

$$p(I) = \text{const} \times \exp(-I/\langle I \rangle),$$

then the average is

$$\langle I^N \rangle = N! \langle I \rangle^N.$$

If, however, the laser is operated on a single highly stabilised mode, there are no fluctuations in the intensity and $\langle I^N \rangle = \langle I \rangle^N$. For the same average intensity $\langle I \rangle$, a fluctuating multimode laser field will produce $N!$ more final states. At extremely high intensities ($\simeq 10^{12} \, \text{W cm}^{-2}$), even higher-order multiphoton processes will compete with lower orders. For example, excitation need not cease after the minimum number of photons necessary for ionisation has been absorbed. If the photon flux is high enough, further photons can be absorbed in exciting more continuum electron states before the electron finally leaves the influence of this ionic core and becomes truly free. These continuum–continuum transitions are called 'above-threshold ionisation' (ATI) transitions and are important in electron heating in multiphoton ionisation. They are observed in experiments in which the ejected photoelectron kinetic energy is measured. In N-photon ionisation by radiation of frequency ω, the photoelectron kinetic energy is

$$E = N\hbar\omega - E_{\text{IP}},$$

where E_{IP} is the ionisation potential. Further absorptions of up to S photons will lead to $(S+1)$ photoelectron energy peaks at $E_S = (N+S)\hbar\omega - E_{\text{IP}}$, separated by $\hbar\omega$. Initially the ATI rates R_{N+S} are proportional to $I^{(N+S)}$, and higher-order terms become compatible with lowest order ones at high intensities. At low intensities, only one peak is observed in the photoelectron energy spectra ($S = 0$). As the intensity is increased, more peaks are observed, and the lowest energy peaks are shifted and diminished by a mechanism thought to involve higher-order transitions (saturation) and the shifting of ionisation potentials by the intense fields. In figure 10.17 we show the photoelectron spectra obtained by Kruit and co-workers at the FOM Institute in Amsterdam from the $(N+S)$ photon ionisation of xenon atoms. A high-intensity multimode neodymium YAG laser is used which produces pulses of $1.06 \, \mu\text{m}$ radiation with a duration of $20 \, \text{ns}$ and intensities of $10^{12} \, \text{W cm}^{-2}$. Twelve photons are normally required at this wavelength to ionise xenon. The first peak in figure 10.17 corresponds to this minimum absorption, and is supplemented by peaks from up to seven further absorptions. Each peak is broadened in part by the acceleration of the free electrons in the focused laser electric field.

Similar deviations from the lowest order multiphoton behaviour are observed whenever one or more intermediate states are resonant with some multiple of the laser frequency. Just as the atom can absorb more photons after it reaches the original final state, it can emit and absorb photons following excitation of the near-resonant intermediate state. Energy levels may be shifted by the intense field and lifetimes shortened by induced emission and absorption, making it impossible to separate target from flux by producing a 'dressed atom' target whose structure depends intimately and nonlinearly on the intensity. This means that R_N is not simply written as $\sigma_N \langle I^N \rangle$. Perturbation theory is still applicable, but higher-order processes must be included, and in some cases summed to infinite order. The excited state can also be stimulated to return to the ground before it has a chance to ionise, and then be re-excited, leading ultimately to a coherent Rabi population oscillation between the resonant states. Under these circumstances, it may not be possible to define a time-independent transition rate, or, in consequence, a cross-section.

The resonant ionisation rate R is now a nonlinear function of intensity which depends on how close to intermediate resonance we tune the laser. The index of nonlinearity K is no longer equal to N except far from resonance. This has been observed in a striking way in a study of three-photon-resonant four-photon ionisation of caesium by the Saclay group where $K = 4$ far from three-photon resonance but rises to 30 on resonance, reflecting large intensity-dependent Stark shifts and induced lifetimes. Resonant multiphoton processes respond to fluctuations in the radiation field in a very complicated way, providing a test of the theory of such 'multiplicative stochastic processes'.

The study of multiphoton excitation has widened and intensified in recent years, so that only a flavour of the field has been given above. Multiphoton excitation has been used extensively in atomic and molecular spectroscopy to excite high-resolution Doppler-free spectra and to excite high-lying levels which cannot be reached by traditional ultraviolet one-photon absorption from the ground state because of dipole selection rules.

Other resonances reached by multiphoton (and of course one weak ultraviolet photon) excitation are autoionising states, states of many-electron atoms in which more than one electron is excited from the normal ground state and which lie in energy above the ionisation threshold for the configuration in which only one electron is excited. The Coulomb interaction between electrons mixes the configurations and allows the two-electron excited state to decay by spontaneously emitting an electron. These autoionising resonances can be fairly long lived and are of great interest in atomic physics and in laser physics where it is hoped they can be exploited as storage levels in potential vacuum ultraviolet lasers. Their behaviour in strong radiation fields typical of normal pulsed-laser excitation is currently under investigation. If the coherent excitation rate into the autoionising state is large, the population of the structured continuum can be induced to return to the ground state periodically, to result in the coherent population Rabi oscillations. If the excitation rate is chosen to be just equal to the autoionisation width, some of the population (up to one-half) can be trapped in a coherent superposition state quite immune to further ionisation. We have the remarkable result that shining light on to an autoionising resonance can in part stop it from ionising!

Multiphoton excitation has become increasingly useful as a diagnostic tool. Resonant multiphoton ionisation is a very sensitive way of detecting small numbers of atoms, and a single-atom detector using such ideas has been pioneered by a group at the Oak Ridge National Laboratory in the United States. Resonant ionisation spectroscopy can be used to detect even a single atom present in mixtures of atoms or molecules of another type using quite modest commercially available lasers. This has obvious implications for studies of pollution and radioactive waste and can be used in a host of applications from the measurement of sodium contaminants in silicon of importance to microelectronics, to the detection of small numbers of atoms created in solar neutrino-induced nuclear reactions. The application of multiphoton processes to generate up-converted vacuum ultraviolet provides a direct laboratory-scale rival to synchrotron sources for the study of highly excited atoms and molecules and of chemical surface and solid-state phenomena in the ultraviolet and soft X-ray régime with ultra-fast time resolution.

The state-selective excitation of atoms and molecules by intense resonant laser fields, their ionisation and collection forms the basis of Atomic Vapour Laser Isotope Separation (AVLIS). AVLIS is the first use of lasers on a large industrial scale with total expenditure planned at the billion-dollar level. Already, laser isotope separation accounts for more than one-eighth of the total continuous dye-laser market with additional major investments in high-repetition rate large-bore copper vapour lasers. AVLIS has now been demonstrated to offer economic advantages over advanced gas centrifuges, and centrifuge work in the USA has now been supplanted by the Lawrence Livermore National Laboratory (LLNL) laser-driven method.

The nuclear industry contributes a substantial part of the generated electrical capacity of the world (and, as we shall see, consumes a substantial part). For example, in the USA, over 5% of its electrical power (15 000 MW) was produced by light water reactors (LWRs) by 1973. This proportion has now increased substantially. A 1 GW LWR consumes 10^5 SWU year^{-1} of uranium fuel. (A separative work unit, or SWU, is a value function related to the free energy of separation required to separate a feed stream of assay into enriched product and depleted tails.) In the case of LWR fuel, the assay contains 0.7% U^{235} feed, 3.2% product and 0.25% tails. Isotope separation of civilian and military product makes up about one-third of the fuel cycle cost (approximately 18 000 MWe, or over

6% of the USA's total electrical capacity). At present, the major separative scheme employed is gas diffusion, which relies on lighter molecules getting through a membrane faster than heavier molecules. This process is extremely inefficient as it depends on the ratio of the square roots of the uranium hexafluoride (UF_6) molecular weights for the different isotopes, and this ratio is only $(1.0042)^{\frac{1}{2}}$. As a result, a huge number of connected cascades are required, feeding through the only utilisable uranium molecule, UF_6, which is gaseous (and then only above 65°C). The total production by gas diffusion of U^{235} ($< 17\,200\,000$ SWU year^{-1}) in the USA consumes more electricity than San Francisco plus Philadelphia plus Denver. The typical cost of an actinide is in excess of 1000\$ kg^{-1}. Gas diffusion uses 3 MeV energy per separated U^{235} atom, whereas normal chemical processes cost 6 eV (a typical ionisation potential).

Figure 10.17. Photoelectron energy spectra produced by multiphoton ionisation of xenon atoms by a neodymium laser at a wavelength of 1.06 μm at an average intensity of 10^{12} W cm^{-2} observed by Kruit and coworkers in Amsterdam. The spectrum is made up of a decreasing set of peaks separated by $\hbar\omega$, the incident photon energy. Each extra peak indicates that another photon has been absorbed by the photoelectron during its intense field ionisation.

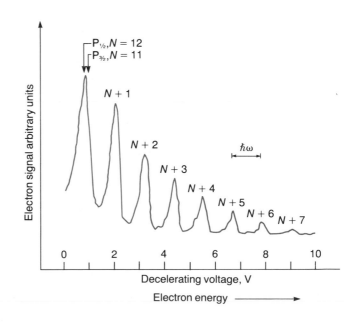

Gas centrifuges are an economic alternative to gas diffusion. A British–West German–Dutch consortium run three plants with a production greater than 10^5 SWU year^{-1}, and contain more than 10 000 centrifuges. If all gas diffusion plants were to be superseded by centrifuges, more than 15 000 rotors would have to be manufactured per week. The USA Department of Energy now believe that AVLIS unit costs are 10% of current projections of those for gas centrifuges, and have suspended centrifuge work in favour of laser isotope separation based on the atomic vapour method. Laser isotope separation is now making the transition from physics feasibility studies to engineering test modules. In 1984/5, a special isotope separation test module (i.e. plutonium AVLIS) was built and operated in 1985 at LLNL, and a development module for uranium enrichment has been designed for joint development by LLNL and Oak Ridge.

In all photoionisation AVLIS methods under consideration, the selective excitation by lasers of well defined frequency of one isotope rather than another is followed by a collection stage. This is shown schematically in figure 10.18. Atomic energy levels depend to a small extent on the isotopic nature of the nucleus and are changed by a small amount by nuclear volume and reduced masses (by perhaps one part in 10^5 in the optical wavelength of the transition shown). A laser of bandwidth much less than this isotope shift will preferentially and resonantly excite isotope A and not isotope B. Of course, the Doppler distribution of resonance frequencies must also not allow species B to come into resonance (this Doppler effect is minimised by the high atomic weights and by using jets of atoms illuminated nearly transversely). Once species A is excited, photoionisation or other collection methods can be used to separate product from tails. LLNL has demonstrated that this ionisation is so efficient that 90% of the uranium atoms end up as ions in the collector.

In the LLNL scheme, the isotope of interest is ionised by absorbing two photons from an intense dye-laser radiation field. The pump lasers for the dyes are high-repetition rate copper vapour lasers which pump a set of dye oscillators, each tuned to particular uranium transition. These are amplified in dye amplifiers. A redundant oscillator–amplifier chain is provided and runs hot ready to be switched in if laser failure occurs. This prevents plant shut-down during repair and maintenance. The U^{235} is generated by a vaporiser at 2500–3000 K in a jet which intersects the laser beams at nearly normal incidence. The product is collected electromagnetically, and both tails and product are removed in liquid form. The copper vapour master-oscillator power-amplifier (MOPA) chains in the largest system developed so far (the LLNL VENUS system) contain thirty discharge heated copper vapour lasers: 3 cm small-bore oscillators and 8 cm large-bore amplifiers. Each amplifier contributes about 100 W, and the VENUS power amplifier contains thirty MOPA chains. The dye MOPA chains have diameters of about 45 cm, running at about 6 kHz repetition rate, with a large volume dye pump to ensure each pulse sees fresh dye. The

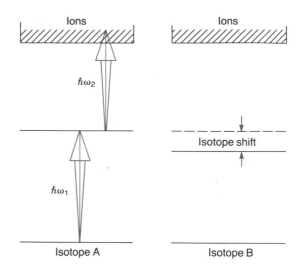

Figure 10.18. Selective excitation of isotopes. Isotope A is separated from B by two-photon excitation by lasers tuned precisely to atomic transitions. Isotope A has resonance frequencies which coincide with the laser frequencies and is excited, ionised and the ion collected electromagnetically. Isotope B has slightly different transition frequencies, so it remains out of resonance with the lasers and unexcited.

dye needs to be recirculated to prevent thermal damage effects. The LLNL AVLIS test module contains forty copper vapour lasers, five master oscillators and seven pre-amps, running thirteen hours each day to excite a low throughput separator named REGULIS and a high throughput separator MARS. It is estimated that 100–1000 of these units will be required for a 10^6 SWU year^{-1} plant. Such a large-scale AVLIS plant will require 10^5 W of dye-laser power. The market for dyes, copper vapour lasers, etc. seems ensured!

10.6   Phase conjugation

A major application of nonlinear optics is the production, through nonlinear frequency mixing, of a polarisation which radiates a signal which is a *phase conjugate* version of an input. A phase conjugate mixing device has remarkable optical properties: perfect retroreflection, the complete cancellation of optical aberrations and inhomogeneities in wavefronts, and the ability to reverse an optical beam and return it to its source.

Three optical waves of electric field strength E_1, E_2 and E_3 mix in a nonlinear medium to produce a nonlinear polarisation

$$P \simeq \chi^{(3)} E^3, \tag{10.7}$$

where $\chi^{(3)}$ is a third-order susceptibility, and $E = E_1 + E_2 + E_3$.

We can write the optical field E_i in terms of its spatially varying amplitude $A_i(r)$ and factor out its gross time and plane-wave behaviour as

$$E_i(r) = \tfrac{1}{2}A_i(r)e^{i\omega_i t - i\mathbf{k}_i \cdot \mathbf{r}} + \text{complex conjugate}. \quad (10.8)$$

Then the polarisation in equation (10.8) is made up of many frequencies $(3\omega_1, 3\omega_2, \dots,$ and so on down to $\omega_1 - \omega_2 - \omega_3)$. One which is particularly interesting in phase conjugation is the difference polarisation which oscillates at $\omega = \omega_1 + \omega_2 - \omega_3$,

$$P(\omega = \omega_1 + \omega_2 - \omega_3) \simeq \chi^{(3)} E_1 E_2 E_3^*, \quad (10.9)$$

where E_3^* is the complex conjugate of the signal wave E_3 ($i = \sqrt{(-1)}$ is replaced by $-i$).

The three waves produce this polarisation which radiates a fourth wave at frequency ω, which is why this process is referred to as four-wave mixing. The frequencies ω_1, ω_2 and ω_3 could be equal, but the three beams distinguished by different wave-vectors (photon momenta) $\mathbf{k}_i$. We imagine the geometry in figure 10.19. The two counter-propagating waves E_1 and E_2 are intense and usually referred to as pump waves. Note that A_4 is shown as retroreflected: the nonlinear polarisation $P(\omega)$ depends upon the photon momenta $\mathbf{k}$ in a strange way which guarantees this. For the geometry shown, $\mathbf{k}_1 + \mathbf{k}_2 = 0$. The signal wave A_3 has a photon momentum $\mathbf{k}_3 = k$ in the z-direction. Then the polarisation is

$$P(\omega) = \chi^{(3)} A_1 A_2 A_3^* e^{i\omega t + ikz}, \quad (10.10)$$

which will radiate a field E_4 with a similar factor $\exp(i\omega t + ikz)$, with a $\mathbf{k}$-vector opposed to that of the signal $\mathbf{k}_3$. Compare this with a one-photon induced polarisation $P = \chi^{(1)} A_1 \exp(i\omega t - ikz)$, where the $\mathbf{k}$-vector is not reversed. In this sense, the radiated wave is like a time-reversal of the incident wave.

The dependence of the radiated field E_4 on the complex conjugate of the signal wave is the origin of the name 'phase conjugation'. The reversal of the photon momentum means that a nonlinear medium pumped by two intense waves E_1 and E_2 acts as a perfect retroreflector for the signal E_3, which is returned along a precisely reversed path (figure 10.20) so that no matter how the source of E_3 is moved, the reflected wave E_4 will always trace its way back to this source. The reflectivity is $|(E_4/E_3)| = \chi^{(3)} A_1 A_2$, which for intense pump beams A_1 and A_2 can be greater than unity: the phase conjugate reflector can *amplify* the incident wave, deriving its gain from the pump fields.

A spherical wave derived from some source S would reflect off a conventional mirror as shown in figure 10.21. A phase conjugate mirror reverses the wavefront, which on reflection (figure 10.22) must retrace its path, converging back to S. Such behaviour makes a phase conjugate mirror invaluable in guaranteeing targetting: any light scattered from the source S may be amplified and retroreflected precisely to S. Applications of this idea to laboratory-scale targetting of laser-fusion beams, and of star-wars interception, are obvious.

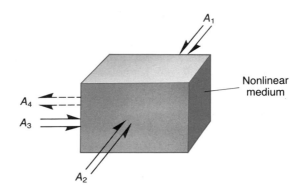

Figure 10.19. Nonlinear four-wave mixing to generate a phase conjugate reflected wave A_4. The two strong pump waves A_1 and A_2 counterpropagate into the nonlinear medium. The signal wave A_3 travels into the medium in the z-direction and generates a fourth phase conjugate wave A_4 in the opposite direction.

Figure 10.20. Perfect retroreflection is provided by phase conjugate four-wave mixing. The reflected wave E_4 always retraces the path of the signal wave E_3. As the angle of incidence θ is varied, the reflected wave will always faithfully follow, never deviating from this retrodirection.

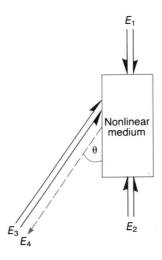

The subject of phase conjugation is rapidly growing, with applications in photolithography, adaptive optics, optical correlation image processing, and so on. Laser resonators can be made with one mirror (and even the active gain element) replaced by a phase conjugate reflector.

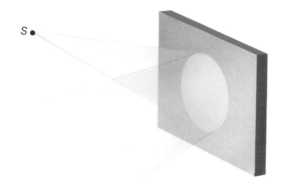

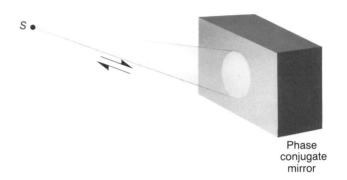

Figure 10.21. Reflection from a normal mirror continues beam divergence. Waves from a source diverge as they propagate outwards so that the beam diameter increases. Upon reflection from a normal mirror this process continues and the beam continues to increase in width.

Figure 10.22. Retracing optical wavefronts in phase conjugate reflection. The waves from the source expand as always as they propagate outwards. When they interact with the phase conjugate mirror, their direction of motion is reversed so the expanding beam turns round, retracing its path, and contracts back to its initial width.

10.7 Ultrashort optical pulses

Pulsed lasers do not naturally produce short optical pulses. The ruby laser first operated by Maiman in 1960 gave pulses which lasted a few ten-thousandths of a second. To make shorter pulses, we need to manipulate the laser action either by shutter techniques called Q-switching, or by correlating all the frequency components in a method called modelocking. To achieve the shortest optical pulses so far made, with durations of eight femtoseconds (where one femtosecond, 1 fs, is 10^{-15} s) and containing only four optical cycles, we resort to nonlinear optical pulse-compression techniques. The progress towards ever shorter optical pulses is shown in figure 10.23.

The reduction of laser pulse durations to 1 ns used Q-switching. The loss in a cavity is parametrised by the time for an electric field to decay to $(1/e)$ of its original value, and this time is Q/ω, where Q is the quality factor and ω is the field frequency. In Q-switching, a loss element in the laser cavity (for example a shutter opened by a fast electrical pulse) prevents the laser gain from exceeding the losses and generating laser action, while the population inversion is still increasing. At the critical moment when the population inversion is greatest, the shutter is suddenly opened, reducing the losses and allowing the large stored energy to be emitted in a giant pulse.

Shorter optical pulses are produced by 'mode-locking'. Lasers can allow gain at many different mode frequencies, each separated by a frequency spacing $c/2L$ so that an integer number of half-wavelengths of each of these mode frequencies are able to fit within the laser mirrors separated by a distance L. Provided each of these modes sees a gain greater than the losses, its energy will grow by stimulated emission. If the radiation from each of these many modes is uncorrelated in phase, the total laser field wildly fluctuates as the different mode components interfere constructively and destructively. But if the modes can be correlated and share a common phase, they will interfere to produce a regular pulse train as the modes periodically interfere in step. The duration of each of the pulses depends inversely on the number of oscillating modes locked in phase. In the past ten years dramatic pulse shortenings from a few picoseconds (1 ps $= 10^{-12}$ s) down to a few tens of femtoseconds have been produced by mode-locking.

The spectral bandwidth Δv of the laser radiation controls the shortest duration T of the optical pulses:

$$T \geqslant (\Delta v)^{-1}. \qquad (10.11)$$

To obtain 100 fs pulses, we require a bandwidth $\Delta v > 10^{13}$ Hz. If the cavity length $L = 1$ m, the mode spacing $c/2L$ is 150 MHz and we need to lock in phase 7×10^4 modes. Clearly a large gain bandwidth is necessary for ultrashort pulse generation, and this is why dye lasers with a huge gain across much of the optical wavelength region are employed for these purposes.

One way of mode-locking a laser is to actively modulate either the gain of the laser or its phase by periodic driving of any cavity parameter at an intermode frequency $c/2L$. For example, the mirror spacing can be modulated by a piezoelectric drive attached to one mirror which alternates periods of high gain and loss. Alternatively, the gain can be directly modulated. In each case, the modulation generates sidebands on each mode; the sidebands coincide in frequency with next-neighbour mode frequencies and share out the energy coherently amongst the modes until a common phase is established. Then the pulse separation is given by the inverse of the mode spacing, i.e. by $2L/c$.

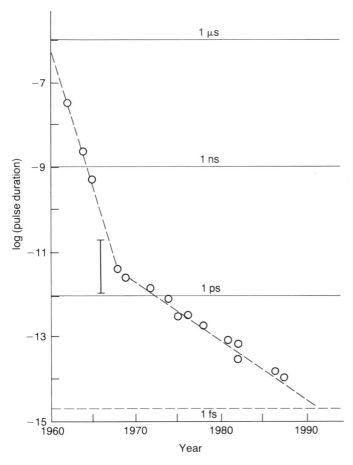

Figure 10.23. Progress in generating ultrashort optical pulses: before the laser, short pulses were extremely difficult to produce. Within a few years of the first laser, submicrosecond pulses were routinely generated. Reduction first to nanosecond and then to picosecond duration was achieved by controlling the laser emission dynamics, principally by mode-locking. Since 1970 progress towards ever shorter pulses has been slower but has resulted in pulses so short that they contain only a handful of optical cycles.

A simpler, yet very powerful, mode-locking technique is to insert an absorber into the laser cavity which will 'passively' mode-lock the laser light. The absorber, usually an organic dye in solution in a cell, preferentially absorbs low intensity parts of the laser field in the laser cavity. High intensity parts are so intense that they saturate the dye by a bleaching which allows most of the intense light through. In this way as the light reflects to and fro in the cavity an intense pulse builds up, with lower intensity parts seeing a greatly reduced total gain. This super-intense pulse takes a time $2L/c$ to make a complete round trip in the cavity. Part is transmitted by the output coupling mirror on each bounce: from outside, we see a mode-locked

pulse train with pulses separated as before by $2L/c$. The passively mode-locked dye laser most used has the dye Rhodamine 6G as the active lasing agent, and the dye DODCI as the saturable absorber. The shortest mode-locked laser pulses have been produced in a colliding pulse geometry where a jet of saturable absorber dye sees counter-propagating pulses which overlap at the jet to saturate the absorption and synchronise the circulating pulses. This technique proved to be a major advance in the production of ultrashort pulses and resulted in a balanced, or 'compensated', colliding pulse mode-locked pulse train, with pulse durations of 27 fs with a repetition rate at 100 MHz.

Nonlinear pulse compression techniques have been developed to produce much shorter optical pulses than those formed by mode-locking. The limitation on pulse lengths is the gain bandwidth Δv, which needs to be increased if shorter pulses are to be formed. The pulse frequency spectrum can be stretched by 'chirping' the light: for example, it could be up-chirped with higher optical frequencies at the front of the pulse compared with the back. Some nonlinear optical processes can produce substantial pulse chirping sufficient to broaden the spectrum: for example, nonlinear propagation of intense laser light in a single-mode glass fibre. The refractive index n of the glass is altered by an amount which depends on the laser intensity I:

$$n = n_0 + n_2 I, \qquad (10.12)$$

where n_0 and n_2 are the linear and nonlinear refractive indices, respectively; n_2 is small and usually positive. The light from a mode-locked laser focused into the small diameter (about 10^{-3} cm) core of the fibre gives internal intensities I large enough for nonlinear effects to be important. A further advantage in using a single-mode fibre is that the entire spatial transverse structure is equally affected by the same refractive index and is all chirped in an identical way. The nonlinear refractive index produces a phase-shift which depends on the instantaneous intensity. The variable phase-shift increases until the peak of the pulse arrives, and then it diminishes. The instantaneous frequency shift (the chirp) is given by the rate of change of the phase-shift, so that frequencies in the leading part of the optical pulse are lowered while those in the trailing edge are raised. The size of this frequency chirp increases directly with the length of the fibre (figure 10.24).

Once the light pulse bandwidth has been stretched by this self-phase modulation, the pulse is compressed by a pair of diffraction gratings (figure 10.25). The angle through which a light beam is diffracted depends on its wavelength (figure 10.26). For the direction α' the diffraction grating gives constructive interference (paths from adjacent rulings differ by integer number of wavelengths) when

$$d\sin\alpha + d\sin\alpha' = n\lambda, \qquad (10.13)$$

where d is the spacing between rulings and λ is the wavelength of the light. For fixed input α, different wavelengths λ are

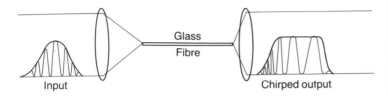

Figure 10.24. Chirping of laser radiation in a nonlinear optical
fibre. The input pulse has the same frequency throughout its pulse
length. When it is focused into a thin glass fibre, nonlinear
refractive index effects inside the fibre in the strong laser field
'chirp' the pulse so that the pulse frequency modulates from lower
to higher frequencies from the front edge to the trailing edge of
the output.

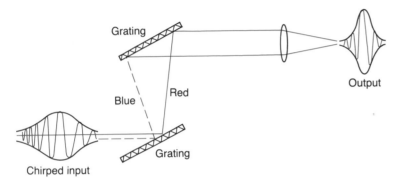

Figure 10.25. Use of diffraction gratings to compress a chirped
optical pulse. The diffraction grating deflects light at an angle
which depends on its wavelength. The chirped input is then
spatially separated into its constituent frequencies which are then
deflected by the second grating. The geometry is arranged for the
constituent frequencies to catch up with each other and compress
the output pulse.

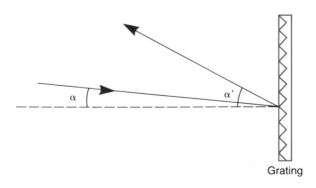

Figure 10.26. Diffraction of a light beam by a grating. The
incident light comes into the ruled grating and is deflected at an
angle α' which depends on its wavelength. The grating consists of
a very closely ruled set of lines or grooves which diffract the light
only at specific angles.

diffracted through different angles α'. In the grating pair, the
leading edge (with longer chirped wavelength) has further to go
in the pair whereas the trailing shorter wavelength edge has a
shorter optical path length and catches up! The chirp stretching followed by grating compression generates an output pulse
much shorter than the original unchirped input with an output
pulsewidth given by the reciprocal of the chirp bandwidth. This
pulse compression method has been borrowed from chirp radar.
Grischkowsky and coworkers at IBM were the first to employ it.
It was further developed by Ippen's group at MIT, who used a
colliding-pulse mode-locked dye laser to produce a train of
pulses with a central wavelength 620 nm each of 65 fs
duration. Single pulses are selected and amplified to energies of
5 kJ, collimated and focused into a short length (about 8 mm) of
fibre. This broadens the pulse spectrum by a factor of four by
self-phase modulation. A grating pair compresses the chirped
pulse to its final length of eight optical cycles (16 fs). Most
recently a Bell Telephone group led by Shank compressed 40 fs
amplified pulses from a compensated colliding pulse mode-
locked laser to 8 fs (only four optical cycles) at a 5 kHz repetition
rate. Tunable femtosecond laser pulses are now available to
study ultra-fast phenomena in atoms, molecules and solids.

A competition between linear dispersion caused by the n_0
term independent of intensity in equation (10.12), and nonlinear self-phase modulation caused by the nonlinear refractive
index n_2 is responsible for the formation of optical solitons in
glass fibres. A soliton is a stable wave which propagates without
change in shape, never spreading or dispersing. In the soliton
laser developed at Bell Labs by Mollenauer and Stolen, a
wavelength range in excess of $1.3 \mu m$ is chosen where the glass
in the optical fibre is negatively dispersive: the group velocity v
of an optical wavepacket centred on a wavelength λ depends on
the refractive index at λ *and* on its rate of change with λ:

$$(c/v) = (n - \lambda dn/d\lambda) \qquad (10.14)$$

and different frequency components within the packet will
disperse in time. If the rate of change of $(dn/d\lambda)$ is negative, low
frequencies are delayed compared with high frequencies (negative group velocity dispersion). If the intensity is low, this
linear dispersion can lead to a rapid broadening of a short input
pulse. A 10 ps input pulse at $\lambda = 1.5 \mu m$ travelling through
30 km of fibre would be lengthened to 150 ps by this mechanism and spoil its usefulness in optical telecommunications. But
if the pulse intensity is increased until nonlinear self-phase
modulation is important, we note from above that the nonlinear chirp works in the opposite direction, with high frequencies delayed over low frequencies. Solitons are steady state
pulses for which the two competing influences balance (figure
10.27). Mollenauer and coworkers observed twenty-seven-fold
pulselength compression from 7 ps to 260 fs in 100 m of fibre.
They used a solid-state laser to generate the light pulses in
which the active optical elements involved in the stimulated
emission are crystal defects called colour centres. When the
fibre is used *inside* a colour-centre laser cavity, it generates a

soliton laser with pulses from 130 fs to 2 ps, depending on fibre length. The soliton laser is clearly of interest in long distance optical communications.

In a femtosecond, light travels one-third of a micrometre (this is 1% of the thickness of a human hair). Such ultrashort optical pulses can be used to study a wide range of fast phenomena previously unresolvable. In condensed matter physics, such pulses can be used to study transport processes in semiconductors, where hot electrons thermalise in about 200 fs, and electron momentum distributions relax in about 20 fs. In photochemistry, direct observations of energy dissipation, transport and mode energy relaxation are possible. Electron transport in biological systems has a 1 ps timescale. All of these are being actively studied with ultrashort optical pulses, so that the rapidly changing processes can be interrogated before they have time to completely disappear.

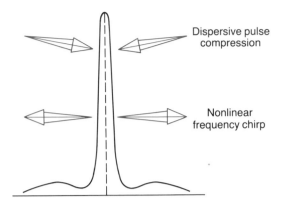

Figure 10.27. Pulse shape as a function of time of an optical soliton generated in a glass fibre. Two competing effects balance to produce the soliton. Dispersion acts to compress the pulse, whilst the nonlinear frequency chirp caused by the intense optical field inside the fibre acts to broaden the pulse. The balance between the two generates a narrow central peak with secondary sidepeaks for this simplest soliton.

10.8   Nonlinearities in optics and chaos

So far in our discussion of the nonlinear response of atoms and molecules to strong light fields, we have concentrated on the steady state behaviour after initial transients die out. This approach assumes that a dynamical equilibrium exists where the radiation forces on the physical system are balanced by dissipative interactions responsible for damping. Steady states are often achieved, but are not inevitable when nonlinearities are present. Nonlinear response can have very complicated dynamics with a wide variety of possible evolutions. Stable solutions, classified as 'fixed points' by mathematicians, can exist, but so too can oscillatory solutions known as limit cycles. More recently it has been discovered that a nonlinear physical system can degenerate into a chaotic, unpredictable time-development apparently indistinguishable from random noise although the system is perfectly deterministic in the sense of being described by well-behaved equations of motion with no stochastic, fluctuating ingredients. Mathematicians and theoretical physicists have made great progress in understanding this transition to chaos (see Chapter 12 by Ford). Here we will describe such phenomena in lasers and in nonlinear optical systems.

To understand the central concepts in the transition to chaos, we turn to a very simple model which contains many of the phenomena governing quantum optical chaos. Imagine a pulse of light travelling round a laser ring cavity made up of highly reflecting mirrors and containing a thin cell of excited atoms which amplify the pulse of light by stimulated emission. The pulse energy will grow through this amplification, but as it grows the pulse will deplete the energy in the amplifying excited atoms. This depletion, or saturation, clearly depends on how intense the pulse is as it enters the amplifying cell. If x_n is the pulse energy (essentially the number of photons in the pulse circulating round the cavity) after the nth transit, then, after one more passage through the cell, the pulse energy is changed to x_{n+1}. The growth can be characterised by a rate r so that x_{n+1} will be related to x_n by the growth rate r. However, the growth cannot continue indefinitely and we must take into account the depletion of available energy. We do this in a simple way by multiplying r by a factor $(1 - x_n)$, which contributes nothing extra for small x_n, but as x_n grows decreases the effect of r to account for this saturation. Then the pulse energy after the $(n+1)$th transit is related to its earlier value by the recurrence relation, or 'map'

$$x_{n+1} = rx_n(1 - x_n). \qquad (10.15)$$

This is called the 'logistic map' and has been much studied by ecologists and mathematical biologists as an example of a birth–death, or predator–prey evolution. Here it is an oversimplified and crude model of photon amplification but contains much of interest. The 'dynamics' of successive iterations of the logistic map round the ring cavity is governed by the single nonlinearity parameter r, related to the ability of the gain medium to amplify light. We will imagine r to be variable as a control parameter which we can vary at will (for example by providing more pumping power to create a higher density of excited amplifier atoms). There are limits on the size of r: it must be greater than zero to be interesting, and because we do not want to get involved with units we have normalised 'pulse energies' x_n which are positive but less than one. This restricts r to be less than 4 (x_{n+1} is less than one and x_n could equal $\frac{1}{2}$, for example).

If we fix the size of our amplifier gain r, we can iterate the logistic map many times and see how the pulse of light evolves from some initial value. For example if $r = 0.9$, an initially large x_n just dies away to zero: the gain is insufficient for the pulse to sustain itself in the cavity. When the gain is set to $r = \frac{3}{2}$, the pulse energy rapidly settles down to a steady state (a fixed point) where $x = \frac{1}{3}$ after each transit. This crudely mimics the transition of the laser to a stable operating point above the threshold where gain balances loss. As r increases, initial relaxation oscillations break out but a steady state is reached. Again this behaviour has been seen in lasers driven well above threshold where the population in the excited state far exceeds that in the lower state so that optical gain hugely exceeds the photon losses. When $r = 3.05$ a new behaviour sets in: the pulse energy oscillates between two stable positions, $x = 0.5$ and $x = 0.7$, forever. Increasing r further uncovers an unexpected diversity of different oscillations, culminating in a completely chaotic evolution in which all order has been eradicated (figure 10.28).

The behaviour of the simple logistic map as we iterate can be seen best by making a graph of the function $f(x) = rx(1 - x)$ for various choices of r (figure 10.29a–f). We see that r describes how steeply $f(x)$ rises. We can then use these graphs to iterate the map. First we pick an initial value of our pulse energy, x_1. After the first passage through the amplifying cell, the pulse energy grows to $x_2 = rx_1(1 - x_1)$. This value is obtained by drawing a vertical line at x_1 to the curve $y = rx(1 - x)$. This is then our new output value, so we set $y = x$ to obtain x_2 by drawing a horizontal line from the curve to the line $y = x$. The next iteration is obtained by a further vertical line to the parabola $y = f(x)$ and so on. If $r < 1$, figure 10.29(a), the iterations draw a sequence of steps which home in on the origin, which is our steady-state fixed point. If the gain parameter r is increased above unity, figure 10.29(b), the iterates home in not on the origin but on a nonzero steady state x (called a simple attractor). When the gain factor r is increased, this simple attractor becomes unstable and successive iterations wind outwards from what has become a repellor. We note that, at this critical value of r, $y = 3x(1 - x)$ crosses $y = x$ at $x = \frac{2}{3}$. In fact, the slope of y at this point $x = \frac{2}{3}$ is equal to -1 and crosses the line $y = x$ at right angles. If $r < 3$, this angle is less than $90°$, and if $r > 3$ the angle is greater than $90°$ and it is this angle which determines whether the iterates spiral into a stable fixed attractor or outwards from a repellor. The curious behaviour at $r = 3$ is best explained using two iterations because we know for $r > 3$ there are two attractors visited in turn, so two iterations are needed to return to a particular attractor:

$$x_3 = rx_2(1 - x_2) = r[rx_1(1 - x_1)][1 - rx_1(1 - x_1)].$$

This is a quartic polynomial, sketched in figure 10.30. When the gain $r < 3$, there is only a single attracting point where the line $y = x$ crosses, and when it exceeds three there are *three* crossing points: two stable attractors and an unstable repellor,

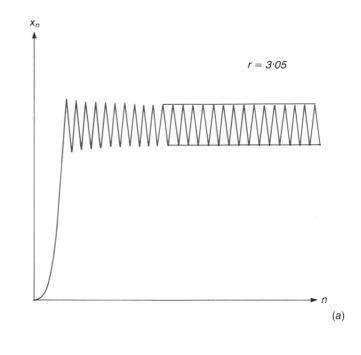

(a)

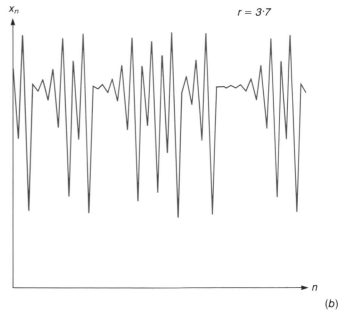

(b)

Figure 10.28. Relaxation oscillations and chaotic output in a simple iterative model of optical amplification. The output light after n transits through a thin region of gain is shown for each transit. (a) When the optical gain $r = 3.05$ the output evolves to an oscillation between two stable values on successive iteration. (b) Increased optical gain of $r = 3.7$ generates a chaotic output in which the next iteration yields a new output effectively unrelated to the last value.

so the system alternates between the two attractors. At $r = 3$, the single attractor is a tangent attractor and our original single stable solution *bifurcates* into an alternation between two solutions, called a two-cycle. This is called period doubling: it takes twice the number of iterations to return (figure 10.31). Remarkably, as the gain parameter is increased, this bifurcation happens again and again. By the time the gain has increased beyond $r = 3.57$ this bifurcation has occurred infinitely many times, and at $r = 4$ there are no periodic attractors at all (figure 10.32): the whole interval from $x = 0$ to 1 is a 'strange attractor'. If there are exceedingly many attractors which the system visits in turn before returning to its starting point, this n-cycle will look quite complicated and as n increases to infinity, it is as if the motion is aperiodic. This is the origin of deterministic chaos and this period-doubling which we have described is one generic route to chaos. In our ring laser model, the output would consist of a sequence of irregular pulses.

Feigenbaum, using a hand calculator, discovered that there are magic, universal numbers in any such bifurcation series. For example, if stable attractors bifurcate at a value $r = r_n$ to an n-cycle, then

$$(r_n - r_{n-1})/(r_{n+1} - r_n)$$

tends to $4.669\ 201\ 660\ 910\ 299\ldots$ as $n \to \infty$, and this number is *not* particular to our simple logistic map. More complicated nonlinear evolutions lead to the same period-doubling route to chaos.

Many quantum optical systems have nonlinear features which lead to a period-doubling sequence to a chaotic evolution. Other routes to chaos which do not involve period-doubling are possible, involving a two-frequency interaction and incommensuracy. Many earlier and unexplained instabilities in laser systems are now known to involve not uncontrollable noise but interesting generic features of basic nonlinear physics. As the gain increases, lasers can show pulsations and chaos. Passive devices such as optically nonlinear crystals inside cavities formed by reflecting mirrors can exhibit two or more different output intensities for the same input, and such optically bistable systems reveal period-doubling instabilities and chaos.

One signature of chaos is the broad-band nature of the power spectrum of the emitted light. Regular oscillation would result in single discrete frequency spectra; as the nonlinearities increase and period-doubling occurs, the spectrum acquires more and more lines. A further indication of chaos is the extreme sensitivity of the output to the initial conditions. Two chaotic systems which start with nearly identical initial conditions develop along exponentially diverging paths until they are quite unrelated. This is in stark contrast to linear systems which start with similar initial conditions but rapidly approach a common long-time behaviour. Indeed the applied mathematicians have seen a close relationship between the chaotic development of simple nonlinear systems and real

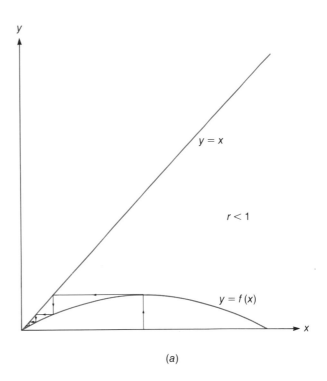

(a)

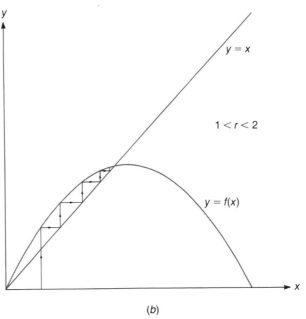

(b)

Figure 10.29. The iteration of the logistic map is illustrated by two curves $y = rx(1 - x)$ and $y = x$. An input value x_1 is chosen and the output $y = rx_1(1 - x_1)$ computed by drawing a vertical line to meet the curve y. This output is then used as the new input x_2 by drawing a horizontal line to meet $y = x$. This is repeated again and again. In graphs (a)–(f), different values of the gain r are chosen to illustrate the sensitivity of the iterations to the size of this gain.

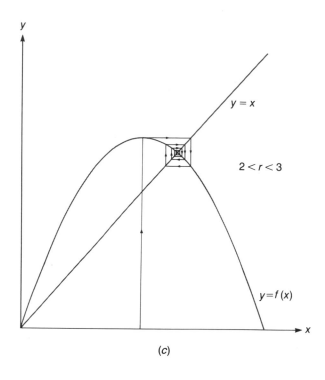

$y = x$

$2 < r < 3$

$y = f(x)$

(c)

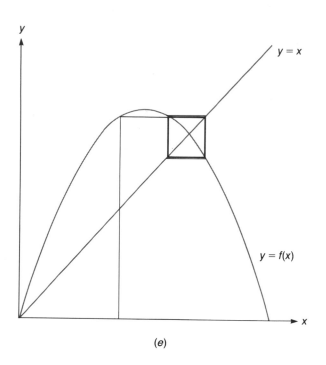

$y = x$

$y = f(x)$

(e)

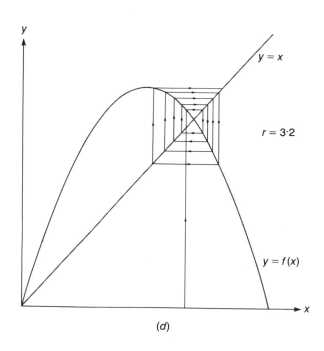

$y = x$

$r = 3\cdot2$

$y = f(x)$

(d)

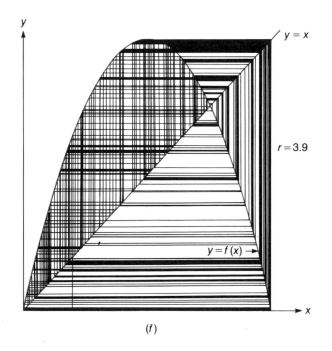

$y = x$

$r = 3.9$

$y = f(x)$ →

(f)

turbulence in complicated fluid flow. The nonlinear equations of turbulent flow, suitably simplified, were shown in 1963 by Lorenz to exhibit strange attractors and chaos. The same equations were shown by Haken and Graham to be equivalent to the basic single mode laser equations describing the development of the laser electric field by an atomic medium with an optical polarisation and inversion controlled by an outside pump mechanism. Whereas the instabilities in fluid flow may take hours or days to develop and study, the closely related instabilities in lasers may be studied in a fraction of a second. Indeed optical chaos seems ubiquitous and stable laser emission is often an exception.

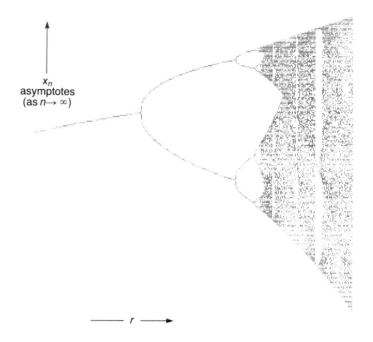

Figure 10.30. Double iteration of the logistic map reveals origin of period doubling. For $r < 3$ only a single crossing of the two curves $y = x$ and $y = r[rx(1-x)]\{1 - [rx(1-x)]\}$ (which represents two iterations) occurs. This implies a one-cycle, where one always gets the same result. At $r = 3$ we find a tangential intersection, whereas for $r > 3$ there are three intersections representing two stable attractors and an unstable repelling point.

Figure 10.32. Cascade of period-doubling bifurcations in the logistic map. As the control parameter r is increased, the stable solutions repeatedly split into two. The range of values of r in which an n-cycle exists decreases rapidly as n increases until an infinite number of bifurcations occur to result in a chaotic evolution.

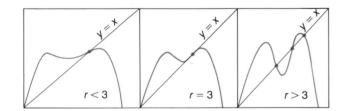

Figure 10.31. Bifurcation from a period one to a period two solution of the logistic map as control parameter r increases. For small r the system evolves to a single stable solution. At a critical value of r the steady-state solution bifurcates into two possible stable solutions which are visited in turn in the evolution as a 'two-cycle'.

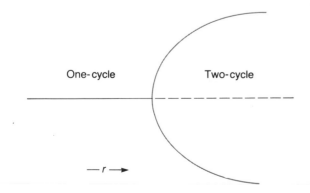

10.9 Where next?

The central concern in quantum optics is the establishment of coherence and order in light against an opposing tendency to randomness and disorder created by the fluctuating quantum environment interacting with the light. An example of current work in progress illustrates this concern in a study of the suppression of quantum fluctuations in so-called 'squeezed' states of light.

Squeezed states of light are nonclassical states of light with phase-dependent quantum noise properties which can be less than those of the vacuum state of the electromagnetic field. Quantum fluctuations of the electromagnetic field are an unavoidable consequence of the quantisation procedures and reflect our inability to measure precisely the electric and magnetic field of light. Even the vacuum state is beset by these random fluctuations, which are intimately linked to the zero-point energy of the light field. The vacuum fluctuations are completely phase insensitive. Yet nonlinear optical interactions can generate squeezed quantum fluctuations which depend strongly on the phase (relative to that of the intense coherent light in the nonlinear interaction generating the squeezing).

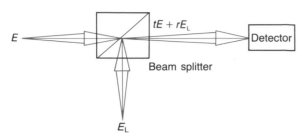

Figure 10.33. Homodyne detection of a signal E by combining with a field E_L from a local oscillator and detecting the beat note. The beam splitter is imagined to be a partially silvered mirror which combines the signal and local oscillator. A fraction t of the signal is transmitted and a fraction r of E_L is reflected to form the detected field.

The phase properties of noisy signals are not revealed by measuring the light intensity, which just averages over the phase. A technique called homodyne detection is used (figure 10.33) to detect phase-sensitive signals. The optical signal E is combined with the light from a coherent local source E_L at a beam splitter. A photomultiplier sees the total field $(tE + rE_L)$, where t is the transmission fraction of the beam splitter and r the reflectivity. The detected intensity is $I = |(tE + rE_L)|^2$, or

$$I = t^2|E|^2 + r^2|E_L|^2 + rt(EE_L^* + E_LE^*),$$

where asterisks denote complex conjugates. We imagine our local oscillator is much stronger than the signal field and pick out the beat signal which depends on both local oscillator field and signal field (this is how a super-heterodyne radio receiver works). If the phase of the local oscillator field is varied, we pick out the components in the beat signal (which depends on the signal E) at that phase, and in this way map out the phase properties of the signal E. A coherent field, or even a vacuum field has completely phase-insensitive noise. A squeezed state, generated by some nonlinear optical processes, has less phase noise in one phase quadrature of the light than that in the quadrature 90° out of phase with it. If we write the electric field of the signal in terms of these quadratures,

$$E = E_0(X_1\cos\omega t + X_2\sin\omega t)$$

then X_1 and X_2 are the quadrature field operators. They obey Heisenberg's uncertainty relation $\Delta X_1 \Delta X_2 > \frac{1}{4}$ for the uncertainties ΔX_1 and ΔX_2 (the standard deviations from the mean values of X_1 and X_2). A coherent field, or a vacuum field, has just the minimum zero-point energy fluctuations permitted by the uncertainty principle, so that $\Delta X_1 \Delta X_2 = \frac{1}{4}$, and the standard deviations are equal: $\Delta X_1 = \Delta X_2 = \frac{1}{2}$. A squeezed state can have one of these standard deviations much less than the vacuum noise $\Delta X_i = \frac{1}{2}$ for one of the quadratures provided the

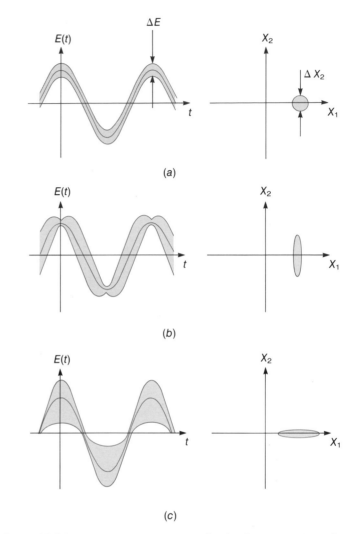

Figure 10.34. Time-variation or equivalently phase-variation of a radiation field. The shaded area's width at any time indicates the size of the field uncertainty ΔE for (a) a coherent field, (b) a squeezed state with reduced amplitude fluctuations (X_1 squeezed), and (c) a squeezed state with reduced phase fluctuations (X_2 squeezed). On the right hand side the quadrature standard deviations ΔX_1 and ΔX_2 for these states are shown. (After C. M. Caves who first squeezed states in this diagrammatic way; Caves, 1981.)

complementary quadrature has an increased noise standard deviation. In figure 10.34 the quantum noise in an optical electric field is sketched as a function of time (or equivalently against the local oscillator phase). As the phase varies, the quadratures X_1 and X_2 contribute varying amounts to the total electric field standard deviation. The origin of the phrase 'squeezed state' is obvious from figure 10.34 when the standard deviations in X_1 and X_2 are shown. The circular 'error' contour for a coherent signal can be squeezed into an ellipse with minor radius ΔX_1 less than the vacuum limit of $\frac{1}{4}$, or with ΔX_2 reduced, provided the noise in the other quadrature is enhanced.

Several groups of experimentalists have succeeded in generating squeezed light with quadrature noise below the vacuum level, some by more than a factor of two. To produce squeezing we require an interaction which nonlinearly couples optical fields together in a phase-sensitive fashion. A whole range of nonlinear optical couplings will do this, including four-wave mixing and a two-photon technique known as parametric down conversion in which laser photons of frequency ω split into subharmonics of frequency $\omega/2$ (the inverse of frequency doubling). Squeezed light sources are a good example of the coherent manipulation of light beams, this time to produce radiation with noise properties less even than the shot noise of a perfectly coherent radiation field. To demonstrate the existence of squeezed light required the development of technologically complex stable lasers, nonlinear optical devices and detectors.

Squeezed light is of great interest to the physicist because it provides new insights into the fundamental quantum fluctuations at the heart of field theory. At the same time an efficient source of intense squeezed light is of more than just theoretical interest. In optical communications the quantum fluctuations of the light field used to carry messages (for example in an optical fibre) contribute significant quantum noise to the transmitted signal. Fibre optic telecommunications engineers are already trying to use squeezed states of light in an attempt to minimise this noise in phase-sensitive transmission schemes. In other areas of physics, squeezed light with its decreased amplitude or phase fluctuations can be used to sense tiny effects which otherwise would be masked by the quantum jitter induced by normal light. One example is the detection of gravity waves produced by remote astrophysical events. Such gravity waves will produce minute strains in detectors on earth. These strains could be detected using light to sense the motion of the apparatus, provided the light fluctuations themselves do not disturb the apparatus by an amount which exceeds that produced by the gravity waves. Squeezed states of light can be used to minimise this disturbance.

Quantum optics is making a major contribution to information science. Ultrashort pulses can be used in communication systems, and coherent light is a central part of imaging and optical signal processing. More recently nonlinear optics has been employed in optical switches, memory devices and parallel processors with great potential for the construction of an all-optical computer characterised by speed and massive parallelism. The key idea in this form of optical computer is optical bistability. A nonlinear medium (for example gallium arsenide or indium antimonide crystals) contained within a reflecting optical cavity can exhibit two different steady-state optical transmission states for the same input intensity. Whether an optical cavity transmits or reflects light depends upon whether an integer number of half-wavelengths of the light fit into the length of the cavity. The effective length of the cavity seen by the light depends on the refractive index of the material inside. This refractive index can be altered by the high intensity provided by a laser so that a low transmission cavity can be changed to a high transmission device merely by increasing the applied laser intensity.

In an optically bistable experiment a cavity containing a nonlinear element is chosen to weakly transmit light at low intensities. The light just cannot fit inside the cavity. When the input light intensity is increased the refractive index increases and changes the effective optical length. Eventually the light is able to penetrate the cavity as the wavelength matches the new apparent cavity length and the cavity switches to a high transmission state. If we reverse the process the large internal cavity field is held onto by the medium until the decreasing input intensity is unable to maintain the nonlinearity and the device switches to the low output state. This hysteresis and switching forms the basis of optical transistors, discriminators, switches and logic elements. Optically bistable elements have been made from solid-state materials with high nonlinearity and integrated into microscopic logic elements. The optical computer is one of the welcome but unexpected bonuses derived from a study of fundamental nonlinear optics.

At the heart of quantum optics is a concern for fundamental ideas (quantum fluctuations, coherence, order–disorder transitions and so on). Yet at the same time quantum optics and its applied sister quantum electronics (in which the quantum optical physics we have described are utilised in practical devices) is having a major impact on technology through optical communications, laser materials, processing techniques such as AVLIS, and a host of other applications.

Further reading

Caves, C. M. (1981). *Phys. Rev. D* **23**, 1693.

Heavens, O. (1970). *Lasers*. Duckworth.
  Although dated, this is an excellent account of the basic ideas of laser physics for the layman.

Knight, P. L. and Allen, L. (1983). *Concepts of Quantum Optics*. Pergamon Press, Oxford.
  This book has an account of photons and coherence, followed by a selection of original papers in this field.

Loudon, R. (1983). *The Quantum Theory of Light*, 2nd edn. Oxford University Press.
  This is the best textbook account of quantum optics.

11 Physics of far-from-equilibrium systems and self-organisation

Gregoire Nicolis

11.1 Introduction

For the vast majority of scientists physics is a marvellous algorithm explaining natural phenomena in terms of the building blocks of the universe and their interactions. Planetary motion; the structure of genetic material, of molecules, atoms or nuclei; the diffraction pattern of a crystalline body; superconductivity; the explanation of the compressibility, elasticity, surface tension or thermal conductivity of a material, are only a few among the innumerable examples illustrating the immense success of this view, which presided over the most impressive breakthroughs that have so far marked the development of modern science since Newton.

Implicit in the classical view, according to which physical phenomena are reducible to a few fundamental interactions, is the idea that under well-defined conditions a system governed by a given set of laws will follow a unique course, and that a slight change in the causes will likewise produce a slight change in the effects. But, since the 1960s, an increasing amount of experimental data challenging this idea has become available, and this imposes a new attitude concerning the description of nature. Such ordinary systems as a layer of fluid or a mixture of chemical products can generate, under appropriate conditions, a multitude of *self-organisation phenomena* on a macroscopic scale – a scale orders of magnitude larger than the range of fundamental interactions – in the form of spatial patterns or temporal rhythms. States of matter capable of evolving (states for which order, complexity, regulation, information and other concepts usually absent from the vocabulary of the physicist become the natural mode of description) are, all of a sudden, emerging in the laboratory. These states suggest that the gap between 'simple' and 'complex', and between 'disorder' and 'order', is much narrower than previously thought. They also provide the natural archetypes for understanding a large body of phenomena in branches which traditionally were outside the realm of physics, such as turbulence, the circulation of the atmosphere and the oceans, plate tectonics, glaciations, and other forces that shape

our natural environment; or, even, the emergence of self-replicating systems capable of storing and generating information, embryonic development, the electrical activity of the brain, or the behaviour of populations in an ecosystem or in an economic environment.

The principal goal of this chapter will be to analyse the status of self-organisation phenomena from the standpoint of the physical sciences. Nonlinear dynamics and the presence of constraints maintaining the system far from equilibrium will turn out to be the basic mechanisms involved in the emergence of these phenomena.

We shall first describe (in Sections 11.2 to 11.4) some particularly representative experiments, which will serve continuously as reference throughout this chapter. In Section 11.5 we shall summarise the 'take home' lessons suggested by these examples. This will allow us to identify the key concepts of nonlinearity, irreversibility, stability, bifurcation and symmetry-breaking, which will be further analysed in Sections 11.6 to 11.9. Finally, we shall illustrate (in Section 11.10) how the self-organisation paradigm allows us to model problems outside the traditional realm of the physical sciences.

11.2 Thermal convection, a prototype of self-organisation phenomena in physics

We shall first be concerned with the bulk motions of fluids under the effect of temperature inhomogeneities known as *thermal convection*.

The study of these motions, even beyond the specific subject that will receive our attention, is far from academic. Without counting the numerous technological applications, thermal convection is the basis of several important and spectacular phenomena on our planet. One example is the circulation of the atmosphere and the oceans (determining to a large extent short- and medium-term weather changes) and the continental drift (the motion of continental plates induced by large-scale motions in the mantle). A little further from us, at the basis of transfer of heat and matter in the sun, is convection, which affects considerably the sun's activity. In the laboratory we may study the mechanism of thermal convection on a setup of more modest dimensions. The following simple experiment, realised for the first time by Bénard, leads to the observation of a number of astonishing properties.

Imagine a layer of fluid (say water) limited by two horizontal parallel plates whose lateral dimensions are much longer than the width of the layer. Left to itself, the fluid will rapidly tend to a homogeneous state in which, statistically speaking, all its parts will be identical. For instance, a minute observer will, on the sole basis of observations of his environment, be unable to tell whether he is within the small volume V_A or the small volume V_B of the fluid (figure 11.1). All volumes that one could define arbitrarily within the fluid will thus be indistinguishable, and the knowledge of the state of one of them would suffice to know the state of them all, independently of their form and their size. In other words, from the standpoint of our observer, the position he occupies makes no difference. Alternatively, there exists no intrinsic way enabling him to perceive the notion of space.

The homogeneity of this sytem extends of course to all its properties and, in particular, to its temperature, which will be the same at all parts of the fluid and equal to the temperature of the limiting plates or, alternatively, to the temperature of the 'external world'.

All these properties are characteristic of a system in a particular state, the state of equilibrium, for which there is neither a bulk motion nor a temperature difference with the outside world. We can express this property in a more quantitative manner as follows. We denote by T_1 and T_2 the temperatures of plates 1 and 2, respectively; then at equilibrium we will necessarily have

$$\Delta T_e = T_2 - T_1 = 0. \qquad (11.1)$$

Imagine now that somebody places a finger on one plate for a moment. The temperature at this part of the plate will be momentarily modified (for instance from 20°C to the human body's temperature 36.9°C). An incident like this, which takes place by chance in a system and modifies locally (and generally weakly) some of its properties, is called a *perturbation*. For our system at equilibrium this temperature perturbation will have no influence, since the temperature will rapidly become uniform again and will equal its initial value. In other words the perturbation dies out; the system keeps no track of it. When a system is in a state such that the perturbations acting on it die out more or less quickly in time, we say that the state is *asymptotically stable*.

From the standpoint of our minute observer, not only the homogeneity of the fluid makes it impossible for him to develop an intrinsic conception of space, but, in addition, the stability of the state of equilibrium eventually makes all instants to be identical. It is therefore also impossible for him to develop an

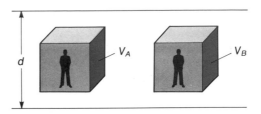

Figure 11.1. A minute observer contemplating the state of the volume elements V_A, V_B in a layer of liquid in equilibrium finds them indistinguishable and concludes that the fluid displays translational invariance along the horizontal direction.

intrinsic conception of time. One can hardly speak of 'behaviour' for a system in such a simple situation. So let us move away from it.

One can increase the complexity of the system, for instance, by heating the fluid layer from below. In doing this, we communicate energy to our system in the form of heat. Moreover, as the temperature T_2 of the lower plate will be higher than T_1, the equilibrium condition equation (11.1) will be violated ($\Delta T > 0$). In other words, by applying an *external constraint* to the system, we do not permit the system to reach equilibrium. Note that in the present example external constraint implies energy flux and vice versa.

Suppose first that the constraint is weak (ΔT small). Our system will again adopt a simple and unique state, in which the only process going on will be a transport of heat from the lower to the upper plate, from which heat will be evacuated to the external world. The only difference from the state of equilibrium will be that temperature, and consequently density and pressure, will no longer be uniform. They will vary, in practically linear fashion, from warm regions (below) to cold ones (above). This phenomenon is known as *thermal conduction*. In this new state that the system has reached to respond to the constraint, stability will prevail again, and the behaviour will eventually be as 'simple' as at equilibrium.

By removing the system from equilibrium further and further, through an increase in ΔT, we observe that, suddenly, at a value of ΔT that we will call *critical*, ΔT_c, matter begins to perform a bulk movement which is visualised in figure 11.2(*a*). Moreover, this movement is far from random: the fluid is structured in a series of small 'cells' (figure 11.2*b*) known as Bénard cells. This is the regime of thermal convection defined at the beginning of this section.

Figure 11.3 outlines a qualitative explanation of the phenomenon. Owing to thermal expansion the layer becomes stratified; the fluid closer to the lower plate being characterised by a lower density than that nearer the upper plate. This gives rise to a gradient of density which opposes the force of gravity. One can easily see that this configuration is potentially unstable. Consider a small volume of the fluid near the lower plate, and imagine that this volume element is weakly displaced upward by a perturbation. Being now in a colder, and hence denser, region, it will experience an upward Archimedes force, which will tend to amplify the ascending movement further. If, on the other hand, a small droplet initially close to the upper plate is displaced downward, it will penetrate an environment of lower density, and the Archimedes force will tend to amplify the initial descent further. We see therefore that, in principle, the fluid can generate ascending and descending currents like those observed in the experiment. The reason why these currents do not appear as soon as ΔT is not strictly zero, as the above argument would suggest, is that the destabilising effects are counteracted by the stabilising effects of the viscosity of the fluid, which generates an internal friction opposing movement, as well as by thermal conduction, which tends to smear out the temperature difference between the displaced droplet and its environment. This explains the existence of a critical threshold, ΔT_c, as observed in the experiment. Figure 11.2(c) further shows the *complexity* of the movements: at one point the fluid goes upward, moves along plate 1, then goes downward, moves along plate 2, goes upward again, etc. The cells unfold, along the horizontal axis, adopting successively a right-handed (R) or left-handed (L) rotation.

This is, you might say, a modest complexity compared to that of the humblest bacterium. But let us turn once again to our minute observer!

At his level, his universe has been totally transformed. For instance, he can now decide where he is and where he is not by observing the sense of rotation of the cell he occupies. Moreover, by counting the number of cells he goes through, he can acquire a quite efficient notion of space. We call this emergence of the notion of space in a system in which, till then, this notion could not be perceived in an intrinsic manner *symmetry-breaking*. In a way, symmetry-breaking brings us from a static, geometrical view of space to a view whereby the space is shaped by the functions going on in the system.

Perhaps the most remarkable feature to be stressed in this sudden transition from simple to complex behaviour is the *order* and *coherence* of this system. When ΔT was below the critical value, ΔT_c, the homogeneity of the fluid in the horizontal direction was rendering its different parts independent of each other. It was totally without importance (figure 11.1) to permute volumes V_A and V_B. In volume V_C, placed between V_A and V_B, no modification of the observable properties would be detected by the fact that volume A is now on its right rather than on its left. In contrast, beyond the threshold ΔT_c, everything happens as if each volume element were 'watching' the behaviour of its neighbours and was 'taking it into account' to play its role adequately and to participate in the overall pattern. This suggests the existence of *correlations*, that is to say,

(a)

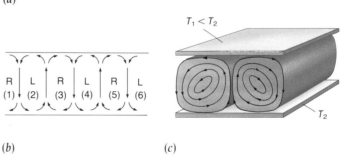

(b)   (c)

Figure 11.2. (a) Top view of convection (Bénard) cells. (b), (c) Schematic views demonstrating the details of the flow. Notice the opposite direction of rotation of two adjacent cells. (L) = left-handed rotation; (R) = right-handed rotation.

of statistically reproducible relations between distant parts of the system. We arrive at this key conclusion here just by observing the phenomenon.

Nevertheless, one should, even at this early stage of our analysis, oppose the long-range character of these correlations with the short range of intermolecular forces. The characteristic space dimension of a Bénard cell in usual laboratory conditions is in the millimetre range, whereas the characteristic space scale of the intermolecular forces is in the angstrom (10^{-10} m) range. Alternatively, a single Bénard cell comprises something like $(10^7)^3 \sim 10^{21}$ molecules. That this huge number of particles can behave in a coherent fashion despite the random thermal motion executed by each of them is one of the principal properties characterising self-organisation and the emergence of complex behaviour.

Notice that throughout the last few paragraphs we have been using a vocabulary including notions such as coherence, complexity and order, which have been an integral part of biology for a long time, but which, until recently, were outside the mainstream of physics. The possibility to describe, through these fundamental concepts, the behaviour of both living beings and of quite ordinary physical systems is a major development that could not possibly have been forecast a few years ago.

But the Bénard cells reserve more surprises for us. On the one hand, the experiment is perfectly reproducible since, by realising the same experimental conditions, one will see always the convection pattern appear at the same threshold value ΔT_c. But, on the other hand, as we saw in figure 11.2 matter is structured in cells that are alternatively right-handed or left-handed. Once a direction of rotation is established, it will remain so in each cell. Still, however sophisticated the control of the experimental setup might be, two qualitatively different situations can be realised just after the critical threshold ΔT_c.* In figure 11.2(*b*) cell 1 is right-handed (and cell 2 left-handed, 3 right-handed, etc.), but it could have been otherwise, and cell 2 would then be right-handed with corresponding changes for the other ones. As soon as ΔT slightly exceeds ΔT_c, we know perfectly well that the cells will appear: this phenomenon is therefore subjected to a strict determinism. In contrast, the direction of rotation of cells is unpredictable and uncontrollable. Only chance, in the form of the particular perturbation that may have prevailed at the moment of the experiment, will decide whether a given cell is right- or left-handed. We thus arrive at a remarkable cooperation between chance and determinism, which is again reminiscent of a similar duality familiar in biology since Darwin's era (mutation–natural selection), and which has so far been limited in the physical sciences to the quantum description of phenomena going on at a microscopic scale. Figure 11.4 summarises the situation.

Figure 11.3. Qualitative explanation of the origin of thermal convection. If a parcel of warm fluid near the bottom of the layer is displaced upward slightly, it enters a region of greater average density and therefore experiences an upward buoyancy force. Similarly, if a parcel of cool fluid near the top of the layer is displaced downward, it becomes heavier than its surroundings and tends to sink.

We see that far from equilibrium, that is when the constraint is sufficiently strong, the system can adjust to its environment in several different ways, or, to be less anthropomorphic, that several solutions are possible for the same parameter values. Chance alone will decide which of these solutions will be realised. The fact that, among many choices, only one has been retained confers to the system an *historical dimension*, some sort of 'memory' of a past event which took place at a critical moment and which will affect its further evolution.

What happens when the thermal constraint increases well beyond the first threshold of structuration? For some range of values the Bénard cells will be maintained globally, but some of their specific characteristics will be modified. And then, suddenly, beyond another critical value $\Delta T'_c$, we witness a new, forceful manifestation of randomness: the structure will become fuzzy, and a regime characterised by an erratic dependence of the variables in time will emerge. This is the precursor of what engineers and fluid dynamicists have, for the past one-hundred years, been calling *turbulence*. More generally, it now appears that turbulence is one aspect of a general trend of several classes of systems to evolve in a *chaotic* fashion under certain conditions. We discuss this point further in Sections 11.7 and 11.8.

To summarise, we have seen that nonequilibrium has enabled the system to transform part of the energy communicated from the environment into an ordered behaviour of a new type, the *dissipative structure*: a regime characterised by a symmetry-breaking, multiple choices and correlations of a macroscopic range. We can therefore say that we have literally witnessed the birth of complexity through self-organisation. True, the type of complexity achieved is rather modest, but nevertheless it presents characteristics which were usually ascribed exclusively to biological systems. More importantly, far from challenging the laws of physics, complexity appears to be an inevitable consequence of them when suitable conditions are fulfilled.

---

*As a matter of fact, one should say that there is an *infinity* of possibilities, since for a large system the whole structure can be translated by any length (breaking of a *continuous symmetry group*). However, the most characteristic manifestation of this multiplicity remains the direction of rotation of the cell, and for this reason we use this criterion throughout this section.

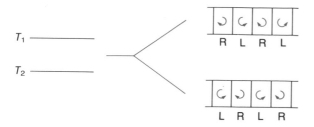

Figure 11.4. Multiplicity of solutions beyond the threshold of the thermal convection instability. Depending on the conditions, a given region of space may be part of a left-handed (L) or a right-handed (R) cell.

11.3   Self-organisation phenomena in chemistry

We have, according to tradition, regarded Bénard cells as part of physics, principally because the chemical nature of the substances constituting the layer remains unchanged during the phenomenon. We shall now consider chemical reactions, namely processes involving modifications of the identity of the constituting molecules. Such phenomena are of quite general concern. Much of the chemical industry is based on heterogeneous catalysis, whereby some of the steps necessary for the synthesis of a product are accelerated by the presence of a surface reacting with the bulk medium. For instance, the oxidation of ammonia is usually carried out in the presence of a platinum catalyst, whereas in the decomposition of N_2O, a catalytic surface of copper oxide is used. Combustion, for instance the burning of hydrocarbons, thanks to which heat engines function, is another class of important chemical transformations. In addition, several manifestations of biological order are amenable to chemical reactions involving special catalysts, the enzymes.

In a typical chemical reaction a molecule of species A (say the iodine molecule I_2) can combine with a molecule of species B (say molecular hydrogen H_2) to produce one molecule of species C and one molecule of species D (or two molecules of the same species like hydrogen iodide, 2HI, in our example). This process is symbolised by

$$A+B \xrightarrow{k} C+D, \qquad (11.2a)$$

in which k is the rate constant, generally a function of temperature and pressure. On the left-hand side, the *reactants* A and B combine and disappear in the course of time, whereas on the right-hand side the *products* C and D are formed and appear as the reaction advances. Nevertheless, in an isolated system it is observed that, even after a very long time, reactants A and B never disppear completely. Moreover, the amounts of the coexisting constituents A, B, C and D are characterised by a certain fixed ratio after a sufficient lapse of time. When this fixed value is attained we say that the system is at *chemical equilibrium*,

and the value of the ratio is the *equilibrium constant*. This is the analogue of the homogeneous state of rest in the Bénard problem. It is, of course, understood that the values of the physical parameters, like pressure, temperature, etc., remain constant.

How can one reconcile this result with reaction (11.2*a*), according to which A and B are bound to disappear? Experiment shows the existence of the reverse transformation of reaction (11.2*a*),

$$C+D \xrightarrow{k'} A+B. \qquad (11.2b)$$

At equilibrium both reactions occur with exactly the same velocity. This fundamental property of nature, known as *detailed balance*, is responsible for most of the properties characterising this state of matter. We represent a reversible reaction as

$$A+B \underset{k'}{\overset{k}{\rightleftharpoons}} C+D. \qquad (11.2c)$$

In the Bénard problem we were able to move away from equilibrium by introducing an energy flux to our system. The most obvious chemical analogue of this procedure is to submit the system to a mass flow from (or toward) the surroundings, thus realising what will be referred to hereafter as an *open system*. For instance, one can eliminate C or D from the reaction vessel when their concentration becomes larger than a prescribed value; or one can feed the system by a flow of a mixture rich in A and B, which goes through the reaction vessel and is eventually evacuated or recycled. Figure 11.5 depicts the way in which an open system functions at a laboratory scale.

By combining appropriately the rates of inflow and outflow of matter, one can create conditions for the system to attain a state in which the concentrations of A, B, C and D remain constant in time, while at the same time the ratio of their products is no longer given by the equilibrium constant. Mathematically, this will be reflected as a vanishing rate of change of the concentrations in time. To express this properly we introduce the time derivatives of the concentration (C) and write

$$\frac{dC_A}{dt} = \frac{dC_B}{dt} = \ldots = 0. \qquad (11.3)$$

We call the state described above a *stationary nonequilibrium state*.

Can we expect from an open system in such a nonequilibrium state a behaviour similar to the Bénard problem? For one thing, in such a system detailed balance no longer holds. Suppose now that, momentarily, one particular process, say the forward step of a reaction, becomes enhanced compared to the backward step. At equilibrium, detailed balance would tend to re-establish the initial state of affairs. Away from equilibrium, however, this is no longer so. Moreover, if some parts of the chemical mechanism could enable the system to capture and further amplify the above-mentioned enhancement, one would have a potentially unstable situation similar to the one depicted for the Bénard problem in figure 11.3. Such

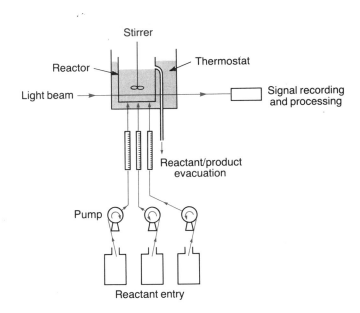

Figure 11.5. Experimental realisation of an open system. Three peristaltic pumps feed the reagents into the reactor. A vigorous stirring mixes the feed streams, giving rise to a practically uniform solution within which the reaction takes place. When the total volume exceeds a limit, the surplus is removed from the reactor. The reaction is monitored by an optical device and the recorded signal is adequately processed.

mechanisms are known to exist in chemistry, and their most striking manifestation is *autocatalysis*. For instance, the presence of a product may enhance the rate of its own production. As a matter of fact, this seemingly exotic phenomenon happens routinely in any combustion process, owing to the presence of free radicals, those extremely reactive substances containing one unpaired electron, which, by reacting with other molecules, give rise to further amounts of free radicals and thus to a self-accelerating process. In addition, *self-reproduction*, one of the most characteristic properties of life, is basically the result of an autocatalytic cycle, whereby the genetic material is replicated thanks to the intervention of specific proteins, themselves synthesised through the instructions contained in the genetic material.

For a long time chemists thought that a homogeneous, time-independent state similar to equilibrium should eventually emerge from any chemical transformation. Any deviation from this rule could only be the result of an artifice, or an amusing curiosity arising from the interference with a phenomenon that does not belong to the realm of chemistry.

We shall now describe a chemical reaction known as the Belousov–Zhabotinski reaction (hereafter referred to as BZ), which under certain conditions of nonequilibrium presents a whole spectrum of fascinating and unexpected behaviours.

There is nothing special about the reactants of the BZ reaction. A typical preparation consists of cerium sulphate, $Ce_2(SO_4)_3$, or another cerium salt, malonic acid, $CH_2(COOH)_2$ and potassium bromate, $KBrO_3$, dissolved in sulphuric acid. The evolution of the system can be followed by the naked eye by the use of a colouring substance such as, for instance, ferroin, which gives a red colour when there is an excess of ions of Fe^{2+} and a blue one when there is an excess of Fe^{3+}. More sophisticated means of observation are provided through specific electrodes, or by spectroscopic measurements of optical absorption caused by one particular substance.

Let us review the types of behaviour exhibited by this system under different experimental conditions, all at ordinary temperature.

BZ reaction in a well-stirred system: chemical clock and chaos

Suppose that the reaction is carried out first under the conditions shown in figure 11.5. Because of the efficient transfer of matter ensured by stirring, the system remains practically homogeneous in space at each moment. This experimental setup also allows one to control quite easily the distance of the system from equilibrium. It suffices to change the rates of pumping of the chemical into (or out of) the system, thus realising different *residence times* of a given substance within the reaction vessel. A very long residence time essentially amounts to realising a closed system, and under these conditions we expect to reach an equilibrium-like behaviour characterised by detailed balance. By decreasing this time we are not allowing a full equilibration of forward and backward steps, and under these conditions we expect that the system can manifest nonequilibrium behaviour. This is precisely what the experiment shows. For very large residence times a homogeneous steady state is reached, in which the concentrations of the chemicals remain time-independent. This is the typical state to which the laboratory chemist is accustomed, and it shares all the qualitative properties of chemical equilibrium. It is the analogue of the state of thermal conduction realised in the Bénard system when a weak temperature difference is applied across the plates.

If we now diminish the residence time, we encounter an altogether different pattern of behaviour. All of a sudden a blue colour (if ferroin is used), indicating an excess of Fe^{3+} (or Ce^{4+}) ions, invades the system. A few minutes (or even a fraction of a minute) later, the blue colour is replaced by a red one, indicating an excess of Fe^{2+} (or Ce^{3+}) ions. And the process goes on, blue, red, blue, red, etc., in a rhythmic manner with a perfectly regular period and amplitude which depend only on the experimental parameters and are thus intrinsic to the system. This oscillation, which measures time through an internally generated dynamics, constitutes a *chemical clock*. Figure 11.6 describes a typical clock behaviour.

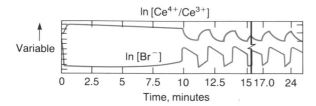

Figure 11.6. Potentiometric traces of $\ln[\mathrm{Br}^-]$ and $\ln[\mathrm{Ce}^{4+}/\mathrm{Ce}^{3+}]$ versus time during the BZ reaction. Initial concentrations: $CH_2(COOH)_2 = 0.032$ M, $KBrO_3 = 0.063$ M, $Ce(NH_4)_2(NO_3)_5 = 0.01$ M, $H_2SO_4 = 0.8$ M and $KBr = 1.5 \times 10^{-5}$ M. After an initial 'induction' period the oscillation starts. If a perturbation is applied to the system, a short transient is developed, and subsequently the oscillatory regime is recovered.

At this stage the lucid observer will raise a natural question. Why is there this bewilderment before rhythmic behaviour? Are we not observing a similar phenomenon when we watch a pendulum? After all, any mechanics course starts with the perfect frictionless pendulum and its oscillatory behaviour as a prototype of such fundamental principles as Newton's second law and the conservation of energy or angular momentum!

Figure 11.7 outlines an explanation of the deep differences between the two kinds of oscillations. Figure 11.7(a) shows a frictionless pendulum vibrating around the vertical direction with a maximum angle of opening θ_1 (the amplitude of the periodic function $\theta(t)$ describing the instantaneous value of the angle) and a period T_1. It also shows the time variation of the concentration of a chemical in the BZ reagent, characterised by an amplitude A and a period T.

We now momentarily disturb this state of affairs by displacing the pendulum at a higher angle θ_2 from the vertical $(\theta_2 > \theta_1)$, and by applying a slight temperature or concentration pulse in the BZ reagent (for instance, we could inject a few millimoles of potassium bromate or touch the vessel for a few seconds). Figure 11.7(b) describes the response of the systems to this disturbing action. The pendulum will again perform an oscillation, but its amplitude will now be equal to θ_2 rather than θ_1, and its period will be slightly larger. In other words, this system will keep the memory of the disturbance forever. In contrast, after a transient the BZ reagent will reset in an oscillatory mode of exactly the same amplitude and period as before. This is the property of *asymptotic stability* to which we referred in the previous section. It is directly related to a ubiquitous property of most of the phenomena observed in nature, namely *irreversibility*, and it is essentially responsible for the reproducibility of events. On the other hand, systems like the pendulum lack this property because their dynamics is invariant under time reversal. They are therefore at the complete mercy of the perturbations that may act on them and,

inasmuch as these perturbations are essentially unpredictable, they are bound to show, sooner or later, an erratic behaviour.

A very familiar oscillator which we have all borne since our embryonic life is our heart. It beats more or less regularly (around seventy or eighty beats per minute for the average individual), but it can also become deregulated. A particularly dramatic form of deregulation is fibrillation, the inability to perform an oscillation encompassing the entire heart. In fact, whatever the ultimate cause of death of an individual, death will always involve a state of fibrillation of the heart. We can now realise that if the heart were functioning as a pendulum, fibrillation would have occurred long before our birth. That the heartbeat – and indeed any other reproducible rhythmic phenomenon observed in nature, from circadian rhythms and the cell division cycle to the change of luminosity of variable stars, the cepheids – should belong to the same realm as the oscillations in the BZ reagent shows the tremendous importance of irreversibility and dissipative systems.

Let us now come back to our chemical clock. In the vocabulary introduced already in Section 11.2, one can say that in the regime of uniform steady state (which is also asymptotically stable!) the system ignores time. But, once in the periodic regime, it all of a sudden 'discovers' time in the phase of the periodic motion and in the fact that the maxima of the different concentrations follow each other in a prescribed order. We can refer to this as *breaking of temporal symmetry*.

From an even more fundamental viewpoint, the maintenance of sustained oscillatory behaviour encompassing the entire system implies that its different parts act in a concerted fashion by maintaining sharp phase relationships between them; otherwise destructive interference would wipe out the oscillatory behaviour. In other words, we again expect, just like in the Bénard problem, the emergence of *long-range correlations* induced by the nonequilibrium constraints.

Figure 11.7. Illustration of the difference between sustained oscillations in conservative systems like the pendulum (left) and in dissipative systems like the BZ reagent (right).

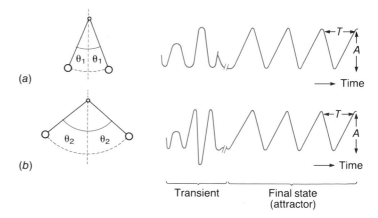

But we have not yet exhausted the list of surprises hidden in the BZ reaction! Detailed experimental analysis shows that, for values of residence times which are intermediate between two kinds of oscillatory regimes, complex nonperiodic behaviour is observed. The appearance of this *chemical turbulence* illustrates once again the tendency of many natural systems to evolve under certain conditions in a chaotic way. It also brings out the following very important property of chemical systems: in fluid mechanics, and in most other physical examples, complex behaviour is invariably associated with spatial inhomogeneities. But in chemistry even a spatially homogeneous system can show complex behaviour in time. The reason has in fact already been mentioned at the beginning of this section: chemical systems are endowed by mechanisms like autocatalysis, which are due to the peculiar molecular structure and reactivity of certain constituents and which enable them to evolve to new states by amplifying (or repressing) the effect of slight perturbations.

BZ reaction in a nonuniform system: spatial patterns

Suppose now that the reaction is carried out without stirring, thus allowing for the possible development of spatial inhomogeneities. One then observes regular spatio-temporal patterns in the form of propagating wavefronts, whose necessarily static representation (figure 11.8) gives only a faint idea of the beauty and activity involved. The waves seen in the figure are created in a thin layer of reagent. They appear in two different forms: circular fronts (figure 11.8a), displaying a roughly cylindrical symmetry around an axis perpendicular to the layer, usually referred to as *target patterns*; and spiral fronts (figure 11.8b) rotating in space clockwise or counterclockwise. It is also possible to obtain, though under rather exceptional conditions, the multi-armed spirals shown in figure 11.8(c). In either case the wavefronts propagate over macroscopic distances of space without distortion and at a prescribed speed, the 'message' released by the chemistry at the centre from which the whole pattern emanates. Once again, we witness the birth of complexity.

As in the Bénard problem, we can associate the formation of wavefronts to a space symmetry-breaking. The symmetry broken by the target patterns of figure 11.8(a) is quite similar to that broken in the Bénard system: essentially, the system is no longer invariant to translations along a particular direction of space. The symmetry-breaking involved in the formation of spirals (figure 11.8b, c) is, on the other hand, quite different, as it is associated to the notion of *chirality* or rotation. This type of asymmetry of matter has always exerted a particular fascination. Louis Pasteur, the founder of modern biochemistry, repeatedly expressed his bewilderment over the optical asymmetry of biomolecules, which is manifested by the rotation of the plane of polarisation of light in a preferred direction. He regarded this property as one of the basic aspects of life.

Moreover, the observation of the morphological asymmetry of adult organisms has introduced into human thinking the notions of 'right' and 'left' which marked philosophers and writers ever since Plato. It is amazing to see these deep notions emerging quite naturally through the intrinsic dynamics of a modest, ordinary-looking physico-chemical system!

The above-described behaviour is not an exclusive feature of the BZ system. It is shared by a host of other phenomena going on in homogeneous phase and involving equally simple chemicals, a partial list of which is given in table 11.1 Characteristically, in all these situations the domain of parameter values for which oscillations are observed is quite close to the domain in which another interesting phenomenon occurs, namely *bistability*. Specifically, two (or sometimes several) simultaneously stable stationary states coexist under exactly the same experimental conditions. The particular state chosen by the system will depend on the experimental conditions. This is illustrated in figure 11.9, in which we plot the value of a variable (e.g. the concentration of a chemical) versus a characteristic control parameter (e.g. the residence time). Suppose that an experiment is performed for a value $\lambda = \lambda'$ of this parameter, for which only one stable state (branch a) is available. By gradually increasing λ we enter in the region of multiple states ($\lambda_1 < \lambda < \lambda_2$). However, the system remains on branch (a) until λ exceeds the value λ_2. At this moment it jumps to branch (b) and remains there. If a variation of λ in the opposite direction is now imposed, starting, say, from value λ'' and going to value λ', the system will remain on branch (b) till the value λ_1 of the parameter. In other words, the system describes different patterns according to its past history. We call this phenomenon *hysteresis*.

In all cases known so far it turns out that the same chemical mechanism can account for bistability, oscillations and waves. As already mentioned, at least one autocatalytic step is usually involved. In the BZ system this step ensures the production of two moles of a substance appearing as an intermediate in the reaction, the bromous acid ($HBrO_2$), out of one molecule of the same substance according to

$$HBrO_2 + BrO_3^- + 3H^+ + 2Ce^{3+} \rightarrow 2HBrO_2 + 2Ce^{4+} + H_2O.$$
$$(11.4)$$

Direct autocatalysis is not, however, the only mechanism capable of generating complex behaviour in chemistry. For instance, in the phenomenon of heterogeneous catalysis, to which we referred at the beginning of this section, some of the chemical steps usually release energy. (They are called, for this reason, *exothermic* reactions.) This heats the medium. Now, as mentioned already, the rate constant of a chemical reaction is temperature dependent. It is in fact an increasing function of temperature, described to a good approximation by the Arrhenius law:

$$k(T) = k_0 e^{-E_0/k_B T},$$
$$(11.5)$$

in which E_0 is referred to as the activation energy and k_B is a

(a)

(b)

Figure 11.8. Wave propagation in a two-dimensional unstirred layer of BZ reagent. (*a*) Target patterns. They predominate when the reagent is in a parameter range compatible with the onset of bulk oscillations. (*b*) Successive stages of formation of rotating spiral waves. These patterns do not arise spontaneously, and must be induced by a suitable initial condition. Notice the annihilation of the wave upon collision of two fronts. (*c*) Multi-armed spirals.

universal constant known as Boltzmann's constant. This dependence can be understood qualitatively as follows (cf. figure 11.10). A reactive transformation involves the breaking of chemical bonds; in other words, it has to overcome an 'energy barrier' corresponding to the energy of the bond. This amount is provided by the kinetic energy of translational motion of the colliding molecules. If the medium is heated the mean kinetic energy of the molecules will increase, and thus a larger number of pairs of colliding molecules will have a sufficient amount of kinetic energy to overcome the barrier. In other words the reaction will be accelerated, in agreement with equation (11.5).

Suppose now that we are dealing with an exothermic reaction. If such a process is accelerated, more heat will be released, the temperature of the medium will further increase, and the reaction will be further accelerated. This provides therefore a potentially destabilising element capable of inducing transitions to new types of behaviour. And, indeed, in many catalytic reactions of industrial interest, patterns of behaviour quite similar to those described for the BZ system are observed. These affect considerably the local state of the catalyst and play therefore a very important role in the course of the reaction and in the yield and efficiency of the chemical plant.

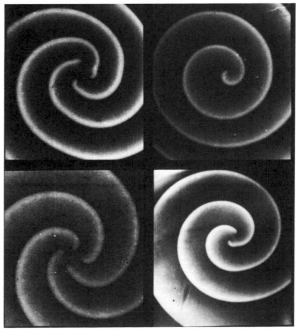

(c)

Table 11.1(*a*). *Chemical oscillators in a continuous stirred tank reactor (CSTR)*

Main species	Additional species
I^-	IO_3^-, MnO_4^- or $Cr_2O_7^-$
I^-	malonic acid
IO^{3-}	H_3AsO_3
IO^{3-}	$Fe(CN)_6^{4-}$, SO_3^{2-}, ascorbic acid or $CH_2O.SO_2$
I_2	$Fe(CN)_6^{4-}$, SO_3^{2-}
IO^{3-}	I^-, malonic acid
IO^{3-}	I^-, H_3AsO_3
I^-	BrO_3^-
BrO_3^-	SO_3^{2-}, $Fe(CN)_6^{4-}$, H_3AsO_3 or Sn^{2+}
I^-	I_2, $S_2O_3^{2-}$

(*b*). *Chemical oscillators in surface catalysis*

Main species	Additional species (catalysts)
$CO + O_2$	Pt, Pd, CuO, Ir
$H_2 + O_2$	Pt, Pd, Ni
$NH_3 + O_2$	Pt
$C_2H_4 + O_2$	Pt
$C_3H_6 + O_2$	Pt
$C_6H_{12} + O_2$	NaY (zeolite)
N_2O (decomposition)	CuO

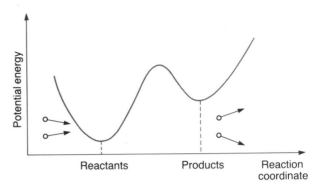

Figure 11.10. Schematic representation of a chemical reaction in terms of the motion of reactants in a double well potential, in which the initial and final states are separated by a barrier.

11.4   Biological systems

By now we have encountered three basic modes of self-organisation of matter giving rise to complex behaviour: bistability and hysteresis, oscillations (both periodic and non-periodic), and space patterns. We have seen these phenomena emerge in two 'case studies' involving bulk flows under a thermal constraint and open systems undergoing catalytic chemical reactions. Before we go on in our survey, it is natural to spend some time discussing biological complexity from the standpoint of self-organisation.

Living beings are undoubtedly the most complex and or-ganised objects found in nature, both in view of their mor-phology and their functioning. They are literally historical structures, since they have the ability to preserve the memory of forms and functions acquired in the past, during the long period of biological evolution. Moreover, living systems func-tion definitely under conditions far away from equilibrium. An organism as a whole continuously receives fluxes of energy (e.g. the solar influx used by plants for photosynthesis) and of matter (in the form of nutrients), which it transforms into quite different waste products evacuated to the environment. At the cellular level, strong inhomogeneities are also observed. For instance, the concentration of potassium ions, K^+, inside the cells of the nervous system, the neurons, is higher than in the outside environment, while the opposite is true for the sodium ions, Na^+. Such inequalities, which imply highly nonequilib-rium states, are at the origin of processes like the conduction of the nerve impulse, which play an important role in life. They are maintained thanks to active transport and bioenergetic reactions like glycolysis or respiration.

Being convinced by now that ordinary physico-chemical systems can show complex behaviour presenting many of the characteristics usually ascribed to life, it is legitimate to inquire whether some of the above features of biological systems can be

Figure 11.9. Illustration of the phenomena of bistability and hysteresis. The system is started on branch (*a*) with an initial value of parameter λ equal to λ'. Next, λ is increased. The system first remains on (*a*), but beyond the limit point value $\lambda = \lambda_2$, it jumps on to branch (*b*). When on the other hand λ is decreased below the value λ'', the system remains on branch (*b*) until λ becomes λ_1, whereupon it jumps on to (*a*). In other words, the specific path of states followed depends on the system's past history. This is known as *hysteresis*. Notice that in the range $\lambda_1 < \lambda < \lambda_2$ the system enjoys *bistability*, i.e. the possibility to evolve, for given parameter values, to more than one stable state, depending on initial preparation.

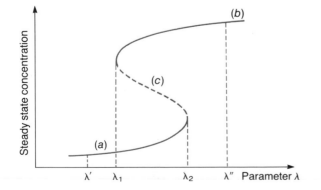

attributed to transitions induced by nonequilibrium constraints and appropriate destabilising mechanisms similar to chemical autocatalysis. This is probably one of the most fundamental questions that can be raised in science. No exhaustive answer can be claimed, but one can mention some examples in which the connection between physico-chemical self-organisation and biological order is especially striking. The particular problem on which we focus here is the control of embryonic development.

Embryonic development is the sequence of events leading from a unique cell, the fertilised egg, to a complete organism. Nature provides us with an unlimited number of illustrations of such processes. Certainly the simplest case is that of bacteria, whose development reduces to a sequence of cellular divisions. At the other end of the spectrum are advanced organisms like mammals, where development leads to a pluricellular body in which the cells form specialised tissues and organs that may comprise 10^{12} cells or more.

At present, it is out of the question to arrive at a detailed understanding of how such processes take place and, particularly, of how they are coordinated with the fantastic precision that allows each cell to fulfil its role at the right time and in the right place. Rather, we shall discuss living systems whose development is characterised by an 'intermediate' level of complexity, like the amoebae of the species *Dictyostelium discoideum* (a slime mould). Here development reduces essentially to a transition phenomenon, very similar to that observed in the BZ reagent, marking the passage from the unicellular to the pluricellular stage of life.

Figure 11.11 describes the life cycle of this species. In figure 11.11(*a*) the amoebae are at the unicellular stage. They are moving in the medium, feeding on such nutrients as bacteria, and they proliferate by cell division. Globally speaking they constitute a uniform system, inasmuch as their density (number of cells per square centimetre) is essentially constant. Suppose now that the amoebae are subjected to starvation (in the laboratory this is induced deliberately; in nature it may happen because of less favourable ambient conditions). This is the analogue of applying a 'constraint' in a physical or chemical experiment. One observes that individual cells do not die. Rather, they respond to the constraint by aggregating toward a centre of attraction (figure 11.11*b*). The initial homogeneity is broken; space becomes structured. The resulting pluricellular body (plasmodium) is capable of moving, presumably to seek for more favourable conditions of temperature and moisture (figure 11.11*c*). During this migration it differentiates and gives rise to two kinds of cells, one of which constitutes the stalk and the other a fruiting body within which spores are formed (figure 11.11*d*). Eventually the spores are disseminated in the ambient medium (figure 11.11*e*), and, if the conditions are favourable, they germinate to become amoebae, and the life cycle begins again.

Let us investigate in more detail the aggregation stage. The following phenomena are observed during this process. First,

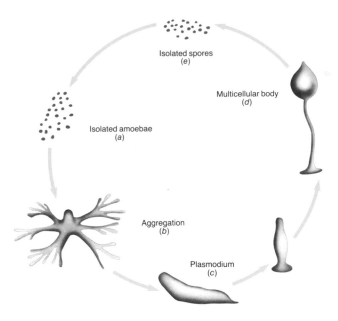

Figure 11.11. Life cycle of the amoeba *Dictyostelium discoideum*.

after starvation, some of the cells begin to synthesise and release signals of a chemical substance known as cyclic adenosine monophosphate (cAMP) in the extracellular medium. The synthesis and emission are periodic, just like in the chemical clock of the BZ system, with a well-defined period for given experimental conditions. By diffusing in the extracellular medium, the cAMP emitted by the 'pioneer' cells reaches the surface of the neighbouring cells. Two types of events are then switched on. First, these cells perform an oriented movement toward the regions of higher concentration of cAMP, referred to as *chemotaxis*. This motion gives rise to patterns of the density of cells (see figure 11.12) looking very much like the wave patterns in the BZ reagent (see figure 11.8). On the other hand, the process of aggregation is accelerated by the ability of sensitised cells to amplify the signal and to *relay* it in the medium. This enables the organism to control a large territory and form multicellular bodies comprising some 10^5 cells.

To sum up, the response to the starvation constraint gives rise to a new level of organisation, implying the concerted behaviour of a large number of cells and enabling the organism to respond flexibly to a hostile environment. What are the mechanisms mediating this transition? Let us first observe that the process of chemotaxis leads to an amplification of the heterogeneity formed initially, when the 'pioneer' cells begin to emit pulses of cAMP. Indeed, by enhancing the density of cells near the emission centre it contributes, thanks to the phenomenon of relay, to an increased release of cAMP from this

As it turns out, a second feedback mechanism is present in *D. discoideum* which operates at the subcellular level and is responsible for both the periodic emission of cAMP and the relay of the chemotactic signal. This mechanism is related to the synthesis of cAMP by the cell. This substance arises from the transformation of another important cellular constituent, the adenosine triphosphate, ATP, which is one of the principal carriers of energy within living cells due to its high energy phosphate bond. The ATP→cAMP transformation is not spontaneous, however; a catalyst is needed to accelerate it to a level compatible with vital requirements. In biological systems the role of catalysis is ensured by special molecules, the *enzymes*. These molecules usually contain several atoms and have a rather high molecular weight – hence the term *macromolecule* by which they are usually referred to. Some enzymes have a single active site which the reactants must hit in order to transform into products. But in many cases one deals with *cooperative enzymes*, which display several sites, some of which are catalytic and some others which are regulatory. When special effector molecules bind to these latter sites, the catalytic function is considerably affected. In some cases the molecules reacting with or produced from the catalytic site may also act as effector molecules. This will switch on a feedback loop, which will be positive (activation) if the result is the enhancement of the rate of catalysis, or negative (inhibition) otherwise.

Let us come back to the ATP→cAMP conversion. The enzyme catalysing this process is called adenylate cyclase, and it is fixed at the interior of the cell membrane. It interacts with a receptor fixed in the exterior phase of the membrane in a cooperative fashion, whose details are not completely elucidated. The cAMP produced diffuses in the extracellular medium through the cell membrane and, with a certain probability, can bind to the receptor and activate it (figure 11.13). In this way it enhances its own production, thereby giving rise to a feedback loop capable of amplifying signals and of inducing oscillatory behaviour.

Figure 11.12. Concentric and spiral waves of aggregating cell populations of *Dictyostelium discoideum* on an agar surface. The bands of amoebae moving toward the centre appear bright and the stationary amoebae appear dark.

Figure 11.13. Mechanism of oscillatory synthesis of cAMP in slime mould *Dictyostelium discoideum*. R and C denote, respectively, the receptor and the enzyme adenylate cyclase. C catalyses the (intracellular) conversion of ATP into cAMP. The extracellular cAMP arising, in part, from diffusion of cAMP produced inside the cell, activates R, and this switches on a positive feedback loop.

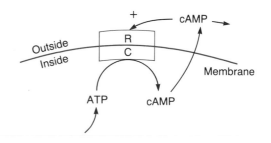

region, thereby enhancing the chemotactic movement of the other cells toward it. This constitutes what one usually calls a *feedback loop*, very similar to chemical autocatalysis or to the self-acceleration of an exothermic reaction encountered in Section 11.3.

Simple as they may be, the processes underlying the development of *D. discoideum* are prototypical to a large class of more sophisticated developmental phenomena. Indeed, it seems that the breaking of spatial symmetry and the concomitant compartmentalisation of cells explain many experimental data on insect morphogenesis. More generally, in developing tissues one frequently observes gradients of a variety of substances like ions or metabolites. It is natural to conjecture that these gradients provide the tissue with a 'coordinate system' conveying *positional information* to the individual cells, thanks to which they can recognise their position with respect to their partners. It is therefore likely that transitions mediated by chemical substances and leading to symmetry-breaking are one of the key features of life amenable to physico-chemical self-organisation phenomena. This astounding idea was enunciated for the first time in 1952 by the British mathematician Alan Turing, and has been ever since a constant source of inspiration for physicists and biologists alike.

11.5   Forces versus correlations

A piece of iron at a temperature above 1044 K presents no detectable magnetic properties. But when we cool it below this 'critical' temperature, the material becomes magnetised. This is a typical example of an important class of natural phenomena known as *phase transitions* (see Chapter 8 by Bruce and Wallace). Above the critical temperature the material is isotropic, in the sense that none of its observable properties is characterised by a preferred direction. But below this temperature a magnetisation emerges, which is a vector quantity pointing in a certain direction of space. The material becomes anisotropic and, in more technical terms, the *rotational symmetry* characterising the state of no magnetisation is broken.

There are many other examples. A liquid is a state of matter in which the molecules move in all possible directions and do not recognise each other over distances longer than a few-hundred-millionths of a centimetre. It can therefore be regarded as a homogeneous material in which all points of space are equivalent. We now cool this system slowly and uniformly. Below a characteristic temperature (e.g. 0°C for pure water at ordinary pressure), we obtain a crystal lattice, a new, solid phase of matter. Its various properties like, for example, the density, are no longer identical as we move along a certain direction in space; in other words, the *translational symmetry* characterising the liquid is broken.

In both cases the breaking of symmetry is concomitant with the appearance of new properties which prompt us to characterise the material as *ordered*. For instance, in the crystal lattice the molecules perform small vibrations around regularly arranged spatial positions which, depending on the case, may lie on the vertices of a cube, the vertices of a regular hexagonal prism and the centres of their hexagonal bases, and so on.

What is happening here? Did we simply forget to include these phenomena in our long list of transitions to self-organisation and complex behaviour?

In the absence of interactions, the molecules of a material move freely in all directions, and this motion is entirely characterised by the molecules' individual kinetic energies $\frac{1}{2}m_i v_i^2$ (m_i, v_i being the molecular masses and velocities, respectively) which remain invariant in time. We know, however, that in any physical system the molecules interact, essentially by forces of electromagnetic origin. A typical example is the van der Waals forces prevailing in an electrically neutral fluid; these are depicted in figure 11.14(*a*). These forces are *short ranged*, that is to say their intensity drops abruptly to very small values beyond an interparticle separation of a few molecular diameters.

When the system is dilute or when the temperature is very high, intermolecular forces are not very efficient: kinetic energy dominates, and the material behaves in a disordered fashion. But when we lower the temperature (or compress the system), the roles of kinetic energy and of intermolecular forces tend to be reversed. Eventually the system is dominated by the interactions and adopts a configuration in which the corresponding potential energy is as small as possible. This is very much like the familiar motion of a body falling under the effect of gravity toward the deepest point of a valley and subsequently oscillating slightly around this position of *mechanical equilibrium*. The spatial structure which results is precisely what we see in a phase transition. Being mediated entirely by the intermolecular interactions, it displays a characteristic length – like the lattice constant – which is microscopic (a few angstroms), comparable to the range of these interactions. Moreover, the behaviour of the material is time-independent if the environment itself is stationary. In this sense one can say that the order associated to a phase transition leads to 'fossil' objects.

Both properties must be contrasted with those characterising the transition phenomena surveyed in the preceding sections of this chapter. Indeed, the space scales characterising the states emerging beyond these latter transitions are *macroscopic*, comparable to or larger than the characteristic scales of biological systems. Moreover, in addition to spatial patterns, a variety of time-dependent phenomena can arise and be sustained, whose characteristic scales are again macroscopic.

The reason behind this basic difference is that a new organising factor not amenable to intermolecular interactions is now at work; the nonequilibrium constraint. Because it encompasses parts of the system containing huge numbers of molecules, new phenomena associated to states of matter enjoying long-range correlations are born. It is instructive to realise that in a dendritic structure like a snowflake (figure 11.14*b*) these two kinds of order are superimposed and can thus be easily distinguished and opposed. Indeed, the underlying crystal lattice has little to do with the emergence of dendrites, their size and their spacing, which are orders of magnitude

(a)

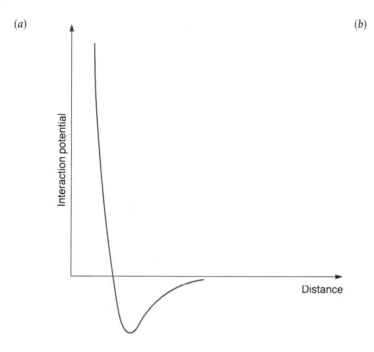

(b)

Figure 11.14. (a) A typical form of intermolecular interaction potential in a neutral gas. The distance corresponding to the minimum of the curve is of a few angstroms, and the value of the potential energy at this point is about $k_B T_c$, where k_B is Boltzmann's constant and T_c is the critical temperature of the liquid–vapour transition. (b) While the basic crystal structure of a snowflake is determined by the intermolecular interactions in (a), its overall shape carries the memory of the nonequilibrium constraints presiding over its formation.

larger than any crystallographic length. In short, the complex behaviour we are interested in here can be thought of as a phase transition of a new type, in which the lowering of temperature is replaced by the application and progressive increase of suitable nonequilibrium constraints. For this reason the term 'nonequilibrium phase transition' will sometimes be used in connection with self-organisation phenomena leading to complex behaviour.

Force-mediated symmetry breaking phenomena similar in some respects to phase transitions appear also at the level of fundamental interactions. One of the most striking examples is the broken symmetry between matter and antimatter. As is well known, to each charged particle quantum relativistic physics associates an antiparticle of the same mass but opposite charge: we have electrons and antielectrons (positrons), protons and antiprotons, etc. Experimentally antiparticles have been observed under special conditions. Still, although the basic equations of physics are symmetric in matter and antimatter, the proportion of antimatter in the observable universe is negligible with respect to matter. How did this asymmetry come about?

One possible answer comes from the grand unified theories of the strong, electromagnetic and weak interactions of elementary particles, which predict how matter behaves at very high temperatures – those prevailing some 10^{-35} s after the Big Bang (see Chapter 15 by Georgi), for example. In particular, it appears that matter can undergo one or perhaps several transitions at 'critical temperatures' as it cools from the initial exceedingly hot state, owing to the expansion of the universe. According to the grand unified theories, matter above the first transition temperature was in a very symmetric state. Quarks, the constituents of protons and neutrons, and leptons, including electrons, neutrinos and their antiparticles, all behaved identically. Below the transition point, however, the differences were made manifest through symmetry-breaking. Eventually these differentiated particles became the raw material that makes up stars, planets and living beings. In this sense, therefore, a transition mediated by the above three exceedingly short-ranged fundamental interactions has reached a macrosopic scale encompassing the present universe, which appears to be a 'relic' of an event produced in a remote past. There is no contradiction, however, between this statement and our previous effort to oppose ordinary phase transitions and nonequilibrium phase transitions on the basis of their widely different ranges. Indeed, the history of the early universe and, in particular, its expansion, is marked by the passage from a state of thermal equilibrium to a state in which the equilibrium between different constituents of matter as well as between matter and radiation is broken. In such a nonequilibrium environment large-scale symmetry-breaking transitions of the kind mentioned above can take place. In this respect, therefore, differentiated matter as we observe it today can be viewed as the outcome of a primordial nonequilibrium.

However, an adequate description of the history of the early universe requires the use of general relativity. This brings on to the stage gravity, the fourth fundamental force of nature, which is well known for its *long-range* character.

Gravity should therefore be regarded as a basic organising factor in the universe mediating the passage from equilibrium to nonequilibrium and enabling in this way microscopic events to manifest themselves at a global scale. In a way, because of gravity and expansion, transitions in the early universe share features of both equilibrium and nonequilibrium phase transitions, as we see them at the more modest scale of our everyday experience.

11.6   Quantitative formulation

In the previous sections we gave an 'operational' definition of equilibrium states, nonequilibrium constraints, stability and nonlinearity. We now adopt a more dynamical point of view, aiming at the establishment of a relation between these concepts and the laws governing the evolution in time.

Consider a system embedded in an environment with which it communicates through the exchange of certain properties, which we call *fluxes* (see figure 11.15). As a result of these exchanges the variables describing the instantaneous state, $\{X_i\}$, vary in time and attain values which are typically different from those characterising the state of the environment, $\{X_{i,e}\}$. In a physico-chemical system $\{X_i\}$ may denote temperature, hydrodynamic velocity, chemical composition, electric polarisation, and so forth. In a biological system they may describe the density of cells in a nutrient medium or the electric potential across the membrane of a neuron. One may likewise apply the picture of figure 11.15 to problems arising outside the domain of strict applicability of the physical sciences. Thus, in a human society X_i may represent the populations of workers exerting different kinds of economic activities, the price of certain goods, and so forth.

Whatever the detailed interpretation of X_i might be, the evolution can often be cast in the following general form:

$$\frac{\mathrm{d}X_i}{\mathrm{d}t} = F_i(X_1, \ldots, X_n; \lambda_1, \ldots, \lambda_m), \quad (i = 1, \ldots, n),$$

$$(11.6)$$

in which F_i denote the rate laws and $\lambda_1, \ldots, \lambda_m$ are a set of parameters present in the problem, which can be modified by the external world. We call these quantities *control parameters*.

Now, a very characteristic feature of the vast majority of systems encountered in nature is that the F's are complicated *nonlinear functions* of X's. In a fluid in motion this has to do with the fact that the transport of its properties like, for instance, energy, is carried out by the motion itself, whose velocity is one of the variables to be determined. In chemical reactions or in biology it has to do with the ability of certain kinds of molecules to perform autocatalytic and other regulatory functions. And, in an animal or human population, nonlinearity may reflect the processes of communication, competition, growth, or information exchange. In short, the equations of evolution of all these systems should admit under certain conditions several solutions, since by definition multiplicity of solutions is the most typical feature of a nonlinear equation. Our basic working assumption will be that these solutions represent the various modes of behaviour of the underlying system, some examples of which were given in Sections 11.2–11.4.

Before we proceed to the classification of these states, let us discuss briefly the particular regime in which properties X_i attain a constant value throughout the system and become equal to environmental value $X_{i,e}$. This state, which was referred to in Section 11.2 as *thermodynamic equilibrium*, is characterised by detailed balance:

> probability of a 'direct' process =
> probability of a 'reverse' process.

We can easily understand that in such a state any attempt at a diversification and self-organisation will be smeared out immediately. Equilibrium is therefore a state of full homeostasis, characterised both by uniqueness and by strong stability properties.

Figure 11.15. Schematic representation of an *open system*, communicating with the environment through the exchange of such properties as mass, energy, momentum, or information. The amount transported per unit surface and unit time is the *flux* of the corresponding property across the system.

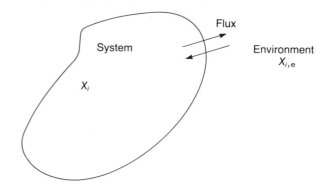

Figure 11.16 presents three alternative views of equilibrium. In figure 11.16(a) a state variable X_i initially far from its equilibrium level $X_{i,e}$ tends to it *monotonously*, performing at most a few undulations before the final decay. In the mathematical literature this property is known as *asymptotic stability*. Its physical origin is to be sought in the celebrated second law of thermodynamics, according to which equilibrium is attained in an *irreversible* fashion and corresponds to the extremum of a quantity known as state function or *thermodynamic potential*: maximum of entropy S in a system at constant energy, or minimum of free energy F (figure 11.16b) in a system at constant temperature. Experiment has shown that all physical systems containing a large number of degrees of freedom obey this law. We shall refer to them subsequently as *dissipative dynamical systems*. On the other hand, certain simple systems encountered in mechanics like the pendulum or the two body problem remain invariant under time reversal, and as a result they cannot enjoy the property of asymptotic stability. The connection between these *conservative systems* and the dissipative dynamical systems is a deep problem of statistical mechanics which is outside the scope of the present chapter.

Figure 11.16(c) presents a still different view of equilibrium, which will turn out to be the most useful one for our subsequent discussion. We embed the evolution of the system in a space spanned by the state variables which we call the *phase space*. An instantaneous state is represented in this space by a point (and vice versa), whereas a succession of such states defines a curve, the phase space trajectory. This representation of the time evolution of a system was originally adopted in classical mechanics, where the state variables involved are the positions and velocities of a set of particles moving under the influence of external forces and their own interaction. Its advantage is to allow for the simultaneous visualisation of both positions and velocities, contrary to the usual description in terms of real space trajectories in which only the positions can be plotted. In a conservative system, time reversal invariance implies that there are not privileged parts of phase space. But, in a dissipative dynamical system, as time grows the phase space trajectory will tend to an object representative of the regime reached by this system when all transients will die out. We call this regime the *attractor*. According to our previous discussion the attractor representing equilibrium will have to be unique and will describe a time-independent situation. Obviously this will give in phase space a point toward which all possible histories have to converge monotonously. In other words, the state of equilibrium is a *universal point attractor*.

We are now in the position to define more sharply the goal of a theory of self-organisation, as the search for new attractors arising when a system is driven away from its state of equilibrium. We proceed to this important point in the next section. We close our present discussion by a simple example showing in an explicit manner how nonequilibrium can indeed affect deeply the behaviour of a system.

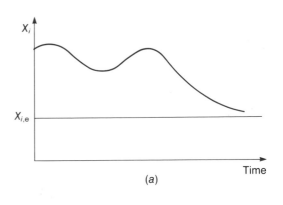

(a)

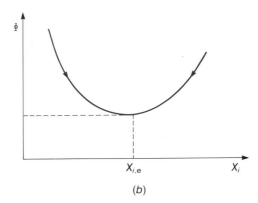

(b)

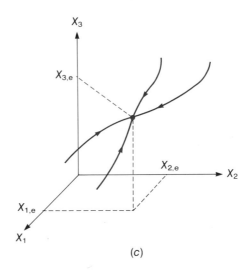

(c)

Figure 11.16. Three views of the state of equilibrium: (a) monotonic approach of state variable X_i toward its final value as $t \rightarrow \infty$; (b) the state of equilibrium $X_{i,e}$ is the extremum of the thermodynamic potential Φ; (c) equilibrium is a universal point attractor in the system's phase space.

Consider the two coupled chemical reactions

$$A + 2X \underset{k_2}{\overset{k_1}{\rightleftarrows}} 3X,$$

and

$$X \underset{k_4}{\overset{k_3}{\rightleftarrows}} B. \tag{11.7}$$

Here the concentration x of product X is taken to be the only state variable, being understood that A and B are continuously supplied from or removed to the outside to maintain fixed concentrations. At equilibrium, detailed balance implies that

$$k_1 a x^2 = k_2 x^3$$

(equilibrium of forward and backward steps
in the first reaction)

$$k_3 x = k_4 b$$

(equilibrium of second reaction).

These relations fix the equilibrium value x_e of x uniquely and impose, in addition, a condition on the concentrations a and b of constituents A and B:

$$x_e = \frac{k_4 b_e}{k_3} = \frac{k_1 a_e}{k_2}$$

$$\left(\frac{b}{a}\right)_e = \frac{k_1 k_3}{k_2 k_4}. \tag{11.8}$$

On the other hand, in a stationary state far from equilibrium it is not necessary for each individual reaction to balance in both directions. Cancelling the overall effect of the two forward reactions by that of the backward reactions is sufficient, and this yields

$$-k_2 x_s^3 + k_1 a x_s^2 - k_3 x_s + k_4 b = 0. \tag{11.9}$$

This is a cubic equation in x_s and can have up to three solutions for certain values of a and b, just like in the case of figure 11.9 and unlike the equilibrium case. We can therefore say that nonequilibrium reveals the potentialities hidden in the nonlinearities, which remain 'dormant' at or near equilibrium.

11.7   Nonequilibrium attractors; instability and bifurcation

The monotonic character of the approach to the state of equilibrium stressed in the preceding section implies that the evolution laws, equation (11.6), should obey, in the neighbourhood of equilibrium, some very particular conditions. Introducing the deviations of X_i from the equilibrium values, $X_{i,e}$,

$$x_i = X_i - X_{i,\,e}, \tag{11.10}$$

one can indeed show that the evolution of $\{x_i\}$ can be cast in the form

$$\frac{dx_i}{dt} = -\sum_j \Gamma_{ij} \left(\frac{\partial \Phi}{\partial x_j}\right), \tag{11.11}$$

where Φ is a thermodynamic potential taking its minimum at equilibrium (cf. Section 11.6) and $\{\Gamma_{ij}\}$ is a symmetric, positive-definite matrix. As shown by Onsager, this symmetry can be traced back to the property of detailed balance or, alternatively, to the invariance of the equilibrium state under time reversal. This remarkable result can be extended to the evolution around nonequilibrium steady states maintained close to equilibrium, because of the theorem of minimum entropy production due to Prigogine.

The search for a generalised thermodynamic potential in the nonlinear range has attracted a great deal of attention, but these efforts finally failed. Typically, therefore, beyond the *linear domain* of irreversible processes equations (11.11) are expected to break down. A first consequence is that the steady state (point) attractor, extrapolating the state of equilibrium as the distance from equilibrium is increased, can now be approached through damped oscillations (figure 11.17a). This behaviour heralds a still more interesting possibility, depicted in figure 11.17(b), in which the oscillation eventually becomes sustained, Topologically, this implies the emergence of a new, *one-dimensional attractor* in phase space, known as a *limit cycle*. Such attractors should therefore constitute the natural models of the rhythmic behaviour observed in nature (cf. Sections 11.2–11.4).

By allowing the intrinsic nonlinearities to be manifested beyond the regime of detailed balance, nonequilibrium can also lead to the coexistence of multiple attractors. This immediately leads to a topological problem, namely how to delimit the relative *basins of attraction* (the set of initial states in phase space that will evolve to either of the attractors). The solution of this problem, depicted in figure 11.18, involves necessarily an intermediate unstable state as well as a family of orbits remaining invariant under the flow known as *separatrices*. Clearly, the coexistence of multiple attractors constitutes the natural model of systems capable of showing adaptive behaviour and of performing regulatory tasks (cf. also figure 11.9).

The existence of one-dimensional attractors suggests the possibility of higher-dimensional attracting objects whose cross section along different phase space coordinates would give a limit cycle. Such objects can indeed be shown to exist and to have the topology of a *torus*. They model multiperiodic behaviour, which is observed under appropriate experimental conditions. We do not discuss these attractors in detail. Instead, we jump directly to what is undoubtedly the most complex and challenging attracting object known to date, namely a *chaotic attractor*. An example constructed from a model system involving three variables is shown in figure 11.19.

We observe two opposing trends. On the one side (see horizontal arrow in figure 11.19) an instability of the motion tending to remove the phase space trajectory away from the steady state solution, which turns out to be the state $X = Y = Z = 0$; and on the other side (see vertical arrow) the

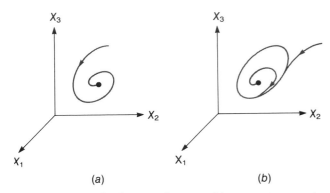

(a)   *(b)*

Figure 11.17. As the distance from equilibrium is increased a system can, *(a)*, evolve toward a steady state in a nonmonotonic fashion or, *(b)*, evolve toward a periodic attractor represented by a closed curve in phase space.

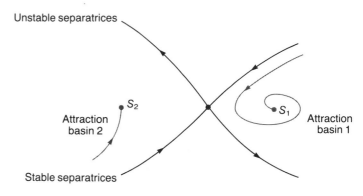

Figure 11.18. Coexistence of multiple attractors in a dissipative system far from equilibrium. The set of initial conditions attracted by each of the stable fixed points S_1 and S_2 is determined by the configurations of two pairs of invariant orbits emanating from or merging to an unstable fixed point, known as separatrices.

bending of the outgoing trajectories followed by their reinjection back to the vicinity of the steady state. Two ingenious tricks of nature allow us to reconcile these contradictory tendencies. One is the adoption of a *fractal geometry*, that is to say the existence of attracting objects of noninteger dimensionality, d_F, which are intermediate between a surface and a volume. And the other is the sensitivity of the trajectories on the attractor to small changes in the initial conditions, as a result of which two nearby initial states can diverge, momentarily, in an exponential fashion. Quantitatively this divergence is measured by the *Lyapounov exponents*, σ_L. To be specific, let d_0 be a (small) initial separation of two trajectories in phase space. Under the action of the dynamics this distance will take a value d_t at time t. If d_t/d_0 varies exponentially with time during this interval, the expression $\sigma = (1/t)\ln|d_t/d_0|$ will exist and be finite. More significant is the limit of this expression for large t and small d_0, as it provides the *mean* rate of this exponential variation. By definition, the Lyapounov exponents σ_L are the values taken by this limiting σ when the distance d_t, viewed now as a vector in phase space, is projected successively along the unit vectors of a coordinate system. There are, therefore, as many Lyapounov exponents as phase space dimensions. If the motion is unstable and presents sensitivity to initial conditions, at least one of them will be positive, indicating an exponential rate of divergence along some direction in phase space. This will be so, in particular, for chaotic attractors. Clearly then, chaotic attractors provide the archetype of natural phenomena characterised by a *limited predictability*. Some obvious examples are turbulence and climatic variability. We come back to this point in Section 11.10.

One might be tempted to deduce, from the above discussion, that a given system can only be modelled by a particular type of attractor. This is not so, however. As a matter of fact the most

Figure 11.19. Chaotic attractor constructed from a model system involving three variables and a single nonlinearity of second degree. The horizontal arrow indicates the instability of the motion of the trajectories spiralling away from the fixed point at the origin. The vertical arrow indicates the reinjection of the trajectories back to the vicinity of the origin.

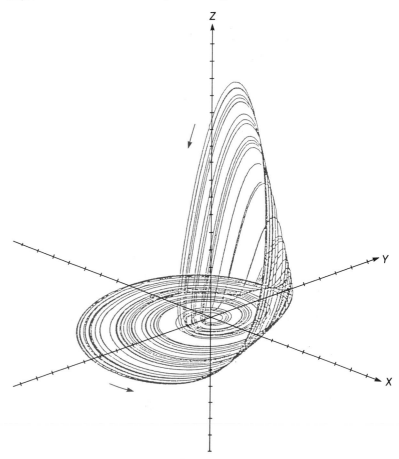

exciting aspect of nonequilibrium phenomena is that the same physical system can show a great variety of behaviours, each corresponding to a different attractor. The mechanism which is at the origin of this diversification is the *instability* of a 'reference' state and the subsequent *bifurcation* of new branches of states as the parameters $\lambda_1, \ldots, \lambda_m$ built in the system are varied.

The simplest bifurcation is depicted in figure 11.20. We represent in a graph the way a state variable of the system X (vertical component of the velocity of the flow at a certain point in the Bénard problem, amplitude of oscillation of concentration of a chemical in the BZ reaction, etc.) is affected by a single control parameter λ (the thermal gradient in the Bénard problem, the inverse of residence time in an open reactor, etc.). We obtain in this way a *bifurcation diagram*.

For small values of λ only one solution – the state of rest in the Bénard experiment, or the steady state in a chemical system – is accessible. It is the direct extrapolation of equilibrium, and shares with it the important property of asymptotic stability, since in this range the system is capable of damping internal fluctuations or external disturbances. For this reason we call this branch of states the *thermodynamic* branch. But beyond a critical value, denoted by λ_c in figure 11.20, we find that the states on this branch become unstable: the effect of fluctuations or of small external perturbations is no longer damped. The system acts like an amplifier; it moves away from the reference state and evolves to a new regime (the state of convection in the case of the Bénard experiment). The two regimes coalesce at $\lambda = \lambda_c$, but are differentiated for $\lambda \neq \lambda_c$. This is the phenomenon of *bifurcation*. We can easily understand why this phenomenon should be associated with catastrophical changes and conflicts. Indeed, at the crucial moment of transition (vicinity of $\lambda = \lambda_c$), the system has to perform a critical choice which, in the Bénard problem, is associated with the appearance of a right- or left-handed cell in a certain region of space (figure 11.20, branches b_1 or b_2). Nothing in the description of the experimental setup permits the observer to assign beforehand the state that will be chosen. Only chance will decide, through the dynamics of fluctuations. The system will scan the 'ground', will make a few attempts, perhaps unsuccessfully at the beginning, and finally a particular fluctuation will take over. By stabilising it, the system will become a historical object in the sense that its subsequent evolution will depend on this critical choice. We have thus succeeded in formulating, in abstract terms, the remarkable interplay between chance and constraint, between fluctuations and irreversibility, which was underlying most of the phenomena surveyed in Sections 11.2–11.4.

Figure 11.21 represents a mechanical analogue of the phenomenon. A ball moves in a valley (figure 11.20, branch a), which at a particular point becomes branched and leads to either of two new valleys (figure 11.20, branches b_1 and b_2) separated by a hill. Although it is always perilous to draw analogies and extrapolations, it is nevertheless useful to insist on the similarity between these ideas and the notion of

mutation and selection familiar from biological evolution. As a matter of fact, one can say that fluctuations are the physical counterpart of mutation, whereas the search for stability plays the role of selection. Even the very structure of a bifurcation diagram is reminiscent of the phylogenetic trees employed abundantly in biology!

Figure 11.20. Bifurcation diagram showing how a state variable X is affected when the control parameter varies. A unique solution (a), the thermodynamic branch, loses its stability at λ_c. At this value of the control parameter new branches of solutions (b_1, b_2), which are stable in the example chosen, are generated.

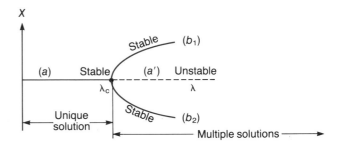

Figure 11.21. Mechanical illustration of the phenomenon of bifurcation.

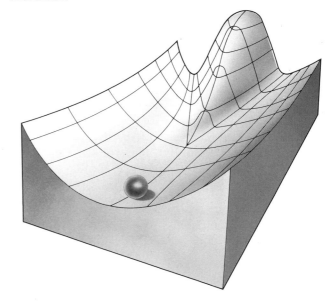

In large classes of systems the diversification of behaviour through bifurcation does not stop only at the first transition – referred to as primary bifurcation – depicted in figure 11.20. One observes, instead, complicated *bifurcation cascades* leading to secondary, tertiary etc. branches and culminating in certain cases to the bifurcation of chaotic attractors. This raises naturally the question of the universal structure of bifurcation diagrams and of the possibility of providing a full list of the attractors of a given dynamical system. We discuss this question in Section 11.8.

11.8   The normal form of the bifurcation equations and the limits of universality

Consider a system described by a set of evolution laws of the form of equation (11.6). In a typical natural phenomenon the number of variables n is expected to be very high, and this will complicate considerably the search on all possible solutions. Suppose, however, that by experiment or by intuition we know one of these solutions, X_s, because for instance it is particularly simple and symmetric. By a standard method, known as *linear stability analysis*, one can then determine the parameter values λ for which this solution, regarded as the *reference state*, switches from asymptotic stability to instability.

As discussed earlier, stability is essentially determined by the response of the system to perturbations. It is therefore natural to transform the dynamical laws, Equation (11.6), in a form in which the perturbations appear explicitly. Setting (cf. equation 11.10)

$$X_i(t) = X_{i,s} + x_i(t), \qquad (11.12)$$

substituting into equation (11.6) and taking into account that $X_{i,s}$ is also a solution of these equations, we arrive at

$$\frac{dx_i}{dt} = F_i(\{X_{i,s} + x_i\}, \lambda) - F_i(\{X_{i,s}\}, \lambda).$$

These equations are homogeneous in the sense that the right-hand side vanishes if all $x_i = 0$. To obtain a clearer form of this homogeneous system we expand $F_i(\{X_{i,s} + x_i\}, \lambda)$ around $X_{i,s}$, and write explicitly the part of the result that is linear in x_j, plus a nonlinear correction whose structure need not be specified at this stage:

$$\frac{dx_i}{dt} = \sum_j L_{ij}(\lambda)x_j + h_i(\{x_j\}, \lambda) \ (i = 1, \ldots, n). \qquad (11.13)$$

L_{ij} are the coefficients of the linear part and h_i are the nonlinear contributions. The set of L_{ij} defines an *operator* ($n \times n$ matrix in our case), depending on the reference state X_s and on the parameters λ.

Now a basic result of stability theory establishes that the asymptotic stability of the reference state ($X = X_s$ or $x = 0$) of system (11.13) is identical to those of its linearised part,

$$\frac{dx_i}{dt} = \sum_j L_{ij}(\lambda)x_j \ (i = 1, \ldots, n). \qquad (11.14)$$

Stability reduces in this way to a linear problem, which is soluble by methods of elementary calculus.

Figure 11.22 summarises the typical outcome of a stability analysis carried out according to this procedure. What is achieved is the computation of the rate of growth of the perturbations γ as a function of one (or several) control parameter(s). If $\gamma < 0$ (as it happens in figure 11.22 branch *a* if $\lambda < \lambda_c$) the reference state is asymptotically stable, and if $\gamma > 0$ ($\lambda > \lambda_c$ for branch *a* in figure 11.22) it is unstable. At $\lambda = \lambda_c$ one has a state of marginal stability, the frontier between asymptotic stability and instability.

In general a multivariable system gives rise to a whole spectrum of γ, just like a crystal has a multitude of vibration modes. (In technical terms, these γ's will be related to the eigenvalues of the operator L.) One will have therefore several γ versus λ curves in figure 11.22. Suppose first that, of all these curves, only one (figure 11.22*a*) crosses the λ axis, while all others are below it. Under well-defined mild conditions, one can then show that at $\lambda = \lambda_c$ a bifurcation of new branches of solutions takes place. Two cases can be distinguished:

(i) If at $\lambda = \lambda_c$ the perturbations are nonoscillatory, the bifurcating branches will correspond to steady-state solutions.

(ii) If at $\lambda = \lambda_c$ the perturbations are oscillatory, the bifurcating branches will correspond to time-periodic solutions in the form of limit cycles.

Figure 11.22. Rate of growth of perturbations, γ, as a function of the control parameter, λ, deduced from linear stability analysis (cf. equation (11.14)). (*a*) The reference state is asymptotically stable for $\lambda < \lambda_c$ and becomes unstable for $\lambda > \lambda_c$, where λ_c is the critical value of marginal stability. (*b*) The reference state remains asymptotically stable for all values of λ.

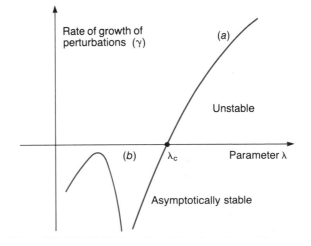

In either case a most remarkable point is that a suitable set of quantities can be defined which obey a closed set of equations having a universal structure, referred to as *normal form*, if the parameters remain close to their critical values λ_c. In case (i) there is only *one* such quantity, which measures the amplitude of the bifurcating branches. In case (ii), on the other hand, there are *two* such quantities characterising both the amplitude and the phase of the oscillation. Effectively therefore the original multivariable dynamics (equations 11.13) is decoupled into a single equation or a pair of equations giving information on bifurcation, and $n-1$ or $n-2$ equations which turn out to be 'irrelevant' as far as bifurcation is concerned. We call *order parameters*, z, the quantities satisfying the bifurcation equations. The structure of the normal form equations turns out to be *for case* (i):

$$\frac{dz}{dt} = (\lambda - \lambda_c) - uz^2$$

or

$$\frac{dz}{dt} = (\lambda - \lambda_c)z - uz^3$$

or

$$\frac{dz}{dt} = (\lambda - \lambda_c)z - uz^2, \tag{11.15}$$

where u is a certain coefficient; and *for case* (ii):

$$\frac{dz}{dt} = [(\lambda - \lambda_c) + i\omega_0]z - uz|z|^2, \tag{11.16}$$

where z and u are now complex valued, $|z|^2 = zz^*$, z^* being the complex conjugate of z, and ω_0 is the frequency of oscillation of the perturbations at the bifurcation point. This powerful result was in fact known to Poincaré, the great mathematical genius of the turn of the century, but it was the American mathematician G. Birkhoff who developed it further for the study of bifurcations in conservative systems. For dissipative systems the idea of reduction to a few order parameters is frequently associated to the name of the Soviet physicist L. Landau, in connection with his theory of phase transitions. In the mathematical literature it is frequently referred to as the Lyapounov–Schmidt procedure and, more recently, as the *centre manifold theory*.

More intricate situations can also be envisaged in which several branches cross the λ axis in figure 11.22. This leads to the interaction between bifurcating solutions generating secondary, tertiary or even higher bifurcation phenomena. The above results carry through, in the sense that one can guarantee that the part of the dynamics that gives information on the bifurcating branches takes place in a phase space of reduced dimensionality. The explicit construction of the normal forms becomes, however, much more involved. Moreover, in contrast to the simple bifurcations discussed above, the normal forms obtained by keeping the lowest order terms of perturbation theory cannot be guaranteed to be universal in the vicinity of criticality. In particular, the effect of

higher order terms or of additional parameters cannot be fully assessed. In more technical terms, one can say that the complete *stable unfolding* of the problem remains an open question. Another characteristic feature of interacting bifurcations is that, in addition to fixed points, limit cycles or tori, new attractors are unexpectedly generated and give rise to *global* bifurcation phenomena in the normal form. If the latter contains at least three coupled order parameters these global bifurcations may lead to chaotic dynamics, a flavour of which was already given in Section 11.7. Further aspects of chaos are discussed in Chapter 12 by Ford.

It should be noticed that the reduction of the dynamics to a normal form remains possible in certain classes of bifurcations leading to space symmetry-breaking in spatially distributed systems. In such cases the order parameters represent combinations of amplitudes of the dominant modes appearing in an expansion of the state variables in Fourier series or, more generally, in a series of linearly independent functions compatible with the symmetry properties of the system and the boundary conditions.

11.9   The microscopic basis of self-organisation

One of the main conclusions emerging from the preceding sections is that self-organisation rests on the ability of nonlinear far-from-equilibrium dynamical systems to create and sustain states of matter displaying regulatory and other remarkable properties, which would be exceedingly improbable under equilibrium conditions. We have already emphasised that such states imply the existence of correlations of macroscopic range, which have little to do with the intermolecular interaction forces. In this section we give a qualitative explanation of how this rather extraordinary coherence can arise. To this end we have to turn to a finer description of nonequilibrium systems compared to the macroscopic description adopted so far in this chapter.

The observation of the state of a physical system involves, as a rule, an averaging of the instantaneous values of the pertinent variables, either over time or over a volume of supermolecular dimensions. For instance, if we put some 0.33×10^{23} molecules of water in a container of one cubic centimetre at ambient temperature and pressure, we would conclude that we have a liquid whose number density is 0.33×10^{23} molecules cm^{-3} and whose mass density is $0.33 \times 10^{23} \times$ (mass of H_2O molecule) $= 1\,g\,cm^{-3}$. These are the sort of variables that we have been dealing with in the preceding sections. On the other hand, the number density in small volume elements of the order of, say, ten cubic angstroms, will be subjected continuously to deviations from this macroscopic value: molecules will cross the boundaries of these volumes and, because of the random character of their motion, the number of particles contained at each small volume will be essentially unpredictable. We call the deviations generated by this mechanism *fluctuations*. Thanks to the fluctuations, physico-chemical systems become capable of exploring continuously the state space and of performing excursions around the state predicted by the solution of phenomenological, deterministic equations.

The natural approach to the problem of fluctuations is thus in terms of probability theory. For instance, in a system undergoing chemical reactions the quantities of interest are the probabilities $P(X_\alpha, \Delta V, t)$ of having, at time t, X_α particles of species α in a volume element ΔV. More complete information will be contained in the multivariate probabilities $P(X_{\alpha j}, \Delta V_j; X_{\beta k}, \Delta V_k; t)$ of simultaneous occupations of various volume elements j, k, by, say, the chemical species $\alpha, \beta, \ldots$.

To compute the form of the probability distribution one sets up a balance equation counting the processes leading the system to a certain state Q, and the processes removing it from this state. Obviously,

$$\text{prob}(Q, t) =$$
(number of transitions to state Q per unit time)
$- $ (number of transitions from Q per unit time)
$$\equiv R_+(Q) - R_-(Q), \qquad (11.17)$$

and the problem now reduces to the determination of the transition rates R_+ and R_-. Whatever the explicit form of these quantities may turn out to be, it will have to be compatible with certain constraints. For instance, it is known that at thermodynamic equilibrium physico-chemical systems satisfy the conditions of *detailed balance*, to which we alluded repeatedly before. Thus, if one decomposes R_+ and R_- into the rates $r_{k,\pm}$ of the elementary processes k taking place in the system,

$$R_\pm = \sum_k r_{k,\pm},$$

the following condition must be satisfied:

$$(r_{k,+})_e = (r_{k,-})_e. \qquad (11.18)$$

This relation must in turn be compatible with the form of the probability distribution in the state of equilibrium, which is known from statistical mechanics. A limiting case of such a distribution is the Poissonian. More generally, as shown first by Einstein, at equilibrium the probability of a fluctuation is determined entirely by thermodynamic quantities. In an isolated system, the inversion of the famous Boltzmann formula $S = k_B \ln$ (number of molecular arrangements compatible with a given value of energy) yields:

$$P_e \sim \exp\frac{1}{k_B}\Delta S, \qquad (11.19)$$

where ΔS is the change of entropy due to the fluctuation,

$$\Delta S = S(Q) - S(Q_e).$$

Another constraint which equation (11.17) should satisfy is to reduce, in a certain limiting sense, to the evolution laws one deals with in the deterministic description, like the equations of fluid mechanics and of chemical kinetics. One expects that the macroscopic observation of a physico-chemical system will yield values representative of its most probable state. We must require therefore that the peaks of $P(Q, t)$ be solutions of the deterministic equations. If the distribution is unimodal (figure 11.23a), that is to say it has a single peak, this guarantees that the equation for the mean value is close to the deterministic equation, the correction being essentially proportional to an inverse power of the size of the system.

Despite their interest, however, the above constraints do not suffice to fix the form of equation (11.17) in a unique fashion. We shall therefore appeal now to some intuitive arguments. We have already stressed that fluctuations arise from the random motion of the molecules, which is due to the intermolecular interactions and to the large number of the particles involved in the dynamics of most physico-chemical systems of interest. They are therefore essentially localised events, whose characteristic space and time scales are extremely short. Under these circumstances one expects that the transition rates appearing in equation (11.17) will depend only on the state Q and on the states that can be connected to Q through a single fluctuation. In other words, we suppose that the memory of successive transitions arising by successive fluctuations is lost beyond the first transition. This condition defines an extremely important class of phenomena known as *Markovian processes* whose study is one of the main preoccupations of probability theory.

A very simple illustration of a Markovian process is shown in figure 11.24. The system considered has five states, I to V. I and V are 'absorbing boundaries' in the sense that the system stays therein once it reaches them. On the other hand, once in states II, III and IV, the system jumps immediately to the left or to the right with probabilities q and $1-q$, respectively. In the limit where the number of states is very large, this 'random walk' becomes a realistic representation of the motion of a particle in a host fluid and leads to the familiar Fick law of diffusion.

In systems presenting complex behaviour, one of course deals with more intricate situations because of the nonlinearities of the kinetics. Nevertheless, an explicit form of equation (11.17), known as the *master equation*, can be written out. It expresses the rate $R_+(Q)$ as the product of the transition probability per unit time of going from state Q' to Q times the probability of being in the state Q' at time t in the first place, summed over all states Q', which can lead to Q in a single step by virtue of the elementary dynamical processes going on in the system. Similarly, $R_-(Q)$ is the product of the probability of being in state Q at time t in the first place times the sum of the transition probabilities per unit time from Q to all states Q' accessible from Q. We write this balance in the following form:

$$\frac{dP(Q,t)}{dt} = \sum_{Q' \neq Q} [W(Q'|Q)P(Q',t) - W(Q|Q')P(Q,t)], \quad (11.20a)$$

where the transition probability per unit time $W(Q'|Q)$ is non-negative for any $Q' \neq Q$. Moreover, as $P(Q,t)$ must remain

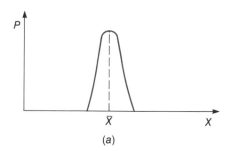 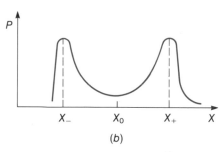 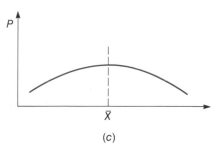

Figure 11.23. Stochastic analogue of bifurcation. As the system crosses the critical value λ_c of a control parameter λ, the probability function switches from a unimodal form peaked sharply on a unique attractor ($\lambda < \lambda_c$, (*a*)) to a multihumped distribution whose maxima coincide with the new attractors emerging beyond bifurcation ($\lambda > \lambda_c$, (*b*)). When λ is in the immediate vicinity of λ_c the probability distribution becomes considerably flattened, a fact that shows up as an enhancement of the fluctuations, (*c*).

normalised to unity for all times, W must satisfy the condition

$$\sum_Q W(Q'|Q) = 0, \qquad (11.20b)$$

expressing the intuitively clear fact that, starting from state Q', all the system can do is either to stay in Q' or perform a transition to the accessible states $Q \neq Q'$.

Equation (11.20*a*) is linear with respect to the unknown function $P(Q,t)$, but this does not mean that its solution is straightforward. In fact, the complexity of the dynamical laws of evolution (for instance, the feedbacks present in a chemical or biological system) is reflected by the presence of non-linearities in the transition rates W. Near equilibrium these feedbacks are ineffective because of detailed balance, but far from equilibrium they are fully manifested. As a result the complete analysis of equation (11.20) constitutes an open problem. Still, by carrying out perturbative calculations, by looking at simple exactly soluble limits, and by carrying out computer simulations one arrives at a general picture of the laws governing the fluctuations in nonequilibrium systems. The main results are presented below. We will be appealing continuously to both the analogies and the differences between self-organisation and the phenomenon of phase transition at equilibrium.

As emphasised in Section 11.5 an equilibrium phase transition like freezing or spontaneous magnetisation is the result of competition between the intermolecular interaction forces, which tend to order the system, and the random thermal motion of the molecules, which has the opposite effect. This is why equilibrium transitions are induced, typically, by increasing the pressure (which favours the effect of molecular interactions), or by decreasing the temperature (which diminishes the effect of thermal noise). As a matter of fact, even far

from a phase transition, intermolecular forces confer to a relatively dense system a certain degree of order whose range, however, is short, as it is entirely conditioned by the inter-molecular forces. A useful visualisation of this situation is provided by figure 11.25.

Consider a 'reference system' at thermodynamic equilibrium in which the effect of intermolecular interactions can be neglected, like for instance, a perfect gas or an ideal solution (figure 11.25*b*). In such a system the fluctuations in a given volume element are Poissonian, a property which reflects the complete disorder that prevails as a result of the thermal motion of the constituting particles. Suppose now that attractive interactions are switched on. The particles will tend to form clusters (figure 11.25*c*), and this will result in larger deviations of their spatial distribution from the average value. If, on the other hand, the interactions are repulsive, the particles will be distributed more efficiently within the volume (figure 11.25*a*), and this will diminish the deviation from the average. In both cases the spatial 'order' associated with the non-Poissonian behaviour remains short ranged, unless the system is driven near a critical point of phase transition: the spatial correlation then acquires a long tail, reflecting the ability to undergo a change encompassing the entire system.

Consider again an ideal system, in which the effect of intermolecular interactions can be neglected, capable of functioning far from as well as close to equilibrium. For such a system one expects that the effects of clustering and critical behaviour described above will not take place under any circumstances. In equilibrium this can be checked explicitly: the probability distribution is Poissonian, and as a result spatial correlations do not exist. But what happens if the system is driven far from equilibrium by means of appropriately applied constraints? To be specific, consider the reaction $A \rightleftharpoons 2X$ and

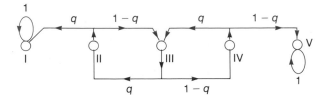

Figure 11.24. Markovian process describing transitions between states I to V. States I and V are absorbing boundaries, whereas starting from states II, III, IV the system can jump to the left or to the right with probabilities q and $1 - q$.

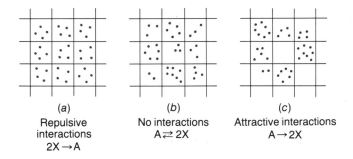

(a)	(b)	(c)
Repulsive interactions $2X \rightarrow A$	No interactions $A \rightleftarrows 2X$	Attractive interactions $A \rightarrow 2X$

Figure 11.25. Origin of spatial correlations of macroscopic range in a nonequilibrium ideal system. In equilibrium, (*b*), detailed balance precludes accumulation of X in preferred regions of space. X is thus distributed randomly (Poisson statistics), just as one would expect in a system of noninteracting particles. Far-from-equilibrium detailed balance is not valid. Pairs of X can thus be destroyed (*a*) or created (*c*) in a correlated fashion. These correlations will be maintained on a distance of the order of the diffusion length (equations (11.22)–(11.23)). The system, though ideal, will thus behave like a system of particles interacting through repulsive (*a*) or attractive (*c*) long-range forces.

suppose first that A is in large excess, so that the step $2X \rightarrow A$ can be neglected. To avoid accumulation of X to infinite concentrations, we also introduce a reaction removing it from the medium. This leads us therefore to the scheme

$$A \xrightarrow{k_1} 2X, \qquad X \xrightarrow{k_2} B. \qquad (11.21)$$

Clearly, the first step amounts to creating a cluster of particles of the species X in the system. In other words, the nonlinear chemical kinetics gives rise to an effect reminiscent of attractive interactions. In equilibrium, the property of detailed balance ensures that any reaction, such as $A \rightarrow 2X$, will be on average as frequenc as its reverse, $2X \rightarrow A$. The above-described clustering will therefore be counteracted by the fact that $2X \rightarrow A$ depletes the system from pairs of particles, thus playing a role analogous to repulsive interactions. In this way the familiar Poissonian behaviour will be recovered. Similarly, if all reaction steps were linear, particles would be created or destroyed at random places in space, and as a result Poissonian behaviour

would still prevail. The presence of both *nonlinear kinetics* and *nonequilibrium constraints* will disturb this balance, and as a result a systematic deviation from the Poisson law will be sustained. How can one relate these deviations to the existence of spatial correlations? In the step $A \rightarrow 2X$ the two X particles are obviously correlated, as they are created together. Thanks to diffusion, this correlation will not remain localised, but will extend in space. Eventually, however, at least one of the particles becomes inactivated through the second step of scheme (11.21), and as a result the correlation will die out. This will happen, on average, at a rate equal to k_2. In the light of this discussion, the correlation length, l_{corr} can be estimated as follows: the rate of inactivation of X through the second step in scheme (11.21) is approximately equal to the diffusion rate over a distance equal to the correlation length, i.e.

$$k_2 x \simeq \frac{D}{l^2_{corr}} x$$

or finally

$$l_{corr} \simeq \left(\frac{D}{k_2}\right)^{\frac{1}{2}}, \qquad (11.22)$$

in which D is Fick's diffusion coefficient of species X.

Note that if we denote by $F(X)$ the rate law corresponding to scheme (11.21), $F(X) = k_1 A - k_2 X$, k_2 is simply equal to $|F'(X_s)|$, which according to Section 11.8, gives the rate of decay, γ, of the perturbations around the steady state $X_s = k_1 A / k_2$. This conclusion is in fact quite general. In any system undergoing a transport process whose rate is described by a coefficient D and a local 'chemical-like' process whose slowest mode is characterised by the decay rate γ, the correlation length is given by

$$l_{corr} \simeq \left(\frac{D}{\gamma}\right)^{\frac{1}{2}}. \qquad (11.23)$$

As we saw in Section 11.8, in the vicinity of a bifurcation point γ tends to zero, and as a result the correlation length diverges. This in turn indicates that coherence tends to encompass the entire volume of the system, which thus becomes able to undergo collectively a transition to a new state (figure 11.23*b,c*). Note, however, that because of the 'critical slowing down' of the evolution near the bifurcation point it becomes impossible to argue in terms of individual modes and their decay rates. A more complete theory taking into account the nonlinear coupling of modes is needed. Such an approach has been worked out using the full master equation. The main qualitative conclusions are that near the bifurcation point the law of divergence of the correlation length is modified with respect to equation (11.23) because of the mode coupling, and that the transition may even be suppressed altogether.

To understand this latter, most surprising, point, it is necessary to introduce the concept of *critical dimensionality*. Remember that we are dealing with systems in which a local process (like, for instance, an autocatalytic chemical reaction) coexists with a transport process coupling among themselves neighbouring volume elements. Because of fluctuations, the 'chemical-like' process will tend to produce deviations from spatial homogeneity. In the extreme case of no coupling between volume elements, this would lead to a random superposition of localised states: even if within each element the parameters are close to their bifurcation values, the system will not 'see' the bifurcation because it will be dominated by noise. On the other hand, in the opposite case in which the space cells are coupled but the chemical-like process is absent, spatial homogeneity will be secured but order will again be absent since there will be simply no 'message' to be relayed and sustained. We conclude therefore that both types of processes are necessary and should be of comparable importance or, in more concrete terms, that they should both have comparable characteristic times. Now, the characteristic time for smearing an inhomogeneity through transport over the length L of a spatial cell is simply

$$\frac{1}{\tau_d} \simeq \frac{D}{L^2},$$

or, using the relation $L^d \simeq \Delta V$, where d is the number of space dimensions,

$$\frac{1}{\tau_d} \simeq D\Delta V^{-2/d}. \tag{11.24}$$

If spatial coherence is established over a macroscopic range, ΔV will be large, and hence diffusion will be slow. Moreover, this slowing down will be more important as the dimensionality becomes smaller. On the other hand, as we recalled previously, close to a bifurcation point the chemical-like process will experience a critical slowing down, as a result of which its characteristic time τ_{ch} will also become large. In contrast to τ_d the slowing down will be independent of the dimensionality, because of the local character of the process. In fact, for a given cell ΔV, it will be given by the inverse of the decay rate, γ, of the slowest mode and will thus only depend on the parameters and the degree of the dominant nonlinearities. By requiring that both τ_d and τ_{ch} are of the same order, one thus arrives at a relation linking the value of dimensionality to the intrinsic properties of the system: this is precisely the critical dimensionality d_c. For $d>d_c$, diffusion will be efficient, and as a result long-range spatial correlations will be sustained. But in a low-dimensional system (for instance, $d=1$ or 2), diffusion will not be able to correlate sufficiently well the various spatial cells, and bifurcation will be 'missed'.

The specific value of d_c depends rather weakly on the detailed structure of the system; what seems to matter are some quite general features like the degree of the dominant nonlinearity. One can in this way define 'universality classes' constituted of quite different systems, which nevertheless, show similar qualitative behaviour. For instance, in the autocatalytic model of equation (11.7) the critical dimensionality for certain bifurcations turns out to be $d_c=4$. Note that far from the bifurcation the dimensionality ceases to play an important role.

For instance, the estimation of equation (11.23) for the correlation length remains valid whatever the value of d.

To go beyond these qualitative results we have to construct the explicit form of the solutions of the master equation, equation (11.20). We do not discuss the technical aspects of this problem here. Let us outline some of the highlights:

(i) As soon as a system is not strictly at equilibrium, correlations of a macroscopic range suddenly arise and can be sustained indefinitely. In systems in which a chemical-like mode is coupled to a transport-like mode, the range is intrinsic and is given essentially by equation (11.23). However, the amplitude is proportional to the strength of the nonequilibrium constraint. In a sense, this amplitude can be considered as an 'order parameter' characterising the 'transition' from an uncorrelated state of matter (equilibrium) to a correlated one (nonequilibrium). On the other hand, in systems in which there is no coupling between different modes, the correlation length may be extrinsic. For instance, in a heat conducting medium at rest subject to a temperature difference, one finds that the correlation length is essentially given by the size of the system. The amplitude, however, remains related to the strength of the non-equilibrium constraint (here the energy flux through the system) and is playing again the role of 'order parameter'.

(ii) When a bifurcation point is approached, the correlation length diverges, as already mentioned above, provided that the dimensionality is close to d_c. Correlations extend therefore throughout the system, even though their length before bifurcation was intrinsically determined. This gives rise to a qualitative change of the properties of the underlying probability distribution, which is somewhat similar to the change between figure 11.23a and b, the difference being that one now deals with multivariate distributions depending on the values of the stochastic variables X at each point of space $\mathbf{r}$. Specifically, in an infinitely extended system like the autocatalytic model of equation (11.7), undergoing a bifurcation to multiple steady states without spatial symmetry-breaking, it turns out that the stationary probability distribution near the singularity has the form

$$P_s(\{x_r\}) \sim \exp\left\{-\frac{2\Delta V}{Q_s}\sum_r\left[-\frac{\lambda-\lambda_c}{2}(x_r-x_s)^2 + \frac{1}{4}(x_r-x_s)^4 + \frac{\mathscr{D}}{8d}\sum_l(x_{r+l}-x_r)^2\right]\right\}. \tag{11.25}$$

Here $\lambda-\lambda_c$ is the distance from bifurcation, $\mathscr{D}$ is the jump rate between neighbouring volume elements ΔV (related to Fick's diffusion coefficient D), Q_s is a certain positive combination of the system's parameters, x_r is the local value of the state variable X_r reduced by the volume element $\Delta V(x_r=X_r/\Delta V)$. The index l labels the neighbours of cells $\mathbf{r}$ to which the exchange of X is limited through the transport processes present in the system.

The non-Gaussian character of this distribution reflects the coherence associated with bifurcation: the system can no longer be partitioned into a collection of weakly coupled subsystems. Similar results are obtained for bifurcations leading to limit cycles and to spatial structures associated to symmetry-breaking.

It should be mentioned that the exponential in equation (11.25), known as the Landau–Ginzburg potential, is very familiar from the theory of phase transitions at equilibrium. Using this analogy one can compute many relevant properties of the system, such as the range of correlations, the variance of the fluctuations, etc., using the technique of renormalisation group developed by Wilson.

In short, we arrive at a simple, appealing picture of how order can emerge in a system. In somewhat anthropomorphic terms, order appears to be a compromise between two antagonists: the nonlinear chemical-like process, which through fluctuations sends continuously but incoherently 'innovating signals' to the system, and the transport-like process which captures, relays and stabilises them. Disturbing the delicate balance between these two competing 'actors' leads to such qualitative changes as an erratic state in which each element of the system acts on its own, or, on the contrary, a 'homeostatic' fossil-like state in which fluctuations are crushed and a full uniformity is imposed. Complexity and self-organisation appear therefore to be limited on both sides by two different kinds of states of disorder.

11.10   The self-organisation paradigm and the modelling of complex systems

We have seen that nonlinear physics of far-from-equilibrium systems is the physics of unstable motions, of bifurcations, of probabilistic behaviour, of multiple choices, and of self-organisation. As pointed out repeatedly throughout this chapter, nonequilibrium constraints and nonlinear dynamics are ubiquitous in real-world phenomena. It is therefore legitimate to expect that these new concepts should provide the natural framework within which certain key features of our natural environment will have to be investigated. In this section we illustrate this possibility by a few selected examples.

Complexity and information

A significant aspect of nonequilibrium physics and self-organisation is the emergence of *new levels of description* brought out by the underlying dynamics. For instance, the state of a nonequilibrium system is described most naturally in terms of the spatial correlations between macroscopically separated elements rather than in terms of the intermolecular forces. Similarly, as we saw in Section 11.8, in the vicinity of a bifurcation point a considerable reduction of description can be achieved owing to the emergence of collective variables – the

order parameters – obeying the normal form equations (11.15) and (11.16). In other words, the phenomenon of bifurcation is best understood by introducing the appropriate order parameter, rather than by arguing in terms of the entire set of variables present in the problem. In this subsection we show that in certain classes of dynamical systems it becomes natural to introduce a still higher level of abstraction, and we speak of symbols, codes, complexity and information.

Suppose that we want to communicate to an observer a set of instructions that would enable him to reproduce the main features of a physical phenomenon (like e.g. the Bénard convection cells, see section 11.2) without performing the experiment again. The communication with such an observer will involve, perforce, a sequence of symbols belonging to a finite 'alphabet', like for instance a sequence of binary digits 1 and 0. We imagine that the cost of communication with the observer is very high; this forces then upon us the additional constraint to make the instructions as short as possible, without compromising of course the feasibility of the original goal.

Consider first a regular phenomenon, say a periodic motion. A characteristic feature of this type of behaviour is that two variables, say X and Y, keep a fixed phase lag between them. If we therefore monitor the times of intersection (with prescribed slope), of their trajectories with preassigned levels L_X, L_Y (figure 11.26) we will find the repetitive sequence

$$XY\ XY \ldots XY \ldots, \qquad (11.26a)$$

which may subsequently be regarded as the symbolic description of all periodic motions. This description makes clear two further properties: the full predictability of the behaviour, as well as the independence of the content on the direction of reading.

Figure 11.26. Symbolic description of a dynamical system constructed by monitoring the successive crossings of prescribed thresholds L_X, L_Y, . . . by the values of the state variables X, Y, ...

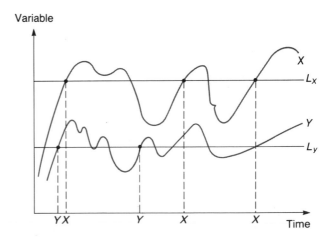

Let us now consider another extreme case, namely random (white) noise. This time the level crossings of X and Y would be distributed randomly, and one would be led to a sequence like

$$XXYXXX \ldots YYYYYX \ldots, \qquad (11.26b)$$

which could be generated equally well by tossing a coin Although this sequence remains independent of the direction of reading, the full predictability of the preceding case has now been replaced by full unpredictability, in the following sense: if we monitor another pair of white noise variables we will find a sequence which, typically, will have nothing to do with sequence (11.26b) or, better, a sequence which could in no way be inferred from this sequence. Two elegant (and related) mathematical concepts that allow us to express this feature in a compact way are *algorithmic complexity* and *information*. Algorithmic complexity measures the length of the shortest description of a given (finite) sequence. In this sense therefore sequence (11.26b) has the maximum possible complexity, essentially equal to its length, whereas sequence (11.26a) has the minimum possible complexity since a single instruction (the initial state XY) suffices to reproduce it fully. Similarly information is considered to be maximum in sequence (11.26b), since the realisation of one particular sequence out of the enormous number of random sequences amounts to localising the system very sharply in the state space. It is zero in sequence (11.26a), since the outcome could be predicted on the basis of the initial state (XY).

It should be clear from the above discussion that the complexity of natural objects should lie somewhere between the cases in sequences (11.26a) and (11.26b). Take, for instance, the genetic code. As Schrödinger put it so lucidly, the DNA is an 'aperiodic crystal': it can therefore not be of the form in sequence (11.26a). If, on the other hand, all random sequences of the four naturally occurring nucleotides (whose number is the order of 4^N for a nucleic acid of N nucleotides long) could be equally good candidates for the genetic material, life would amount to selecting one unique event out of a tremendously large number of possibilities. The *a priori* probability of such a selection would be completely negligible. What is needed, therefore, is a process capable of producing with high probability a complex, information-rich aperiodic sequence of states. Moreover, we want this edifice to be stable (in the sense of reproducibility) and asymmetric (in the sense of a well-defined direction of reading, as observed in present-day DNA). A similar reasoning clearly holds for brain activity, the structure of a language and, most probably, for music and other forms of art.

Now, the self-organised states of matter allowed by non-equilibrium physics provide us with models of precisely this sort of complexity. Most important among these states ranks, for our present purposes, chaotic dynamics. Indeed, the instability of motion associated with chaos allows the system to explore continuously its state space, thereby creating information and complexity in the form of aperiodic sequences of symbols. On the other hand, being the result of a physical mechanism, these sequences are produced within probability one: the selection of a particular sequence out of a very large number of *a priori* equiprobable ones simply does not arise. In a way, the dynamical system generating chaos acts as an efficient selector rejecting the vast majority of random sequences and keeping only those compatible with the underlying rate laws. Equally importantly, perhaps, the irreversibility incorporated in these laws gives rise to a preferred direction of reading and allows for the existence of attractors enjoying asymptotic stability and thus reproducibility.

The insertion of the symbolic concepts of complexity and information into physics achieved by chaotic dynamics establishes a highly interesting link between physical sciences on the one side, and cognitive sciences on the other. This remarkable synthesis is likely to give rise to important advances in such areas as biological evolution or the development of computing devices.

Probabilistic behaviour and adaptive strategies in animal and human societies

The phenomenon of bifurcation, in which a transition between different modes of behaviour is achieved by a cooperation between the deterministic laws of evolution and the fluctuations arising from the system's variability, finds a most natural application in systems involving interacting populations. We begin our discussion with some aspects of the behaviour of social insects.

The usual view of an insect society holds that nest construction, food searching and other collective activities are the prototypes of a deterministic world in which individual insects are small, reliable automatons obeying a strictly established genetic program. Today, this picture of absolute rigidity is fading, and a new paradigm is gradually emerging in which random elements from the environment and a plasticity of the individual behaviour begin to play an important role.

What is most striking in an insect society is the existence of two scales: one at the level of the individual, characterised by a highly probabilistic behaviour, and another at the level of the society as a whole, where, despite the inefficiency and unpredictability of the individuals, coherent patterns characteristic of the species develop at the scale of an entire colony. Let us see how these two aspects are tied up to ensure the overall organisation of the society. We choose the example of food searching by ants, and compare the course of this activity in two different situations: (i) a predictable food source, like a colony of tinier insect species (say aphids) whose lifetime is of the order of several months, exists in the vicinity of the nest; and (ii) an unpredictable source, like a dead bird, becomes suddenly available.

In the first case one observes the formation of stable routes from the nest to the nearby colony, and each route has its own specialised users. Moreover, few ants are found outside these regions. Clearly, under the existing circumstances, it is beneficial for the society to develop permanent stable structures with a low noise level.

In contrast, a permanent structure in an unpredictable environment may well compromise the plasticity of the colony and bring it to a suboptimal regime. A reaction toward such an environment is thus to maintain a high rate of exploration and the ability to develop rapidly temporary structures suitable for exploring any favourable occasion that might arise. In other words, it would seem that randomness presents an adaptive value in the organisation of the society. This appears to be supported by the experimental data. For instance, let two food sources be proposed to an ant colony of species in which recruitment – the assembly of great numbers of individuals around the food – is not very accurate. Some ants quickly discover the first source and subsequently recruit other ants. However, as recruitment is not very rigorous, a large number of ants lose the recruitment trail and explore the foraging area. They quickly discover the second source, even when recruitment toward the first source was already well established. Typically, when the two sources are of equal quality, one is exploited maximally until it is exhausted. Thereafter, the second source is fully colonised, and its exploitation is intensified, so that no interruption in food-collection rate is recorded. When the second food source is more concentrated in glucose than the first, the ants shift their collective efforts toward the most rewarding source, without however completely abandoning the first one. This behaviour is strongly reminiscent of the bifurcation of two simultaneously stable states. Moreover, we see clearly how randomness allows the system to switch between these two modes of behaviour.

We now ask the following question. Ants losing the trail because of errors are able to discover new food sources, but they do not exploit the already known ones. What is the best balance between fluctuations, allowing discoveries and innovations, and accurate determinism, allowing immediate exploitation? The following simple model provides some clues.

We first describe how global recruitment works in the presence of a single food source. Let X be the number of workers at the food source. The mean flux of ants arriving at the same source, j_+, may reasonably be set to be proportional to the number of encounters between the X individuals and the $N-X$ remaining ones, where N is the total number of ants able to participate in the recruitment. Thus:

$$j_+ = aX(N-X), \qquad (11.27a)$$

where a is the recruitment rate per individual. The flow of departure from the food source, j_-, is, on the other hand, merely

$$j_- = -bX, \qquad (11.27b)$$

where b is the inverse of the mean time spent to stay near the food and come back to the nest. When the food source is exhausted, this last term alone governs the evolution of X. But

when the source still contains food the overall rate of change of X is given by:

$$\frac{dX}{dt} = aX(N-X) - bX \text{ (food not exhausted).}$$
$$(11.28)$$

This relation is known as the *logistic equation*, and it has a widespread use in a large variety of population problems. The threshold value N is referred to as the carrying capacity. If $X < N$, j_+ is positive, and this expresses the positive feedback of X population into itself, albeit with the slowing down saturation factor expressed by $N-X$. But, if $X > N$, the above feedback disappears as the rate j_+ becomes negative.

Suppose now that the colony is confronted with K identical, regularly distributed food sources which are equidistant from the nest. Recruitment is initiated toward the median source, and it is assumed that recruited ants are distributed around the source according to a normal distribution. Ants hitting a source are supposed to initiate recruitment toward it.

A straightforward generalisation of equations (11.27*a*, *b*) and (11.28) yields:

$$\frac{dX_i}{dt} = \sum_{j=1}^{K} a_{ji} X_j (N - X_1 - \ldots - X_K) - bX_i \qquad (11.29a)$$

(food source not exhausted)

$$\frac{dX_i}{dt} = -bX_i \qquad (11.29b)$$

(food source exhausted)

where $X_1, \ldots, X_K$ now represent the number of recruited ants toward the source $1, \ldots, K$. The coefficient a_{ji}, $j \neq i$, is the recruitment rate for the ants recruited to j to reach by error source i. It can be estimated as:

$$a_{ji} = \frac{1}{S} \int d\mathbf{r} \exp\left[-\frac{(\mathbf{r}_j - \mathbf{r})^2}{2\sigma^2} \right],$$
$$|\mathbf{r}_i - \mathbf{r}| \leq d \qquad (11.30)$$

in which S is a suitable normalisation factor, $2d$ is the dimension of the food source, and σ^2 is the variance of the Gaussian probability distribution. The smaller σ is the more accurate the recruitment.

Figure 11.27 describes the results of a numerical solution in which the time needed to exploit fully a given quantity of food is plotted against the error, as expressed by the variance of the statistical distribution, for given numbers of food sources. As expected, when only one source is present the best recruitment strategy is the one that functions without mistake. However, when food is parcelled there is one 'optimal' value of the error which minimises the time of exploration. This defines the level of randomness during communication which can be advantageous by increasing the possibility of discoveries allowing the society to concentrate its activities on the most rewarding resources, and by promoting the colonisation of resources which will be fully exploited later.

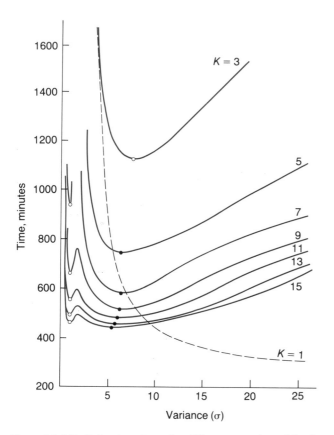

Figure 11.27. Collection times for different number of food sources (*K*). Value of parameters: $N = 650$, $a = 10^{-3}$, $b = 1.6 \times 10^{-3}$, $2d = 1.4$ cm.

Adaptation and plasticity, two basic features of nonlinear dynamical systems, also rank among the most conspicuous characteristics of human societies. It is therefore natural to expect that dynamical models allowing for evolution and change should be the most adequate ones for social systems.

A dynamical model of a human society begins with the realisation that, in addition to its internal structure, the system is firmly embedded in an environment with which it exchanges matter, energy and information. Think, for instance, of a town in which raw materials and agricultural products arrive continuously, finished goods are exported, while mass media and professional communication keep the various groups aware of the present situation and of the immediate trends.

The evolution of such a system is an interplay between the behaviour of its actors and the constraints imposed by the environment. It is here that the human system finds its unique specificity. Contrary to the molecules, the 'actors' of a physico-chemical system, or even the ants or the members of any other animal society, human beings develop individual *projects* and *desires*. Some of these stem from anticipations about how the future might reasonably look and from guesses concerning the desires of the other actors. The difference between desired and actual behaviour acts therefore as a constraint of a new type which, together with the environment, shapes the dynamics. A basic question that can be raised is whether, under those circumstances, the overall evolution is capable of leading to some kind of global optimum, or, on the contrary, whether each human system constitutes a unique realisation of a complex process whose rules can in no way be designed in advance. In other words, is past experience sufficient for predicting the future, or, on the contrary, isn't a high degree of unpredictability of the future the essence of human adventure, be it at the individual level of learning or at the collective level of history making? The developments outlined in the present chapter suggest that the answer to this question should lean toward the second alternative. Extensive mathematical modelling has been carried out, which permits us to justify this intuitive feeling and, at the same time, to specify more sharply the nature of the unpredictability of the human system.

Reconstruction of the dynamics of natural systems from time-series data

In addition to providing ideas and tools for the modelling of complex systems, nonlinear physics of far-from-equilibrium phenomena also allows us to gain a deeper understanding of such systems *independent* of any modelling.

The quantitative study of a system occurring in nature is based on the observation of a variable, or of a limited set of variables, during a sufficiently long period of time. Figures 11.28 and 11.29 provide two important examples of such *time series* pertaining, respectively, to *climatic variability* and to the *electrical activity of the brain*, the two phenomena we choose in this subsection to illustrate the main ideas.

In figure 11.28 the global ice volume present on the planet Earth is plotted against time for the last million years. Its variation, which can be inferred from the study of the oxygen isotope record of ice cores or deep sea core sediments, reveals a number of peaks associated to the glaciations, the most dramatic climatic episodes of the quaternary era. Both the position and strength of these peaks are rather irregular, although on average a time scale of 10^5 years is clearly emerging. It is therefore natural to inquire whether one deals with a well-defined dynamical phenomenon or, on the contrary, with a random process which essentially reflects the variability of the environment to which our planet is subjected. Mathematical modelling alone cannot settle the issue, since for both alternatives one can produce scenarios fitting the record in a reasonably satisfactory way. A new argument is therefore needed.

Figures 11.29(*a*) and (*b*) refer to the electroencephalogram (EEG), a record of the sum of elemental sustained neuronal activities of relatively low frequency (0.5 to 40 Hz) emanating from small volumes of cortical tissue just underneath the scalp. Figure 11.29(*a*) describes the EEG of a normal human subject in the stage of deep sleep, whereas figure 11.29(b) describes the EEG of a patient during an epileptic seizure. Both records show an irregular succession of peaks, although (b) looks definitely more 'coherent' than (a). Is it possible to characterise this coherence more sharply? Do we deal with a well-defined dynamics or, rather, with a random course of events impossible to control? Below, we show how nonlinear physics can provide us with clues to these important questions.

The first step is to embed the dynamics of the system under study in phase space. This means, in particular, that one should go beyond the 'one-dimensional' view afforded by a time series of a single variable, $X_0(t)$. It can be shown that a phase space satisfying all requirements of dynamical systems theory is the phase space generated by $X_0(t)$ and its successive lags $X_0(t+\tau), \ldots, X_0(t+(n-1)\tau)$. It suffices to choose τ in such a way that these different functions are linearly independent.

Next, in each of the above defined phase spaces (that is to say, for each choice of value of n), one tries to identify the nature of the set of data points, viewed as a geometrical object in an n-dimensional space. Again, dynamical systems theory (cf. Sections 11.6–11.8) provides algorithms for accomplishing this. One particular quantity which can be identified in this way is the dimensionality, $d_F(n)$, of our data set. Once $d_F(n)$ is determined one can also obtain information on dynamical properties like, for instance, the largest positive Lyapounov exponent $\sigma_L(d_F(n),n)$, if any.

Finally, $d_F(n)$ and $\sigma_L(d_F(n),n)$ are plotted against n for increasing values of the embedding dimensionality. If these dependencies are saturated beyond some relatively small n, the system represented by the time series should be a deterministic dynamical system possessing an attractor. The saturation values of d_F and σ_L will be the dimensionality and the largest

Figure 11.28. Variation of global ice volume during the last 1 000 000 years, inferred from the isotope record of deep sea cores.

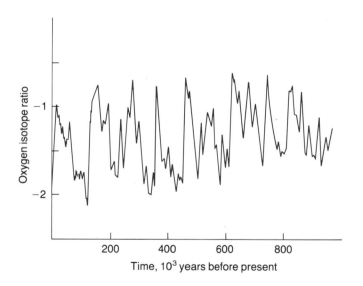

Figure 11.29. (a) EEG record during the deep sleep stage of a normal human subject; (b) EEG record during a 5 s epileptic seizure of a human subject.

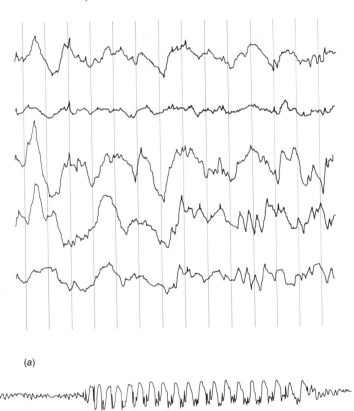

(a)

(b)

Lyapounov exponent of the attractor. The value of n beyond which saturation is observed provides the minimum number of variables necessary to model the behaviour represented by the attractor. If on the other hand there is no saturation trend, the conclusion will be that the system described by the time series evolves in a random way.

This procedure, applied to the climatic data of figure 11.28, indicates a fractal dimensionality of about $d_F \simeq 3.1$, a positive Lyapounov exponent corresponding to a characteristic predictability time of about 30 000 years and a minimum value of $n=4$. We are thus allowed to conclude that long-term climatic change can be viewed as a deterministic, unstable dynamics possessing a chaotic attractor. This provides us with a natural way to understand the well-known variability of the climatic system. It also allows us to calculate, *quantitatively*, the limits beyond which predictions about the future become meaningless.

Applied to figures 11.29(*a*) and (*b*), the procedure leads to two strikingly different predictions. In both cases a chaotic attractor seems to exist. However, for figure 11.29(*a*) the dimensionality turns out to be slightly above four, whereas for figure 11.29(*b*) it drops to a value slightly above two. This provides us with a quantitative measure of the coherence associated to the pathological state of the brain. Moreover, as pointed out at the beginning of the present section, chaotic attractors are potential information-generating devices. We are thus led to a tantalising picture of how information, one of the most conspicuous attributes of the human brain, can be linked to, and even emerge from, its dynamical activity.

The reconstruction of the dynamics of a host of other complex systems occurring in nature can be undertaken along similar lines. Two particularly important examples are short term weather variability and economic activity. All in all, these developments contribute to the emergence of a novel picture of our natural environment, more complex, more realistic, and undoubtedly more challenging, than the idealised archetype proposed by classical science.

Further reading

Physical aspects
Presentations of the physical ideas underlying self-organisation addressed to the nonspecialist reader having a general scientific background can be found in

Babloyantz, A. (1986). *Molecules, Dynamics and Life*. Wiley, New York.
Nicolis, G. and Prigogine, I. (1987). *Exploring Complexity*. Piper, Munich.
   Large parts of Sections 11.1–11.5 and 11.9 are inspired from this book.
Peacocke, A. (1983). *The Physical Chemistry of Biological Organization*. Clarendon, Oxford.
Prigogine, I. (1980). *From Being to Becoming*. Freeman, San Francisco.

Self-organisation phenomena in physics and chemistry
Velarde, M. and Normant, C. (1980). *Sci. Am.* July, p. 78.
   The problem of thermal convection treated in Section 11.2 of the present chapter is addressed.
Winfree, A. (1972). *Science* **175**, 634, and
Müller, S., Plesser, T. and Hess, B. (1986). *Naturwissenschaften* **73**, 165.
Ross, J., Müller, S. C. and Vidal, C. (1988). *Science* **240**, 460.
   In these reviews pattern formation in chemistry, treated in Section 11.3 of the present chapter, is beautifully illustrated.

Biological systems
Meinhardt, H. (1982), *Models of biological pattern formation*, Academic, London.
   The concepts of positional information and symmetry breaking are discussed from the standpoint of developmental biology.
Martiel, J. L. and Goldbeter, A. (1987), *Biophys. J.* **52**, 807.
   Develops mathematical models of the control of development in *Dictyostelium discoiteum*.

More technical presentations
Ebeling, W. and Feistel, R. (1982). *Physik der Selbstorganisation und Evolution*. Akademie-Verlag, Berlin.
   Emphasis is placed on various problems related to biological evolution, in connection with nonlinear dynamics and information theory.
Glansdorff, P. and Prigogine, I. (1971). *Thermodynamics of Structure, Stability and Fluctuations*. Wiley, London.
   In this book there is emphasis on the search of thermodynamic criteria of evolution and instability.
Haken, H. (1977). *Synergetics*. Springer, Berlin.
   Here an alternative presentation of bifurcation phenomena and fluctuations is provided. Applications deal mostly with hydrodynamics and quantum optics.
Nicolis, G. and Prigogine, I. (1977). *Self-organization in Nonequilibrium Systems*. Wiley, New York.
   The nonlinear phenomena associated with bifurcation, the role of fluctuations, and applications in chemistry and biology, are discussed in detail.

Mathematical aspects
Presentations of the mathematics of nonlinear systems connected with the topics developed in this chapter can be found in

Guckenheimer, J. and Holmes, P. (1983). *Nonlinear Oscillations, Dynamical Systems and Bifurcations of Vector Fields*. Springer, Berlin.
   The technically oriented reader will find here a rewarding complement of the material of Sections 11.7 and 11.8.
Mandelbrot, B. (1977). *Fractals: Form, Chance, Dimension*. Freeman, San Francisco.
   An inspired presentation of the fractal geometry and of its role in the understanding of natural phenomena. A useful complement for the material of Section 11.7.
Schuster, H. (1984). *Deterministic Chaos*. Physik-Verlag, Weinheim.
   This is an informative presentation of chaotic dynamics with technicalities limited to the minimum required. It is a useful complement of the material of Sections 11.7 and 11.10.

The concept of complexity is discussed in

Chaitin, G. (1975). *Sci. Am.* May, p. 47.
   The reading of this article is strongly recommended as a complement to Section 11.10.

Modelling of complex systems

The role of chaos in information processing and cognition, discussed briefly in Section 11.10, is also developed in

Nicolis, J. S. (1986). *Dynamics of Hierarchical Systems*, Springer, Berlin.

Shaw, R. (1981). *Z. Naturf.* **30a**, 80.

The modelling of social phenomena from the standpoint of dynamical systems, to which we alluded in Section 11.10, is inspired from

Allen, P. (1982). *Environment & Planning* **B9**, 95.

Deneubourg, J. L., Pasteels, J. and Verhaege, J. C. (1983). *J. Theor. Biol.* **105**, 259.

Alternative views of the quantitative formulation of social phenomena can be found in

Montroll, E. and Badger, W. (1974). *Quantitative Aspects of Social Phenomena*. Gordon and Breach, London.

The reconstruction of the dynamics of the climatic system and of the electrical activity in the brain, described in Section 11.10, is based on

Babloyantz, A., Salazar, M. and Nicolis, C. (1985). *Phys. Lett.* **111A**, 152.

Nicolis, C. and Nicolis, G. (1984). *Nature* **311**, 529.

Babloyantz, A. and Destexhe, A. (1986). *Proc. Nat. Acad. Sci. (USA)* **83**, 3513.

The methods followed in these papers are described in

Grassberger, P. and Procaccia, I. (1983). *Phys. Rev. Lett.* **50**, 346.

Packard, N., Crutchfield, J., Farmer, J. and Shaw, R. (1980). *Phys. Rev. Lett.* **45**, 712.

Eckmann, J. P. and Ruelle, D. (1985). *Rev. Mod. Phys.* **57**, 617.

12   What is chaos, that we should be mindful of it?*

Joseph Ford

There are more things in Heaven and earth, Horatio,
Than are dreamt of in your philosophy.
                                    Hamlet: Act I, Scene V

12.1   In the beginning . . .

Earliest man perceived the world as total chaos and called his most primeval gods by that name. But, as generations passed, the regularity of the solar day, the lunar month, and the sidereal year introduced man to the notion of predictability and order. It was a heady notion which expanded and eventually led him to believe in a completely ordered and predictable universe. Specifically, as emphasised by Laplace, the Newtonian dynamics which governs all the macroscopic world is strictly deterministic, meaning that past and future states of a Newtonian system uniquely follow from its present state. Moreover, quantum mechanics is also deterministic because here again present system state uniquely determines past and future state. To emphasise, regardless of their dissimilar definitions of state, Schrödinger's equation and Newton's equations are equally deterministic. In consequence, mid twentieth century man could believe in the triumph of determinism at both the microscopic and macroscopic levels.

But despite all his devotion to and deep belief in determinism, man has for millennia regarded certain Newtonian systems as unpredictably chaotic – dice and roulette wheels for example. Indeed, a caveman croupier or his more contemporary Las Vegas counterpart would welcome us to test our deterministic schemes at his gambling tables. Of greater scientific interest, however, is the problem of turbulence in all its forms. Orbital behaviour for turbulent systems – beautifully pictured in the September 1984 issues of *Discover* magazine and reproduced in part here in figures 12.1–12.4 – exhibits a chaos having structure on every scale, sometimes self-similar and sometimes not. In regard to the latter, can even modern man maintain that non-self-similar structure on every scale is deterministically predictable? But gambling games and turbulence aside, perhaps the most startling revelation of contemporary dynamics is the discovery that Newtonian mechanics can yield orbital chaos in systems having only two degrees of freedom. Moreover, there is now substantial though nonrigorous evidence that chaotic orbits are the rule in Newtonian dynamics rather than the exception. But this means we have come full circle. Cavemen perceived their world as largely chaotic, and now so do we! There is a difference, however. The caveman viewed nature as indifferently rolling unbiased dice; modern man recognises that nature's dice are only slightly but nonetheless purposefully loaded. Our scientific task is thus to determine the loading and the purpose.

*Should the above title induce a slight feeling of déjà vu, consult Psalms 8:4.

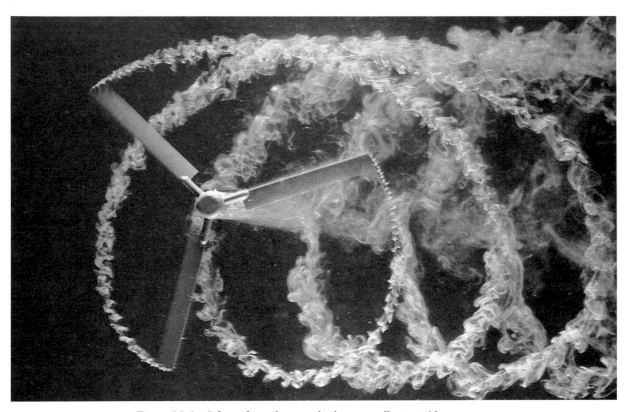

Figure 12.1. Coloured smoke reveals the normally invisible
turbulent flow trailing behind the tips of a helicopter blade.
(Reproduced with permission from the Office National d'Etudes et
de Recherche Aérospatiales (ONERA). This photograph appeared in
the September 1984 issue of *Discover* magazine.)

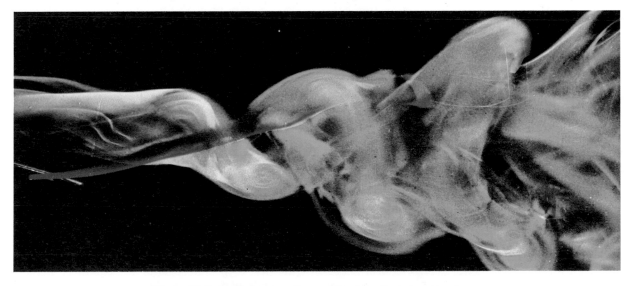

Figure 12.2. Turbulence pattern created by the wing tip of an
aircraft. (Reproduced with permission from ONERA. This
photograph appeared in the September 1984 issue of *Discover*
magazine.)

Figure 12.3. A ghost-like vortex pattern generated by water
flowing around the cylinder half seen at the far left of the figure.
(Reproduced with permission from Prof. Sadatoshi Taneda, Kyushu
University. This photograph appeared in the September 1984 issue
of *Discover* magazine.)

12.2   What is chaos?

In the foregoing, we have asserted first that ours is a strictly
deterministic world and second that ours is mostly a chaotic
world, an interesting but perhaps confusing juxtaposition of
assertions. Upon reflection, a reader pondering whether determinism and chaos are contradictory terms will surely
welcome a definition of chaos more precise than that conveyed
by the vague words 'erratic', 'irregular', 'disordered', or 'seemingly unpredictable', so commonly used in the current literature. Greater sophistication might seek to correlate chaos with
precise mathematical concepts – positive Liapunov exponents,
positive metric or topological entropy, or fractal attractors –
whose definitions need not concern us here, but even these
notions do not convey the simplicity or the generality of the full
truth. Specifically, in its strictest technical sense, chaos is
merely a synonym for randomness as the algorithmic complexity theory of Andrei Kolmogorov, Gregory Chaitin, and Ray
Solomonov so clearly reveals; this theory considers the deterministic computation of a physical quantity Q specified as a
digit sequence S and asserts that S, hence Q, is random provided
the shortest computer program which prints S is the copy
program 'PRINT S'. Thus, no matter the program used to
deterministically compute a truly random sequence, our CRAY
I computers become nothing more than photocopiers. The
central point here is that the term 'deterministic randomness' is
not self-contradictory and that, at the macroscopic level, we
can indeed live in both a determinate and a largely random
world. Determinate because of existence and uniqueness for
classical orbits, and random because chaotic orbits cannot be
computed by any algorithmic rule appreciably simpler or
shorter than the copy program. Alternatively stated, Newtonian determinism assures us that chaotic orbits exist and are
unique, but they are nonetheless so complex that they are
humanly indistinguishable from realisations of truly random
processes. In the above, the word 'random' is not meant to
imply a uniform probability distribution only; chaos does not
exclude loaded dice.

Let us now seek to further reconcile the notions of determinism and randomness by describing a few more details of the
algorithmic complexity viewpoint. Analytically solvable systems having highly ordered motions like the harmonic oscillator or the Kepler problem have orbits which require only a
relatively small amount of input information to reap a large
quantity of output orbital information; here, 'amount of
information' means the length of the binary digit string which
encodes this information. Thus, these analytically solvable
systems are not only deterministic, they are also reasonably
predictable and nonrandom. However, when the Newtonian

Figure 12.4. The familiar turbulent wake which trails a boat moving through water. In addition to the turbulent flow, a symmetrically ordered pattern of waves is seen to angle off each side. (Reproduced with permission from Prof. J. N. Newman, MIT. This photograph appeared in the September 1984 issue of *Discover* magazine.)

system becomes sufficiently complex – a hard sphere gas or a roulette wheel, for example – our orbital computation is found to require just as much informational input as it provides in informational output. This means that our computations have now ceased computing or predicting anything because the output orbital data are so chaotic, so unpredictably random that our input information must, of necessity, be equivalent to

a copy of the output. Alternatively stated, our orbital computation serves primarily as a dictionary with which to translate an 'input word' into a synonym of equal length called the 'output word'. For chaotic systems, determinism, existence-uniqueness remain, as they must, but predictability is lost. Let us now briefly examine deterministic randomness from the viewpoint of computational complexity rather than algorithmic; that is, let us inquire about the time required to run a program rather than about its size. The clock time required to numerically compute along an analytically derivable orbit is proportional to $\log(t)$, where t is system time. On the other hand, the clock time required to compute along a chaotic orbit is proportional to system time itself. As before, an analytically derivable orbit is deterministic and essentially predictable, since we can announce future system state long before it occurs. Contrariwise, a chaotic orbit is deterministic but not predictable, since we cannot predict the future appreciably before it arrives. To summarise, a chaotic orbit is its own briefest description and its own fastest computer; it is both determinate and random.

Finally turning now to the microscopic level, there is currently only confusion. Although few observers doubt that chaos is ubiquitous throughout nature, quantum mechanics has not been shown to exhibit chaos, meaning, for example, randomness in the determinate time evolution of the quantum state. We shall return to this quantum anomaly in a later section.

12.3 What does chaos mean?

It is instructive to regard the source of chaos as missing information, for chaos is what humans (cavemen or modern) observe when they lack the information to perceive the underlying order. In this regard, one must not assume that the needed information is missing because of some temporary theoretical ignorance or experimental ineptitude which will be overcome in a future epoch, any more than one is permitted to assume that some future theory will predict the precise influence of a position measurement upon a particle's momentum. In both cases, we are dealing with a fundamental uncertainty, an ultimate limit on man's abilities. Chaos shall always be a mystery, perhaps the ultimate, all-encompassing mystery. It is a paradox hidden inside a puzzle shrouded by an enigma. It is visible proof of existence and uniqueness without predictability.

But precisely what is the missing information which places chaos beyond human comprehension? Don't Newton's equations provide us with a relatively simple computational rule for determining each orbit once the initial state S_o is given? They do indeed, but who provides the initial state S_o? Recall that S_o is shorthand for a set of numbers specifying all system coordinates – positions and velocities, for example. In turn, each of these numbers is, in decimal notation, given by an infinite sequence of digits, e.g. $\frac{1}{3} = 0.3333\ldots$ or $\sqrt{3} = 1.732$ .... Thus, if we wish to specify any possible S_0, we must be able to specify any possible infinite digit sequence. But providing this specification is, in fact, humanly impossible, as we now indicate. First, consider the collection of decimal digit sequences corresponding to all numbers on the unit interval $(0,1)$. Here, each digit sequence defines a number and every possible digit sequence appears in the collection. Now shift gears and regard each decimal digit sequence as specifying not a number but the outcome of an infinite sequence of random roulette wheel spins – a special roulette wheel bearing only the integers $0, \ldots, 9$ equally spaced around the wheel. Here again, each decimal digit sequence defines a possible roulette wheel sequence and all possible roulette wheel sequences occur in the collection. Thus, we now perceive that there is a one-to-one correspondence between number associated decimal digit sequences and random roulette wheel sequences. In consequence, the decimal digit sequences for almost all numbers on the interval $(0,1)$ – or the interval $(0,-1)$ as well – are seen to be both unpredictable and uncomputable, for they are as random as sequences of roulette wheel spins. Our argument is now completed by noting that randomness in a digit sequence for a number outside the interval $(-1,1)$ cannot possibly depend on the position of its decimal point and that any number can be brought to the interval $(-1,1)$ by a shift of its decimal. In summary, most numbers and most initial states S_0 are specified by random digit strings which are unpredictable, uncomputable and undefinable. Clearly then, only a god can provide the initial state S_0. Thus, we have now tracked our missing information to the real number system whose individual members are, in general, beyond man's ability even to define, much less specify or compute.

However, for analytically solvable Newtonian systems, which are the heart and soul of traditional mechanics texts, this randomness of S_0 real number digit strings is, in general, of no practical consequence for orbit calculations extending over reasonable system time intervals. Specifically, a bundle of analytically derivable orbits emerging from closely spaced initial states surrounding a specified initial state S_0 will spread apart only linearly with system time. This relatively slow separation of initially close orbits means that knowledge of a precise S_0 is generally unneeded for practical orbital predictability in analytically solvable Newtonian systems. However, as Max Born and others have pointed out, this linear separation of orbits eventually makes even analytically derivable orbits unpredictable, though not random, if one seeks to know very distant future states. But if analytically derivable as well as chaotic orbits are unpredictable, one begins to suspect that Newtonian dynamics is but a theorist's fanciful description of a perfect world inhabited by perfect observers. Rather than pursue this point here, however, let us examine missing information in chaotic orbits.

The Newtonian algorithms (computation rules) for chaotic orbits transform the missing information of uncomputable,

random S_0 digit strings directly into the missing information of uncomputable random orbits; here orbital random variables are simply functions of the S_0 random variables. Strictly speaking, analytically derivable orbits are subject to this same transformation of randomness, but for them the informationally incompressible, random S_0 digit strings become so diluted with the redundant information injected by their Newtonian algorithms that their orbital output digit strings are not random. In contrast to analytically derivable orbits, however, a bundle of initially close chaotic orbits spread apart exponentially with system time; this means that a small initial error, 10^{-8} say, propagates and grows each second approximately like 10^{-8+k}, where k is integer time in system seconds. In essence, one decimal digit of orbital accuracy is being lost each system second. To regain this lost digit of accuracy in the output, we must add one digit of accuracy to the input S_0. Thus, if we seek to accurately integrate a chaotic orbit, we must put in just as much information as we get out. Moreover, since we are putting randomness into the computation via the S_0 digit strings, we must get it back again in the emerging orbital output digit strings. Finally, this exponential loss of accuracy with time, this propagation of missing information from random input to random output means that chaotic orbits are, humanly speaking, uncomputable. Our lack of knowledge about and our lack of ability to deal with real numbers are the ultimate culprits, and, this side of Heaven, there is no help for it. Let us now turn to a laboratory view of missing information.

Consider figure 12.5, which schematically depicts a part of the world called the local universe. In order to accurately predict and measure the time evolution of this local universe, Newtonian dynamics postulates universal and perfect clocks and metre sticks which have either negligible or controllable interaction with the local universe. But this postulate was clearly made in ignorance of chaos. Since chaos is ubiquitous at the macroscopic level, the clock and the metre stick as well as the local universe are each chaotic, in general. Interactions between chaotic systems create effects which locally magnify exponentially with time; they cannot, therefore, be neglected no matter how small they may be. Moreover, because chaos means random, the effect of these interactions is uncontrollable, unpredictable and uncomputable. Indeed, the effect of one chaotic system weakly acting upon another appears as random noise. A weak interaction among several chaotic systems thus introduces a low-level random noise throughout the composite system. Hence, when we are missing the information supplied by noninteracting, universal clocks and metre sticks and when nature forces us to use weakly interacting but chaotic substitutes, the act of measurement introduces a small and uncontrollable noise (error) into the quantity being measured. The analogy here with quantum measurement theory needs no amplification. However, in order to make the classical notions discussed in this paragraph concrete, let us examine the three-particle oscillator system dramatised in figure 12.6.

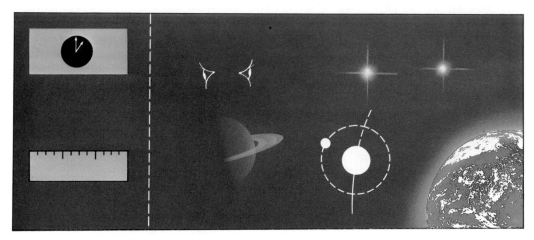

Figure 12.5. An imaginative view of the local universe showing the clock and metre stick with which measurements are made. The eyes denote the observer; the dotted line permits a weak interaction between the local universe and the measuring devices.

Figure 12.6 shows a massive, non-rotating Earth and two particles, each of which is coupled to the Earth by a spring. The human observer at the North Pole is assumed not to affect the motion; the ideal metre sticks attached to the Earth permit perfect position measurements; and the motion of this three-particle oscillator system is assumed to occur in one dimension. Kronos, Greek god of time, watches from a great distance and hides his perfect, universal clock. The human observer now seeks to measure the motion of the leftmost particle in figure 12.6, but, having no universal clock, he is forced to use the oscillatory motion of the rightmost particle in figure 12.6 as his clock for making time measurements. Noting the human's intent, Kronos sets out to predict what the human will observe.

Kronos, who knows Newton quite well, immediately perceives that the springs are linear and in consequence that this three-particle system is nothing more than a set of coupled harmonic oscillators. He therefore concludes that the displacement D_L of the left mass point from the surface of the Earth, aside from an ignorable uniform translation, is merely a sum of two elementary trig functions; symbolically, he writes $D_L = D_L(T)$, where T is his universal, perfect time. Now regardless of how the human observer elects to measure his Earth time t using his rightmost oscillator, Kronos can determine the functional relationship $T = G(t)$ between T and t. Then, Kronos recognises that, for the displacement D_L, the human will observe $D_L = D_L[G(t)]$, a type of frequency modulated oscillation typified by the simple $D_L = \sin[at + b\sin(ft)]$ of FM radio. Had the human access to the universal clock, he would, like the god, observe the simple $D_L = D_L(T)$, a function containing only two frequencies; but, being denied the universal clock and forced to use his rightmost particle as clock, he actually observes not $D_L = D_L(T)$ but $D_L = D_L[G(t)]$, a function known to contain an infinity of frequencies. The human will, therefore, observe an incredibly complex motion in his simple oscillator system. In summary, when the Earth observer is missing the information supplied by the universal clock, his observations will contain a type of noise as mentioned earlier, but, even worse, the more the human increases the accuracy of his measurements, the more he observes things which are not there (frequencies, in this example). One suspects there may be a moral here for those seeking to probe the shortest or longest intervals of time or distance.

Finally, it is enlightening, though somewhat audacious, to view chaos as the beginning of the third scientific revolution in this century. If the aeons have taught man anything, it is that phenomena lying outside the range of his everyday experience may contain fundamental surprises whose explanation requires totally new theories. An object moving far in excess of everyday speeds requires relativity. An object far smaller and less massive than everyday macroscopic objects requires quantum theory. Both these theories were revolutionary, among other reasons because they revealed a previously unexpected 'discontinuity' in continuous speed and mass variables and because each placed a new and fundamental bound on man's abilities. Taking its turn, chaos now implies a complexity beyond everyday human experience, indeed beyond human understanding. Deterministic objects so complex they are uncomputable, unpredictable, undefinable and random; so complex they imply another limitation of man. As we have seen, the human limitation exposed by chaos extends far back into the foundations of our number system – to the question of whether or not most real numbers have been or humanly can

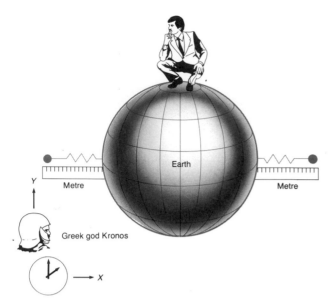

Figure 12.6. A three-particle oscillator system with the Earth as middle particle. The noninteracting human observer stands at the North Pole of the Earth. Kronos, son of Chaos and god of time, observes from a great distance and protects his universal clock.

chaos exposes the clay feet of Newtonian dynamics as it stands in that final nonrelativistic, nonquantal domain which serves as its last refuge. Newtonian dynamics is now seen to be useful only for the nonrandom orbits of slowly moving, macroscopic objects, a collection of orbits whose relative size is perilously close to zero. It is thus highly appropriate that Newtonian dynamics should at last be returned to the gods from whom it was stolen for it is only they who can supply that missing information which is beyond human grasp. In its place, we must erect a theory compatible with human capabilities. In this regard, James Clerk Maxwell, who perceived these issues with great clarity, sought to leave us a hint when he wrote, 'The true logic of this world is in the calculus of probabilities.' If Maxwell has correctly pointed the way, then the new theory now standing off stage will provide us with a probability calculus appropriate to finite digit strings, finite because the real number continuum is also a possession of the gods which should be returned. The detailed structure of this new theory is presently only dimly seen, but even now the set designers of chaos are decorating its stage aided by the propmen of cellular automata theory (see *Discover* magazine, August 1984, p. 81) and the stagehands of algorithmic and computational complexity theory (see the *International Journal of Theoretical Physics*, vol. 21, issues 3–4, 6–7 and 12).

be properly defined. But at an even deeper level, chaos is forcing us to recognise the physical significance of the long-neglected theorem due to Kurt Gödel which in everyday language asserts: a one-pound theory can no more produce a ten-pound prediction than a pregnant, one-hundred pound woman can give birth to a two-hundred pound child. Before Gödel, humans assumed that simple questions had simple answers, that any object known to exist and be unique could be found and exhibited. Alas, Gregory Chaitin has proven that even simple questions can have answers so complicated that they contain more information than man's entire logical system, and nonlinear dynamics has proven the existence and uniqueness of chaotic orbits which no human can ever exhibit. Those trying to deterministically predict a random digit string or a chaotic orbit would do well to bear Gödel and Chaitin in mind.

To summarise, no man can go faster than the speed of light; no man can make simultaneous measurements of two conjugate variables with infinite precision; and no man can compute or measure any continuum variable precisely. In consequence, we can no longer disguise the fact that deterministic Newtonian dynamics has been dealt a lethal blow. Relativity eliminated the Newtonian illusion of absolute space and time; quantum theory eliminated the Newtonian dream of a controllable measurement process; and chaos eliminates the Laplacian fantasy of deterministic predictability. But of the three, chaos deals perhaps the most devastating blow because

12.4   Chaos: who needs it?

Over the centuries, chaos has been indicted in every disaster from riots in the street to the heat death of the universe. In truth, uncontrolled chaos can indeed be a fearsome thing. Nonetheless, chaos wears yet another face having great charm and fascination – the enchanting unpredictability of flames dancing over a burning log, the mesmerising variation of visual patterns presented by waves breaking on a seashore, the incredible variety of shapes which fuel the imagination when white clouds drift upon an azure sky. Chaos is dynamics freed from the shackles of order and predictability. It permits systems to randomly explore their every dynamical possibility. It is exciting variety, richness of choice, a cornucopia of opportunities. But how can we harvest this desirable richness without also reaping disaster's dreaded hemlock? Dare we dream of utilising the randomness of chaos to generate order, probability one? Let us look to nature for an answer.

In setting up the scheme humans call evolution, nature wished to insure in perpetuity the survival of life forms against every possible variety of natural catastrophe; in addition, nature wished to encourage the expansion of life into every possible ecological niche no matter how harsh or specialised. In principle, nature could have written a deterministic program to cope with the temporal unfolding of exceedingly complex, almost random patterns of life affecting events; instead, nature chose a highly effective technique which uses randomness to defend against the unexpected. Specifically, nature uses random mutations to provide the wide variety of life forms needed to meet the demands of natural selection. In essence, evolution is chaos with feedback. Random mutations alone would correspond to nature indifferently rolling unbiased dice, but the added feedback of natural selection and survival of the fittest, in effect, biases the dice so that, over many rolls, life forms not only survive they improve, probability one. In this article, we shall say that an event occurs 'probability one' whenever its occurrence is overwhelmingly likely.

Chaos serves a similar role in one theory of the human autoimmune response. Here, the body must be prepared to repel microscopic invaders of whatever stripe: germs, microbes, bacteria, or viruses. A deterministic program written for this purpose would not only have to catalogue all currently existing invaders but all possible mutants, natural or man-made, which might arise in the future. Rather than develop such an horrendous, deterministic program whose storage might require a finite fraction of the body, nature elected to fight fire with fire and provided us with an economical program based on chaos.

Briefly, the body responds to the presence of any microscopic invader by activating an internal manufacturing plant which can produce randomly shaped molecules. Molecules having a marvellous variety of shapes are made and then carried throughout the body. When one of the randomly generated molecules is, at last, seen to attach itself conformally to an invader, a feedback message is immediately sent to the manufacturing plant ordering it to cease random production and to churn out a half-zillion copies of the successful molecule, Model #39. Very soon, each invader is tagged by a conformally fitting Model #39 molecule and, using Model #39 as a 'handle', the circulatory defence system carries the invaders to waste elimination centres – bowels, kidneys, etc. Finally, the manufacturing plant is told to shut down operation and all is quiet again. When nature must correctly respond to a sequence of events whose nature and arrival time are essentially random, then nature uses the richness of chaos to nondeterministically solve its problem, probability one. Using direct or indirect feedback, nature uses and controls chaos to achieve its goals. But should not this fact embolden us to accept the challenge: 'What nature can do; man can do better'? Indeed, man has already begun to use and control chaos, as we now discuss.

Consider the prime numbers. To date, there is no known deterministic computer program for locating primes whose run time does not depend exponentially on program size; in essence, no known deterministic program is appreciably better than a simple search through all possibilities, a procedure which rapidly becomes humanly intractable as the number of possibilities grows. Similar remarks hold for determining the shortest route a salesman can take when he visits a number of cities. This so-called travelling salesman problem may sound quite trivial to the uninitiated, but it is actually – see the Garey and Johnson reference in our bibliography – a member of a whole class of equivalent, very hard problems all of which appear to require run times exponential in program size. However, both these problems can be 'quickly' solved, run time polynomial in program size, provided one permits computer programs which contain chaos – a random number generator, for example. By introducing chaos into the computer program, one obtains answers which, with probability close to one, either are or are very close to the correct answer. Additional problems solvable using chaos but apparently intractable otherwise are discussed on page 917 of the 2 March 1984 issue of *Science* magazine. Here we discuss only one of these: the dining philosophers' problem.

A group of philosophers sit around a dining table which is set with one fork to the right of each philosopher and with one large bowl of spaghetti at its centre. Each philosopher requires two forks to eat the spaghetti; thus, the philosophers cannot all eat simultaneously. Various deterministic schemes exist which, if followed, would permit all philosophers to eat, e.g. electing a benevolent philosopher-king who dictates the eating order. But if we insist on finding a deterministic solution which treats all philosophers equally and which makes deadlock impossible, then we, and determinism as well, face a formidable challenge indeed. On the other hand, Michael Rabin and Daniel Lehmann of the Hebrew University have discovered a quite simple solution based on randomness, although the proof that their solution always works is not simple. Specifically, when any philosopher is hungry, he tosses a random coin, seeking his right fork if heads and his left fork if tails. Suppose his coin came up heads; the philosopher then vigilantly waits for his right fork to become available. When it does, he picks it up and quickly looks to his left. If the left fork is available, he picks it up, eats, and returns both forks to their respective places. Should the left fork not be available after he obtains his right fork, he immediately returns his right fork to its place and again flips his coin. Without deadlock and with probability one, this strategy can be shown to permit all the hungry to eat. On the practical side, the dining philosophers represent processes in a computer which must be treated equally and which must, probability one, be able to communicate pairwise without deadlock.

We now consider the possibility that chaos can actually be used to predict chaos as proposed by Michael Barnsley and co-workers at Georgia Tech. Let us begin by considering contractive mappings, $X_{n+1} = f(X_n)$, of the unit interval $(0, 1)$ upon itself, where contractive means that iteration via the function $f(X)$ causes all lengths dX_n to sequentially decrease toward zero. This contractive property implies that, under iteration, the interval $(0,1)$ having unit length must tend asymptotically to a finite or infinite collection of points having zero length. These finite or infinite collections of points are called attractors. For such contractive maps, all the points of the interval $(0,1)$ can asymptotically approach a single point, called a point attractor, a finite set of points called a periodic attractor, or, most interesting of all, an infinite set of points having zero length called a chaotic or strange attractor. Almost all initially close

Figure 12.7. Each object in this delicate 'Japanese print' is the chaotic attractor of some nondeterministic mapping similar to those discussed in the text. Here, the beauty and 'naturalness' of nature is revealed to be nothing more than the beauty and 'naturalness' of chaos. Indeed, this pretty landscape testifies to the uniquity of chaos in nature. (Professor Michael Barnsley (Georgia Tech) designed and prepared this computer generated scene; Vincent Mallette (Georgia Tech) prepared the final photograph.)

Figure 12.8. Professor Michael Barnsley's bedroom window on a frosty Atlanta morning. The boundary of the frost is the chaotic attractor of one of nature's chaotic mappings. (Original slide courtesy Michael Barnsley; final photograph courtesy Vincent Mallette.)

points on a chaotic attractor are found to exponentially separate upon iteration yielding that much heralded 'sensitive dependence upon initial conditions'. It is precisely this sensitive dependence upon initial conditions in contractive systems which leads many experts to assert that not even the total wealth of the United States can buy accurate weather predictions extending appreciably beyond two weeks into the future. But does this mean that the situation is hopeless? Barnsley and co-workers think not.

These investigators asked themselves if, given a chaotic attractor obtained by experimental or analytical means, they could find a contractive mapping whose chaotic attractor is precisely the given attractor. As a test case, they considered the limiting collection of points obtained by first deleting the middle third $(\frac{1}{3}, \frac{2}{3})$ from the unit interval, by then deleting the middle third from each remaining interval $(0, \frac{1}{3})$ and $(\frac{2}{3}, 1)$, and finally by repeatedly deleting the middle third from each then remaining segment of the original unit interval. It is straightforward to show that, in the limit, the resulting collection of points has zero length; it is less straightforward to verify that this collection of points contains just as many points as the original interval $(0, 1)$. But now, what mapping has this exotic collection of points as its chaotic attractor? To find a deterministic mapping meeting this goal could be difficult and might yield a computationally intractable mapping. Thus, as no surprise at this point in our narrative, we now expose a random scheme which works remarkably well. Let $M_1 : X_{n+1} = \frac{1}{3} X_n$ and $M_2 : X_{n+1} = \frac{1}{3} X_n + \frac{2}{3}$ denote two contractive maps M_1 and M_2 of the unit interval upon itself. Also let there be a weighted coin having probability P_1 for heads and $(1 - P_1)$ for tails. Starting with any initial iterate X_0 lying in the unit interval, we first toss our coin and then apply M_1 or M_2 as the coin is heads or tails, thus obtaining our first iterate X_1. Thence, we again flip our coin applying M_1 or M_2 as the coin is heads or tails to obtain iterate X_2, and so on. At every stage of iteration, this composite, nondeterministic mapping is a contractive map. Moreover, we may prove that all mapping iterates asymptotically approach and move along the collection of points formed by sequentially deleting middle thirds from the interval $(0, 1)$ and, surprisingly, that this approach does not depend on the specific value used for the probability P_1. In addition, one may prove that this 'deleted third' collection of points, call it C, is such that $C = M_1(C) \cup M_2(C)$, where $M_1(C)$ or $M_2(C)$ is the collection of points which results when the mapping M_1 or M_2 acts on all points in the collection C and where $M_1(C) \cup M_2(C)$ denotes the total sum of the points in either $M_1(C)$ or $M_2(C)$. A reader who finds this last sentence confusing may interpret it as simply saying that the collection of points C is a type of invariant under the composite mapping. Turning now to the more general case, if $M_1, \ldots, M_n$ denote n contractive maps and if $P_1, \ldots, P_n$ denote probabilities associated with a loaded, n-sided die, then the random mapping generated using the $[M_k, P_k]$ set as before yields a chaotic attractor A satisfying $A = M_1(A) \cup \ldots \cup M_n(A)$. Examples of such chaotic attractors are shown in figures 12.7–12.9. Additional, quite remarkable photographs of computer generated chaos appear in figures 12.10–12.13.

Now let us be given a digital or graphical recording of some noisy output signal from a chaotic system. Thence, let us regard this output record as the chaotic attractor of some non-deterministic mapping, as above. Our problem then is to find a mapping set $[M_k, P_k]$ which yields the specified attractor. As a first step, Barnsley and co-workers sought an approximate mapping whose chaotic attractor precisely matched the noisy output record only at a specified set of points. Let us now omit discussion of the intermediate steps and leap straight ahead to the heart of the matter; in essence, one is here engaged in curve fitting using chaotic 'curves' rather than straight lines or polynomial curves. The great surprise is the striking agreement between experimental and theoretical chaotic 'curves', both within and without the fitted interval. This work suggests that one may detect statistical regularities in chaos provided one uses chaos as a probe.

In all our discussions thus far, we have examined only the taming of chaos by chaos. Thus, let us now investigate the possibility that order can also be used to control chaos. To this end, recall that thermodynamics assumes all many-body systems to be chaotic, implying that most system motions are so complex they are not controllable by any macroscopic means. In consequence, when the macroscopically observable total energy U of a many-body system is found to increase by an amount δU, in general only part of this increase can be reckoned as macroscopic mechanical work δW done on the system; typically δW represents an external force causing an increase in a system position variable or a slight increase in outside pressure causing a decrease of system volume. The remaining part of the increase δU represents energy absorbed by the macroscopically uncontrollable chaotic system motions; the only thermodynamic way to determine this 'hidden' energy transfer is by taking the difference $(\delta U - \delta W)$. Here, $(\delta U - \delta W)$ represents a transfer of energy $\delta Q = (\delta U - \delta W)$ by nonmechanical means, δQ is called heat; typical heat transfers involve energy exchange due to a temperature difference between system and environment or due to absorption by the system of radiant energy such as sunlight. The equation $\delta Q = \delta U - \delta W$ is called the first law of thermodynamics, and it provides a very general statement of the conservation of energy at the macroscopic level. Not all energy transfers permitted by the first law actually occur in nature, however. The transfer $\delta W \Rightarrow \delta U$ followed by the transfer $\delta U \Rightarrow \delta Q$, that is, 'ordered energy' residing in controllable coordinates being transformed into the 'chaotic energy' of uncontrollable coordinates, is always allowed. However, the inverse sequence $\delta Q \Rightarrow \delta U$ followed by $\delta U \Rightarrow \delta W$, that is, chaos to order, is never observed. In this second sequence, energy is first placed into uncontrollable system motions; it cannot thence be totally regained from the system as work, adding ordered energy to the environment, because this would imply that a process in the macroscopic

Figure 12.9. Enlargement of a section of figure 12.8 showing that each small frosty 'branch' in the boundary of figure 12.8 also has 'branches', which in turn themselves have 'branches', etc. Chaos has structure on every scale. (Original slide courtesy Michael Barnsley; final photograph courtesy Vincent Mallette.)

Figure 12.10. This remarkable picture of chaos is reproduced with permission from *The Beauty of Fractals* (Springer-Verlag, Berlin, 1986), by H.-O. Peitgen and P. H. Richter, Center for Complex Dynamics, University of Bremen. Chaotic attractors frequently have a fractal, self-similar structure analogous to the 'deleted third' point set mentioned in the body of this paper. Specifically, we here see the magnificently fierce fractal dragon having self-similar structure on every scale – some see instead the heraldic British Lion, but these underprivileged few were never read fairy tales as children. This is the first of four animals we shall exhibit from the fractal zoo; all derive from chaotic mappings discussed by Peitgen and Richter.

Figure 12.11. The awesomely deadly, fractal scorpion who carries a lethal poison in each self-similar tail. (Reproduced with permission from Peitgen, H.-O. and Richter, P. H. (1986). *The Beauty of Fractals*. Springer-Verlag, Berlin.)

environment can control the uncontrollable system motions, a logical contradiction. Lord Kelvin stated this impossibility as follows: there is no thermodynamic process, no engine operating in a cycle, whose net effect on the universe is to convert heat δQ completely into work δW; this is Kelvin's statement of the second law of thermodynamics. The above remarks are only part of a larger body of thermodynamics theory which has met with such frequent and widespread success that many theorists tacitly regard it as universally valid. But prejudice to the contrary notwithstanding, analytically solvable many-body systems exhibiting highly ordered motions do exist, however sparsely, in the chaotic majority, and they most certainly do not obey thermodynamic laws; for example, electromagnetic radiation confined within an ideal superconducting cavity is an analytically solvable 'many-body' problem which lies outside thermodynamic law. All internal motions of such analytically solvable systems are macroscopically controllable; all their energy transfers may be treated as work; and all lie outside the range of validity of the second law of thermodynamics.

Suppose now that a brief interaction permits energy transfer from a chaotic to an analytically solvable system, before and after which both systems are isolated. In general, some of the energy transferred left the chaotic body as heat and, as internal energy of the analytically solvable system, is now completely available for doing useful work. No matter how thermodynamically heretical the previous sentence may sound, it is theoretically quite correct. Experimentally however, finding a suitable analytically solvable system may be extremely difficult. No

Figure 12.12. The comedic, fractal dog. Note that his face repeats
on every scale but that the repetitions are *not* self-similar!
(Reproduced with permission from Peitgen, H.-O. and Richter,
P. H. (1986). *The Beauty of Fractals*. Springer-Verlag, Berlin.)

Figure 12.13. The many-tentacled, fractal octopus from which there can be no escape. We have now seen but four of the many denizens residing in the fractal zoo; a more comprehensive pictorial catalogue appears in the opus by Peitgen and Richter. (Reproduced with permission from Peitgen, H.-O. and Richter, P. H. (1986). *The Beauty of Fractals*. Springer-Verlag, Berlin.)

system is truly isolated, and an analytically solvable system becomes at least weakly chaotic when coupled to the outside world. The essential point here is that the second law can always enter through the walls of a nonisolated system, although perhaps slowly if the coupling with the outside is weak. Indeed, even the physical pendulum, that paradigm of dynamical order, does not actually have the isolated, one degree of freedom motion so frequently assumed; specifically, the stately swing of a long, massive pendulum very slowly comes to rest as its internal degrees of freedom absorb most of the energy originally residing in its macroscopic oscillation. In summary, our entire problem now reduces to the study of relaxation times. Can the second law be kept at bay long enough to extract a 'forbidden' amount of energy as work. Nonlinear dynamics suggests that we need only find a suitable

analytically solvable system to attain our dream. Heretofore, over the centuries, those seeking to violate the second law have been quite silly. They have sought to construct a perpetual motion machine using systems which obey thermodynamics. Logically speaking, this is the equivalent of starting with the premise '2+2=4' and deriving the conclusion '2+2=5'. Nonlinear dynamics now offers us an escape from this historical and logical trap by revealing the existence of analytically solvable systems which do not obey the second law of thermodynamics. Figure 12.14 dramatises the confrontation between nonlinear dynamics and thermodynamics in facetious terms; however, the possible control of chaos offered by analytically solvable systems having ordered motions is truly no laughing matter.

Kolmogorov   Arnold   Moser

VS.

Clausius   Carnot   Kelvin

CAN THE 2nd LAW
SURVIVE 10 ROUNDS ?

Figure 12.14. A facetious dramatisation of the factual confrontation which exists between nonlinear dynamics and thermodynamics. The noted mathematicians, Kolmogorov, Arnol'd and Moser, enter the ring for nonlinear dynamics while Clausius, Carnot and Kelvin champion the second law of thermodynamics. (This composite was created by Franco Vivaldi (University of London), artist, with the help of Vincent Mallette (Georgia Tech), photographer.)

12.5   Chaos: an illustrative example

Few equations have received more attention in recent years than the discrete logistic equation

$$Y_{n+1} = \alpha Y_n(1 - Y_n), \qquad (12.1)$$

which maps the unit interval $(0,1)$ upon itself when $0 \leq \alpha \leq 4$. Equation (12.1) can be used to illustrate much contemporary pioneering work: the period doubling to chaos publicised by Robert May, the universal numbers and renormalisation theory of Mitchell Feigenbaum, and the period three implies chaos of Tien-Yien Li and James Yorke – for definitions and

details consult the review articles in the bibliography. In equation (12.1), as α varies there occurs a wondrously discontinuous alternation between types of order and chaos, but of greatest interest to us here is the full turbulence on the unit interval which occurs at $\alpha = 4$. In order to make this fully developed chaos transparently obvious, let us follow Stanislaw Ulam and John von Neumann in changing the dependent variable via the equation

$$Y_n = \sin^2 \pi X_n. \qquad (12.2)$$

Simple algebra then shows that, at $\alpha = 4$, equation (12.1) is equivalent to

$$X_{n+1} = 2X_n \ (\text{mod } 1), \qquad (12.3)$$

where (mod 1) means drop the integer part of each X_n. Like equation (12.1), equation (12.3) maps the unit interval upon itself. Equation (12.3) is a linear difference equation for which existence and uniqueness proofs are trivial. This equation thus has a unique solution passing through each initial iterate X_0. Indeed, the analytic form of this solution reads

$$X_n = 2^n X_0 \ (\text{mod } 1). \qquad (12.4)$$

A more imaginative form of the solution is found by writing X_0 as a binary digit string of zeros and ones. For example,

$$X_0 = 0.1110000100110111 \ldots . \qquad (12.5)$$

One then notes that future iterates of equation (12.3) may be obtained simply by sequentially moving the binary point to the right in equation (12.5) and dropping the integer part, since multiplying X_n by two in equation (12.3) is equivalent to moving the binary point one place to the right in equation (12.5). However, it is absolutely crucial to emphasise here that the obvious determinism, existence-uniqueness of solutions (12.4) and (12.5), becomes meaningful in human terms if and only if X_0 is specified through some auxiliary experimental or theoretical means. But is it, in fact, humanly possible to measure or compute the binary digit string for an X_0, in general? Upon recalling our earlier discussion of roulette wheel sequences, we immediately realise that the answer to this question is no; but for emphasis, we now repeat the old argument in terms of base two digit sequences. Specifically, glance again at the binary digit string for the X_0 of equation (12.5), but now regard this digit string as describing a particular infinite coin toss sequence in which one means heads and zero means tails. Viewed in this light, the set of all X_0 digit strings on the unit interval may be regarded as being identical with the set of all possible random coin toss sequences. In short, almost all X_0 digit strings are truly random and cannot be deterministically predicted by experiment or deterministically computed by theory. Equation (12.3) is now seen to generate sequences of truly random numbers, a fact which follows from the randomness of the X_0 digit strings. The deterministic algorithm of equation (12.3) thus merely transforms the randomness (missing information) of equation (12.5) into the randomness (missing information) of the 'orbital' X_n set. The full solution to equation (12.3) exists and is unique, but humans can never gain full knowledge of it. Surely, the silvery lustre of determinism, existence-uniqueness, and formal analytic solution is now visibly tarnished.

At this point in the discussion, an unwary reader may be starting to form a false conclusion; be warned that it is not merely some academic question regarding infinite digit strings or infinite precision which is crucial here. Suppose we initially specify the X_0 of equation (12.4) to one binary digit accuracy and suppose we then require only one binary digit accuracy in all future X_n iterates. From equation (12.5), we immediately see that we will have to add one additional binary digit (one bit of input information) to X_0 for each additional iterate X_n (one bit of output information) we obtain. Quite clearly, our output information is identical to our input information. Our algorithm does not compute, it copies (or translates). Hence, we cannot know the 'orbit' in whole or in part unless we are given the 'orbit' in whole or in part beforehand. Deterministic chaos is a mystery, perhaps the ultimate mystery which only the gods can understand.

Let us close this section by seeking to eliminate one final confusion which sometimes arises when a reader critically reviews our discussion of equation (12.3). We have seen that randomness of the X_n sequence generated by the difference equation (12.3) arises directly from the innate randomness of their X_0 digit strings. But randomness of initial condition digit strings occurs for all difference equations; why then aren't all difference equations chaotic? The answer is that a difference equation yields ordered or chaotic solutions depending on how it manipulates its random X_n digit strings. Some difference equations pass the initial condition chaos directly on to their X_n solution iterates while others do not. Equation (12.3) transfers the randomness while the difference equation

$$X_{n+1} = X_n + \beta \pmod 1 \qquad (12.6)$$

does not. Nonetheless, the solution iterates of equation (12.6) do pass certain, though not all, tests for randomness. Specifically, when β is irrational, it may be shown that the equation (12.6) difference equation is ergodic. Here, ergodicity means that the X_n solution iterates for each X_0 are densely and uniformly distributed on the interval (0, 1) and, more important, that

$$\lim_{N \to \infty} N^{-1} \sum_n f(X_n) = \int_0^1 dX f(X), \qquad (12.7)$$

where f denotes any well-behaved function. In words, equation (12.7) states that the 'time' or n-average of f on the left, where the sum runs from $n=1$ to $n=N$, equals the position or X-average of f on the right; over the decades, ergodicity has been regarded as a rather strong statistical property. However, the solution iterates of equation (12.6) are easily shown to be nonrandom. Let X_0' and X_0'' be two close initial conditions

generating the solution sequences X_n' and X_n''. Then, using equation (12.6), we may immediately establish that

$$[X_n'' - X_n'] = [X_0'' - X_0']. \qquad (12.8)$$

Thus, the initially small separation distance between X_0'' and X_0' is perpetually maintained by each (X_n'', X_n') pair. Alternatively stated, each small interval $(X_n'' - X_n')$ moves under iteration like a very short rigid rod; hence, the digital randomness associated with the multitudinous points inside the interval $(X_n'' - X_n')$ is clearly not being transferred into 'orbital' randomness. Briefly, equation (12.3) yields random behaviour; equation (12.6) does not, and thereby hangs a tale.

12.6 Quantum chaos: is there any?

Let us begin by recalling that quantum mechanics is widely regarded as the most general and fundamental description of nature yet devised by man. In this sweeping theoretical development, classical dynamics is perceived as merely a natural consequence – a limiting case – of quantum mechanics, a concept usually referred to as the correspondence principle. This notion obviously implies that, since chaos exists at the classical level, it must also be contained in the underlying quantum mechanics. Consequently, most investigators have regarded the existence of quantum chaos as an indisputable reality, despite the fact they could give no canonical definition of it. But permitting the definition of chaos to be 'up for grabs' is surely a form of madness, and blessed with such insanity how would we recognise chaos even if we met it? In the beginning, this lack of definition may have served the useful purpose of encouraging exploration and innovation; but alas, it now provides only an anarchy in which every man contributes and defends his private definition in the literature. Order should be brought to this definitional chaos, and indeed it can be. For classical chaos, there is one definition which subsumes all others and which survives the transition to quantum mechanics without alteration: chaos is merely a synonym for randomness. We adopt this meaning for chaos throughout this article. With this definition in hand, we may legitimately inquire if there is any chaos (randomness) in quantum dynamics beyond that of the type contained in the probability density $\psi^* \psi$. If we exclude from consideration the innate randomness of $\psi^* \psi$, and the like, what opportunities for chaos remain? There are two: randomness in the Schrödinger time evolution or flow of the wave function ψ or randomness in the quantum eigenvalues, eigenfunctions and matrices.

As we turn to these matters, let us first seek to motivate our treatment of them. We have already mentioned that chaos may be the beginning of the third scientific revolution of this century. In this section, we go further and suggest that the first major battle of this revolution is even now being waged over the existence of quantum chaos. Should chaos not be found in quantum mechanics, then an earthquake in the foundations of physics appears inevitable, say about magnitude twenty on the Richter scale. Thus, we are here addressing extremely fundamental, 'high stakes' issues. Unfortunately, from the expositor's viewpoint, these questions are as complex and subtle as they are vital. Hence, if we are to provide a meaningful summary which does not short-change the reader, we must here permit discussions to be a bit longer and to contain a few more technical 'nuts and bolts' of the subject than earlier sections. For this, we ask the reader's indulgence and perseverance. Let us now review the case for and against chaos in the time evolution or flow of the wave function ψ.

For finite particle number, spacially bounded, undriven (time independent) quantum systems, the general solution $\psi(x,t)$ to Schrödinger's equation reads

$$\psi(x,t) = \Sigma A_n U_n(x) e^{-2\pi i E_n t/h}, \qquad (12.9)$$

where x stands for all space variables, t is the time variable, the A_n are expansion coefficients, the $U_n(x)$ are the energy eigenfunctions, the E_n are discrete energy eigenvalues, h is Planck's constant, and the sum is over all allowed integer values of n. Because the E_n are discrete, the $\psi(x,t)$ of equation (12.9) is an almost periodic function of time meaning that $\psi(x,t)$, having achieved a value, makes near returns to this value throughout all time; $[\sin t + \sin t\sqrt{2}]$ is a simple but typical example of an almost periodic function. Because of this recurrent behaviour, $\psi(x,t)$ cannot approach and remain at thermodynamic equilibrium. Again, because the E_n are discrete, the time evolution of $\psi(x,t)$ is, like the evolution of equation (12.6), no more than ergodic which means that correlations do not decay. In consequence, the time evolution of $\psi(x,t)$ for finite, bounded, undriven quantum systems exhibits no chaos despite the fact that most of their corresponding classical systems do. But worse, it may be shown that the time flow of $\psi(x,t)$ remains ergodic and nonchaotic even when h is arbitrarily small but not zero. Thus, if equation (12.9) is a quantum solution for a classically chaotic system, letting h tend to zero cannot convert ergodic quantum motion into random classical motion as seems required by the correspondence principle. Similar remarks are valid when particle number or spacial size tend to infinity. However, in the limit as h^{-1}, system size, or system particle number tends to infinity, the energy spectrum of equation (12.9) is found to be continuous rather than discrete, and the almost periodic behaviour of equation (12.9) is lost. Here, the time evolution or flow of $\psi(x,t)$ can be shown to be mixing, where, intuitively speaking, a flow is mixing provided almost all close initial states swirl apart in time as if being stirred by an egg beater. For a mixing flow, as opposed to an ergodic one, all correlations tend to zero with time and an approach to equilibrium can occur. However, the correlations for a mixing flow need approach zero only as some inverse power of the time, whereas, for a chaotic flow, correlations decay exponentially. In short, a continuous spectrum in a quantum system guarantees mixing but asserts absolutely nothing definite about randomness. The fact that randomness does not occur at all in the flow of $\psi(x,t)$ for a finite, bounded, time independent quantum system, and the fact that randomness cannot be assured even in the limit as h^{-1}, system size, or particle number tend to infinity is profoundly troubling. Moreover, these facts cast a heavy pall over our basic assumption that classical mechanics is a natural consequence (or limit) of quantum mechanics.

Let us now continue our search for quantum chaos by examining time driven systems. The flow of $\psi(x,t)$ for a finite particle number, spacially bounded, time driven quantum system can have a continuous spectrum and can therefore be mixing. Is it possible that here, at last, a continuous spectrum and mixing will be found to go hand-in-hand with randomness? Unfortunately, gaining insight into this question is a nontrivial matter, but the problem is eased slightly by the fact that we are here asking a quite generic question whose answer may be insensitive to the particular system being studied. Nonetheless, care is still required. Analytic attack on even a simple time dependent partial differential equation is quite difficult in general, and a brute force numerical attack fares little better. To circumvent such adversity, one needs a gimmick, a trick, or more important, a cleverly selected system which can be analysed. Some years ago, Boris Chirikov discovered just such a time driven system. Specifically, he first considered the classical motion and then the quantal motion of a plane rigid rotator subjected to periodically applied, impulsive kicks of such brief duration they may be regarded as instantaneous. Hence, at almost all times, this system is merely the analytically solvable, undriven, free rotator; each impulsive kick serves only to instantaneously change the state of the free rotator. It is now straightforward to derive simple difference equations which relate system states just before a kick to system states just before the previous kick. Numerical integration then does the rest.

Chirikov first showed that the classical kicked rotator exhibits a transition to chaos as a certain system parameter increases above a critical value. For our purposes, the most significant feature of kicked rotator chaos is the rotator's diffusive absorption of energy from the periodic driving kicks, detected as an energy growth which (on average) is proportional to system time. One now needs to recall that a diffusing quantity, energy in this case, must satisfy the so-called diffusion equation, a partial differential equation which has been shown to be the continuum limit of a discrete random walk. The meaning is now clear: diffusion implies randomness. Thus, Chirikov next sought to observe diffusive energy absorption in the quantum kicked rotator. For suitable system parameter values, he did in fact find the rotator energy growing proportional to system time, apparently mimicking the chaotic behaviour of its classical twin. But then came the real shock. Diffusive energy absorption continued only for a brief time t_B, called the break time, following which the growth of energy with system time essentially ceased. Moreover, upon integrating the original quantum 'orbit' from $t=t_B$ back to $t=0$, one easily regained the initial state. This result is as contradictory to the notion of diffusion as it would be to find a random walker who always proceeds backward through each step of his original walk when he is turned around after the nth step. In essence, these two facts definitively establish that no diffusive energy absorption and no randomness have been observed in the quantum rotator. One might seek to explain the lack of chaos here by supposing that the kicked rotator spectrum is, in some sense, not continuous, but alas, Giulio Casati and Italo Guarneri have proven that rotator models of this type do, under suitable circumstances, exhibit a continuous spectrum. Finally, there is no *a priori* reason to suspect that Chirikov's kicked quantum rotator is atypical of the general case. Taken in concert, these findings considerably dim the hope that chaos accompanies mixing in finite, driven quantum systems. In consequence, hope also dims for finding chaos in infinite, undriven systems, since a driven, finite system can be regarded as a 'driven', finite part of the infinite system. Moreover, when these conclusions are added to the earlier negative results for finite, undriven systems, chaos appears unlikely to occur in any Schrödinger equation flow! For some, an appealing way to reconcile this issue is to assert that the lack of chaotic flows in quantum mechanics merely reflects the lack of chaotic flows in nature at the microscopic level. But if this assertion be true, how do we dismiss quantum statistical mechanics which states that chaotic flows do occur in nature at the microscopic level? Whatever one's bias in these matters, the existence of significant, profound and unsettled questions regarding quantum chaos is difficult to deny.

Having found no demonstrable chaos in the time flows of quantum systems, let us now consider the remaining opportunities for chaos: eigenvalues, eigenfunctions and matrices. In this context, chaos would mean that one or more of these quantities is a random variable. We begin by examining the evidence regarding chaos in quantum eigenvalues. If the energy eigenvalues for a system are random, they would surely be highly sensitive to and rapidly change with a small variation of perturbation on the system. Ian Percival and Neil Pomphrey numerically investigated this possibility for a quantum system composed of two harmonic oscillators weakly coupled by very simple nonlinear forces. In the parameter range for which the corresponding classical system was chaotic, Percival and Pomphrey observed that energy eigenvalues were indeed highly sensitive to slight variation in nonlinear coupling, thus lending support to the existence of quantum chaos. A decade or so later, however, William Reinhardt and colleagues straightforwardly computed many of these sensitive energy eigenvalues using semi-classical perturbation theory. Eigenvalues are not random or chaotic when they can be straightforwardly computed, sensitivity to perturbations notwithstanding; thus, no quantum chaos is apparent here. But let us continue the search.

If the energy eigenvalues E_n for a given quantum system were not only random but independent, meaning that knowledge of one E_n says nothing about the value of another, then we would expect the observed relative frequency of occurrence $F(S)$ for nearest neighbour energy level spacings $S=(E_n-E_{n-1})$ to be the probability distribution for the random variable S. The above notion can be made less abstract by likening it to the toss of a random coin. For a brief moment, regard the variable S as taking on only the values ± 1, corresponding to heads or tails for the toss of a random but biased coin. Then, by observationally measuring the relative frequency of occurrence of heads or tails during long sequences of tosses, we thereby determine the unequal probabilities for heads or tails. Returning now to energy level spacings, this relative frequency of occurrence $F(S)$ has been extensively investigated numerically and/or analytically for a variety of quantum systems whose classical counterparts were either analytically solvable or else chaotic. The intent here was to establish the existence of quantum chaos and to show that classical and quantum chaos were intimately linked.

Turning now to specifics, quantum systems which classically are analytically solvable have been shown to yield an $F(S)$ for energy level spacings given by

$$F(S) \simeq e^{-aS}, \qquad (12.10)$$

indicative of a Poisson probability distribution; energy level spacings for analytically solvable systems thus appear to be as random as radioactive decay. On the other hand, quantum systems which classically are chaotic yield

$$F(S) \simeq S^n e^{-bS^2}, \qquad (12.11)$$

similar to a skewed Gaussian or Wigner probability distribution. These results have frequently been offered as proof of chaos in the quantum eigenvalue spectrum. Moreover, when a classical system undergoes the transition order to chaos as some system parameter varies, its quantum counterpart undergoes the transition Poisson to skewed Gaussian in its frequency of occurrence $F(S)$. This parallel between classical and quantum behaviour led some to believe that the historically vague notion of quantum chaos had, at last, been given a firm foundation. Would that nature were so simple.

The fly in the above soothing ointment is that $f(S)$ is not always a probability distribution. To illustrate, consider the digit string C given by

$$C=0.01234567890123456789012345678901234567\overline{89} \\ \cdots \cdots \qquad (12.12)$$

Quite clearly, each digit $0,\ldots,9$ in this sequence appears with relative frequency 10^{-1}. Moreover, the relative frequency of occurrence of digits in equation (12.12) is identical to that of a digit string having a uniform probability distribution. But, in fact, the sequence in equation (12.12) is obviously periodic; it is, therefore, highly predictable and quite nonrandom. Specifically, the $F(S)$ for digits in equation (12.12) is most certainly not the probability distribution for the digits of equation (12.12). As an example having greater physical relevance, consider a point particle freely moving inside a plane rectangular box. Its quantum energy levels are given by

$$E_{mn} \simeq \alpha m^2 + n^2. \qquad (12.13)$$

Casati and Guarneri numerically computed $F(S)$ for this analytically solvable quantum system using 10^5 energy levels and, as expected, obtained the Poisson distribution $F(S) \simeq e^{-aS}$. But then, using algorithmic complexity theory, these investigators proceeded to prove that the E_{mn}-sequence is, in fact, not random at all. Unfortunately, comparable rigour has not been applied to the quantum energy level spacings for a system which is classically chaotic. Here, analytic formulas for the spacings are not available, and numerical schemes strain to obtain 10^3 levels. It would, therefore, be highly premature to assert that quantum chaos appears in such eigenvalue data. Nonetheless, these results do reveal an interesting fact: when a classical system makes the transition, order to chaos, the corresponding quantum system also makes a transition, but not necessarily to chaos.

Let us now direct our attention to randomness in quantum energy eigenfunctions. First, recall that each orbit for an isolated, chaotic classical system in general densely and ergodically covers its energy surface, implying that states of equal energy are equally likely. Thus, if randomness occurs in an energy eigenfunction $U_n(x)$ for an isolated quantum system whose classical equivalent yields chaotic orbits, then one expects the probability, derived from $U_n(x)$, for measuring a specified value of the position–momentum pair (q,p) is essentially zero unless this (q,p) value corresponds to a point on the classical energy surface $E(q,p) = E_n$, where $E(q,p)$ is the classical energy expression and where E_n is the quantum energy eigenvalue associated with $U_n(x)$. The agreement here between classical and quantal behaviour is expected to improve as the quantum number n increases. After this prelude, we can announce that A. I. Shnirelman has verified the above expectation for the energy eigenfunctions of a quantum particle moving freely on a billiard table having the shape of a stadium, a system whose classical motion is known to be fully chaotic. Michael Berry and colleagues have used semi-classical quantum mechanics to extend Shnirelman's results to arbitrary quantum systems whose classical counterparts are chaotic. Charting his course along a different tack directed toward a frontal assault upon the problem, Boris Chirikov invokes the fact that an eigenfunction $U_n(x)$ is random if, when $U_n(x)$ is expressed as a series having the form $U_n(x) = \Sigma B_{nk}\Phi_k(x)$ analogous to a Fourier series, its expansion coefficients B_{nk} are random; Chirikov then presents evidence that randomness of the B_{nk} can actually be observed in a specific atomic system. Along related lines, Moshe Shapiro and co-workers argue that energy eigenfunctions will be random if the values of $U_n(x)$ at different spacial points rapidly become less and less correlated as the distance between points increases. This notion is formalised in terms of spacial correlation functions. Shapiro also presents supporting evidence for his views.

The works of Shnirelman, Berry, Chirikov and Shapiro have one feature in common; each proposes a test for randomness which is necessary but not sufficient. Specifically, the tests proposed by Shnirelman and Berry prove only ergodicity, a necessary but not sufficient condition for randomness as we have seen earlier; this work is certainly significant but it says nothing definitive about quantum chaos. The test proposed by Shapiro is not without merit, but it is currently so lacking in discriminatory power that it indicates randomness even in analytically solvable quantum systems. At first glance, Chirikov's test appears both necessary and sufficient. It is certainly true that an eigenfunction $U_n(x)$ is random if and only if its expansion coefficients B_{nk} are random. But this only shifts the question of randomness from the $U_n(x)$ to the B_{nk}; thus, what evidence does Chirikov present indicating that the B_{nk} are, in fact, random? Why, evidence that involves a necessary but not sufficient condition, of course. The fault here lies not with the investigator but with the delicate subtleties of proving randomness. If an eigenfunction is not random, then humans are usually able to find a proof. But if it is random, even certified to be such by a god, then humans can do no better than accept the god's word for it; for man is incapable of proving that a given object is random, be it sequence or function – more details on this will be presented later. But fortunately or unfortunately, depending on one's bias, the eigenfunctions of Schrödinger quantum mechanics appear not to involve such delicate issues; Eric Heller has recently discovered that the quantum eigenfunctions for a particular classically chaotic system are, in general, not even ergodic, much less random. No known evidence indicates that Heller's results are atypical of the general situation. Thus, as with quantum eigenvalues, it would be highly premature to suggest that quantum eigenfunctions exhibit chaos. Moreover, should eigenfunctions and eigenvalues not be chaotic, as seems likely, then neither are quantum matrices.

This completes our rummage through the more conspicuous 'nuts and bolts' of quantum chaos. As an aid to evaluating claim and counterclaim, let us now pause to review the bidding. From its inception to this very day, the search for quantum chaos has borne a disfiguring birthmark seldom mentioned in polite society; quantum chaos may exist but a generally accepted definition does not. Lacking a precise definition, those seeking to discover chaos in quantum mechanics have adopted the same intuitive terms – 'erratic', 'irregular', 'disordered' or 'seemingly unpredictable' used earlier in classical chaos. Necessity, not choice, dictated this selection, for most of the classically precise terms are meaningless in quantum mechanics. Privately, many would admit that chaos, classical or quantal, had to mean randomness, but none wished to take on the task of stating in print how mathematical randomness could arise in strictly deterministic quantum systems. Their fears were groundless; the determinate–random question was answered some years ago by algorithmic complexity theory, but this fact is still not widely known. Again, we shall return to this point in a moment. Actually, in the deterministic time evolution of Schrödinger's equation for the wave function ψ, the question of deterministic randomness has, strictly speaking, never arisen. Finite, bounded quantum systems, whether driven or undriven, whether small or large in particle number or spacial size, simply give no hint of making a proper approach to equilibrium. This lack in the quantum flows encounters a conflict with laboratory reality long before the question of deterministic randomness can be raised. Eventually, our ability to analytically treat increasingly sophisticated quantum systems will force an encounter with at least the border of randomness, but that lies in our future.

On the other hand, the definitional confusion regarding quantum chaos has had a quite deleterious impact on researchers studying the question of chaos in the quantum eigenfunction–eigenvalue problem, forcing them, in effect, to attack the problem 'sideways'. Specifically, they have focused their attention on properties a presumed 'erratic', 'irregular', 'disordered', 'seemingly unpredictable' quantum system might possess. Then, if any quantum systems could be shown to exhibit such properties, these properties were immediately nominated as the defining properties of quantum chaos. Unfortunately, this methodology is equivalent to asserting that, since a duck has a broad yellow bill, web feet and oily white feathers, then any fowl having these properties is a duck. Quite obviously, however, a fowl is a duck if and only if it exhibits all the properties of a duck, just as a quantum system is random (chaotic) if and only if it exhibits all the properties of randomness. But if we adopt such stringent criteria, will we ever be able to prove randomness or its lack in the quantum eigenfunction–eigenvalue problem? The answer lies in algorithmic complexity theory, which asserts that the well-known deterministic procedure for computing quantum eigenfunctions and eigenvalues will pass all tests for randomness provided that the bit length of input, program plus data, is *of necessity* approximately the same as that of the output. Here, a bit is one digit in the binary encoded input or output. This algorithmic criterion is precisely the one previously used to definitively establish chaos or its lack for classical orbits. Quite generally, if the bit length of input is appreciably smaller than that of the output, then the quantity or object being computed is not random. However, if the bit lengths of input and output are approximately equal, randomness is not guaranteed; we must still prove the necessity of this equality; that is, we must prove that no shorter input algorithm exists which computes the same output, a difficult task indeed. In particular, even though the randomness of the quantity or object being computed be certified by a god, actually proving that input cannot be significantly less than output is a Gödel–Chaitin question whose answer lies beyond human capability. Nonetheless, there is a conclusion, well within human reach, that is even better! In particular, we are capable of proving the generic

statement: this quantity or object is random, probability one. For example, we may easily prove that almost all digit sequences are random despite the fact that we cannot prove any individual sequence to be random; in short, a given sequence is random, probability one. Quite frequently, proving the generic, rather than the specific, is not only easier but also more significant. Returning now to the quantum eigenfunction–eigenvalue problem, the complexity theory analysis is not yet completed, but no evidence supporting quantum chaos has appeared thus far.

Wherever one looks, a cleansing flood tide sweeps over quantum mechanics. Should quantum chaos not stand revealed at ebb tide, how are we to explain the historical development of a quantum theory containing no chaos in its time flows, its eigenfunctions, its eigenvalues, or its matrices? First, let us recall that the founding fathers of quantum theory were struggling to explain analytically solvable systems such as the hydrogen atom and the harmonic oscillator; insofar as they thought about 'quantum chaos' at all, they relegated it to what is now called quantum statistical mechanics. Groping to find a description of an unseen and unseeable microscopic world and using the correspondence principle as guide, the founders quite understandably selected the simplest theory which met the constraints. Moreover, since examples of classically chaotic systems were essentially nonexistent at the time, the pioneers could not have checked their theory against classical chaos had they wished. In retrospect, is it not reasonable to suppose that the founders picked a theory too simple to contain chaos? A paradigm for classical chaos is the simple difference equation $X_{n+1} = 2X_n$ (mod 1); a paradigm for quantum mechanics may well be the equally simple difference equation $X_{n+1} = X_n + \beta$ (mod 1). The randomness of real number digit strings is available to both these models; but as we know, only one uses it. Thus far, Schrödinger's equation has been found to mimic only nonrandom mixing and/or ergodic difference equations. Note here that the linearity of Schrödinger's partial differential equation is not the source of the problem; Liouville's equation, the partial differential equation equivalent of Newton's equations, is also linear but nonetheless exhibits chaos. If Schrödinger's equation is too 'simple', one fault must reside in the fact that, for undriven systems, many product wave function solutions having the form $\psi = F(x)G(t)$ always exist. Indeed, the complete solution can always be written as a sum of these product solutions – see equation (12.9), for example. This analytic simplicity has long been regarded as a virtue of quantum mechanics, and for analytically solvable systems it most certainly is. But as our horizons expand to include chaos, this virtue becomes a vice. Last but not least, the fact that the ∇^2 operator in Schrödinger's equation is a 'smoothing operator' is another virtue turned vice; in this context, smoothing means that the value taken by $\nabla^2 F(x,y,z)$ at the point (x,y,z) is an average over the values taken by $F(x,y,z)$ in some neighbourhood of the point (x,y,z). To fully appreciate this point, recall that the solutions $F(x,y,z)$ to Laplace's equation $\nabla^2 F(x,y,z) = 0$

are so analytically smooth that they possess derivatives of every order. The fabric of chaos, on the other hand, is grainier, is more robust, and has greater resistance to the wear and tear of an abrasive world.

Although the founders of quantum mechanics relegated chaos to quantum statistical mechanics, they firmly believed that, with sufficient ingenuity, chaos and quantum statistical mechanics as well could be derived from the underlying quantum mechanics. Most especially, they had no confusion regarding the concept we now call quantum chaos, for they knew quite well that the basic question of chaos and/or statistical mechanics, whether classical or quantal, asks how deterministic laws can ever yield true randomness. Thus, the seemingly new definition of quantum chaos presented in this article is actually decades old; Maxwell, Boltzmann, Gibbs, Fermi and Pauli would immediately recognise an old friend: chaos is merely a synonym for randomness.

But would they so easily agree that there might be no chaos in quantum mechanics and therefore that quantum mechanics is not random enough? Quite doubtfully; however, they would nonetheless appreciate the deep significance of and eagerly await results from ongoing theoretical and laboratory investigations concerning the ionisation of hydrogen by microwave radiation. Here, the hydrogenic electron is initiated in a highly excited energy state E_n with $n \simeq 40$. Then microwave photons, with energy much less than that required to cause the first upward electronic transition, irradiate the hydrogen atom. Intuitively, one anticipates that the electron will random walk or diffuse up the energy ladder toward ionisation. Indeed, over a certain parameter range, both laboratory experiment and semiclassical quantum calculations indicate that diffusive energy absorption may be occurring. Moreover, in some but not all parameter ranges, numerical solution of Schrödinger's equation provides similar results.

But life is never simple, for all these Schrödinger wave function calculations exhibit the same fatal defects as did those of Chirikov's kicked rotator. Although these time driven quantum systems can mimic diffusive energy growth over some time interval, this diffusive absorption sometimes ceases at a break time t_B. But much worse, starting with the wave function $\psi(x,t)$ at any time t, Schrödinger's equation can be integrated backward in time, easily regaining $\psi(x,0)$. As mentioned earlier, a diffusive random walker does not in general return to his initial position in n steps after being turned around at the nth step. Thus the lines of battle over quantum chaos (if any) have been drawn; combat has begun. As cold fact without melodrama, the fate of modern physics may hang in the balance, for might it not be possible that the microwave ionisation of hydrogen is the twentieth century's equivalent of the nineteenth century's electromagnetic radiation in a black body cavity? These then are the 'high stakes' issues which led us to describe selected 'nuts and bolts' of chaos with care. It is hoped that the reader now shares with us the drama and the excitement of our time.

At present, the evidence weighs heavily against quantum chaos, but no matter how damning, it is only circumstantial. Pending a definitive proof, a fair and impartial jury would be forced to render the Scottish verdict: *not proven*. Therefore, suppose we now assume that all forms of chaos have been found to exist in quantum mechanics; would the third scientific revolution then have to be called off for lack of a quorum? Most definitely not, for quantum mechanics still suffers the same terminal illnesses as Newtonian dynamics. Specifically, quantum mechanics is firmly rooted in the uncomputable number continuum and firmly supports the god-like notion that any single continuum variable can, in principle, be computed and/or measured to arbitrary accuracy. It assumes the existence of essentially noninteracting, perfect clocks and metre sticks. Indeed, according to certain theories of quantum measurement, the quantum clocks and metre sticks are identical to the Newtonian ones. Regardless, if chaos is ubiquitous in nature at the microscopic level, whether or not in quantum mechanics, then there is an uncomputable and uncontrollable noise level on the smallest scale. Intuitively here, one dimly perceives the existence of a small, irreducible noise level in any measurement implying a generalised uncertainty principle $\delta A / A \geqslant \alpha$ for each single observable A, where δA is the uncertainty in A and where α is a new fundamental, dimensionless constant of nature. The winds of change push us toward this uncertain future.

12.7   That we should be mindful of it

The present manuscript is based on an elementary review talk presented to general audiences around the world. While the flavour and the simplicity of the original talk have, for the most part, been retained, this written version has taken on an independent existence and has metamorphosed into an open letter addressed not only to those untutored in chaos but to the experts as well. Specifically, I here seek to rally the amateurs, the dilettantes, the connoisseurs, the aficionados and the disciples of chaos to the challenge facing us all. Chaos now presages the future as none will gainsay. But to accept the future, we must renounce much of the past, a formidable challenge indeed. For as Leo Tolstoy poignantly recognised, even brilliant scientists can seldom accept the simplest and most obvious truths if they be such as to contradict principles learned as children, taught as professors, and revered throughout life as sacred ancestral treasures. This is the dichotomy chaos presents. Should we cling to the past and present truths, contributing to a gradual and comfortable evolutionary chaos, or should we iconoclastically seek out and prove those dimly seen future truths toward which the revolutionary chaos points. As the evangelist of chaos – some say bishop, some say guru – I am personally committed to chaos as revolution. In consequence, all these words have been but an evangelist's invocation, a prelude to the future, assuring the faithful that the roots of tomorrow visibly entwine the hours of today. Even so, straight is the gate and narrow is the way; through the ages, the human mind has ever been asked to solve the unsolvable, predict the unpredictable. But rejoice, under such circumstances would nature have evolved a human brain containing no chaos to ease the burden? As ever, literature anticipates science. Nietzsche speaks:

> Yea verily, I say unto you
> a man must have chaos yet within him
> to birth a dancing star.

Figure 12.15. Proposed logo for the Third International Conference on Quantum Chaos held in Como, Italy, 22–26 September 1986, emphasising that existence, or its lack, of quantum chaos is currently the burning issue of this subject area. Unfortunately the idea was dropped. (Franco Vivaldi (University of London) created the logo; Vincent Mallette (Georgia Tech) provided the photograph.)

Further reading

Alekseev, V. M. and Yakobson, M. V. (1981). *Phys. Rep.* **75**, 287.
Although much of this article is written at the highest level of mathematical sophistication, its first six pages provide a readable introduction to the various 'measures' of classical chaos, including algorithmic complexity theory and randomness.

Bayfield, J. E. and Koch, P. M. (1974). *Phys. Rev. Lett.* **33**, 258.
This progress report of Bayfield and Koch as well as the report of Casati *et al.* (1984) seek to determine if a microwave excited hydrogenic electron 'diffuses' up to the continuum and ionises. This ongoing work constitutes the first sharp test of whether or not chaos (randomness) exists in quantum mechanics.

Berry, M. (1978). *Regular and Irregular Motion in Topics in Nonlinear Dynamics, AIP Conference Proceedings*, Vol. 46. American Institute of Physics, New York.
This magnificently written, elementary review of analytically solvable, nearly solvable and chaotic systems is widely regarded as the best introduction to conservative chaos now available.

Casati, G. (1985), ed. *Chaotic Behavior in Quantum Systems*. Plenum Press, New York.
These are the conference proceedings of the first conference restricted to quantum chaos. They present an interesting picture of a subject area at its birth.

Casati, G., Chirikov, B. V. and Guarneri, I. (1985). *Phys. Rev. Lett.* **54**, 1350.
This readable paper deflates the notion that analytically solvable systems have random energy spectra.

Casati, G., Chirikov, B. V. and Shepelyansky, D. L. (1984). *Phys. Rev. Lett.* **53**, 2525.

Casati, G. and Ford, J. (1979), eds. *Stochastic Behavior in Classical and Quantum Hamiltonian Systems, Lecture Notes in Physics*, Vol. 93. Springer Verlag, New York.
These are the proceedings of the very first conference on chaos. Much of the work described here is still under active investigation.

Chaitin, G. J. (1982). *Intl. J. Theor. Phys.* **22**, 941.
This paper provides a new and understandable meaning for Gödel's theorem on undecidable propositions and exposes its link to algorithmic complexity theory.

Chaitin, G. J. (1987). *Algorithmic Information Theory*, Cambridge University Press, Cambridge.
This remarkable text reveals that arithmetic contains a countable infinity of independent undecidable propositions. Moreover, randomness is shown to occur at a level no higher than number theory.

Ford, J. (1983). *Phys. Today* **36**, (4), 40.
This elementary review resolves the seeming paradox between determinism and randomness using algorithmic complexity theory.

Ford, J. (1985). In *Chaotic Dynamics and Fractals*, Barnsley, M. F. and Demko, S. G., eds. Academic Press, New York.
This review presents practical, physically interesting applications of algorithmic complexity theory. It seeks to bring the mathematically abstract notions of Gödel's theorem and algorithmic complexity theory into the mainstream of theoretical physics.

Garey, M. R. and Johnson, D. S. (1979). *Computers and Intractability*. W. H. Freeman, New York.
This book is an elementary introduction to the startling theory – computational complexity theory – which arises from asking how much clock time a computer needs to run a program to completion.

Helleman, R. H. G. (1980). In *Fundamental Problems in Statistical Mechanics*, Cohen, E. G. D., ed., Vol. V. North-Holland, Amsterdam.
This excellent, elementary review of dissipative chaos nicely complements the Berry (1978) article; moreover, the exhaustive reference list appended to this paper is without parallel.

Heller, E. J. (1984). *Phys. Rev. Lett.* 53, 1515.
Heller here deflates the notion that quantum eigenfunctions for classically chaotic systems are random.

Mehta, M. L. (1967). *Random Matrices*. Academic Press, New York.
This book and the following one by Porter describe the theoretical effort to characterise complicated energy level spectra when the underlying fundamental theory is unknown.

Porter, C. E. (1965), ed. *Statistical Theories of Spectra: Fluctuations*. Academic Press, New York.

13 Conceptual foundations of quantum mechanics

Abner Shimony

13.1 Plan

Quantum mechanics is a framework theory which, according to prevailing present opinion, applies to every kind of physical system. In order to use quantum mechanics for the purpose of predicting or explaining physical phenomena, it is necessary to specify the compositions and the interactions of the systems involved in the phenomena. Most of the striking implications of quantum mechanics – such as tunnelling, electron diffraction, superfluidity, and vacuum fluctuations – cannot be derived without such information about the systems involved. Except for illustrations and citations of crucial experimental evidence, however, our concern will not be with special systems, but rather with the radical innovations of the quantum mechanical framework itself.

Section 13.2 will summarise these radical innovations: objective indefiniteness, objective chance, objective probability, potentiality, entanglement, and quantum nonlocality. The words which have just been used will be explained and related to the standard quantum mechanical formalism. Some classically baffling phenomena will be seen in Section 13.3 to be quite natural in the light of these innovations.

In Section 13.4 we shall assess proposals that the quantum mechanical formalism, with its almost miraculous predictive power, can be retained without abandoning the general conceptual framework of classical physics. The most important of these proposals is the family of hidden variables theories, which assert that the quantum state of a physical system is an incomplete description, in need of supplementation. The experimental failure of the most important subclass of hidden variables theories – those which are 'local', in a sense which will be spelled out – provides the strongest evidence that the conceptual innovations of quantum mechanics are inseparable from its predictive power.

In Section 13.5 we shall examine the problem of 'the actualisation of potentialities', also known as the problem of 'the reduction of the wave packet'. This problem arises from the linearity of quantum dynamics, and it is deeply embedded in the present structure of quantum mechanics. It has often been claimed that this problem is specious, arising from a narrow or inadequate representation of the measuring process. We shall argue, however, that the problem is a fundamental anomaly, which cannot be lightly dismissed. Some proposals for solving this problem by postulating nonlinear or stochastic modifications of quantum dynamics will be considered, and the difficulties of these modifications will be discussed. We shall also discuss the suggestions for solving this problem which are contained in Everett's and Bohr's interpretations of quantum mechanics.

Our conclusion is that the main conceptual innovations of quantum mechanics are probably embedded permanently in physical theory, but that some further radical innovation will probably have to be made.

13.2 The conceptual innovations of quantum mechanics

The concepts of indefiniteness, chance, probability, and entanglement

When the general constitution of a physical system is given, there remain many contingencies concerning the system which are not yet specified: e.g. where the system is located relative to a frame of reference, what its velocity is relative to the frame, what its energy is, what the relative motions are of its parts (if it has parts). Some of these contingencies have very many possible values, and even a continuum of possible values – for instance, the position. Some contingencies, however, have only two possible values, which can be labelled 'truth' and 'falsity'. For example, it is either true or false that the centre of mass of the system is located within a specified region R. Our discussion will be greatly simplified by restricting our attention for the most part to the two-valued contingencies. The suggestive word 'eventuality' has been proposed to refer to a two-valued contingency, and we shall adopt this usage (although other words, such as 'proposition', are also in common use).

According to classical physics, the specification of the state of a system determines the truth or falsity of each of its eventualities (or, more precisely, of those eventualities which pertain to it alone and do not concern its relations to other systems). By contrast, quantum mechanics holds that a maximal specification of a system – its state – does not assign a definite truth or falsity to each of its eventualities. When any state of the system is given, there exist eventualities which do not have a definite truth or falsity, not because of ignorance on the part of some or all human beings, but because they are *objectively indefinite*. (It should be noted that we use the term 'state' to refer to a maximum specification of the system, which cannot be supplemented without contradiction. Sometimes the expression 'pure state' is used to refer to a maximal specification and 'mixed state' refers to less than maximum specifications, as in the descriptions which are common in statistical mechanics.)

Suppose now that an eventuality *e* is indefinite in the state *S*, and that the system is subjected to a procedure – possibly an experimental procedure, but conceivably a natural process not involving human observers – such that a definite truth value is elicited. According to quantum mechanics the outcome obtained is a matter of *objective chance*. Not only is the outcome not determined by *S*, but it is not determined by *S* in conjunction with the states of all other systems entering into the procedure. The chance character of the outcome, furthermore, is not a matter of ignorance on the part of one or all observers; it is a property of the physical situation itself. The concept of objective chance just mentioned presupposes the concept of objective indefiniteness, and therefore it cannot even be formulated in classical physics. There is, to be sure, a concept of chance in some parts of classical physics, especially classical statistical mechanics, but it always arises from ignorance of or incomplete specification of the microscopic properties of a system.

Suppose now that the eventuality *e* is indefinite when the system $\mathscr{S}$ is in the quantum mechanical state *S*, and suppose that $\mathscr{S}$ is subjected to a procedure which elicits a definite result regarding *e*. Then, according to quantum mechanics, there is a definite probability of finding *e* to be true and a definite probability of finding it to be false. These probabilities depend upon *e* and *S*, not upon the knowledge or beliefs of the observer, and in this sense they are objective. If all members of a large population of systems of the same kind are prepared in the state *S*, and in each case truth or falsity is elicited by an appropriate procedure, then the frequency of results can be expected to approximate these objective probabilities. It would not be correct to say that we have here an initially heterogeneous ensemble, in which there are definite frequencies of true and false *e* prior to the elicitation procedure. If that were the case, then *S* would not be the state in the sense of a maximum

specification of each system, for in each case there would be some objective property not contained in *S*.

It will be convenient (following Heisenberg) to introduce a single word to characterise an eventuality *e* which is indefinite when the system is in the state *S*, but which has definite probabilities of the two possible outcomes when a truth value is elicited: we say that *e* is *potential* in this case. If a procedure has the consequence of yielding definite truth or falsity for an eventuality *e* which initially is indefinite, then we speak of the *actualisation of a potentiality*. A quantum state is constituted not only by a specification of the truth or falsity of some of the eventualities, but also by the specification of the probabilities of finding truth or falsity upon actualisation of all the other eventualities. Thus a quantum state is a network of potentialities.

In classical physics the state of a composite system $1 + 2$ at an instant *t* is completely specified by the states of 1 and 2 separately, provided, of course, that the state of 1 contains information about the position and orientation of 1 relative to a frame of reference, and similarly for system 2. If 1 and 2 interact, then the temporal evolutions of the states of each are coupled, but at any instant the state of each can be characterised without reference to the state of the other. In quantum mechanics, however, it is possible for the state of $1 + 2$ to be *entangled* at an instant *t*: neither 1 by itself nor 2 by itself is in a definite state, yet the two together are in a definite state. Such a conception is incomprehensible from the standpoint of classical physics, but it is comprehensible from the standpoint of quantum mechanics, which regards the state of a system as a network of potentialities. In an entangled state the potentialities of 1 and of 2 are so correlated that there are pairs of eventualities, e_1 of system 1 and e_2 of system 2, which are both indefinite, but are such that when they are actualised by appropriate procedures then either both turn out to be true or both turn out to be false.

The language which we have employed for describing the conceptual innovations of quantum mechanics is quite philosophical. We have no apology for this language, because we consider it to be appropriate to the subject. We do not regard philosophy as an autonomous discipline, with a subject matter distinct from that of other disciplines, but rather as the general investigation of foundations questions and the general search for perspective. The change of framework of physics from classical to quantum mechanical is clearly a fundamental transformation of the conception of nature, and hence is a philosophical matter according to our usage of the term. It may be objected that the philosophical terminology which has been employed above introduces some unnecessary unclarity into physics. To this objection we reply that a highly formal exposition of quantum mechanics is unclear concerning interpretation even though it is clear concerning structure, whereas the formal and philosophical expositions in combination may supplement each other and achieve a fuller clarification. In the final part of this section we shall give a summary of the formal principles of quantum mechanics, and it will be seen that concepts of indefiniteness, chance, probability, and entanglement are implicit in these principles. In order to prepare for our exposition of the formal principles of quantum mechanics, however, it will be useful to review some relevant mathematical tools.

Some mathematical preliminaries

According to quantum mechanics, the state of a physical system is represented by a vector in a *linear vector space* $\mathscr{V}$ associated with the system. The concept of a linear vector space is not difficult, but it is somewhat abstract, and therefore we shall first present a special case which is easy to visualise, namely a two-dimensional real vector space, which we shall call $\mathscr{V}_2$. Thereafter, we shall indicate generalisations.

The essential features of $\mathscr{V}_2$ are exhibited in figures 13.1(a)–(e). Each vector in $\mathscr{V}_2$ can be visualised as a directed line segment in a plane, originating from a common origin. In figure 13.1(a) two vectors u_1 and u_2 are shown, which differ from each other both in direction and length. An arrow-head is placed at the end of the vector away from the common point of origination, in order to emphasise that a vector has a direction. The length of a vector u is a real number, which is designated as $\|u\|$. Two distinct vectors may have the same length but different directions, as have u_1 and u_3 in figure 13.1(b); and two different vectors may have the same direction but different lengths, as have u_2 and u_4 in figure 13.1(b). More specifically, the relation between u_2 and u_4 can be expressed by

$$u_4 = 5u_2,$$

and the length of u_4 is five times that of u_2. We shall assume that if u is any vector in $\mathscr{V}_2$ and r is any real number, then ru is also a vector in $\mathscr{V}_2$. The meaning of ru is intuitively clear if r is a positive real number, but if r is negative or zero some explanation is necessary. If r is negative, then $\|ru\|/\|u\| = |r|$, where $|r|$ is the absolute value of r; but because r is negative ru points in the opposite sense from u, as indicated in figure 13.1(b). If $r = 0$, then ru is a vector of zero length, which we shall call the *null vector* and denote by the numeral 0. (The context will determine whether 0 stands for number zero or for the null vector.) The null vector is hard to represent diagrammatically, since its tail and its head coincide.

The sum $u_1 + u_2$ of two vectors is exhibited diagrammatically in figure 13.1(c) by placing the tail of u_2 upon the head of u_1. Then the directed line segment from the origin to the transported head of u_2 represents $u_1 + u_2$. It is geometrically obvious that if the order is reversed, and the tail of u_1 is placed upon the head of u_2, the same result will be obtained. In other words,

$$u_1 + u_2 = u_2 + u_1.$$

Clearly we can add three or more vectors. For example, $u_1 + u_2 + u_3$ can be defined either as the sum of u_1 and $u_2 + u_3$, or

the sum of $u_1 + u_2$ and u_3, and the results of summing in either of these two ways are the same.

Figure 13.1(d) exhibits that some pairs of vectors in $\mathscr{V}_2$, like u_1 and u_2, and like u_1' and u_2', are *orthogonal* to each other. In $\mathscr{V}_2$ there do not exist sets of three vectors which are all non-null and mutually orthogonal, because $\mathscr{V}_2$ is two dimensional. Figure 13.1(d) also illustrates that an arbitrary vector v can be expressed uniquely as the sum of two vectors, one of which is a multiple of u_1 and the other a multiple of u_2. We shall call the first of these two vectors 'the projection of v upon u_1' and denote it by $P_{u_1}v$; and we shall call the second 'the projection of v upon u_2' and denote it by $P_{u_2}v$. We can therefore write

$$v = P_{u_1}v + P_{u_2}v.$$

Clearly, we can equally well project v upon the directions of u_1' and u_2' and write

$$v = P_{u_1'}v + P_{u_2'}v.$$

It is easy to generalise everything that we have said about $\mathscr{V}_2$ to an n-dimensional vector space $\mathscr{V}_n$, where n is an integer greater than or equal to unity, even though diagrams are difficult if n is greater than three. All the foregoing statements about multiplying a vector u by a real number, about the length $\|u\|$ of a vector u, and about adding two or more vectors, hold in $\mathscr{V}_n$ exactly as in $\mathscr{V}_2$. The statements about orthogonality have to be modified somewhat. In $\mathscr{V}_n$ it is possible to find n non-null vectors $u_1, \ldots, u_n$ which are mutually orthogonal, but it is impossible to find $n+1$ mutually orthogonal non-null vectors. An arbitrary vector v in $\mathscr{V}_n$ can be written as a unique sum of projections upon the directions of n mutually orthogonal non-null vectors:

$$v = P_{u_1}v + \ldots + P_{u_n}v.$$

A useful concept is that of a subspace of $\mathscr{V}_n$. $\mathscr{E}$ is a subspace of $\mathscr{V}_n$ if it is a collection of vectors of $\mathscr{V}_n$ with the following properties: if u belongs to $\mathscr{E}$, so does ru for any real number r, and if u_1 and u_2 belong to $\mathscr{E}$, so does $u_1 + u_2$. An example of a subspace is the set of all vectors which can be expressed in the form $r_1u_1 + r_2u_2$, where u_1 and u_2 are two vectors of $\mathscr{V}_n$, neither of which is a multiple of the other; this subspace can be called the *plane* determined by u_1 and u_2 (see figure 13.1e).

The concept of projection can be generalised to *projection upon a subspace*. Let $\mathscr{E}$ be a subspace of $\mathscr{V}_n$, and let $\mathscr{E}^\perp$ be the set of all vectors orthogonal to every member of $\mathscr{E}$. Then every vector v in $\mathscr{V}_n$ can be expressed uniquely as the sum of a vector in $\mathscr{E}$ and one in $\mathscr{E}^\perp$. We call the first 'the projection of v upon $\mathscr{E}$' and denote it by $P_{\mathscr{E}}v$; and we call the second 'the projection of v upon $\mathscr{E}^\perp$' and denote it by $P_{\mathscr{E}^\perp}v$. Then

$$v = P_{\mathscr{E}}v + P_{\mathscr{E}^\perp}v.$$

An illustration is given in figure 13.1(e), where n is taken to be three and $\mathscr{E}$ is the plane determined by the orthogonal vectors u_1 and u_2, while $\mathscr{E}^\perp$ consists of all vectors orthogonal to both u_1 and u_2.

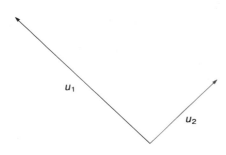

(a)

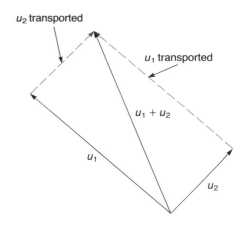

(c)

Figure 13.1. (a) All vectors are represented by arrows proceeding from a common origin. A vector is uniquely determined by its direction and length. The two vectors u_1 and u_2 represented here differ both in direction and length.

Figure 13.1. (c) $u_1 + u_2$ is constructed by transporting u_2 so that its tail falls upon the head of u_1, and then an arrow is drawn from the origin to the transported head of u_2. $u_2 + u_1$ is constructed analogously. The constructions indicate that $u_1 + u_2$ is the same vector as $u_2 + u_1$.

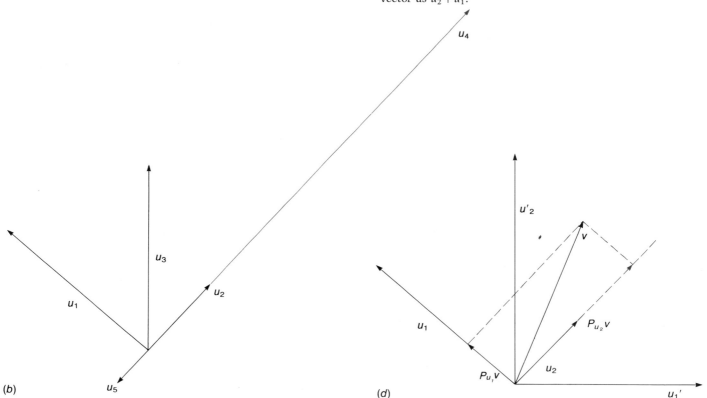

(b)

(d)

Figure 13.1. (b) The length of a vector u is denoted by $\|u\|$. Here $\|u_1\| = \|u_3\|$, i.e. u_1 and u_3 have the same length. The directions of u_2 and u_4 are the same, but their lengths are different; in fact, $u_4 = 5u_2$, and therefore $\|u_4\| = 5\|u_2\|$. The directions of u_2 and u_5 are the same, but their senses are opposite; in fact, $u_5 = -\frac{1}{2}u_2$, $\|u_5\| = \frac{1}{2}\|u_2\|$.

Figure 13.1. (d) u_1 and u_2 are orthogonal to each other. $P_{u_1}v$ is the projection of v upon u_1, and $P_{u_2}v$ is the projection of v upon u_2. The construction indicates that v is the same vector as $P_{u_1}v + P_{u_2}v$. Another pair of orthogonal vectors is u_1' and u_2'. The projections of v upon u_1' and u_2' could also be constructed, and v also can be expressed as $v = P_{u_1'}v + P_{u_2'}v$.

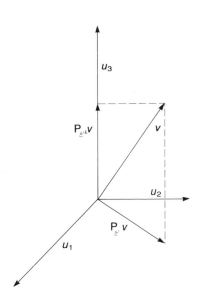

Figure 13.1. (e) u_1, u_2, u_3 are mutually orthogonal in $\mathscr{V}_3$. $\mathscr{E}$ is the set of all vectors of the form $r_1 u_1 + r_2 u_2$, where r_1 and r_2 are real numbers. $\mathscr{E}^\perp$ is the set of all vectors orthogonal to every vector in $\mathscr{E}$, i.e. $\mathscr{E}^\perp$ consists of all vectors of the form $r u_3$, where r is a real number. $P_\mathscr{E} v$ is the projection of v upon $\mathscr{E}$. $P_\mathscr{E}{}^\perp v$ is the projection of v upon $\mathscr{E}^\perp$. The construction indicates that $v = P_\mathscr{E} v + P_\mathscr{E}{}^\perp v$.

One further mathematical concept will be important for our exposition. If $\mathscr{V}'$ and $\mathscr{V}''$ are two complex linear vector spaces of dimensions n' and n'', respectively, we can form from them a new complex linear vector space called the *tensor product* of $\mathscr{V}'$ and $\mathscr{V}''$, denoted by $\mathscr{V}' \otimes \mathscr{V}''$, having dimension $n'n''$. We shall construct the tensor product space in the following way. Let $u_1, \ldots, u_{n'}$ be n' mutually orthogonal vectors in $\mathscr{V}'$, each of unit length, and let $v_1, \ldots, v_{n''}$ be n'' mutually orthogonal vectors in $\mathscr{V}''$, each of unit length. Then we form $n'n''$ product vectors $u_i \otimes v_j$, where i is any integer from 1 to n' and j is any integer from 1 to n''. Each product $u_i \otimes v_j$ is determined by the ordered pair consisting of u_i and v_j, even though it is not accurate simply to identify the product with an ordered pair. Each $u_i \otimes v_j$ is a vector in the new vector space $\mathscr{V}' \otimes \mathscr{V}''$, and each has unit length in this vector space. If either $i \neq k$ or $j \neq l$, then $u_i \otimes v_j$ is orthogonal to $u_k \otimes v_l$. The space $\mathscr{V}' \otimes \mathscr{V}''$ consists of all vectors of the form

$$\psi = \sum_{i=1}^{n'} \sum_{j=1}^{n''} c_{ij} u_i \otimes v_j,$$

where c_{ij} are arbitrary complex numbers. There are a few additonal rules governing the tensor product spaces, but we shall not need them in our exposition. We may also note that a real tensor product space can be defined, just as well as a complex one. The one peculiarity which we wish to emphasise, since it turns out to have fundamental significance for quantum mechanics, is that ψ cannot be expressed in the form

$$\psi = w \otimes z,$$

where w is a vector of $\mathscr{V}'$ and z is a vector of $\mathscr{V}''$, for some choices of the c_{ij}.

The formal principles of quantum mechanics

The discovery of the principles of quantum mechanics has an intricate history, which lies beyond the scope of this chapter. It will suffice to say here that the main body of physical phenomena from which the principles were extracted concerned the absorption and emission of light by atoms; that the decisive proposals were made in very different ways by W. Heisenberg in 1925 and E. Schrödinger in 1926; that they made essential use of earlier work of M. Planck, A. Einstein, N. Bohr, L. de Broglie and others, and that a number of physicists and mathematicians, including M. Born, P. Jordan, W. Pauli, P. Dirac, and J. von Neumann, made crucial clarifications.

Our procedure here will be to summarise succinctly the general principles of quantum mechanics in a simple way, by using the language of linear vector space theory. The statement of each principle will be followed by some comments, for the sake of explicitness and emphasis. In Section 13.3 it will be shown that these principles provide a straightforward and elegant explanation of a number of phenomena concerning the

For a complete exposition of quantum mechanics two further generalisations of the concept of linear vector space are required. The first is that we permit a vector to be multiplied not only by a real number r, but by a complex number c (c having the form $r_1 + r_2 i$, where r_1 and r_2 are real numbers and i is the square root of minus one). The second generalisation is that we permit the dimension of the vector space $\mathscr{V}$ to be either a finite integer n or infinite. Although complex and infinite-dimensional vector spaces differ in certain respects from real finite-dimensional ones, the differences will not be crucial for any of the ideas or illustrations of quantum mechanics which we shall introduce in the remainder of this chapter, and hence they need not be studied here.

We shall occasionally use the concept of a *linear operator* on the vector space $\mathscr{V}$. A is a linear operator if it assigns to every vector in a subspace $\mathscr{D}_A$ of $\mathscr{V}$ (called the *domain* of A) a definite vector, denoted by Av, such that

$$A(c_1 u_1 + c_2 u_2) = c_1 A u_1 + c_2 A u_2,$$

for every pair of vectors u_1, u_2 in the domain and every pair of complex numbers. (In all the examples which we shall give, $\mathscr{D}_A$ will be the whole of $\mathscr{V}$.) The projection operators P_{u_i} and $P_\mathscr{E}$ which were defined above are instances of linear operators.

transmission of polarised light through Polaroid sheets. These phenomena form a small sample of the vast range of physical phenomena which have been understood successfully in terms of quantum mechanics and which therefore provide empirical support for its principles.

Principle 13.1: Associated with every physical system is a complex linear vector space $\mathscr{V}$, such that each vector of unit length represents a state of the system.

Principle 13.1 does not prescribe how a vector in $\mathscr{V}$ represents a state. We previously characterised a state as a maximal specification of the system. Hence the state should determine which eventualities of the system are true, which are false, and which are indefinite, and it should assign probabilities to those which are indefinite. But this information is not given by Principle 13.1, which asserts only that a maximal amount of information about the system is 'encoded' in a unit vector. How this information is encoded and how it can be extracted when the vector is given are the content of Principles 13.2 and 13.3. What is informative about Principle 13.1 is the structure which it attributes to the space of states, namely a structure which mirrors the mathematical structure of a linear vector space.

It should be noted that Principle 13.1 does not assert a one-to-one correspondence between vectors of unit length and states of the system. In fact, if u is a vector of unit length and $u' = cu$, where c is a complex number of absolute value unity, then u' also has unit length, and it will be evident from Principles 13.2 and 13.3 that u' represents the same state as u. (The absolute value $|c|$ of a complex number written in the form $r_1 + r_2 i$ is $|c| = (r_1{}^2 + r_2{}^2)^{\frac{1}{2}}$.) Finally, we note that Principle 13.1 could be generalised to permit any non-null vector to represent a state, but then the rules for extracting information about eventualities would have to be made more complicated than those stated in Principles 13.2 and 13.3, and we prefer to avoid the complication.

Principle 13.2: There is a one-to-one correspondence between the set of eventualities concerning the system and the set of subspaces of the linear vector space associated with the system, such that if e is an eventuality and $\mathscr{E}$ is the subspace which corresponds to it, then e is true in a state S of the system if and only if any vector v which represents S belongs to $\mathscr{E}$; and e is false in the state S if and only if v belongs to $\mathscr{E}^{\perp}$.

Principle 13.2 provides a physical interpretation of the mathematical formalism in those cases in which eventualities are definitely true or false. Given any state S of the system, the Principle specifies precisely which eventualities are true and which are false in that state. However, the Principle does not say that every eventuality is definitely true or definitely false in a given state. Except in the trivial cases in which $\mathscr{E}$ is either the

entire vector space $\mathscr{V}$ or is the null subspace $\mathscr{V}^{\perp}$ there is a vector v of unit length which belongs neither to $\mathscr{E}$ nor to $\mathscr{E}^{\perp}$; or, equivalently, the projection of v upon $\mathscr{E}$ is non-null, and likewise the projection of v upon $\mathscr{E}^{\perp}$ is non-null; and therefore the eventuality e is neither definitely true nor definitely false in the state represented by v.

In a fuller exposition of the principles of quantum mechanics than we are able to present here, more would be said about the correspondence between the set of eventualities and the set of subspaces. Both of these sets have definite structures which must be related to each other. The details of this relation are not needed for our purposes, but we shall make two points informally in order to indicate what remains to be filled out. If e is an eventuality, there is a unique negation e' which is false whenever e is true and true whenever e is false. Since e corresponds to a subspace $\mathscr{E}$, it is natural to let e' correspond to $\mathscr{E}^{\perp}$, for this assignment guarantees that, for every state which is represented by a vector belonging either to $\mathscr{E}$ or to $\mathscr{E}^{\perp}$, the eventualities e and e' are opposite with respect to truth and falsity, as required by the concept of negation. The second point is that the subspace associated with the necessarily true eventuality $\mathbf{1}$ is the whole vector space $\mathscr{V}$, since every vector belongs to $\mathscr{V}$, and hence by Principle 13.2 the eventuality $\mathbf{1}$ is true in every state, as required.

Principle 13.3: If S is a state and e is an eventuality corresponding to the subspace $\mathscr{E}$, then the probability that e will turn out to be true if initially the system is in S and an operation is performed to actualise e, is

$$\mathrm{prob}_S(e) = \|\mathrm{P}_{\mathscr{E}} v\|^2,$$

where v is a unit vector representing S.

In the equation just written, we have used some notation previously explained. $\mathrm{P}_{\mathscr{E}} v$ is the projection of v upon the subspace $\mathscr{E}$ and $\|\mathrm{P}_{\mathscr{E}} v\|$ is the length of this projection. The projection of a vector v upon any subspace has a length which is at most equal to $\|v\|$ and at least equal to zero. It follows that $\|\mathrm{P}_{\mathscr{E}} v\|$ is between zero and unity, consistent with the standard convention that represents probabilities by real numbers between zero and unity. If v belongs to $\mathscr{E}$, then $\mathrm{P}_{\mathscr{E}} v = v$, and therefore $\mathrm{prob}_S(e) = 1$. On the other hand, if v belongs to $\mathscr{E}^{\perp}$, then $\mathrm{P}_{\mathscr{E}} v$ is the null vector, and hence $\mathrm{prob}_S(e) = 0$. Consequently, in these two extreme cases Principle 13.3 agrees with Principle 13.2, which asserts that e is true if v belongs to $\mathscr{E}$ and that e is false if v belongs to $\mathscr{E}^{\perp}$.

An example in which v belongs neither to $\mathscr{E}$ nor to $\mathscr{E}^{\perp}$ will be illuminating at this juncture. Suppose that $\mathrm{P}_{\mathscr{E}} v = v_1$ and $\mathrm{P}_{\mathscr{E}^{\perp}} v = v_2$, and let the lengths of v_1 and v_2, respectively, be r_1 and r_2. It can easily be shown by the rules of the theory of linear vector spaces that $r_1{}^2 + r_2{}^2 = 1$; in fact, this equation concerning the lengths of projections is a straightforward generalisation of the Pythagorean theorem. From Principle 13.3 and the comment on Principle 13.2 it follows that $\mathrm{prob}_S(e) = r_1{}^2$ and

$\text{prob}_S(e') = r_2{}^2$. Hence, the probabilities of the two mutually exclusive and exhaustive outcomes once e is actualised, namely truth and falsity, sum to unity, in agreement with standard usage in probability theory.

Implicit in all three Principles is the famous Superposition Principle, which asserts that any two states can be combined (actually in infinitely many ways) to form states which have characteristics intermediate between those of the two which are combined. The Superposition Principle is derived in the following way. Let u_1 and u_2 be two vectors of unit length so that by Principle 13.1 each represents a state, and let them be orthogonal to each other. Then there is a subspace $\mathscr{E}$ to which u_1 belongs, while u_2 belongs to $\mathscr{E}^\perp$. By Principle 13.2 the eventuality e corresponding to $\mathscr{E}$ is true in the state represented by u_1 and is false in the state represented by u_2. Now construct a new vector u as

$$u = c_1 u_1 + c_2 u_2,$$

where neither c_1 nor c_2 is zero, and the sum of $|c_1|^2$ and $|c_2|^2$ is unity. Then u is a vector of unit length, again by the generalisation of the Pythagorean theorem, and hence represents a possible state of the system. The projections of u upon $\mathscr{E}$ and $\mathscr{E}^\perp$ are both non-null, for

$$P_{\mathscr{E}} u = c_1 u_1, \text{ and hence } \|P_{\mathscr{E}} u\| = c_1,$$
$$P_{\mathscr{E}^\perp} u = c_2 u_2, \text{ and hence } \|P_{\mathscr{E}^\perp} u\| = c_2.$$

We have therefore constructed a state which is intermediate between the states represented by u_1 and u_2, for in the state represented by u the eventuality e is neither true as in the former nor false as in the latter, but is indefinite. Furthermore, by Principle 13.3, the probability that e will turn out to be true upon actualisation is $|c_1|^2$ and that it will turn out to be false is $|c_2|^2$. Clearly, then, the closer c_1 is to having absolute value unity, the more does the state represented by u_1 dominate in the superposition, and the closer c_2 is to having absolute value unity, the more does u_2 dominate.

We may summarise the results so far by saying that Principles 13.1, 13.2 and 13.3 suffice to show that quantum mechanics exhibits the first three conceptual innovations listed earlier: objective indefiniteness, objective chance and objective probability. Furthermore, the three Principles add mathematical structure to these conceptual innovations.

Principle 13.4: If 1 and 2 are two physical systems, with which the vector spaces $\mathscr{V}^1$ and $\mathscr{V}^2$ are associated, then the composite system $1 + 2$ consisting of 1 and 2 is associated with the tensor product space $\mathscr{V}^1 \otimes \mathscr{V}^2$.

Some of the states of $1 + 2$ which are postulated to exist by Principle 13.4 are completely in accord with common sense and the conceptions of classical physics. Thus, if u_1 and u_2 are non-null vectors in $\mathscr{V}^1$, and v_1 and v_2 are non-null vectors in $\mathscr{V}^2$, then $u_1 \otimes v_1$ represents the state which can be characterised by saying 'system 1 is in u_1 and system 2 is in v_1'; and likewise with $u_2 \otimes v_2$. In both cases the state of the composite system $1 + 2$ is specified simply by specifying the states of the two components 1 and 2 separately. Now suppose that u_1 and u_2 are orthogonal and that v_1 and v_2 are orthogonal, and for simplicity suppose that u_1, u_2, v_1, v_2 are vectors of unit length. Then in $\mathscr{V}^1 \otimes \mathscr{V}^2$

$$\Psi = \frac{1}{\sqrt{2}}(u_1 \otimes v_1 + u_2 \otimes v_2)$$

is a vector of unit length and hence it represents a state of $1 + 2$. We may ask what state 1 is in by itself, and what state 2 is in by itself, when $1 + 2$ is in the state Ψ. The answer is remarkable. Neither 1 nor 2 is in a definite state! It is straightforward to show, in fact, that there are no vectors w and z of $\mathscr{V}^1$ and $\mathscr{V}^2$, respectively such that $\Psi = w \otimes z$. Another way to put the matter is to say that the above expression for Ψ applies the Superposition Principle to the states $u_1 \otimes v_1$ and $u_2 \otimes v_2$, both of which have a commonsensical character, in order to generate the noncommonsensical state Ψ, which is *entangled*. The connection between entanglement and objective indefiniteness is very close. Suppose that e_1 is an eventuality of system 1 which is definitely true in the state u_1 and definitely false in the state u_2, and e_2 is an eventuality which is definitely true in the state v_1 and definitely false in the state v_2. Then e_1 and e_2 are both indefinite in the state Ψ of $1 + 2$, but indefinite in a correlated manner. If e_1 and e_2 are both actualised, there is probability $\frac{1}{2}$ that both will be true, probability $\frac{1}{2}$ that both will be false, and probability 0 that one will be true and the other false. Entanglement of states would be completely baffling if a state of a physical system were nothing more than a compendium of its actual properties; but if we stretch our imaginations to accept the concept of potentiality, as quantum mechanics apparently requires us to do, then entanglement becomes comprehensible.

There is a corollary to entanglement which we barely mention now but shall discuss in detail in Section 13.4; that is, that if the systems 1 and 2 are spatially well separated, then the entanglement of the state of $1 + 2$ constitutes a kind of *nonlocality*, and the correlated actualisation of potentialities of 1 and 2 appears to be a kind of nonlocal process.

The fifth principle concerns the time development of the quantum state for a special class of systems, namely those in an environment which is unaffected by the behaviour of the system of interest – which we shall call a *nonreactive environment*. It is an idealisation to regard an environment as nonreactive, but in a large class of problems this idealisation is not a serious departure from the literal truth. A nonreactive environment may change, for example, because of a periodic electromagnetic field, but the changes are not due to feedback from the system of interest.

Principle 13.5: If a system is in a nonreactive environment between times 0 and t, then there is a linear operator U(t) such that U(t)v represents the state of the system at time t if v represents the state at time 0. Furthermore, U(t) is length-preserving, i.e. $\|U(t)v\| = \|v\|$ *for all v in* $\mathscr{V}$.

In the statement of Principle 13.5 t is an arbitrary time, which may be positive (later than the initial time) or negative (earlier than the initial time) or zero.

Although Principle 13.5 cannot be used directly to infer the temporal development of the quantum state of a system in a reactive environment, it can often provide some important information indirectly. Suppose that the system of interest, which we shall now refer to as system 1, has non-negligible effect only upon system 2, which is a part of the environment. Then attention can be focused upon the composite system 1 + 2, the states of which are represented in a tensor product space, according to Principle 13.4. If the environment of this composite system is to a good approximation nonreactive, then Principle 13.5 can be used to follow its temporal development, and some important information can then be extracted concerning system 1 by itself.

A very interesting instance of a reactive environment is one which contains a measuring apparatus for measuring some property of the system under consideration. In order to perform as a measuring apparatus, a part of the environment must be sensitive to the values of the property, e.g. to the truth or falsity of an eventuality e. It must react one way if e is true and another way if e is false. Consequently, Principle 13.5 cannot be used directly to follow the temporal evolution of a system while it is being subjected to a measurement. It is possible, however, that the system together with the apparatus of measurement constitute a composite system 1 + 2 with a nonreactive environment, and then, as remarked above, Principle 13.5 is applicable to 1 + 2. Our analysis of the measurement process in Section 13.5 will be based upon this idea. We shall see in that section that there is some tension between Principle 13.5 and Principle 13.3.

13.3   Example: linearly polarised photons

The first indications that electromagnetic radiation has a granular structure were the discoveries by M. Planck in 1901 concerning the spectrum of black-body radiation and by A. Einstein in 1905 concerning the photo-electric effect. The least unit of light of a given frequency is called the photon. The subsequent development of atomic physics and quantum optics has overwhelmingly supported the photon theory of radiation. Consequently, when we study linearly polarised light normally incident upon a sheet of Polaroid, we may consider it to be a beam of photons, which under ordinary circumstances are uncorrelated. In this section we shall discuss the quantum mechanical treatment of the polarisation states of a photon and show how naturally quantum mechanics accounts for the phenomena of partial transmission of beams of photons through sheets of Polaroid. We shall assume that the reader is acquainted with the appearance and behaviour of Polaroid sheets, since they are very commonly used in sunglasses and other products, and we shall not discuss the mechanism of operation of Polaroid.

If the propagation direction of a photon is along the z-axis, then the direction of its linear polarisation is in the x–y plane, perpendicular to the z-axis. The vector space associated by quantum mechanics (in accordance with Principle 13.1) with the polarisation of a photon is a two-dimensional space, which we shall call $\mathscr{V}_{\text{pol}}$. Every vector in $\mathscr{V}_{\text{pol}}$ can be expressed in the form $c_1 u_1 + c_2 u_2$, where u_1 and u_2 are any two orthogonal non-null vectors of $\mathscr{V}_{\text{pol}}$. One interesting choice of u_1, u_2 is u_x, u_y, each of unit length, representing, respectively, states of polarisation along the x- and y-axes. Another interesting choice would be $u_{x'}$, $u_{y'}$, which have analogous meanings, representing states of polarisation along the x'- and y'-axes, where x'–y' is a pair of axes in the plane perpendicular to the z-axis, obtained by rotating the x–y-axes through an angle θ. If a beam of photons is prepared to be linearly polarised along the x-axis and is made to impinge upon a sheet of Polaroid which is oriented so that its optical axis is along the x'-axis, then a proportion $\cos^2\theta$ of the initial beam passes through the sheet (see figure 13.2). This phenomenon is an instance of the law of Malus, discovered long before the photon theory of light.

The quantum mechanical explanation of this and related phenomena uses the fact that u_x and u_y can be expressed in terms of $u_{x'}$ and $u_{y'}$ as follows:

$$u_x = \cos\theta u_{x'} + \sin\theta u_{y'},$$
$$u_y = -\sin\theta u_{x'} + \cos\theta u_{y'}.$$

Let $\mathscr{E}_x$ be the subspace of $\mathscr{V}_{\text{pol}}$ consisting of all vectors proportional to u_x, and let $\mathscr{E}_y$, $\mathscr{E}_{x'}$, $\mathscr{E}_{y'}$ be similarly related to the vectors u_y, $u_{x'}$, $u_{y'}$. The eventuality e_x corresponding to $\mathscr{E}_x$ according to Principle 13.2 asserts that the photon is linearly polarised along the x-axis. The eventualities e_y, $e_{x'}$ and $e_{y'}$ have similar significance. When a photon impinges upon the Polaroid sheet, which is oriented so that its optical axis is along x', then it is reasonable to suppose that the eventuality $e_{x'}$ is actualised. Consequently, Principle 13.3 can be used to compute the probability that $e_{x'}$ turns out to be true upon actualisation: it is the square of the length of the projection of u_x upon the subspace $\mathscr{E}_{x'}$ which corresponds to the eventuality $e_{x'}$. But by the above expression of u_x in terms of $u_{x'}$ and $u_{y'}$, the projection of u_x upon $\mathscr{E}_{x'}$ is $P_{\mathscr{E}_{x'}} u_x = \cos\theta u_{x'}$, and therefore the probability that the outcome is the truth of $e_{x'}$ is $\cos^2\theta$. But this outcome is the same as the passage of the photon through the Polaroid sheet. Hence, the probability of the passage of the photon through the sheet is $\cos^2\theta$, and therefore if there is a beam of similarly prepared photons, all in the state represented by u_x, then a proportion $\cos^2\theta$ of the beam will pass through the Polaroid sheet.

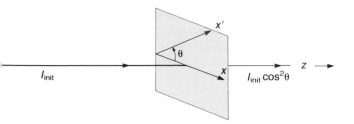

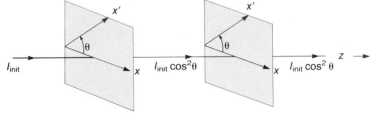

Figure 13.2. The initial beam of photons propagates in the z direction with intensity I_{init} and polarisation along the x-axis. A polarisation analyser (e.g. a sheet of Polaroid) lies in the x–y-plane, perpendicular to the incident beam, and the optical axis of the analyser is along the x'-axis, the angle between x and x' being θ. Then the intensity of the transmitted beam is $I_{\text{init}}\cos^2\theta$.

Figure 13.3. The intensity of the beam is not further diminished by passage through a second polarisation analyser with an optical axis in the same direction as that of the first analyser. The analysers are assumed to be ideal.

A further experimental result allows us to conclude that those photons which have passed through the Polaroid sheet oriented along x' are in the polarisation state represented by $u_{x'}$. If these photons are allowed to impinge upon another sheet also oriented along x', then it is found that all of them pass through the second sheet (correction being made for the nonideality of the Polaroid). This experiment is pictured in figure 13.3. This result is exactly what is expected if the state of a photon after passage through the first Polaroid is $u_{x'}$, for the projection of $u_{x'}$ upon $\mathscr{E}_{x'}$ is just $u_{x'}$ itself, which has unit length, and therefore the probability that the photon will pass through the second Polaroid sheet is one (either by Principle 13.3 or by Principle 13.2). If, however, the state of the photon after passage through the first Polaroid were not represented by a vector proportional to $u_{x'}$, then the probability of passage through the second Polaroid sheet would be less than one, contrary to the experimental result.

One further experiment is particularly striking and instructive. An initial beam of light polarised along the x-axis, with intensity I_{init}, impinges upon a pair of crossed polarisers consisting of one sheet of Polaroid with optical axis in the x direction followed by a sheet with optical axis in the y direction (figure 13.4a), and the result is that no photons emerge. This result is predicted by quantum mechanics, since after emergence from the first sheet the photons are in the state u_x, and the probability then of passage through the second sheet is $\|P_{\mathscr{E}_y}u_x\|^2$, which is zero. Now, however, let another sheet of Polaroid, oriented along x', be inserted between the crossed polarisers (figure 13.4b). The intensity of the beam which emerges from the series of three sheets of Polaroid is not zero, but rather

$$I_{\text{final}} = I_{\text{init}}\cos^2\theta\sin^2\theta,$$

which can be checked photometrically by varying the angle θ. This phenomenon is uncanny to the uninitiated and exhilarating even to the initiated. The opaque pair of crossed polarisers is

made partially transparent by inserting an intermediate sheet, even though from a commonsense point of view this additional sheet constitutes a further obstacle to the photons. Quantum mechanically, however, everything falls into place. The photons emerging from the first sheet are in state u_x, and their probability of passage through the intermediate sheet is

$$\|P_{\mathscr{E}_{x'}}u_x\|^2 = \|\cos\theta u_{x'}\|^2 = \cos^2\theta;$$

but any photon that emerges from the intermediate sheet is in state $u_{x'}$, and its probability of passing through the second of the crossed polarisers is

$$\|P_{\mathscr{E}_y}u_{x'}\|^2 = \sin^2\theta.$$

Consequently, the probability of passing through the series of three sheets is $\cos^2\theta\sin^2\theta$, in agreement with the observed intensity I_{final}.

If a theory of photons were formulated within a classical framework (which is quite a different matter from retreating to the classical electromagnetic theory of light), then a state or maximal specification of a photon would presumably determine whether the photon would pass or would not pass through each ideal polarisation filter. In our notation, the state would specify the truth value of each eventuality of the form e_n, where n is an axis perpendicular to the direction of propagation. The transparency of the series of sheets of Polaroid in figure 13.4(b) is very difficult to understand from the standpoint of this classical picture. Difficult is not impossible, however, and one could imagine that impingement of a photon which has passed through the first sheet (oriented along x) upon the intermediate sheet (oriented along x') has the effect of changing the state of the photon in a way that depends upon the detailed constitution of the intermediate sheet, and this change might permit the photon to pass through the final sheet (oriented along y). Hidden variables theories make proposals of this kind in order to fit the experimental data within a classical framework. We postpone the assessment of hidden variables theories to Section 13.4.

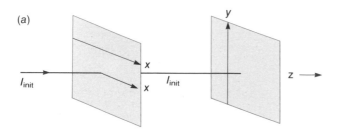

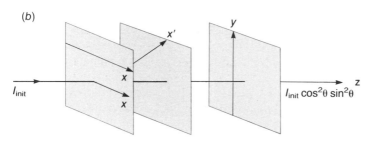

Figure 13.4. (*a*) If the initial beam is polarised along the *x*-axis, it passes without diminution through an analyser having an optical axis along the *x*-axis, but it is not transmitted at all through an analyser oriented along the *y*-axis. (*b*) The insertion of an analyser oriented along the *x'*-axis between two analysers oriented, respectively, along the *x*-axis and the *y*-axis enables partial transmission through the series of three analysers. This phenomenon is mysterious from a classical point of view but comprehensible quantum mechanically.

13.4   Hidden variables theories

Types and motivations

All hidden variables theories deny that the quantum mechanical description of a physical system is a state in the sense of a maximal specification. When we discuss hidden variables theories it will therefore be convenient to introduce some additional terminology and refer to the *complete state*, which is the true maximal description, as distinguished from the *quantum state*, which is postulated by quantum mechanics to be maximal. For the most part, hidden variables theories acknowledge that a system prepared in a given way is correctly described by one quantum state and not by others, but that even the correct one gives only an incomplete description; just as a bottle of gas in thermodynamic equilibrium is correctly described by one thermodynamic state and not by any other, but the correct thermodynamic state fails to specify the microscopic constitution of the gas. The expression 'hidden

variables' refers to those properties which are specified by the complete state but not by the quantum state. In this usage, there is no intention to say that the hidden variables are in principle inaccessible to experimental investigation. Often this intention has been imputed by critics, who then unjustly argue that the postulation of a hidden variables theory is tantamount to the introduction of a radically nonempirical element in physics.

There are several types of hidden variables theories, but at this point we shall only make preliminary remarks about the main distinctions among them. A *simple* hidden variables theory (sometimes called 'noncontextual') is one in which the complete state λ of the system assigns a definite truth value to each of the quantum mechanically recognised eventualities, i.e. to each one corresponding to a projection operator. A *contextual* hidden variables theory is one in which the truth value of each eventuality is not determined by a complete state λ, but the outcome of each experimental test of an eventuality is determined by λ together with some features of the experimental arrangement. In a *stochastic* hidden variables theory the outcomes of such experimental tests need not be determined by λ together with the features of the experimental arrangement, but the probability of each possible outcome is determined by these factors. A distinction which will be very important in our later discussion is between *local* and *nonlocal* hidden variables theories, but we shall not explain these terms at this point.

Historically there have been three main motivations for proposing hidden variables theories. The first motivation is an unwillingness to accept the radical conceptual innovations of the quantum mechanical framework. A theorist thus motivated may try to interpret the quantum mechanical formalism so as to eliminate objective indefiniteness, objective chance, objective probability and entanglement, and nevertheless preserve the formalism for the purpose of predicting physical phenomena. It is not important for our purposes to inquire whether the tenacious preference of the theorist for the framework of classical physics is rational or irrational; we shall only be concerned with whether the experimental evidence and analytical considerations decisively require the conceptual innovations summarised above. The second motivation is the argument of Einstein, Podolsky and Rosen (hereafter abbreviated as EPR), which is so significant that we shall devote the next subsection to it. A third motivation is to avoid the problem of the actualisation of potentialities, which arises if the quantum state is taken to be a complete state while also the dynamical evolution of the state of an isolated system is governed by the linear law of Principle 13.5. One way to prevent this problem from arising is to deny the premise that the quantum state is a maximal specification of the system.

The type of hidden variables theory which would be acceptable to a theorist clearly depends upon the theorist's motivation. Thus, someone opposed to objective chance and objective probability would presumably not consider a stochastic hidden variables theory to be a substantial improvement over quantum mechanics. And it will be clear that the kind of hidden variables theory envisaged by EPR must be local.

For the most part, our survey of hidden variables theories will consist of negative results. There is a rigorous mathematical argument against nontrivial simple hidden variables theories. There is very strong experimental evidence against local hidden variables theories, both contextual and stochastic, although some loopholes – in our opinion unattractive – remain open. Nonlocal hidden variables theories remain live options, but only at the price of accepting some modifications of relativistic space–time structure. These results do not definitively discredit the entire programme of hidden variables theories, but they do succeed in making it very unappealing. In our opinion, the strongest argument for the permanent establishment of the radical conceptual innovations of quantum mechanics is the preponderance of negative evidence against the programme of hidden variables theories.

The argument of EPR

We shall use a pair of photons propagating, respectively, in the z and $-z$ directions in order to present the essential content of the argument of EPR, even though they studied a different pair of systems and formulated their premises somewhat differently. The notation of Section 13.3 will be employed with slight modifications: $u_x(1)$ is a vector in the vector space $\mathscr{V}_{pol}(1)$ representing polarisation states of photon 1, and the subscript x indicates that the photon is polarised along the x-axis; $u_x(2)$ has an analogous meaning for photon 2. We shall suppose that the photon pair is prepared in the quantum state

$$\Psi = \frac{1}{\sqrt{2}} \left[u_x(1) \otimes u_x(2) + u_y(1) \otimes u_y(2) \right]$$
$$= \frac{1}{\sqrt{2}} \left[u_{x'}(1) \otimes u_{x'}(2) + u_{y'}(1) \otimes u_{y'}(2) \right]$$

(see figure 13.5). The equality of the two different expressions for Ψ, employing different bases, can be checked algebraically, but the physical reason for the equality is that Ψ represents a state of zero polarisation angular momentum, and this is invariant under rotation about the common line of propagation of the two photons; consequently, Ψ has the same form with respect to any pair of perpendicular axes in the plane normal to the line of propagation. In the state represented by $u_x(1) \otimes u_x(2)$ both photons have linear polarisation along the x axis, and similarly for $u_y(1) \otimes u_y(2)$, but when these two states are superposed to form the state Ψ, then the linear polarisations are indefinite. Furthermore, Ψ is an entangled state, which cannot be expressed in the form $w \otimes z$, where w is a state of photon 1 and z of photon 2. The entanglement is constituted by the strict correlations of the indefinite linear polarisations of the two photons. If their polarisations with respect to the x–y-axes are made definite by a suitable procedure, then with certainty it

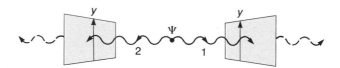

Figure 13.5. The photon pair $1+2$ is prepared in the entangled state Ψ. If the optical axes of the two polarisation analysers are parallel to each other, then either both photon 1 and photon 2 pass through their respective analysers or both fail to pass through them.

would be found either that both photons are polarised along x or both along y, though which of these outcomes would occur is a matter of chance. Exactly the same kind of correlation would be found if the polarisations were made definite with respect to the x'–y'-axes. That Ψ represents a possible quantum state is guaranteed by Principle 13.4, but it should be remarked that to a very good approximation this state has been experimentally realised.

An argument along the lines of EPR, concluding that Ψ does not represent a complete state of the pair $1+2$, uses two premises.

(*a*) If an experimenter has the option of performing an operation which permits a prediction with certainty of the truth value of an eventuality e of a system without in any way disturbing the system, then e has a definite truth value whether or not the operation is performed. (Comment: In order to know that the operation has the property mentioned in premise (*a*), one could perform it repeatedly so as to make the prediction concerning each of a large ensemble of replicas of the system of interest, and then make a direct test on each member of the ensemble in order to see whether the prediction is correct. Those checks would indeed disturb the members of the test ensemble. But if in each case the prediction is correct, then we may reliably infer by the usual canons of inductive logic that the operation permits a prediction with certainty concerning the system of interest, which is not subject to check and hence is not disturbed.)

(*b*) For a photon pair in the quantum state Ψ, there is an operation which can be performed so as to determine the truth or falsity of an eventuality of the form $e_n(1)$ (for any direction n perpendicular to z) without disturbing photon 2; and likewise, with 1 and 2 interchanged. (Comment: In order to carry out the argument of EPR it is irrelevant whether this premise is true because of the relativistic limit upon the speed of signals or because of insulating one photon from the effects of operations performed on the other or because of some other reason. Later, however, we shall return to the question of the grounds for premise b.)

The EPR argument now proceeds as follows. One option of the experimenter is to test the truth value of $e_x(1)$, which is the eventuality that photon 1 is linearly polarised along x, and when this is done the truth value of $e_y(1)$ is automatically obtained. By premise (*b*) there is an operation for testing $e_x(1)$ without disturbing photon 2. But, because of the correlation between the polarisations of photons 1 and 2 that is implicit in Ψ, it follows that this operation also yields the truth values of $e_x(2)$ and $e_y(2)$. Then by premise (*a*) the truth values of $e_x(2)$ and $e_y(2)$ are definite. But another option of the experimenter is to test the truth values of $e_{x'}(1)$ and $e_{y'}(1)$, where the x'–y'-axes are obtained from the x–y-axes by rotation through an arbitrary angle θ. The same reasoning as before shows that $e_{x'}(2)$ and $e_{y'}(2)$ are definite. Therefore the truth values of eventualities concerning linear polarisation of photon 2 along all axes in the plane normal to the propagation direction are definite. A similar conclusion can obviously also be derived for photon 1. Consequently the quantum state Ψ is far from a complete state of $1+2$. In fact, the complete state attributes to each photon more simultaneously definite eventualities than any quantum states admit.

In our opinion the logic of EPR is correct and therefore if the conclusion is false, as the experiments to be discussed later strongly indicate, then either premise (*a*) or premise (*b*) is at fault. We shall return later to the assessment of these premises.

The impossibility of a simple hidden variables theory

One of the most striking results concerning hidden variables theories is obtained by pure mathematics. It is the direct consequence of a mathematical theorem about the structure of the subspaces of a vector space of dimension greater than two, which we shall merely state without proof.

Theorem: If the dimension of a linear vector space $\mathscr{V}$ is greater than two, there is no function m which assigns to every subspace $\mathscr{E}$ of $\mathscr{V}$ a number zero or unity in such a way that

$$m(\mathscr{V}) = 1,$$

and

$$m(\mathscr{E}) = m(\mathscr{E}_1) + m(\mathscr{E}_2),$$

if $\mathscr{E}_1$ and $\mathscr{E}_2$ are orthogonal subspaces and $\mathscr{E}$ is the smallest subspace containing both $\mathscr{E}_1$ and $\mathscr{E}_2$.

This theorem is a corollary of an intricate theorem of A. Gleason in 1957, but it was later obtained more simply by J. S. Bell and by S. Kochen and E. Specker. The theorem has an immediate interpretation concerning the eventualities of a physical system, because of the correspondence between subspaces and eventualities postulated in Principle 13.2. The interpretation takes the numbers 1 and 0 to represent truth and falsity, respectively. Then if the dimension of $\mathscr{V}$ is greater than two, it is impossible to assign truth or falsity to all the eventualities in such a way that (i) the necessary eventuality **1** (which

corresponds to $\mathscr{V}$ itself) is assigned the value 'true', and (ii) the disjunction of two mutually exclusive eventualities is true if and only if one of them is true. But (i) and (ii) are obligatory if the hidden variables theory is to be a satisfactory treatment of eventualities from the standpoint of ordinary logic.

The theorem just stated is of fundamental importance for the understanding of the formalism of quantum mechanics. It says that, in all but the simplest cases, the structure of the set of eventualities (which by Principle 13.2 is mirrored in the structure of the set of subspaces of a vector space) precludes a simultaneous assignment of truth or falsity in a way that is compatible with standard logic. Hence, the earlier assertion in Section 13.2 that in every state there are some indefinite eventualities is inseparable from the formal structure of quantum mechanics. It is impossible to maintain that the states recognised by quantum mechanics are not in fact maximal specifications of the system but can be further refined to yield states which simultaneously determine truth or falsity of all the eventualities.

There is, however, an avenue of escape from this theorem for those who find the conceptual innovations of quantum mechanics unpalatable. The conception of a hidden variables theory can be broadened. Instead of assigning definite truth or falsity to each eventuality when a state of the system is given, the theory can say that the state together with relevant features of the environment are required in order to determine whether an eventuality is true or false. In a given environment only a subset of the eventualities are true or false (notably, those being measured), but these outcomes are not a matter of chance.

Yet another avenue of escape is a stochastic hidden variables theory, which maintains that the outcome of a measurement may be a matter of chance, even when the complete state of the system and that of the environment are both given. This avenue would surely not satisfy those who object to the indeterminism of quantum mechanics, but it may escape other innovations of quantum mechanics, such as entanglement.

In the following subsection we shall see, however, that neither of these avenues of escape will succeed, if one respects the reasonable demand that the hidden variables theory be local.

Bell's inequalities and the experimental disconfirmation of local hidden variables theories

In his papers of 1964 and 1971, J. S. Bell showed that every local hidden variables theory implied under certain conditions an inequality (known as 'Bell's Inequality' or, because of the proliferation of similar results, 'an inequality of Bell's type'). He then showed that there are circumstances under which the predictions of quantum mechanics violate the inequality. There is thus a discrepancy between quantum mechanics and a whole family of hidden variables theories, encompassing all physically attractive theories. This discrepancy, known as 'Bell's Theorem', is the strongest result yet obtained, showing that the formalism of quantum mechanics cannot be maintained for the purpose of predicting phenomena without accepting also a radical conceptual framework. In addition, the discrepancy makes possible an experimental test between quantum mechanics and the entire family of local hidden variables theories.

In order to define 'local' it is necessary to consider a class of systems, of which EPR's was the prototype. Each system of this class consists of two spatially well separated parts 1 and 2, and each part is subjected to bivalent experimental tests (with outcomes labelled + and −), the test on 1 being parametrised by a variable X which takes on at least two distinct values, and the tests on 2 being parametrised by a variable Y which takes on at least two distinct values. A local hidden variables theory for this set of tests is one for which every complete state λ of $1 + 2$ assigns a definite probability $p_\lambda^1(X)$ to the + outcome of the test on 1 when the parameter has value X, independently of what test is performed on 2 and independently of the outcome of that test; and likewise it assigns probability $p_\lambda^2(Y)$ to the + outcome of the test on 2 when the parameter has value Y, independently of the choice of X or the outcome of the experiment on 1 (see figure 13.6). We shall call these independence conditions 'Bell locality'.

Comment (i): By standard probability theory the probability of a − outcome of the experiment on 1 is $1 - p_\lambda^1(X)$, and the probability of a − outcome of the experiment on 2 is $1 - p_\lambda^2(Y)$. Because of the independence conditions the probability of a joint + + outcome in an X-test on 1 and a Y-test on 2 is $p_\lambda^1(X)p_\lambda^2(Y)$.

Comment (ii): The definition of a local hidden variables theory makes no reference at all to the quantum mechanical characterisation of parts 1 and 2, specifically not to the vector spaces associated with 1 and 2 and the projection operators on these spaces. The definition merely refers to two families of bivalent empirical tests. In order to make comparisons with the quantum mechanical predictions, however, it is essential to associate a subspace $\mathscr{E}_X$ with the test on 1 parametrised by X, and a subspace $\mathscr{E}_Y$ with the test on 2 parametrised by Y. In the local hidden variables theory the complete state λ and the parameters X and Y (which may contain information about the contexts of the measurement) may not determine the outcomes of the tests, but may only fix the probabilities of the possible outcomes. Hence the local hidden variables theories are allowed to be *stochastic*.

Comment (iii): No indication has yet been given about the circumstances under which the independence conditions are expected to hold. In the experiments to be discussed below, especially that of A. Aspect, J. Dalibard and G. Roger, the independence conditions appear to be justified by considerations of relativistic space–time structure.

Figure 13.6. A composite system consists of part 1 and part 2. Part 1 impinges upon an apparatus with an adjustable parameter X, and one of two outcomes (labelled + and −) then occurs. Part 2 impinges upon an apparatus with an adjustable parameter Y, and one of two outcomes (again labelled + and −) occurs.

We shall now give in a few steps an outline of a simple derivation of Bell's Inequality invented by J. Clauser and M. Horne. First, it is a fact of elementary algebra that any four real numbers r', r, s', s lying between 0 and 1 satisfy the following inequalities:

$$-1 \leqslant r's' + r's + rs' - rs - r' - s' \leqslant 0.$$

If x' and x are two possible values of the parameter X in the foregoing definition of a local hidden variables theory, and y' and y are two possible values of the parameter Y, then $p_\lambda^1(x')$, $p_\lambda^1(x)$, $p_\lambda^2(y')$, $p_\lambda^2(y)$ are probabilities lying between 0 and 1, which may therefore be substituted for r', r, s', s, respectively. As noted in Comment (i), the probability of the joint + + outcome is $p_\lambda(X,Y) = p_\lambda^1(X)p_\lambda^2(Y)$, whence the algebraic inequality yields

$$-1 \leqslant p_\lambda(x',y') + p_\lambda(x',y) + p_\lambda(x,y') - p_\lambda(x,y) - p_\lambda^1(x') - p_\lambda^2(y') \leqslant 0.$$

In an experimental situation it is usually impossible to control the choice of the complete state λ of the pair $1 + 2$, but it may be assumed that the experimental arrangement determines a probability distribution ρ over the space of complete states (just as the thermodynamic parameters determine the Gibbs distribution over a phase space in equilibrium statistical mechanics). The averages of $p_\lambda^1(X)$, $p_\lambda^2(Y)$ and $p_\lambda(X,Y)$ using the distribution ρ will be denoted, respectively, by $p^1(X)$, $p^2(Y)$ and $p(X, Y)$, and these are the probabilities of finding in the laboratory a + outcome for an X-test on 1, a + outcome for a Y-test on 2, and a joint + + outcome in a joint X-test and Y-test. Consequently, averaging with the use of the distribution ρ in the foregoing inequality yields the following inequality governing the laboratory probabilities:

$$-1 \leqslant p(x',y') + p(x',y) + p(x,y') - p(x,y) - p^1(x') - p^2(y') \leqslant 0.$$

This is the inequality of Bell's type which we promised.

We wish to emphasise the generality of the argument. No quantum mechanical assumptions were made. No restriction was placed upon the character of the complete states λ or upon the distribution ρ or upon the nature of the parts 1 and 2. The crucial assumption is Bell locality (the independence conditions). Finally, although the concept of complete states is used in the course of the derivation, there is an averaging over the space of these states, so that the inequality is formulated in terms of probabilities which are susceptible to laboratory testing.

A laboratory situation in which the quantum mechanical predictions violate the Bell's Inequality was designed in 1969 by J. C. Clauser, M. Horne, A. Shimony, and R. Holt (figure 13.7). They considered a pair of photons 1 and 2 emitted from an atom in a cascade, with the initial and final atomic states having total angular momentum zero. The quantum mechanical polarisation state is almost exactly represented by the Ψ that we met earlier. The quantum mechanical probabilities for $+$ outcomes in an X-test on 1 and in a Y-test on 2 (i.e. of passage through polarisation analysers oriented in the X and Y directions, respectively) are easily calculated to be

$$p_\Psi^1(X) = p_\Psi^2(Y) = \tfrac{1}{2},$$

and the quantum mechanical probability for joint passage through the two polarisation analysers is

$$p_\Psi(X,Y) = \tfrac{1}{2}\cos^2(Y-X),$$

where Y and X are taken to be the angles which the optical axes of the two polarisation analysers make with a fixed axis in the plane orthogonal to the propagation direction of the photons. If the values x', x, y', y of the parameters are, respectively, chosen to be $45°$, $0°$, $22\tfrac{1}{2}°$ and $67\tfrac{1}{2}°$, then

$$p_\Psi(x',y') + p_\Psi(x',y) + p_\Psi(x,y') - p_\Psi(x,y) - p_\Psi^1(x') - p_\Psi^2(y') = 0.207,$$

which conflicts with the upper limit in Bell's Inequality.

The polarisation correlation test of Bell's Inequality was first performed in 1972 by S. Freedman and J. C. Clauser, and with variations has been performed eight times since then. In all but two instances the results agreed with the quantum mechanical predictions and violated Bell's Inequality, and in these two instances there are concrete reasons for suspecting systematic errors. Whether the preponderance of violations of Bell's Inequality is of fundamental significance, however, depends upon whether the experimental situations were such that Bell's Inequality could be expected to hold. In every experiment but one, the parameters of the two polarisation analysers were kept at fixed values during time intervals of the order of a minute. Because the analysers were metres apart, the commonsense judgment of the experimenters and of most commentators was that the choice of the value of Y and the outcome of the Y-test did not affect the probability $p_\lambda^1(X)$, and likewise the choice of X and the outcome of the X-test did not affect the probability $p_\lambda^2(Y)$. Nevertheless, there was a loophole for the defenders of local hidden variables theories, who could say that a minute is enormously long compared to the time needed for a signal travelling at the velocity of light to go from one polarisation analyser to the other, possibly conveying to the 'conspiratorial' hidden variables of each analyser some crucial information about the test being performed with the other analyser, and that conspiratorial decisions of the hidden variables accounted for the failure of Bell's Inequality. This loophole is blocked, however, by the spectacular experiment of Aspect, Dalibard and Roger (1982), in which the choice of the values of X and Y is made by electro-acoustical devices within time intervals of 10 ns, whereas the time required for a signal at light velocity to connect the two analysers is about 40 ns. Some criticisms have

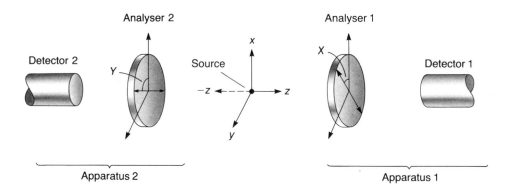

Figure 13.7. Photons 1 and 2 from an atomic cascade at the source impinge, respectively, on polarisation analysers oriented at angles X and Y to the vertical direction. Coincident detections are counted. The apparatus represented here is a specialisation of the generic apparatus indicated in figure 13.6. By making two appropriate choices of the angle X and two appropriate choices of the angle Y the predictions of quantum mechanics conflict with Bell's Inequality.

been made of the design of this experiment, but most commentators agree that Aspect *et al.* achieved an experimental arrangement in which Bell's independence conditions hold unless there is some kind of action at a distance. But the data of Aspect *et al.* conflict with Bell's Inequality and confirm the predictions of quantum mechanics. Consequently, Bell's independence conditions must fail, in spite of the space-like separation of the tests on photons 1 and 2.

Discussion of consequences

Several questions must be discussed in the light of the experimental results summarised above. (i) What assessments should be made of the premises of the EPR argument? (ii) Is the kind of action at a distance exhibited in the experiment of Aspect *et al.* in conflict with the special theory of relativity, and, more generally, is there a discrepancy between quantum mechanical nonlocality and special relativity theory? (iii) Is there any promise in nonlocal hidden variables theories?

In answer to (i), EPR provided a powerful argument for the incompleteness of the quantum mechanical description of physical systems, provided that their premises (*a*) and (*b*) are correct. The experimental results violating Bell's Inequality, especially in the experiment of Aspect *et al.*, throw grave doubt upon premise (*b*). If Bell's independence conditions fail, then it is hard to see how 'an operation can be performed concerning the linear polarisation of photon 1 without disturbing photon 2'. But in what sense can there be a disturbance in the experiment of Aspect *et al.*, where the photons are well separated when the choices of experimental tests are made? In order to throw some light on this question, it is reasonable to re-examine the quantum mechanical treatment of the phenomenon, since quantum mechanics does correctly predict the results in the experiment in question. The quantum state represented by

$$\Psi = \frac{1}{\sqrt{2}}\left[u_x(1)\otimes u_x(2)+u_y(1)\otimes u_y(2)\right]$$
$$= \frac{1}{\sqrt{2}}\left[u_{x'}(1)\otimes u_{x'}(2)+u_{y'}(1)\otimes u_{y'}(2)\right]$$

is such that the eventualities $e_x(1)$, $e_x(2)$, etc. all have indefinite truth values, but if a test is performed on photon 1 yielding a result concerning $e_x(1)$ and $e_y(1)$, then one of the two terms $u_x(1)\otimes u_x(2)$ or $u_y(1)\otimes u_y(2)$ is singled out from the superposition; and whichever of these alternatives occurs has the consequence of making $e_x(2)$ and $e_y(2)$ definite; similarly, if the test performed on photon 1 makes $e_{x'}(1)$ and $e_{y'}(1)$ definite; and likewise, if the test is performed on photon 2. An actualisation of a potentiality for one of the two photons is *ipso facto* an actualisation of the corresponding potentiality of the other one. Hence an operation on one of the photons does disturb the other, and the disturbance is in *some sense* a causal process,

even though the cause (a test on a photon) and the effect (the actualisation of a potentiality of the other photon) are events with space-like separation.

A common objection to this causal interpretation is that the test concerning photon 1 merely conveys *knowledge* about photon 2, by making an inference from the known initial correlation between the polarisations. An analogy is made to a situation in ordinary life, in which a known amount of a conserved quantity is partitioned in an unknown way between two vessels, so that the measurement of the amount in one vessel permits an inference with certainty of the amount in the other. The intention of this analogy is to dispense with a causal explanation of the agreement of truth values of $e_n(1)$ and $e_n(2)$, and incidentally to explain the agreement within a commonsense framework. This explanation fails because it relies upon a tacit premise that the pair $1 + 2$ is such as to fix the truth or falsity of all eventualities $e_n(1)$ and $e_n(2)$, and in particular to fix them so that they agree. With this tacit premise, the discovery of the truth or falsity of $e_n(1)$ reveals enough about the complete state to infer the truth or falsity of $e_n(2)$. But such a tacit premise is a commitment to a hidden variables theory, in fact to a fragment of a simple hidden variables theory, to use our previous terminology. It is easy to see that Bell's independence conditions would be satisfied by such a hidden variables theory, from which his Inequality follows. Consequently, the violation by experiment of the Inequality rules out this commonsensical attempt to evade a causal explanation of the joint truth or joint falsity of the eventualities of photons 1 and 2.

Clearly there is tension between the theory of relativity and the causal interpretation of correlated actualisations of potentialities, as we shall discuss forthwith. For the present, however, our main concern was to have undermined the solidity of premise (*b*) of the EPR argument, and with it the whole train of reasoning which led to the conclusion that hidden variables are needed to supplement the quantum mechanical description of a physical system.

In order to explore further the tension between quantum mechanics and relativity theory, let us consider an experimental arrangement in which $e_x(1)$ and $e_x(2)$ are tested by observers at rest in different inertial frames, and suppose that the tests are events of space-like separation. If the reduction of Ψ is to be interpreted causally, then which of the events is the cause and which is the effect? There is obviously no relativistically invariant way to answer this question. It could happen that in one frame of reference the testing of $e_x(1)$ is earlier than the testing of $e_x(2)$, and in the other frame the converse is the case. Should we relativise the identifications of cause and effect to the frames of reference? The wiser course is to say that quantum mechanics presents us with a kind of causal connection which is generically different from anything that could be characterised classically, since the causal connection cannot be unequivocally analysed into a cause and an effect. Similarly, if the two tests concern polarisation along different axes, e.g. $e_x(1)$ and $e_{x'}(2)$, we cannot say in a frame-invariant way that one outcome occurred first, thereby changing the probabilities of outcomes of the other. The quantum mechanical probabilities of outcomes for $e_x(1)$ and $e_{x'}(2)$ are not independent of each other (since then the Bell independence conditions would hold), but they are due to reciprocal influence, without singling out one event as the cause and one as the effect. This kind of causal connectedness between two events with space-like separation has no classical analogue, and no classical analogue should be expected, since quantum mechanical potentiality has essentially broadened the concept of an event.

It may be asked, however, whether the tension between relativity theory and quantum mechanical nonlocality is not more serious than we have admitted, in that a clever arrangement would permit faster than light communication by using an entangled state as a medium of communication. There are, however, several demonstrations of great generality – by R. Peierls, D. Page, G. C. Ghirardi *et al.* and J. Jarrett – that superluminal communication in this manner is impossible. We shall not recapitulate the general arguments, but instead shall study one particular simple arrangement, which at first seems promising for communication, in order to see why communication does not occur.

We shall take an ensemble of photon pairs $1+2$, all prepared in the quantum state represented by Ψ, and all emitted from a common source within a very short time interval. Two experimenters are well separated on the z-axis, one with a linear polarisation analyser for observing the 1 member of each pair (i.e., the one propagating in the $+z$ direction) and the other with an analyser for observing the 2 member of each pair (i.e., the one propagating in the $-z$ direction). The observer of the 1's makes a binary choice between orienting the optical axis of the analyser of the 1's along the x axis or along the x' axis, and the observer of the 2's knows that these are the only options. If the observer of the 2's can infer from the statistics of photons passing or not passing the analyser of the 2's whether the first observer chose x or x', then one bit of information has been communicated; and if the entire set of observations on the 1's constitutes an event of space-like separation from the entire set of observations on the 2's, then the bit is communicated faster than light (see figure 13.8). Simple calculations show, however, that no information is communicated in this way. If the choice is x, then in half the cases the term $u_x(1) \otimes u_x(2)$ is singled out by chance from the superposition Ψ and in half the cases the term $u_y(1) \otimes u_y(2)$ is singled out. If the observer of the 2's chooses to test $e_x(2)$, it will be found that in half of the cases $e_x(2)$ will be true and in half it will be false – i.e. half of the 2's will pass through the analyser. If the choice of the observer of the 1's is x', then in half the cases $u_{x'}(1) \otimes u_{x'}(2)$ is singled out by chance from Ψ and in half the cases $u_{y'}(1) \otimes u_{y'}(2)$ will be singled out. By the law of Malus, a proportion $\cos^2\theta$ of the first group of 2's will pass an analyser oriented in the x direction and a proportion $\sin^2\theta$ of the second group will pass, where θ is the angle between the x-axis and the x'-axis. Again, a total of half of the 2's will pass through the

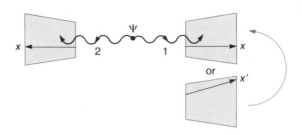

Figure 13.8. An ensemble of pairs $1+2$ is prepared, all in the quantum state Ψ. An observer on the right decides whether to observe the 1's with a polarisation analyser oriented along the x-axis or with one oriented along the x'-axis. The observer on the left tries to infer which decision was made by observing the statistics of passage of the 2's through a fixed polarisation analyser.

analyser. The statistics are exactly the same, independently of the choice of the first observer. No bit of information is communicated. In other words, the quantum mechanical probabilities are perfectly fitted to preclude the possibility of capitalising upon quantum nonlocality for the purpose of sending a signal faster than light. In this sense there is peaceful coexistence between the space–time structure of special relativity (SR) theory and the nonlocality of quantum mechanics (QM). We may use the peace sign ☮, which is well known in political action, to express the theoretical situation schematically:

QM ☮ SR.

In answer to our third question, one possible reaction to the experimental disconfirmation of the family of local hidden variables theories is to take seriously nonlocal theories. Might it not be possible to keep one of the radical conceptual innovations of quantum mechanics – namely, nonlocality – and nevertheless devise alternatives to quantum mechanics which dispense with objective indefiniteness, objective chance and objective probability? Suggestions of this kind have been made by D. Bohm and J.-P. Vigier. Both, in different ways, postulate subquantum models in which pulses like shock waves propagate superluminally in a kind of ether and establish correlations among separated systems in a way that yields statistical predictions agreeing with those of quantum mechanics. Both authors claim that the instability of the postulated medium precludes controllable communication faster than light. We are not convinced by their arguments. It is hard to see how pulses can propagate 13.5 m in the postulated ether in such a way as to establish precisely the correlations observed in the experiment of Aspect *et al.*, and nevertheless not be susceptible to enough control to transmit an

SOS signal. It is not clear either that the nonlocal hidden variables theories are fully in agreement with the crucial experiments or that they peacefully coexist with relativity theory.

13.5   The problem of actualising potentialities

The strength and the weakness of the quantum mechanical framework

Since much of our analysis has been devoted to the possibility of accounting for all the predictions of quantum mechanics within the conceptual framework of classical physics, the impression may have been conveyed that the radical conceptual innovations of quantum mechanics are philosophically or aesthetically distasteful to the author and were stoically accepted only after the failure of all reasonable hidden variables theories. Quite the contrary, the author finds the quantum mechanical framework to be strong, beautiful, and even intuitive. Two examples will illustrate the virtues of this framework.

It is well known that symmetry principles and their systematic study by the mathematical theory of groups play a much more prominent role in quantum than in classical physics. A major reason for this fact is that the indefiniteness of eventualities permits the existence of quantum states which exhibit a much higher degree of symmetry than is possible with classical states. For a classical particle moving in a spherically symmetrical field of force about a point P, the invariance of the laws of motion under rotations about P has the consequence that if $x(t)$ is a possible motion of the particle, so also is the trajectory which results by subjecting $x(t)$ to any rotation about P. Any actual trajectory $x(t)$, which is the sequence of classical states of the particle through time, obviously breaks the symmetry. Potentialities, however, can exhibit more symmetry than actualities, and the quantum state is a network of potentialities. The spherical symmetry of the field of force permits the existence of quantum states which are spherically symmetrical and stationary in time: at all times the expected value of the component of velocity in any direction is zero, and at all times the angular distribution of the position of the particle is the same in all directions. Even more remarkable, this spherical symmetry is not achieved at the price of complete quiescence, for the kinetic energy of the particle is not zero.

As a second example, we note that the power of quantum mechanics to explain the cohesion of matter, which was inexplicable in classical physics, makes essential use of the concept of entanglement of states. Striking examples are the covalent bond of the hydrogen molecule, the stability of the benzene ring, and the tensile strength of metallic crystals. In all these cases calculations show that no nonentangled state of the electrons of the systems can explain the tightness of binding which is found experimentally.

In spite of triumphs such as these, there remains one crucial weakness in the framework of quantum mechanics. The conditions for the actualisation of indefinite eventualities are not made explicit in the framework, and, as we shall see in the next subsection, the dynamical law of quantum mechanics is a serious obstacle to understanding actualisation. But this difficulty undermines the concepts of the framework themselves. In our presentation of the concept of objective chance, we considered an eventuality e which is indefinite in the state S of the system, but is such that definite truth or falsity is elicited when the system is subjected to an appropriate procedure. If actualisation of e never occurs, then it is vacuous to inquire whether it occurs by chance or deterministically. Similarly, it is vacuous to attribute objective probability to the possible outcomes if definite outcomes do not occur. Furthermore, the linearity of the dynamical law of quantum mechanics implies a difficulty in understanding how potentialities are actualised in situations where, on a practical level, we know them to be actualised. Consequently, there is not only an anomaly within quantum mechanics, but also tension between quantum mechanics and some of our practical knowledge of the physical world.

Linear dynamics and measurement procedures

We shall present the problem of the actualisation of potentialities in a specialised way, by looking at measurement procedures. Our reason for doing this is not that we think the problem concerns only human interventions in physical systems for the purpose of obtaining knowledge. It is rather that in a measurement procedure there is an unusual degree of control over the preparation of the system and the relevant parts of the environment with which it interacts, with the consequence that we can say with confidence both what the predictions of quantum mechanics are and what practical knowledge we have of the possible outcomes of the process. The so-called 'quantum measurement problem' is just a special case of the problem of the actualisation of potentialities, but it is a particularly well sharpened case. Another term which is a near synonym is 'the problem of the reduction of the wave packet', the locution arising from the fact that eventualities of interest in measurement often concern the position of a particle, and the initial state is one in which the position is indefinite and the wave-function (which is the position representation of the state) is a spread-out packet. Yet another near synonym is 'the problem of the reduction of superpositions', which is self-explanatory.

We shall consider a composite system $1 + 2$ in which system 1 is a microscopic object of interest and system 2 is the measuring apparatus. Let e be an eventuality of system 1. According to Principle 13.2, e corresponds to a subspace of the vector space $\mathscr{V}^1$ associated with system 1. Any vector of $\mathscr{V}^1$

can be expressed as the sum of a vector in $\mathscr{E}$ and one in $\mathscr{E}^\perp$. The eventuality e is definitely true if the state of system 1 is represented by a vector in $\mathscr{E}$, it is definitely false if the state is represented by a vector in $\mathscr{E}^\perp$, and it is indefinite if the state is represented by a vector which has non-null projections in both $\mathscr{E}$ and $\mathscr{E}^\perp$. We assume that system 2 is prepared at time $t = 0$ in a *neutral state*, in which the apparatus is ready to register subsequent to interaction system 1 whether e is true or not. The neutral state is represented by the vector v_0. The system $1 + 2$ is assumed to be in a nonreactive environment, so that its temporal evolution is governed, according to Principle 13.5, by the family of linear operators $U(t)$. Finally, we assume that there are two orthogonal subspaces $\mathscr{F}_+$ and $\mathscr{F}_-$ of $\mathscr{V}^2$, such that the following conditions are satisfied:

if u belongs to $\mathscr{E}$, then $U(t_f)(u \otimes v_0)$ belongs to $\mathscr{V}^1 \otimes \mathscr{F}_+$
if u belongs to $\mathscr{E}^\perp$, then $U(t_f)(u \otimes v_0)$ belongs to $\mathscr{V}^1 \otimes \mathscr{F}_-$,

for a specified t_f, which is the time of completing the measurement.

Comment (i): What we have just described is clearly a schematic measuring procedure for the eventuality e of the object. Since $\mathscr{F}_+$ and $\mathscr{F}_-$ are mutually orthogonal subspaces of the vector space associated with the apparatus, there is an eventuality f of the apparatus which is true for any apparatus state represented by a vector in $\mathscr{F}_+$ and false for any apparatus state represented by a vector in $\mathscr{F}_-$. It follows that an observer who knows that initially e either is definitely true or definitely false, but does not know which, can determine which of these alternatives is the case by examining the eventuality f of the apparatus. We refer to f as an 'indexical eventuality', since its truth value is a reliable index for the truth value of e (see figure 13.9).

Figure 13.9. This is an idealised measuring apparatus for determining whether eventuality e is true or false. In both the upper and the lower parts of the figure the initial neutral state of the apparatus is represented. In the upper part it is shown that if e is initially true, then at the conclusion of the measurement process the eventuality f of the apparatus is true. The lower part shows that if initially e is false, then at the conclusion of the measurement process f is false.

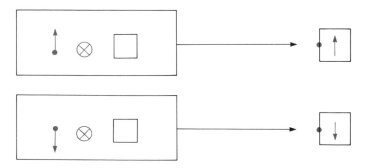

Comment (ii): It is straightforward to generalise the above procedure so as to measure a dynamical variable $\mathscr{A}$ with a discrete set of possible values. All that is needed is to replace $\mathscr{E}$ and $\mathscr{E}^\perp$ by a finite or denumerably infinite set of orthogonal subspaces $\mathscr{E}_i$, each associated with a specific value of $\mathscr{A}$; and to replace $\mathscr{F}_+$ and $\mathscr{F}_-$ by a corresponding set of orthogonal subspaces of $\mathscr{V}^2$.

Comment (iii): In several respects the scheme of measurement which has been presented is more general than usual quantum mechanical treatments of measurement. System 1 is not assumed to be in the same state at t_f as at $t = 0$. Also, there is no guarantee that a remeasurement of e will yield the same result, since e is not even assumed to be true (or false) at t_f if it was at $t = 0$. There are, however, yet more general schemes of measurement, which take into account the impossibility of laboratory preparation of a definite quantum state of a large system, like a piece of apparatus, and therefore describe the initial 'neutral state' as a mixed state, giving only probabilities of finding the system in each of a set of pure quantum states. We shall not consider in detail this gesture towards realism, because it does not change our conclusion that actualisation of potentialities is problematic in a measurement process governed by quantum dynamics, even though the argument required is more complicated than the one which will be given. For our purposes it suffices to think of v_0 as representing the correct though unknown initial quantum state of the apparatus. We could also consider schemes of measurement in which the truth or falsity of the apparatus eventuality f is not a perfectly reliable index of the truth or falsity of the object eventuality e, but again the conclusion would not be changed.

The scheme of measurement presented above generates an anomaly in a straightforward way. Let u belong to $\mathscr{E}$ and u' belong to $\mathscr{E}^\perp$, both of unit length, and consider the superposition $cu + c'u'$, where neither c nor c' is zero, and $|c|^2 + |c'|^2 = 1$. The initial state of object + apparatus is then represented by $(cu + c'u') \otimes v_0$. Then the state of object + apparatus at the final time t_f is represented by

$$\Phi_f = U(t_f)[cu + c'u') \otimes v_0] = cU(t_f)(u \otimes v_0) + c'U(t_f)(u' \otimes v_0),$$

where use has been made of the linearity of $U(t_f)$. But in the sum on the right the first term lies in $\mathscr{F}_+$ and the second in $\mathscr{F}_-$, and hence in the state represented by Φ_f the eventuality f of the apparatus is indefinite. Here is a disturbing conclusion, because f is an eventuality of a macroscopic apparatus and therefore could very well be decidable by direct inspection. Thus, objective indefiniteness of eventualities is not confined to microscopic systems, where our commonsense preconceptions might be suspected not to hold, but is also exhibited in the macroscopic domain, where conventional wisdom holds that classical descriptions are adequate. There is a further and more serious problem. An observer who looks at the apparatus always finds after the foregoing procedure that eventuality f is definitely true or false (e.g. a light is definitely on or definitely off), and the indefiniteness inferred from the dynamical law of quantum mechanics is not exhibited. Consequently, there is a conflict between the literal implications of quantum mechanics and the practical observation report. The most dramatic example of this conflict is supplied by Schrödinger's cat paradox, in which the apparatus includes a cat which is neither definitely alive nor definitely dead in the state Φ_f (see figure 13.10). (It must be emphasised that Schrödinger did not believe in the indefiniteness of the eventuality of the cat's being alive, but only reached this conclusion as a *reductio ad absurdum* argument against the literal correctness of quantum mechanics.)

The conflict which we have just noted depends upon interpreting Φ_f as the objectively correct description of system $1 + 2$, rather than as a compendium of knowledge. Much of the literature on the quantum measurement problem presents solutions which either tacitly or explicitly interpret quantum mechanical indefiniteness as ignorance on the part of the observer. Our long analysis of hidden variables theories in Section 13.4 had the net effect of confirming that superpositions like Φ_f signify objective indefiniteness, thereby precluding easy solutions to the problem of measurement.

Conjectures on the revision of quantum dynamics

The derivation above of the objective indefiniteness of the apparatus eventuality f at the conclusion of the measuring process rested upon the existence and the linearity of the time evolution operators $U(t)$. Revision of the quantum mechanical dynamics has been suggested as the way to prevent the problem of measurement from arising in the first place. Advocates of such revision have in general sought to preserve intact the time-independent part of the quantum mechanical formalism, together with the interpretation of the quantum state as a network of potentialities. The modified dynamics is to have the property of assuring the evolution of a state of a macroscopic system represented by $\Sigma c_i \phi_i$, in which a macroscopic dynamical variable $\mathscr{A}$ has an indefinite value, into one represented by one of the ϕ_k, in which $\mathscr{A}$ has a definite value. Possibly the adjective 'macroscopic' can be dropped in a wide range of cases, but we have stated only a minimum demand upon the modified dynamics, since the conflict between the linear quantum dynamics and laboratory practice is most evident when $\mathscr{A}$ is a dynamical variable of the apparatus.

This programme is endangered from the onset by a mathematical difficulty which was explored by G. Mackey, B. Simon and others: that the time-independent parts of quantum mechanics together with some very moderate assumptions about temporal evolution suffice to infer the linear quantum dynamics. There are, however, ways of evading this difficulty, by weakening the moderate assumptions about temporal evolution or by omitting part of time-independent quantum mechanics. Consequently, some hope remains of a successful

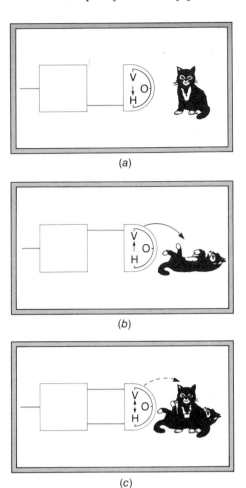

Figure 13.10. Schrödinger's cat. If a photon comes out through the horizontal channel of the polarisation analyser the cat is unaffected and remains alive (*a*), but if it comes through the vertical channel a lethal device is triggered and the cat is killed (*b*). Does quantum physics imply that, until the box is opened and its state measured, the cat is neither alive nor dead?

replacement of the usual time-dependent Schrödinger equation either by a nonlinear differential equation or by a stochastic equation. These options have been explored by B. Mielnick, I. Bialynicki-Birula and J. Mycielski, C. Piron, P. Pearle, N. Gisin, G. C. Ghirardi, and others, but much work both of surveying the mathematical possibilities and assessing them experimentally remains to be done.

In our opinion, the prospects of a successful nonlinear variant of quantum dynamics are not good, because reasonable desiderata for such a theory pull in opposing directions. On the one hand, the deviation of the nonlinear theory from the usual quantum dynamics must be small, at least for systems with few degrees of freedom, because of the excellent confirmations of quantum dynamics for these systems. (In particular, two neutron optics experiments by C. Shull *et al.* and by R. Gähler *et al.* place extremely low upper bounds upon the magnitude of the nonlinear term in one important family of nonlinear dynamical equations.) On the other hand, the magnitude of the nonlinear term must sometimes be large for some systems with many degrees of freedom, because the nonlinearity must account for the apparatus's transition from a superposition of the form $\Sigma c_i \phi_i$ to a single term ϕ_k. Furthermore, if the apparatus is actually to have a definite value of the macroscopic dynamical variable $\mathscr{A}$ at the conclusion of the measurement process, the final state cannot be a superposition of various of the ϕ_i with one large coefficient and other coefficients small but nonzero; it is a matter of principle that the dynamical evolution of the initial superposition must annihilate completely every term but one. Since practical measurements, yielding definite values of macroscopic dynamical variables, are often accomplished in time intervals of the order of nanoseconds, the annihilation of competing terms must be accomplished very quickly, and surely cannot be allowed to linger until t goes to infinity. Thus stringent requirements are placed upon the form of the putative nonlinear dynamical equation and upon the magnitude of its nonlinearity. In spite of the obvious tension among the various desiderata for a nonlinear theory, we cannot say that they are mathematically incompatible. It is conceivable that a well designed nonlinear equation will have the property that the magnitude of the nonlinearity increases rapidly with the number of degrees of freedom and with the mass of the system, thereby permitting all the desiderata to be satisfied, and incidentally providing a bridge between the micro-level and the macro-level. Nevertheless, none of the work done so far on nonlinear dynamical theories provides grounds for optimism about this programme.

Somewhat more promising is the programme of replacing the time-dependent Schrödinger equation by a stochastic equation. A rough model is provided by Langevin's equation for the evolution of the classical state of a Brownian particle, using a stochastic term to represent the random forces of molecules colliding with the particle. One fascinating possibility is that the source of the stochastic modification of the Schrödinger equation is the space–time structure itself, which cannot be abstracted away from the dynamical behaviour of the system of interest even if it is isolated from other material systems. An argument could be considered along the following lines. First, the general relativistic idea is accepted that the matter–energy distribution provides source terms for the field equations governing space–time structure. Second, it is supposed that each of the values a_k of the dynamical variable of the system is associated with a different source term for the field equations – e.g., each a_k is the location of the centre of mass of a large piece of apparatus. Third, it is assumed (contrary to programmes for quantising general relativity) that the Superposition Principle

does not hold for states of space–time itself, so that the indefiniteness of the dynamical variable $\mathscr{A}$ cannot give rise to an indefiniteness of the space–time metric. The resistance of space–time to objective indefiniteness is then assumed in effect to trigger a stochastic choice on the part of the system, thereby actualising the indefinite dynamical variable $\mathscr{A}$. A speculation of this kind has been advanced by F. Károlyházy, A. Frenkel and B. Lukács, and they have even proposed an experimental test of their theory. Whether or not the experiment is performed, and regardless of the outcome, their proposal is valuable as a suggestion for a physical source of a stochastic modification of the Schrödinger equation.

It should be pointed out that the peaceful coexistence between quantum mechanics and special relativity theory may not be maintained by a modified quantum mechanics. The failure of peaceful coexistence would not *ipso facto* be grounds for dismissing a revision of quantum mechanics. After all, the best classical space–time theory we now possess is not special but general relativity, and the quantisation of general relativity remains a major unsolved problem of contemporary theoretical physics. (See the chapter by Isham.) Quite apart from nonlocal quantum correlations, there are difficulties in understanding quantum fluctuations of the space–time metric. It is not inconceivable that both kinds of difficulties will be resolvable only if relativity and quantum mechanics are modified in tandem. Needless to say, some remarkable experimental confirmations would be needed in order to justify major innovations on two fronts of theoretical physics simultaneously.

The 'many-worlds' interpretation

H. Everett daringly proposed to deal with the problem of explaining how quantum mechanical potentialities are actualised by denying that actualisation occurs. He claims that his proposal is analogous to dealing with Newton's problem of determining which among the inertial frames is objectively the rest frame by denying that there is such a thing as an objective rest frame. According to Everett, any closed system – and specifically the universe as a whole, which is the only good candidate for a system which is unequivocally closed – is at any time in a pure quantum state, the temporal development of which is governed by a group of unitary operators U(t). The dynamical development of the quantum state $\Psi(t)$ of the universe may readily make a macroscopic dynamical variable $\mathscr{A}$ have an indefinite value. No intrusion of some superposition-reducing dynamical process, supplementary to the time-dependent Schrödinger equation, is postulated for the purpose of singling out one value a_k as the actual value of $\mathscr{A}$. Of course, $\Psi(t)$ can be written as a superposition of states, in each of which $\mathscr{A}$ has a sharp value. Let $\mathscr{A}$ be a dynamical variable of system I (a subsystem of the universe as a whole), and let ϕ_i $(i=1,2,3,..)$ be vectors of unit length in the vector space $\mathscr{V}^{\mathrm{I}}$, such that $\mathscr{A}$ has the value a_k in ϕ_k; and let system II consist of all of the universe other than I. Vectors χ_k in the vector space $\mathscr{V}^{II}$ can be found such that

$$\Psi(t) = \Sigma c_i \phi_i \otimes \chi_i.$$

Every term in this superposition with a nonvanishing coefficient is a 'branch' of the universe as a whole, which is as real as any other branch. When a state of I is chosen, in any manner whatever, the choice together with $\Psi(t)$ itself uniquely determines a *relative state* of II. If, for example, ϕ_k is chosen for I, then the relative state of II is by χ_k. The advantage of the relative state conception, according to Everett, is the emphasis given to correlations. For example, if II contains a number of independent observers as subsystems, each of whom makes readings of $\mathscr{A}$, then upon assumption that their readings are accurate they will all agree that $\mathscr{A}$ has the value a_k in the branch labelled by k. In this way the intersubjective agreement among physicists is understandable quantum mechanically, even though from a global point of view $\mathscr{A}$ does not have a definite value. There is no attribution of the power to reduce superpositions to observers, for there is a relative state of each observer, in each branch, with sensori-neural apparatus and presumably also the corresponding consciousness exhibiting the properties appropriate to that branch.

Because of the thesis that all the branches in the superposition $\Sigma c_i \phi_i \otimes \chi_i$ are equally real, Everett's theory is sometimes called 'the many-worlds interpretation of quantum mechanics', though he himself seemed to prefer to call it 'the theory of the universal wave-function'. His solution of the problem of the actualisation of potentialities, or more accurately his dismissal of the problem, has attracted a quite large following. Some followers are also enthusiastic about a feature of the theory which we have not yet mentioned: his programme of deriving a frequency interpretation of the coefficients c_i by looking at the statistics of repeated events in typical branches, thus maintaining the independent postulate of probabilistic interpretation of the coefficients. Conversations with Everett's followers suggest that they may be resonating to the theme of parallel histories of the world, which has captivated the imagination of readers and viewers of science fiction. There are, however, numerous authors, including B. d'Espagnat, J. S. Bell, and H. Stein, whose imaginations have resisted this theme and who have presented some powerful criticisms of Everett's interpretation.

One line of criticism rests upon the observation that in any vector space of dimension greater than one there are infinitely many different ways to choose a set of mutually orthogonal vectors, and each way provides the means for expressing an arbitrary vector of the space as a superposition. Consider, in addition to the vectors ϕ_i mentioned above, another set of mutually orthogonal vectors in $\mathscr{V}^I$, namely $\phi_1', \phi_2', \ldots$, each representing a state in which $\mathscr{A}$ has an indefinite value, and write

$$\Psi(t) = \Sigma c_i' \phi_i' \otimes \chi_i',$$

where χ_i' is a vector of $\mathscr{V}^{II}$ determined by ϕ_i'. Then the universe characterised by the same wave-function $\Psi(t)$ as before ramifies into branches, but now not in a way that is congenial to our imaginations, for each of the equally real branches exhibits an indefinite value of the macroscopic dynamical variable $\mathscr{A}$, and furthermore the relative state of II in each branch makes each observer reflect the indefiniteness of $\mathscr{A}$. But branches of this kind are alien to our experience. As observers we never do see a pointer on a dial somehow suspended between pointing up and pointing down (which presumably is a different experience from perceptual vagueness). Everett's interpretation appealed to a properly disposed imagination because he focused his attention upon a set of branches for each of which $\mathscr{A}$ has a definite value. There is, however, a democracy of such sets from the standpoint of the vector space formalism itself, and if some are to be preferred to others it can only be for reasons which have not yet been made explicit. It could be conjectured, for example (though Everett himself seems to be silent on this point), that human consciousness is restricted to branches in which macroscopic variables have definite values. Branches which do not satisfy this condition would be as real as any others, but they would be automatically censored by human consciousness and would not contribute to the empirical evidence. This conjecture should not be dismissed out of hand, and yet it cannot be regarded as a serious proposal unless an appropriate psychophysics is worked out within the framework of quantum mechanics. If Everett's interpretation is not coherent without the accomplishment of such an ambitious programme, then it loses the simplicity which at first sight is one of its charms.

It has also been objected that the many-worlds interpretation violates the methodological rule that entities ought not be multiplied beyond necessity. The only necessity which has been adduced for maintaining the equal reality of an infinity of branches is unwillingness to curtail the range of validity of standard quantum dynamics. But in view of the fact that the experimental confirmation of that dynamics is almost entirely based upon the behaviour of microscopic systems – the only macroscopic evidence being provided by special systems like superconductors – it is very hazardous indeed to extrapolate the validity of that dynamics to the universe as a whole.

Finally, we must reflect that equal reality of all branches wipes out the distinction between potentiality and actuality, which is central to decision making, to ethical choice, and to all practical activity. Bell has said, 'If such a theory were taken seriously it would hardly be possible to take anything else seriously. So much for the social implications.'

The philosophy of Bohr

More than any other person, Niels Bohr formulated a defence of the intelligibility of quantum mechanics within a few years after its discovery and laid to rest the philosophical scruples of most of a generation of physicists. His principle of complementarity provided a point of view within which the wave–particle duality and the Heisenberg uncertainty relations could be understood, and his answer to the argument of EPR was generally accepted by the community of physicists as a vindication of the completeness of quantum mechanical descriptions.

Bohr maintained, as a general principle of interpretation in physics, that theoretical concepts, including assertions of the reality of entities or of their properties, cannot be used unambiguously without careful reference to the experimental arrangement in which the concepts are applied. This principle introduces no relativity to the individual observer, because the irreversible registration that occurs at the conclusion of a measurement is accessible to many observers, and the information obtained can be communicated without ambiguity. What quantum mechanics adds to this general principle is the indivisible quantum of action, which implies that the interaction between the object and the apparatus cannot be made negligible or entirely compensated for. Consequently, properties of an object which are measured by incompatible experimental arrangements must not be regarded as simultaneously real, as was possible in classical physics. The experimental arrangements exhibit a relationship of complementarity to each other, and the resulting limitation upon the simultaneous realisability of properties of a specified object is the principle of complementarity. Furthermore, the phenomena which result from the interaction of an object and an apparatus cannot be subdivided in such a way that the contributions of the two physical systems can be discerned, because 'any attempt at a well-defined subdivision would demand a change in the experimental arrangement incompatible with the definition of the phenomena under investigation'. Bohr's conception of complementarity thus has two distinguishable elements. One, the indivisibility of the quantum, is purely physical. The other, concerning the condition of unambiguous use of concepts, is more generally philosophical, making it possible for Bohr to extend his principle of complementarity to other domains, especially biology and psychology, in which an appropriate surrogate for the indivisibility of the quantum can be found.

Bohr believed that the apparent anomalies of quantum mechanics are consequences of misinterpretation, resulting from the use of theoretical concepts without detailed attention to the experimental arrangement and from neglect of the fact that the arrangement must be described in the language of classical physics. The use of expressions like 'creation of physical attributes of objects by observation' are manifestations of such misinterpretation, and we conjecture that he would object equally strenuously against the phrase 'actualisation of potentialities', which has been used repeatedly in this chapter.

Can the anomalies of quantum mechanics be resolved or, more accurately, prevented from arising, in the way that Bohr suggests? To do justice to this question would require a longer excursion into philosophy and a more detailed examination of Bohr's subtle and condensed assertions than space permits. All

that can be said briefly is that an answer depends crucially upon the choice between two radically different conceptions of legitimate demands upon a theory of knowledge. It is essential to Bohr's theory of knowledge that the ordinary human elements in experience be accepted without challenge or revision; he wrote 'we must not forget that, in spite of their limitation, we can by no means dispense with those forms of perception which colour our whole language and in terms of which all experience must ultimately be expressed'. In ordinary experience there are objects with definite macroscopic properties which are perceivable by all properly endowed observers. It is illegitimate to demand a complete explanation of how these definite perceptions are permitted by the basic laws of nature, because the definiteness is the presupposition of the whole scientific enterprise of systematising experience. The enterprise requires that there are objects with definite macroscopic properties which are perceivable by all properly endowed observers. Bohr wrote: 'Without entering into metaphysical speculations, I may perhaps add that an analysis of the very concept of explanation would, naturally, begin and end with a renunciation as to explaining our own conscious activity.' The theme of renunciation and submission to the unavoidable limitations of the human condition is recurrent in Bohr's writings, and places him, more perhaps than he realised himself, in a philosophical tradition of renunciation of excessive claims to human knowledge, including Hume and Kant. The latter systematically maintained that human beings have no knowledge of 'things in themselves', but only of the objects of experience.

There is, however, another philosophical tradition, according to which the presuppositions of human knowledge are open to full rational investigation. A coherent philosophy, according to this tradition, has not one but two starting points. One starting point is what is given in ordinary human experience, on the basis of which inferences are made about the constitution of the world beyond us. The other starting point consists of the fundamental principles of the constitution of the world, among which are the principles of physics. According to this philosophical tradition, a coherent philosophical system shows how the two starting points are compatible and connected. In particular, the cognitive power of the knowing subject is a legitimate object of investigation, and there is no 'renunciation as to explaining our own conscious activity'. Of the very diverse philosophers belonging to this tradition (including Aristotle, Locke, Leibniz, and perhaps Einstein), not many would claim that a coherent philosophical system has been achieved, but all would regard coherence in the sense defined as a philosophical desideratum, and they discern no limitation of human faculties which in principle precludes its achievement.

The problem of the actualisation of potentialities serves as an illustration of a challenge to the achievement of philosophical coherence. Suppose, as Bohr insists, that definite results of observations employing macroscopic apparatus provide the indispensable data to which physical theories refer. Then, according to the second tradition, the principles governing the interactions of microscopic objects and macroscopic apparatus must be such as to guarantee the definiteness of the physicists' observations. We cannot be content with Bohr's assertion that the definiteness of these observations is the starting point of scientific investigation, but should be able to explain in terms of natural principles how this definiteness occurs. It would follow, then, that if the unlimited validity of quantum mechanics throughout the physical world prevents the explanation of the definiteness of measurement results, some modification or limitation of quantum mechanics is needed.

The debate for nearly three decades between Bohr and Einstein concerning human knowledge and the foundations of quantum mechanics can be understood as a conflict between the two philosophical traditions which have just been cited. It should be obvious that the bulk of the present paper was written with sympathy for the second tradition, which aims at a 'coherent' philosophy in which the two distinct starting points are connected. The problem of the actualisation of potentialities has turned out to be a serious obstacle to this philosophical programme. There remain avenues to be explored, and it is surely premature to say that the programme as a whole is a failure. Nevertheless, the prospects of the programme are less favourable than they were a generation ago, and Bohr's philosophical views on limitations of explanation and the inexplicability of the starting point of human knowledge have been favoured by the outcomes of analysis and experimentation. It should be emphasised, however, that even if in the long run Bohr's general theses continue to be supported by scientific developments, a major task will remain of expanding his condensed dicta into a detailed account of human knowledge.

13.6   Conclusions

Intensive research on the foundations of quantum mechanics has yielded quite firm answers to some questions, but has left others undecided. The experimental disconfirmation of local hidden variables theories is unlikely to be reversed by further experimentation. Consequently, it does not seem feasible to interpret quantum mechanical indefiniteness, chance, probability, entanglement and nonlocality merely as features of the observer's knowledge of a physical system. Rather, they seem to be objective features of the systems themselves. Thus the conceptual innovations of quantum mechanics are likely to remain a permanent part of the physical world view. To be sure, the family of nonlocal hidden variables theories has not been refuted, but the prospects for such theories are not at present promising, and, even if one turns out to be successful, it would still have the consequence of recognising nonlocality as a feature of the physical world.

There is, however, no consensus among students of the foundations of quantum mechanics concerning a solution to the problem of the actualisation of potentialities, or, to its special case, the measurement problem. Possibly a solution will be obtained by yet another radical modification of physics, such as a change of the dynamical law of quantum mechanics. Another possibility is a further radical modification of the conception of physical reality: for instance, the elimination of the dichotomy between potentiality and actuality proposed by Everett, or the attribution to consciousness of the power to actualise potentialities. By contrast with these radical proposals, Bohr's treatment of the problem of measurement may seem conservative, but that judgment would be inaccurate, for Bohr himself recognised that he was proposing radical limitations on the scope of human knowledge.

Whatever the outcome of our present uncertainties may be, it is sure to be philosophically significant. Those who have deplored the rift between science and philosophy which began to develop in the eighteenth century may take comfort in the mutual relevance of these disciplines exhibited in the foundations of quantum mechanics. They will find here a vindication of the old sense of 'Natural Philosophy'.

Further reading

By far the best anthology on the foundations of quantum mechanics is *Quantum Theory and Measurement*, edited by John A. Wheeler and Wojciech H. Zurek (Princeton University Press, 1983). This anthology contains many of the papers of the Bohr–Einstein debate and important papers of Schrödinger, Heisenberg, Wigner, Bohm, Bell, Aspect and Everett. The annotated bibliography is a good guide to further literature, and it contains references to many of the works which were mentioned in the text of this chapter only by author. References not given in this bibliography can almost all be found in the bibliographies of one of the following:

Bell, J. S. (1983). *Foundations of Physics* **12**, 989–99, reprinted in Bell, J. S. (1987). *Speakable and Unspeakable in Quantum Mechanics*. Cambridge University Press, pp. 159–68.

d'Espagnat, B. (1984). *Physics Reports* **110**, (4), 202–64.

Kamefuchi, S., Ezawa, H., Murayama, Y., Namiki, M., Nomura, S., Ohnuki, Y. and Yajima, T. (1984). eds. *Foundations of Quantum Mechanics in the Light of New Technology*. Physical Society of Japan, Tokyo.

Shimony, A. (1986). In *Quantum Concepts of Space and Time*, C. Isham and R. Penrose, eds., pp. 182–203. Oxford University Press.

A valuable exposition of Bohr's philosophical viewpoint is

Petersen, A. (1968). *Quantum Physics and the Philosophical Tradition*. MIT Press, Cambridge, MA.

Two fine popular expositions are

Polkinghorne, J. C. (1985). *The Quantum World*. Longman; and

Rae, A. (1986). *Quantum Physics: Illusion or Reality?* Cambridge University Press.

The very rapid switching between settings of the polarisation analyser, which was described in Section 13.4, is reported in

Aspect, A., Dalibard, J. and Roger, G. (1982). *Phys. Rev. Lett.* **49**, 1804–7.

14 The quark structure of matter

Frank Close

14.1 Why quarks?

For more than two millenia philosophers and scientists have asked the question 'What are things made of?' The ancient Greeks believed that matter is ultimately built from small, indivisible pieces which they called 'atoms'. Even so, some wise people criticised this belief and warned that some day you would have to face the question 'What are atoms made of?'

The obvious way to answer these questions is to go and look at things in fine detail. To appreciate why this task is not as simple as it seems at first sight, we should realise what is involved in seeing things.

There are three essential ingredients. First you need a source of illumination, some radiation which shines on the object in question. This radiation then bounces off the object and enters a detector. For example, the radiation could be light from the sun, the object could be this page and the detector could be your eye.

Atoms are much too small to be seen with the naked eye and the ancients had not developed tools with which they could look at matter in fine detail. Today we have microscopes that can resolve minute structures, far smaller than the naked eye can discern. People tend to think that the power of a microscope is its ability to magnify objects, whereas in reality it is the ability to distinguish objects that are close together. If we build more powerful microscopes we can resolve smaller and smaller distances until we encounter an essential limitation imposed by Nature. Beams of light (or of matter) act like waves, and to resolve small objects needs a beam whose wavelength is smaller or comparable to the structure of interest.

The human eye only responds to electromagnetic radiations whose wavelengths lie in a rather small range. This 'visible light' cannot resolve objects smaller than about a thousandth of a millimetre. To resolve details on a finer scale than this requires more powerful illuminating radiations. In devices known as electron microscopes one uses beams of electrons in place of visible light and bounces these off the object of interest. By applying large voltages we can accelerate the electron beam and consequently improve the resolving power at will. The better the resolution we want, so the more energy we must first pump into the beam and in turn this requires ever larger 'microscopes'.

With an electron microscope on the laboratory bench one can resolve structures on the scale of molecules, but barely perceive individual atoms. To resolve structures within the atom needs electron accelerators that are several metres long. Indeed we now know that atoms are not the smallest pieces of matter but have a detailed inner structure of their own. They contain electrons, which are particles carrying negative electrical charge, surrounding a compact nucleus that carries positive electrical charge.

Accelerators that are several miles in size can produce beams of electrons that are powerful enough to resolve the structure within the nucleus and show that atomic nuclei consist of two types of particle, a positively charged proton and a neutral neutron. And even this is not the end. In recent years we have found that these neutrons and protons are in turn built from yet smaller particles called quarks.

This is the current limit that we can resolve. As far as we know electrons and quarks are fundamental particles from which the matter in us and around us is built. Put these few types of particle together in different combinations and you build the infinite varieties of forms, substances and even life itself.

It is possible that in discovering the existence of quarks we have found the ultimate seeds of nuclear matter. However, it may be that there is a layer of reality beyond quarks, entities which cluster to build up quarks as composite structures, and which we are currently unaware of through lack of powerful enough microscopes to resolve them.

Whether this is the case or not, there is no doubt that quarks are an essential layer of reality. Their behaviour governs the properties of bulk matter. There is an elegance in the way Nature operates at the quark level which is often hidden at the atomic, let alone in the macroscopic, domain. So, irrespective of what may lie beyond, it is important to understand the behaviour of quarks for they are the deepest that we have yet managed to probe Nature's secrets.

In this article I shall describe how we became aware of the quarks, how we study their behaviour and what we have learned about them. I will also look for clues from history that may help us make an informed guess as to whether or not quarks are indeed fundamental or instead made from more basic units. But before entering into the more detailed exposition it may be helpful to make a preview of the material that will be covered in this chapter.

Structured systems have the common feature that the constituents can be rearranged relative to one another. Quantum mechanics limits these states to a discrete set each with a characteristic energy, and as a result the system can exhibit a discrete spectrum of energy levels. Familiar examples include molecules, where the constituent atoms vibrate against each other; atoms, where the electrons can be excited into various orbits; and atomic nuclei, whose neutrons and protons vibrate and orbit in various configurations.

The smaller the system is, so the greater is the scale of excitation energy. For atoms the excitations are of the order of an electron-volt (eV). For molecules the scale is a fraction of this, whereas for nuclei the scale is a factor of thousands (keV) or even millions (MeV) greater.

During the 1960s similar excitation spectra were identified for the hadrons (strongly interacting particles like neutrons and protons). When a proton is hit by a beam of particles, it may be excited into a short lived 'resonance' state, much as atoms, molecules or nuclei are. The only essential difference is one of scale: the proton excitations involve tens or hundreds of MeV.

The reason for these excitations is much the same as before. The proton is a composite system built from quarks, and it is the different quantum states adopted by the quarks that generate the resonance states.

The proton is the lightest example of a system formed by three quarks. To every particle variety of matter there exists a corresponding form of antimatter whose mass is the same but whose electrical charge is opposite in sign to that of the corresponding particle of matter. Thus as the proton is built from three quarks, so is the antiproton built from three antiquarks. A single quark and a single antiquark form the 'meson' family of hadrons whose lightest example (quark and antiquark in the lowest energy state – the 'ground state') is the pion. No other combinations of one, two or three quarks and antiquarks have ever been seen. We will see later why these particular combinations and not others occur.

In all composite systems, the excited states tend to be short lived, decaying to the stable ground state and emitting energy in the form of electromagnetic radiation. For molecules and atoms we can literally see the evidence for this in colours all around us.

The existence of excited states in nuclei may be crucial to our existence. The carbon nucleus contains an excited level that is in fortunate near-coincidence to the collision energy of three alpha particles. As a result the chance of three alphas fusing in a star is not small and so the heavy elements essential to life are seeded. The carbon nucleus is formed in an excited state and then emits energy and falls to the stable ground state. Similarly the excited states (resonances) of individual protons and neutrons are short lived and unfamiliar in everyday life. They decay to the ground state configurations. In the case of three quark systems this is the proton or neutron (two different varieties of quark, the up and down, generate these *two* independent combinations) which are the stuff of atomic nuclei. Similarly, a system of quark and antiquark will restore to a pion, but only for so long as the quark and antiquark do not mutually annihilate. The pion is unstable.

These 'spectra' hint that the supposedly simple objects are in fact structured. However, this alone is not sufficient to tell you the nature of the constituents. Clues about these fundamental pieces come from another direction.

There has been a recurring theme during the march to the fundamental constituents of matter. At any given moment in history there are candidates for 'THE' basic elements or fundamental units, such as the atomic elements in the nineteenth century, or the electron, proton and neutron some fifty years ago. Initially these are few in number and there is optimism that simplicity and order has arrived. There is the promise that the many combinations of these few units will generate the richness and variety previously discerned; diversity is equated with complexity and lack of fundamentality.

However, this simplicity is a myth. As we probe the fundamental units in more detail we find that not only do they have excitation spectra but there are many more species of them than at first we thought. For example, the handful of elements known in the eighteenth century were joined by more and more varieties, until over a hundred are now known. Furthermore, these elements are not all independent; several of them appear to be systematically related in their properties. This is the origin of the famous periodic table of the elements due to Dmitri Mendeleev. This is the first major step towards identifying the nature of the constituents.

(i) A pattern

Mendeleev's periodic table of the atomic elements was clear cut but contained gaps. Not all of the elements had been discovered. The position of the gaps in the pattern enabled him to predict that further elements existed with specified properties. Later they were found with the expected properties. This is how it was for Mendeleev: gallium, germanium and scandium being notable predictions that were subsequently discovered in France, Germany and Scandinavia.

In the 1960s a similar train of events occurred with the so-called 'elementary particles'. The proton and neutron are but two of a multitude of particles that respond to the strong nuclear force. A pattern (the 'Eightfold Way', described on p. 401) was discerned and, analogous to Mendeleev's earlier success, gaps were noted. New particles were predicted to fill the gaps and within a year or two had been discovered in the laboratory.

Predicting the existence of new forms of 'fundamental' matter from the behaviour of other forms confirms that the scheme is correct. The richness of the Eightfold Way pattern provided clues as to the properties of the fundamental constituents, the quarks, whose various combinations gave rise to the pattern. For example, it was clear that they carried electrical charges that were fractions $\frac{2}{3}$ or $-\frac{1}{3}$ of a proton's charge. This was a startling result since no evidence for an isolated charge smaller than that carried by a proton had ever been seen.

The detailed description of the Eightfold Way, of hadron spectroscopy and the way that quarks underwrite the properties of hadrons is described in Sections 14.2 and 14.3.

(ii) Scattering 'sees' the constituents

As we noted earlier, you can resolve objects only if you irradiate them with a beam of radiation whose wavelength is comparable to or smaller than the size of the object. To see inside atoms and particles we must use radiations whose resolving power is much better than that of ordinary visible light.

For example, in the years leading up to 1920, Rutherford's group at Manchester fired alpha particles at atoms. These beams had good enough resolving power to penetrate the atom and reveal its inner structure. He was surprised to learn that the alpha particles were thrown back in their tracks when they hit a thin sheet of gold leaf. The most powerful magnetic fields then available could only deflect alphas slightly, yet in the atom there had to be huge forces at work sufficient to turn the alphas through a full 180 degrees. It was from this discovery of violent collisions that Rutherford proposed his model of the atom which has survived the test of seventy years.

To Rutherford the nucleus appeared to be a point. The alpha particles available to him could not resolve the inner structure of the nucleus. Today we have beams of high energy electrons, powerful enough to resolve the inner structure of nuclei and even see inside the proton and neutron.

When this was first done, in the latter half of the 1960s, a parallel sequence of events occurred with the proton and neutron. The beam of electrons suffered violent collisions when it met these nuclear particles, suggesting that the proton's charge is concentrated on quarks within. I shall describe these experiments in more detail in Section 14.4.

(iii) Ionisation produces the constituents

When you irradiate a target with enough energy to resolve its inner structure it becomes likely that you will eject the constituents from within the system. This is known as ionisation. Thus, in the case of atoms we get electrons, electronics, PacMan etc.

We can remove electrons from within atoms and eject protons and neutrons from within nuclei. However, this step has not been achieved at the quark level. Isolated free quarks have not been produced and the theoreticians believe that Nature forbids quarks to exist as free particles; we say that quarks are permanently confined (see Chapter 17 by Taylor). Even so we have no doubt that quarks exist, albeit permanently confined within clusters of twos or threes forming particles such as the pion and proton. Much of this article will survey the evidence for this.

(iv) The force that binds the quarks

Having identified a new layer of reality, there are important questions such as what laws govern the behaviour of the new elementary particles, what forces cluster them together to build up the complex structures that had previously been thought of as the elements?

In the atomic case the force had been known for centuries. The binding of atoms involves the electrostatic attraction of opposite electrical charges, negatively charged electron and positively charged nucleus. The motion of these charges gives rise to magnetic effects too, so the force is more correctly called the 'electromagnetic' force. When electrical charges are disturbed they may emit electromagnetic radiation, such as light. The theory of the electromagnetic force and the interaction of electrical charges and light is called 'quantum electrodynamics' or QED for short (see Chapter 16 by Georgi).

There are many subtle effects in QED manifested in the behaviour and properties of atoms. We shall see that there are tantalising parallels between these and the way that quarks behave.

Quarks carry electrical charge and so feel the electromagnetic force. But in addition they possess a new form of charge called 'colour'. This acts in many ways like electrical charge but, as we shall see later, with some subtle and far reaching differences. As electrostatics and electromagnetism are to electrical charge, so are chromostatics and chromomagnetism to colour. Much as electrical charges attract and neutralise in atoms, so do colour charges attract and neutralise. This is how quarks mutually attract and form the colour-neutral or 'colourless' clusters that we know of as proton and neutron for example.

In the same way that neutral atoms are built from electrically charged constituents, so do colourless protons and neutrons contain coloured quarks. Two colourless protons feel practically no colour forces when they are far apart from one another. However, when they are very near to one another, the coloured quarks in the one can feel the colour force exuded by the coloured quarks in the other. Complicated short-range attractions and repulsions result and give rise to a large variety of nuclear phenomena. This is the source of the strong nuclear forces between neutrons and protons.

Chromostatics is similar to electrostatics. Combined with relativity and the quantum theory we have 'quantum chromodynamics' by analogy with quantum electrodynamics. This theory of quark interactions has survived a decade of intense scrutiny so far. I will give more details in Section 14.3, and the foundations of the theory are described at more length in Taylor's chapter, 'Gauge theories in particle physics'.

(v) The menu

As far as we know, all matter is made of quarks and leptons. These particles behave as if they are spinning.

We are all familiar with the concept of momentum as a particle moves through space; if a particle rotates then we say that it has angular momentum. Quantum mechanics restricts the allowed magnitudes to be integer multiples of a basic amount known as Planck's quantum. This rule applies to particles in motion around one another or around some common centre. However, the electron and the quarks themselves possess an intrinsic angular momentum, as if they are spinning, and the amount of this spin angular momentum is one-half of Planck's quantum. We say that they have 'spin one-half' (spin $\frac{1}{2}$). They can spin in either of two senses, which you may visualise as clockwise or anticlockwise if you wish.

There is a profound principle of Nature that constrains the behaviour of particles whose spins are $\frac{1}{2}$. This is known as the Pauli exclusion principle after its discoverer, Wolfgang Pauli. His principle forbids more than one spin $\frac{1}{2}$ particle to occupy the same quantum state. Electrons and quarks have spin $\frac{1}{2}$, and in consequence are constrained by the Pauli exclusion principle.

The most stable systems of electrons or of quarks arise when these particles have no rotary motion relative to one another. We say that they have no net angular momentum. This preferred configuration is called the 'S-state'. Now, although they are not rotating about one another, they still have their own intrinsic spinning behaviour. But the directions of their spins must conspire such that the Pauli principle is satisfied. Thus in an S-state one electron can spin in one direction, the other one in the opposite sense and that is the lot. It is impossible to have three electrons in the S-state.

It is hard to overrate the importance of the Pauli principle. It restricts the number of possible structures that can occur. Moreover, it generates a regularity among those that do occur. In the case of electrons orbiting atomic nuclei it generates the periodicity among the atomic elements as in Mendeleev's periodic table. In the case of quarks it generates a systematic pattern among the quark clusters (the nuclear particles such as protons and neutrons) that was the first clue to the existence of the deeper layer of reality. That Nature limits the available structures, and forms a regular pattern rather than a disordered mess, may be crucial for our ability to have discerned order in the scheme of things and to have made any progress in science at all.

There are believed to be six kinds of quarks named up, down, charmed, strange, top and bottom, each type having spin $\frac{1}{2}$. They are denoted by the letters u, d, c, s, t, b. (The top and bottom are sometimes called truth and beauty.) Whereas five of these varieties are well established, evidence for the 'top' quark is controversial. Each quark can carry any one of three varieties of the colour charge. There are also six varieties of lepton – colourless particles such as the electron which do not exert colour forces and do not respond to the strong nuclear force. In addition to the well known electron there is the neutrino, which always accompanies the electron in radioactive beta-decay processes. There are two heavier relatives of the negatively charged electron, each of which is negatively charged: these are the muon and the tau. In turn these each have their own variety of neutrino. This gives a total of six species of lepton which are denoted by the letters e, ν_e (electron and its neutrino); μ, ν_μ muon and its neutrino); τ, ν_τ (tau and its neutrino).

The fundamental electromagnetic, colour and weak forces act on quarks and leptons. Electromagnetic forces act on the electrical charges; the colour or strong force acts on the colour charges and the weak force has its most familiar manifestation in radioactive beta-decays.

These forces are transmitted by a family of particles known as gauge bosons (see the aforementioned chapters by Georgi and Taylor). The most familiar example is the electromagnetic force whose gauge boson is the photon, the quantum bundle of electromagnetic radiation. The analogous transmitters of the strong colour force are gluons. The weak force is transmitted by W and Z bosons.

Whereas we have known of the photon for decades, we have only recently become aware of the gluons, W and Z bosons. And their discoveries have been direct results of our ability to recognise the existence of quarks and to put those quarks to use. This has given us new insights into the possible unity underlying these apparently disparate forces.

The way that quarks and leptons respond to these forces specifies these particles' properties. If we define the proton charge as our basic unit of electric charge then there are three leptons with an electric charge of minus-one unit: the electron, muon and tau. In addition there are three neutrinos that have no electrical charge at all. There are three quarks that have electrical charge $\frac{2}{3}$ (up, charm and top) and three with electric charge $-\frac{1}{3}$ (down, strange and bottom).

As far as we know the world is built from up and down quarks together with electrons and their neutrinos. For some unknown reason, Nature has repeated this pattern (at least) twice. Each of these sets is known as a 'generation' of quarks and leptons. Thus we know of three generations of fundamental particles of matter.

In any generation there are quarks and leptons whose properties are identical to those in another generation but for their masses. The concept of quark mass is rather a vague one since the quarks appear to be permanently confined in clusters and are not freely available. Even so the relative magnitudes of their masses are more readily defined and we find that the charmed quark is some five times heavier than the up, while the top is at least thirty times heavier than the charmed variety, but otherwise these quarks seem to respond in precisely the same way to electromagnetic, colour and weak forces. Similar remarks apply to the three quarks with charge $-\frac{1}{3}$; the strange quark is somewhat heavier than the down and the bottom is an order of magnitude heavier than the strange, while in other respects these quarks appear to act in the same ways.

For each type of quark or lepton there also exists an antiparticle. The mass of an antiparticle is identical to that of the corresponding particle; their electrical and colour charges are opposite in sign but equal in amount to that carried by their particle counterpart (see table 14.1). These regularities are very striking and may be the first clue that there is a pattern among the quarks and leptons hinting that they are not truly fundamental. We will return to this at the end of the article and first concentrate on the evidence for quarks and the way that they behave.

In this chapter I will develop the above sequence of events one by one. First we will review the way that coloured quarks build up the pattern of observed strongly interacting particles (the hadrons) and show the crucial role of the Pauli principle in this. Next we will see how the quarks were detected in scattering experiments, the modern analogue of Geiger–Marsden and Rutherford's work on the nuclear atom.

The lessons learned will then be applied. We can fire beams of quarks at one another even though we cannot get individual quarks out of their parent hadrons. Protons are swarms of quarks and so beams of protons are in effect beams of quarks. Once we know how the quarks are distributed inside the protons we can correct for this and interpret a beam of one billion protons as a beam of three billion quarks, say. This is the rationale behind recent experiments at CERN, the European Centre for Nuclear Research in Geneva, where protons and antiprotons (quarks and antiquarks) are collided head-on.

Table 14.1. *The fundamental particles of matter – quarks and leptons*

	Quarks		Leptons	
	$\frac{2}{3}$	$-\frac{1}{3}$	0	-1
Charge				
	u	d	ν_e	e
	c	s	ν_μ	μ
	(t)	b	ν_τ	τ

Antiparticles have the same masses and spins, but the charges are opposite in sign.

From the annihilation of matter and antimatter new forms of radiation and matter emerge. Thus were discovered the W and Z bosons (carriers of the force responsible for beta-radioactivity). In order to design and interpret these new experiments, it was necessary first to have detailed knowledge on the quark constituency of protons.

Finally we will raise questions that are currently being debated. Are there new forms of hadronic matter, such as glueballs, as suggested by quantum chromodynamics? Are quarks really the end of the road or is there a level of reality beyond the quark?

14.2   Quarks and hadron spectroscopy

In this section we will voyage into the atomic nucleus and see its structure revealed. Initially this will be in terms of protons and neutrons. In addition to the familiar proton and neutron many unfamiliar ephemeral particles have been discovered, which are also built of quarks. We will see how the idea of quarks was first developed and show how all of the known strongly interacting particles are explained as clusters of quarks.

When Rutherford first realised that atoms contain a central nucleus, he saw the nucleus only as a point object. The alpha particles, emitted in radioactive processes, were unable to penetrate heavy nuclei and show their structure.

Having found that atoms consist of electrons in the peripheral regions with a compact and complex nucleus deep in the centre, the next question was 'What is the nucleus made of?'. By the 1930s the answer was becoming known. Nuclei are built from two varieties of heavy particles whose masses are almost identical. One is electrically neutral, the neutron, the other is positively charged, the proton.

These advances were made by studying light nuclei, that is those that could be disrupted by alpha particles, aided by the development of particle accelerators. By accelerating charged particles artificially, one could produce beams of particles with enough energy to shatter nuclei and study their structure in some detail. Later, very high energy beams in cosmic rays and modern accelerators revealed a whole menagerie of short-lived particles.

The first sense of order emerged with the realisation that there are two distinct types of matter. There are particles like the electron that do not experience the strong nuclear force; these are called *leptons* (from the Greek for lightweight, thin). Then there are particles that do feel the strong force; these are called *hadrons* (from the Greek for bulky).

Today we know of six varieties of lepton: the electron, the muon and the tau, and three corresponding varieties of neutral neutrino. As far as we know these are truly elementary particles. The really exciting results were in the family of hadrons. Over 200 of these are now known – exceeding even the number of atomic 'elements'!

It will be useful to divide the hadrons into two subspecies called *baryons* and *mesons*. The proton and neutron are the lightest and best known members of the baryon family. Baryons are particles of spin $\frac{1}{2}$ that feel the strong force and that are unstable and decay, ultimately returning to a proton. (Protons might be permanently stable. If they are unstable their half life is at least 10^{32} years – see Chapter 15 by Georgi.) The mesons, on the other hand, are integer spin particles that feel the strong force, are unstable and ultimately decay to electrons, photons and neutrinos. The lightest and best known examples of mesons are the pions and kaons. Pions transmit nuclear forces and hold nuclei together. Kaons are 'strange' particles, containing a strange quark. Kaons were first seen in cosmic rays in the 1940s and were the first heralds of stuff 'beyond ordinary matter'.

Today we recognise that all of these mesons and baryons possess one or more of six profound attributes. They are charged or neutral, some have a property called strangeness, some are charmed and some have beauty (sometimes called bottomness). Many have several of these properties, e.g. charge and strangeness (as the K^+), or charge, strangeness and charm (as the F^+), and so on. The search is on for particles predicted to possess a sixth attribute, called 'topness', or 'truth'. These whimsical names are codewords that we use to describe deep properties of nature. What they *are* is not well understood. For example, what *is* electrical charge? We know it by its effects but that does not tell us what it is. Similarly we recognise that other bizarre properties occur and so we have to call them something – such as strangeness or charm.

A pattern – the Eightfold Way

This proliferation of 'fundamental' particles has already been commented upon as being the first hint of deeper structure. Historically this is what happened with the atomic elements; now we see a modern analogue with the hadrons.

The next essential step soon followed when a *pattern* was discerned (known as the Eightfold Way, discovered in the early 1960s). This led to the proposal that the strongly interacting particles, the hadrons, are not fundamental but are built from quarks. Six varieties, or 'flavours', of quarks give rise to the six previously mentioned attributes of hadrons.

When the Eightfold Way patterns were first deduced, only three of the attributes of matter were known – charge, neutrality and strangeness. These were sufficient to point the way forward; the subsequent discoveries of charm and bottom and top matter generalises the patterns into more complicated forms. Indeed, it is fortunate that we discovered the various forms of matter sequentially – it enabled us to see the wood rather than the trees. Charm and bottom matter were not found until the 1970s, and the first claims for sighting top matter only emerged in 1985 (but have subsequently been questioned; the proof of the existence of top is still awaited).

If you collect together the proton and other spin $\frac{1}{2}$ baryons and plot them on a diagram according to their strangeness and electrical charge you obtain the pattern shown in figure 14.1. The pattern becomes significant when you make a similar plot of the pion and kaons (spinless mesons whose quantum wavefunctions change sign in a mirror, the so-called 'pseudoscalar mesons'; figure 14.2). The same pattern qualitatively emerges for the pseudoscalar mesons as for the spin $\frac{1}{2}$ baryons.

Figure 14.1. Families of eight baryons, each with spin $\frac{1}{2}$, and ten baryons each with spin $\frac{3}{2}$. The top line in each case has zero strangeness; the Σ and Λ members have strangeness -1; Ξ and Ξ^* have strangeness -2; Ω has strangeness -3. Their electrical charges are shown by superscripts $+$, $-$ or 0. Their quark constituencies are in parentheses.

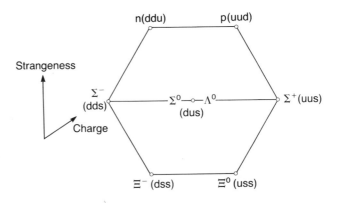

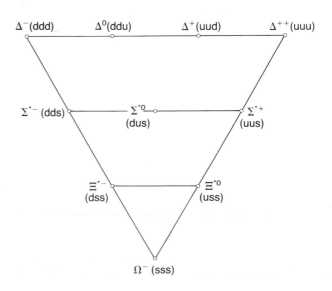

Although this similarity proved suggestive in 1962, we shall see that the patterns arise for rather different reasons. A corresponding pattern emerges for the spin 1 'vector' mesons (ρ, ω, K*, ϕ) while the $J=\frac{3}{2}$ baryons, (Δ, Σ^*, Ξ^*, Ω^-) form an enlarged triangle.

When these patterns were first noticed in 1962, the omega-minus (Ω^-) particle was unknown. Gell-Mann predicted its existence and what properties it must have if the pattern were to be completed. The subsequent discovery of the 'omega-minus' confirmed the validity of the scheme and directed attention to the next stage: why are the patterns as they are?

These particular hadrons are built from three varieties, or flavours, of quark; the up, down and strange. These are sufficient to build all the hadrons discovered before 1974.

The quarks have charge and strangeness as in figure 14.3(a) and so form a triangle on a strangeness–charge plot. The antiquarks have the opposite strangeness and charge to their quark counterparts and form a triangle that is inverted relative to that of the quarks (figure 14.3b).

The quarks and antiquarks all have spin $\frac{1}{2}$. So three quarks clustered together, with no orbital motion relative to one another (known as the 'S-wave' state), have a net amount of spin $J=\frac{1}{2}$ or $\frac{3}{2}$. The properties of these configurations are the same as those found in the family of spin $\frac{1}{2}$ particles that includes the proton or the family of spin $\frac{3}{2}$ particles such as the omega-minus. A pair, in particular quark and antiquark, will form $J=0$ or 1 (like the spinless 'pseudoscalar' pion and kaon or the family of spin 1 'vector' mesons). Now let's see how the patterns emerge.

Two quarks form a total of six possible combinations as shown in figure 14.4. All these states have fractional electric charge and none are seen in nature. But put three together and you get the large triangle of states in figure 14.1. These have integer charges and strangeness from 0 to -3. They correspond precisely with the ten members of the family including the delta and omega-minus. If we lop off the corners (the meaning of this will emerge when we examine the constraints of the Pauli principle) we obtain the pattern of the $J=\frac{1}{2}$ states including the proton and neutron. The charges and strangeness agree with the empirically observed states.

When we combine two quarks we obtain fractionally charged, unseen, states. But if we combine a quark and an antiquark we obtain integer charged, 'real' states corresponding to the known mesons. This is shown in figure 14.5.

The patterns of mesons and baryons appear to be the same, but closer examination shows that they are not. There are octets and decuplets (tens) of baryons, whereas mesons come in nonets. This basic pattern has been verified over and over as more hadrons with higher spins have been uncovered during the last twenty-five years. In all cases, hadrons containing strange quarks are slightly heavier than their nonstrange counterparts. The up and down quarks have nearly identical masses, whereas the strange quark at rest has some 150 Mev more energy or mass.

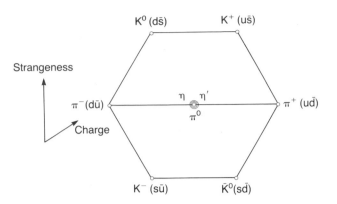

Figure 14.2. Family of nine spinless mesons exhibiting a similar pattern to figure 14.1. However, here the top line has strangeness $+1$, the middle has zero strangeness and the bottom line has strangeness -1. The quark and antiquark constituencies are in parantheses. (Antiquarks are denoted by $\bar{u}$, $\bar{d}$, $\bar{s}$.)

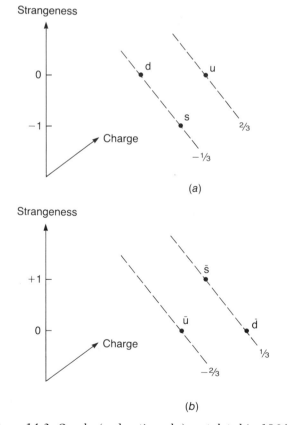

Figure 14.3. Quarks (and antiquarks) postulated in 1964 as constituents of strange and nonstrange hadrons. They form a basic triangular pattern in a two-dimensional space of strangeness and charge. The up, down and strange quarks are denoted by u, d, s. Their corresponding antiquarks are denoted by $\bar{u}$, $\bar{d}$, $\bar{s}$. Their properties are summarised in table 14.1.

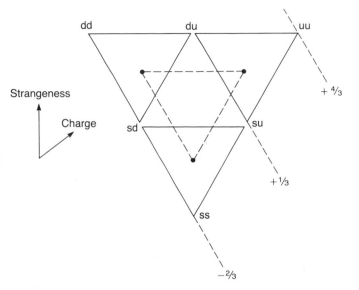

Figure 14.4. Pattern resulting from pairs of quarks in strangeness–charge space. The u, d, s quarks form an inverted triangle as in figure 14.3 (the dashed triangle). The addition of a second quark generates a total of six possible combinations of charge and strangeness. Diagrammatically this appears as a set of inverted triangles centred on the vertices of the original dashed triangle.

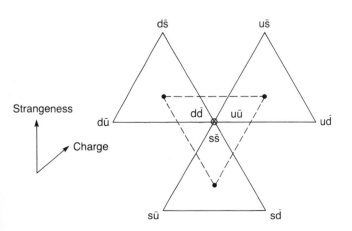

Figure 14.5. Pattern resulting from quark and antiquark in strangeness–charge space. The u, d, s quarks form an inverted triangle (dashed) as in figure 14.4. The addition of an antiquark generates nine possible combinations of charge and strangeness. These appear as a set of triangles (compare with figure 14.3 for antiquarks) centred on the vertices of the original dashed triangle.

All of these hadrons can be explained if we regard the quarks and antiquarks as dynamical objects that interact much as electrons and nuclei in atoms. So there will be a set of excited states with the quark spins coupled to their relative orbital angular momenta. First we illustrate this for mesons; a quark orbiting an antiquark.

Mesons

There are so many hadrons known today that no one can remember all their names, let alone their properties. A physicist once commented that 'If I had to remember all their names I might as well have become a botanist'. You don't have to! First, data on them are collected and updated every two years in a publication known as the *Particle Data Tables*. In these data tables you will see the mesons classified by properties such as parity (*P*) (behaviour under spatial reflection) and charge conjugation (*C*) (particle–antiparticle symmetry). Second, they are all composed of quarks and/or antiquarks whose spins and orbital angular momenta have coupled in various ways, like atomic physics. We don't expect students to remember all the atomic energy levels, let alone give them names. Similarly, it is sufficient to be able to work out the allowed combinations of spin and other quantum numbers by coupling the quarks together. This turns out to describe the pattern of hadrons precisely and provides compelling evidence for the quark model of hadrons. A summary of the long lived particles is contained in the appendix to this chapter.

A meson consists of a quark and an antiquark. If we were to look in a mirror we would see a quark and antiquark still, but left and right have swapped. If the quark was on the left of the meson, it will be on the right of the centrepoint in the mirror image, and vice versa. We still have the same meson, and in quantum mechanics we can ask how the state in the mirror relates to that in the real world.

The quark–antiquark system is described by a wavefunction – a mathematical expression related to the chance of finding a quark at position $+r$ and the antiquark at $-r$ relative to their centre of mass. In the mirror the quark will be at $-r$ and the antiquark at $+r$. The relation between the meson as seen in the mirror and the real world is contained in the mathematical wavefunction – how it behaves as we replace $+r$ by $-r$ everywhere. This is called parity. It turns out that the rules of quantum mechanics imply that for a quark and antiquark with L units of angular momentum, the parity is

$$P = (-1)^{L+1}.$$

Quantum mechanics restricts L to integer values, either even or odd. So if the pair have no orbital angular momentum, $L = 0$, or any even amount, the parity is negative – the wavefunction changes sign from positive to negative in going to the mirror world. If L is an odd number then the parity will be positive.

Analogous to this we can imagine a magic mirror that interchanges matter and antimatter. This sends quark to antiquark and vice versa. So a quark–antiquark pair remains an antiquark–quark pair; a meson remains a meson but, again, quantum mechanics shows how the mathematical wavefunction behaves under this operation (called 'charge conjugation', denoted by C). Whereas the parity, P, depended only on the relative orbital angular momentum between the two objects, C depends on both the orbital angular momentum, L, and the combined spin state, S, of the quark and antiquark.

Quantum mechanics constrains both the amount of orbital angular momentum, L, and the spin S of the quark and antiquark system to be an integer. If the sum of the magnitudes of $L + S$ is an even number, the wavefunction is unchanged when charge conjugation is performed; if it is an odd integer, then the wavefunction changes sign.

One can therefore work out the set of allowed $J(PC)$ for a quark and antiquark in net spin 0 and 1 coupled in orbital angular momentum L, and total spin $\mathbf{J} = \mathbf{L} + \mathbf{S}$ (the vector sum of $\mathbf{L}$ and $\mathbf{S}$). The lowest few are shown in table 14.2. Note that there are no states with $J(PC)$ (odd) $- +$ nor (even) $+ -$. Such states (in particular $1 - +$) are exotics within the framework of the quark model, where the states consist of a single quark and an antiquark. If such an exotic were found it would show that states exist beyond this simplest configuration. None have yet been seen. This supports the quark model of mesons.

Some members of these multiplets and their masses are shown in figure 14.6. Exciting a quark from the ground state into a configuration where it rotates around its fellows costs energy, and correspondingly gives the system higher mass overall. Exciting a quark to each successive level costs about 0.5 GeV energy. A similar price is seen for baryons and also for heavy states made from charmed and bottom quarks.

The discovery of mesons containing charmed quarks in 1974 provided a perfect example of nonrelativistic bound states where the constituents orbit and spin as in atoms. A bound state of a charmed quark and a charmed antiquark forms mesons which have 'hidden charm'. The charmed antiquark has the opposite amount of charm as a charmed quark, so in a system of a charmed quark and charmed antiquark the net charm is zero, even though there is charm 'hidden' within.

The lightest examples of such mesons have masses of some 3 GeV, about three times as much as a proton. They are known as 'charmonium' by analogy with the atomic system 'positronium'. The latter is a bound state of electron and positron bound by the electromagnetic force. The charmed quark and charmed antiquark are bound by a powerful 'colour' force – the quarks carry a new form of charge called colour (see Section 14.3). Analogous to hydrogen or positronium, the charmonium system has a series of energy levels. The ground state is when the quarks are in the S-wave and couple their spins to total zero or one. There are both orbital excitations (P,D by analogy with the standard nomenclature in atomic physics) and radial excitations (1S, 2S, etc.).

The same effects occur for 'bottomium' – the bound states of a bottom quark and bottom antiquark. These have masses in excess of 9 GeV. Four radial excitations have been identified so far in the S-wave and two in the P-wave. The separation in energy between successive levels in bottomium is almost identical to those in charmonium, even though the mass scale is some three times greater. If you interpret these states as evidence for quarks interacting through some potential, then the mass systematics suggest that the potential varies roughly as the logarithm of the distance (some 0.5×10^{-15} m) separating the quarks in these states.

There should also exist states consisting of charm or bottom in association with up, down or strange antiquarks. Several of these have been identified.

Baryons

If we restrict our attention to the states made from three of up, down or strange quarks we see that the mesons occur in families of nine for each combination of spin and orbital angular momentum. For baryons, however, it is more subtle. Baryons contain three quarks. These can be identical in flavour and the Pauli exclusion principle places important constraints on the possible quantum states that they occupy, leading to a rich and detailed spectroscopy.

The essential details of the Pauli principle have been mentioned in Section 14.1.

In quantum theory it implies that the wavefunction of a system of quarks must change sign when any two of the quarks

Table 14.2. *The possible spins of mesons formed from a single quark* (q) *and an antiquark* ($\bar{q}$)

	$S = 0$	$S = 1$
$L = 3$	3^{+-}	4^{++} 3^{++} 2^{++}
$L = 2$	2^{-+}	3^{--} 2^{--} 1^{--}
$L = 1$	1^{+-}	2^{++} 1^{++} 0^{++}
$L = 0$	0^{-+}	1^{--}

The $q\bar{q}$ couple their individual spins to $S = 0$ or $S = 1$. Their orbital angular momentum about one another is denoted by $L = 0,1,2,3 \ldots$ in units of $\hbar$. Then L and S couple together giving the total. The superscripts denote the affect on the wavefunction's sign under parity and charge conjugation operations. L can have any integer value, but only a few are listed here. Compare with figure 14.6.

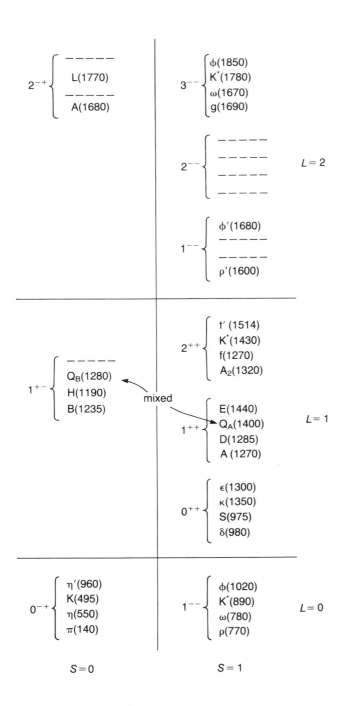

$$2^{-+} \left\{ \begin{array}{c} \text{-----} \\ \text{L}(1770) \\ \text{-----} \\ \text{A}(1680) \end{array} \right.$$

$$3^{--} \left\{ \begin{array}{c} \phi(1850) \\ \text{K}^*(1780) \\ \omega(1670) \\ \text{g}(1690) \end{array} \right.$$

$$2^{--} \left\{ \begin{array}{c} \text{-----} \\ \text{-----} \\ \text{-----} \\ \text{-----} \end{array} \right. \quad L = 2$$

$$1^{--} \left\{ \begin{array}{c} \phi'(1680) \\ \text{-----} \\ \text{-----} \\ \rho'(1600) \end{array} \right.$$

$$2^{++} \left\{ \begin{array}{c} \text{f}'(1514) \\ \text{K}^*(1430) \\ \text{f}(1270) \\ \text{A}_2(1320) \end{array} \right.$$

$$1^{+-} \left\{ \begin{array}{c} \text{-----} \\ \text{Q}_B(1280) \\ \text{H}(1190) \\ \text{B}(1235) \end{array} \right. \quad \xrightarrow{\text{mixed}}$$

$$1^{++} \left\{ \begin{array}{c} \text{E}(1440) \\ \text{Q}_A(1400) \\ \text{D}(1285) \\ \text{A}(1270) \end{array} \right. \quad L = 1$$

$$0^{++} \left\{ \begin{array}{c} \epsilon(1300) \\ \kappa(1350) \\ \text{S}(975) \\ \delta(980) \end{array} \right.$$

$$0^{-+} \left\{ \begin{array}{c} \eta'(960) \\ \text{K}(495) \\ \eta(550) \\ \pi(140) \end{array} \right.$$

$$1^{--} \left\{ \begin{array}{c} \phi(1020) \\ \text{K}^*(890) \\ \omega(780) \\ \rho(770) \end{array} \right. \quad L = 0$$

$$S = 0 \qquad\qquad S = 1$$

Figure 14.6. Meson states built from a quark and an antiquark. The quark and antiquark couple their spins to a total of spin 0 or 1. Their relative orbital angular momentum is denoted by *L* and can have integer values only. The total angular momentum of the system (**J**) is the vector sum of **L** and **S**. The states are denoted J^{PC}(N(mass)). The superscripts *PC* refer to the wavefunction's behaviour under parity and charge conjugation. N is the name of the resulting meson and its mass is given in units of MeV.

are interchanged. We say that the wavefunction must be 'antisymmetric'.

The power of the Pauli principle is evident already when the quarks are in the ground state, the S-wave. They couple their individual spins of $\frac{1}{2}$ to a total of $\frac{1}{2}$ or $\frac{3}{2}$. These will be the total spin of the system, and indeed the nucleon belongs to an octet of spin $\frac{1}{2}$ baryons, the next family up the mass spectrum being the spin $\frac{3}{2}$ family of ten (decuplet) that includes the delta and omega-minus states. The decuplet contains states which are manifestly symmetric under interchange of flavour, such as the omega-minus that consists of three identical strange quarks. Their spins have coupled to a total of $\frac{3}{2}$, which is symmetric in spin too. Being in the ground state the spatial wavefunction is trivially symmetric, and so we appear to have a system that is symmetric under exchange of its constituents – something that is in conflict with the Pauli principle which requires *antisymmetric* states where spin $\frac{1}{2}$ constituents are concerned.

There is an extra ingredient, however, that we have not yet discussed. The quarks carry a new form of charge called colour. This is the source of the forces that cluster the quarks together forming the hadrons. We will see how this operates later, but for the moment it is sufficient to note that there are three available colour labels for the quarks. The Pauli principle is satisfied if the three quarks are antisymmetric under interchange of their colour labels.

The way that the colour forces act appears to attract quarks to one another only if their overall colour state is antisymmetric. The Pauli principle then requires that the three quarks are *symmetric* under interchange of flavour, spin and spatial quantum numbers; the *antisymmetric* colour configuration takes care of the overall *antisymmetry*.

Once this subtlety has been allowed for, all of the known baryons are fitted with the spectrum expected in the quark model. If we excite a quark to the P-wave then we have added one unit of orbital angular momentum to the net spin of $\frac{1}{2}$ or $\frac{3}{2}$. This yields total spins of $\frac{1}{2}$, $\frac{3}{2}$ or $\frac{5}{2}$. States exist with these spins and masses about 0.5 GeV more than the ground state proton. This is in line with what we saw already for mesons.

Summary

All particles that are built of quarks belong to one of two families. Mesons consist of a single quark and a single antiquark; baryons consist of three quarks. No one has yet detected a particle consisting of two quarks or two quarks and an antiquark, nor yet a single isolated quark.

These regularities must be an important clue to the nature of the forces that cluster the quarks together. We do not yet understand these forces fully, but some of the main features have been identified.

Next we will survey our knowledge of these forces and show why the mesons and baryons are the preferred combinations. In turn this will given us new insights into the strong force that builds up the nuclei at the heart of atoms.

14.3   Colour: a new force acting on quarks

Quarks carry electrical charge and also have another form of charge called 'colour'. This is the source of the powerful forces that bind the quarks to one another and build up the mesons and baryons. In this section we will see how the colour forces act on the quarks, building the mesons and baryons and determining their properties. We will learn that the theory of colour implies that new forms of matter made from 'gluons' should exist. The search for these is underway but there is, as yet, no clear sighting of a 'glueball'.

The colour charge is in many ways similar to electrical charge. As we have some familiarity with the properties of electrical charges, it may be helpful to present colour by analogy with them.

Whereas electrical charges are simply positive or negative, there are three different varieties of colour: positive or negative of either red, green or blue variety. These have nothing to do with real colours of course; 'colour' is simply a name to distinguish them. If quarks carry positive colour charge, then their antimatter partners, the antiquarks, carry negative colour charge. The rules of chromostatics are rather similar to those of electrostatics: like colours repel, opposite colours attract. Thus the attraction of opposites, such as red quark and red antiquark, forms the familiar mesons.

Because there are three forms of positive colour charges, the possible attractions are richer than we are used to for electromagnetic forces. In addition to 'opposite' charges we have also 'unlike' charges, such as red and green. The attraction of opposites generalises to 'attraction of unlike charges', but with a slight subtlety. We have to take into account the quantum state that the quarks are in. What does this mean? Suppose we have two quarks, one coloured red (R) and one blue (B). If we arbitrarily call the position of the red '1' and the blue '2' then we denote the pair as R_1B_2. Conversely if they were swapped around we would write B_1R_2. In quantum theory we do not distinguish between these; the physical state is a 50:50 mixture which is described as either the sum or the difference of the two, i.e.

$$R_1B_2 + B_1R_2$$

or

$$R_1B_2 - B_1R_2.$$

If we swap the position labels 1 and 2 then the first state above stays the same (symmetric) whereas the second state changes sign (antisymmetric).

The physical significance of this has to do with the nature of the forces that the quarks mutually experience. It turns out that if the pair are in the symmetric quantum state they will mutually repel one another; if they are in the antisymmetric state they will mutually attract.

So much for the circumstances in which quarks can mutually attract. You can immediately see that this rule succinctly summarises the rules of repulsion too. Two red quarks, say, are clearly symmetric under interchange of the red labels. By the symmetry rule they must repel. This is what we know from our electromagnetic experience: like (colour) charges repel.

Two different colours can form an antisymmetric state and attract one another. A third will be strongly attracted by a pair only if its colour differs from the initial pair's colours and the quantum state is antisymmetric under the interchange of any pair's colour labels. Thus red–green–blue clusters form: the familiar baryons exist.

Notice how the way that the attractions and repulsions work has necessarily forced three quarks each of different colour to be present in the baryon. Moreover, any pair is antisymmetric under interchange of their colour labels. The Pauli principle requires that the state be antisymmetric under exchange of any pair of quarks. The colour labels alone take care of this, and so, if we forget about colour, the state must be *symmetric* under exchange of quark flavours and spins.

This has important consequences for the internal structure of the neutron and proton, in particular is gives the uncharged neutron a finite magnetic moment: the neutron contains charged spinning quarks, and although their charges add up to zero their magnetic effects do not – the up and down quarks are constrained by Pauli to spin in preferred directions giving an overall magnetic moment to the neutron.

It is instructive to carry through the discussion of the colour attractions and repulsions to its logical conclusion. We have seen that three quarks of different colours will attract one another to form a baryon. What happens if a fourth quark is brought up to this trio?

A fourth quark will necessarily have the same colour as one of the three already present. Its repulsion is exactly balanced by its attraction for the other pair because the attraction only operates in the antisymmetric mode and there is only a 50% chance that this is present; as there are two quarks attracting then, with the 50% taken into account, there is an exact balance of forces.

The hadrons have no net colour but feel the strong forces because of their coloured constituents. This is in its way analogous to the way that covalent or van der Waals forces arise between electrically neutral atoms. The strong nuclear force is but a remnant of the more powerful forces between quarks. This powerful colour force is neutralised within protons and neutrons with the result that its remnant acts only over short distances – the nucleus.

So we see that colour naturally explains the combinations of quarks and antiquarks that cluster to form mesons and baryons. It controls, via the Pauli principle, the way the quarks behave inside the baryons, and in turn it is responsible for the properties of those baryons, such as magnetic moments. It also plays an important role in generating the masses of the hadrons, giving rise to subtle fine structure effects in the spectrum of particle masses which are colour analogues of the familiar electromagnetic fine structure effects in atoms.

If we recap briefly the atomic case, then we will see the colour analogue immediately.

An electron has an intrinsic spin and hence a magnetic moment. In atomic spectroscopy there are important effects arising from the magnetic interactions among the constituents. Most famous of these is the hyperfine splitting of the ground state of the hydrogen atom. The electron and proton can spin parallel or antiparallel. In these two configurations their magnetic interactions differ, in one case attractive, in the other repulsive. This causes the ground state energy level of hydrogen to be split in two depending on the relative orientations of the electron and proton spins.

When their net spins add to one ('spin-triplet') the energy is slightly higher than in the spin 0 ('spin singlet') combination. This is called the 'hyperfine splitting' phenomenon. Quarks have an intrinsic spin identical to the electron's. Consequently quarks have magnetic moments. However, the fact that they in addition have colour charge causes them to have colour-magnetism as well as simply electromagnetism. This causes colour force analogues of hyperfine splitting to occur in the hadron spectrum.

The lightest meson arises when a quark and an antiquark are in the ground state energy level of the colour force (analogous to the ground state of the electromagnetic force in hydrogen atoms). When they are in overall spin 0 we have the hadronic state called the pion. This is the lightest of all mesons. Somewhat heavier is the rho-meson. This consists of the same quark and antiquark but in an overall state of spin 1.

Just as the spin 1 level has more energy than the spin 0 in hydrogen (an electric system), so does the spin 1 meson have more energy or mass than the spin 0 counterpart in the colour system. In hydrogen the hyperfine splitting is less than an electron-volt. In the mesons it is of the order of hundreds of millions of electron-volts. This is due to the colour forces between quarks in the pion and rho-mesons being much stronger than the electromagnetic forces in atoms. This phenomenon of hyperfine splitting of meson energy levels is seen to arise whatever flavour of quarks are in the meson.

Magnetic moments are inversely proportional to the mass of the relevant particle. The same is true of colour magnetism. Thus the colour magnetic moment of heavy quarks is smaller than that of light quarks. This implies that the size of the colour-magnetic hyperfine splitting is less for mesons containing heavy charmed, or bottom, quarks than for their analogues built from strange or up and down flavours. This is indeed what is seen in practice. Whereas the pion and rho, built from light up and down quarks, are separated by hundreds of millions of electron-volts (hundreds of MeV) the psi and eta-charm, which consist of a massive charmed quark and charmed antiquark, are separated by only some tens of MeV.

The hyperfine splitting is not restricted simply to mesons. The same thing happens in the baryon spectrum. The three quarks in the ground state can couple their spins to a total of $\frac{1}{2}$ or $\frac{3}{2}$. These two states would have identical masses but for the colour-magnetic forces. The hyperfine splitting pushes the spin $\frac{1}{2}$ state down in energy and the $\frac{3}{2}$ state up.

This is precisely as seen in Nature. The spin $\frac{3}{2}$ state is the short-lived delta particle which is some 300 MeV heavier than the spin $\frac{1}{2}$ proton. The hyperfine splitting, rooted in the colour-magnetism, is thus responsible for ensuring that protons are the lightest baryons, and hence they are (relatively) stable and able to build nuclei – the seeds of atoms, biology and life.

Thus we owe our existence to the bizarre colour force.

Coloured gluons, glueballs and how hadrons decay

The colour charges generate forces that are similar in origin to the familiar electrical forces. We have seen how chromostatics is similar to electrostatics, and how chromomagnetic effects parallel electromagnetic ones in the hyperfine splitting of energy levels in atoms and quark systems. Ultimately we have a similarity in the fully fledged theory that arises when relativity and quantum field theory are applied to the force fields generated by the respective charges. In the case of electromagnetism one has quantum electrodynamics, QED; in the colour case one has quantum chromodynamics, QCD (see Chapter 17 by Taylor).

When the motion of electrically charged particles is disturbed, they emit electromagnetic radiation. Quantum theory shows that this is not a smooth legato wave but rather comes in staccato bundles known as photons. Analogous phenomena arise when coloured particles are disturbed. The radiation in this case comes in quantum bundles; QCD's analogue of the photon is known as the gluon.

It is an oversimplification to refer to 'the' gluon. Whereas in QED there is only one type of photon, in QCD there are eight varieties of gluon. This richness arises from the fact that there are three independent varieties of colours that the quarks can possess, and this allows eight different colour combinations to be carried by the gluons (for more details, such as why eight and not nine combinations are allowed, see the chapter by Taylor). The gluons carry colour charges and so attract quarks and other gluons by means of the same colour forces that attract quarks to one another.

As a result of this, theorists predict that there should be a whole new spectroscopy of hadrons in Nature. States that consist of pure glue are called 'glueballs'. Combinations might occur where the colours of the quarks do not mutually cancel out among themselves but attract gluons to form an overall colour neutral state containing both quarks and gluons. These hybrid states have been called hermaphrodites or 'meiktons' from the Greek for mixture. These rich varieties of matter appear to be a natural consequence of colour forces, and there is currently much debate about their properties: how might we best produce them, what will be their distinctive signatures, what do they decay into? There is as yet no conclusive evidence for their existence.

To motivate ways of producing these gluonic hadrons we need first to understand how conventional quark matter is produced and decays. First let's draw a comparison with atomic systems.

The excitation energies of atomic systems are but a few electron-volts. So an excited atom can radiate excess energy as photons and relax to the ground state. That is all that it can do. If the photon had energy more than twice the rest mass of an electron then it could materialise as an electron–positron pair. This requires an energy of a million electron-volts, quite beyond the atomic energy scale.

Quark systems, the hadrons, have excitation energies of hundreds of MeV. Excited levels can decay by emitting photons or by creating quarks and antiquark pairs from their excess energy and relaxing to their ground state, such as the proton in the case of baryons.

There are two ways that the new quark and antiquark can depart from the production site as the excited hadron decays. These are illustrated in figure 14.7(a) and (b). In these diagrams we represent quarks by lines with an arrow pointing to the right; antiquarks are similarly denoted but with left pointing arrow; and gluons are the curly lines. Thus in figure 14.7(a), for instance, a meson, represented as a solid blob, consists of a quark and an antiquark. The quark radiates a gluon which does not escape but instead converts into matter and antimatter – a quark and antiquark. The antiquark then combines with the original quark to make one meson (the upper one) while the new quark combines with the original antiquark to make a second meson (lower meson). The mode shown in figure 14.7(b) is where the new quark and antiquark bind together and move off as a hadron. It transpires that this is less likely than the alternate mode, figure 14.7(a). This one is where the new antiquark drags one of the pre-existing quarks with it to form one hadron and the new quark combines with the other pre-existing quarks or antiquarks to form other hadrons.

The succinct way of distinguishing figure 14.7(b) from figure 14.7(a) is by topology. Figure 14.7(b) can be cut in two without severing any quark or antiquark lines (see the dotted line in the figure); figure 14.7(a) cannot.

Empirically it appears that any process where new quarks and antiquarks emerge in a way that can be cut, like figure 14.7(b), are suppressed relative to processes of the figure 14.7(a) topology. This observation has been elevated to an *ad hoc* rule, known as the Zweig rule or OZI rule after Okubo, Zweig and Iizuka, three physicists who formulated it independently in slightly different ways. A deep understanding of why the rule works, and the details of its operation, are still not to hand though various partial insights are accepted.

When a new quark and antiquark emerge in motion out of the vacuum, a force must have been applied. Forces are

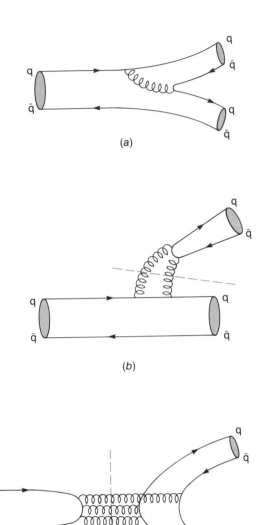

Figure 14.7. (a), (b) Two ways that a meson (q$\bar{\text{q}}$) can decay (time elapses from left to right). Quarks or antiquarks can radiate gluons, which in turn generate further quarks and antiquarks. By this sequence, a new quark and antiquark (q and $\bar{\text{q}}$) are created. Solid lines denote quark ⟶ and antiquark ⟵; gluons are represented by ⟿. The dotted line in (b) shows that it can be cut in two without severing any quark or antiquark lines. (c) If the original meson contains quark and antiquark with identical flavours, such as charm and anticharm, then the constituents can mutually annihilate. New mesons can emerge from the gluons in the intermediate step.

transmitted by quanta, such as photons in the case of electromagnetic forces and gluons for the forces acting on coloured quarks. The quark and antiquark in figure 14.7(a) and (b) are produced by the action of one or more gluons. Figure 14.7(a) can occur with only a single gluon being needed to produce the new quarks. Figure 14.7(b) is more complicated. Suppose we were looking upstream in figure 14.7(b) along the new quark and antiquark towards their point of creation. The quark and antiquark have formed a hadron with no net colour. So the gluons that created them must have no overall colour. Yet individual gluons *do* have colour. Therefore there must be two or more gluons involved, whose colours conspire to give an overall colourless combination. The chance of this conspiracy is rather small, at least it is less likely than that a single gluon will be involved. When a single coloured gluon is involved, the emergence of overall colourless systems can only happen if the quarks and antiquarks pair off as in figure 14.7(a). To make figure 14.7(b) several *conspiring* gluons are needed, which is relatively improbable. This seems to be at the root of the Zweig rule. Another important example of the Zweig rule at work is in the decays of mesons built from heavy quarks and antiquarks, such as charmonium.

If the charmonium system is in an excited state, whose mass exceeds 4 GeV, it can shed energy which materialises as a light quark and antiquark. This can occur as in figure 14.7(b), and so form two 'charmed mesons' – a charmed quark and light antiquark, or vice versa. This decay occurs without inhibition, the lifetime being some 10^{-22} s.

If the charmonium is in its ground state then this form of decay is closed to it. The way it decays is for the charmed quark and charmed antiquark to mutually annihilate. Their mass–energy is transformed into gluons and/or photons which then can rematerialise as hadrons built from light quarks and antiquarks. This is illustrated in figure 14.7(c); on comparing this with figure 14.7(a) we see that this is Zweig forbidden and so should be less probable. This is indeed the case; the lifetime of the psi, the ground state configuration of charmonium, is 1000 times greater than the heavier charmonium states that decay by the allowed processes, figure 14.7(a).

The charmonium system has no overall colour, so when it decays it has to produce gluons that are overall colour neutral. The quantum numbers of the psi particles, the spin 1 charmonium state, place further constraints on the gluons. It turns out that the minimum number of gluons that emerge is three, if gluons alone are involved, or two if accompanied by a photon.

This latter decay path, into a photon and two gluons, is predicted by QCD to occur in about one in seven of all psi decays. This rate is confirmed by data and gives us some confidence that the description of the decay is correct: the psi decays into a photon and two gluons.

The individual gluons carry colour but the pair is overall colour neutral. This implies that the two gluons' colours balance exactly and so they should attract each other strongly, much as coloured quarks do. As a result there is a good chance that glueballs – bound states of pure glue – might form in this channel. The glueballs are only a transitory phase and rapidly decay to conventional hadrons built of quarks and antiquarks, so the end result is that the psi decays to a photon and several hadrons. It requires some detective work to see if these hadrons are the progeny of glueballs.

There has been a lot of work during recent years studying the decays of psi particles into a photon and hadrons and looking for evidence of glueballs. There is good evidence that some of the hadrons are the progeny of a new particle, the iota, that had not been clearly seen before. This is the first candidate for a glueball.

The iota weighs about 50% more than a proton, which fits in with the theorists' expectations for the mass range of the lightest glueballs. Moreover it appears in the right way. Historically, experiments have tended to involve beams of quarks and antiquarks which collide and produce particles made from quarks and antiquarks. There was no clear sign of the iota in such experiments. Contrast this with psi decay. Here the original quark and antiquark are mutually destroyed and gluons are directly involved. New hadrons here are indeed *prima facie* candidates for glueballs.

The iota may be the first sighting of a glueball, but there is much argument as to whether it is, or is not, such a beast. The question of whether or not glueballs exist is a crucial one for the validation of QCD. Consequently we require much further study of the iota, and also need to find other states like it, before we can agree that glueballs exist.

Psi decay is one of the best places to look for glueballs. Other processes include proton and antiproton annihilation at low energies. This is possible at CERN in the machine LEAR (Low Energy Antiproton Ring). Here too the original beams (consisting of protons and antiprotons built from quarks and antiquarks, respectively) can mutually annihilate and form glue. The fact that several quarks and antiquarks are present in the initial beams makes this a difficult process to interpret. However, one looks to see if the same 'new' objects turn up here as in psi decay. If they do this will add greatly to the glueball interpretation of them. This is currently under scrutiny.

Summary

We have seen how the quarks build up the mesons and baryons. The forces acting on the quarks are due to the colour charges on the quarks. But the theory implies that gluons exist and that there will be new forms of matter made from them. None has yet been conclusively seen.

This is as far as we can currently reach by concentrating on structures – the neutron, proton and as yet undiscovered glueballs – whose dimensions are of the order of 10^{-15} m (1 fermi). If we resolve distances of the order of 1 fermi, then we are seeing the quarks acting collectively. To see the quarks acting as individuals we must resolve distances less than 1 fermi.

We now encounter the important principle of Nature that the smaller a thing is so the shorter wavelength your probing radiation must have to resolve the object. To produce short wavelength radiation requires sources of high energy. Thus to see quarks individually, inside their 1 fermi prisons, brings us to the realm of very high energy interactions. The higher the energy of the beam, so the better the resolution and the more details we will see.

14.4   Quarks in high energy interactions

The substructure of atoms was first hinted at by Mendeleev's discovery of the periodic table of the atomic elements. We have seen the analogue of this for the hadrons, where the discovery of the Eightfold Way pattern received its explanation in the quark model of hadron structure.

The electronic and nuclear atom was confirmed by seeing the constituents directly. High energy interactions involving hadrons have provided similar evidence for the quarks within hadrons.

We see things by bouncing light off them. Sunlight bouncing from trees shows us trees, but visible light cannot resolve the molecules and atoms that the tree is built from. It is no use looking at the tree with a microscope, or taking a picture and enlarging it. The power of a microscope is not its ability to make things larger; rather it is the ability to distinguish things that are close together – to tell things apart. The smaller a thing is, the more energy you need to resolve it, and visible light simply does not have the power. To see atoms and their constituents we need beams with billions of times as much energy as visible light and special types of 'microscopes' to deal with them.

Visible light is shaken from atoms, and each photon carries a few electron-volts energy at most. To resolve atomic structure requires energies of millions of electron-volts, and resolving the nuclear structure requires up to billions of electron-volts.

In 1911 the scattering of low energy alpha particles on atoms revealed the atomic nucleus. The beam was powerful enough to resolve subatomic dimensions but was not able to penetrate the nucleus itself. The nucleus appeared to be a point charge.

More powerful beams, of electrons in particular, reveal the inner structure of nuclei. We can bounce the electron beam off nuclei and make what is in effect a radar sweep of the target. The resolution improves with the increasing momentum of the electron beam. We begin by seeing the whole nucleus, and then we progressively resolve the neutrons and protons and finally the quarks within.

What we resolve depends on the violence with which the nucleus is struck. We can vary this by changing the incident energy or the angle at which we watch for scattered electrons. We then measure the probability that electrons scatter at chosen angles with particular amounts of energy transferred to the target. The distribution of events as a function of scattering angle tells us about the spin of the constituents.

The results imply that the electron beam is scattered from objects that have spin $\frac{1}{2}$ and which are in motion aligned with the target. This is as expected for quarks.

The modern convention is to plot the data against a particular combination of the energy lost by the electron and the amount of sideways kick – or momentum change relative to its original line of flight. The definition of this quantity (usually denoted by x) is less important for us than is its physical significance. Imagine we are travelling with the electron beam and see the target coming towards us with some high momentum. It is like a swarm of constituents, protons and neutrons, or even quarks, and its momentum is shared among them. The electron scatters from one, or a collection, of these constituents. The quantity x is the fraction of the total momentum that is carried by the relevant constituent(s) that scatter the electron, and can have any value from 0 to 1.

At SLAC and CERN in 1968–72 electrons and neutrinos with energies above 10 GeV were fired at protons and neutrons and resolved the quarks within. The resulting x-distribution is summarised in figure 14.8.

The x-distribution vanishes as x approaches unity. This means that there is little chance of finding a single quark carrying all of the proton's momentum. This is no surprise as there are three quarks within the proton, and on average each will carry one-third of the total momentum – the average x will be $\frac{1}{3}$. However, the x-distribution rises up even when x is less than $\frac{1}{3}$ and has a substantial contribution as x approaches zero. This is due to a profound property of Nature.

The interactions of coloured quarks are in some ways similar to those of electrically charged electrons. As electrons are surrounded by electric field, virtual photons and pairs of electrons and positrons, so are the quarks surrounded by gluons, quarks and antiquarks. The intrinsic quark is called the 'valence' quark; its surrounding cloud is the 'sea'. The sea doesn't care what the variety of valence quark is. Thus the valence quarks determine whether we have a proton or a neutron overall; the sea is common to both.

When the x-distributions of proton and neutron targets are compared they are different. This is because they contain different valence quarks. So the difference between the two is entirely due to their different valence quarks. When the difference of the two is extracted from the individual data sets it shows that the x-distribution of each of the three valence quarks peaks near $x = \frac{1}{3}$, in accord with expectation.

This is what is found for the x-distributions when the proton is probed at momenta between 1 and 10 GeV. As we increase the probing momentum up to 100 GeV we gradually discern a slow change in the x-distribution (figure 14.9). There is a leftwards shift in the distribution as the resolving power increases. At poor resolution we pick up the quark and its sea but cannot separate them. At higher resolution we discern the individual members. Each individual carries only a fraction of the whole momentum. Thus the perceived increase in low momentum quarks with improving resolution is due to the fractal nature of the quark: the closer we look, the more detailed structure we see, and so we perceive a sharing of the momentum among more and more participants. The average per member thus goes down, and the average value of x decreases.

This behaviour is expected in QCD. As the coupling between quarks and gluons is rather feeble at high resolution, the quarks act as nearly free particles when responding to the high momentum probe. QCD perturbation theory predicts that the quarks and gluons interact and give subtle changes to the x-distribution as the resolution improves. The observed variation of the x-distribution with resolution appears to be as required by QCD.

If the quarks move around laterally in the target, they will deflect the electrons more or less depending on whether they are moving towards or away from the beam at the moment of interaction. This will smear the angular distribution of the scattered electrons. This is a rather small effect, however, and is hard to extract from the data with certainty. There is some suggestion that occasionally the electron hits a quark that has been deflected sideways as if by radiating a gluon. However, this is by no means certain. This is a precision measurement at the limit of present techniques. In addition, the theoretical interpretation is not straightforward and requires inputs that we cannot yet analyse within QCD: we do not understand the mathematical subtleties of QCD well enough.

The comparison of electron and neutrino interactions gives important information on the presence of quarks versus antiquarks in the target. The electrons interact electromagnetically. This appears the same when viewed in a mirror – electromagnetic interactions 'conserve parity'. Left and right are equivalent, and this affects the form of the angular distributions. The electron scatters and moves off at some angle relative to its incoming direction. The chance that it emerges at some given angle is determined by looking at millions of events. You find that the distribution in angle can be described mathematically as the sum of various *even* powers of sine or cosine of the angle.

Weak interactions, which control the behaviour of neutrinos, do not respect mirror symmetry – they are said to 'violate parity'. This causes characteristic asymmetries in the angular distributions: *even and odd* powers of the sine or cosine are needed to describe the scattering distribution. Neutrino beams spin left handed and antineutrino beams corkscrew in a right handed sense. Quarks and antiquarks respond in different ways to these different probes. Quarks respond to left handed probes as antiquarks do to right handed ones. So by using neutrino beams or antineutrino beams we can distinguish between quarks and antiquarks in the target; we can isolate quarks and antiquarks separately by careful study of the angular distributions.

When neutrinos scatter from quarks the angular distribution turns out to be given by the sum of even and odd powers of the cosine of the angle. Exactly the same distribution arises when antineutrinos scatter from *antiquarks*. However, if we scatter neutrinos from *antiquarks* (or antineutrinos from quarks) we see a different distribution. Instead of a *sum* of even and odd you find the same set of even powers as before, and the same set of odd powers too, but, instead of a sum, this time you see a *minus* sign between the set of even and the set of odd terms. So by measuring the angular distributions and fitting them to a set of even and odd powers of cosine you can weigh the relative amount of even and odd powers required and in turn determine the relative importance of quarks and antiquarks that scattered the beam.

If quarks and antiquarks are equally important then the odd powers cancel out: the angular distribution is a sum of even powers only. However, if only quarks or only antiquarks are responsible then the even and odd contributions are equally important. You measure the angular distributions for the case where the beam has some chosen energy and the scattered beam has some other chosen energy. You can do it all over again with different energies. By selectively changing the energies or momenta of the incoming beams you can filter out the momentum distribution of the quarks or antiquarks that scatter them. By measuring the angular distribution in each case you can determine the relative importance of quarks and of antiquarks as a function of x: you measure the momentum distribution of quarks and of antiquarks separately.

This requires many very detailed and precise measurements. It has taken many years to perform and is still being refined. But we have already learned a great deal about how the quarks and antiquarks are distributed in the proton and neutron.

Antiquarks occur in the sea. So measurement of the antiquark distribution shows us the sea directly and proves that it is indeed concentrated in the region of x near to zero. Compare this with the data where electrons scatter from protons, where sea and valence quarks are both contributing. By removing the known sea distribution from the total you can deduce the valence distribution. We find that this agrees with the results found by the independent method of comparing the neutron and proton target data and so confirms that there is overall consistency in the interpretation.

Glue in the proton

A proton contains both up and down flavours of quark. By comparing the response of electron, neutrino and antineutrino beams to proton targets we can isolate the contributions of the up and down quarks individually.

Electrons couple to all electrically charged pieces of the proton, in particular to both up and down quarks. Neutrinos and antineutrinos, on the other hand, are selective. Their most noticeable interactions are when they scatter and change into electrically charged electrons and positrons or muons. In these circumstances neutrinos tend to filter out down flavours, antineutrinos filter out the up quarks.

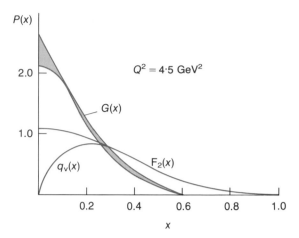

Figure 14.8. Qualitative representation of $P(x)$, the probability distribution of finding constituents carrying a fraction x of a proton's momentum. $q_v(x)$ is the distribution of valence quarks and peaks when $x \simeq \frac{1}{3}$. $F_2(x)$ is the net average momentum distribution for all electrically charged objects: quarks and antiquarks. The total area under $F_2(x)$ is a measure of the fraction of the target's momentum that is carried by quarks and antiquarks. It is about 0.55, which implies that electrically neutral objects must carry the remaining 45%. These are the gluons and their distribution is denoted by $G(x)$. The measured distributions depend upon the resolution (momentum transfer) of the probing beam. In this figure the square of the momentum transfer is denoted Q^2 and equals 4.5 GeV^2.

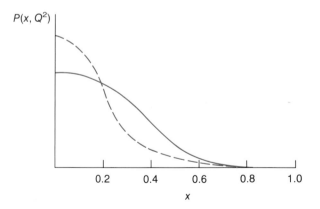

Figure 14.9. The momentum distribution $F_2(x)$ in figure 14.8 was measured at moderate resolution, corresponding to a momentum transfer (Q^2) of 4.5 GeV^2. This is shown as the solid curve in this figure. As Q^2 increases, resolution improves. The momentum carried by the quark is seen to be shared among a cloud of quarks, antiquarks and gluons surrounding the quark. On average these will carry less momentum than the poorly resolved quark. The probability to find a low momentum quark increases at the expense of high momentum ones; the distribution shifts to the dotted curve by $Q^2 = 50$ GeV^2. The shift is predicted by QCD and is seen in the data.

These experiments reveal that each up quark on average carries more momentum than the down quark in the proton. This is a subtle consequence of the colour forces at work within the proton. As electrically charged electrons and protons feel magnetic forces within hydrogen atoms, so do coloured quarks feel colour-magnetism in the proton. We already commented that, collectively, this colour-magnetism separates the masses of the proton and the delta resonance. Within the proton it energises the up and down in a discriminatory manner. So in the down and up quark momentum asymmetry we see colour-magnetism working on individual quarks.

Having extracted the detailed behaviours of the up and down quark distributions, we can add them up and see if they yield the total energy of the parent proton. They do not! A small amount of energy, 5%, is carried by the sea of quarks and antiquarks, but the 'valence' up and down quarks only carry about 50% of the remainder. So nearly half of the energy is carried by constituents that do not respond to electron nor to neutrino and antineutrino probes.

These are the gluons. By careful study of the quarks we can deduce what the gluons must be doing to 'balance the books' – that is, to conserve energy and momentum. Details of QCD imply that the perceived sharing of energy–momentum between quarks and gluons depends on how closely we look. Whereas at poor resolution we perceive the momentum carried by a misty quark, at high resolution we see that the sharp quark carries only a part – the mist is due to a cloud of gluons which carry some of the momentum. And this is indeed what we find. At high resolution, we see that as much as 50% of the total energy is in the glue.

Now that we know the momentum distribution of the gluons as well as the quarks within the proton, we can design new experiments. This has been put to use recently in experiments at CERN. The quark and antiquark distributions in a proton are identical to the *anti*quark and quark distributions, respectively, in an *anti*proton. The gluon distribution is the same for both. So when we fire protons and antiprotons head-on at one another we can interpret the results as if we were firing bags of quarks, antiquarks and gluons together. At ultra-high energies, such as at CERN, the proton and antiproton beams are like well collimated rays of quarks and gluons. The astonishing results of those experiments will be described later.

Quarks in nuclei

Atomic nuclei are built of protons and neutrons which in turn are made of quarks. So if you look at a nucleus in great detail you will see a swarm of quarks in motion much as you find them inside individual protons or neutrons. At least, that is what people expected, but it is not quite as things have turned out.

For more than half a century we have been familiar with the notion of neutrons and protons inside atomic nuclei. Only in the last fifteen years have we had lepton beams powerful enough to study the behaviour of quarks in detail within those protons and neutrons. These early experiments concentrated on the simplest nuclei, hydrogen (a single proton) and deuterium (one proton and one neutron), in order to obtain the clearest pictures of protons or neutrons.

Then around 1982 a group of physicists at CERN, Geneva, were making similar experiments but with iron nuclei as targets. According to the theorists, quarks in iron behave much the same as in free protons or neutrons; but when the CERN team examined their data they found that the theorists were quite wrong. When iron and free protons are viewed under the same microscopic resolution, the quarks in iron are seen to be moving slower than in protons. No one had expected this radical result, and at first the CERN experimentalists suspected that some error had occurred in their analysis. They double checked everything but could find nothing wrong. They finally published their discovery a year later in 1983.

Initially there was a slow response to the announcement; its meaning was unclear, and the easiest explanation seemed to be that some quirk of the experiment was responsible, not that a genuine effect had been seen. This scepticism disappeared in early Spring 1983 when a team at Stanford announced that they had found similar anomalies. They irradiated aluminium targets instead of iron, but here too the quarks refused to behave. The way that the Stanford team obtained their results is interesting.

In the classical experiments of the early 1970s, when quarks were discovered in protons, a target of hydrogen and deuterium contained in canisters of aluminium was used. Sometimes the electron beam scattered from the hydrogen and sometimes from the container. As the experimentalists were primarily interested at that time in the hydrogen data, they removed the unwanted events where the scattering took place on the container. Now suddenly these were interesting in light of the CERN discovery.

They searched and found the computer tapes on which the old data had been stored, but in the intervening years the computer system had been modernised and the tapes could no longer be loaded. Luckily the Argonne National Laboratory near Chicago still had a suitable computer system. They successfully loaded the tapes, extracted and analysed the aluminium data, and the same sort of anomaly emerged as had been found by the CERN team. The effect was real. Theorists could ignore it no longer.

There is an intriguing phenomenon in these data which is currently being pursued both by theorists and experimentalists in a desire to understand what is going on.

The paradox had been that the high resolution pictures of quarks in protons and nuclei did not match. However, there is an almost perfect match between portraits of a nucleus at one resolution and that of a proton taken at a different, much higher, resolution. What does this mean?

High resolution pictures reveal the quarks inside the proton or nucleus. The displaced matching implied that images of quarks are in effect clearer in nuclei than in free protons or neutrons; showing up at moderate rather than high resolution. Since big objects are easier to resolve than small ones, this suggests that the quarks might be forming bigger bunches in nuclei than in individual protons and hence be more easily identifiable.

This also agrees with the observed sluggishness of quarks in iron. The subtle interplay between momentum and spatial position that is at the heart of quantum mechanics implies that quarks in a small prison will move faster than in a large one; quarks in big bunches, as in iron, move slower than in the smaller free proton. It appears that protons in nuclei are effectively enlarged or that some unexpected fusing of protons and neutrons is taking place.

It is the different behaviours of quarks when in nuclei and free protons that have shown us this new detail. Whereas quarks are trapped within a small prison when inside free protons, in nuclear matter their perimeter fence is pushed back slightly. This extra freedom is only 10% in radius but changes the nuclear make-up considerably; nuclei are seemingly not simply built of the same neutrons and protons that we find in hydrogen and deuterium. This small extra freedom affects the motion of the quarks quite considerably, providing the tell-tale clue by which we have discovered some new insights into nuclear complexity. But it is still an open question as to what precisely causes this.

Electron–positron annihilation

Protons, pions and indeed all hadrons are made of quarks and/or antiquarks. When we fire electrons at targets of protons and neutrons we see the quarks inside those nuclear particles. When we fire electrons at electrons or positrons we do not see any quarks because there are none in the beam or target: electrons and positrons are, so far as we can tell, fundamental particles.

However, this does not mean that we learn nothing about quarks from such experiments. Far from it. Electrons colliding with positrons provide one of the most exciting ways of learning about bizarre varieties of matter made from quarks.

The key feature is that positrons are the antiparticles of electrons. When matter and antimatter meet, they can mutually annihilate. Mass has been converted into energy: $E = mc^2$ at work.

What is the point of this?

Destroying the electrons and positrons is just the start. The aim is to watch what happens when their energy recongeals into new forms of matter and antimatter. It can return whence it came, into electron and positron, but more interestingly it may produce new forms of matter with their corresponding antimatter. The hunt is on for those occasions when new forms of matter, not previously seen on Earth, emerge from the encounter. Exotic forms of matter can occur fleetingly in the heat of stars, and when we temporarily simulate that heat on Earth, so can we capture these new varieties in earthbound laboratories.

This continuous destruction and recreation of matter and antimatter was common in the brief heat of the primordial Big Bang. By annihilating electrons and positrons in the laboratory we are reproducing conditions similar to those that occurred within a split second after the Big Bang. We can create matter and antimatter, built of quarks and antiquarks, to order.

This is very exciting because it enables us to break away from the familiar universe of neutrons and protons built from up and down quarks. We can materialise natural phenomena that have never before occurred on Earth.

Examples include the discovery of charmed and bottom matter during the last decade. Matter on Earth comprises up and down quarks; strange quarks arrive in the cosmic rays. But charm and bottom are much more ephemeral. By concentrating enough energy into the electron–positron annihilation, teams at Stanford congealed charmed matter in 1976.

A total energy of 4 GeV is enough to make charmed particles pour out from collisons. By 1977 more powerful machines were being built, capable of colliding electrons and positrons with a total energy of more than 10 GeV. At these energies you can produce bottom matter, massive particles containing the bottom quark – a fifth variety of matter in the universe. By the end of the 1980s we hope to have machines at Stanford and at CERN that will collide electrons and positrons at energies up to 100 GeV. Under these circumstances we hope to produce the sixth variety of quark – the top quark – in copious quantities. (There are tentative hints of its presence in debris from proton–antiproton annihilations at CERN, but this is unlikely to be totally settled until the new electron–positron collider, LEP, begins operation at CERN in the 1990s.)

Electron–positron annihilation is very useful for producing new flavours of quark. It is also ideal for studying the behaviours of freshly produced quarks, how they behave as they separate from the antiquark that is born alongside.

The minimum energy needed to create a particle of mass m is mc^2. We cannot produce matter alone in electron–positron annihilation; it emerges with its corresponding antimatter. So the minimum total energy needed is $2mc^2$. This will be sufficient to produce the pair at rest. They will soon meet and annihilate. But if we put in more energy at the start, then we can produce the matter and antimatter (quark and antiquark) in motion. They fly apart and escape each other's clutches.

There are close parallels between the production of quark and antiquark in the electron–positron annihilation and the process of muons being produced.

When electron and positron annihilate it is possible that the energy will rematerialise as a positively charged muon together with a negatively charged muon. This happens very frequently and is a bench mark against which other processes, such as quark and antiquark production, can be compared.

The muons and electrons are, as far as we can tell, structureless fundamental particles. As a result, the probability for producing these in pairs has a characteristic behaviour as the energy of the incident beams is varied. If you double the incident energy, the chance, or more precisely the 'cross section', for a muon pair to emerge dies out fourfold. Increase the energy threefold and the cross section drops ninefold. In general, the cross section varies as the inverse square of the energy.

The cross section to produce extended objects falls away faster with energy; the size of the object introduces an extra dependence on length, and hence energy, into the description of the process relative to the case with point particles.

So the cross section for producing quarks will have the same energy dependence as that for producing muons if the quarks are point particles like the muons. There is a problem in that individual quarks do not emerge free in Nature; instead they cluster and form extended hadrons. However, quantum mechanics tells us that if we add together the individual cross sections for producing any set of hadrons then the sum total cross section for electron and positron to annihilate and produce hadrons is described *as if* free quarks were produced. The cross section is equal to that corresponding to the sum of cross sections for producing any flavours of quark possible at the energy of interest.

For example, at low energies the cross section is given by the sum of the cross sections for up, down and strange flavours. Above 4 GeV we have to add in the charmed contribution, and so the ratio of hadron to muon cross sections rises due to this extra piece. Above 10 GeV we add in the contribution from the bulky bottom quarks and so on.

Empirically the cross section to produce hadrons does have the same energy dependence as that to produce muons, confirming the pointlike nature of the quarks at presently accessible energies. Moreover, the ratio (figure 14.10) of the hadron and muon cross sections tells us about the varieties and properties of the quarks that are responsible for building the hadrons. Theory predicts that the ratio is given by the sum of the squares of the charges of each variety of flavour contributing.

We can see what this is by first supposing that only the up flavour could be produced. The up quark has charge $\frac{2}{3}$ times that of a muon, and so the ratio of the hadron production via up quarks to the cross section for muons will be $\frac{4}{9}$. The down quark, with an amount $\frac{1}{3}$, will add a further $\frac{1}{9}$ to this; the strange will likewise add another $\frac{1}{9}$ and so a total ratio of $\frac{2}{3}$ should result.

In 1970 the first data emerged that could test this idea. The data came from the Italian machine at Frascati near Rome which could annihilate electrons and positrons at energies up to about 3 GeV. The data were not very precise but showed that

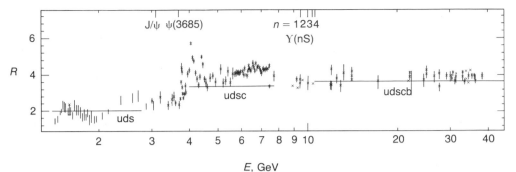

Figure 14.10. R denotes the ratio of the cross sections for electron–positron annihilation producing hadrons as compared to muons. This is predicted to be equal to the sum of the squares of the electrical charges of the various types of quarks that can contribute. Up to a collision energy of some 4 GeV, up, down and strange quarks alone contribute and the ratio should be equal to 2. Above 4 GeV charmed quarks contribute and increase the ratio to $3\frac{1}{3}$, and finally above 10 GeV the presence of bottom quarks brings it to $3\frac{2}{3}$. These values are indicated by straight lines. The slight excess in the data is expected from QCD. At certain energies there are pronounced peaks due to the production of ψ and Υ particles. These have been removed and the locations indicated.

the ratio appeared to be a constant as expected but its magnitude was much larger, more consistent with 2 than $\frac{2}{3}$. Within three years the more powerful machine, SPEAR, at Stanford had started to operate, and this showed clearly that the ratio was 2, within errors, at energies up to about 4 GeV. Further work revealed that the ratio then rose by a step of about $\frac{4}{3}$ as the threshold for producing charmed hadrons was crossed.

So the ratio was behaving as expected for pointlike quarks, but for the fact that everything was a factor of three bigger than had been predicted.

The reason for this is that each flavour of quark has any of three colours. The electron–positron annihilation will produce a red quark and a red antiquark as often as a blue pair or a green pair. So the cross section for producing a pair of red up quarks will be $\frac{4}{9}$ that of a (colourless) pair of muons; there is a further factor of $\frac{4}{9}$ for green and $\frac{4}{9}$ for blue. This gives a grand total of $\frac{4}{3}$ for the ratio.

The discovery of these large cross sections for hadron production in electron–positron annihilation proved to be one of the first dramatic pieces of evidence for the existence of colour. This provided the impetus to develop gauge theories of colour, leading to the modern successful theory of QCD described in Taylor's chapter.

One of the features of QCD is that it naturally justifies the arguments above, namely that the cross section for producing hadrons is given to high accuracy by the sum of the cross sections for producing quasi-free quarks. However, it also shows that it is not the whole story. There are obvious structures when resonances are produced at certain energies, but away from these QCD predicts that the ratio is not exactly given by the sum of the squares of the quarks' charges (amplified by three for colour). There is an additional contribution that dies away slowly with energy and is about a 5 to 10% correction at presently available energies. This extra contribution arises from detailed arguments about the interactions between quarks and gluons that are inherent in QCD. The data do show some hint of this anticipated 10% effect.

This is all consistent with expectation but is not yet a very precise test of QCD. A cleaner measure of the interaction between quarks and gluons comes from looking at the outcome of the electron–positron annihilation in more detail.

According to QCD the quarks are produced at a point and move apart from one another. If the electron and positron had a lot of energy, the quark and antiquark will move away from one another rapidly. As they separate, the potential energy between them grows, probably proportional to their spatial separation. This growth is soon enough to materialise further quarks and antiquarks from the vacuum. These pop out with very little kinetic energy and the original fast moving quarks attract them and sweep them along. Two jets of hadrons emerge. One jet follows the direction of the initial quark, the other jet follows the original antiquark, see figure 14.11(*a*).

This phenomenon is very noticeable in the data. The hadrons are not diffusely spread around everywhere; two distinct jets appear instead. This is the nearest that we get to seeing the quarks in electron–positron annihilation. They are like yeti. The quarks have gone; their footprints, the two jets of hadrons, remain.

However, QCD predicts that about 10% of the events should have not two but three distinct jets. The reason for this is due to a similarity between QCD and QED.

As QED is for electrical charge so is QCD for colour. When electric charge is accelerated, photons are radiated. When colour is accelerated so are gluons radiated. The chance that this happens depends upon the quark–gluon coupling (see Taylor's chapter), the analogue of QED's 1/137, somewhere about $\frac{1}{10}$ in QCD at present energies. The quark and antiquark are produced with considerable kinetic energy;

colour charges have appeared from nowhere and shot off rapidly. This sudden acceleration of colour causes gluons to be radiated. Most likely is that they travel along with the quarks and are not resolved. However, there is a chance that a gluon will be emitted at a large enough angle relative to the flight of the quark and antiquark that it will be noticeable (figure 14.11*b*).

Just as quarks are permanently confined within hadrons, so do gluons appear not to exist free. The same set of circumstances that convert the quarks and antiquarks into jets of comoving hadrons also cause a jet to emerge along the direction of the gluon. So there will be a sample of events where three jets emerge (figure 14.12) (and a tiny sample of four or five jets, though these are very hard to resolve at present accuracy).

The relative abundance of three jets as against two jets tells us the magnitude of the quark–gluon interaction strength. The angular distribution of the two jet and the three jet events relative to the direction of the incoming electron and positron confirm that the two jet events are indeed initiated by quarks and are consistent with the hypothesis that the third jet comes from a gluon whose spin is the same as that of a photon – as required in QCD. There is currently much interest in studying the hadrons that emerge in events with three jets to see if the gluon leaves its mark by producing glueballs in addition to the many conventional quark hadrons that pour out from these events.

Figure 14.11. Electron and positron enter left and annihilate into a photon (ᴍ). This rematerialises into a fast quark and antiquark. These do not escape as free particles but generate jets of hadrons denoted by the arrows ⇨. In (*a*) two jets emerge; in (*b*) one quark first radiates a fast gluon (ᴍ) which generates its own 'gluon jet' of hadrons. Thus in (*b*) we have a three-jet event.

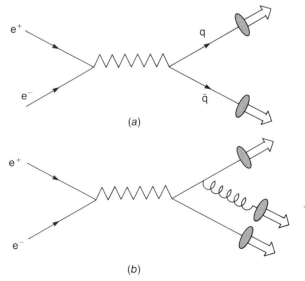

How quarks produce hadrons

To establish whether there are new phenomena in gluon jets that contrast those in quark jets we need to study the hadrons that comprise the jets. In the events containing two jets we are sure that one jet is initiated by a quark and the other by an antiquark. Within a jet there will be one hadron that carries the most energy; this is called the leading hadron. Then there will be a next-to-leading hadron and so on down to slow hadrons.

The details of jet formation are not well understood by theory. It seems likely that the electron–positron annihilation produces an initial quark and antiquark which quickly separate from each other. This generates further quarks and antiquarks from the vacuum which then cluster with the original pair and form the hadrons, bound states of quarks (baryons) or quark and antiquark (mesons) that are detected.

Although current theory is not well enough developed to describe all the details of jet formation, there are some features of the data that theory explains well. This concerns the nature of the fastest hadrons emerging from the production site. The chance of these being positive or negative pions, for example, can be described and related to the chance of finding these same hadrons in other processes. This follows from their quark content and how well it matches the quarks in the beam or target.

A proton contains more up flavours than down and, when an electron beam hits, it is four times as likely to interact with one of the up than the down varieties. This is because the up quark has twice as much electrical charge as the down and hence four times as much probability to contribute to cross sections. So to a good approximation the proton is viewed as a source of up quarks when seen by incoming electrons.

Suppose that the interaction throws one of these quarks violently forwards. It cannot appear as a free quark in the detector, instead it picks up antiquarks and quarks from the vacuum and dresses itself up into a hadron. What hadrons are more likely than others?

The most likely quark to be thrown forward is the up quark, as we outlined above. So it is most likely that the fastest forward moving hadron contains that up quark. A negative pion is built from a down quark and an up *antiquark* and so does not contain the crucial up quark. A positive pion, by contrast, consists of an up quark and a down antiquark. So a positive pion can be expected to emerge fast forward, due to the projected up quark having dragged a down antiquark from the vacuum, attracted mutually and formed a positive pion.

To form a negative pion requires two sequential steps. We just saw how the up quark dresses itself into a positive pion – it drags a down antiquark from the vacuum. So a down quark must have also emerged from the vacuum. If this in turn can drag an up quark and antiquark out, then it can combine with the up antiquark and form a negative pion. This is illustrated in figure 14.13. Clearly the positive pion is formed from a fast quark whereas the negative pion contains slower quarks. This

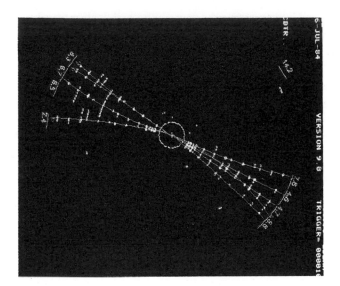 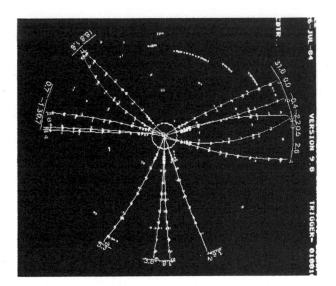

Figure 14.12. Two- and three-jet events. (*a*) An example of a quark–antiquark back to back jet event seen at the PETRA electron–positron collider in Hamburg, Germany. Most of the tracks are made by pions. This event was observed in the TASSO detector. (*b*) Electron–positron annihilation at PETRA sometimes gives rise to three-jet events as in this event seen in the TASSO detector. Such events are believed to be due to the fragmentation into ordinary hadrons of a quark and an antiquark, together with a gluon.

naturally leads us to expect that positive pions will, on the average, be faster than negative ones. This is indeed as seen empirically.

Neutrinos select out down quarks, antineutrinos select out up quarks and so the chance of finding positive or negative pions in these processes can be related to the electromagnetic interaction by the same line of arguments as above. By combining information from these complementary processes it is possible to extract detailed information on the chance that an up quark, say, will end up in a positive or negative pion that is moving fast, medium fast or slow. This corresponds to asking 'How much is the original quark slowed down in the act of picking up the antiquark?'

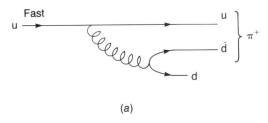

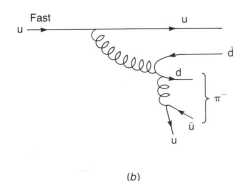

Figure 14.13. Fragmentation of quarks into hadrons. The fast quarks in figure 14.11 fragment into hadrons in stages. Here a fast up quark (u) enters left and the colour force field (denoted ⚡) generates a quark and antiquark (a down (d) and anti-down (d̄) in this example). (*a*) The ud̄ form a π^+. The colour force field surrounding the down quark in turn similarly induces production of further quark–antiquark pairs. (*b*) If the original quarks are up and anti-up (uū) then a π^- can emerge. Note that two stages are needed to produce the π^- from an initial up quark. Thus the π^- will have less momentum than π^+ on average when an up quark is the source.

There is now detailed information on how this varies with the mass of the initiating quark. This comes from studying the emergence of charmed or bottom hadrons from electron–positron annihilation. These necessarily contain a bulky charmed or bottom quark which is almost certainly created in the original electron–positron collision – there simply isn't enough concentrated energy in the vacuum for such bulky objects to appear spontaneously. So the distribution in momentum of the charmed or bottom hadron tells us how much the charmed or bottom quark slows as it picks up the light antiquark from the vacuum.

What we find is that inertia plays an important role. The energy of a heavy quark is transmitted to the hadron containing the heavy quark. The attached light quark doesn't decelerate it. We can compare the effect for light up and down quarks, the slightly heavier strange quark, the massive charm and very bulky bottom quarks as initiators. The trend is continuous: the heavier the initial quark, the less it slows, the more chance that the hadron carries a lot of energy in the debris of the collision. This suggests that when we have enough energy to produce the top quark, the most massive of all, the top quark will be decelerated hardly at all. This punch-through property of top hadrons may prove to be a decisive signature that will enable them to be identified in the debris of complicated high energy reactions in the future.

Drell–Yan annihilation: the route to the W and Z bosons

Protons contain three valence quarks surrounded by a cloud of glue and further quarks and antiquarks in pairs. When two protons collide violently there is a chance that a quark in one proton will hit and annihilate with an antiquark in the other proton. Only a small amount of the proton's momentum is carried by antiquarks, and so a high energy collision between two protons will produce a lot of debris due to their quarks, while the quark–antiquark annihilation will occur with only a small fraction of the total initial energy.

However, if we collide protons with pions then conditions can be very different. A pion contains a valence antiquark which carries a significant fraction of the pion's momentum. Hence a violent annihilation between quark and antiquark can occur here. The most extreme case is when beams of protons and antiprotons collide since here we have valence quarks in a proton meeting valence antiquarks in the antiproton. Protons and antiprotons counter-rotate in the same tunnel at CERN. Their head-on collisions were at the highest energy in the world until 1985 and they provided the highest energy quark–antiquark annihilations. (In autumn 1985, Fermilab near Chicago matched these conditions.)

The annihilation of quark and antiquark can provide gluons which rematerialise as glueballs or as new quarks and anti-quarks and hadrons. Another possibility is that they annihilate and convert into an electron and a positron. This is like electron–positron annihilation, but in reverse. It is known as the Drell–Yan process after the two theorists, Sidney Drell and Tung-Mow Yan, who first realised its interesting applications. In particular it provides a clean signal that a quark and antiquark have annihilated and enables identification of these rare and interesting events.

It was in the late 1960s when Drell and Yan first thought about this process. At that time the idea of quarks was still rather new and tests for their presence were eagerly sought. The Drell–Yan process was particularly useful in this respect. It helped to identify quarks and antiquarks in hadrons; it revealed an important dynamical process and showed that proton collisions are in effect quark collisions. Whereas in 1970 the Drell–Yan process might have been categorised as an esoteric line of research, by 1985 it had become an essential tool being put to use in producing the W and Z bosons, the carriers of the weak force.

The chance that two colliding hadrons undergo the Drell–Yan process has a characteristic energy dependence if the quarks and antiquarks are point particles. This was the original motivation, and its expectations have been confirmed giving yet more proof of the presence of quarks and antiquarks in protons.

Once the Drell–Yan process had been seen to work in practice rather than just in principle, a lot of interest developed in applying it to learn new things. For example, the quark and antiquark momentum distributions in protons and neutrons were known but how do the quarks and antiquarks behave in a pion? The pion is unstable so you can't easily fire electrons at targets of pions and measure the quark distributions inside pions. However, the Drell–Yan process made the measurement possible.

The technique involved firing beams of pions into nuclear or proton targets. If an electron and a positron emerge with high momentum then it is likely that they are the result of an annihilation involving an antiquark from the pion beam and a quark from the target. The energy and direction of motion of the electron and positron enable you to compute the momenta of the initial quark and antiquark. Thousands of events are accumulated, and from them a moment distribution can be obtained for the constituents of the proton and pion. However, the separate distributions are still convoluted together. But we already know the momentum distributions for quarks inside protons and so we can disentangle new information from the Drell–Yan data, namely the momentum distribution of the antiquark in the pion.

A nice illustration of the Drell–Yan phenomenon is given by comparing positive and negatively charged pions incident on nuclei containing the same number of protons and neutrons (for example, carbon). These nuclei contain an equal abundance of up and down quarks. Now, a negative pion is built from a down quark and an up antiquark, whereas the positive pion is the other way round, namely an up quark and a down antiquark. It is the antiquark that annihilates with a quark in the target nucleus, and produces a photon. Thus, in the case of a positive pion, it is the down antiquark that is involved, whereas for the negative pion it will be the up antiquark.

The probability that Drell–Yan annihilation occurs is proportional to the squared magnitude of the electrical charge of the participating quark and antiquark and also depends on the chance of finding the required quarks in the target. In carbon the chance of finding up quarks is the same as finding down quarks and so the rate for annihilation is controlled entirely by the electrical charges. Thus the relative rate for positive pion beams will be four times greater than that with negative beams: the squared charge of up quarks is four times that of down.

This fourfold favouring of positive pions over negative ones in the ability to instigate Drell–Yan annihilation is confirmed by experiment. This is a clear indicator that quarks and antiquarks are responsible and that the magnitudes of their charges are in the ratio 2:1.

With this clear proof that Drell–Yan annihilation indeed proceeds by quark–antiquark annihilation one can confidently extend the experiments to learn about the quark constituency of other hadrons. Similar experiments have been performed with kaon beams replacing pions and the distributions of strange quarks compared with the up and down quarks. From this one learns what role the mass of the quark plays in determining the momentum distribution. The conclusions are not yet totally clear as data are hard to come by and, in the kaon case in particular, are subject to considerable uncertainties.

Such uncertainties do not plague protons and antiprotons. These particles are being put to use at CERN in experiments that have discovered the W and Z particles; the gauge bosons of the electroweak force (see the chapter by Taylor).

The essential key is that the symmetry between matter and antimatter implies that the way antiquarks behave inside antiprotons must be identical to the way that quarks behave inside protons. The momentum distribution of the up and down quarks in protons has been determined very well from lepton scattering experiments with proton targets. Consequently we immediately know also the distributions of up and down antiquarks in antiprotons. From these you can calculate what is the chance of an up quark in the proton annihilating with a down antiquark in the antiproton and so producing a positively charged W. Similarly you can compute the chance for a down quark and an up antiquark to produce a negatively charged W, and for like flavours, such as an up quark and an up antiquark, to annihilate into a neutral Z.

There is a beautiful result from these experiments showing the intrinsic violation of parity (mirror symmetry) inherent to the weak interactions. For over thirty years, since the work of T. D. Lee and C. N. Yang, we have known that the beta decay of neutrons produces electrons with an asymmetric distribution in space. Thus the distribution seen in a mirror differs from that in the real experiment. Study of the distribution has revealed that the emitted electron is polarised 'left handed' (that is, its spin axis is, conventionally, antialigned to its direction of motion. Beta decay tends not to produce right handed electrons).

This violation of parity arises because beta decay is caused by the weak interaction. The modern theory of weak interactions predicts that they are caused by the intermediate agency of W bosons. These are quanta of the weak force much as photons are the quanta of the electromagnetic force. Drell–Yan annihilation can produce photons by electromagnetic annihilation of quark and antiquark, and so should analogously produce W bosons when quark and antiquarks of different flavours annihilate under the action of the weak force. As the weak interaction violates parity symmetry there should be interesting asymmetries in the process.

The W is produced when a quark polarised left handed annihilates with an antiquark polarised right handed. The two beams collide head-on from opposite directions. As they are moving in opposite directions, and also have opposite spin polarisations, the net effect is that both spins are pointing the same way (see figure 14.14). If we call the collision axis the z-axis, then the quark and the antiquark spins are each $J_z = +\frac{1}{2}$ and so the W is produced with $J_z = +1$. This is parity violation at work; it is never produced with $J_z = -1$.

The W then decays and this spin polarisation shows up in the direction that the electron (if W^-) or positron (if W^+) are ejected.

Weak interactions generate left handed electrons and right handed antineutrinos. These are the decay products of a W^-. As the W^- has $J_z = +1$ and angular momentum must be preserved, the electron must be ejected in the same direction as the initial proton beam. This is most easily seen by referring to the diagram.

In the case of a W^+ the whole story is reversed. Weak interactions generate right handed positrons and left handed neutrinos. So this time the W^+ ejects the neutrino in the direction of the proton and it is the positron that follows the antiproton's direction. The net effect is that the charged *particles* (or antiparticles) preserve their sense of direction: proton throws electron forward, antiproton throws anti-electron forward. This is illustrated by the diagram (figure 14.14), where thin arrows indicate the direction of motion and thick arrows denote spin orientation.

In practice, the electron or positron are not thrown exactly forward and backward but are distributed in angle relative to the beams. A distribution of $(1 + \cos\theta)^2$ is predicted and beautifully realised in the data (θ is the angle between the electron and the direction of the incident proton, or between the positron and antiproton). If parity had been a good symmetry then the distribution would have been $(1 + \cos\theta)^2$ and $(1 - \cos\theta)^2$ in equal measure, i.e. $(1 + \cos^2\theta)$ overall. This

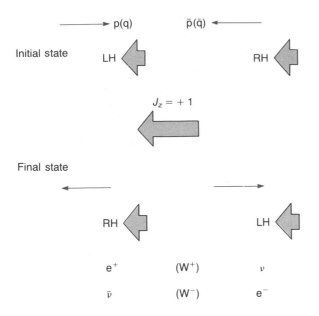

Figure 14.14. A quark in a proton annihilates with an antiquark in an antiproton and produces a W boson (either a W$^+$ or W$^-$ depending on the particular varieties of initiating quarks). The z-direction is from left to right. Small arrows denote motion; large arrows denote the z projection of spin, J_z. At the top we see a left handed quark meet a right handed antiquark (left, right, mean that J_z is aligned opposite or parallel to the direction of flight). The total $J_z = +1$ is indicated, and is identical to the spin projection of the W$^\pm$. The W$^+$ decays into e$^+\nu$; the W$^-$ decays into e$^-\nu$. Spin (angular momentum) conservation requires that each of these final particles has $J_z = -\frac{1}{2}$; thus one is right handed and the other left handed (denoted RH and LH). Weak interaction theory only allows right handed e$^+$ or ν, left handed ν or e$^-$. Thus the W$^-$ and W$^+$ will eject the e$^-$ and e$^+$ into opposite hemispheres. The e$^-$ follows the direction of the original proton; the e$^+$ follows that of the original antiproton.

latter is what is seen in Drell–Yan annihilation induced by electromagnetic annihilations, where parity is a good symmetry.

The W is already being put to use as a tool by which further discoveries are appearing. It can decay into any lepton or quark flavour pairing with equal strength, enhanced by three for colour in the case of quarks relative to leptons (red pairs, green pairs and blue pairs come out with equal probability and we have to sum over all of them). Thus the W decays to up and down quarks (or, more precisely, hadrons built from these quarks) three times as often as it decays into an electron and a neutrino. It will decay to a muon and its neutrino or to a tau and its neutrino as often as to the electron and its neutrino. It will also decay to charm and strangeness as often as to up and

down. Finally, top and bottom quarks should emerge at a similar rate to the other quarks (though these quarks are rather massive and their chance of being produced in W decays is cut down somewhat).

In 1984 the first reports emerged of possible sightings of the top flavour being produced in W decays, when the W decayed into top and bottom. However, it now appears that these claims were premature and evidence for the top quark is still being sought.

So the menu of quarks appears to be complete apart from the top quark. We see them directly in high energy experiments, and for twenty-six years we have watched them collectively in our studies of hadron spectroscopy. We understand why the mesons and baryons occur: they are the combinations of quarks and antiquarks where the colour forces saturate. Nothing in this says that quarks cannot exist alone. After all, the electrical forces saturate in atoms but ions can also exist, so why not for colour too – why no coloured analogues of ions such as free colour carrying quarks?

How sure are we that free quarks do not exist? And if they do exist, then what opportunities might they offer us in the future? These are the questions we now address.

14.5 Free quarks and fusion

Looking for free quarks

All particles that have ever been seen in isolation have integer charges. The fractional charges of quarks, $\frac{2}{3}$ or $-\frac{1}{3}$ relative to a proton, are their most distinctive signature. You would notice at once if a fractionally charged particle passed through your detector, and yet no one has ever conclusively seen a single quark in isolation.

In 1968 Brian McCusker found a track in a cloud chamber picture that appeared to show that a fractionally charged particle had passed through the chamber. The density of the track was markedly less than would be expected for a 'conventionally' charged particle such as a pion or proton. However, some technical doubts were expressed about the circumstances under which the picture was taken, and the majority of scientists were, and remain, sceptical. Certainly no other such tracks have been seen in the subsequent decades, and millions of photos have been taken in this time.

One of the first things that physicists do at each new breed of accelerator is to search for free quarks. One possibility is that, although quarks are apparently lightweight when inside the proton, they are very massive when free. Accelerators might simply be too feeble to have produced free quarks yet, and so access to new higher energy regimes offers the promise of the first free quark.

Although we have not yet produced free quarks in our 'feeble' accelerators on Earth, there is the possibility that massive free quarks were produced in the huge energies attained during the Big Bang. This raises the possibility that quarks might arrive in the cosmic radiation. If they do, then they must be extremely rare to have escaped detection.

Mankind has had apparatus capable of watching cosmic rays for less than a century. Although free quarks are rare, it is still possible that several of them have landed during the Earth's history and accumulated on or near the surface in rocks. If the atmosphere is opaque to quarks then there is the possibility that quarks might still be found in moonrock. Astronauts have brought back rocks from the moon; no quarks were found. Large volumes of substances have been dredged and examined on Earth without success. A problem is that no one is sure which substances are the most 'quark friendly', so everything has to be examined.

William Fairbank at Stanford might have found evidence for fractionally charged particles residing on niobium coated tungsten balls! His technique is to do an analogue of Millikan's oil drop experiment, balancing gravity and electromagnetic forces and so deducing the charge on the object of interest. Niobium becomes a superconducting material at low temperatures. Fairbank cooled 0.25 mm diameter balls and then levitated them magnetically. Then he tickled the balanced ball with a small electrical force. An uncharged ball won't react to the force; a charged ball will. If the ball reacted, electrons or positrons were added from a radioactive source until the niobium was neutralised. If the ball contained fractional charge, then it can never be neutralised.

In a few cases, balls apparently had a charge of $\frac{2}{3}$ or $-\frac{1}{3}$, as would be the case if a free quark was sitting on them. However, no one else has managed to reproduce this effect in similar experiments. It is very hard to remove all uncertainties, and many physicists believe that some unknown source is responsible. So, although there is no agreed explanation as to the source of Fairbank's signal, there is doubt that it is evidence for free quarks. More experiments are in progress and their results are eagerly awaited.

So the situation at present is that theorists believe that quarks are permanently confined in colourless clusters: the hadrons. However, at present, that is still an unproved assumption, and it is amusing to contemplate what the possible consequences of free quarks could be.

Using free quarks to initiate fusion

One of the quark species must be stable. The key reason behind this is the bizarre property of quarks – that they have fractional electrical charges, i.e. $\frac{1}{3}$ or $\frac{2}{3}$ as much as a proton.

Now, it is an important property of Nature that electrical charge is neither created nor destroyed. If you create a positive charge, then a negative one must accompany it so that overall the sums are balanced at zero. Imagine you are watching a quark decaying into a shower of particles. The quark has a charge of $\frac{1}{3}$ or $\frac{2}{3}$. The shower that it turns into must have the same net charge as the quark that gave birth to it: $\frac{1}{3}$ or $\frac{2}{3}$. If 'quark' means 'something with fractional charge' then the shower must contain a quark. Once a spare quark somewhere, always a spare quark.

So, after death, any given quark leads to at least one sequence of quarks along with other integrally charged or neutral 'conventional' particles.

If the original quark is heavy then it contains a lot of energy. This energy is shared among its progeny. The new quark that is born in the shower must have less energy than the original quark because the other members of the shower have used up some of the energy themselves. In turn, when this quark decays, it creates another shower, containing a quark of even less energy, and so on. The sequence eventually terminates when the lightest of all quarks is reached.

Now suppose that you were moving along with this quark so that it appeared to be at rest. If it decays it must share its rest energy, mc^2, among the decay products. It can only do this if these newborn particles are lighter than their parent. This is clearly impossible since by hypothesis the parent is the lightest quark of all and we can no longer balance both energy and charge. The lightest quark of all must be stable. The only way that it can disappear is to meet an antiquark and be annihilated.

A stable fractionally charged object could have bizarre applications. A quarkonics industry might be built around it. Fractionally charged atoms could be used as tracers in a variety of chemical, physical and biological processes. The most significant impact might be in initiating fusion, converting nuclear mass into energy.

The essential idea of fusion is that two deuterium nuclei (one proton and one neutron makes one deuteron) combine. They will make either a nucleus of helium-3 (two protons and one neutron) which leaves a free neutron, or alternatively they will make tritium (hydrogen-3; two neutrons and a proton) and leave a free proton spare. In both cases several millions of volts of energy are released from the nucleus.

So if you could fire two deuterons at one another, and they stayed together long enough to stick (fuse), you could help to solve the world's energy supplies by using the energy released. There is a major research effort attempting to harness fusion. A difficulty is that the two deuterons mutually repel due to their positive electrical charges. As the two nuclei approach they feel an intense repulsive force which slows them and they bounce away instead of merging.

A trick that can overcome this is to make atoms where negatively charged electrons orbit around the positively charged nuclei. The atom is electrically neutral at distances greater than 10^{-10} m. Two atoms, this distance apart, feel no intense disruption. Unfortunately this is still an enormous distance compared to that where neutron and proton will fuse. We have to get the nuclei to within 10^{-15} m of each other; some 10 000 times nearer than the range that atoms usually approach.

If we could make smaller atoms then it might be easier to get the nuclei closer together. Heavy objects moving slowly in a small circle have the same angular momentum as light ones moving fast round a large circle. The light electron moves at a large distance round the atomic nucleus. But, replace it with the heavy electron-like particle called the muon, and you have a small atom: the heavy muon orbits nearer to the nucleus than the light electron does.

We call this a 'muonic' atom to distinguish it from the usual 'electronic' atom. The muon's negative charge shields the positively charged nucleus allowing atoms to approach within 10^{-13} m; one-hundred times closer than before. This is near enough that the nuclei have a chance to fuse. The catch is that the muon is unstable; the atom survives for less than one millionth of a second, and in this short time fusion hardly has the chance to happen. This is where a stable massive quark could be useful.

The lightest quark of all is the up quark with charge $\frac{2}{3}$; the up antiquark has charge $-\frac{2}{3}$. So we could make an 'atom' where the positively charged proton is encircled by a stable negatively charged up antiquark, which shields the proton's positive charge. It will orbit very close to the nucleus because it has a strong attraction for it: nuclei are made of quarks that bind together tightly, not loosely like electromagnetically attracted atoms. Under these conditions fusion would be rapid. The antiquark is not used up, but lives on to catalyse another reaction.

To bind two deuterons together needs at least two up antiquarks, net charge $-\frac{4}{3}$. The two quarks will bind the two deuterons in an analogous way to that where two electrons bind two protons in a hydrogen molecule. The two deuterons can then have a good chance to fuse together and release energy.

Some of the practical details have already been evaluated and are as follows.

A pair of antiquarks with charge $-\frac{4}{3}$ come to rest in hydrogen or deuterium at pressure of 2000 p.s.i. $(1.4 \times 10^{7} \, \mathrm{N \, m^{-2}})$. They capture a molecule and form a system with charge $-\frac{1}{3}$. The resulting 'quark-molecule' is very excited. It takes a short time to settle down, 'de-excite', and in this time it captures another nucleus, of charge $+1$, and fuses. A quark is released to go on and do the job again. The process can be summarised by the sequence

$$Q + [D(pn) + D(pn)] \rightarrow [3H(pnn) + p] + (\text{energy}) + Q,$$

where Q = quark, D = deuteron and $3H$ = tritium.

The new quark moves on to the next nucleus and can initiate further fusion. The reaction rate is one per second at 2000 p.s.i. pressure. For deuterium this releases 3.65 MV energy per fusion. A few grammes of quarks could catalyse fusion and generate 10% of the total USA energy consumption!

There appears to be only one weak link in the whole enterprise: no one has yet found isolated free quarks!

14.6 Quarks and leptons: what next?

As already discussed at the beginning of this chapter, at various stages during our voyage into matter we have convinced ourselves that we have identified the basic building blocks. There is a psychologically based belief that Nature is economical, that there should be only a few elementary particles at the root of the infinite variety about us.

We are periodically encouraged in this belief when we discover an underlying economy that describes diversity. Nearly one-hundred atomic elements represents what I mean by diversity. Then the inner structure of atoms was revealed and all of these varieties of element were seen to be built from a few more elementary objects.

Around 1935 physicists believed that all matter was made of protons, neutrons, electrons and neutrinos. This was extremely economical, and we have seen what happened next: by the 1960s nearly as many varieties of hadron had been identified as there had been atomic elements in the last century. The hint that we had discovered economy in Nature was misunderstood. It merely showed that we had not looked hard enough. We uncovered a deeper layer of reality, which appeared at first glance to be the philosopher's stone. Detailed study then revealed that extensive structures occurred at this new layer; initially we had merely uncovered a few, and mistakenly thought that we had seen it all.

Today we have the quarks and leptons as the basic building blocks of matter along with various force-mediating particles, the photon, W, Z and gluons. Initially there was the promise of economy here too. Three varieties of quark, the up, down and strange, proved sufficient to build up the many hadrons that had confused us. Two neutrinos, an electron and muon completed the lepton half of the menu.

The discovery of charmed quarks was the first hint that we had glimpsed only the low energy extremes of a richer tapestry. As we extended our investigations into new higher energy regions we found further leptons, the tau and its neutrino, and also bottom quarks. These latter have masses orders of magnitude greater than the up and down quarks and the electron that build up our homely environment. Moreover, the quarks carry colour and so are triplicated; the up quark occurs in red, blue or green form as do all the other quarks. List them all and you see that the total number of species is quite large. If we believe in a 'few' basic building blocks of matter, then this proliferation may prove to be the first hint that there is physics beyond the quark.

We have accumulated data on the properties of the various quarks and leptons which must be explained by any ultimate theory. Each of these particles has its own, seemingly random, mass value. So in addition to the proliferating building blocks we also have about a score of arbitrary unexplained parameters in the theory. These include the masses of the quarks and leptons, the strengths of the various forces, the various mixing angles relating the ways that quarks in different generations respond to the weak interactions, and the bizarre relations between their electrical charges.

This latter point is extremely profound and so much part of our existence that we hardly think about it. Atoms are neutral because the electrical charge of the proton exactly balances the negative charge on the electron. There is no explanation of this in the 'standard model' of weak and electromagnetic interactions. It is put in by hand. To explain it requires going beyond the standard model and constructing theories which relate quarks and leptons. Grand Unified Theories (GUTs) offer this promise and are discussed in Chapter 15 by Georgi.

Whether the way forward is via GUTs or whether we have the first hints of a reality beyond quarks and leptons, there is no doubt that we have not yet tapped the final truths. There are regularities among the quarks and leptons which demand explanation.

The most noticeable of these is the existence of generations. Each generation consists of two leptons and two quarks each with spin $\frac{1}{2}$. The quarks and leptons in a generation respond to weak interactions in identical ways and to electromagnetic forces with strength scaled by their electrical charges but otherwise the same. This whole pattern is then triplicated by Nature. There must be a reason for this and even the GUTs have no compelling explanation of this yet.

Many questions come to mind of which I offer a selection.

Why are all integer charged particles colourless and all fractionally charged particles coloured? Is there a connection between threefold colour and third fractional electrical charge? If the quarks and leptons are composed of smaller particles, then why do they all have spin $\frac{1}{2}$; are there spin $\frac{3}{2}$ examples to be found? Are there examples with charge $\frac{4}{3}$ or $\frac{5}{3}$? Is it simply a case again that we have looked at Nature at 'low' energies and convinced ourselves that we now have the whole picture?

There is no doubt that the standard model gives a very good description of phenomena at distances above 10^{-17} m. The challenge now is to probe to shorter distances, or equivalently higher energies, and hopefully learn why Nature is as it is at the larger distances. That is how things have been in the past. If there are new constituents and new forces to be discovered they must be restricted to these minute distances. It will require extreme energies to examine them.

If there are particles that are the common seeds of the quarks and leptons, then the relations among the masses and properties of the present 'fundamental particles' will follow from the charges and other quantum numbers of the new 'elementary' particles. This question is currently only answerable by experiment. We do not have enough information to constrain flights of fancy as to the number and properties of the new elementary 'prequarks'. Large though the menu and parameters of the standard model may be, they are not extensive enough to be a clear guide. If the generations are excited states of composite systems, then there will presumably be more generations to be found. This at least is an exciting prospect for new machines, such as LEP, to examine.

Although there is almost total freedom in inventing possible composite models of the quarks and leptons, there is one profound problem that advocates of this route have to overcome, and which hints that things are different this time around.

Composite systems, such as atoms, nuclei or protons, can be characterised by two important scales: their mass and their spatial size or 'radius'. The uncertainty principle of quantum mechanics relates the physical size of objects and the typical momenta of components moving within. The smaller a system is, so the larger are the momenta and energies of its constituent parts.

Atoms and nuclei are more massive than the energies carried by their constituents. Protons are built from quarks, and the whole mass of the proton might arise from the energy of the quarks. This is in accord with the uncertainty principle; the mass of the proton is in accord with the energy scale associated with its intrinsic size. So all of the known composite systems 'make sense'. If quarks and leptons are composite then something essentially new is taking place. If we confined objects within the size of an electron then we would expect the composite system to have a mass of the order of hundreds of GeV. The electron, by contrast, is at least a million times lighter than this. The 'composite' electron, neutrino and quarks are almost massless on the scale of their constituents' energies.

This is the problem that any advocates of composite quarks and leptons have to solve. It may be the hint that the quarks and leptons are truly fundamental. On the other hand, it may be the first hint of paradox that will lead to essential new insights about Nature's universal law.

Table of long-lived particles

Family groups	Particle	Symbol	Mass	Charge states	Spin	Strangeness	Average lifetime	Principal decay products
	Graviton	g	0	0	2	0	∞	Stable
	Photon	γ	0	0	1	0	∞	Stable
Neutrinos		$\nu_e \bar{\nu}_e$	0(?)	0	1/2	0	∞	Stable
		$\nu_\mu \bar{\nu}_\mu$	0(?)	0	1/2	0	?	Stable (?)
		$\nu_\tau \bar{\nu}_\tau$	?	0	1/2	0	?	?
	Electron	$e^+ e^-$	0.51100	$+1\ -1$	1/2	0	∞	Stable
	Muon	$\mu^+ \mu^-$	105.66	$+1\ -1$	1/2	0	2.2×10^{-6}	$e\nu\bar{\nu}$
	Tauon	$\tau^+ \tau^-$	1784.2	$+1\ -1$	1/2	0	3.4×10^{13}	Hadrons
	Pion	$\pi^+ \pi^-$	139.57	$+1\ -1$	0	0	2.6×10^{-8}	$\mu\nu$
		π^0	134.96	0	0	0	0.8×10^{-16}	2γ
	Kaon	$K^+ K^-$	493.67	$+1\ -1$	0	$+1\ -1$	1.2×10^{-8}	$\mu\nu\pi\pi^0,\ \pi-\pi^+\pi^-$
		$K^0 \bar{K}^0$	497.67	0	0	$+1\ -1$	0.9×10^{-10}	$2\pi^0,\ 3\pi^0,\ \pi^+\pi^-$
	Eta	η	548.8	0	0	0	2.5×10^{-19}	$2\gamma,\ 3\pi^0,\ \pi^0\pi^+\pi^-$
	Proton	$p\bar{p}$	938.28	$+1\ -1$	1/2	0	$> 10^{39}$	Stable(?)
	Neutron	$n\bar{n}$	939.57	0	1/2	0	898	$pe\nu$
	Lambda	$\Lambda\bar{\Lambda}$	1115.60	0	1/2	$-1\ +1$	2.6×10^{-10}	$p\pi^-,\ n\pi^0$
Sigma		$\Sigma^+ \bar{\Sigma}^+$	1189.36	$+1\ -1$	1/2	$-1\ +1$	0.8×10^{-10}	$p\pi^0,\ n\pi^-$
		$\Sigma^0 \bar{\Sigma}^0$	1192.46	0	1/2	$-1\ +1$	5.8×10^{-20}	$\Lambda\gamma$
		$\Sigma^- \bar{\Sigma}^-$	1197.34	$-1\ +1$	1/2	$-1\ +1$	1.5×10^{-10}	$n\pi-$
Xi		$\Xi^0 \bar{\Xi}^0$	1314.9	0	1/2	$-2\ +2$	2.9×10^{-10}	$\Lambda\pi^0$
		$\Xi^- \bar{\Xi}^-$	1321.3	$-1\ +1$	1/2	$-2\ +2$	1.6×10^{-10}	$\Lambda\pi^-$
	Omega	$\Omega^- \bar{\Omega}^-$	1672.5	$-1\ +1$	3/2	$-3\ +3$	0.8×10^{-10}	$\Xi^0\pi^-,\ \Xi^-\pi^0,\ \Lambda K^-$

'Long-lived' here means with lifetime much greater than 10^{-23} s.
Antiparticle symbol is shown after particle (if different). Mass is in MeV, charge in units of the proton charge and lifetimes in seconds. This list is not exhaustive.

15 Grand unified theories

H. M. Georgi

15.1 The rules of the game

Theoretical particle physicists are a fortunate breed. We are privileged to be players in the most interesting game we know. The game is simple. We try to determine the rules according to which the world works, making use of the clues provided for us by our experimental colleagues. We carry around in our heads bits and pieces of a majestic jigsaw puzzle, the ever evolving quantitative model of our physical world. One evening, in the Winter of 1973, I had an experience that is rare and wonderful even for a particle theorist. Some of my pieces fitted together in a particularly surprising and beautiful way. The results were the first 'Grand Unified Theories', simple unified descriptions of the interactions we had already seen that made dramatic predictions for new processes, including the decay of the proton, and thus of all neutral matter. The jury is still out on these theories. Experimental physicists have not seen evidence for proton decay, despite a heroic effort. But 'GUTs' have had a profound effect, for better or worse, on theoretical particle physics. In this essay, I will try to give you a glimpse into the marvellous game of the particle theorist, to show you how the pieces of a grand unified theory fit together and to trace the consequences of this idea into the 1980s.

15.2 The pieces of the puzzle

Physics is an accumulative science. Our present understanding of particle physics, in particular, has been built up gradually over the years, incorporating the brilliant and painstaking work of thousands of physicists. I cannot begin to explain, or even to list, the hundreds of distinct facts on which our current theory is based. Furthermore, the mathematical language in which the theory is expressed (called relativistic quantum field theory) is not easy to understand. Indeed, the language itself raises many questions that are not understood by anyone and are actively studied both by physicists and by mathematicians. Clearly, I cannot hope to give a detailed account of the theory on which GUTs are based. Various aspects of the theory are described elsewhere in this anthology. Here, I will present a particularly oversimplified version that, I hope, conveys some glimpse into those pieces of theory that are especially important for GUTs.

To approach GUTs, we must go down to very short distances. As we go, we will get deeper and deeper into the world of modern particle physics. I will try to explain the important concepts in the theory of particle interactions when we get to the scale at which they first become important.

We have to start somewhere, so let us take 10^{-2} cm. This is the size of an amoeba. It is about the smallest distance that can be resolved by the unaided eye. At smaller distances the nature of 'seeing' changes.

At first the change is trivial. To see a red blood cell, at 10^{-3} cm, we just need a microscope. But if we go down two more factors of 10 and try to look at a bacterial virus with a diameter of 10^{-5} cm, an ordinary light microscope is not very useful. The trouble is that visible light has a wavelength of a few times 10^{-5} cm. It wiggles back and forth (more or less) in this distance, so it cannot be used to see details of this size. It is not uniform enough.

Visible light is only a small part of the spectrum of electromagnetic radiation. It is theoretically possible to build a microscope that uses electromagnetic waves with a shorter wavelength than visible light, ultraviolet light or X-rays, but it is not very practical. The trouble can be traced to a fundamental fact about the quantum mechanical world. Any wave like light that carries energy and momentum is built out of packets of energy and momentum, particles of light, and the momentum and energy of the individual packets go up as the wavelength goes down. Ultraviolet and X-ray photons, as the particles are called, have a higher energy than the photons of visible light. For various reasons, this makes them difficult (though not impossible) to use in a microscope.

Microscopes need not use light beams. An electron microscope uses a beam of electrons in much the same way that an ordinary microscope uses light. But even for material particles like electrons there is a wavelength that goes down as the momentum and therefore the energy go up. The wavelength, λ, of any particle is related to its momentum, p, by the relation $\lambda = 2\pi\hbar/p$ or $p = 2\pi\hbar/\lambda$, where $\hbar$ is Planck's constant. This is a general property of the world that follows from quantum mechanics: *to see short distances, you need high energies*. It follows that as we try to 'see' shorter and shorter distances with higher and higher energy beams, the process will require more and more elaborate equipment and involve more and more violent collisions between the particles in the beam and the object we are trying to see. But, if we are willing to put up with these difficulties, we can continue on down the distance scale.

At about 10^{-5} cm we see large molecules like haemoglobin. Smaller molecules range in size down to 10^{-8} cm, the size of a typical atom. At some point on the distance scale near 10^{-8} cm, it is reasonable to stop calling our microscopes 'microscopes' and start calling them 'particle accelerators'.

If we look inside the atom, we see that it is a cloud 10^{-8} cm across of electrons surrounding a tiny nucleus with a diameter of 10^{-12} to 10^{-13} cm. The nucleus, in turn, is built out of protons and neutrons, each 10^{-13} cm in diameter. At these small distances, many properties that are familiar from the macroscopic world show up in unfamiliar forms. For example, electric charge, the basic stuff of electricity, in the macroscopic world behaves somewhat like a fluid. It flows through wires and onto and off of objects. It is pushed by generators and batteries and controlled by switches. But at small distances, electric charge is an unchangeable property of each type of particle. Each proton carries the same quantity of positive electric charge, called, naturally, the charge of the proton, and given the symbol e. Each electron carries exactly the same quantity of negative electric charge, $-e$. The photon and neutron carry no electric charge; they are neutral. Charge cannot flow onto or off of any of these particles unless the particle changes its identity. And even then, the total charge before the change and after must be the same. When these particles are bound together into nuclei or atoms, the total charge is just the sum of the charges of the parts.

15.3   Coupling constants

The quantum mechanical theory of the interactions of electrons and photons is called 'quantum electrodynamics'. It is an extremely successful theory in that its predictions have been verified experimentally to great precision (see my article on EFTs, Chapter 16). One reason that precise predictions can be extracted from quantum electrodynamics is that the electromagnetic interaction is fairly weak. As I will show you, it is characterised by a small dimensionless 'coupling constant', α, about equal to $1/137$. The concept of a 'coupling constant' will be central to the idea of unification, so I will describe it in some detail.

The idea starts with Coulomb's law: the force between two charged particles is proportional to the product of their electric charges and inversely proportional to the square of the distance between them. In a particular system of units, the force looks like

$$F = q_x q_y / r^2,$$

where q_x and q_y are the charges and r is the distance between them. This equation defines what we mean by the charges in this system of units. The product of the two charges is equal to the force times the square of the distance between the charges. In particular this means that, in this system of units, the product of the charges is measured in the same units as a force times the square of a distance, which in turn has the same units as a product of an angular momentum and a velocity. This product depends on the units we use to measure angular momentum and velocity. But there are two constants which are apparently built into the structure of the world; Planck's constant, $\hbar$, an angular momentum that characterises the size of quantum mechanical effects, and the speed of light, c, a velocity, that is central to relativistic effects (see my chapter on EFTs). Thus, if we form the combination $q_x q_y / \hbar c$, we have a dimensionless measure of the strength of the interaction between q_x and q_y. It is independent of the units in which we measure force and distance. This would not be very interesting except for the experimental fact that charge is quantised. All particles that can be isolated in the laboratory seem to have an electric charge (as measured by Coulomb's law) that is an integral multiple of the proton charge, e, so it makes sense to use the proton charge as the standard against which all other charges are measured. Thus, taking out factors of the proton charge, we can write

$$q_x q_y / \hbar c = Q_x Q_y e^2 / \hbar c,$$

where Q_x and Q_y are dimensionless (integer) 'charges', the ratios of q_x and q_y to the proton charge, and $e^2/\hbar c$ is a dimensionless measure of the strength of the electromagnetic interactions; this is the coupling constant:

$$\alpha = e^2 / \hbar c.$$

The experimental value of α is much smaller than 1 (as stated previously it is about $1/137$). It is small enough to allow a straightforward treatment of the quantum theory that makes use of 'perturbation theory', a series expansion of the predictions of the theory in increasing powers of the small parameter α.

In the discussion below, I will absorb the proton charge e into the dimensionless coupling constant α and describe the charge of any particle by its ratio, Q_E, to the proton charge. The E suffix stands for electric charge. We need the subscript because we will soon meet some other types of charges, similar to electric charge in some ways but associated with other forces.

I should emphasise that, while the quantisation of electric charge is an experimental fact, the quantum electrodynamics theory does not require it. The theory would still make sense if there were also a particle with electric charge $Q_E = \pi$, for example.

15.4 Antimatter

Another peculiarity that shows up at these short distances is antimatter. Occasionally, in the violent collisions required to probe the interior of an atom, a new kind of particle is produced that has exactly the same mass as an electron, but $Q_E = 1$ instead of $Q_E = -1$. This particle is called an 'antielectron' or a 'positron'. Antielectrons usually do not stay around very long. If an antielectron collides with an electron, the two can cancel each other out and annihilate into a burst of photons.

There are good theoretical reasons to believe that all particles have antiparticles (although some, like the photon, may be their own antiparticles). The antiparticles always have exactly opposite values of electric charge (and any other charges). The antiproton (with $Q_E = -1$) and the antineutron (with $Q_E = 0$) have also been seen in high energy collisions.

Loosely speaking, for any process in which a particle is coming in, there is a related process in which its antiparticle is going out, and vice versa. Consider, for example, a process in which an incoming electron is scattered by a magnet and goes off in another direction. The above dictum implies, for example, that there is a related process in which the magnet creates two particles, one electron and one antielectron. This process is just the reverse of the process in which a positron and electron annihilate each other.

15.5 Virtual particles

In a relativistic quantum theory, forces are associated with exchange of particles. The electromagnetic force between two charged particles (the only force we have talked about so far) is associated with photon exchange. Photons have energy and momentum and (of course) travel at the speed of light. But they have no mass, because there is no way to slow one down and measure its mass at rest. They are emitted and absorbed when charged particles are moved around.

A free electron (one that is not being accelerated) cannot emit or absorb a real photon. The process cannot conserve energy and momentum. But a quantum mechanical system can violate conservation of energy for a very short time dt, so long as the magnitude of the energy imbalance dE and its duration dt satisfy the Heisenberg uncertainty relation that the product of dE and dt is less than Planck's constant. A particle with the wrong energy or momentum is called a 'virtual' particle.

The force between two charged particles results from exchange of virtual photons which carry momentum but not energy. Here it is crucial that the photons have zero mass. If a particle has mass m, Einstein tells us that the energy of a particle when it is at rest should be mc^2. If such a particle exists as a virtual particle with zero energy, the energy imbalance is at least mc^2. Then Heisenberg tells us that this particle can only exist for a time $\hbar/mc^2$. Even travelling at the speed of light, the

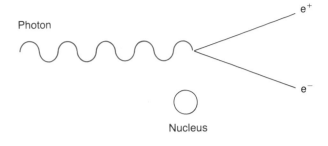

Figure 15.1. A photon splitting into an electron–positron pair in the electric field of a nucleus.

particle can only go a distance $\hbar/mc$ in this time. This quantity $\hbar/mc$ is an important one. It is a length which depends only on the mass of the particle and the fundamental constants $\hbar$ and c. It is called the Compton wavelength of the particle. What we have seen is that the range of a force produced by a particle with a nonzero mass is about equal to its Compton wavelength. The Compton wavelength of the proton, for example, is about 2×10^{-14} cm. Not very large. In fact, the *photon's* rest mass seems to be exactly zero, so that two charged particles can exchange virtual photons of arbitrarily small momentum and energy imbalance giving rise to the Coulomb force which has infinite range. But we will return to the virtual exchange of a massive particle when we discuss the weak force, below.

I will sometimes describe the virtual exchange of particles in a diagram such as figure 15.2 in which the solid lines represent charged particles and the wavy line represents the exchanged virtual photon. In fact, to a theoretical particle physicist such a diagram is more than a picture. They are called 'Feynman diagrams' because Feynman taught us how to associate each such diagram with a definite mathematical expression that can be used to calculate force or a scattering probability.

15.6 Hadrons and quarks

10^{-13} cm is a very small distance. To 'see' things of this size and smaller, to look inside the protons and neutrons, requires a special effort. The microscopes are large particle accelerators that produce beams of very high energy electrons or protons (or sometimes other particles). When such a high energy particle collides with a proton, instead of producing a nice clear picture of the inside of the proton, it usually makes a mess. The mess consists of several new particles instead of or in addition to the particles that collided. The particles that are produced most often are called π (pi) mesons or pions. There are three kinds: π^+, π^0 and π^- with $Q_E = +1$, 0 and -1. The pions are about the same size as the proton and neutron but only about one-seventh as heavy.

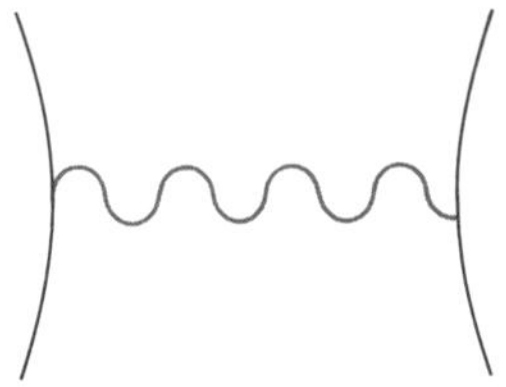

Figure 15.2. At the quantum level the electromagnetic force between two charged particles is ascribed to the exchange of 'virtual' photons ($\sim$). The photon may thus be regarded as a sort of 'messenger' particle, conveying the force between the charged particles and enabling them to act on each other at a distance.

Protons, neutrons, antiprotons, antineutrons and pions all belong to a large class of particles called 'hadrons' that feel the strong interactions that bind protons and neutrons together into complicated nuclei. In the last fifteen years or so, physicists have found better ways of looking inside the hadrons than just slamming two hadrons together. For example, a high energy electron beam can be used to probe the interior of a hadron if you look only at the scattered electron and ignore the hadronic mess of pions and other particles that comes along with it (see the discussion of the parton model in the chapter by Close). These techniques show rather clearly that the hadrons are built out of quarks that are stuck, or confined inside them. There are several types (or flavours) of quarks, but the protons, neutrons and pions are built out of just two, the u quark (u is for 'up') and the d quark (d is for 'down'). The electric charges of the quarks are fractions of the proton charge, $\frac{2}{3}$ for u and $-\frac{1}{3}$ for d. The proton is built out of two u's and one d. The charge works out because $\frac{2}{3} + \frac{2}{3} - \frac{1}{3} = 1$. The neutron is built out of two d's and a u ($-\frac{1}{3} - \frac{1}{3} + \frac{2}{3} = 0$).

Of course, there are corresponding antiparticles. The antiparticle of the u is called the $\bar{u}$ (read u-bar) antiquark. It has electric charge $-\frac{2}{3}$. The antiparticle of the d is the $\bar{d}$ antiquark, with charge $\frac{1}{3}$. The antiproton and antineutron are made out of antiquarks. For example, the antiproton is built out of two $\bar{u}$'s and a $\bar{d}$. The pions are made out of one quark and one antiquark. The π^+, for instance, is a u and a $\bar{d}$ ($\frac{2}{3} + \frac{1}{3} = 1$) and the π^0 is a combination of $u\bar{u}$ and $d\bar{d}$ ($\frac{2}{3} - \frac{2}{3} = -\frac{1}{3} + \frac{1}{3} = 0$).

What is it that binds the quarks together into hadrons? A clue can be obtained just by counting the quarks. Many different kinds of experiments have been done to count the quarks. They all agree that each type of quark (u, d, etc.) comes in three distinct states. We call these three states colours, red, green and blue, for no very good reason. But while the name is not important, the 'threeness' is crucial. It is the reason that a proton is made of three quarks rather than two or five or some other number. We now have good evidence that the forces which bind quarks together are due to an interaction between the colours that is similar to the force between electrically charged particles, but stronger. Colour is a different kind of charge.

The theory of the colour interaction has a great deal of symmetry. All the colours are treated on exactly the same footing. If we interchange the names of red and blue quarks, for example, the theory looks the same. But there is more to it than just passive symmetry. This interchanging of colours is really a crucial part of the dynamics of the colour interaction. You can think of the colour interaction as a process in which the colour of a quark (or antiquark) changes in one place and the change is compensated by an inverse change taking place somewhere else.

There is a very simple geometrical representation of the symmetry of the colour interaction that can be a real help in understanding it. The most symmetrical thing that can be built with three objects is an equilateral triangle. Represent the three colours of quarks as the corners of the triangle. Then if you think of the quarks as weights, the colour interactions tend to bind the quarks together into systems that can be balanced at the centre of the triangle without tipping over. These are called *colourless* combinations. For example, a system of three quarks, one of each colour is colourless. This is the reason that the proton and neutron consist of exactly three quarks; one of each colour.

The antiquarks have colours (or anticolours if you will) that are precisely opposite to the colours of the quarks. In the geometrical picture, this means that the antiquarks are also an equilateral triangle, but it is rotated by $180°$. Comparing the quark and antiquark triangles, you can see that you can make a colourless state by taking one colour quark combined with the corresponding anticolour of antiquark. These states are the mesons.

Electrons and antielectrons and any other particles that do not feel the strong force have no colour. In the geometrical picture, they belong at the centre of the equilateral triangle.

15.7 Colour charges

Just as the electric force between charged particles is related to the exchange of photons, the colour force is produced by the exchange of particles called 'gluons'. There are eight different gluons, and correspondingly eight different kinds of colour charge! The square of the charge measures the probability that a quark which is being accelerated will emit a corresponding gluon. The multiplicity of charges is due to the fact that the colour of the quark can change when a gluon is emitted. We have to think of the colour charges as operations on the colour states of the quarks. The action of the eight charges is shown in table 15.1 (taking out a dimensional factor, e_3, like the proton charge e in electromagnetism, so that the Q's are dimensionless). The reasons why the charges take this form will, I hope, become apparent as we go further.

Table 15.1. *The colour charges*

	Acting on		
	$\lvert \text{red}\rangle$	$\lvert \text{green}\rangle$	$\lvert \text{blue}\rangle$
Q_{c1}	$\frac{1}{2}\lvert \text{green}\rangle$	$\frac{1}{2}\lvert \text{red}\rangle$	0
Q_{c2}	$\frac{1}{2}\lvert \text{green}\rangle$	$-\frac{1}{2}\lvert \text{red}\rangle$	0
Q_{c3}	$\frac{1}{2}\lvert \text{red}\rangle$	$-\frac{1}{2}\lvert \text{green}\rangle$	0
Q_{c4}	$\frac{1}{2}\lvert \text{blue}\rangle$	0	$\frac{1}{2}\lvert \text{red}\rangle$
Q_{c5}	$\frac{1}{2}\lvert \text{blue}\rangle$	0	$-\frac{1}{2}\lvert \text{red}\rangle$
Q_{c6}	0	$\frac{1}{2}\lvert \text{blue}\rangle$	$\frac{1}{2}\lvert \text{green}\rangle$
Q_{c7}	0	$\frac{1}{2}\lvert \text{blue}\rangle$	$-\frac{1}{2}\lvert \text{green}\rangle$
Q_{c8}	0	$\frac{1}{2}\lvert \text{green}\rangle$	$-\frac{1}{2}\lvert \text{blue}\rangle$
$Q_{c8}{}'$	$-\frac{1}{2}\lvert \text{red}\rangle$	0	$\frac{1}{2}\lvert \text{blue}\rangle$

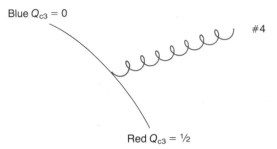

Blue $Q_{c3} = 0$

#4

Red $Q_{c3} = \frac{1}{2}$

Figure 15.3. The figure shows a process in which a red quark emits a virtual gluon and changes into a blue quark. The gluon carries away the appropriate colour charge in order that the overall colour may be conserved.

In table 15.1, the notation $\lvert \ldots \rangle$ refers to the colour state of the quark. A 'state', in quantum mechanics, is a description of the system. But the peculiar thing about quantum mechanical states is that they can be added together or multiplied by constants to get other states. The square of the coefficient of a given colour state in the description of a quark is proportional to the probability that the quark has the corresponding colour. If a quark is described by the state $\lvert \text{red}\rangle$, its colour is known to be red. But if the colour state is, for example, $\lvert \text{red}\rangle + \lvert \text{blue}\rangle$, then a measurement of the quark colour will give red 50% of the time and blue 50% of the time. Thus, in the table, Q_{c3} acting on a green quark state gives $-\frac{1}{2}$ times the same green quark state, which means that the corresponding gluon (gluon #3) can be emitted by the green quark state with probability proportional to $\frac{1}{4}e_3{}^2$ and no change in the quark colour. Q_{c6} acting on a green quark state gives $\frac{1}{2}$ times a blue quark state, which means that gluon #6 can be emitted from a green quark, but in the process the quark colour changes from green to blue. Note that I have listed nine rather than the eight charges I promised you. In fact, however, only eight of these charges are independent. $Q_{c8}{}'$ can be obtained as minus the sum of Q_{c3} and Q_{c8}, so it is not really new. I included it to make the symmetry of this set of charges more apparent.

The colour interactions differ from the electromagnetic interactions in a very important way. The photon has no charge, but the colour gluons carry the colour charges and so they can also emit gluons when they are accelerated. This must be so if the colour charges are to be conserved. Consider the process shown in figure 15.3 in which a red quark emits a #4 gluon and becomes a blue quark. Look, for example, at the charge Q_{c3}. If the total charge is to be the same before and after the emission, since the initial red quark has $Q_{c3} = \frac{1}{2}$ and the final blue quark has $Q_{c3} = 0$, the #4 gluon must carry a $Q_{c3} = \frac{1}{2}$. Thus the fact that the gluons carry charge is related to the fact that the corresponding charges are not just numbers, but change the colour states of the quarks. This has another important consequence. For ordinary numbers, like three and five, the product is the same whether we multiply three by five

or five by three. We say that ordinary numbers 'commute' under multiplication. There is a natural way to multiply two charges, say Q_{c1} and Q_{c2}, together, by acting first on a colour state with Q_{c2} and then on the resulting state with Q_{c1}. This gives, in symbols,

$$Q_{c1}Q_{c2}\lvert \text{red}\rangle = \tfrac{1}{2}Q_{c1}\lvert \text{green}\rangle = \tfrac{1}{4}\lvert \text{red}\rangle$$

But performing these operations in the opposite order gives a different result. Q_{c1} acting on $\lvert \text{red}\rangle$ gives $\frac{1}{2}\lvert \text{green}\rangle$ and Q_{c2} acting on $\frac{1}{2}\lvert \text{green}\rangle$ gives $-\frac{1}{2}\lvert \text{red}\rangle$. Symbolically,

$$Q_{c2}Q_{c1}\lvert \text{red}\rangle = \tfrac{1}{2}Q_{c2}\lvert \text{green}\rangle = -\tfrac{1}{4}\lvert \text{red}\rangle.$$

In fact, the difference between the product of the charges $Q_{c1}Q_{c2}$ acting on any colour state and the product in the other order $Q_{c2}Q_{c1}$ acting on the same state is just twice the charge Q_{c3} acting on the state. Therefore we say that, as an operation,

$$Q_{c1}Q_{c2} - Q_{c2}Q_{c1} = Q_{c3}.$$

This is called a commutation relation. There are similar relations between all the other colour charges. Only two independent charges can be chosen to act like numbers. In our table, these are Q_{c3} and Q_{c8}. The others all involve nontrivial operations on the states. The advantage of the commutation relations is that they are independent of the details of the states on which the charges act. In fact, these particular forms occur in many very different physical and mathematical systems, because they are associated with a simple and ubiquitous symmetry on the states. The charges Q_{c1} through Q_{c8} comprise almost all the essentially different ways of transforming the three states: you can change the sign of one compared to another; you can interchange two; or you can interchange two with a sign change on one state. The only other charge that could act on the three colour states is a charge that always gives you back the same state, perhaps multiplied by a number:

$$Q_{c0}\lvert \text{red}\rangle = \lvert \text{red}\rangle, \quad Q_{c0}\lvert \text{green}\rangle = \lvert \text{green}\rangle, \quad Q_{c0}\lvert \text{blue}\rangle = \lvert \text{blue}\rangle.$$

The charge Q_{c0} is very different from all the others precisely because it acts like a number, not an operation. It commutes with all the other charges and never appears in the commutation relation for any other charges. This is not one of the colour charges. There is no corresponding gluon.

15.8   Group theory

The study of commutation relations such as those described in the previous section belongs to the branch of mathematics known as group theory. The surprising thing is that one can get a great deal of information just from the commutation relations, without knowing exactly what the charges are, or even what the particles are that carry the charges. In the words of Sir Arthur Stanley Eddington*:

> We need a super-mathematics in which the operations are as unknown as the quantities they operate on, and a super-mathematician who does not know what he is doing when he performs these operations. Such a super-mathematics is the Theory of Groups.

The commutation relations of the eight colour charges are associated with a symmetry of the interactions under what is called 'the group SU(3)'. The '3' in SU(3) refers to the three colours that are mixed up under the action of the group. The 'U' stands for the mathematical term 'unitary' that describes the nature of the symmetry. This kind of unitary symmetry is particularly important for quantum mechanical systems. The 'S' in SU(3) stands for 'special', which refers to the fact that the number-like charge Q_{c0} is not included among the colour charges. In fact, the charge Q_{c0} is associated with a different (and rather trivial symmetry of the system called a U(1) symmetry (this is also unitary but the 1 refers to the fact that it acts as a number and does not mix up two or more states).

Group theory tells us many remarkable things about any set of states on which charges with these commutation relations act. The states must fall into families called 'representations of SU(3)'. Any state within such a family can be obtained from any other by repeated operations with the charges. The triplet of quark colours is the simplest such representation. Others include the antitriplet of antiquark anticolours and the octet of eight gluon states.

Physically, the reason that we can extract this kind of information is that any system which carries colours must be able to emit gluons. Then, as we have seen, colour charge conservation puts severe constraints on the colour charges of the system, just as it determines the charges of the gluons themselves. For example, the difference between the Q_{c3} charges of any two states must be a multiple of $\frac{1}{2}$ so that you

can get from one to another by the emission of gluons. In fact, the same sort of result holds for any of the colour charges. They necessarily come in multiples of $\frac{1}{2}$. This colour charge quantisation is an automatic ' consequence of the commutation relations of the charges. We did not have to impose it to account for experimental facts, as we did for electric charge quantisation. Note, however, that charge quantisation would not have been automatic had the colour charges included the charge Q_{c0}. Since this charge just acts like a number, it does not transform the states or participate in the commutation relations. Group theory tells us nothing about it. If there were a gluon corresponding to this charge, it would be neutral.

The numerical values of the colour charges are closely connected with the geometrical representation of the quark colour states I discussed earlier. Because of the nontrivial commutation relations, only two of the colour charges can act like numbers. These could be any two combinations of the charges that commute with one another. In table 15.1 they are Q_{c3} and Q_{c8} (and the dependent $Q_{c8}' = -Q_{c3} - Q_{c8}$). We can plot these two independent charges on a plane, as in figure 15.4. In this figure, the Q_{c3} and Q_{c8} axes are oriented at an angle of 120° to one another in order that Q_{c3}, Q_{c8} and Q_{c8}' are treated symmetrically. The Q_{c3} and Q_{c8} charges of the quarks form the equilateral triangle of the geometrical representation. If we change the signs of all the quark charges, we get the charges of the antiquark triplet. In the same plane, this produces a triangle with the opposite orientation.

There is another way of looking at the antiquark triplet. Group theory tells us that, by taking appropriate combinations of the states in the quark triplet representation, we can build up any other representation. The antiquark representation can be built out of pairs of states from the quark triplet in a very simple way, by identifying anticolour states with pairs of distinct colour states. The correspondence is the following:

$$|\text{red}\rangle * |\text{green}\rangle \quad \rightarrow \quad |\text{antiblue}\rangle$$
$$|\text{green}\rangle * |\text{blue}\rangle \quad \rightarrow \quad |\text{antired}\rangle$$
$$|\text{blue}\rangle * |\text{red}\rangle \quad \rightarrow \quad |\text{antigreen}\rangle.$$

Not only do these pairs have the right colour charges, they give the right transitions between states. For example, a blue quark can emit a #6 gluon and become a green quark. Therefore an antigreen antiquark, with the colour of a state of blue + red can emit a #6 gluon and become an antiblue antiquark with the colour of a state of green + red. The octet of gluons can be described in a similar way. The gluons have the colours of appropriate combinations (determined by group theory) of three quark states. This building process is another way to understand the quantisation of the colour charges. Since any representation can be built out of the triplet, all charges must be multiples of the charges of the triplet.

Electrons and antielectrons, with no colour, have the value zero for all the colour charges. In the geometrical picture, they belong in the centre of the equilateral triangles.

*Eddington, Sir A. S. (1956). In *The World of Mathematics*, J. R. Newman, ed. Simon and Schuster, N.Y.

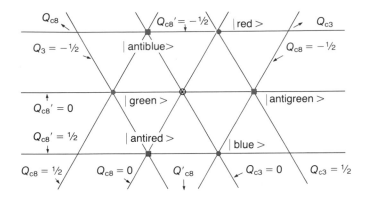

Figure 15.4. If the values of Q_{c3} and Q_{c8} for red, blue and green quarks are plotted on a lattice they form a triangular arrangement. The antiquarks would form a reflected triangle.

Since the colour charge is quantised in units of $e_3/2$, it is useful to define a dimensionless coupling constant that characterises the strength of the colour interactions,

$$\alpha_3 = e_3^2/\hbar c.$$

Experiments show that α_3 is larger than α. The colour SU(3) interactions (which are supposed to explain the 'strong interaction' after all) are stronger than the electromagnetic interactions.

15.9 Asymptotic freedom

The relativistic quantum mechanical vacuum is a lot more than just empty space. It is seething with virtual particles. For example, the same electromagnetic interaction that allows an electron to emit a photon, also allows an electron, an anti-electron and a photon to appear out of nothing. Of course, these cannot be real particles because their energy and momentum cannot be related as they would be for a real particle. There is no energy available to produce real particles. But they can exist as virtual particles for a short time consistent with the uncertainty principle.

If a real charged particle is added to this complicated vacuum, it can polarise the virtual electron–antielectron pairs. If the real particle has a positive charge, the virtual positive charges are pushed away from it slightly while the virtual negative charges are pulled toward it. The net result is that some of the virtual positive charge is pushed far away and the real charge is surrounded by negatively charged vacuum.

You may now suspect that the statements we made earlier about the charges and coupling constants were a little naïve.

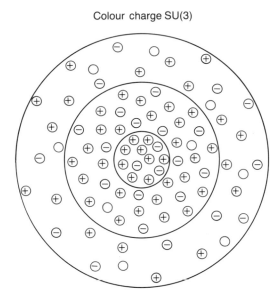

Colour charge SU(3)

○ Colour- neutral gluon
⊕ Colour-positive gluon
⊖ Colour- negative gluon
⊕ Colour-positive quark
⊖ Colour- negative quark

Figure 15.5. Virtual quarks, antiquarks and gluons that inhabit the quantum vacuum surround a central point colour charge. The strong force attracts virtual quarks of the appropriate anticolour and repels virtual quarks of the same colour. This polarises the vacuum, creating a screening cloud around the central charge that reduces its effective magnitude, A similar, but opposite, effect polarises the virtual gluons. When all the virtual particles are taken into account the gluons win. The net effect is 'antiscreening', i.e. the effective colour charge grows larger at short distances.

The charge on a point particle in this complicated vacuum may depend on how you measure it. For example, if we measure the electrical field on the surface of a ball of radius R surrounding the particle, we can infer the charge inside the ball. But this charge will depend on R because it will always include both the positive charge we put in and some of the negatively charged vacuum surrounding it. You might think that we could find the 'bare' charge of the particle by going in very close, making R very small. But that doesn't work. The density of negative charge in the vacuum increases as you get closer to the real charge, so that no matter how small you make the sphere, there is always a significant amount of negative charge in the enclosed vacuum. If you think about this for a while, you will see that the 'bare' positive charge must be infinite and the density of negative charge in the vacuum must go to infinity as R goes to zero. This is one of the infinities in relativistic quantum field theory that I discuss in Chapter 16. It is nothing to worry about, really. It is only a 'real' infinity if you can really make R infinitely small without any change in the rules of quantum field theory. That's silly. But it might as well be an infinity, in the sense that the rules probably continue to work down to distances far beyond any we can probe directly. At any rate, the charge and coupling constant defined as a function of R do make sense, and they are the appropriate ones to use when we discuss physics at distances of order R.

Charge quantisation is still there. The charge in a sphere of radius R around a d quark is $\frac{1}{3}$ the charge in a sphere of the same size around an electron. The dimensionless charge Q_E still takes on integer (or $\frac{1}{3}$ integer for the quarks) values, but the coupling constant α now depends on R. The 1/137 we talked about is the value of α at atomic distances, around 10^{-8} cm and larger.

In quantum electrodynamics, the dependence of α on R is weak because α is small. We usually don't have to worry about it, but it has been measured. In quantum chromodynamics, where α_3 is larger, this effect is very important.

But there is something else that happens in the colour SU(3) interactions. Not only does a positive colour charge polarise the vacuum, but the fact that the gluons carry the colour charges actually allows the central charge to spread out. The mechanism that produces this effect is not at all obvious. It has to do with the special role of the gluons in carrying the colour force. This is discussed in detail in the chapter by Taylor. At any rate, in quantum chromodynamics this effect overwhelms the effect of vacuum polarisation. A positive colour charge appears to be spread out, so the vacuum surrounding the central charge is positively charged. In fact, because the charge is spread out, the charge inside a sphere of radius R decreases (very slowly) to zero as R goes to zero. The colour interactions get weaker at short distances. This property of the colour interaction, which is called *asymptotic freedom*, allows theorists to calculate many properties of the quarks and gluons at short distances. The electromagnetic interaction has the opposite property. It is weak and therefore theoretically tractable at the long distances that we observe easily in low energy experiments. At very short distances (if there were no other change in the physics besides the quantum mechanical effects of vacuum polarisation) electromagnetism would be strong and difficult to understand. But the simplicity of the colour interactions shows up only at short distances and high energies. Only in the last ten years have we been able to reach high enough energies to make our theoretical predictions reasonably reliable. The fact that these predictions are borne out in very high energy scattering experiments that probe the short distance properties of the quarks and gluons gives us confidence that quantum chromodynamics is the correct theory of the strong interactions.

15.10   Weak interactions and decays

We have now seen, in our descent into the abyss of particle physics, two of the three types of interactions that are important at the distance scales that we can probe with today's technology. We have seen the electromagnetic interactions, mediated by photons, and the strong colour interactions, mediated by gluons. The mediators of the third important type of interaction, the weak interactions, are the W and Z particles, which do not show up directly until we get to very small distances, of the order of 10^{-16} cm, one-tenth of 1% of the diameter of the proton. But well before these short distances were directly accessible to high energy machines, theoretical physicists were confident that they understood their properties. The reason for this confidence was the success of a theory of the weak interactions in explaining the slow decays of many particles, including the neutron. Thus we will begin our study of the weak interactions by talking about the process of particle decay.

Most types of particles do not live forever. Instead, they blow up into two or more lighter particles. Because of quantum mechanics, the decay process involves an element of chance. Given a single particle, you cannot say when it will decay. But, if you have lots of identical particles, the average rate at which they decay can be used to define what is called the particle's lifetime. The lifetime is the total number of particles divided by the rate of decay. Suppose that you have 100 000 particles and 1000 of them decay in a year. Then, the lifetime is one hundred years. Since decay is a random process, any given particle would have only one chance in one hundred of decaying in any given year.

In a decay process, some mass is converted into energy. If you stop the products of a decay and weigh them, the result for their total mass is always less than the mass of the original particle. The missing mass is converted into energy that is used to set the decay products flying apart.

Some particles do not decay. They are 'stable'. The electron is stable for an interesting reason. All interactions are believed to conserve electric charge. That means that the total electric charge of the products of a decay must be equal to the charge of the original particle. But the electron and antielectron are the lightest charged particles. Any other combination of particles with the same electric charge as the electron must have a total mass greater than or equal to the electron mass. Thus the electron cannot decay. The antielectron is stable for the same reason. An electron and an antielectron together can annihilate each other into photons because the total electric charge is zero, but neither one, by itself, can decay.

Now, let us look at some particles that can decay. The π^0 is unstable. This should not surprise you. I have said that the π^0 is built out of a combination of $u\bar{u}$ and $d\bar{d}$ quark–antiquark pairs. What this means is that the π^0 state is a sum of a $u\bar{u}$ state and a $d\bar{d}$ state. The π^0 has a 50% probability of being a $u\bar{u}$ and a 50% probability of being a $d\bar{d}$. In either form, the quark and antiquark sometimes annihilate into photons through the electromagnetic interactions. This happens fairly quickly, because it is fairly easy for a particle and its antiparticle to annihilate each other. The lifetime of a π^0 is only about 10^{-16} s.

A neutron is a bit heavier than a proton. A neutron is also unstable. It decays with a lifetime of about 1000 s into a proton, an electron and a very light or massless particle called an 'antineutrino' (because it is the antiparticle of a particle called the neutrino). The neutrino and antineutrino do not feel the strong forces that hold quarks inside hadrons. They also have zero electric charge, so they do not feel electric forces.

Two questions may occur to you about neutron decay. What kind of interaction causes it? And why does it take so long, compared, say, to the π^0 decay? We can give a provisional answer to the second question. Whatever the interaction is that causes neutron decay, it must be much weaker than the interaction that causes π^0 decay, so that the decay is less probable. Historically, this fact gives the force its name. It is called the 'weak interaction'. Weak interactions are also seen in many other radioactive decays of nuclei, and in decays of more exotic hadrons.

The modern theory of the weak interactions is based on a two-fold symmetry associated with the group SU(2) similar to the three-fold symmetry of the colour SU(3) interactions. But to understand how the particles fit into SU(2) representations, we must first discuss the notion of handedness.

All of the particles we have been discussing have spin. They behave like little perpetually spinning tops, carrying angular momentum $\pm\frac{1}{2}\hbar$ for the electron, the neutrino and the quarks and $\pm 1\hbar$ for the photon and gluons.

For a moving particle, the natural axis along which to measure the spin angular momentum is the direction of motion. Particles with positive spin angular momentum along the direction of motion are called right handed (because a top has positive angular momentum along the thumb of the right hand when it is spinning in the direction in which the fingers curl). Particles with spin angular momentum opposite to the direction of motion are called left handed (see figure 15.6).

The handedness of a massive particle can be changed by bringing the particle to rest and accelerating it in the opposite direction without changing its spin. Thus, massive particles have both left handed and right handed components. However, the neutrinos apparently have zero mass or at least a mass so small that it has not been detected. If the neutrinos are actually massless, then like photons they always travel at the speed of light. Since a massless neutrino cannot be brought to rest, its handedness never changes. In fact, experimentally, only left handed neutrinos and right handed antineutrinos have been observed. Their counterparts, right handed neutrinos and left handed antineutrinos are presumed not to exist. It is convenient to think of the handedness changing under the interchange of particles and antiparticles. So, for example, the left handed neutrino is the antiparticle of the right handed antineutrino.

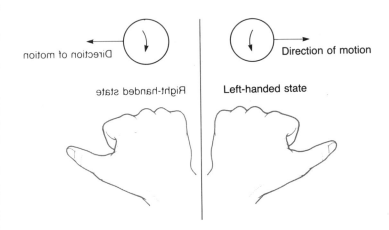

Figure 15.6. The orientation of a particle's intrinsic spin is quantised. The direction that a particle moves defines an axis, and if the spin vector points along it in the sense of a right-handed corkscrew, the particle is said to be in a right-handed state. The mirror image defines a left-handed state.

15.11   Weak SU(2)

Handedness is important because the weak interactions are associated with an SU(2) symmetry that acts differently on the left handed and right handed components of the various particles. The '3' in the SU(3) of colour referred to the three colours in the simplest representation, the triplet of quark colours. The '2' in SU(2) indicates that the simplest nontrivial representation of SU(2) is a doublet. The particles all fall into doublet representations of the SU(2) or singlets with 0 values for the SU(2) charges. The singlets are:

$$e^-_{\text{right}} \quad e^+_{\text{left}} \quad u_{\text{right}} \quad \bar{u}_{\text{left}} \quad d_{\text{right}} \quad \bar{d}_{\text{left}}.$$

The doublets are:

$$\begin{bmatrix} v_{\text{left}} \\ e^-_{\text{left}} \end{bmatrix} \qquad \begin{bmatrix} u_{\text{left}} \\ d_{\text{left}} \end{bmatrix} \qquad \begin{bmatrix} e^+_{\text{right}} \\ \bar{v}_{\text{right}} \end{bmatrix} \qquad \begin{bmatrix} \bar{d}_{\text{right}} \\ \bar{u}_{\text{right}} \end{bmatrix}$$

Here e^- means an electron state and e^+ means an antielectron state. Notice the particle on the top of each doublet has an electric charge, Q_E, larger by one than the particle on the bottom of the same doublet. There are three weak SU(2) charges, Q_{w1}, Q_{w2} and Q_{w3}. The singlets have zero values for all of the weak charges. The doublets have the charges as in table 15.2, where a dimensional factor of e_2 has been taken out.

Table 15.2. *SU(2) properties of quarks and leptons*

Charge	Acting on	
	$\lvert+\rangle$	$\lvert-\rangle$
Q_{w1}	$\frac{1}{2}\lvert-\rangle$	$\frac{1}{2}\lvert+\rangle$
Q_{w2}	$\frac{1}{2}\lvert-\rangle$	$-\frac{1}{2}\lvert+\rangle$
Q_{w3}	$\frac{1}{2}\lvert+\rangle$	$-\frac{1}{2}\lvert-\rangle$

In table 15.2, + and − refer generically to the top and bottom states of a doublet. Thus Q_{w1} acting on a left handed neutrino state gives $\frac{1}{2}$ times a left handed electron state. Q_{w2} acting on a left handed d quark state gives $-\frac{1}{2}$ times a left handed u quark state. Q_{w3} acting on a right handed $\bar{u}$ antiquark state gives $-\frac{1}{2}$ times the right handed $\bar{u}$ antiquark state. All of the SU(2) charges give zero when they act on a singlet state.

The particles that mediate the weak force by their exchange are associated with the weak SU(2) charges, but, because the electric charges of the two members of each doublet are different, some of these particles have electric charge. For example, the combination $Q_{w1} + Q_{w2}$ when acting on a $\lvert+\rangle$ state gives a $\lvert-\rangle$, but it gives zero when acting on $\lvert-\rangle$. This means that the corresponding particle can be emitted by a left handed u quark (for example) which then turns into a left handed d quark. But then Q_E conservation requires that the corresponding particle must have negative charge. It is called the W^-. Similarly, the particle that corresponds to the charge $Q_{w1} - Q_{w2}$ is the W^+ with $Q_E = +1$. The exchange of these two charged particles produces the weak decays.

A typical process is illustrated in figure 15.7. A left handed d quark emits a virtual W^- and becomes a left handed u quark. The W^-, in turn, decays into a left handed electron and a right handed antineutrino (since the left handed neutrino can absorb a W^- and turn into a left handed electron, the process in which the right handed antineutrino is emitted is allowed as well). The net result is that a d quark turns into a u quark, an electron and an antineutrino, which is the transition that takes place in the beta decay of a radioactive nucleus. When this happens to a d quark that is bound to another d quark and a u quark in a neutron, the electron and neutrino escape because they have no colour, but the u quark remains bound to the u and d quarks that did not decay, leaving a proton. This is neutron decay.

15.12   SU(2) is not enough

The W^+ and W^- carry electric charge, so the weak SU(2) has something to do with electromagnetic interactions. But none of the SU(2) charges is exactly equal to the electric charge. Therefore the SU(2) interactions, by themselves, do not include the electromagnetic interactions. In fact, however, the weak

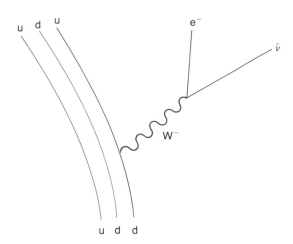

Figure 15.7. Beta-decay of the neutron as viewed in terms of quark transformations. The weak force, which can change quark flavours, turns a d quark into a u quark plus an emitted W^- particle. This changes the neutron into a proton. Meanwhile the W^- decays via the weak force into an electron (e) and an antineutrino (ν).

charge Q_{w3} is related to the electromagnetic charge as follows:

$$Q_E = Q_{w3} + Q_{U(1)},$$

where $Q_{U(1)}$ is a charge that has the same value for each component of each weak SU(2) doublet. It is a U(1) charge, which means that it acts like a number. It commutes with the weak charges (and also with the colour charges). For particles in weak SU(2) singlets, Q_{w3} is zero, so $Q_{U(1)}$ is just equal to the electric charge, Q_E. For particles in weak SU(2) doublets, $Q_{U(1)}$ is the average electric charge of the doublet. For example, the right handed electron has $Q_{U(1)} = -1$, while the left handed electron and the neutrino have $Q_{U(1)} = -\frac{1}{2}$. With both SU(2) and U(1) charges, and their corresponding particles, we can describe both the weak and the electromagnetic interactions.

The charges Q_{w3} and $Q_{U(1)}$ are associated with two neutral particles. Since $Q_E = Q_{w3} + Q_{U(1)}$, we get the usual electromagnetic interactions if the photon is associated with this combination of charges. The other combination is associated with a particle called the Z^0. Like the photon, it has $Q_E = 0$ and can be emitted by a particle without changing the particles identity. But unlike the photon, the Z^0 can be emitted and absorbed by neutrinos. The virtual exchange of the Z^0 gives rise to a weak force between neutrinos and the other particles, as illustrated in figure 15.8. This is called a neutral current weak interaction because the exchanged particle is electrically neutral. In contrast, the classic weak interaction that causes decays is a charged current interaction. The neutral current interaction is much harder to observe because no change in particle identity is involved.

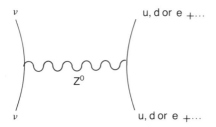

Figure 15.8. An important test of the modern theory of the electroweak force is the existence of neutral currents. This means that weak processes of the type shown can occur, in which an electrically neutral virtual quantum (Z^0) is exchanged between a neutrino and a quark, leaving their identities unchanged. Contrast figure 15.7, where particle transmutations occur.

When two or more commuting sets of charges are involved in the symmetry of a theory, like the SU(2) and U(1) charges in the weak interactions, the names are just strung together with $\times$'s in between to indicate that the corresponding symmetries are there but independent of one another. Thus the weak interactions are described by an SU(2) $\times$ U(1) theory. The full theory of colour strong interaction and weak and electromagnetic interactions is an SU(3) $\times$ SU(2) $\times$ U(1) theory.

15.13   Spontaneous symmetry breaking

Something crucial is missing from the above description of the weak interactions. The weak interactions are observed to be short range and the W and Z are observed to be heavy. What has happened to the symmetry that should relate them to the massless photon? Furthermore, the left handed neutrino and the left handed electron are part of an SU(2) doublet. If the SU(2) symmetry behaved like the SU(3) symmetry of the colour interactions, it would be impossible to tell them apart. But, in fact, they have very different properties. Where is the SU(2) symmetry?

The answer is that the underlying force law is symmetrical, but the vacuum is not. We have already seen, in the discussion of vacuum polarisation, that the vacuum state of a quantum mechanical theory is not the boring empty place that we normally associate with the word 'vacuum'. It is, in fact, the substrate on which all of physics takes place. The vacuum state in which we all live distinguishes the W and Z from the photon, making them heavy but leaving the photon massless. It also allows the electron to get a mass, joining the left handed and right handed components of the electron into a massive particle, even though they have different SU(2) charges. Such a situation is called 'spontaneous symmetry breakdown'. A useful analogy is the breakdown of rotational symmetry in a

crystal. Consider, for example, a grain of salt. It is built up out of sodium ions and chloride ions. The forces between the ions are electromagnetic forces that do not pick out any special directions in space. However, when the ions are packed together they form a cube. If you lived inside a grain of salt, you would find that your world does have some special directions, for example the directions perpendicular to the faces of the cube, with special properties. It is true that a 'giant' living outside the cube of salt could pick it up and rotate it (and you with it) without affecting the physics of your world, but to you inside the cube, the rotational symmetry of the law of physics would not be obvious. The rotational symmetry has been broken by your environment, the salt crystal.

The relativistic quantum mechanical vacuum, in which we all live, is like the salt crystal. The symmetry that is broken spontaneously is not rotational symmetry but the SU(2) $\times$ U(1) symmetry of the weak interactions. This breaking allows the W and Z and the electron and the quarks to have the masses we observe, even though these masses do not respect the SU(2) symmetry.

There is another aspect of the spontaneous breakdown of the SU(2) $\times$ U(1) symmetry that can be understood in the crystal analogue. It is 'graininess'. The salt crystal is grainy on the atomic scale, about 10^{-8} cm. Living inside the crystal, you might well distinguish three different domains of distance, associated with different physics.

(1) *Distances much larger than 10^{-8} cm.* If you do an experiment that probes the structure of your world at these macroscopic distances, you will see the cubic structure of the crystal. There will be special directions with special properties. Rotational symmetry will be quite obviously broken.

(2) *Distances of the order of 10^{-8} cm.* If you do an experiment that probes the structure of your world at typical atomic distances, the result will be very complicated. You will see also the complicated interatomic forces that are responsible for packing the atoms into the crystal.

(3) *Distances much smaller than 10^{-8} cm.* If you do an experiment that probes the structure of your world at distances much smaller than 10^{-8} cm, it doesn't matter very much that you are living in a crystal. You will find that the results of your experiment will be approximately rotationally invariant. This is because the nonuniform electric fields in the crystal that mess up rotation invariance are very weak compared to the kinds of fields you have to generate to probe subatomic distances. Once you are inside the atom, it is the constituents of the atom that matter, not the way the atom is put together with other atoms in the crystal.

Just as a crystal is grainy at an atomic scale, the vacuum of our world is grainy at a scale of about 10^{-16} cm. Again, we can distinguish three distinct regions of distance scale.

(1) *Distances much greater than 10^{-16} cm.* If you do an experiment that probes distances much larger than 10^{-16} cm, you will not see the SU(2) × U(1) symmetry at all. You will not even see W's and Z's directly. You do not have a probe with high enough energy to produce them. You see the photon and the gluons directly, but the heavy W and Z show up only in the short range interactions caused by their virtual exchange.

(2) *Distances of the order of 10^{-16} cm.* If you do an experiment that probes distance of the order of 10^{-16} cm, the results will be very complicated. W's and Z's will show up in your experiments, but they will look very different from photons, because they will be heavy and not moving very fast. You should also see the physics, whatever it is, that produces the spontaneous breaking of the SU(2) × U(1) symmetry. We are now on the threshold of this region in present day experiments. We have seen the W and Z, but cannot yet see these short distances clearly enough to understand the physics of spontaneous symmetry breaking.

(3) *Distances much smaller than 10^{-16} cm.* If you do an experiment that probes the structure of the world at distances much smaller than 10^{-16} cm, you will see SU(2) × U(1) as an explicit (though approximate) symmetry. At such short distances, the masses of the W and Z are negligible compared to the energies involved in your experiment, which must be large compared to the W and Z mass so that the wavelength of your 'microscope beam' will be smaller than 10^{-16} cm. At these high energies, the W and Z have energies large compared to their masses, so they are moving very fast, close to the speed of light, just like photons. Likewise, the masses of the electron and the quarks are irrelevant so that, for example, the left handed neutrino and electron look like identical parts of an SU(2) doublet. Such experiments are still far in the future.

Now that we have a better idea of what the weak interactions are, we can give a more satisfactory answer to a question that we asked earlier. Why are they weak? The answer lies not in the coupling constant of the SU(2) charges but in the short range of the weak force. Because the W and Z are heavy, the forces produced by their virtual exchange operate only at very short distances, smaller than the W and Z Compton wavelength, 10^{-16} cm. Thus, for example, the scattering process neutrino plus d quark into electron plus u quark is weak: it doesn't happen very often until the energies of all the particles are large enough to probe these small distances. In the related neutron decay process, all the energies are very small, so the interaction is very weak and the process is very slow. More precisely, the probabilities for all of these weak processes at low energies are inversely proportional to the fourth power of the W mass. It is the factor of $1/M_W^4$, rather than the coupling constant, α_2, that makes the weak interaction weak. In fact, as we will see later, α_2 is larger than the electromagnetic coupling constant α. The weak interactions would hardly be noticeable except

that they do things that the electromagnetic and strong interactions do not. W exchanges change particle identities. These show up in the particle decays, as we have discussed. Z exchanges cause neutral current interactions of neutrinos. And they all violate parity symmetry (the 'mirror' symmetry that interchanges right and left) because the left and right handed components of the quarks and the electron interact differently.

15.14   Neutral currents and charge quantisation

The spontaneously broken SU(2) × U(1) theory of weak and electromagnetic interactions was worked out in the 1960s by Sheldon Glashow at Harvard, Steven Weinberg at MIT and Abdus Salam at Imperial College, London, and Trieste. Glashow worked out the form of the theory, but did not know how to give mass to the W and Z. Weinberg and Salam worked out the effect of spontaneous symmetry breakdown and produced a consistent theory. Many years later, in 1979, when it was clear that SU(2) × U(1) was the right theory of the weak interactions (although still before the experimental discovery of the W and Z) they received the Nobel prize in physics for their efforts. However, in the late 1960s and early 70s, it was not at all obvious that the SU(2) × U(1) theory was right. When the full theory was first written down by Weinberg in 1967, it seemed to have both theoretical and experimental problems.

The apparent theoretical problem was that the theory involved massive particles with spin 1, the W and Z. At the time, it was thought by most physicists that such theories did not make sense, that they were not renormalisable (see my Chapter 16 on EFTs and Chapter 17 by Taylor). I remember seeing Weinberg's paper in 1967 while I was a graduate student and not being excited because I didn't see how to make sense of it. Most others, I think, had the same reaction. Fortunately, there was a brighter graduate student around. Gerhard 'tHooft, in Utrecht, a graduate student working with Martinus Veltman, had been spending his time trying to make sense of theories of this kind, and in 1971 he announced his proof that theories with massive particles like the W and Z are sensible, so long as the masses come from spontaneous symmetry breaking. At the same time, he more or less independently constructed the SU(2) × U(1) model of Glashow, Weinberg and Salam.

'tHooft's insight solved the theoretical problem, but in 1971 the experimental problem was still there. As we discussed above, the SU(2) × U(1) theory predicts so-called neutral current weak interactions of neutrinos, produced by virtual Z exchange. By 1971, these interactions had been looked for at the level predicted by the SU(2) × U(1) theory, and not seen! In such a situation, the theorist can adopt either of two attitudes. He can choose to wait for better experiments. Modern elementary particle experiments are very hard, and it is easy to make subtle errors, particularly if, as was the case here, you think you know what the answer is going to be. There was, in fact, for irrelevant reasons, a widespread prejudice against neutral current interactions, so that when they were not seen, people were not surprised. If the theory is sufficiently beautiful and compelling, and if new experiments are in the offing, it is often best to wait. Alternatively, he can try to construct a new theory, keeping the successes of the old theory but somehow eliminating the unseen effect.

The $SU(2) \times U(1)$ theory is not particularly beautiful. It is often called a unification of the weak and electromagnetic interactions, but, in fact, the unification is partial, at best. The problem is the $U(1)$ charge. As I have already remarked, this is a charge that commutes with all the other weak and colour charges, so group theory tells us nothing about it. In particular, because of the $U(1)$, the theory gives us no explanation of the striking experimental fact of electric charge quantisation. Further, because of the $U(1)$, there are two separate dimensional charges or coupling constants required to specify the theory, one for the $SU(2)$ charge e_2 and coupling constant α_2 and another for the $U(1)$ charge e_1 and coupling constant α_1. This introduces another unknown parameter into the theory and again reduces its explanatory power.

Thus, it seemed reasonable to look for alternatives to $SU(2) \times U(1)$. Besides, working is more fun than waiting. So Shelley Glashow and I plunged into the model building game in 1971, using the tools that 'tHooft had developed for us, in an attempt to construct an alternative theory of the weak interactions that did not suffer from the 'problem' of neutral currents. The next few years were an interesting and productive time, during which we learned a lot about the structure of the new class of theories that 'tHooft's work had enabled us to study. We built many models. Indeed, sometime in the following three years, a local wag posted a sign on my door that went something like this:

Algorithm for constructing weak interaction models

Step 1: Choose symmetry group.
Step 2: Choose particle representation of symmetry.
Step 3: Break symmetry spontaneously.
Step 4: Calculate particle masses.
Step 5: Identify weak interactions and find neutral
        current effects.
Step 6: Write paper.
Step 7: Go to Step 1.

Actually, of course, we were learning many things about the new theories. Early in 1972, we found a very interesting alternative to the $SU(2) \times U(1)$ theory, a theory based on the charges of the group $SO(3)$. This has to do with the rotation symmetries of a three dimensional space. There are three $SO(3)$ charges and they have the same commutation relations as the three $SU(2)$ charges. In fact, the only difference between $SU(2)$

and $SO(3)$ is that in $SO(3)$ the representations with charges of odd multiples of $\frac{1}{2}$ are not allowed. Only integral charges appear. Thus we could incorporate the electric charge of the electron without adding the extra $U(1)$ charge. To do this we had to postulate the existence of additional heavy particles, a charged particle like a heavy electron and a neutral particle like a heavy neutrino. We had to assume that they were heavy and had not yet been observed. But in this process we eliminated the neutral current interactions. In our model, there was no Z. The three $SO(3)$ charges were associated only with the W^+, the W^-, and the photon.

In the $SO(3)$ theory, weak interactions are really unified with electromagnetic interactions. There is only one coupling constant, the electromagnetic coupling constant α. What is more, the quantisation of electric charge is automatic. Too much so, in fact, because only integral charges are allowed. There is no way to incorporate quarks. This didn't bother us at the time, because it was not at all clear that quarks had anything to do with the strong interactions. Asymptotic freedom had not been discovered, and the crucial experiments that would eventually convince us of the existence of quarks and gluons were still in progress.

15.15 Magnetic monopoles

The quantisation of charge is such a dramatic experimental fact that it was very exciting to have a theoretical explanation. One other explanation had been proposed much earlier by P. A. M. Dirac. Dirac realised that he could understand the quantisation of electric charge if magnetic monopoles exist. Everyone is familiar with the north and south poles of a magnet. In magnets, the north and south poles always go together, because the magnetic force is produced not by poles (like the electric charges that produce electric forces) but by electric currents. In a permanent magnet it is the permanent currents associated with the spinning electrons in the magnet that produce the magnetism. But Dirac imagined that there could exist just a north pole. Dirac showed that in a quantum mechanical world, where there exist both electric charges of strength e and magnetic charges of strength m, the product em is necessarily quantised in units of $\hbar/2$ because it behaves like a quantum mechanical angular momentum. Then the ratio of any two charges e_x and e_y must be a rational number, the ratio of integers $(2me_x/\hbar)/(2me_y/\hbar)$. This is charge quantisation.

It did not occur to us that our explanation of charge quantisation had anything to do with Dirac's. Again, 'tHooft had a better idea. He and the Russian physicist Polyakov (independently) showed that our model predicts the existence of magnetic monopoles. They are not put into the theory. They come out automatically in a subtle and interesting way. When the symmetry is spontaneously broken, it picks out a 'direction' in the three-dimensional 'space' of the $SO(3)$ charges. This has nothing to do with ordinary three-dimensional space. But 'tHooft and Polyakov realised that the two could be put together. The usual vacuum picks out a single fixed direction in the SO(3) space. But they considered a situation in which, around a point in ordinary space, the vacuum is always 'pointing' out in the SO(3) space. Polyakov called this configuration a 'hedgehog', because it reminded him of the outward pointing quills of this porcupine-like animal, a favourite in Russian fairy tales (see figure 15.9). One might think that this peculiar twisted vacuum would simply unwind into the usual constant vacuum. And if the model has a U(1) piece, as in the SU(2) × U(1) theory, that is just what happens. But in the SO(3) theory, the twist in the vacuum cannot unwind. Instead, 'tHooft and Polyakov argued that it acts like a heavy magnetic monopole. Thus the two explanations of charge quantisation, ours and Dirac's, are closely related. If there are no U(1) charges so that Q_E is quantised because of group theory, then monopoles exist, so that one can also see that Q_E is quantised by Dirac's argument.

Alas, this appealing complex of ideas had to be shelved because neutral current weak interactions do exist. Evidence began to pile up by the end of 1972. It had led to some interesting physics, but SO(3) was wrong. Unification of weak and electromagnetic interactions was not so simple.

Figure 15.9. A magnetic monopole acts as a source (or sink) of radial magnetic field lines, reminiscent of the spikes on a hedgehog. In conventional magnetic theory, where monopoles are excluded, magnetic field lines cannot end. Instead, they form closed loops.

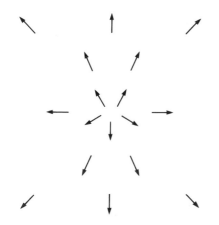

15.16   Unification of strong, weak and electromagnetic interactions

Glashow and I and others continued to search for alternatives to the SU(2) × U(1) theory that would implement a real unification. One group actively involved in the search was the team of Jogesh Pati of the University of Maryland, and Salam. In 1973, they found an interesting partial unification in which the electromagnetic interactions were combined partly with the weak interactions and partly with the colour interactions in the group SU(2) × SU(2) × SU(4) (I will talk more about it later). They still had two coupling constants, a large coupling for the colour interactions and a smaller coupling for the SU(2) interactions. But they did get rid of the U(1) and get automatic charge quantisation. The significance of their model was not widely appreciated because they had the symmetry breaking all wrong. They wanted to break the colour symmetry and give the quarks integral charges. This turned out to be the wrong thing to do. But what was worse, it cluttered up their otherwise attractive partial unification.

Meanwhile, Glashow and I were getting nowhere with our own efforts. The quarks just wouldn't fit. However, important things were happening in other fields. Asymptotic freedom was discovered at the beginning of 1973. This was recognised by many as an important clue to the nature of the strong interactions. But it took some time to really get it straight. At first, most of us thought that we ought to break the colour SU(3) symmetry spontaneously to give mass to the gluons. I am not sure who is most responsible for recognising that this was not only unnecessary, but undesirable. I learned it from Weinberg. At any rate, the idea is that, along with the fact that the colour interactions get weaker at short distances goes the fact that they must get stronger at long distances. It was hoped then and is now thought that the strong forces at long distances permanently confine quarks and gluons within hadrons. This put the strong interactions in a very different light.

When he heard about quark confinement, Shelley Glashow, always eager to expound a crazy idea, suggested that we try a different kind of unification, not just of weak and electromagnetic interactions, but of strong interactions as well. I thought that this was a little strange, because I still felt that, despite confinement, the colour coupling constant, α_3 must be larger than the SU(2) coupling constant, α_2. But Shelley thought that we should give it a try, even though we didn't understand in detail how the couplings would work out. Neither of us had the faintest idea how to find such a unification. We talked inconclusively until it was time to go home. After dinner, I retired to my living room, stretched out in a comfortable reclining chair and tried to think about the question more systematically.

It is very easy in retrospect. It always is. Such is the nature of our field, and probably of many other intellectual endeavours, that when you (or somebody else) finally has the good idea, you feel very stupid for not having seen it sooner. At the time, what was hard about it was putting quarks and antiquarks into the same representation of the symmetry. This was hard, technically, for a reason that I will talk about later.

I knew that I was looking for a set of charges of a single unified group that contained the charges of SU(3), SU(2) and U(1). Any idiot, you might think, would think of SU(5), the symmetry of five identical quantum mechanical states, as the obvious way of combining the SU(3) symmetry of three colour states with the SU(2) symmetry of two weak states. But I didn't. The reason I didn't was that it wasn't obvious what representations to use for the particles. There was no way to do it without putting quarks and antiquarks in the same representation. I was missing the obvious. I had to be led by the nose.

Instead, I thought about the partial unification that I had learned about recently from Pati and Salam. There were two things about it that I didn't like very much. I mentioned earlier their odd asymmetrical treatment of the colour interactions. Later, they were to push this view very hard, as an alternative to confined quarks, forming what they called the 'quark liberation front'. For now, it was merely an inessential complication of their model. I realised that it could be stripped away, leaving a simpler and more attractive theory of confined quarks. I also told you earlier that the right handed neutrino and the left handed antineutrino are not observed. Pati and Salam needed these states in order to associate part of the electric charge with the colour interactions in the group SU(4) and the rest with weak interactions in two sets of SU(2) charges, one (the usual one) acting on the left handed quarks and electron and neutrino, the other on the right handed particles. The nice part of the Pati–Salam model was the representation of the particles. They were very simple.

The simplicity of the Pati–Salam representations had to do with the fact that, together with the right handed neutrino and the left handed antineutrino, they had thirty-two states in all, and thirty-two is a power of two; it is 2^5. The powers of two are associated with a remarkable set of representations of rotation symmetries, the so-called 'spinor' representations. The spinor representation of the rotation group of our three-dimensional world describes the spin of the electron. In fact, Pati and Salam's SU(2) × SU(2) × SU(4) charges are equivalent to the charges of two rotation groups, those for four- and six-dimensional space, and their particle representations are the appropriately combined spinor representations of these groups.

15.17 SO(10)

Now curiously enough, although I couldn't add two and three to get five, I could add four and six to get ten! I considered the charges associated with rotations in a ten-dimensional space,

SO(10). A ten-dimensional space can be broken up into a four-dimensional space and a six-dimensional space (remember, these are not physical dimensions – just mathematical devices for talking about the charges). There were several reasons that this was easier for me. The most important was that it was clear what representations to look at – the spinors were the obvious choice. What made this even easier was the fact that I had recently learned how to construct all these representations in my unsuccessful attempts to unify weak and electromagnetic interactions. So I did it. And it worked beautifully. SO(10) has two different spinor representations, each of which can accommodate sixteen particles. All sixteen of the left handed particles fit into a single spinor representation. All sixteen of the right handed particles fit into the other. But when I say sixteen left handed particles, I am counting both particles and antiparticles, both electron and antielectron and *both quarks and antiquarks*. The group theory had done *for* me something I had been reluctant to do for myself. It had put quarks and antiquarks into the same representation.

I spent some time that evening studying the structure of this SO(10) unified theory. It was obvious that the SO(10) symmetry would have to be spontaneously broken at some small distance (at the time, I thought of small as 'smaller than the W Compton wavelength' without really understanding just how small). Also, the right handed neutrino had to be hidden in some way. It wasn't hard to find ways of doing these things. They were the kind of things I was good at from all my practice with model building. Everything fell into place very naturally.

One thing that pleased me about SO(10) was that the model did not suffer from a disease called 'anomalies'. Sometimes a quantum field theory is less symmetrical than it seems. Because the theory must be defined by a limiting process of 'renormalisation' (see the following chapter and the chapter by Taylor), an apparent symmetry may not survive as a real symmetry of the quantum theory. The symmetry is said to be broken by a quantum mechanical 'anomaly'. Such a quantum mechanical breakdown of symmetry does not mix well with the spontaneous breakdown of the symmetry that is needed to make the theory realistic. In particular, 'tHooft's proof that such theories make sense does not work if anomalies break the symmetry. It was well known (from the work of the European physicists Claude Bouchiat, John Iliopoulos and Phillip Meyer and of Pati and Salam) that there were no anomalies in the SU(2) × U(1) × SU(3) theory of weak, electromagnetic and strong interactions. But it was not clear that a larger unifying symmetry would have the same property. In fact, a naïve unification in which all the left handed particles are put into a single representation of a unitary group (like SU(15)) would have anomalies. But the SO(10) theory is safe. I knew this immediately because Shelley and I had shown a year before that only the unitary groups can suffer from this disease.

15.18　SU(5)

But something was still bothering me about SO(10). I thought that there must be a simpler way. I knew from the group theory that the SO(10) charges contain the charges of SU(5). And I knew that the SU(5) charges, in turn, contained SU(3) charges and SU(2) × U(1) charges. Now that I had the idea, from SO(10), of putting quarks and antiquarks together, I could see that it might be possible to find SU(5) representations with the right properties. In fact, I could have done this entirely from the group theory, just by taking the SU(10) spinor representation apart into its SU(5) pieces. But it was too late in the evening (or too early in the morning) for such mental gymnastics, so I found a simpler way. The two simplest SU(5) representations are the 5 and the 10, describing five and ten states, respectively. The 5 is similar to the 2 of SU(2) and the 3 of SU(3). The SU(5) charges acting on these five states are just all possible charges, except the charge that doesn't change the states at all. There are twenty-four charges in all. I knew that the 5 had to contain both an SU(2) doublet and an SU(3) triplet, so I just labelled the states of the 5 by putting together the SU(2) labels $|+\rangle$ and $|-\rangle$ and the colour SU(3) labels, $|red\rangle$, $|green\rangle$ and $|blue\rangle$, giving five states in all. The SU(2) and SU(3) charges on these states were then exactly the charges we have already discussed. Schematically, I had finally thought of the 5 as $2 + 3$.

The SU(5) charges include four charges that can be taken to act like numbers. Three of these are the weak charge Q_{w3} and the two colour charges Q_{c3} and Q_{c8}. The last one had to be the charge $Q_{U(1)}$ of the SU(2) × U(1) model. The key point is that, for the theory to be a real unification, the $Q_{U(1)}$ charge that is needed to fit the particles we see into the 5 had to be the same $Q_{U(1)}$ charge that was demanded by the group theory. The constraint from group theory is that the sum of the charges of the particles in the 5 must be zero, so that the charge does not involve any piece of the charge that doesn't change the states at all. Also $Q_{U(1)}$ could have only two values in the 5, one on the three colour states and another on the two weak states (otherwise it would not commute with the weak and colour charges). Since the quark charges had to be either $\frac{2}{3}$ or $-\frac{1}{3}$, there were only two possibilities:

	A	or	B

$$Q_{U(1)}|red\rangle = \tfrac{2}{3}|red\rangle \qquad\qquad Q_{U(1)}|red\rangle = -\tfrac{1}{3}|red\rangle$$
$$Q_{U(1)}|green\rangle = \tfrac{2}{3}|green\rangle \qquad\quad Q_{U(1)}|green\rangle = -\tfrac{1}{3}|green\rangle$$
$$Q_{U(1)}|blue\rangle = \tfrac{2}{3}|blue\rangle \qquad\quad Q_{U(1)}|blue\rangle = -\tfrac{1}{3}|blue\rangle$$
$$Q_{U(1)}|+\rangle = -1|+\rangle \qquad\qquad Q_{U(1)}|+\rangle = \tfrac{1}{2}|+\rangle$$
$$Q_{U(1)}|-\rangle = -1|-\rangle \qquad\qquad Q_{U(1)}|-\rangle = \tfrac{1}{2}|-\rangle$$

Possibility A doesn't work, because there is no particle that is an SU(2) doublet with $Q_{U(1)} = -1$. But B works beautifully. The colour triplet must be the three colour states of the right handed d quark. The weak doublet must be the right handed antielectron and antineutrino. If we include the $Q_{U(1)}$ value along with the label, these states look like this:

$$\left.\begin{array}{l}|red,\,-\tfrac{1}{3}\rangle\\ |green,\,-\tfrac{1}{3}\rangle\\ |blue,\,-\tfrac{1}{3}\rangle\end{array}\right\}\ d_{right}$$
$$|+,\tfrac{1}{2}\rangle \qquad e^{+}_{right}$$
$$|-,\tfrac{1}{2}\rangle \qquad \bar{\nu}_{right}$$

There is another representation of SU(5) that also involves five states, but it is simply related to the five we have just discussed by a change in sign for all the charges. Thus, this representation is just right to describe the antiparticles of the particles in the 5, the left handed $\bar{d}$ antiquark and the left handed electron and neutrino.

I was pleased that there was a unique choice of the assignment of the particles into the 5. But now came the real test. Knowing how the charges act on the 5, I could calculate how they act on the 10. If the unification was to succeed, those charges had to be the charges of the rest of the particles (either left handed or right handed). The 10 is obtained from the 5 by putting together distinct pairs of states, just as the antiquark antitriplet in SU(3) is obtained by taking distinct pairs of states from the triplet. When the states are combined in this way, the $Q_{U(1)}$ values just add. Thus the states are as in table 15.3.

Table 15.3. *SU(5) properties of quarks and leptons*

$|green,-\tfrac{1}{3}\rangle$	$*\ \ |blue,-\tfrac{1}{3}\rangle$	$\rightarrow$	$|\overline{red},-\tfrac{2}{3}\rangle$	
$|blue,-\tfrac{1}{3}\rangle$	$*\ \ |red,-\tfrac{1}{3}\rangle$	$\rightarrow$	$|\overline{green},-\tfrac{2}{3}\rangle$	u_{left}
$|red,-\tfrac{1}{3}\rangle$	$*\ \ |green,-\tfrac{1}{3}\rangle$	$\rightarrow$	$|\overline{blue},-\tfrac{2}{3}\rangle$	
$|red,-\tfrac{1}{3}\rangle$	$*\ \ |+,\tfrac{1}{2}\rangle$	$\rightarrow$	$|+,red,\tfrac{1}{6}\rangle$	
$|green,-\tfrac{1}{3}\rangle$	$*\ \ |+,\tfrac{1}{2}\rangle$	$\rightarrow$	$|+,green,\tfrac{1}{6}\rangle$	u_{left}
$|blue,-\tfrac{1}{3}\rangle$	$*\ \ |+,\tfrac{1}{2}\rangle$	$\rightarrow$	$|+,blue,\tfrac{1}{6}\rangle$	
$|red,-\tfrac{1}{3}\rangle$	$*\ \ |-,\tfrac{1}{2}\rangle$	$\rightarrow$	$|-,red,\tfrac{1}{6}\rangle$	
$|green,-\tfrac{1}{3}\rangle$	$*\ \ |-,\tfrac{1}{2}\rangle$	$\rightarrow$	$|-,green,\tfrac{1}{6}\rangle$	d_{left}
$|blue,-\tfrac{1}{3}\rangle$	$*\ \ |-,\tfrac{1}{2}\rangle$	$\rightarrow$	$|-,blue,\tfrac{1}{6}\rangle$	
$|+,\tfrac{1}{2}\rangle$	$*\ \ |-,\tfrac{1}{2}\rangle$	$\rightarrow$	$|1\rangle$	e^{+}_{left}

The right hand side of table 15.3 is the good news. When we combine distinct pairs of the colour states, we get the triplet of anticolour states. Two $-\tfrac{1}{3}Q_{U(1)}$ charges give a $-\tfrac{2}{3}Q_{U(1)}$ charge. Thus the first three states are exactly right to be the left handed $\bar{u}$ antiquark. The next six states combine a colour state with a weak state. This gives six states that are both a colour triplet and a weak doublet. The $Q_{U(1)}$ charge is $\tfrac{1}{6}$, just right to be the left handed u and d quarks. Finally, the tenth state is a $+$ and $-$ combination. This gives a state that has no colour charge (because neither state had any to begin with) and also no weak charge (because the weak charge of the two states cancels out (note, for example, that the Q_{w3} values are $\tfrac{1}{2}$ for the $+$ and $-\tfrac{1}{2}$ for the $-$, so they add to 0). The total $Q_{U(1)}$ charge is 1. This is just right to be a left handed antielectron state. As with the 5, there is another 10 with opposite charges that includes all the right handed antiparticles of the left handed particles in the 10 above.

It all fits. With these SU(5) representations, we can include all the particles that we want, and none that we don't want. We don't need the right handed neutrino. It was necessary in the SO(10) representation, but all the SU(5) charges give zero on the right handed neutrino state, so there is no reason to include it at all.

Again, it was a straightforward matter to break the SU(5) symmetry spontaneously down to the colour SU(3) and SU(2) × U(1) interactions. Like the SO(10) theory, the SU(5) theory has no anomalies to spoil the symmetry. This time the absence of anomalies is not trivial. If only the particles in the 5 were present, the theory would not make sense. But when both the 5 and the 10 are there, the anomalies cancel and the theory is sensible.

15.19   X particles

All through the long evening, my level of excitement had been rising. By the time I had finished constructing the SU(5) theory, the pieces fitted together so well that I was sure that I had found something important. But as I sat, tired and happy, admiring the simplicity and elegance of the thing, I began to worry again about putting quarks and antiquarks together in the same representation. Technically, it was now clear how it worked. The process of taking distinct pairs of states (by which I got the 10 from the 5) made an antiquark if both states were colour states but a quark if only one was a colour state. But the result, I now realised, was very dangerous. There were SU(5) charges that changed a quark state to an electron state in the 5 but an antiquark state to a quark state in the 10. I called the corresponding particles X particles. I was afraid that the exchange of virtual X particles could lead to processes that change the number of quarks. Such processes could cause the proton to decay. I drew the picture which is in figure 15.10.

A right handed red d quark can emit an X particle and become a right handed antielectron. Then a right handed antielectron can certainly absorb the same X particle and become a right handed red d quark. But from table 15.3, you see that a left handed green u quark has the same SU(5) charges as the pair, right handed antielectron and right handed green d quark; while the left handed antiblue ū antiquark has the same colour charges as the pair, right handed red d quark and right handed green d quark. Therefore, a left handed green ū quark can absorb the X particle and become a left handed antiblue ū antiquark. The net result is a process in which two quarks come in and a quark and an antiquark come out. If this happens inside a proton, the result can be proton decay into an antielectron and a π^0 meson, as shown in figure 15.11. I knew that the proton was very stable. I could not imagine any reason

why the symmetry breaking should be so severe that this process would be slow enough. So I went to bed.

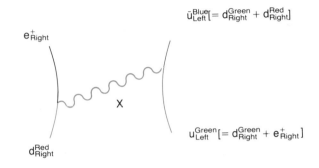

Figure 15.10. In Grand Unified Theories, the exchange of virtual X quanta can change quarks into leptons, and quarks into antiquarks. Here a d quark changes into a positron, and a u quark changes into a ū antiquark.

$$\bar{u}_{Left}^{Blue}[= d_{Right}^{Green} + d_{Right}^{Red}]$$

$$e_{Right}^{+}$$

$$X$$

$$u_{Left}^{Green}[= d_{Right}^{Green} + e_{Right}^{+}]$$

$$d_{Right}^{Red}$$

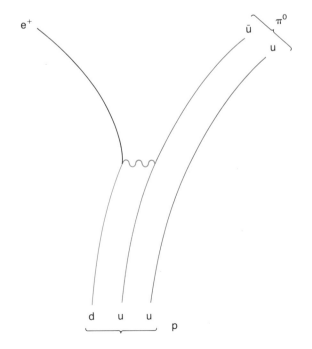

Figure 15.11. The process depicted in figure 15.10 can lead to proton decay. On the left, one of the d quarks inside the proton changes into a positron, which flies away, and an X particle (〰). The X then encounters one of the proton's u quarks, and turns it into an antiquark, ū. The ū then combines with the remaining u quark to make a neutral pion. The net reaction is thus $p \rightarrow e^+ + \pi^0$.

$$e^+ \qquad \bar{u} \quad \pi^0$$
$$u$$
$$d \quad u \quad u \qquad p$$

The next morning, in Glashow's office, I described the beautiful fit of the SU(5) theory and what I thought of as the depressing fact of proton decay. Shelley was not fazed. Instead, he dragged me up to the library where we looked up the most recent data on the lifetime of the proton. It was said to be greater than 10^{29} years. That is a long time! Undaunted, Shelley estimated that if the X particle were 10^{14} times as heavy as a proton, or more, the proton would live long enough. This was a little crazy. These particles were at least a billion times heavier than anything that we had ever heard anyone discuss before. But the fit of the known particles into SU(5) was too beautiful to be ignored. We wrote it up and went on to other things, ignoring the amused reactions of our particle physics colleagues, many of whom thought that *we* had gone a bit crazy.

It may sound slightly peculiar that, having found something as interesting as the SU(5) theory in 1973, we would spend most of the next three years working on other problems. But that is exactly what happened. The reason is that the SU(5) theory was really found before its time. At the end of 1973, the colour SU(3) theory of the strong interactions and the SU(2) × U(1) theory of the weak and electromagnetic interactions were amusing and attractive guesses, but we were far from confident that they had anything to do with the world. The three-year period from 1974 to 1977 was a time of frenetic activity directed primarily towards fleshing out and verifying these guesses. Fortunately, Nature cooperated in a spectacular way, by supplying a hitherto unobserved particle, the charmed quark, exactly on cue (see Chapter 17 by Taylor). Without that helping hand from Nature and our experimental friends, theorists might still be arguing about whether the SU(3) × SU(2) × U(1) theory is the correct description of physics at distances greater than 10^{-16} cm.

15.20   The grand unification scale

Grand Unified Theories were not entirely forgotten during this period. Soon after the discovery of SU(5), Helen Quinn, Weinberg and I did a careful analysis of the couplings in the GUTs. Since all the charges in SU(5) are treated symmetrically, in some sense there ought to be only one coupling constant in the theory. But, as we have seen, that is certainly not the situation at distances of the order of 10^{-16} cm. There the strong coupling α_3 is larger than the SU(2) coupling α_2, which in turn is larger than the U(1) coupling α_1. However, because of the constraint from proton decay, we knew that the SU(5) symmetry had to be spontaneously broken down to SU(3) × SU(2) × U(1) at some very small distance L. It is only for distances smaller than L that the SU(5) symmetry appears as an explicit, approximate symmetry. We realised that the dependence of the coupling constants on the distance scale, that I discussed when I talked about asymptotic freedom, would cause the couplings to split apart at larger distances. The situation is illustrated schematically in figure 15.12. The SU(3)

coupling α_3 is the most asymptotically free, because asymptotic freedom is an effect of the charges of the particles that correspond to the charges and there are more of them in SU(3) than in SU(2). α_3 therefore increases faster than the SU(2) coupling α_2 as the distance at which it is measured increases. The U(1) coupling α_1 is not asymptotically free at all (it is like electromagnetism – the particle corresponding to $Q_{(1)}$ has no $Q_{(1)}$ charge). Thus α_1 decreases as the distance increases.

As you can see from figure 15.12, at distances much larger than L, the couplings have the right form, at least qualitatively. But we can do better than that. At distances smaller than 10^{-16} cm, all of the couplings are rather small, and their dependence on the distance at which they are measured is actually calculable. Therefore, if two of the couplings at 10^{-16} cm are known, they can be followed down to shorter and shorter distances until they meet. This will happen at a distance of the order of L. Furthermore, the third coupling can then be predicted, by following the couplings back up to larger distances. What we actually did was to use α, the combination of α_2 and α_1 that actually occurs in the electromagnetic interactions, and then estimate L and $\sin^2\theta$ ($\sin^2\theta$ is the ratio α/α_2 – it is determined by looking in detail at neutral current weak interactions, or measuring the W mass). The results are that L is about 10^{-29} cm and $\sin^2\theta$ is about 0.21.

In 1974, when Quinn, Weinberg and I worked out this prediction, it looked bad for the SU(5) theory. At the time, the experimental value of $\sin^2\theta$ was about 0.35 (with rather large errors). This didn't bother us too much at the time, because we were far from confident that the SU(3) and SU(2) × U(1) theories on which SU(5) was based were the correct description of physics at 10^{-16} cm. As the years have passed, we have become more and more confident of SU(3) × SU(2) × U(1). But the experimental value of $\sin^2\theta$ has also changed. Today it is in good agreement with the SU(5) prediction.

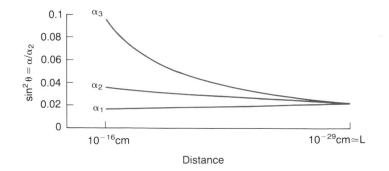

Figure 15.12. The three effective coupling constants α_1, α_2 and α_3 vary with distance (see figure 15.5 for an explanation of this). While α_1, the weakest of the three, grows stronger at short range, the other two grow weaker, in such a way that all three converge to a common value at a distance of about $L = 10^{-31}$ m.

Another consequence of the SU(5) unification that is based on similar ideas was worked out by Mike Chanowitz, Andrzej Buras, John Ellis, Mary Gaillard and Demetres Nanopoulos at CERN in 1977 (some of these same people are responsible for the adjective 'grand' in grand unified theories and the abbreviation 'GUTs'). Shelley Glashow and I had noticed that, in the simplest SU(5) theory, the masses of the d quark and the electron are the same. This relation, while qualitatively reasonable, did not seem to be quantitatively correct. But Chanowitz and company noticed that the same argument could be applied to the b quark and the tau lepton which are heavier copies of the d quark and the electron, respectively. Furthermore, they corrected the naïve relation using arguments similar to those that Quinn, Weinberg and I had used for the coupling constants. They discovered that the b quark should be heavier than the tau, at the long distances at which we observe them, by a factor of between two and three. In fact, the ratio of masses is about $2\frac{1}{2}$. Unfortunately, the corresponding relations still do not seem to work for the lighter particles, the electron and the d and the s quark and the muon, so it is not clear how seriously we should take the success of the SU(5) prediction for the b and tau.

15.21 The search for proton decay

The most spectacular prediction of the SU(5) grand unified theory is the existence of interactions that change the number of quarks, mediated by exchange of the X particles. Like the weak interactions, which are weak because the W and Z are heavy, these interactions are weak because the X particles are heavy. The probability that such an interaction will take place at low energies is proportional to $1/M_X^4$. Only at energies greater than the X mass do these interactions become fairly strong. But M_X is of the order of $\hbar c/L$, where L is the very small distance computed by Quinn, Weinberg and I at which the SU(5) symmetry is spontaneously broken. M_X is an enormous mass, nearly 10^{15} times the proton mass. Simple dimensional analysis suggests that the rate of proton decay from such an interaction will be $m_p^5/(M_X^4/\hbar c^2)$, which gives a lifetime of about 10^{31} years. More refined estimates in the simplest SU(5) theory give rates for the decay of a proton into an antielectron plus a π^0 as large as one every 10^{29} years. If that were correct, it would mean that any given proton has one chance in 10^{29} of decaying into an antielectron and a π^0 in any given year. That doesn't sound like a very high decay rate. The total age of the universe is only something like 10^{10} years. However, there are lots of protons and neutrons in matter (the quark number changing decay of a neutron would be as spectacular as proton decay). In one tonne of matter, there are about 6×10^{29} protons and neutrons. Thus if the lifetime were 10^{29} years, you would expect to see a handful of events per year in one tonne of matter.

Many experimental groups have taken up the challenge of trying to find proton decay. These experiments are done underground, in mines, to minimise the effect of cosmic rays that can mask and sometimes mimic proton decay. There are now six different detectors in operation that look at masses ranging from 60 to 3300 tonnes. The largest of these is the IMB (for Irvine, Michigan and Brookhaven) detector, located in a salt mine under Lake Erie, outside Cleveland. This detector is a 20 m cube hollowed out of the salt, lined with plastic and filled with ultrapure water. A proton decay into $e^+ + \pi^0$ near the centre of the detector would produce a characteristic flash of Cherenkov light in the form of back to back cones. This light is a kind of electromagnetic shock wave produced because the particles into which the proton decays move faster than the speed of light in the water. The Cherenkov light can be picked up by some of the 2048 photoelectric tubes that line the cube, inside the plastic liner. A complex array of electronic monitors look for events of this (or any other interesting) form and record them on magnetic tape, for more detailed analysis by the physicists in the collaboration.

Most of the detectors see events that could be proton decay, but all of these events are such that they might have been produced by neutrinos in cosmic rays. None of the groups see conclusive evidence for proton decay. Their results indicate that the lifetime of the proton is at least 10^{32} years. This is disappointing because it seems to indicate that the simplest SU(5) theory is too simple. This conclusion is not 100% certain, because all of the estimates of the proton lifetime contain large and unknown uncertainties due to the difficulties of dealing with the colour SU(3) theory at large distances (like 10^{-13} cm, the radius of the proton) where the coupling constant α_3 is large. But the likelihood is that the simplest theory must be modified. Unfortunately, it does not take much of a modification to raise the proton lifetime by several orders of magnitude, because the extrapolation from 10^{-16} cm to 10^{-29} cm is so huge that the numerical results are very sensitive to small changes in the parameters of the theory.

Experimenters will continue to try to find proton decay. With several large detectors in place and working well, the bound on the lifetime can be pushed down by another order of magnitude at least. It would be sad if Nature denied us the chance to look through this window into the world of very short distances.

15.22 Superheavy magnetic monopoles

There is one other place where the physics of SU(5) might show up directly. Like the SO(3) theory, SU(5) provides a theoretical explanation for the quantisation of electric charge. In SU(5), the explanation is much more attractive, because it accounts not only for the integral charge of the electron but also for the fractional charges of the quarks. The principle, however, is the same. Charge is quantised because of group theory. All the SU(5) charges participate in nontrivial commutation relations, and therefore all possible representations of these charges, and

Figure 15.13. A view from the bottom of the IMB proton decay detector near Cleveland, Ohio, showing the photoelectric tubes lining the walls. Note the clarity of the water, which is essential to the effectiveness of the experiment. Both this detector, and the KAMIOKA detector in Japan, saw neutrinos from the supernova in the Large Magellenic Cloud.

therefore all particles have charges that are multiples of the charges of the 5. We might expect that the mechanism identified by 'tHooft and Polyakov in the SO(3) theory would operate here as well and imply the existence of magnetic monopoles in this theory. Indeed, that is what happens. The SU(5) vacuum can twist up into the nontrivial configurations that are the magnetic monopoles. But it is only at distances of order L that unification prevents the unwinding of these configurations. Thus the monopoles are very small and very heavy. They are expected to be even heavier than the X particles. If such monopoles exist, it should be possible to recognise them, by their magnetic charges, and perhaps by other peculiar properties associated with their large mass. Many experimental physicists have set up detectors to look for monopoles that may be floating around loose in the universe. On St Valentine's Day, 1983, one such detector, put together by Blas Cabrera at Stanford, recorded an event that looked very much like the passage of a monopole through the apparatus. This caused great excitement at the time. Unfortunately, neither Cabrera, nor anyone else, has seen a monopole since. Most people assume that the St Valentine's Day event was some kind of rare experimental glich, not a real monopole. However, people are continuing to look.

15.23   Physics at the unification scale

Apart from proton decay, monopoles and the relations I have discussed between couplings and masses, the physics of grand unification is down at very small distances of the order of L, the unification scale. This is sad, because we have no way of doing experiments that directly probe such small scales. The energies required are far too high. However, some people are very hopeful that we can learn about grand unification not by experimenting, but by observing the universe. If the standard big-bang cosmology is correct, the universe was once very hot. If we could follow the history of the universe back far enough, the temperature might be so high that the typical energies of the particles bouncing around in the primordial fireball are of the order of M_X. At such ridiculously early times, the interactions of the X particles in SU(5) were as important as anything else. One might then hope to find effects of the SU(5) interactions that could be followed forward in time to give observable features of the universe as we see it today.

The most obvious feature that could be explained in this way is the fact that the universe seems to make much more out of matter than of antimatter. If the quark number changing interactions of SU(5) can destroy protons, they can also, in the right circumstances, create protons. Perhaps the protons out of which our world is built were cooked into existence shortly after the big bang by the interactions in some grand unified theory. This interesting speculation was anticipated by the great Russian physicist Andrei Sahkarov, who suggested that interactions that can create and destroy protons were needed to account for the observed assymetry between matter and antimatter in the universe. In the modern context of grand unified theories, this speculation was first made by Motohiko Yoshimura of the Japanese National High Energy Physics Laboratory, KEK and was subsequently elaborated by many physicists, including Ellis, Gaillard, Nanopoulos, Weinberg, Savas Dimopoulos and Leonard Susskind from Stanford and Sam Treiman and Frank Wilczek from Princeton. They have shown that to produce an excess of protons over antiprotons requires that the quark number changing interactions look different when they are run backwards in time. This condition is satisfied in realistic unified theories; however, it is an open question whether any of these theories can account quantitatively for what we see (see Chapter 3 by Guth and Steinhardt).

My own suspicion is that, while such speculations are great fun, they will probably never yield real hard information about the nature of the interactions at the unification scale. There is a great difference between cosmology or astrophysics and particle physics which, at its root, is the difference between an observational science and an experimental science. The same comments, of course, apply to any sort of physics at the unification scale. Even if we are lucky, all we can do is look, rather than directly manipulating such small distances.

15.24　The artist's hand

Some parts of a puzzle are easier than others. When the colours and textures are rich and distinctive, when there are lots of holes to fill, when the shapes are easy to recognise, the puzzler's efforts are often rewarded. Such was the state of the puzzle of particle physics in the early 1970s. It was a wonderful time, full of challenges and triumphs. But we have completed that easy patch of our puzzle. The surrounding area seems, at least for the moment, rather uniform, without striking features, and, most important, with few obvious holes for us to try to fill. We have answered most of the obvious questions and resolved all the obvious problems.

The only hole that we see clearly is SU(2) × U(1) breaking. We still have no idea what causes the spontaneous breaking of the SU(2) × U(1) symmetry of the weak and electromagnetic interactions. The answer may not be easy, but we have the tools to find it. If particle physicists can make clear the need to build a 'supercollider' that probes distances from 10^{-17} to 10^{-18} cm, we *can* generate new experimental puzzles in this area.

But most of the other issues that motivate theorists today are based on aesthetic criteria rather than puzzling experimental results. These can be broadly divided into two classes. There are issues like the flavour puzzle, the question of the significance of the heavy copies of the electron and the u and d quarks, whose answers may be nearby, or far away. We simply have no idea whether a higher energy machine will address this issue. It might, but it might not. Then there are questions that seem likely to be resolved only at very tiny distances, beyond the reach of manipulative experimental techniques. It is here that grand unified theories have had their greatest impact, for good or ill, focusing theoretical attention on such tiny scales.

One such question is the so-called hierarchy puzzle: Why do there exist two such vastly different scales of spontaneous symmetry breaking as the SU(2) × U(1) breaking scale at 10^{-16} cm and the grand unification scale at 10^{-29} cm (or perhaps even smaller). Some of the attempts to answer this question will be addressed in other articles in this anthology. It was once hoped that supersymmetry would resolve this puzzle. My own preference, at the moment, is for a theory that would produce the SU(2) × U(1) breaking by some dynamical mechanism similar to the dynamics that makes the π mesons light in the quantum chromodynamic theory of strong interactions.

Another question might be called the 'uniqueness' question: Why do the particular particles that we see exist? The SU(5) grand unified theory is a partial answer to this question, in the sense that the striking *fit* of these particles into a simple structure is an indication of their special status. But clearly one can go on asking such questions (Why SU(5)? Why 5 + 10?, etc.) until you can determine the structure of the world by pure thought.

The important thing to understand about such questions is that they do not have to have answers. In fact, they do not even have to be questions. The issues may have changed completely long before we get to such tiny distances.

Unlike a picture puzzle, our puzzle does not come in a box with the entire picture on the outside for comparison. We must generate our pieces as we go, growing outwards from the completed patches as technology, ingenuity and insight allow. We see the hand of the artist not in the big picture, but in the rich texture of every patch. It is as an example of this richness that I most value SU(5). Whatever, its significance at smaller distances, SU(5) is a clear signal to us that, despite the frequent difficulties and frustrations, we should keep working at the puzzle.

16 Effective quantum field theories

H. M. Georgi

In Chapter 15 I discussed the idea of Grand Unified Theories (GUTs) of particle interactions. I still find that subject very interesting, despite the disappointing failure of experiments to find proton decay. But I have mixed emotions. I feel about the present state of GUTs as I imagine that Richard Nixon's parents might have felt had they been around during the final days of the Nixon administration. I am very proud that the grand unification idea has become so important. After all, at first it was something of an ugly duckling, roundly ridiculed by everyone. But proud as I am, I cannot help being very disturbed by the things which GUTs are doing now.

GUTs were motivated by the physics of $SU(2) \times U(1)$ and colour $SU(3)$ and the desire to predict the value of the weak mixing angle (a parameter in the $SU(2) \times U(1)$ model that measures the relative strength of the $SU(2)$ and $U(1)$ interactions) and to explain the quantisation of electrical charge. They were certainly not an attempt to emulate Einstein and produce an elegant geometrical unification of all interactions including gravity, despite the parallels which have been drawn in the semipopular press. Einstein's attempts at unification were rearguard actions which ignored the real physics of quantum mechanical interactions between particles in the name of philosophical and mathematical elegance. Unfortunately, it seems to me that many of my colleagues are repeating Einstein's mistake. It is primarily for this reason that I want to address the larger picture.

The language of relativistic quantum mechanics is called 'quantum field theory'. In a quantum field theory, a field (like the magnetic field surrounding a magnet) is assigned quantum mechanical properties and is seen to be associated with a type of particle (the magnetic field, for example, is associated with the photon, the particle of light). In the last fifteen years, our understanding of quantum field theory has changed considerably. I believe that we have arrived at a mature and satisfying view of the subject. It is this modern view of quantum field theory that I will discuss here. The view of many field theorists today is that the most appropriate description of particle interactions in the language of quantum field theory depends on the energy at which the interactions are studied.

Table 16.1. *Logarithmic energies and related distances*

Energy $(M_p \times c^2)$	Distance (cm)	Associated physics
10^{19}	10^{-33}	Quantum Gravity?
10^{14}	10^{-28}	GUTs?
10^2	10^{-16}	W and Z particles Quarks
10^{-1}	10^{-13}	Nuclei
10^{-6}	10^{-8}	Atoms Molecules
10^{-11}	10^{-3}	Amoeba
10^{-16}	10^2	People

The description is in terms of an 'effective field theory' that contains explicit reference only to those particles that are actually important at the energy being studied. The effective theory is the most useful means of extracting the physics in a limited range of energies, but it changes as the energy changes, to reflect changes in the relative importance of different particles and forces. I should say at the outset that my own role in the development of this view has been minor compared to that of giants like Ken Wilson and Steve Weinberg. But the invention of GUTs was important because it forced us to think seriously about particle physics at extremely short distances. But I am getting ahead of myself. I am going to begin by making some rather obvious statements about the nature of physics.

16.1   Dimensional analysis

Dimensional analysis is one of the oldest and most important physical ideas. The key principle in dimensional analysis is that physics should not depend on the units in which physical parameters are measured. Because of this principle, the dimensions of a physical quantity can tell you a lot about the physics itself. To some degree, this principle is incorporated into our common sense, that uneven distillation of instinct, experience and learning that we carry with us in our study of the limited physical universe of our everyday lives. When we hear that something is measured in square centimetres, we immediately think of an area. Kilometres per hour identifies a velocity. Calories measure energy (often in a tempting form that produces guilt when we eat it). We know from experience that the actual value of each of these quantities depends on comparison with a set of units that has no fundamental significance except perhaps historically. We know that we can convert from one set of units to another, as long as the quantities measured are the same. But while we can compare centimetres and inches, we know that it doesn't make sense to compare centimetres with square inches. Inches and centimetres represent the same dimensional property in different units. Centimetres and square inches are dimensionally different so they always measure physically different things.

Some other units are less familiar. Momentum, which has units of mass times distance over time, grammes times centimetres per second for example, measures the tendency of a moving object to keep going in a straight line. Momentum times distance (for example grammes times square centimetres per second) is the unit of something called angular momentum, the unit of spin. It is the physics of angular momentum that an ice skater uses to perform a rapid revolution. For a fixed angular momentum (and it is fixed unless the skater is given a twist by some outside force such as the push or pull of the skates on the ice) the rate of the skater's turning is inversely proportional to the area over which his mass is spread. When he lifts his arms over his head, this area decreases and his rate of revolution increases, even though his angular momentum stays about the same (see figure 16.1).

We all understand in our bones that dimensional quantities, those with units, have values that depend on the system of units in which they are measured. That is very different from a pure number like 1 or π. This is common sense. But dimensional analysis is much more general than common sense. Common sense applies only in the bounded domain of human experience. Classical physics, the physics of Newton, is built on common sense and quantifies and extends it in a precise mathematical language to explain in great detail such different phenomena as the fall of an apple and the motion of a planet. It works so impressively well in the domain of our everyday experience and at larger, astronomical distances, that in the nineteenth century, many physicists were confident that they knew most of what was worth knowing about the way the world works.

Figure 16.1. Conservation of angular momentum is well illustrated by the spinning ice skater. By the drawing in of the arms, so concentrating the skater's mass closer to the axis of rotation, the rate of rotation increases.

16.2   Biology is not a branch of physics

In fact, however, classical and nonrelativistic physics are only approximate theories which work well for velocities much smaller than the speed of light, c, which is roughly equal to $3 \times 10^{10}\,\mathrm{cm\,s^{-1}}$ and angular momentum much larger than Planck's constant, $\hbar$, which is about $10^{-27}\,\mathrm{cm^2\,s^{-1}}$. c and $\hbar$ are fundamental constants which mark the boundaries between different appropriate descriptions of the world. The word 'appropriate' is crucial here. It is easy to say that classical and nonrelativistic physics have simply been replaced by quantum mechanics and relativistic physics, the theories that we know are needed to understand the world at small angular momenta and large velocities. But it is not true that they have simply been replaced. It is not true in a sense which is similar to the sense in which the statement 'chemistry and biology are branches of physics' is not true. It *is* true that in chemistry and biology one does not encounter any new physical principles. But the systems on which the old principles act differ in such a drastic and qualitative way in the different fields that it is simply not *useful* to regard one as a branch of another. Indeed the systems are so different that 'principles' of new kinds must be developed, and it is the principles which are inherently chemical or biological which are important. In the same way, to study phenomena at velocities much less than *c* and angular momentum much greater than $\hbar$, it is simply not useful to regard them as special cases of phenomena for arbitrary velocity and angular momentum. In fact, we usually put the logic the other way around. The correspondence principle is the statement that our quantum mechanical description of the world must reduce to the simpler classical description in the appropriate domain. We don't need relativity and quantum mechanics for small velocities and large angular momenta. It's just as well, too, because if we had had to discover the laws of relativistic quantum mechanics from the beginning, we probably would never have gone anywhere.

Particle physicists, like me, tend to forget all this, because we are interested almost exclusively in velocities nearly equal to the speed of light, *c*, and angular momenta not much larger than $\hbar$. We don't need the fundamental constants *c* and $\hbar$ as boundaries because we are always in the same domain, so we just set them equal to one and measure all dimensional quantities in units of mass, or whatever. For example, when we set *c* equal to one, a second can be either a time of one second or the distance that light travels in one second. When we set $\hbar$ equal to one, one centimetre can be either a distance of one centimetre or the inverse of the momentum required to produce an angular momentum of $\hbar$ at a distance of one centimetre from the axis. Thus we can measure energy and momentum in units of mass, time and distance in units of inverse mass, force in units of mass squared, etc. This habit is so ingrained that we tend to use these units interchangeably. For example, I will often convert the mass of a particle, *M*, into a distance, $1/M$, called the Compton wavelength of the particle, the length at which we see the particle's quantum mechanical properties. For example, the Compton wavelength of an electron is about 4×10^{-11} cm (40 millionths of a millionth of a centimetre). The proton, which is almost 2000 times heavier, has a Compton wavelength about 2000 times *smaller*, or 2×10^{-14} cm. This is a trivial exercise in dimensional analysis, but it illustrates a general feature of the quantum mechanical world. The heavier a particle is, or the higher its energy, the smaller is the distance at which its quantum effects appear.

Having said these obvious (?) things, I will now proceed with a brief review of the history of quantum field theory.

16.3   Local quantum field theory

Field theory developed in the late 1920s and early 1930s to describe the interactions of electrons and photons. It was the natural synthesis of quantum mechanics and relativistic wave equations like Maxwell's equations (the equations that describe the properties of classical electric and magnetic fields) and the Dirac equation (an attempt to describe the properties of the field associated with the electron). Today we would say that this particular synthesis was more than just natural, it was inevitable. Local quantum field theory is the only way to combine a quantum mechanical theory of particles with special relativity consistent with causality. *Causality* is the general principle that causes should always happen before their effects. The word 'local' here is the crucial one. A local quantum field theory is one in which the interactions which cause scattering or creation or annihilation of particles take place at single space-time points. Locality is important because action at a distance causes trouble with causality in relativistic theories.

Obviously, the assumption of locality is an act of incredible hubris. After all, a 'space-time point' is not a physical thing. It is a mathematical abstraction – infinitely small. To really know how particles interact at a single point you have to understand how the world works down to arbitrarily small distances. That is ridiculous! Only a particle theorist would have the infernal gall to even propose such a thing!

Nevertheless, early quantum field theory yielded many important results, although, from our modern vantage point, the logic often seems confused. The Dirac equation was the first relativistic treatment of electron spin. Goudsmit and Uhlenbeck discovered in the 1930s that the electron behaves like a spinning top. Its angular momentum is exactly half of Planck's constant, $\hbar/2$. A spinning electrically charged particle should act as a magnet. The strength of the magnetism of such a particle is measured by its *g* factor, which is the ratio of the actual strength of the particle's internal magnet to that of a point particle with the same charge and mass moving around a fixed axis to give the same angular momentum. Dirac's theory not only incorporated the spin of the electron in a way which was consistent with Einstein's relativity principle, but it also 'predicted' that the *g* factor of the electron is 2, close to its experimental value. In fact, however, this was really only an aesthetic argument. The arguments which led to the Dirac equation also allow an additional term in the equation called a Pauli term with an arbitrary coefficient, which gives an arbitrary *g* factor. It is really only the 'simplest' Dirac equation which gives $g = 2$.

The 'prediction' of the positron was confused in a different way. The Dirac equation has negative energy solutions. Dirac realised brilliantly that, if nearly all the negative energy states were filled, the few 'holes' would behave like particles with positive charge. But at first he interpreted these as protons! The work of Weyl, Oppenheimer and others convinced Dirac that this interpretation was untenable and that his equations predicted the existence of a genuinely new particle with the same mass as the electron but with positive charge. Then the positron was discovered, and eventually it was recognised by all that it was Dirac's hole particle. Today, of course, we believe that all particles have antiparticles with the same mass and opposite charges (particles with no charge, like the photon, can be their own antiparticles).

Perhaps the most impressive successes of early quantum field theory were the calculations of scattering probabilities for a variety of processes. These were possible because the theory contained a small dimensionless parameter, the 'coupling constant', $e^2/\hbar c = \alpha \simeq 1/137$ (see Chapter 15). Scattering probabilities could be calculated unambiguously to first order in α. This means that the contribution to the probability that is proportional to the small parameter α is calculated while the contributions proportional to α^2 and higher powers of α are ignored. The processes which were studied included (where e^- is an electron and e^+ a positron):

electron–photon scattering	Klein and Nishina, 1929
e^+e^- annihilation into photons	Dirac, 1930
electron–electron scattering	Moller, 1932
$e^- \rightarrow e^- +$ photon in nuclear field	
photon $\rightarrow e^+e^-$ in nuclear field	Bethe and Heitler, 1934
$e^+e^- \rightarrow e^+e^-$	Bhabha, 1936

They were in reasonable agreement with experiment.

16.4   The tragic flaw

But not all was well. For one thing, in the 1930s and 40s, it became clear that a theory of electrons and photons could not be the whole story of particle interactions because there were other particles and interactions: the proton, the neutron, the pion, the muon, the neutrino . . ..

But what was worse was that the formalism of quantum field theory itself seemed to have a tragic flaw. When nontrivial calculations were attempted to higher order in α, the results were infinite! This meant that the theory defined in the most naïve way simply did not make sense. These infinities arose precisely because of the local nature of the interaction. They were, in fact, the punishment imposed on particle theorists for the hubris of locality. And they worried people a lot. The possibility that it might be possible to absorb the infinities by 'renormalisation' of the physical parameters was discussed, but not completely understood.

16.5   Quantum electrodynamics

So matters stood until the theoretical physics community reassembled after World War II. Stimulated by exciting experiments (such as the measurement by Willis Lamb of the small energy shift between two states of the hydrogen atom that were predicted to have the same energy to first order in α), theorists used renormalisation to do finite calculations of quantum corrections to the first order results (such as the g factor of the electron and the Lamb shift). In renormalisation, the theory is defined by a limiting procedure. First the physics at distances shorter than some cut-off length is modified so that the calculations make sense. This involves giving up one or more of the cherished principles which led to local interactions in the first place. Then the parameters in this modified theory are expressed in terms of physical, measurable quantities (masses, scattering probabilities, etc.). Finally, the renormalised theory is defined by taking the cut-off length to zero. Presumably, this restores all the nice properties that a local quantum field theory should have. With the new theoretical tool of Feynman diagrams, it was possible to show that renormalisation was sufficient to absorb all the infinities in a quantum field theory into renormalised physical parameters. However, unless the theory is carefully constructed, the number of parameters required is infinite! The special theories in which only a finite number of physical parameters are required to unambiguously define the physics were called 'renormalisable'. Fortunately, the simplest theory of electrons and photons had the special property of renormalisability. This was exactly the form of the theory in which the electron's g factor came out right. *The local quantum field theory that made sense was also the one that accurately described the world. Quantum electrodynamics had come of age.*

Note that the logic here is a bit peculiar. The infinities in local quantum field theory, which were regarded as a disaster when they were first uncovered, became an asset with the development of renormalisation and quantum electrodynamics. Particle physicists now had another principles, renormalisability – another constraint to impose on their theories. An extra constraint is always very useful because it decreases the number of theories that you have to think about, which, in turn, decreases the amount of work that you have to do. But many physicists were uneasy about it, because this particular asset still had its roots in an apparent disaster. Here is a similar situation. Suppose I need a used car and I go to the only used car dealer in town and find that every car, when I start the engine, makes a horrible scraping noise and, after a minute or two, starts smoking and smelling awful and stops running. But the dealer tells me that there is one car which starts the same way, but if I gun the engine and pound on the dashboard the scraping sound goes away and the car runs beautifully. Well, I try it out and it works! Terrific! I don't even have a decision to make. So I buy the car and it's a great car, just what I need. But somehow I can never stop wondering about what causes the horrible scraping noise.

Nevertheless, the success of renormalisable quantum electrodynamics in the following decade was spectacular. The electron magnetic moment (*g* factor) and other quantities were calculated to incredible accuracy and agreed well with increasingly precise experiments. Quantum electrodynamics became the paradigm of a successful physical theory. It began to look as if hubris were justified.

16.6   New physics

The quantitative success of renormalisable quantum electrodynamics sustained the prestige of quantum field theory for nearly ten years. But in the late 1950s and the 1960s, renormalisable quantum field theory began to run out of steam. It was not that there was anything wrong with quantum electrodynamics. There were just a lot of other things going on. Experimental physics marched forward. Meson and baryon resonances were discovered and studied. By definition, resonances are incredibly short-lived structures, so that it barely makes sense to call them particles at all. They decay as fast as they are produced through so-called strong interactions into more stable things such as pions and protons. Strange particles were also discovered and studied. These are comparatively long-lived particles, produced in particle–antiparticle pairs by the strong interactions but decaying singly much more slowly by so-called weak interactions.

At first, neither the strong nor the weak interaction could be usefully described using renormalisable quantum field theory. The problem with the strong interactions was that they were strong. There was no small dimensionless parameter, like $\alpha = 1/137$ of quantum electrodynamics. Reliable dynamical calculations were impossible. Progress *was* made by using symmetry arguments to extract information which was independent of the dynamical details, but that was all.

The weak interactions as seen in the decays of various particles, on the other hand, could apparently be described very well by a quantum field theory, to lowest order in a parameter, the Fermi constant, G_F, which is about equal to 10^{-5} in units of inverse square of the proton mass. The trouble was that the constant had dimensions of inverse mass squared. As I will show you in a minute, this means that the theory is *not* renormalisable. That in turn means that you must specify an infinite number of physical input parameters to determine the physics unambiguously. Most physicists felt, not unreasonably, that this was an unsatisfactory situation.

16.7   Dimensional parameters and renormalisability

That dimensionless parameters are OK while parameters with dimensions of $1/m^n$ lead to trouble can be seen, more or less, by dimensional analysis. The physical idea which underlines renormalisation is as follows. In the limiting procedure, in which the cut-off length is taken to zero, the physics itself must not depend on the cut-off, but our description of the physics may. In fact, the parameters of the modified, cut-off theory *must* be chosen to depend on the cut-off in order to keep physics at distances much larger than the cut-off length fixed when the cut-off changes. It follows that any dependence on the cut-off that is important for very small cut-off lengths must be associated with some parameter in the theory. To understand the cut-off dependence, we must understand these parameters. The parameters in a quantum field theory are the masses of the particles it describes and the relativistically invariant amplitudes for the various elementary scattering processes which describe the interactions. Because the theory is quantum mechanical, we expect these parameters to determine the quantum mechanical amplitudes whose squares are related to the probabilities for particle scattering.

The essential fact is that the local nature of the interactions requires the mass dimension of the amplitudes that describe the scattering processes to become more negative as the number of particles involved in the scatterings increases. The reason is just Heisenberg's uncertainty relation. In a quantum mechanical theory, there is a limit to the accuracy with which one can determine simultaneously the position of a particle and its momentum. The product of the uncertainties in the position and the momentum of any particle must always be greater than Planck's constant $\hbar$. But if several particles interact at the same point in space, we know their relative positions exactly. At the instant of the interaction, they are all sitting at the same point. Therefore, according to Heisenberg, we can't know anything about their relative momenta. Apart from the difference between the total momentum before and after the scattering (which is fixed by energy and momentum conservation), all the momenta can be varied continuously over the whole kinematically allowed region for the process. That, in turn, implies that the probability that the scattering will take place for any definite value of the momenta is infinitesimally small.

Here is a similar situation. Suppose that I put a penny somewhere at random inside a cubic box of side one metre which is full of sand. Now I can offer to pay you a billion dollars if you guess exactly the position of the centre of the penny in the box. I am completely safe because the probability of the penny being exactly in any given spot is zero. To give you a chance (which I would never do because I can't pay off the bet!) I would have to let you specify a volume in which you guess that the centre of the penny lies. The probability that your guess will be correct is then the ratio of the volume you specify to the total volume of the box ($1 \, \text{m}^3$). This means that the quantity that describes the likelihood that the centre of the penny will be at a given point is dimensional. It has units of inverse cubic metres. I know that because I must multiply it by a volume with units of cubic metres to get a dimensionless probability.

In the same way, to get a finite value for the probability that a local interaction will cause particle scattering, we must multiply the square of the amplitude by a 'phase space' factor that gives the 'volume' or range of momentum, with units of mass cubed. We must do this for each particle momentum. The result is a probability which is dimensionless. That means that the amplitudes, which contain the parameters that describe the scatterings, must have mass dimension which decreases (in this case, becomes more negative) as the number of particles involved in the scattering increases. It turns out that only the simplest scattering processes involving two, three or four particles at a time can be associated with parameters with dimensions of mass to the zero or a positive power. So you see that the parameters are asymmetrical. There are only a finite number of possibilities for parameters with dimensions of mass to the zero or a positive power. But the number of possible parameters with dimensions of mass to a negative power is infinite. The more particles involved, the more negative the mass dimension.

When we calculate quantum corrections in perturbation theory in these parameters, the results can depend on the cut-off length which we will eventually take to zero or alternatively on a cut-off mass which we will eventually take to infinity. We worry about contributions that go to infinity as M goes to infinity. Any dangerous term will involve a product of two or more of the parameters of the quantum field theory times some increasing function of the cut-off mass, M. The increasing function may be either a positive power of M or a logarithm of M. But the effects of the cut-off at very short distances is just to redefine or 'renormalise' the parameters which describe the physics. Thus each correction term proportional to M^k or $\log M$ is associated with some physical parameter with the same mass dimension as the correction. In particular, only those combinations with the dimensions of one of the possible parameters actually appear. We have to worry about only a finite number of M dependent corrections with positive mass dimension, because there are only a finite number of parameters with positive mass dimension for such corrections to renormalise. But there could be an infinite number of dangerous terms with negative mass dimension.

Notice that if all the parameters have positive or zero mass dimension, so will all of the dangerous quantum correction terms. This is precisely the situation in which there are only a finite number of dangerous terms, all of which can be absorbed into a finite number of renormalised physical parameters. Thus, theories in which all the physical parameters have zero or positive mass dimension are renormalisable. But if any of the

The logarithmic function, $\log M$, is something that increases with M more slowly than any power (even more slowly than a fractional power like the square root, $M^{\frac{1}{2}}$). Physicists and mathematicians like to use what is called a natural logarithm that is defined by $e^{\log M} = M$, where e is an irrational number about equal to 2.71828. A graph of this function is shown in figure 16.2. Here is a physical picture of the log function. Suppose that you are trying to stop your car but your brakes are wearing out so that they never bring you to a complete stop. Instead, the speed of the car at time T (measured in hours) is 1 km divided by T (this gives the speed in kilometres per hour). When T is 1 h, your speed is $1 \, \text{km h}^{-1}$. When T is 2 h, your speed is $0.5 \, \text{km h}^{-1}$. When T is 10 h, your speed is $0.1 \, \text{km h}^{-1}$. And so on. How far does your car travel between time T equals 1 h and a later time T? The answer (in kilometres) is $\log T$. From the graph, you can read that by $T = 2$, one hour after $T = 1$, the car has gone about 0.7 km. After two hours at $T = 3$, it has gone about 1.1 km. The car never stops, so the distance it travels is increasing, no matter how large T is. But the rate of increase keeps going down because the car keeps slowing down. For T less than one hour, the distance travelled is negative (as shown in the graph) because the car has not yet got to where it will be at $T = 1$.

Figure 16.2. A graph of the natural logarithmic function.

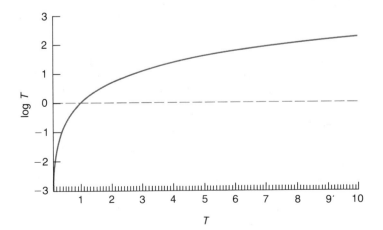

parameters have negative mass dimension, the corrections can have *arbitrarily negative mass dimension* because the parameters with negative mass dimension can appear many times in the product. Thus you can never stop with a finite set of such parameters. The cut-off dependent quantum corrections will always have pieces with more negative mass dimension than any parameter in your set. Only with an infinite number of physical parameters can you absorb all the cut-off dependence. This is why theories with parameters with negative mass dimension are not renormalisable. Thus, amplitudes whose coefficients have dimensions of inverse mass to a power are called 'nonrenormalisable interactions' because they destroy renormalisability.

It is possible to define what you mean by a theory that is not renormalisable by establishing a set of arbitrary rules for dealing with the cut-off dependence. But that doesn't help. If you construct, in this way, a theory with a finite set of parameters, some of which have negative mass dimension, you are just fooling yourself. The theory really depends on an infinite number of physical parameters but you have fixed all but a finite number of them according to an arbitrary prescription. There is no reason to expect such a construction to have anything to do with the world.

16.8   The sociodynamics of particle theory

So far, I have been discussing history which I read about in books, or learned from talking to my older colleagues. In 1947, while quantum electrodynamics was being created, I was busy being born in the post-war baby boom. But, by the mid 1960s, I was an undergraduate at Harvard, getting interested in particle physics myself. By this time, things had deteriorated to the point that, at Harvard, no course in quantum field theory was taught! Julian Schwinger, one of the heros of quantum electrodynamics and soon to be a Nobel prize winner, had given up on quantum field theory in favour of what he felt was a more phenomenological formalism which he taught in a course called 'relativistic quantum mechanics'. He convinced me at the time because he was a masterful lecturer and I was an impressionable undergraduate. But I now believe that he was pulling the wool over my eyes. Relativistic quantum mechanics *is* quantum field theory, properly defined. Schwinger gave up too early.

At any rate, though I didn't realise it at the time and didn't fully appreciate it until 1971, the seeds of the explosive 1970s were being sown all around me during the comparatively boring 1960s.

This may be a good time to tell you my theory about how theoretical particle physics works as a sociological and historical phenomenon. The progress of the field is determined, in the long run, by the progress of experimental particle physics. Theorists are, after all, parasites. Without our experimental friends to do the real work, we might as well be mathematicians

or philosophers. When the science is healthy, theoretical and experimental particle physics track along together, each reinforcing the other. These are the exciting times. But there are often short periods during which one or the other aspect of the field gets way ahead. Then theorists tend to lose contact with reality. This can happen either because there are no really surprising and convincing experimental results being produced (in which case I would say that theory is ahead – this was the situation in the late 1970s and early 1980s, before the discovery of the W and Z) or because the experimental results, while convincing, are completely mysterious (in which case I would say that experiment is ahead – this was the situation during much of the 1960s). During such periods, without experiment to excite them, theorists tend to relax back into their ground states, each doing whatever comes most naturally. As a result, since different theorists have different skills, the field tends to fragment into little subfields. Finally, when the crucial ideas or the crucial experiments come along and the field regains its vitality, most theorists find that they have been doing irrelevant things. But the wonderful thing about physics is that good theorists don't keep doing irrelevant things after experiment has spoken. The useless subfields are pruned away and everyone does more or less the same thing for a while, until the next boring period.

This theory explains, I hope, how I can say that the 1960s were boring despite the fact that many of the pieces of the puzzle of the $SU(3) \times SU(2) \times U(1)$ theory of strong and electroweak interactions were discovered in the 1960s (and even the 1950s). There were people, like Feynman, Gell-Mann, Glashow, Weinberg, Ken Wilson and others, who had many of the right ideas all along, but they were isolated islands in a sea of confusion, unable to convince everyone else that what they were doing was right, and frequently unable even to convince themselves. After all, the original papers of Glashow and Weinberg on the $SU(2) \times U(1)$ theory were pretty much ignored even by the authors themselves until 1971. It wasn't obvious that the theory was renormalisable because it still apparently contained interactions proportional to a dimensional parameter, the inverse of the W mass.

While the decade of the 1960s was a difficult time to live through, it is fascinating to look back on. Other articles in this collection will go into this history in more detail. What I find particularly amusing about it is the peculiar interplay between the attempts to understand the weak interactions and the attempts to understand the strong interactions. These had to be closely connected because many of the particles which decay by weak interactions are produced by, or otherwise participate in, the strong interactions. But, in fact, the interplay was much more subtle. Many of the ideas that were developed in an attempt to understand the strong interactions eventually found their most important application to the weak interactions instead and vice versa. I regard this curious historical fact as significant and mysterious. The crucial concepts that were developed during this period include the renormalisation group, gauge theories, flavour SU(3) symmetry, quarks, charm, colour, spontaneous symmetry breaking, scaling, the parton model, and some others that I will discuss later. All of these are essential components of our present understanding. It is not necessary that you know in more detail what all of these words mean (you can learn more about them in some of the other articles in this collection). But what I do want to convey is the sense that during the 1960s, while we had almost all the pieces of the puzzle of the strong and weak interactions, they were scattered among a myriad of other ideas which would soon be forgotten. Missing were the crucial theoretical insights and experimental results needed to prune away the dead wood.

16.9   Spontaneously broken gauge theories

The missing theoretical idea, provided by 'tHooft in 1971, was the demonstration that spontaneously broken gauge theories can be renormalisable. In particular, he showed the renormalisability of the $SU(2) \times U(1)$ gauge theory written down by Glashow and Weinberg and Salam in which the weak interactions are due to the exchange of W and Z particles. The W and Z are gauge bosons like the photon, but heavy because of spontaneous symmetry breaking. The main reason that this was nontrivial is that, in gauge theories, the symmetry and the dynamics are intertwined in such a way that you need the symmetry to perform the renormalisations. But the physical states in the theory do not respect the symmetry because the vacuum breaks it spontaneously. Thus the absorption of the infinites into the physical parameters does not work in any simple way. But what 'tHooft, Ben Lee and a few other people realised was that the important thing in the renormalisation programme was to get rid of the dependence of physical results on the physics at very short distances. This should have nothing to do with the spontaneous symmetry breaking which gives mass to the W and Z because the spontaneous symmetry breaking affects the structure of the physics only at distances longer than the scale at which the symmetry is broken, in this case the Compton wavelength of the W and Z. With the help of Veltman, 'tHooft was able to disentangle the symmetry from the dynamics. They also introduced a new renormalisation scheme in which the short distance effects are simply thrown away by a completely automatic procedure that makes no explicit reference to the physical parameters. With these, and a few other, bits of cleverness he was able to show that spontaneously broken gauge theories are renormalisable.

There are many nice things about 'tHooft's scheme. One thing I liked very much (when I finally understood it) was the fact that in 'tHooft's procedure the M^n divergences do not even have to be thrown out. They never show up at all. In fact, I think that this is good because these divergences have no physics in them. They can be renormalised away completely, so it is nice that they never appear at all. On the other hand, the logarithmic divergences are interesting because they must be associated with a log of some other dimensional parameter, a momentum or mass. This is dimensional analysis again. The log function is dimensionless. You can find the log of a dimensionless number unambiguously, but the log of a mass depends on the units in which the mass is measured. Thus the logarithmic divergences must have the form:

$$\log (M/m),$$

where m is some physical parameter with the dimensions of mass so that M/m is dimensionless and independent of the choice of units. m may be the mass or energy or momentum of one of the particles involved in the process you are calculating. When the logarithmic divergence is absorbed, the rest of the logarithm remains. Thus renormalisation requires a dimensional parameter to set the scale of the logarithm. This is called the renormalisation scale μ. It is arbitrary in principle, but in fact, for any given calculation, some renormalisation scales are more convenient than others because of the presence of logs of m/μ. If all the masses and momenta in a process are of the same order of magnitude, it pays to choose μ in the same range to minimise the effects of the logs and make the perturbation theory better behaved. It is this logarithmic dependence on the renormalisation scale which is responsible for the renormalisation group dependence of parameters on the distance or momentum scale, first discussed by Gell-Mann and Low. In the renormalisation group, you use the physics at one scale μ to figure out what the physics will look like at a very nearby scale. But by putting together many of these small steps, you can understand how the physics changes under large changes of the scale.

We now know that all this solved the problem of the weak interactions. Fortunately, it wasn't obvious in 1971 because it wasn't clear that the Glashow–Weinberg–Salam $SU(2) \times U(1)$ theory was the right one. So theorists had a good excuse to explore the vast new class of renormalisable theories which 'tHooft had opened up for us. Five years later, the experimental evidence had settled down to the point where we could be confident that $SU(2) \times U(1)$ was right, but, in the mean time, a great deal was learned about the properties of the new kind of theory.

16.10   Scale dependence

One of the most important properties was discovered by Sidney Coleman and Eric Weinberg. They thought about quantum field theories which naïvely have no dimensional parameters and only a single dimensionless parameter, such as quantum electrodynamics with a zero mass electron. They realised that, because the renormalisation scale μ has to be introduced to define the quantum theory, the physics is actually determined by a dimensional parameter instead of a dimensionless one. The point is that the renormalised dimensionless coupling is a function of the renormalisation point, but by dimensional analysis it must therefore be a function of μ/Λ, where Λ is some fixed dimensional parameter. Thus

$$\alpha(\mu) = f(\mu/\Lambda).$$

Furthermore, the μ dependence of $\alpha(\mu)$ is determined by the theory. Therefore f is some fixed, computable function. The Λ is the only thing that can actually be varied. They called this process, in which a dimensionless parameter α is traded for a dimensional parameter Λ, 'dimensional transvestism'. This was too much for the editors of the *Physical Review*, who consider it part of their job to keep the language of particle physics as boring as possible, so the effect is now called 'dimensional transmutation'.

In quantum electrodynamics, the coupling constant is an increasing function of μ, which looks approximately like

$$\alpha(\mu) = b/\log(\Lambda/\mu)$$

for some constant b. Since $\alpha = e^2/\hbar c$ must be greater than zero, this only makes sense for Λ greater than μ where the log is positive. Indeed, the quantum electrodynamics theory probably only makes sense, in principle, for μ less than Λ. This used to worry some people. It no longer bothers us in quantum electrodynamics. Because α is quite small at ordinary scales, the log must be big. That means that Λ is a truly enormous mass. Since we do not believe that quantum electrodynamics is a complete theory of the world, we don't worry too much that it doesn't make sense at energies much larger than anything that we care about. However, it was thought, at one time, that all interesting quantum field theories behave as quantum electrodynamics does. That is they are sick at very short distances. But David Politzer and others showed that this is not true. Gauge theories based on groups such as SU(2) and SU(3) (which are called non-Abelian because their group multiplication laws are not commutative, see my chapter on GUTs) have exactly the opposite property. Asymptotic freedom! In these theories, the coupling constant is again dimensionless, and again dimensional transmutation occurs so that the actual parameter which determines the physics is a mass Λ. But here the coupling decreases with scale:

$$\alpha(\mu) = b/\log(\mu/\Lambda).$$

This makes sense only for μ greater than Λ. For scales much larger than Λ the theory can be simply described in terms of the gauge couplings. But, for scales of order Λ and smaller, the description of the theory in terms of perturbation theory in the coupling α no longer makes sense. In this region, the character of the theory must change in some way.

In the colour SU(3) theory of the strong interactions, quantum chromodynamics, we believe that the change in the character of the theory at large distances is associated with colour and quark confinement. The Λ parameter in quantum chromodynamics is the length scale at which confinement becomes important. At shorter distances, the theory can be described accurately in terms of the interactions between quarks and gluons (the gauge particles of quantum chromodynamics, like the photon in quantum electrodynamics). But at longer distances this picture breaks down. The colour force between quarks, which behaves much like electromagnetism at distances shorter than $1/\Lambda$, does not drop off further for distances longer than $1/\Lambda$. This was not at all obvious before 1973, because most of our understanding of the strong interactions was based on experiments at long distances, where the nature of the physics is dominated by confinement and the underlying dynamics is obscured.

16.11   Grand Unified Theories

Once we understood SU(2) × U(1) and quantum chromodynamics, Grand Unified Theories were a simple step. The motivation for the simplest GUT, SU(5), was not any mystical desire to follow in Einstein's footsteps and unify everything. Shelley Glashow and I were just trying to understand SU(2) × U(1) better. For several years, we had realised that if we could incorporate the SU(2) × U(1) gauge symmetry into a single simple group it would give us some extra information. It would fix the value of the weak mixing angle, a free parameter in the ordinary SU(2) × U(1) theory and it would explain why all the electric charges we see in the world are multiples of the charge of the electron. But we were having great difficulty doing it. The quarks never seemed to fit in properly. When he heard about quantum chromodynamics, Shelley suggested that we might have to incorporate colour to get it to work. When I pointed out to him that the strong interactions are strong, he replied that we only know that they are strong at long distances where confinement is important. At sufficiently short distances, they could be as weak as SU(2) × U(1). With that hint, it was easy for me to find the SU(5) theory, into which SU(3) colour and electroweak SU(2) × U(1) fit very neatly, basically because $2 + 3 = 5$. Of course, this theory also predicted proton decay, but that just meant the scale at which the SU(5) symmetry was broken had to be extremely large.

A few months later, Helen Quinn, Weinberg and I figured out how to actually calculate the scale of SU(5) breaking in the simplest SU(5) model and we discovered that the scale really is very large, about 10^{14-15} GeV.

16.12   Effective field theories

What I want to emphasise about all this is the following. In our understanding of the weak interactions and strong interactions based on SU(2) × U(1) and SU(3) and in the attractive speculation of GUTs, based on SU(5), there is a crucial role played by dimensional parameters, the confinement scale of quantum chromodynamics and the breaking scales of SU(2) × U(1) and SU(5). This was a dramatic change from quantum electrodynamics in which the physics seemed (at least to the naïve observer) to be mostly in the dimensionless coupling constant α. Furthermore, we were getting used to incorporating physics at short distances (such as GUTs) without disturbing our understanding of physics at longer distances. Many physicists began to verbalise and answer a question which had been nagging at them for a long time. If there is all this wonderful stuff going on at short distances, how come quantum electrodynamics worked so well? Of course, the reason is obvious and had been known, in some sense, for a long time. Quantum electrodynamics works extremely well for the electron because the distances at which other stuff is happening are very small compared to the electron's Compton wavelength. It was easy to see this explicitly in theories such as $SU(2) \times U(1)$ in which quantum electrodynamics was embedded in a more complicated but still renormalisable theory at a smaller distance scale. But physicists were slow in appreciating the full power of the idea, which is unleashed only when quantum electrodynamics is thought of as an 'effective field theory', approximately valid at long distances.

The point is this. At distances of the order of the electron Compton wavelength, the only particles we really have to think about are the electron and the photon. All other charged particles are heavier, and, at such large distances, there is not enough energy to produce them, so we do not have to include them in our theory. There are light neutral particles, neutrinos, but at these distances they are so weakly interacting that they don't matter much, so we can ignore them as well. Thus we can describe the electron–photon interaction at these large distances by an effective field theory involving only the electron and the photon. This has to work. With a completely general quantum field theory, we can describe the most general possible interactions consistent with relativity, quantum mechanics and causality. We do not give up any *descriptive* power by throwing out the heavier particles and going to an effective theory.

It might seem, though, that we have given up *predictive* power. After all, an arbitrary effective theory has an infinite number of nonrenormalisable interactions and thus an infinite number of parameters. But this is not quite right for two reasons, one quantitative and one qualitative. Quantitatively, if we know the underlying theory at shorter distances, we can calculate all the nonrenormalisable interactions. Indeed, there is a straightforward and useful technology for performing these calculations. Thus quantitative calculations can be done in the effective theory language.

The qualitative message is even more interesting. All of the nonrenormalisable interactions in the effective theory are due to the heavy particles which we have ignored. Therefore, the dimensional parameters that appear in the nonrenormalisable interactions in the effective theory are of the order of the heavy particle masses. If these masses are all very large compared to the electron mass and the photon and electron energies, the effects of the nonrenormalisable interactions will be small.

They will be suppressed by powers of the small mass or momenta over the large masses.

Thus, not only do we not lose any quantitative information by ignoring the heavy particles and going to an effective field theory language, but we gain an important qualitative insight. When the heavy particle masses are large, the effective theory is approximately renormalisable. It is this feature that explains the success of renormalisable quantum electrodynamics.

To extract the maximum amount of information from the effective theory with the minimum effort, we should renormalise the theory to minimise the logarithms that appear in perturbation theory. We can do this by using 'tHooft's scheme and choosing the renormalisation scale, μ, appropriately. If all the momenta in a process of interest are of order μ, there will be no large logarithms. The standard techniques of the renormalisation group can be used to change from one μ to another as required.

In the extreme version of the effective field theory language, we can associate each elementary particle Compton wavelength with a boundary between two effective theories. For distances larger than its Compton wavelength, the particle is omitted from the theory. For shorter distances, it is included. The connection between the parameters in the two effective theories on either side of the boundary is simple. They must be related so that the description of the physics just below the boundary (where no heavy particles can be produced) is the same in the two effective theories. These relations are called 'matching conditions' for obvious reasons. They are calculated with μ equal to the mass of the boundary particle to eliminate large logs.

If we had a complete renormalisable theory at infinitely short distances, we could work our way up to the effective theory at any larger distance in a totally systematic way. Starting with the mass M of the heaviest particles in the theory, we could set $\mu = M$ and do the matching to find the parameters of the effective theory with the heaviest particles omitted. Then we could use the renormalisation group to scale μ down to the next heaviest mass and repeat the matching calculations to produce the next effective theory. *And so on!* In this way we get a tower of effective theories, each with fewer particles and more small nonrenormalisable interactions than the last. We simply have to continue this procedure until we get to the large distances in which we are interested.

There is another way of looking at it, however, which corresponds more closely to what we actually do in studying physics. We can start at long distances and try to build up each member of the tower of effective theories stretching down to arbitrarily short distances only as it becomes relevant to our understanding of physics. In this view, we do not know what the renormalisable theory at short distances is, or even that it exists at all. In fact, we can dispose of the requirement of renormalisability altogether and replace it with a condition on the nonrenormalisable interactions in the effective theories. The condition is this:

In the effective theory which describes physics at a scale μ, all the nonrenormalisable interactions must have dimensional couplings less than $1/\mu$ to the appropriate power. If there are nonrenormalisable interactions with couplings $1/M$ to a power, for some M greater than μ, there must exist heavy particles with a mass m less than or about equal to M that produce them. In the effective theory including the heavy particles, the nonrenormalisable interactions must disappear.

Note that an effective field theory, like any nonrenormalisable theory, depends on an infinite number of parameters (which are related at shorter distances). But the above condition insures that only a finite number of them are actually important in any physical situation because all the nonrenormalisable interactions are suppressed by powers of μ/M where μ is less than M. Thus, as we go down in distance through the tower of effective field theories, the effects of nonrenormalisable interactions grow and become interesting on the boundaries between theories, at which point they are replaced by renormalisable (or, at least, less nonrenormalisable) interactions involving heavy particles.

This condition on the effective theories is, I believe, a weaker condition than renormalisability. One can imagine, I suppose, that the tower of effective theories goes down to arbitrarily short distances in a kind of infinite regression. This is a peculiar scenario in which there is really no complete theory of physics – just a series of layers without end. More likely, I think, the series does terminate, either because we eventually come to the final renormalisable theory of the world, or (most plausible of all) because, at some very short distance, the laws of relativistic quantum mechanics break down and an effective quantum field theory is no longer adequate to describe the physics.

Renormalisability is still very important. When the gap between two neighbouring mass scales is large, the effective field theory near the lower scale is approximately renormalisable, because the nonrenormalisable interactions have a very small effect. This is the situation in quantum electrodynamics near the electron mass scale. But we no longer have to assume that renormalisability is a fundamental property.

In this picture, the presence of infinities in quantum field theory is neither a disaster, nor an asset. It is simply a reminder of a practical limitation – we do not know what happens at distances much smaller than those we can look at directly.

Whatever happens at short distances, it doesn't affect what we actually *do* to study the theory at the distances we can probe. We have purged ourselves of the hubris of assuming that we understand infinitely short distances. This is the great beauty of the effective field theory language.

16.13   Dollars and direction

Some of you are probably saying to yourselves, by this time, that the whole idea of effective field theories is rather simple and obvious, so why have I subjected you to an article on the subject? One reason is that it makes a difference in dollars. Since our understanding of physics is organised by distance scale (we understand physics at distances greater than 10^{-16} cm, but not at smaller distances), we must push for experimental information at short distances. But short distances mean large energies which mean larger and more expensive accelerators. More specifically, our general understanding of the connection between small nonrenormalisable effects and the heavy particles which produce them is important in the planning of future accelerators.

But I have another reason for talking about effective field theories. As I suggested at the beginning of this chapter, I am somewhat concerned about the present state of particle theory. The problem is, as I mentioned before, that we are in a period during which experiment is not pushing us in any particular direction. As such times, particle physicists must be especially careful.

We now understand the strong, weak and electromagnetic interactions pretty well. Of course, that doesn't mean that there isn't anything left to do in these fields any more than the fact that we understand quantum electrodynamics means that there is nothing left to do in atomic physics. The strong interactions, quantum chromodynamics, in particular will rightly continue to absorb the energies of lots of theorists for many decades to come. But it is no longer frontier particle physics in the sense that it was fifteen years ago.

What then is there to do? If we adopt the effective field theory point of view, we must try to work our way down to short distances from what we know at longer distances, working whenever possible in the effective theory which is *appropriate* to the scale we are studying. We should not try to guess the ultimate theory at infinitely small distances. Even if we could do it, it would probably be about as useful as explaining biology in the language of particle physics. This seems to me to be an extremely important bit of common sense, a useful antidote to the Einstein complex (that is a desire to work on difficult and irrelevant theoretical questions just because Einstein did it) to which most theoretical particle physicists are very susceptible.

Thus, for example, one subject that certainly deserves the attention of all theorists is the question of what causes the spontaneous breaking of the $SU(2) \times U(1)$ symmetry of the electroweak interactions. This is the physics of the *next* effective theory and it *will* be explored by experiment in the near future, if we have the strength and the will to build the superconducting supercollider and push the experimental frontier to the next scale at energies many tens of thousands of times higher than the proton mass!

It is not so obvious that GUTs are interesting things to study. Some years ago, in a panel discussion, Feynman presciently asked me what I would think about SU(5) if proton decay was not observed at the predicted level. In my youthful enthusiasm, I replied that I would believe that it is right anyway. It is too pretty to be wrong. I think that I still believe that. But what I didn't see at the time was that SU(5) or closely related GUTs could be right but not very interesting. If proton decay is actually observed, they become extremely interesting. But until then, apart from a few numbers which express the relations between parameters in our low energy world which follow from unification, their only connection with reality is through cosmology. Cosmology is fun, but it seems unlikely to me that we will know enough about it to extract much quantitative information about physics at very short distances, at least not anytime soon.

I am particularly suspicious of attempts to guess the structure of physics below the Planck length (the length at which quantum gravitational effects are expected to become important, about 10^{-33} cm). If there is any scale at which we might expect quantum field theory to break down, *this is it*, because there is no satisfactory quantum theory of gravity based on conventional relativistic quantum mechanics in ordinary space-time. Indeed, most of the popular theories (such as Kaluza–Klein theories or string theories) assume that physics changes rather dramatically here and that space-time actually has more than four dimensions.

Apparently, the mathematics of these ideas is so appealing that no one is immune. Steve Weinberg, one of the heros of the effective field theory idea, has become so wrapped up in it that he came to Harvard recently to give us a series of talks on differential geometry. I was so moved that I composed the following poem for the occasion:

> Steve Weinberg, returning from Texas
> brings dimensions galore to perplex us.
> But the extra ones all
> are rolled up in a ball
> so tiny it never affects us.

One problem with all this, of course, is that 'it never affects us'. These theories probably have no experimental consequences at all in the practical sense, because we will never probe small enough distances to see their effects. But there is another subtler objection to this kind of speculation. Once you start relaxing the assumptions of relativistic quantum mechanics, where do you stop? In practice, theorists have considered only theories which they happened to know something about for purely accidental historical reasons. That does not seem to me to be a good enough reason to look at them. Theoretical physics must be more than an historical accident.

My personal suspicion is that Nature is much more imaginative than we are. If we theorists approach her study with the proper respect, if we recognise that we *are* parasites who must live on the hard work of our experimental friends, then our field will remain healthy and prosper. But if we allow ourselves to be beguiled by the siren call of the 'ultimate' unification at distances so small that our experimental friends cannot help us, then we are in trouble, because we will lose that crucial process of pruning of irrelevant ideas which distinguishes physics from so many other less interesting human activities.

17 Gauge theories in particle physics

John Taylor

17.1 The weak and strong forces

During the last two decades, two forces of nature have, for the first time, been understood. These are the so-called 'weak' and 'strong' forces. They have been shown to be varieties of a type of force called, rather mysteriously, 'gauge' forces. I must first explain what the weak and strong forces are, and then explain what a gauge theory is.

The best known and understood forces of nature are the gravitational and electric and magnetic ones. The laws of the gravitational force were determined, to a very good approximation, by Newton. A better formulation was contained in Einstein's general theory of relativity (see Chapter 2 by Will). The laws of the electric and magnetic forces (which are inseparably inter-linked) were found by Maxwell. Gravitational forces are most obvious in astronomy, and electromagnetic forces determine most of the properties of atoms, molecules, stable matter and, presumably, living creatures.

But the existence of other forces has been known since the beginnings of nuclear physics. First, there is the force which binds protons and neutrons together in the nuclei of atoms. It is not electric, since all protons have positive charges, which *repel* one another. This force is stronger than electromagnetism, and it acts only over short distances of roughly 10^{-15} m. This is the strong force.

It is now known that protons and neutrons and other particles, like pions, are all made of quarks (see Chapter 14 by Close), so the strong forces must be reducible to the forces between quarks. This is a little like the way in which the forces between atoms and molecules are reducible to the basic forces between electrons and nuclei.

Then there is the force responsible, for example, for radioactivity, and for the transmutation of hydrogen into helium which provides the energy for stars such as the sun. This is the weak force.

A simpler example of the operation of the weak force – though one much more recently discovered – is the deflection of the neutrino in the collision

$$\text{neutrino} + \text{proton} \rightarrow \text{neutrino} + \text{proton}. \quad (17.1)$$

Neutrinos are electrically neutral, and they have no strong interactions either. So the deflection observed in this process must be due to weak interactions (gravitational effects are minute). The probability of deflection is small, and hence the adjective 'weak'. Actually the process is strongly dependent on the neutrino's energy, and the deflection would increase at very high energies; so the term 'weak' is not really a good one.

Another example of the effect of the weak force is the process

$$\text{electron} + \text{proton} \rightarrow \text{neutrino} + \text{neutron}. \quad (17.2)$$

In this process, in addition to a deflection, the nature (including the electric charge) of the particles changes. In fact, charge passes from the hadron (the heavy particle) to the lepton (the light particle). But the effect is also weak. The two processes in equations (17.1) and (17.2) have very similar properties.

Now take the decay process

$$\text{neutron} \rightarrow \text{proton} + \text{electron} + \text{antineutrino}. \quad (17.3)$$

This is obtained from equation (17.2) by shuffling the particles about from one side of the arrow to another (with a judicious change of neutrino into antineutrino). This isn't a deflection process but represents the decay of the neutron. (This can take place because the mass of the neutron is a little more than the combined masses of the proton and electron. I assume the neutrino to have no mass. Whether this is exactly true is a point of current debate, but the mass is certainly very small.)

General principles of relativity and quantum theory allow the rate of the decay in equation (17.3) to be related to the deflection probability in equation (17.2). The mean lifetime of the neutron, before it decays by equation (17.3), is about 15 minutes. This is a very long time by nuclear physics standards, and this is partly due to the weak nature of the force responsible.

The decay in equation (17.3) is the prototype of many modes of radioactive decay of nuclei and elementary particles. The manifestations of the weak force are many and various.

It may seem odd to speak of a 'force' being responsible for transmutations such as equation (17.3); but perhaps it is not so strange if the links between equations (17.1), (17.2) and (17.3) are appreciated.

The gravitational and electromagnetic forces are long-range. They manifest themselves in macroscopic situations, like the solar system and radio waves. This allowed the nature of the forces to be determined by large-scale experiments, without the complications of quantum theory. Both the strong and weak forces are short-range ones, acting over nuclear distances of about 10^{-15} m or less. On these scales, quantum theory can never be neglected. Hence the eludication of the character of these forces is more difficult.

The discovery that has been made in the last twenty years is that, in spite of the obvious differences, the same principle controls strong and weak forces as controls the electromagnetic ones. This principle is called 'gauge invariance'. My main aims in this chapter are to explain what a gauge theory is, and to explain how it is that forces of the same basic character can manifest themselves so differently as the electromagnetic, weak and strong forces do. Most importantly, I must explain how the weak and strong forces can be short-range. (The gravitational force may also, with some generalisation of the term, be said to be a gauge force, but this would take us beyond the scope of this chapter.)

Before attempting to define the term 'gauge theory', we must explain a number of preliminaries. The first of these is the background of special relativity and quantum theory, against which all elementary particle physics is done.

17.2 Forces and the background theory

I want to make a distinction between the forces and the 'background theory'. Two examples should show what I mean.

In Newtonian astronomy, the background theory consisted of such propositions as:

$$\text{mass} \times \text{acceleration} = \text{force}.$$

This equation is true whatever the law of force. But then, in addition, in Newtonian gravity theory the force is prescribed as being inversely proportional to the square of the distance, and so on.

In atomic physics, the background theory is (to a good approximation) nonrelativistic quantum mechanics. Then, in addition, the force is given to be the electrostatic force between electric charges.

In elementary particle physics, although physicists have been groping to find the forces, the background theory has been clear for nearly sixty years. This background theory must be based upon quantum theory *and* special relativity. Any theory which contains both these elements must necessarily be rather abstract. Before explaining why this is so, I remind the reader of the basic ideas of special relativity and quantum theory.

(There was a period, in the 1960s, when some physicists thought that, in elementary particle physics, the constraints of quantum theory plus relativity were so powerful that a law of force might not be required at all. This view did not prevail, but there was probably a grain of truth in it.)

17.3 Special relativity

In relativity, space (with co-ordinates x, y, z) and time (t) are inextricably joined together, to make 'space-time'. Two observers, moving relative to one another at a constant velocity, will assign co-ordinates, x, y, z and t, to space-time in different ways. But they will agree about all the laws of physics, and in particular they will agree about the speed of light, c. This speed, c, is the maximum possible speed at which a particle can move.

Because of the fundamental importance of c, it is sometimes useful to choose a system of units in which $c = 1$. For example, if time is measured in years and distance in light-years, then $c = 1$ light-year/year, by definition. Equally, one might measure distance in metres and time in 'light-metres', where a 'light-metre' is the time it takes light to travel 1 m, i.e. about $\frac{1}{3} \times 10^{-8}$ s. We shall put $c = 1$ in the remainder of this section.

Consider a light-flash emitted at the origin at the zero of time. At a later time t, the light will have reached a sphere in space given by

$$x^2 + y^2 + z^2 = t^2. \tag{17.4}$$

In space-time, this is the equation of a hyper-cone (i.e. the analogue of a cone, but in one dimension more). We can visualise this by leaving out the z co-ordinate and drawing a perspective picture in the (x, y, t) part of space-time. The cone is then as sketched in figure 17.1.

This cone represents the entire history of the light-flash. The apex, O, represents the instant and place of emission, called the 'event' of emission. At a later time, the light-pulse has reached a sphere; but, in the diagram with one dimension missing, this sphere is replaced by the black circle as shown in figure 17.1.

Different observers (moving relative to one another at constant velocity) agree about the light-cone. It is an 'invariant' property of space-time. Actually, different observers do not only agree about the cone described in equation (17.4) they agree about the value of the quantity s^2 defined by

$$s^2 = t^2 - x^2 - y^2 - z^2. \tag{17.5}$$

The quantity s is called the 'interval'. The light-cone is the special case $s = 0$.

When special importance is attached to the interval s, space-time is called Minkowski space-time. The interval plays the same sort of role in Minkowski space-time that ordinary distance plays in ordinary three-dimensional space. But there is a crucial difference. Ordinary distances r are given, by Pythagoras's theorem, as $r^2 = x^2 + y^2 + z^2$. In equation (17.5), on the other hand, there are some terms with positive signs and some with negative signs. So, be warned, the distances we see in the diagrams we draw on paper do not represent the intervals. For example, in figure 17.2, the hyperbolic 'bowl' represents points in Minkowski space-time for which s^2 has a constant (positive) value. So OC and OD have the same value of the interval, although they are lines of different lengths in the page. In the same way, in figure 17.1, OA and OB each have *zero* interval (since they lie on the light-cone, $s=0$).

The entire history of a moving particle is represented by a curve in Minkowski space-time. An example is shown in figure 17.3. (This is for a particle moving with constant speed in a circle.) Such a curve is called the 'world-line' of the particle ('world-curve' would have been a better name).

The physical law that the speed of the particle must be everywhere less than 1 (the speed of light in our units) is expressed by saying that the world-line must lie within the light-cone at any point on it. An example of the light-cone is shown at point E in figure 17.3.

In ordinary space, with ordinary distances, one might define a right-angled triangle to be one for which Pythagoras's theorem applies to the lengths of its sides. In Minkowski space-time, one can define a right-angled triangle to be one for which Pythagoras's theorem applies to the *intervals* corresponding to its sides. This gives us a notion of 'perpendicular' in Minkowski space-time. But, just as the lengths we see on the paper in the space-time diagrams don't give us a good idea of the intervals, so the angles we see in the diagrams don't give us a good idea of perpendicularity. For example, in figure 17.4, OE and OF are perpendicular to Oy, as they look, but OE and OF are perpendicular (in the Minkowski sense) to each other (provided they are drawn equally inclined to the light-cone in the diagram). In fact, a line *on* the light-cone, like OA in figure 17.1, is perpendicular (in the Minkowski sense) to itself!

I said that two observers moving relative to one another at a constant speed attach different co-ordinates to space-time. This is illustrated a little more in figure 17.5, where we leave out y *and* z to avoid complications. Here, Ox and Ot are one observer's co-ordinate axes and Ox' and Ot' are the other's. Note that Ox' and Ot' are perpendicular (in the Minkowski sense), just as Ox and Ot are. (If the second observer were drawing this diagram, he would draw Ox' and Ot' *looking* perpendicular and Ox and Ot *not looking* perpendicular. The relationship between Ox, Ot and Ox', Ot' is in reality a symmetrical one.) The speed, u, of one observer relative to the other is related to the angle θ shown in figure 17.5. (Actually, for those who are familiar with hyperbolic trigonometric functions, the relation is $u=\tanh\theta$.)

In ordinary three-dimensional space, a 'vector' is a quantity with direction as well as magnitude. For example, velocities and electric field strengths are vectors. Given a cartesian co-ordinate system, a vector is represented by three 'components', e.g. (x, y, z) for a position or (p_x, p_y, p_z) for a momentum.

This notion generalises to Minkowski space-time, and the corresponding type of quantity is called a '4-vector'. It has four components. For example, the co-ordinates $(x, y, z; t)$ of an

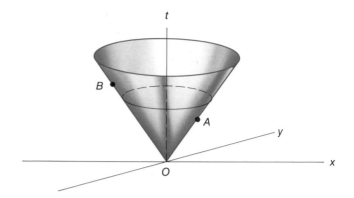

Figure 17.1. Space-time and the light-cone. Space and time are plotted together (the z-axis of space has perforce to be omitted). The light-cone shows the successive positions of a light flash emitted at O (i.e. at time $t=0$ from the position $x=y=z=0$). For instance, the black circle represents the sphere the light-flash has reached after a certain time. OA and OB are examples of lines lying in the light-cone. They have zero value for the interval in Minkowski space-time.

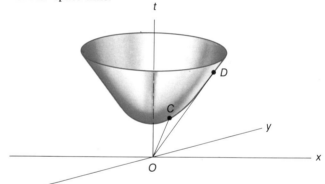

Figure 17.2. Space-time, showing a surface of events whose interval from O is constant. The surface has a hyperbolic shape. OC and OD have the same interval (although their lengths as lines in the diagram are quite different).

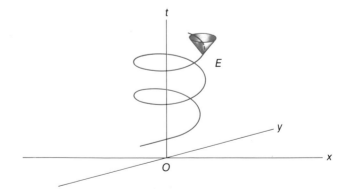

Figure 17.3. The world-line of a particle moving in a circle at constant speed. To find the position of the particle at a time t, slice the diagram with a plane at height t above the x,y-axes. The light-cone at an event E on the world-line is shown. The world-line must point inside the light-cone at that event.

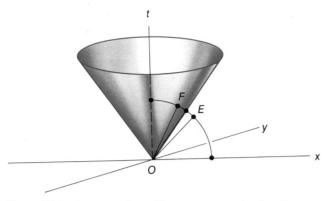

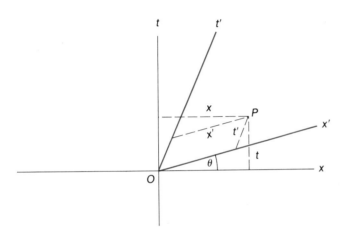

Figure 17.4. An example to illustrate perpendicular directions in the Minkowski sense. The lines *OE* and *OF* are each perpendicular to the *y*-axis, as they look. But *OE* and *OF* are actually perpendicular, in the Minkowski sense, to each other. This is because, in the diagram, they are equally inclined to the light-cone but *OF* is inside and *OE* is outside it.

Figure 17.5. How different observers divide up space-time in different ways. One observer uses *x* and *t* in this space-time diagram, and another uses *x'* and *t'*. The axes *Ox'* and *Ot'* (like *Ox* and *Ot*) are perpendicular to one another in the Minkowski sense. The event *P* is called (*x,t*) by one observer and (*x',t'*) by the other. (The *y*- and *z*-axes are omitted.)

event constitute a 4-vector. (The semi-colon in (*x*, *y*, *z*; *t*) is a reminder that time is not the *same* as space: the minus signs in $s^2 = t^2 - x^2 - y^2 - z^2$ show us this.)

It turns out that, for a moving particle, the momentum (p_x, p_y, p_z) and the 'total' energy *E* make up together another 4-vector (p_x, p_y, p_z; *E*). ('Total energy' will be defined in a moment.) Analogously to the interval $s^2 = t^2 - x^2 - y^2 - z^2$, the combination

$$m^2 = E^2 - p_x^2 - p_y^2 - p_z^2 \qquad (17.6)$$

is an invariant for all observers, and should have some special physical significance. It is in fact the 'rest-mass' of the particle. Physicists usually call this just the 'mass' – it is a fixed property of the particle, and doesn't vary with the speed or anything like that.

If the particle is at rest (relative to the observer whose co-ordinate system we are using), the momentum must be zero, so equation (17.6) gives (assuming *E* and *m* are positive)

$$E = m.$$

If we had used units in which *c* was not equal to 1, this equation would have been

$$E = mc^2,$$

Einstein's famous equation.

When the momentum is small, there is an approximate solution to equation (17.6)

$$E \simeq mc^2 + (p_x^2 + p_y^2 + p_z^2)/(2m).$$

(One can check this by putting it back into equation 17.6.) This contains the Einstein 'rest-energy' mc^2, and the remainder of the right side is the ordinary nonrelativistic kinetic energy. This is the sense in which *E* is the 'total' energy. If you leave out the rest-energy, you no longer have a component of a 4-vector.

Like (*x*, *y*, *z*; *t*), (p_x, p_y, p_z; *E*) can be represented in a Minkowski diagram. For example, in figure 17.6, *OP* and *OQ* represent two energy–momentum 4-vectors. The hyperbolic 'bowl' surface shown represents equation (17.6); so the two 4-vectors have the same mass *m*. In the case of *OP*, the momentum is zero, $p_x = p_y = p_z = 0$, so it corresponds to a particle at rest. *OQ* represents the energy–momentum for a particle of the same mass, but in motion.

We know that light sometimes exhibits particle-like properties, and the 'particles' are called photons. They carry momentum and energy, so there is a corresponding 4-vector. What is the appropriate value of *m*? The answer is, $m = 0$. The reason is as follows.

Consider an ordinary matter particle (like an electron) moving with a constant speed *u* (relative to some observer *Q*). It is always possible to describe the motion of the particle from the point of view of a new observer *Q'*, with respect to whom the particle is at rest. (For example, if the particle's world-line was *Ot'* in figure 17.5, the new observer *Q'* would use *Ot'*, *Ox'* as co-ordinate axes.)

If (p'_x, p'_y, p'_z; *E'*) is the energy–momentum 4-vector as described by the new observer *Q'*, then, since the particle is at rest, $p'_x = p'_y = p'_z = 0$. So, by equation (17.6) applied in the *Q'* frame of reference (this equation is true for *any* observer), $E' = m$.

Now suppose the particle was a photon. By the fundamental laws of relativity, it moves with speed *c* relative to *any* observer. So, in this case, it is *not* possible to find an observer for whom the photon is at rest. What, then, prevents us choosing a co-ordinate system in which $p'_x = p'_y = p'_z = 0$, as we did for the ordinary particle? The answer is that this is impossible if, and only if, $m = 0$. For then the energy–momentum 4-vector lies on the light-cone, as in figure 17.7. The light-cone is the same for any observer, and no change of observer can bring OR into a rest position like OP in figure 17.6. All that a change of observer can do is bring OR to another position on the light-cone, like OR' in figure 17.7.

Figure 17.6. An energy–momentum diagram. The energy E of a particle is plotted vertically, and its momentum (p_x, p_y) is plotted in the horizontal plane (p_z is omitted). The surface shown is for a particle of fixed mass. OP represents the energy and momentum when the particle is at rest (so that $p_x = p_y = p_z = 0$ and $E = m$). OQ corresponds to the same particle in motion.

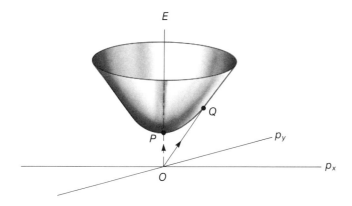

Figure 17.7. An energy–momentum diagram with the cone for massless particles. OR and OR' lie on the cone and are examples of energy–momentum 4-vectors for massless particles.

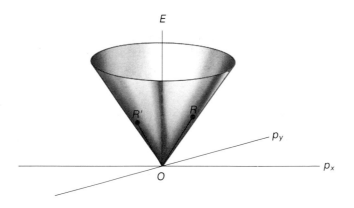

In conclusion, I list the important points about special relativity which we shall require in this chapter:

(*a*) Space and time have to be treated together as space-time.

(*b*) Different observers choose space-time co-ordinate axes in space-time in different ways.

(*c*) All observers agree on the light-cone and the speed of light, which is the maximum possible speed of a particle.

(*d*) Energy and momentum make up a 4-vector, which can also be represented in diagrams like the space-time diagrams.

(*e*) Photons have zero rest-mass, $m = 0$, and their energy–momentum 4-vectors lie on the light-cone.

(*f*) There is a (strange looking) notion of perpendicularity in space-time. (This last point will only be used once, so the reader who finds it perplexing need not worry too much.)

17.4   Quantum theory

In quantum theory, quantities like position co-ordinates (x, y, z) and momentum components (p_x, p_y, p_z) cannot in general have definite numerical values assigned to them. In fact, Heisenberg's uncertainty principle states that

$$(\text{uncertainty in } x) \times (\text{uncertainty in } p_x) \gtrsim \hbar, \quad (17.7)$$

where $\hbar$ is Planck's constant. (More precisely, $\hbar = h/2\pi$, where h is Planck's original constant, and the 2π has been adopted to reduce the number of explicit 2π factors in equations.) The symbol $\gtrsim$ is a little vague here. It means 'greater than or of the order of'. To make it more precise, we would have to define 'uncertainty in', but we need not bother with this.

There is a similar uncertainty relation

$$(\text{uncertainty in } t) \times (\text{uncertainty in } E) \gtrsim \hbar, \quad (17.8)$$

where t is the time and E the energy. There are actually some caveats about this equation, but it will do for our purposes.

The constant $\hbar$ has the value

$$\hbar = 1.05 \times 10^{-34} \, \text{kg m s}^{-1}.$$

Just as c is the fundamental constant in relativity, $\hbar$ is the fundamental one in quantum theory. The ratio

$$\hbar/c = 3.5 \times 10^{-43} \, \text{kg m}.$$

Just as c allowed us to measure time in 'light-metres', so $\hbar/c$ allows us to measure mass in units which one might call

$$(\text{quantum light-metres})^{-1} = 3.5 \times 10^{-43} \, \text{kg},$$

that is, units in terms of which $\hbar = c = 1$. So only one man-made unit remains, which we may choose to be the metre (or, alternatively, the kilogramme or the second).

Instead of *certain* values of *x* etc., quantum theory gives us probabilities. The probabilities are found from a 'wave-function', often called by the Greek letter ψ (psi).

The wave-function ψ is in general a complex number. This is quite a remarkable fact about quantum theory. All of classical (i.e. nonquantum) physics is formulated in terms of real numbers (complex numbers are occasionally used as a convenient device, but they are not essential).

Let me remind the reader that a complex number may be represented by a directed line segment in a plane. Thus it can be specified by two things: its length and the angle it makes with some reference line. The latter is called the 'phase' of the complex number. In figure 17.8, *OB* represents a general complex number, with phase θ; *OA* is the special case of a real and positive one; *OD* is real and negative (it has phase 180°); *OC* is pure imaginary (phase 90°).

In quantum theory, probability (of, for example, observing a certain value of *x*) is given by the square of the length of ψ. So the phase of ψ has no direct physical significance. However, if ψ_1 and ψ_2 are two possible wave-functions for a system (e.g. a hydrogen atom in two different states),

$$\psi_3 = \psi_1 + \psi_2$$

is also a possible wave-function. Addition of complex numbers is defined by a triangle rule, as illustrated in figure 17.9. Thus the length of ψ_3 depends not only upon the lengths of ψ_1 and ψ_2 but also upon the *difference* of the phase, $\theta_2 - \theta_1$. The probability associated with ψ_3 may be greater or less than the sum of the probabilities associated with ψ_1 and ψ_2, according as $\theta_2 - \theta_1$ is greater or less than 90°. Thus, although a single phase is not directly measurable, the phases play an essential role, and cannot be got rid of from quantum theory.

An example where the quantum mechanical phase-angle plays a crucial role is an interference experiment, such as is shown schematically in figure 17.10. A beam of particles is split into two parts which go through, say, two slits *D* and *D'*. The number of particles arriving at a particular point B_1 is measured. We may think of ψ_1 and ψ_2 above as representing the two beams. The intensity measured at *B* is given by the length of ψ_3, and this depends upon the value of $\theta_2 - \theta_1$ and varies for different positions of B_1. For example, if $\theta_2 - \theta_1 = 180°$, *no* particles are measured at B_1.

How can the phases θ_1 and θ_2 be calculated? Let us suppose that the lengths of ψ_1 and of ψ_2 are each constant along the beams, so that all that can vary along them are the phases. It turns out that, for electrically neutral particles, quantum mechanics gives:

$$\text{(rate of change of } \theta \text{ along path)} = \frac{\hbar}{m} \times \text{(velocity of particle).}$$
$$(17.9)$$

(Note that it is consistent with the uncertainty principle, equation (17.7), for the particle to have a definite velocity, since its position along the beam is uncertain.) Starting from the

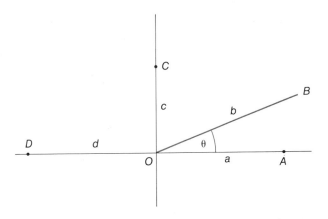

Figure 17.8. The Argand diagram for complex numbers. *a* is real and positive, *d* is real and negative, *c* is pure imaginary and *b* has phase θ.

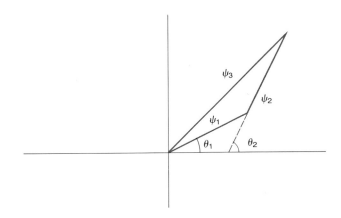

Figure 17.9. The addition of complex numbers: $\psi_1 + \psi_2 = \psi_3$. θ_1 and θ_2 are the phases of ψ_1 and ψ_2.

common value of the phase-angle at point *A*, relation (17.9) can be used to predict the phase-difference $\theta_1 - \theta_2$ at B_1, and hence the number of particles detected at *B*. For example, if the speed is constant, $\theta_1 - \theta_2$ is just proportional to the difference of the two distances ADB_1 and $AD'B_1$.

In general, relations like equation (17.9) give a meaning to the value of the phase-angle at any point of space, provided it is chosen arbitrarily at any one point (like the point *A* in figure 17.10). We shall see in Section 17.7 that there is an important difference when we come to consider electrically charged particles, and equation (17.9) no longer applies.

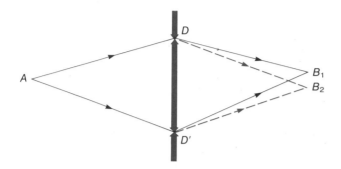

Figure 17.10. An interference experiment. Particles going from A to B_1 or B_2 can go through either slit D or slit D'. In quantum theory, the number of particles detected at, say, B_1 depends upon the relative phase of the two beams. This in turn depends upon the difference of the distances along the two paths ADB_1 and $AD'B_1$. Thus the intensity at B_1 might be a maximum, but at a neighbouring point B_2 it might be a minimum, because of the slight alteration in the path lengths.

The final thing about quantum theory that we must emphasise is the duality between particle-like properties on the one hand and wave-like or field-like properties on the other. This duality is a consequence of uncertainty relations. Which set of properties gives the more appropriate picture in given circumstances depends upon which uncertainties are small on the left-hand side of relations like equations (17.7) or (17.8).

The wave-function ψ tells us about the probability of finding a *particle* at a certain position. So ψ is something which varies from point to point. In this respect, it reminds us of a *field*, like the electric or magnetic field.

17.5   Relativity plus quantum theory

Elementary particle physics is a domain in which *neither* special relativity *nor* quantum theory may be neglected. Important consequences flow from this.

In order to point the contrast, consider first atomic physics, where relativity has only a small effect (because electrons in atoms move at speeds much less than the speed of light). Here, we may decide to study, say, a hydrogen atom. This contains just one electron and one proton, and so has a definite, fixed number of 'degrees of freedom' – six in this case, made up of three position co-ordinates for the electron and three for the proton.

Now take a relativistic case. Suppose we thought we had a two-particle system, say an electron and a positron approaching one another at speeds comparable with the speed of light. Let us enquire about the state of this system at some fairly

definite time while the collision is taking place. According to the uncertainty principle, equation (17.8), at a rather definite time the energy is very uncertain. Therefore, we cannot be sure that there is not considerable extra energy available.

But special relativity tells us, by Einstein's equation $E = mc^2$, that energy may appear as mass. So the system, at this definite time, may contain *extra* particles, e.g. an extra electron and positron, or two electrons, two positrons and three photons, or anything. Extra particles, which are present for very short times like this, are called 'virtual' particles. What do we mean by 'short times' here? We can guess the rough answer to this question by forming a quantity with dimensions of time out of $\hbar$, c and a mass m. Using $E = mc^2$ and the uncertainty relation in equation (17.8), we see that the required combination is

$$\hbar/mc^2. \qquad (17.10)$$

For an electron mass, this gives 1.3×10^{-21} s.

In relativistic quantum mechanics, because of the presence of virtual particles, we *cannot* beforehand limit the number of particles we are talking about. We cannot limit the number of degrees of freedom. We must always be ready to take into account more and more virtual particles. We need a formalism powerful enough to handle this.

We can see how to do this if we think about photons. We know that photons are a quantum aspect of light, i.e. of electromagnetic waves. In other words, if we take Maxwell's equations of electromagnetism and treat them by quantum theory, we find ourselves talking about photons. 'Treating by quantum theory' means, among other things, subjecting the *fields* themselves to uncertainty relations. In so far as the values of the fields are well-defined, the number of photons is uncertain. In so far as the number of photons is well-defined, the values of the fields are uncertain.

In a like manner, for any type of particle, we can find a field which, when treated by quantum theory, gives those particles, and allows us to talk about as many of them as we like. This is what is needed in order to do relativistic quantum theory.

In classical (i.e. nonquantum) physics, fields are associated with forces. For example, the electric field is associated with the electrostatic, inverse-square-law force between charges. Now, for any type of particle, we have a field, and we can ask how the mass of that particle determines the character of the force associated with the field. The answer is that the *mass* of the *particle* determines the *range* of the *force*. We can argue this as follows.

In order for the particle to have an effect, it has to be present as a virtual particle for a short time, where the short time in question was estimated as $\hbar/mc^2$ in equation (17.10). In this time, the particle can move a distance of, at most,

$$\frac{\hbar}{mc^2} \times c = \frac{\hbar}{mc}, \qquad (17.11)$$

and this is our estimate for the range of the force. For example, if m is the mass of an electron, this formula gives a range of about

10^{-13} m. Outside this distance, it turns out that the force falls off exponentially.

Because the photon has zero mass, equation (17.11) gives an infinite range for the force connected with the photon, e.g. the electrostatic force. The meaning of this is that the force just falls off according to the inverse-square of the distance (which, as we shall see in Section 17.7, is for a geometrical reason) and not exponentially. So we may call electromagnetic forces 'long-range' or 'infinite-range'.

To summarise, relativistic quantum mechanics gives us the following scheme:

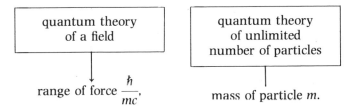

According to this scheme, there should be a field for the electron, and so a force caused by virtual electrons. In fact, we may think of this as giving a force between an electron and a photon (the electrons are appearing in two roles here: as experiencing a force and as causing a force). It is an odd sort of force which exchanges the electron and photon when they collide, but we have already seen examples of forces which change the character of the particles in equation (17.2).

There is one other point about relativistic quantum mechanics which I must stress. This is to do with the physicist's mental picture of the vacuum, i.e. of empty space. Because virtual particles can always, momentarily, be present, physicists tend to think of the vacuum like some sort of rich and complex 'ether' shimmering with activity. Such a picture must be treated with caution. The virtual particles only become apparent if the vacuum is probed in some way. And there must be nothing in the vacuum to give a preferred direction or a preferred standard of rest.

Nevertheless, the picture of the vacuum as complicated has led people to compare it with other complicated media, like metals and dielectrics for example. These analogies (and they are no more than that) have inspired some fruitful thinking in elementary particle physics. I will use some such analogies in later parts of this chapter.

17.6 Spin in quantum mechanics

Elementary particles have a property, called spin, which is subtly different for particles with zero rest-mass and for those with nonzero rest-mass. This difference will be crucial to some of our later discussions.

Spin is the word used for the angular momentum of a particle as it turns about its own centre. But the easiest sort of angular momentum to understand is that of a particle orbiting about

some point outside it. For example, if a particle of mass m moves with speed u in a circle of radius r, it has angular momentum mru.

Angular momentum is important because it is often constant in time. For example, if there is no torque tending to change the angular momentum, a decrease in r must be compensated by an increase in u – a fact made use of by dancers, who can increase their turning speed by bringing their arms in closer to their bodies (see figure 16.1 from the previous chapter).

Angular momentum is a directional quantity, i.e. it is represented by a vector. In the above simple example, the direction of the angular momentum is perpendicular to the plane of the circle in which the particle is moving. If a body is spinning about an axis of symmetry, it has angular momentum directed along that axis; this is illustrated by the familiar motion of a spinning top (see figure 17.11). (For an unsymmetrical body spinning about a general axis, the direction of the angular momentum is less obvious.) Since angular momentum is a vector, it has three 'components' referred to a given set of axes.

The spin of a particle is defined to be the angular momentum of that particle when its centre is at rest. It is important to take the centre to be at rest, so as not to muddle the spin with angular momentum due to the motion of the centre.

In quantum mechanics, angular momentum has some special properties. There is an uncertainty principle which says that

$$\text{(uncertainty in angular momentum)} \times \text{(uncertainty in angle)} \gtrsim \hbar. \quad (17.12)$$

The angle referred to here is an angle which determines how far the body has rotated from some standard position.

Figure 17.11. An example of angular momentum. Angular momentum is a quantity with a direction associated with it. If the top is spinning steadily about its axis, the direction is that shown by the arrow.

Axis of symmetry

There is an obvious analogy between the ordinary uncertainty relation in equation (17.7) and equation (17.12). But
there is an important difference: whereas x can take any value,
an angle is limited to a range of 360° (or 2π radians). It turns
out that this implies that each component of angular momentum can only change by multiples of $\hbar$.

There is a further uncertainty relation which has in its left-
hand side the product of the uncertainties of two different
components of the angular momentum (say the x- and y-
components). Because of this and equation (17.12), it is usual
to consider states of a spinning body in which *one* component,
say the z-component, of the angular momentum has a definite
value. Then neither the angle nor the other components have
definite values. Also, the z-component of spin can only change
by integral multiples of $\hbar$. (Of course, for macroscopic objects, $\hbar$
is very small by comparison, and these uncertainties have no
practical effect.)

Now consider a particle, like an electron, quark, proton, W
particle, etc. By experiment, it is known that these particles
have spin. We may surreptitiously imagine them as tiny
spinning spheres, but quantum mechanics strictly forbids any
such detailed picture. It is intuitively clear that, if a particle has
a nonzero value of spin, we may 'turn the particle', so that its
angular momentum vector points in a different direction. So we
expect a particle to be able to exist in states with different
directions of angular momentum. These states will, in general,
have different values of a component, say that z-component, of
the spin vector. Let us call the z-component of the spin vector s_z.
From the rule about changes in angular momentum being in
multiples of $\hbar$, possible values of s_z must differ from each other
by multiples of $\hbar$. Also, intuitively, we expect that, if s_z is one
possible value, then so is $-s_z$ another possible value, corresponding to 'turning the particle right round'.

It is observed that the number of different values of s_z for
elementary particles is quite small. For electrons, there are just
two values:

$$-\tfrac{1}{2}\hbar, \ +\tfrac{1}{2}\hbar \tag{17.13}$$

(differing by $\hbar$, of course). The next simplest possibility is

$$-\hbar, \ 0, \ +\hbar. \tag{17.14}$$

It will turn out that this is what is predicted (and observed) for
the $W^\pm$ and Z^0 particles which cause the weak force.

So far, we have defined spin for a particle (whose centre is) at
rest. But, in special relativity theory, it must be possible to re-
express the definition for a particle moving at constant velocity
relative to the observer. To see how to do this, we must think in a
four-dimensional space-time way. In figure 17.12, the line L
represents the world-line of a particle at rest. We have to
concentrate on one component of the spin, conventionally the z-
component. This is the component in the direction of the arrow
in the figure, but, clearly, the x- or y-component would have done
as well, or, more generally, the component in any direction
perpendicular to the world-line L.

The italicised phrase has told us how to deal with a particle
which is moving (in our frame of reference). The world-line of
such a particle is shown as L' in figure 17.13. We must use
components of the spin along directions perpendicular to L'
now. Two such directions are the y-axis and the z-axis (omitted
in the figure). A third direction is indicated by the arrow n in the
figure. This is perpendicular to L' in the Minkowski sense of
perpendicular, as explained in figure 17.5.

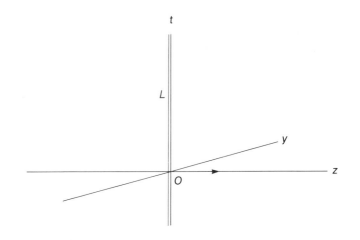

Figure 17.12. A space-time diagram, showing the world-line L for
a particle at rest at $x=y=z=0$. The arrow shows a possible
direction for the spin of the particle.

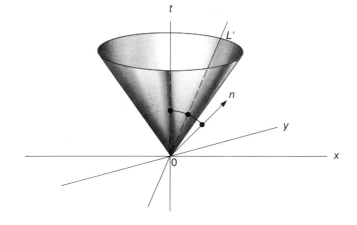

Figure 17.13. Possible directions for the spin of particles moving in
the x-direction. The world-line of the particle is L' (inside the light-
cone). The spin has to be perpendicular, in the Minkowski sense,
to L'. Two obvious directions are the y- and z-axes. But a third
direction is On. It lies in the same plane as Ot, Ox and L.

For a moving particle, and for a given observer, there is always a direction *n* defined like this. The projection of the spin in the *n*-direction is called the 'helicity' of the particle. To understand this name, note that the helicity is, roughly speaking, the spin in the direction of motion. An example of an object which spins about its direction of motion is a corkscrew, which has a helical form (see figure 17.14). For helicity, just as for s_z, possible values are those in (17.13) and (17.14).

Helicity is sometimes a convenient quantity to use, because it doesn't depend upon the choice of an arbitrary spatial direction as s_z depends upon the *z*-axis. However, helicity does depend upon the observer defining it. One way to see this is to go to the special case of an observer with respect to whom the particle is at rest (as in figure 17.12). Then, since *Ot* coincides with the world-line of the particle, there is no unique choice of *n* anymore.

So far we have dealt with a particle of nonzero mass, because we started our discussion of spin with the particle at rest. In the case of a zero mass particle, like the photon, there is no observer for whom the particle is at rest. So we have to reconsider what we mean by spin. What we can do is to try to take the limit of the definition of helicity as the mass becomes smaller and smaller. In this limit, the world line *L'* in figure 17.13 moves towards the light-cone, and so *n* approaches the light-cone from the other side (*L'* and *n* are equally inclined to the light-cone, in order to be perpendicular in the Minkowski sense). In the limit, the directions of *L'* and *n* coincide.

Thus, for a zero mass particle, the helicity is the spin in the direction of the world-line of the particle itself. But now there is a vital difference. Helicity in this case is independent of the choice of observer. It is *intrinsic* to the particle itself. No change of axes or change of observer changes the helicity of a massless particle.

This means that, for massless particles, there is no *a priori* reason for helicities to come in groups like (17.13) or (17.14). In fact, neutrinos (which are probably massless) all have helicity $-\frac{1}{2}\hbar$ (so far as is known). It is the *antineutrinos* which have helicity $+\frac{1}{2}\hbar$. (This statement would be empty if the helicity was the *only* thing distinguishing antineutrino from neutrino. But they have other distinguishing properties: neutrinos can convert into electrons under the action of the weak force, but antineutrinos convert into positrons. There was an example in equation 17.2.)

What about photons? There are photons with helicity $\hbar$. The corresponding light-waves are circularly polarised in a right-handed sense. (Again the appropriateness of the word 'helicity' is clear if one considers the helical, screw-like motion of the tip of the electric field vector in a circularly polarised wave. This is illustrated in figure 17.15.) But Maxwell's equations have the property that, for any solution, there is another obtained by reflection in a mirror. The reflection of a right circularly polarised wave is a left circularly polarised one. So helicity $-\hbar$ must exist also. Thus photons have helicity

$$-\hbar, \ +\hbar. \tag{17.15}$$

The value 0 in (17.14) does *not* appear in this case.

Thus the set of helicity values in (17.14) for a particle with mass are qualitatively different from the set in (17.15) for a massless particle.

We may now put together the conclusion of the preceding section with the conclusion of this to say that:

  (i) *infinite*-range forces correspond to *massless* particles with *two* helicity states; but
  (ii) *finite*-range forces correspond to particles with *mass* which have *three* helicity states.

One of our aims in this chapter is to show how, in spite of this contrast, infinite- and finite-range forces can be unified into a single type of theory.

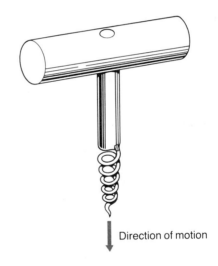

Figure 17.14. An example of an object which rotates about its direction of motion (which direction is indicated by the arrow).

Direction of motion

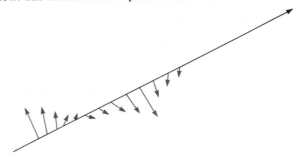

Figure 17.15. The electric vector (grey) in a circularly polarised light-ray. The direction of the light is represented by the black arrow. The electric vector spirals around the direction of the ray.

17.7   Electromagnetism

We are at last ready to explain what are the gauge theories which are claimed to do the wonderful things mentioned above. They are generalisations of electromagnetism, so we must first formulate electromagnetism in a way which will open the way to a generalisation.

In electromagnetism, we have, at each point of space, values of the electric and magnetic fields, E and B, each of which are vectors, i.e. quantities with magnitude and direction (in ordinary three-dimensional space). There are equations, Maxwell's equations, which relate E and B to the electric charges which are present and to the electric currents which are flowing. Also, the rate of change of E with time influences B, and vice versa. But the details of these equations need not concern us.

Some of the content of Maxwell's equations can be visualised in terms of 'lines of force'. Consider E for definiteness. The direction of the electric lines of force at any point is the direction of E. The magnitude of E is proportional to the number of lines of force crossing unit area (when the area is perpendicular to the lines). With these definitions, one of Maxwell's equations says that electric lines of force do not end except on electric charges, and the number emerging from a charge is proportional to the value of that charge.

For a single point positive charge at rest, the lines of force look like figure 17.16. The direction of E is everywhere radially outwards. An area S is shown, and a number of lines of force cross it. If we take a complete sphere centred at the origin, the same number of lines of force goes through it, independently of its area. Therefore, the strength of E at some point is inversely proportional to the area of the sphere through that point, that is inversely proportional to the square of the radius to that point.

It is for this reason that electric forces are said to have 'infinite range' (or sometimes 'long range'). The force decreases with distance of course, but only in the obvious geometrical way. There is no special length in the theory, beyond which forces fall off exponentially, as happens with a 'short-range' force.

Magnetic lines of force can be defined similarly. No isolated magnetic pole has ever been found, so the magnetic lines of force, unlike the electric ones, do not end *anywhere*. For a magnet, for example, the lines of force, which go into the magnet at one end, run along inside it and emerge from the other end. We will, for the moment, concentrate on magnetic, rather than electric, lines of force.

There is another, less obvious, way of characterising the magnetic field. Take any closed curve C in space and count the number of lines of force which thread through the area enclosed by it. This quantity is called the 'flux' through C, and we shall denote it as $F(C)$. The magnetic field is completely determined if we know $F(C)$ for *all* possible curves C.

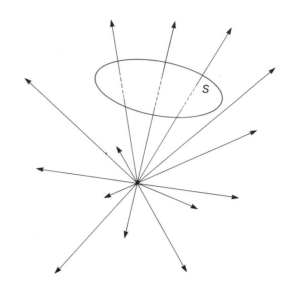

Figure 17.16. Lines of force emerging from an electric charge, some of which go through a surface S. The lines of force diverge outwards equally in all directions.

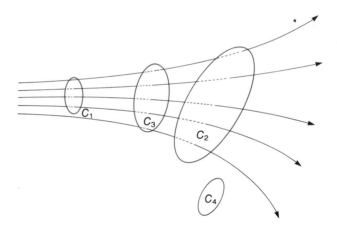

Figure 17.17. A pattern of lines of force and some closed curves C_1, C_2, C_3, C_4. The flux through one of these is defined by the number of lines of force threading through it.

Figure 17.17 shows an example, with lines of force and four examples of curves C_1, C_2, C_3 and C_4. $F(C_1) = F(C_3)$ and indeed F has the same value for any sufficiently large curve encircling all the lines of force. $F(C_4) = 0$, and the same is true for any closed curve lying entirely outside the region where the field is.

This may seem a perverse way of characterising the field, and indeed it is not often used in classical (as opposed to quantum) theory.

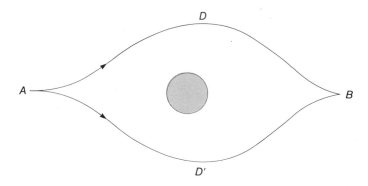

Figure 17.18. The Bohm–Aharanov experiment. Magnetic field intersects the paper in the red-shaded region. *ADB* and *AD'B* are electron beams, which diverge at *A*. The electron intensity is measured at *B*.

In quantum theory, though, the quantity $F(C)$ attains a remarkable new significance. This is shown by an experimental result called the Bohm–Aharanov effect. (It was predicted by Bohm and Aharanov in 1959, and first observed in 1960.) The essentials are shown in figure 17.18. A beam of electrons is split into two beams at *A*. The beams are made to follow different routes *ADB* and *AD'B*, meeting again at *B* where there is a detector. In the shaded region in the centre, there is a magnetic field perpendicular to the plane of the paper (in practice, contained in a very thin 'whisker' of magnetic material). The effect is also mentioned in section 9.4. The strength of this field can be varied.

Apart from the presence of the magnetic field, this is like the interference experiment illustrated in figure 17.10. Again, in quantum theory, the particles in the two beams are represented by wave-functions ψ_1 and ψ_2 and a measurement of particle intensity at *B* measures the length of $\psi_3 = \psi_1 + \psi_2$. This, in turn, depends upon the phase-angle difference $\theta_2 - \theta_1$, as in figure 17.9. It is found in the experiment that the intensity at *B* depends upon the strength of the magnetic field, so $\theta_1 - \theta_2$ must depend upon that field. This would be quite impossible if equation (17.9) applied to charged particles, because then it could be used to compute θ_1 and θ_2 without any reference to the magnetic field at all. (Note that the beams do not enter the region where the magnetic field is!)

The results of the experiment (as predicted by Bohm and Aharanov) are consistent with the relation

$$\theta_2 - \theta_1 = (\text{result computed from equation (17.9)}) + \frac{e}{\hbar} F(C),$$
(17.16)

(where e is the charge of an electron) and this is what replaces equation (17.9) for charged particles. For neutral particles, we just put the charge $e = 0$ in equation (17.16), and then of course it is consistent with equation (17.9).

Equation (17.16) gives less information than equation (17.9), because the former only tells us how to compare two paths *ADB* and *AD'B̄*, whereas the latter tells us about a single open path *ADB*. One may think of equation (17.16) as applying to the *closed* loop *ADBD'A*.

For electrically neutral particles, one may use equation (17.9) to relate the phases of the wave-function at different points of space. For charged particles, however, where equation (17.9) does not apply, the phase seems to have no physical significance. One may' choose it in an arbitrary way, but a change of the phase anywhere in space has no physical significance. Equation (17.16) does not relate phases at *different* points, but tells us what happens on going round the closed loop *C* and *returning* to the starting point.

Thus the quantum theory of electrically charged particles has what should best be called 'local phase-angle independence'. In fact, it is universally called 'local gauge invariance' or just 'gauge invariance'. (The name is due to an historical accident. In 1921, Herman Weyl proposed a theory of charged particles in which *lengths* could be changed arbitrarily from point to point. The theory was quite untenable, but the word 'gauge', meaning 'scale', has stuck.)

Above, we have concentrated upon the magnetic field. The electric field can be included provided that we generalise the closed curve *C* in space to become a closed 'curve' in four-dimensional Minkowski space-time. Since the curves then have much more freedom, the extra information about the electric field can be encoded in $F(C)$. Just as space and time are intimately linked in relativity theory, so are the electric and magnetic fields.

Let us summarise the features of the quantum theory of charged particles which we have just described. They are:

(*a*) The phase-angle of the wave-function of an electrically charged particle has no physical meaning and can be chosen arbitrarily at any point of space-time.

(*b*) On traversing a closed loop *C*, the phase-angle is increased by $(e/\hbar)F(C)$, where $F(C)$ is the flux through *C*. That is, there is a *rotation* through an angle $(e/\hbar)F(C)$ in the Argand diagram.

These are the two ideas we will use to construct more general gauge theories by generalising the type of rotation involved.

Before going on to generalised gauge theories, there is one further fact about electromagnetism we need to recall. This concerns the magnitude of the charge e on an electron. The dimensions of e are such that, in suitable electromagnetic units, the ratio

$$e^2/4\pi\hbar c$$

is a dimensionless number. We can see this by defining two lengths, r_1 and r_2, in terms of e, $\hbar$, c and a mass m:

$$e^2/r_1 = mc^2,$$
$$r_2 = \hbar/mc.$$

The first of these equations is dimensionally correct, because each side is an energy (an electrostatic potential energy and the Einstein rest energy). r_2 was the range defined in equation (17.11). Then the stated dimensionless number is just the ratio r_2/r_1. The actual value of this number is, to a good approximation 1/137. Since it is proportional to the square of electric charge, this is a natural measure of the strength of the electromagnetic interactions in relativistic quantum mechanics (when c and $\hbar$ are relevant).

This number, 1/137, is of enormous importance in physics. The fact that it is small relative to 1 (a meaningful attribute, since it is dimensionless) makes all sorts of approximation schemes viable. In more detail, it turns out that the average speed of an electron in a light atom is of the order $(1/137)c$, and this fixes all sorts of properties of atomic and molecular physics (and molecular biology).

All elementary particles have charges which are small multiples of e (the quark charges are $+\frac{2}{3}e$ and $-\frac{1}{3}e$), so that the smallness of 1/137 is equally relevant to all of them.

17.8　Generalised gauge theories

It turns out that the weak and strong forces are each derived from theories in which the gauge aspect of electromagnetism is generalised.

To get a clue how this might be done, let us concentrate on the term $(e/\hbar)F(C)$ in equation (17.16). Call this quantity α, so that α is the extra piece in the phase-angle difference $\theta_2 - \theta_1$. An increase in a phase-angle *of a complex number* by α means a rotation through α in the Argand diagram (see figure 17.19).

So, in electromagnetism, gauge invariance uses rotations in a plane.

A possible generalisation is to use rotations in a three-dimensional space. Of course, this 'space' has nothing to do with the space in which we live, any more than the Argand diagram does. It is just a mathematical construction; but fortunately we can use our knowledge of 'real space' as a guide.

We shall see that the gauge theories used by nature actually involve more complicated transformations than rotations in three dimensions (which I will henceforth call '3-rotations'). But the 3-rotations illustrate the principles perfectly well, and we will just indicate some of the extra complications later.

Take a set of axes in our three-dimensional space, and call them 1, 2, 3. They are just like the x-, y-, z-axes of 'real space', but we give them different names to emphasise that 'gauge space' has no immediate physical significance. Where we had α above, we will now use the letter R to denote a general 3-rotation. A random example of a particular R is: a rotation through 105° about an axis inclined at 45° to the 1-axis and to the 2-axis. We won't require any general rule for specifying a general rotation R in terms of numbers or angles or wot-not.

Since R is more complicated than just an angle α, the wavefunctions ψ must each now be more complicated than just

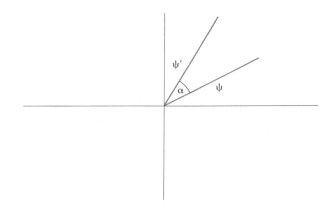

Figure 17.19. The Argand diagram, showing how a complex number ψ is rotated to ψ' when its phase is changed by α.

complex numbers. One of the simplest possibilities is to allow ψ to have three real components (ψ_1, ψ_2, ψ_3), corresponding to the three dimensions in which R operates.

Any rotation can be built up out of lots of very small rotations, and this can be done even if the very small rotations are restricted to three particular kinds: very small rotations in the 2,3-plane in the 3,1-plane, and in the 1,2-plane. (The 1,2-plane is the plane containing the 1-axis and the 2-axis. Rotations in it are determined by just one angle, like α.) In this sense, there are just three independent rotations.

Of course, if the quantity $F(C)$ is generalised, then the notion of electric and magnetic fields must also be generalised, because $F(C)$ is just a way of encoding information about these fields.

In the previous section, the rotation through α was associated with particles carrying electric charge. In the generalised theory, there are three independent kinds of 'charge', corresponding to three independent kinds of very small rotation. We call these kinds of charge (2,3), (3,1) and (1,2). They are all on a footing, but they are each different. Each type of charge will have its own type of 'electric' and 'magnetic' fields. So, for example, electric lines of force of type (2,3) are emitted from a charge of type (2,3). (Note: in electromagnetism, positive and negative charges are not different types of charge in this sense. They just correspond to clockwise as against anticlockwise rotations.)

It is an important geometrical property of rotations that the different types of rotation are inextricably related. A simple example of this is illustrated in figure 17.20, where it is shown that if you do two rotations about different axes and then do the opposite rotations, but in the wrong order, you don't get back where you started from.

Now comes a vital point. *Anything* that is altered by a rotation carries the corresponding charge. Consider a rotation in the 1,2-plane. It does not move the 1,2-plane, but it rotates the 3,1-plane and the 2,3-plane, as indicated in figure 17.21. So it mixes the fields of type (3,1) and (2,3). Therefore, the *fields* of these two types each carry *charge* of type (1,2). That is, in generalised gauge theories, the *fields are charged*.

This means that we have a complicated 'nonlinear' theory. A (2,3)-type field is charged, and is the source of, for example, (1,2)-type lines of force. These in turn are charged, and are the source of (3,1)-type lines of force, and so on.

Why are the extra complications of a generalised gauge theory required for the weak and strong forces? We will answer this question for each in turn.

First, in weak interactions, we saw in Section 17.1 examples of two kinds of weak process. In equation (17.1) no electric charge was transferred from the leptons (the neutrinos in this case) to the hadrons (the proton in this case). This is an example of what is called a 'neutral-current' interaction. In equation (17.2), a unit of electric charge is transformed from the hadron to the lepton – a 'charged-current' interaction.

If there is a field responsible for the force, and there are particles associated with this field, then we require three kinds of particles, one neutral, one positively (electrically) charged and one negatively charged. Let us call these X^0, W^+, W^- particles. The $W^\pm$ particles were discovered at CERN during 1983–4. The X^0 we will leave for the moment, because we have oversimplified.

These three types of field correspond to the three kinds of 3-rotation which we had in our model. For example, the (1,2)-rotations could correspond to the X^0, and the $W^\pm$ could be two different (complex) combinations of the (2,3)- and (3,1)-type fields.

The 'nonlinear' nature of the generalised gauge theory results in mutual interactions between W^+, W^- and X^0. For example, the W^+ particle is a source of the X^0 field. Similarly, the X^0 particle can convert into a W^+ particle and give rise to a W^- field, and so on. The details don't matter. The mutual interaction between $W^\pm$ and X^0 is the important thing.

We must now correct for an oversimplification which we have made. It turns out that we cannot discuss the weak force isolated from the electromagnetic one (the reason for this will become clear in the next section). To include electric charge as well as the three kinds of weak charge, we must construct a generalised gauge theory based upon two types of rotations: the 3-rotations introduced at the beginning of this section, and also 2-rotations (i.e. rotations in a two-dimensional plane, mathematically just like the electromagnetic gauge transformations of the previous section). We call the gauge field associated with the 2-rotations Y^0.

Just as the electric charge strength e appears in equation (17.16), there will be two charge strengths, g and g', associated with the 3-rotations and 2-rotations respectively.

I will now state the physical interpretation of X^0 and Y^0, without justifying it. It turns out that the two known, electrically neutral, gauge particles, the photon and the Z^0 are *mixtures* of X^0 and Y^0. Also, the electric charge strength e is derived from g and g', and is given by

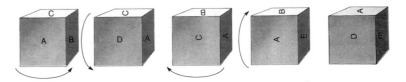

Figure 17.20. The inter-linked property of rotations in three-dimensions. The cube in the first position is rotated through 90° in the sense shown by its arrow. The result is the second position, and so on. The third rotation is the opposite of the first, and the fourth is the opposite of the second. But the cube does not end up in its starting position. (It would have done so if the fourth rotation had preceded the third.)

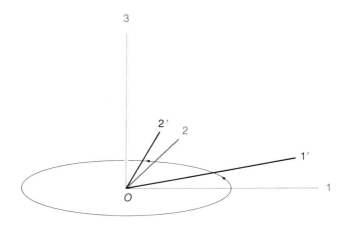

Figure 17.21. A rotation about the 3-axis does not move the 1,2-plane, but it rotates the 1,3- and the 2,3-planes.

$$\frac{1}{e^2} = \frac{1}{g^2} + \frac{1}{g'^2}$$

The dimensionless magnitudes come out, experimentally, to be about

$$\frac{g^2}{4\pi\hbar c} \simeq \frac{1}{30}, \; \frac{g'^2}{4\pi\hbar c} \simeq \frac{1}{107}. \qquad (17.17)$$

These are, of course, consistent with $e^2/4\pi\hbar c \simeq 1/137$.

So g is quite small, but not as small as e. Whence, then, the weakness of 'weak' forces? It turns out that the weakness is mainly due to the very short range of the weak forces. A better name than 'weak forces' would really be 'very-short-range forces'.

As explained in Section 17.5, range of forces is linked to mass of particles. But the whole question of the mass and range associated with the weak gauge fields has been studiously avoided up to now. We will face this question in Section 17.10.

So much for weak forces for the moment. Why do strong forces require a generalised gauge theory? The reason here is connected with colour. As explained in Chapter 14 by Close, quarks come in three colours. This extra degree of freedom doesn't have any obvious physical manifestation in the sense that one can recognise which colour is which. It is just required to give the right counting of numbers of multiquark–antiquark states.

It is a reasonable guess that one should make the colour-transformations into gauge-transformations. The nature of the resulting set of transformations is described in a little more detail in Section 17.12, and also in Chapter 15 by Georgi. But it is clearly going to be a generalised gauge theory, i.e. one with more than just a *single* type of transformation.

17.9   Gauge fields interacting with leptons and quarks

We have seen that generalised gauge fields are automatically self-interacting. But the weak interactions easily observed (e.g. the examples in Section 17.1) depend upon the weak forces acting on leptons and quarks (and hence on hadrons made of quarks). What is the structure of such interactions?

In the case of electromagnetism, we just have to specify the electric charge of any particle, and then the interaction of that particle with the electromagnetic field is determined. How is this generalised? The weak gauge theory involves 3-rotations and 2-rotations, so we must specify how each lepton and quark transforms under each of these. To do this, we must group the particles into sets which transform into each other.

The simplest such set of particles would be a triplet, corresponding to the three dimensions in which 3-rotations act. Mathematically, this would have been possible, but experimentally the leptons and quarks group themselves into doublets, not triplets, so far as the weak force is concerned. For example, an electron and a neutrino are involved in the weak process, equation (17.2), and these have to be put into a doublet. Actually, experiment shows that it is the helicity $-\frac{1}{2}\hbar$ spin states which are involved.

Let us denote the fields of the electron and neutrino as e and v (the Greek letter 'nu' for neutrino). To emphasise that we are concerned with the helicity $-\frac{1}{2}\hbar$ components, we will add a suffix L: e_L, v_L. L stands for 'left-handed', because a negative helicity particle is like a left-handed screw. These have to be complex quantities, not real. One way to see this is to remember that the electron carries electric charge, and so has to admit a phase-change under electromagnetic gauge transformations. This is only possible if it is a complex number.

Can one find a consistent rule for associating transformations of a complex doublet under the 3-rotations and 2-

rotations of the weak gauge theory? The answer is yes. I will indicate what the rules are, without showing why they are consistent. The rules are:

(a) Under a rotation through an angle α in the 1,2-plane, v_L changes phase by $\frac{1}{2}\alpha$ and e_L changes phase by $-\frac{1}{2}\alpha$.

(b) Under rotations in the 2,3-plane or 3,1-plane, e_L and v_L mix with one another. (The mixing is determined by a real number in the case of the 2,3-plane and by an imaginary number in the case of the 3,1-plane.)

(c) Under a 2-rotation through an angle β, both the v_L and e_L change phase by an angle $+\frac{1}{2}\beta$.

Because of the half-angles appearing here, this is a sort of 'square-root' of a rotation.

We can now ask how the electric properties of the neutrino and electron are contained in all this. The electromagnetic phase transformations are made up out of a combination of a particular 3-rotation and a 2-rotation. In fact, for an electromagnetic phase transformation we do a rotation through α in the 1,2-plane, and simultaneously a 2-rotation through the same angle α. From the rules 1(a) and (c) above, the overall effect is to change the phase of e_L by α but to do nothing to v_L. This is just right, considering that the electron is charged and the neutrino is neutral.

There is also the helicity $+\frac{1}{2}\hbar$ state of the electron, e_R. Electromagnetism is reflection-invariant and so does not distinguish between left- and right-handedness. On the other hand, as a matter of observations, the helicity $+\frac{1}{2}\hbar$ spin state of the neutrino doesn't undergo weak interactions and perhaps doesn't exist at all. All these things are correctly taken account of if we add one more rule:

(d) e_R does not change under 3-rotations, but under a 2-rotation through β it changes phase by β (not $\frac{1}{2}\beta$).

From these four rules, it may be checked that e_L and e_R couple to photons in the same way, as they should since they have the same charge. Their couplings to Z^0 are more complicated.

For the purposes of weak interactions, quarks with helicity $-\frac{1}{2}\hbar$ are also grouped into doublets. One such doublet is made up of the up quark (charge $+\frac{2}{3}e$) and of the down quark with a smallish admixture of the strange quark (each of which have charge $-\frac{1}{3}e$). This doublet transforms according to rules (a) and (b), and a slight modification of (c). The helicity $+\frac{1}{2}\hbar$ partners of these quarks do not transform under 3-rotations, but each transforms under 2-rotations in an appropriate way.

Then there are heavier leptons, muon and tau, each with its own neutrino partner, and heavier quarks (charm, bottom and perhaps top), which also fall into doublets. The partner of the charm quark is the strange quark with a small admixture of the down quark. The existence of the charm quark was predicted because of the need to fill in this pattern of doublets (which couldn't easily be done with *three* quarks). The top quark is predicted now for the same sort of reason.

17.10   Short-range forces from gauge theories

We must now face the fundamental problem: how can gauge theories generate short-range forces? The apparent insolubility of this problem held up the application of gauge theories for some ten years.

According to our exposition in Sections 17.7 and 17.8, the idea of lines of force is fundamental to gauge theories. Also, according to the discussion of figure 17.16, lines of force seem inevitably to lead to infinite-range forces (i.e. the inverse-square law).

The only escape from this conclusion is if the lines of force can somehow terminate in 'empty space', in the vacuum. This could happen if there were charges in the vacuum which normally neutralise each other, but which could respond to a field so as to produce a local nonzero density of charge.

An analogue from another branch of physics suggests how this might be possible. (As I mentioned at the end of Section 17.5, elementary particle physicists have learnt to think of the vacuum as being, in some ways, like a rich and complicated medium.)

The analogue we will take is that of a plasma: free negative charges in (say for simplicity) a fixed positively charged background, so that the total charge is zero. In equilibrium, the negative charges distribute themselves evenly, so that the average charge density is zero everywhere.

But, charge density oscillations can be excited, in which there are local regions of compression or rarefaction of the negative charge density. Let us estimate (very crudely) the frequency of such 'plasma oscillations'. Suppose that the negative charges which normally occupy a sphere of radius r are compressed, so that they are in a slightly smaller sphere of radius $r - \delta$ (δ being much less than r); see figure 17.22. Then there is an excess negative charge within the sphere proportional to $enr^2\delta$, where e is the value of each charge and n is the number per unit volume (in equilibrium). This excess charge produces a radial field at the surface of the sphere proportional to $en\delta$. So there is a force $e^2n\delta$ on each negative charge, tending to push it out. In these circumstances, when there is a restoring force proportional to the displacement δ, one gets a vibration with frequency ω_p given by

$$\omega_p^2 \propto e^2 n/m,$$

where m is the mass of one of the negative charges. (Alternatively, we could have derived this formula by a dimensional argument, assuming ω_p to depend only upon e, n and m.)

The plasma frequency, ω_p, turns out to be the minimum frequency of any electromagnetic wave in the plasma. A slower one is just damped out by the movement of the charges, but for a faster one the charges can't respond quickly enough.

Of course, this analogy – between a plasma and the vacuum – cannot be pushed too far. In a plasma, there are different sorts of

waves with different speeds. In the vacuum, by the laws of special relativity, the speed must always be c.

Nevertheless, let us translate this minimum frequency into quantum terms. The quantum of a wave with frequency ω_p has energy $\hbar\omega_p$ (this is a sort of example of the uncertainty principle, equation 17.8). So there is a minimum energy of amount $\hbar\omega_p$. But the minimum energy of a particle in relativity is, according to equation (17.6), the Einstein rest-energy mc^2. So the existence of a minimum frequency mimics the effect of a nonzero mass which, as explained in equation (17.11), is associated with a finite-range force c/ω_p.

Thus, in this model, roughly speaking, photons acquire a mass and electric forces get a finite range.

Now let us try to use this model to inspire a mechanism for making the weak forces have a short range, and, concomitantly, giving the $W^\pm$ and Z^0 particles masses.

We first give an oversimplified discussion, which ignores the Z^0, but which contains the essence of the matter. The shortcoming will be removed later.

We need an analogue of the free charges in the plasma. To make this, we have to postulate the existence of a new type of field (other than the fields of leptons, quarks, photons, $W^\pm$, Z^0, etc.). So far as 3-rotations go, let it have three (real number) components (ϕ_1, ϕ_2, ϕ_3). Suppose that, in the vacuum, this field has a nonzero value. Why this should be is not immediately obvious, and at first sight it is contrary to one's notion of the 'vacuum'. But we press on.

This is a gauge theory, so we have the freedom to carry out a 3-rotation, R, at every point of space, without changing anything physical. We can use this freedom to do such a rotation so that any nonzero part of (ϕ_1, ϕ_2, ϕ_3) points along the 3-axis. Thus, the field in the vacuum can be *chosen* to have the form (0, 0, ϕ_3). Having done this, for convenience, we have used up most of the freedom given to us by gauge invariance, and we must take care not to appeal to it again.

Figure 17.22. A disturbance in an electrical plasma. In equilibrium, the positive and negative charges (red and grey dots) are each uniformly distributed. In the disturbances, the negative charges have been locally compressed, so that the number which previously filled a sphere of radius r now fill one of the smaller radius $r - \delta$. Consequently, there is a net negative charge within this sphere, and the repulsion between like charges tends to push out the negative charge again.

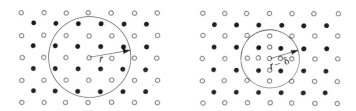

One might ask why the vacuum 'full' of ϕ_3 should be called a vacuum at all. Why isn't it full of weak charge and therefore immediately observable? The answer is that weak charge is connected with rotations in the 1,2,3-space, and therefore with transitions from one component of (ϕ_1, ϕ_2, ϕ_3) to another. When this field is fixed in the form $(0, 0, \phi_3)$, there are no such transitions. So the vacuum does have zero weak charge.

Now suppose that a weak gauge field is applied, say the (3,1) field. The ϕ-field in the vacuum can respond to this, producing locally a small ϕ_1 component. Now there is a ϕ_1 part as well as the ϕ_3 part, and so there is a nonzero (3,1) charge. This charge, in turn, can produce a new contribution to the (3,1) field, modifying the original applied field.

This is analogous to the response of the free charges in a plasma to an applied electric field. The effect in the present case can be described as giving a mass to the particles ($W^\pm$) of the weak field, or giving a short range to the weak force, just as in the plasma model.

The mass turns out to be proportional to the vacuum value of ϕ_3 and to the weak charge strength g (just as the plasma frequency ω_p was proportional to e). The mass of the $W^\pm$ is measured to be $85\,m_p$, where m_p is the mass of a proton. Given the value of g in equation (17.17), we can conclude that the magnitude of ϕ_3 in the vacuum is

$$260\,m_p. \tag{17.18}$$

The rates of most weak processes turn out to be inversely proportional to the fourth power of this constant. So they are very small except at energies of several-hundred proton rest-energies.

The above mechanism works to give mass to the (3,1)- and (2,3)-type gauge particles (i.e. to $W^\pm$). But, since rotations in the 1,2-plane do not have any effect on the vacuum field $(0, 0, \phi_3)$, the (1,2)-type gauge particle does *not* get a mass. This particle, which we called the X^0, *might* have been the photon. But nature is a little more complicated. We have neglected the fact that there is a Z^0 as well as a photon, and that these particles couple to different helicity states of the leptons and quarks than do the $W^\pm$. I will briefly indicate how the above account has to be modified.

Instead of taking for ϕ a triplet of real fields, we take a doublet of complex ones, which I will denote as (ϕ, ψ).

Under the weak gauge-transformations, this doublet transforms in the same sort of ways as do the lepton and quark doublets, discussed in Section 17.9. In the vacuum, the only nonzero thing is the real part of the complex field ϕ. The electromagnetic gauge-transformations are precisely the ones which don't alter ϕ, so the photon does not get a mass. Each of the other weak gauge-transformations either mixes ϕ and ψ or changes the phase of ϕ; so all three of $W^\pm$, Z^0 get a mass.

This is all very well, but what is the physical interpretation of these new fields (ϕ, ψ)? Do they correspond to new types of particle, and if so of what sort? The answer is that one, and only one, electrically neutral particle is predicted to exist.

The reason that there is not more than one particle is because we may always use the freedom provided by local gauge invariance to make ψ zero and ϕ real. Does this mean that three degrees of freedom (the imaginary part of ϕ and the real and imaginary parts of ψ) have somehow been spirited away? This is not the case. By choosing to 'gauge away' these fields we have lost some of the gauge arbitrariness of the gauge fields. So the gauge fields $W^\pm$, Z^0 have each *gained* a degree of freedom.

If we remember what was said in Section 17.7 about the number of helicity states for particles with and without mass, it is not difficult to guess where these degrees of freedom have gone. They have gone to provide the extra helicity zero in (17.14) (the nonzero mass case) as compared with (17.15) (the zero mass case). Thus the trick is complete.

The new ϕ particle is electrically neutral. Because of its role in producing mass for other particles, it turns out that it has to interact with all particles which have mass – the bigger the mass the stronger the interaction. It also interacts with itself, since it has mass. Unfortunately, the theory does not predict how big this mass should be, so that all that experimenters can do is keep on looking for it as new energies are attained.

What can be said is that the particle cannot be *too* heavy (not much more than about $1000\,m_p$). If it were, its self-interactions would be so strong that the theory would be very complicated. Then it would be difficult to understand why the predictions about the $W^\pm$ and Z^0 have worked out so well.

The ϕ particle would certainly be unstable and would decay most often into heavy particles. For example, if it were heavy enough, it would often decay into a W^+, W^- pair, and it might be detected this way.

The mechanism described here for giving mass to gauge particles was discovered by P. W. Higgs in Edinburgh and T. W. B. Kibble in London. It was applied to the physics of weak interactions by A. Salam in London and S. Weinberg in Harvard. The ϕ particle is called the 'Higgs particle'. Its discovery would finally clinch the theory. If, at high enough energies, it does not turn up, there is something we don't understand.

17.11   The high-energy limit

Suppose we calculate, in some approximation, a dimensionless physical quantity, concerned with a collision between particles at energy E. Let m be a 'typical' mass of a relevant particle, say the $W^\pm$ mass. In units chosen so that $\hbar$ and c are 1, the quantity can depend upon the dimensionless ratio E/m. For example, there might be a dependence something like

$$A + B\frac{m^2}{E^2} + C\frac{E^2}{m^2}, \tag{17.19}$$

where A, B and C are constants.

Now suppose we consider the very-high-energy limit in which E/m is much greater than 1 (a limit not yet, unfortunately, attained by accelerators). If $C = 0$ in equation (17.19), there is a smooth limit. If C is nonzero, things are more tricky. There is a grave risk that the approximation in which we calculated equation (17.19) has broken down. There is thus a great methodological advantage in being able to construct theories in which terms like C are all zero. The rationale of this is discussed more fully in Chapter 16 by Georgi.

This requirement becomes very important if there are particles with helicity $\pm \hbar$ (rather than the $\pm \frac{1}{2}\hbar$ of leptons and quarks). For then, according to (17.14) and (17.15), the situation when $m = 0$ is qualitatively different from that with $m \neq 0$. If, then, we take just any old theory describing particles with helicities $\pm \hbar, 0$ when $m \neq 0$, there is likely to be singular behaviour as m approaches zero, i.e. as E/m becomes large. Indeed, in general, terms like C in equation (17.19) do turn up.

There is a unique exception to this. That is when the particles are generated from a gauge theory by the Higgs–Kibble mechanism, as described in the preceding section. Then there is a smooth limit from $m \neq 0$ to $m = 0$. The nature of that limit is this: the Higgs field ϕ ceases to be nonzero in the vacuum, the gauge fields lose their mass, and the helicity-zero degrees of freedom in the gauge particles are returned to the (ϕ, ψ) fields. The four real fields in (ϕ, ψ) (the real and imaginary part of each), *each* represents a real physical particle in the limit.

'Degree of freedom counting', like this, might have led people to discover the Higgs–Kibble effect (but it didn't).

What isn't so obvious is that, when the masses are nonzero, *one* of the Higgs fields has to be left over when the others are absorbed as the helicity-zero part of the gauge fields.

17.12   Strong interactions

At first sight, the strong force seems very different from the weak one. The latter is responsible for a great variety of different processes, which have been studied in great detail. The particles ($W^\pm, Z^0$) associated with gauge fields have been found, and calculations can be made with good accuracy, thanks to the smallness of the coupling strengths g and g' (see equation 17.17). The strong force, on the other hand, is only rather indirectly known. Neither the particles (gluons) associated with the strong field nor even the quarks can be produced as free particles. Calculations are very difficult, and can usually only be made with an error of 25% or more. Nevertheless, the gauge principle is thought to underlie the strong force just as it does the weak one, albeit manifested in a different mode.

The first clue is that all quarks come in three 'colours'. In addition to the different 'flavours' of quarks (which are distinguished by their masses, electric charges, strangeness, etc.), each flavour of quark has three 'colour' types, which are thought to be indistinguishable by any directly observable physical property. (The words 'flavour' and 'colour' are, of course, purely arbitrary – and mildly facetious – words, to call to mind the way that the particles come in sets; with different varieties within each set. In a similar way the phrase 'isotopic spin' was used to describe the way in which nucleons come in two similar kinds, neutrons and protons.)

Let q stand for any particular flavour of quark (e.g., the up-quark). Denote its three colour types by (q_r, q_b, q_g) (for **red**, **blue**, **green**). The fact that these are physically indistinguishable is expressed by saying that the theory is unchanged by transformations which mix these three fields, one with another. One might think that these transformations should constitute another group of 3-rotations, like those in Section 17.8. But actually each q is a complex quantity (because the quarks carry electric charge), and the transformations we require are like a generalisation of the transformations on the (complex) lepton doublet (v, e) in Section 17.9, but there are now three complex fields instead of two.

The structure of this group of transformations (which has no simple geometrical interpretation this time) is described in Chapter 15. These turn out to be eight independent transformations.

Any set of transformations of this type can be used to generate a gauge theory, just as the 3-rotations did in Section 17.8. In the present case, the transformations must be mathematical objects capable of acting on the complex triplet (q_r, q_b, q_g) (i.e. complex 3×3 matrices of a certain kind). Then, since there are eight independent transformations, there will be eight charges and eight gauge fields.

This set of transformations is less easy to visualise than the 3-rotations of Section 17.8. But the same general ideas apply. The eight gauge fields are themselves affected by the transformations, and so they themselves carry 'colour charge'. Thus the eight gauge fields are self-interacting, just like the $W^\pm$, X^0 fields of Section 17.8.

The colour gauge theory is called quantum chromodynamics, or QCD for short. The eight gauge particles are called 'gluons'.

As usual, a gauge force is naturally an infinite-range one, and its quanta are naturally massless particles. How then can the short-range strong force be produced by QCD? This is the question that was presented by the weak forces, but the problem with strong forces is not really the same. This is because no particle with nonzero colour charge, whether it be quark or gluon or anything else, has ever been seen as a truly free particle (as opposed to existing for a very short time – there is direct evidence for quarks in this sense, as explained in Chapter 14). The principle called 'confinement' is therefore postulated: that, for some physical reason, the existence of any free particle with nonzero colour charge is forbidden.

If confinement occurs, there is no problem with infinite-range forces, because one cannot separate quarks from each other to the long distances where they would feel the infinite-range forces. Similarly, there is no problem with the masslessness of gluons, since free gluons do not even exist.

The strong forces that *are* observed are those between, say, nucleons, i.e. between colourless compounds of quarks. These are somewhat analogous to the forces between neutral atoms, which are consequences of electric forces but are more indirect than the forces between electrically charged particles. Similarly, the forces between nucleons are presumably a complicated, secondary matter; and the first thing to do is to understand the inter-quark forces.

Thus the crucial problem with QCD is to explain confinement. This is a harder problem than any arising in the electro-weak gauge theory, and the Higgs–Kibble mechanism is not relevant.

There is no definitive theory of confinement. Nevertheless, many physicists believe that they have a partial, intuitive understanding of it. There has been much activity in trying to find approximate, numerical solutions to the equations of QCD, by using powerful computers. The results of these attempts do not contradict the intuitive understanding of confinement.

In Section 17.15 we shall describe a model which indicates how confinement *might* come about. But first we must explain a very important property which *can* be derived from QCD. This property is called 'asymptotic freedom'. As a preliminary, I shall recall some properties of ordinary matter which I shall later use as analogies for the properties of the vacuum in QCD.

17.13   A digression on the electric and magnetic properties of matter

As I have already mentioned, it seems to be fruitful to think of the vacuum as being, in some ways, like a complicated medium, particularly in the way it responds to fields. The properties of ordinary matter may therefore serve as a source of analogies and ideas to help understand the properties of the vacuum; though, as with all analogies, one must watch carefully not to stretch them too far. For this reason, it will be useful to recall some simple facts about the behaviour of matter in electric and magnetic fields, beginning with the electric case.

All insulating substances are, to a greater or lesser degree, dielectrics – they are polarised by an electric field. That is to say, the electrons in the molecules move a little bit relative to the atomic nuclei. This is illustrated, crudely, in figure 17.23, in which the positively charged nuclei are represented as red dots and the negatively charged electronic clouds are shown as grey. The lines of force outside the material are shown (inside, they would look very complicated). These are pulled in towards the material, because there is a net positive charge on the surface at the left and a negative one at the right.

Figure 17.24 illustrates what happens if a small point negative charge is introduced into the dielectric (let us think of a gas for ease of visualisation). A sphere drawn about the point charge will, on average, contain a slight excess of positive charge from the atoms. The introduced negative charge is thus, to some extent, 'shielded' by a surrounding positive charge. The

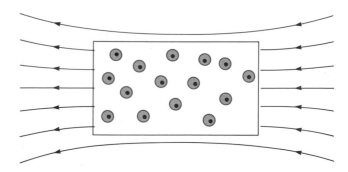

Figure 17.23. A dielectric material in an electric field. Red dots are positively charged nuclei and grey circles are negatively charged electron clouds. The electric field displaces the positive charges relative to the negative ones, causing the material to be polarised.

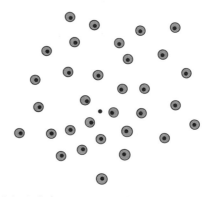

Figure 17.24. A dielectric material with a point negative charge (black dot) inserted into it. The positive charges in the material are attracted towards the point charge, tending to shield its charge.

magnitude of this effect is independent of the size of the sphere which we draw: for larger spheres the polarising field is less but the number of relevant atoms at the surface of the sphere is greater.

From this, it is fairly clear that, if two charges are introduced into the dielectric, the force between them is reduced by a fixed factor. It is

$$\frac{e^2}{\varepsilon r^2}$$

instead of

$$\frac{e^2}{r^2},$$

where r is the separation and ε is the 'dielectric constant' of the material. By the above argument, ε is always greater than 1, though its actual value depends strongly upon the material. Materials with large values of ε (say about 5 to 10) are used in some capacitors to increase their capacitance.

In some materials, the molecules are permanently polarised, i.e. the centre of their negative charge is displaced relative to the centre of their positive charge. When no field is applied, such molecules are randomly oriented. An applied field tends to orient them, so that the final result is roughly the same as in figure 17.23. Again there is a dielectric constant ε which is greater than 1.

Neither of these two mechanisms can ever give $\varepsilon < 1$, because they are both based upon the movement of electric charges under the influence of the applied field.

The above descriptions apply over scales which are large compared with the distances between molecules, for it is only then that effects of the molecules average out in a simple way. Very near to the point charge, in the empty space between the molecules, there is of course no shielding. So we can think, roughly, of a sort of r-dependent dielectric constant, $\hat{\varepsilon}(r)$, where r is the distance from the point charge. The force between two charges is

$$\frac{e^2}{r^2\hat{\varepsilon}(r)},$$

and $\hat{\varepsilon}$ approaches 1 for very small r and approaches the constant ε for large r ('small' and 'large' meaning compared to the molecular spacing). The symbol $\hat{\varepsilon}(r)$ is introduced to distinguish this r-dependent quantity from the large-distance value ε.

Now we turn to the magnetic case, which is more complicated. Some molecules are always magnetic, like tiny permanent magnets. In most substances (ferromagnetic materials like iron excepted) the molecules are normally oriented at random, so the bulk magnetisation is zero. An applied magnetic field tends to orient the molecules. The effect is just like in the electric case with electrically polarised molecules. There is a constant μ, analogous to ε, and $\mu > 1$. Such a material (like oxygen or aluminium) is called paramagnetic. It pulls magnetic field into it, and it is attracted towards regions of stronger magnetic field.

When the individual molecules are not permanently magnetised, it is possible in some cases to have $\mu < 1$. Such a material (like hydrogen or bismuth) is called diamagnetic. It tends to expel magnetic field, and is repelled from regions of stronger magnetic field. (The names are confusing. *Para*magnetic is analagous to *dielectric*. *Dia*magnetic is the opposite!)

It is not possible to give a simple argument why diamagnetism can occur. It is strictly speaking a quantum effect. However, one can see that there *might* be diamagnetic tendencies if electric currents can flow within the molecules. An increasing magnetic field always tends to induce currents to flow in such a way as to tend to prevent the increase of field. This is (at least temporarily) a diamagnetic kind of effect.

Thus the case where $\mu < 1$ is connected with the flow of *electric* charges in a *magnetic* field. There is no analogous case with electric fields since there isolated magnetic poles do not, so far as is known, exist.

17.14 Vacuum polarisation and asymptotic freedom

How far does the vacuum resemble a dielectric? In quantum electrodynamics, the analogy is quite good. As explained in Section 17.5, quantum theory and relativity together imply that, for very short times, there is the possibility of finding pairs of oppositely charged particles present in the vacuum. If there is an electric field present, the average position of the positive charge of such a pair will be displaced relative to the negative one. The result is then somewhat similar to the situation in a dielectric, depicted in figures 17.23 and 17.24. Again there is shielding, and again the force between two charges has the form

$$\frac{e^2}{r^2\hat{\varepsilon}(r)},$$

where $\varepsilon(r)$ *increases* with r. The scale of the r-dependence of $\hat{\varepsilon}$ is now set by the length $\hbar/mc$ associated with the mass m of the particle pairs. (For electrons, this is about 10^{-13} m.)

There is, however, a difference from the case of a material dielectric, in the way that $\hat{\varepsilon}(r)$ is conventionally defined. The charge on a particle is *defined* in terms of the force at macroscopic distances. So, by the definition of e, $\hat{\varepsilon}(r)$ approaches 1 for large r. What the dielectric properties of the vacuum tell us, then, is that $\hat{\varepsilon}(r)$ becomes *less* than 1 for very small r. What is really happening at such small r is that we are seeing the 'bare', unshielded charge, whereas normally in macroscopic experiments we only see the shielded one.

This tendency of $\hat{\varepsilon}(r)$ to decrease at very small distances is a real physical effect which influences, for example, the spectrum of the hydrogen atom. The dependence upon r is only a mild logarithmic one. Nonetheless, at small enough r, $e^2/\hat{\varepsilon}(r)$ ceases to be small (in the sense that $e^2/4\pi\hbar c \simeq 1/137$). At these extreme distances, existing methods of calculation break down, and what really happens is unknown.

But now we must reveal a crucial way in which the analogy between the vacuum and ordinary materials breaks down. The speed of light in a medium is given by

$$c/(\varepsilon\mu)^{\frac{1}{2}}.$$

In the vacuum, special relativity says that the speed is c. Therefore, the vacuum must have the property that

$$\varepsilon\mu = 1.$$

(In matter, the speed of propagation is not, in general, c, because what is observed is a sort of effective speed, as the light is scattered from atom to atom.)

This means that if the vacuum is dielectric ($\hat{\varepsilon}$ increasing with r), then it is diamagnetic ($\hat{\mu}$ decreasing with r). There is nothing surprising about this, since diamagnetism is known to occur.

The above statements are all true, without qualification, if the virtual particles in the vacuum have no spin. In practice in quantum electrodynamics, however, we are most concerned with electrons and positrons. These do have spin (helicities $\pm \frac{1}{2}\hbar$) and are also magnetic. We might therefore expect the vacuum to be paramagnetic, in contradiction to what we said above. The true state of affairs is, in fact, rather complicated (partly by the fact that electrons obey the Pauli exclusion principle). The result of calculation is that the vacuum *is* dielectric and diamagnetic as described above.

Now let us turn to QCD. Here calculation reveals the *opposite* effect to quantum electrodynamics. In the vacuum: $\hat{\varepsilon}(r)$ decreases with r and $\hat{\mu}(r)$ increases with r. It is paramagnetic and has the opposite electrical properties to a dielectric – there is 'antishielding' of charges. (Electric and magnetic here mean 'colour-electric' and 'colour-magnetic'.)

In the calculation, the reason for this can be traced to the self-interaction of gluons, i.e. to the virtual pairs of gluons themselves. Gluons have spin (helicities $\pm \hbar$) and are colour-magnetic, so it is quite natural to have a paramagnetic effect. However, intuitively, one would also expect the vacuum to be dielectric. Because of the relativistic relation $\varepsilon\mu = 1$, these two intuitions contradict one another. The calculation reveals that both effects are present, but the paramagnetic tendency dominates. I do not believe there is any simple argument to show this.

There are also virtual quarks in the vacuum as well as virtual gluons. The quarks have exactly the same effect as the electrons in quantum electrodynamics – they tend towards shielding. But the 'antishielding' effect of the gluons dominates, provided there are not more than sixteen quark flavours.

In electro-weak theory, the self-interaction of $W^\pm$ and Z^0 tends to produce antishielding, but the Higgs fields (ϕ, ψ) have the opposite effect. It can be shown that, whenever gauge particles get masses by the Higgs–Kibble mechanism, shielding results.

Thus QCD, uniquely, has the property of antishielding – at short distances, the forces become *weaker*. This effect is called 'asymptotic freedom', because at 'asymptotically' short distances (or, by the uncertainty principle, at asymptotically large momenta) the quarks become almost *free*. This has the fortunate consequence that good simple approximations are possible at short distances, starting with the simplification of free quarks.

The simplest example of this occurs in experiments in which a beam of electrons collides with a beam of positrons at high momentum. Sometimes, an electron and positron annihilate with one another, and then electromagnetism allows their total energy to reappear in a pair of any other electrically charged particles, for instance a quark–antiquark pair. When a quark and antiquark are first produced, they are close together, and so almost free. The probability of this happening is easily calculated, and so is the distribution of directions of the quark.

The pair then move apart, QCD forces become stronger (because of 'antishielding') and detailed calculation becomes more difficult. One expects that, somehow, the quark and antiquark each tear further pairs out of the vacuum, which combine together to form several pions (or other hadrons), all moving in a 'jet' roughly in the direction of the parent quark. In spite of the complications of the later stages of the process, the probability of this happening and the distribution of jet directions are simply predicted.

There is a subtlety about this. The quark has colour charge, but, if confinement is right, each jet must have zero colour charge. This is thought to be possible because there are some low-momentum quarks or antiquarks which belong to neither jet. They don't affect the jet momentum much, but they allow colour charge to trickle across from one jet to another. The idea is represented in figure 17.25.

At large (greater than about 10^{-15} m) distances, QCD forces become strong, and no simple way of calculating exists. *If* the forces increased fast *enough* with distances, one would have a simple explanation of confinement: the attraction between quark and antiquark would be so strong that they could not be pulled apart. I think, however, that this is probably an oversimplified picture.

17.15   Ideas about confinement

Free quarks, free gluons, or indeed free particles with any nonzero colour charge, have never been seen. The evidence for quarks and gluons is all indirect (see the chapter by Close) – for quarks bound into hadrons, or for quarks or gluons living for a very short time and manifesting themselves as jets. Hence, it is assumed that QCD somehow prohibits colour charge being free. This postulated property is called confinement. It has not been proved, but many physicists believe that they have a vague idea how it is caused, and there are some hints that they are on the right track. In this section, I describe what I believe to be the key ideas. But I emphasise that we are venturing into an area where hunches and unjustified approximations take the place of reliable calculations.

In previous sections I have used analogies, but only so as not to trouble the nonspecialist reader with technical details. In this section, I use analogies because there is nothing much better.

So the question is: why can a hydrogen atom be ionised (i.e. split into an electron and a proton), while a pion cannot, apparently, be split into quark and antiquark?

In the hydrogen atom, the force is the electric inverse-square law force; so the energy required to ionise is given by the integral of this force from some characteristic radius of the atom to infinity. That is, by an area like that shown in figure 17.26. This is a finite area. If the force fell off less fast than $1/r^2$, the energy required to ionise might be infinite, which could be interpreted as confinement.

Let us remember the cause of the $1/r^2$ law (Section 17.7). It is because the electric lines of force do not terminate, but spread out evenly over spheres whose areas grow as r^2. Thus the pattern of lines of force between two opposite electric charges is as shown in figure 17.27.

Suppose that the colour lines of force could somehow be induced to remain in a thin tube, as in figure 17.28, with the cross-sectional area of the tube fixed (at least roughly) at some value A. Then the force would be e_c^2/A instead of e_c^2/r^2, and would thus be independent of the separation r (where e_c is, in some sense, the colour charge). In this case, the energy required to separate the charges to a large distance would be essentially infinite – we would have confinement.

Before wondering how tubes of lines of force could be caused, let us look at the physics of this 'infinite energy' a little more closely. What would actually happen if one tried to 'pull apart' the quark and antiquark in a pion? The energy stored in the tube would increase. At some stage, the energy would exceed the rest-energy (mc^2) of a pion. Then, a new quark–antiquark pair could be torn out of the vacuum, splitting the tube into two pieces, as shown in figure 17.29. Rather than pull the quark and antiquark apart, all we would have done is create another pion. The process of jet formation illustrated in figure 17.22 may be thought of as just such a process of tubes stretching and tearing into fragments in this manner.

It is salutary, also, to reflect on the orders of magnitude concerned. Suppose that the radius of the tube is roughly a typical nuclear distance, about 10^{-15} m. Then the order of magnitude of the tension in the tube is, simply on dimensional grounds,

$$\hbar c/A = \frac{1.1 \times 10^{-34} \times 3 \times 10^8}{10^{-30}} = 33\,000\,\text{N} = 3 \text{ tonne weight.}$$

How could tubes of lines of force be caused? I cannot answer this question. All I can do is point to a partially analogous physical situation in which such tubes are known to form. This arises in the physics of superconductors (see section 9.4 Leggett's Chapter), where electric currents can flow without resistance. If we attempt to apply a magnetic field to a superconductor, induced electric currents flow in such a way as to keep the magnetic field zero within the superconductor. A superconductor expels magnetic field. We may describe it as being infinitely diamagnetic, i.e. $\mu = 0$ (see Section 17.13 for a definition of diamagnetic).

What happens, then, if one continues to increase the magnetic field applied to a superconductor? Eventually a critical value of the magnetic field strength is reached beyond which the material ceases to be a superconductor and becomes an ordinary metal (just as it does if its temperature is raised). Then, of course, the magnetic lines of force can penetrate into it.

In some types of superconductor, this breakdown of the superconducting properties does not occur throughout the material, but along thin filaments, and the magnetic lines of force are trapped inside these filaments. In other words, the bulk of the material is in the superconducting state, with no magnetic field. Through it run thin tubes of material in the normal, nonsuperconducting state, along which lines of force are threaded.

Thus it *is* possible for lines of force to be concentrated into tubes. However, all we have is an analogy concerning *magnetic* field, whereas we want the colour-*electric* field.

If we think of the superconductor as being infinitely diamagnetic, what we require in the QCD case is for the vacuum to have a zero (colour) dielectric constant ε. This is, at least, consistent with asymptotic freedom. As pointed out in the preceding section, asymptotic freedom means an r-dependent (colour) dielectric constant $\hat\varepsilon(r)$ which decreases with r. The present suggestion for confinement would require $\hat\varepsilon$ to be *zero* beyond a certain range for r.

But the present picture contains more than (an extreme case of) asymptotic freedom. It suggests that the vacuum should be capable of two states. The analogue of the superconducting state has $\varepsilon = 0$, and colour-electric lines of force are excluded. In the presence of strong enough colour fields, there should be another state of the vacuum (the analogue of the nonsuperconducting state) with nonzero ε (but $\varepsilon < 1$), in which colour lines of force can run.

The second state of the vacuum would occur within the tubes (like figure 17.21), and also within a certain distance (of order 10^{-15} m) of any quark. The transition between the two states of the vacuum would not be a precisely sharp one.

Let us, finally, consider a system in which some of the above mechanisms can, perhaps, be seen at work. The system is the bound state of a charmed quark and antiquark, called 'charmonium'. The first state of charmonium to be discovered was one of the ones produced directly from the annihilation of an electron and a positron, and called by one of its discoverers J and by the other ψ (and since called, diplomatically, J/ψ). This was in fact, the first confirmation of the existence of charmed quarks, as required by the electro-weak gauge theory.

Since the charmed quark is quite heavy (about $1.5\,m_p$), the size of the charmonium system (of order $\hbar/mc$) is quite small. Therefore, in the inner part of the structure, asymptotic freedom tells us that the colour force is not too strong. The nonlinear (gluon–gluon interaction) nature of QCD may not be too important there, and the force looks roughly like the inverse-square law of electricity. In so far as all this is true, the system closely resembles positronium (the bound state of an electron and a positron), with appropriate changes of scale because of the different masses.

On the other hand, in the outer regions of the charmonium system, the tendency for colour lines of force to concentrate into tubes should show up. This would make the force at these distances decrease less fast than $1/r^2$.

By fitting together a short-range $1/r^2$ force with a long-range, stronger force in this way, people have obtained fits to the energy levels of charmonium (of which ten are known). With the parameters obtained from these fits, the structure of the energy levels of the bound state of the bottom quark and antiquark (called Υ, upsilon) is successfully predicted. The theory should be rather more reliable for bottom quarks, since they are heavier (about $5\,m_p$) and so the system is smaller.

Thus, with a little optimism, one may hope that a better understanding of QCD is not so far off. The best hope of progress in the near future seems to lie with numerical solutions of the equations of QCD using large computers. But perhaps someone will find some new physical insight into the structure of the vacuum in QCD, just as there are now insights into, for example, the physics of superconductors (see Chapter 9 by Leggett).

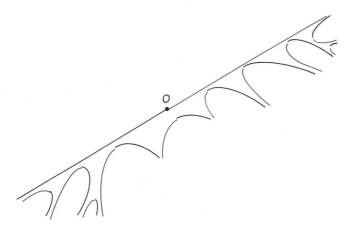

Figure 17.25. The formation of a jet. The quark and antiquark are produced at O. Other quarks (red) and antiquarks (black) are torn out of the vacuum, to make pions (quark and antiquark moving together). The pions nearest to O do not belong unambiguously to either jet.

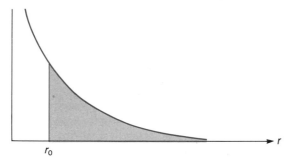

Figure 17.26. The electric attraction between oppositely charged particles. The $1/r^2$ law of force is plotted. The shaded area represents the (finite) energy required to separate from an initial distance r_0 to large distances.

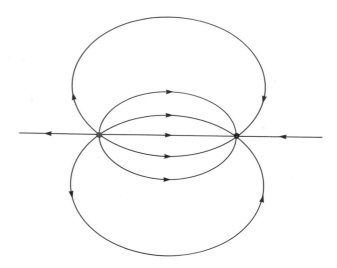

Figure 17.27. Electric lines of force produced by a positive and negative point charge. The lines of force spread out, and the inverse square law of attraction results.

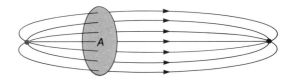

Figure 17.28. Two opposite charges, with the lines of force as they would be if they were confined to a tube, instead of spreading out as in figure 17.27. A is the, roughly constant, area of the tube. In this configuration, the attraction would be independent of the separation.

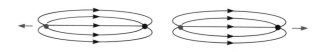

Figure 17.29. A flux tube breaking into two, by the creation of an extra quark–antiquark pair (the black and red dots nearest the centre). The outcome is the production of a pair of pions moving apart as shown by the arrows.

18 Overview of particle physics

Abdus Salam

18.1 Introduction

This chapter is concerned with an overview of the situation in particle physics and a prognosis of its future. Particle physics is a quest for the ultimate building blocks of matter, and the fundamental forces that operate to control and shape matter. It continues a tradition started in Ancient Greece with the early Atomists, and is surely one of the greatest intellectual endeavours of mankind. In this context, perhaps it is in order to remark on the sociological significance of physics as a whole. Physics is an incredibly rich discipline: it not only provides us with the basic understanding of the laws of nature, it also provides the basis of most of modern high technology. Because of this intimate connection with important sections of high technology, physics is the 'science of wealth creation' *par excellence*. This is even in contrast to chemistry and biology which, though as important for development, may perhaps be classed as 'survival sciences', in the sense that chemistry in application is concerned with fertilisers, pesticides, etc., while biology is concerned with medical sciences. Thus, together, chemistry and medical sciences provide the survival basis of food production and pharmaceutical expertise. Physics takes over at the next level of sophistication. If a nation wants to become wealthy, in the conditions of today, it must acquire a high degree of expertise in physics, both pure and applied.

18.2 Overview of particle physics

In the past, particle physics was driven by a troika which consisted of (1) theory, (2) experiment, and (3) accelerator and detection-devices technology. To this troika have been added two more horses. Particle physics is now synonymous with (4) early cosmology (from 10^{-43} s up to the end of the first three minutes of the universe's life) and it is strongly interacting with (5) pure mathematics. One may recall Res Jost who made the statement (towards the end of the 1950s) that all the mathematics which a particle physicist needed to know was a rudimentary knowledge of Latin and Greek alphabets so that one can populate one's equations with indices. This is no longer true.

The situation in this regard has changed so drastically that a theoretical particle physicist must now know algebraic geometry, topology, Riemann surface theory, index theorems and the like. The more mathematics that one knows, the deeper the insights that one may aspire to.

In the last decade or so, in particle physics, we have been experiencing an age of great syntheses and of great vitality. This is in contrast to when I started research (in the late 1940s and early 1950s) when we had ever-increasing quantities of undigested experimental data, and theoretical vignettes of great beauty and power, but little *coherent* corpus of concepts.

At the same time, this is an age of great danger for the future of the subject. Much of particle physics requires the use of giant accelerator machines to produce highly energetic collisions between subatomic particles. These impressive technological marvels are very costly, and future progress depends on the availability of bigger and bigger accelerators to achieve still greater energies, as well as somewhat less costly nonaccelerator and passive experiments. Such devices will, in the future, take an injection of funds of hundreds of millions of dollars, as well as longer experimentation times, for discovering new phenomena or for testing the truth or the inadequacy of theoretical concepts.

18.3 Three types of ideas

I shall divide my remarks into three topics: (1) ideas which have been tested or will soon be tested with the accelerators which are presently being constructed; (2) theoretical ideas whose time has not yet come, so far as the availability of accelerators to test them goes; and (3) passive experiments which have tested – but not conclusively so far – some of the theories of the 1970s. To give a brief summary, I consider each of these three topics in turn (for more detailed explanations of nomenclature and concepts, see the later sections).

(1) *Ideas which have been tested or will soon be tested*

(i) A common mathematical description of the electromagnetic force and the weak and strong nuclear forces is in terms of certain so-called gauge groups. The now standard model of particle physics elaborated in the late 1960s and early 1970s, and based on the SU(3) group for the strong force together with SU(2) × U(1) for the unified electromagnetic–weak (electroweak) force, has no known discrepancy with experiment.

(ii) The existence of a new type of particle, called a Higgs scalar, which is predicted by the unified electroweak theory, may be discovered in the Stanford Linear Accelerator (commissioned in 1987) or at the LEP accelerator at CERN when it is commissioned in 1989.

(2) *Theoretical ideas whose time has not yet come*

Accelerators are not yet being specifically constructed to test certain ideas. These ideas include

(i) theories known technically as $N=1$ supersymmetry and $N=1$ supergravity. These theories predict that each of the known subatomic particles will be matched in nature by yet-to-be-discovered partners whose properties are related to the known articles by a powerful mathematical framework known as supersymmetry. The lower limit on the masses for the supersymmetric partners of presently known particles may now be as large as 50 GeV. However, persuasive theoretical arguments lead us to expect that such partners of quarks and leptons may exist with masses below 1 TeV. To find these new particles we shall need new machines, such as the LHC (large hadron collider using the LEP tunnel), or SSC (superconducting supercollider being considered in the USA), or an electron–positron collider with energy in the TeV range. There has been no sanction from the European or US Governments for any of these accelerators, which may not arrive before the year 2000.

Other ideas in this category which also need higher collision energies to test are

(ii) the possible existence of a type of weak nuclear force with the opposite 'handedness' to the familiar weak force (i.e. a mirror reflection);

(iii) the possibility that quarks and leptons, which are currently accepted as fundamental particles, may in fact themselves be composite bodies made up of still smaller entities, called preons;

(iv) theories which involve extending the supersymmetry concept to include a still wider range of superpartners to the known particles;

(v) the very recent so-called superstring theories which describe both matter and forces in terms of the oscillations of string-like entities.

(3) *The set of ideas for which nonaccelerator and passive experiments have been mounted*

These are mostly concerned with grand unification of electroweak and strong forces in its multifarious ramifications and include (i) proton decays, (ii) transmutation of neutrons into antineutrons, (iii) the measurement of neutrino masses and their oscillations, in a speculated identity between different neutrino species (neutrino oscillations, in particular $v_e \rightarrow v_\mu$ or v_τ, have recently been invoked to explain the negative results of the celebrated Davis experiment, which concerns measuring flux of solar neutrinos of the v_e variety), (iv) magnetic monopoles, (v) various varieties of dark or invisible matter, in the form of very weakly interacting particles that could permeate the universe yet still remain undetected. A number of experiments to find these effects have been tried, but not with much success so far.

Let us now turn to each of these topics in turn.

18.4   Ideas which have been tested or will soon be tested

Since in this context we shall be concerned with the early availability of suitable particle accelerators, we shall start with table 18.1, which gives a list of already existing, soon to be commissioned, or proposed accelerators.

While we are discussing the availability of future accelerators, one must remember the following. The highest *electric field* gradients (which determine the size of an accelerator) achievable with today's technology, are at most around $0.1 \, \text{GeV} \, \text{m}^{-1}$. Twenty years hence, when we may have perfected the technology of laser beat-wave plasma accelerators, this gradient may go up by a factor of 1000, i.e. to $0.1 \, \text{TeV} \, \text{m}^{-1}$. This may mean that a 30 km long accelerator may produce energies of the order of $3 \times 10^4 \, \text{TeV}$. An accelerator circling the moon may generate $10^6 \, \text{TeV}$*.

An accelerator circling the earth – as Fermi once conceived – may be capable of achieving a centre-of-mass energy of about $10^7 \, \text{TeV}$, while an accelerator extending from earth to the sun would be capable of $10^{11} \, \text{TeV}$. In the same crazy strain, for an accelerator to be capable of generating $10^{16} \, \text{TeV}$ (theoretically, the ultimate 'Planck' energy) one would need an accelerator 10 light years in length; clearly one must eventually fall back on

---

*In discussing accelerator energies, account must be taken of the fact that the kinetic energy relative to the laboratory is not really the relevant quantity. A particle which collides with a stationary target will cause a recoil, thus wasting some of its energy. On the other hand, in some accelerators counter-rotating beams of particles that collide can avoid this making all their energy available for reactions. The relevant quantity is thus the energy in the centre-of-mass reference frame. Because of relativistic effects this centre-of-mass energy can be very different from the laboratory energy.

Table 18.1. *New and planned accelerators*

Year	Machine	$\sqrt{s}$ (GeV)	Constituent $\sqrt{s}$ (GeV)	Luminosity	Locality
1986	SpS	900	100–200	10^{30}	CERN
1987	Tevatron	2000	200–600	10^{31}	FERMILAB
1987	TRISTAN (e^+e^-)	60	60	10^{32}	Japan
1987	SLC (e^+e^-)	100	100	10^{30}	Stanford
1987	Bepc (e^+e^-)	4	4		Beijing
1989	LEP (I) (e^+e^-)	100	100	10^{31}	CERN
1995	LEP (II) (e^+e^-)	200	200	10^{31}	CERN
1995	UNK	3000	300–900	10^{32}	Serpukhov
1991	HERA (ep)	320	100–170	10^{31}	Hamburg
?	LHC	16 000	1600–5000		CERN
?	SSC	40 000	4000–10 000	10^{33}	USA
?	e^+e^-	4000	4000	10^{33}	Serpukhov

natural sources of high energy particles – the highest energy cosmic rays – to study, for example, the likes of the recently discovered high energy muon signals apparently associated with the astronomical object Cygnus X3 in the constellation of the Swan (see figure 18.2). This X-ray source, discovered in 1966, is some 37 000 light years distant from us, and has a duty cycle of 4.8 h and an integrated luminosity of 10^5 suns. Recently it has been claimed that Cygnus X3 is beaming to us particles of a new kind (one of the supersymmetry partners?) with energy as large as 10^4 TeV.

If this experimental evidence is taken at its face value, how is the radiation beamed at us by Cygnus X3 generated? One speculative idea is that the Cygnus system may consist of a binary star – a conventional main sequence star plus a pulsar or a black hole. Matter from the conventional star accretes around the compact pulsar or the black hole, forming a disc. The protons thus accelerated (up to maximum energies of 10^5 TeV) go into a beam dump, wherein is created the mysterious radiation, which hits our atmosphere and which creates the observed muons.

One sad aspect of the situation is that the Swan may be dying (figure 18.3) – the emitted flux seems to be decreasing at the rate of a decade over three years, much as if the beam dump was being blown away.

The standard model and the light Higgs particle

The standard model of today's particle physics describes three families of quarks and leptons. The first family consists of the so-called up and down quarks. These particles possess intrinsic spin, and may be divided into two classes according to their 'handedness' or 'chirality', denoted (u_L, d_L) and (u_R, d_R) (L and R stand for left-handedness and right-handedness). Each quark comes in three varieties, known whimsically as colours: red, yellow and blue. There are, in addition, three light objects, the 'leptons', (e_L, ν_L) and (e_R). Thus this family has twelve quarks and three leptons (i.e. fifteen two-component objects); see Chapter 14 by Close.

The second family is a near replica of the first. New labels, called charm and strangeness, replace the up and down for the quarks, while the electron and its neutrino are replaced by the

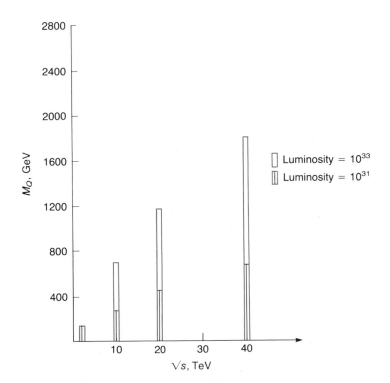

Figure 18.1. Windows for heavy quarks, squarks or leptoquarks.

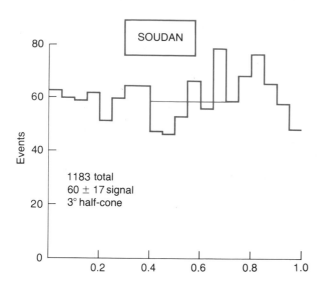

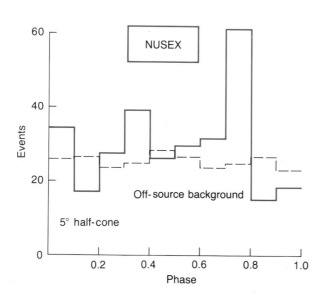

Figure 18.2. Phase structure of Cygnus X-3 underground muon signals.

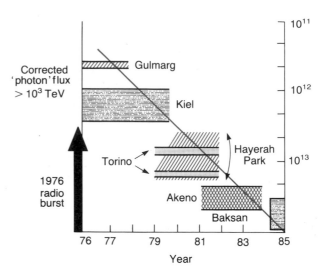

Figure 18.3. Time-dependence of the phase flux from Cygnus X-3.

muon and its neutrino. Like the first family, there are altogether fifteen two-component objects. The third family likewise consists of the so-called top and bottom quarks plus the tauon and its neutrino. An interesting question (see below) concerns whether there are still more families yet to be discovered. In addition to these $45 = 3 \times 15$ objects with intrinsic spin of $\frac{1}{2}$ units there are the twelve spin one objects: the photon (γ^0), W^+, W^-, Z^0 and eight gluons (the superscript gives the electric charge). Nine of these (γ and eight gluons) are

massless. These twelve objects are all 'messengers' associated with conveying the variety of forces between quarks and leptons. In addition, theory predicts one additional special particle, the spin-zero Higgs H^0, giving a total of 118 ($118 = 3 \times 15 \times 2 + 9 \times 2 + 3 \times 3 + 1$) degrees of freedom for the particles in the standard model. All particles except the Higgs in this list have been discovered and their masses and spins determined (though the discovery of the top quark is still disputed). In this context it is worth remarking that CERN data has confirmed the *theoretical expectation* of W^+, W^-, Z^0 masses to within 1%.

The spinless Higgs particle must exist if current ideas about unification of the electromagnetic and weak forces are correct. The Higgs is associated with the crucial field that couples to the other particles and provides them with a mass. However, theory does not specify the mass of the Higgs itself.

Defining a *light Higgs* as an object with a mass 300 GeV (and a heavy Higgs as an object with a mass beyond, up to 1 TeV), one may remark that a heavy Higgs would have a large uncertainty in its mass due to the effects of Heisenberg's uncertainty principle. Thus the concept of sharply defined 'particle' would essentially be lost. Also, it would cause the W and Z to interact strongly. One would then expect a new spectrum of bound states which would modify the properties of the known W and Z. No one likes this possibility, but it could happen.

Figure 18.4 (due to G. Kane) shows the possible signals of the standard model Higgs particles. As one can see, beyond a mass of 60 GeV, one would need the LEP II accelerator to detect these and eventually the SSC supercollider if the mass is in the TeV region.

One of the measurements which was carried out during 1985, relevant to the number of families in the standard model, was the estimate of the number of *light* neutrinos which may couple to the Z^0 particle and thus affect its decay rate. By

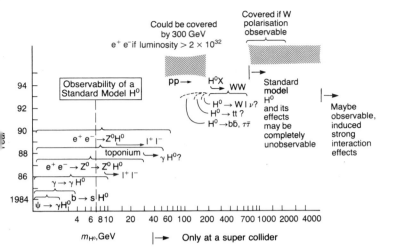

Figure 18.4. The possible signals of the standard model Higgs. Beyond a mass of 60 GeV one would need LEP II and eventually the SSC supercollider. (G. Kane.)

tradition, neutrinos had been considered to have zero rest-mass and to travel at the speed of light, but this is now open to question. Experiment seems to show that, at least for electron and muonneutrinos, the masses are very small, but not zero. The number of light neutrino species has been estimated from the collider measurements of the decay rate Z^0 to be 5.4 (± 1).

Remarkably, this sort of number is confirmed by an intriguing cosmological argument. Neutrinos produced in the big bang permeate the universe and contribute to the total mass of its constituents. The greater the number of neutrino families, the larger the mass of the universe would be. Now the total mass of the matter in the universe affects the rate at which it expands, and this in turn affects the details of nuclear reactions that took place during the first few minutes of the universe's existence. From the known cosmological abundances of the elements He^3 and He^4, thought to have been manufactured in the primeval phase of the universe, tight limits on the number of neutrino species can be derived. Cosmologists have suggested that cosmology may be consistent with three or four light neutrinos. No longer can one say with Landau 'Cosmologists are seldom right, but never in doubt'. They could even be right this time.

An important set of experiments which could be carried out at SLC or LEP concerns higher order quantum field effects in the unified electroweak theory in which virtual particles are emitted and reabsorbed. Careful measurements of these effects in electroweak processes could provide indirect information about the masses of the Higgs particle and the top quark, or the existence of hitherto unseen species of quarks. The hope is that by detecting small deviations from the predictions of the standard model, one may discover how to *extend* the standard model.

18.5 Ideas whose time has not yet come

Supersymmetry

The most important ideas in this category go by the names of $N = 1$ supersymmetry and $N = 1$ supergravity (see the chapter by Isham). The starting point of supersymmetry is the concept of intrinsic spin. A particle such as an electron is found to possess a type of intrinsic rotation reminiscent of a body spinning on its axis. The quantity of spin is always the same: ($\frac{1}{2}\hbar$, where $\hbar$ is Planck's constant divided by 2π). All quarks and leptons have spin $\frac{1}{2}$ in units of $\hbar$. Messenger particles have integral spin, like $\hbar$, e.g., for the photon, or $2\hbar$ (the graviton), while the hypothetical Higgs particle has zero spin.

All known particles possess half-integral number of units of spin ($\frac{1}{2}$ or $\frac{3}{2}$) or an integral number (0, 1, 2). Integral or half-integral spin particles have distinct physical properties, e.g. the latter obey the Pauli Exclusion Principle – they are individualists in the sense that no two of them can occupy the same state. Particles with integral spins, on the contrary, are gregarious, in the sense that they like to crowd into the same state. The name fermion (after Enrico Fermi) is given to particles with half-integral spin, and bosons (after Satyendra Bose) to the integral spin particles.

Because fermions and bosons have such different physical properties they are generally treated separately by theorists. Supersymmetry, however, is a more embracing symmetry that attempts to unite fermions and bosons in a common mathematical framework. This unification relates particles of different spins by placing them in families ('supermultiplets') interconnected by the 'supersymmetry' operation. In fact, it is hoped that all the known particles derive from such a fundamental supermultiplet. In this way all the fundamental bosons and fermions are just different supersymmetric manifestations of a single 'superparticle'. Supersymmetry is the ultimate proposal for a complete unification of all particles.

There are several different supersymmetry theories. The simplest, called $N = 1$, proposes that a spin $\frac{1}{2}$ particle must be accompanied by a spin zero particle, a massless spin 2 graviton must be accompanied by one massless spin $\frac{3}{2}$ particle called the gravitino, and so forth. The designation N refers to the maximum number of supersymmetry operations required to connect any particle in a supermultiplet with every other particle. For $N = 2$ extended supersymmetry, one would group together into a single multiplet two spin 0, two spin $\frac{1}{2}$'s and one spin 1 object. Such a theory would contain *two* gravitinos in the same supermultiplet with the graviton.

The largest supersymmetric structure is $N = 8$ extended supersymmetry, where there is just one multiplet containing one spin 2, accompanied by eight spin $\frac{3}{2}$ gravitinos, twenty-eight spin 1 gauginos (so called because they are the quanta of the gauge fields that mediate the forces such as electromagnetism), fifty-six spin $\frac{1}{2}$ and seventy spin 0 states.

Supersymmetry is an incredibly beautiful theory – a compelling theory if there ever is one, even though there is yet no physical evidence of the existence of any supersymmetry partners to the known particles – at least for masses up to 30 GeV (and even perhaps up to 50 GeV).

One aspect of its compulsion lies in its superior mathematical properties, particularly in relation to the long-standing problems of infinite terms that arise in all quantum field theories (see the section on renormalisation in Chapter 5 by Isham). An additional attraction is the possibility that supersymmetry may open up our understanding of how the large numbers which occur in particle physics could arise 'naturally'.

Consider as an example one of the more fundamental large numbers in physics, the dimensionless ratio $m_P/m_W \simeq 10^{17}$, where m_W is the mass of the W particle associated with the electroweak force and m_P is the so-called Planck mass, defined as $m_P = (hc/G)^{\frac{1}{2}}$. (Here G is Newton's gravitational constant and c is the speed of light.) The Planck mass, 'm_P', corresponding to about 10^{19} GeV, is many orders of magnitude larger than other mass scales in particle physics. It represents the energy at which quantum gravitational effects start to be important, and so it occurs naturally in theories involving gravitons.

The mass–energy scale m_W can be thought of as the regime at which the electromagnetic and weak forces merge in identity, while m_P would mark the regime at which gravity would merge with the other forces of nature. Theorists would like to be able to compute the ratio m_P/m_W, and in particular to know why it is such a large number. Only in supersymmetric theories will the value of this number be unaffected by higher order quantum field effects due to emission and reabsorption of virtual quanta (so-called radiative corrections). This important virtue hints that an understanding of the mysterious ratio m_P/m_W might be obtained through supersymmetric theories.

In spite of the aesthetic and physical attractiveness of supersymmetry, the real world is *not* supersymmetric. However, nature possesses many symmetries at a fundamental level that are broken in the real world. For example, the rotational symmetry of interatomic forces is broken in a ferromagnet in which there is a preferred magnetic alignment. In the same way, it could be that nature is supersymmetric in its underlying dynamics, even though the actual state of the world breaks supersymmetry – and breaks it fairly strongly.

Theorists would like to know at which mass–energy scale supersymmetry is broken. This in turn would tell us at what mass the missing supersymmetry partners of quarks, leptons, photons, W^+ and Z^0 lie. The theoretical expectation seems to be that the relevant mass of supersymmetrical partners could be below an upper limit of 1 TeV, *if supersymmetry is relevant to electroweak phenomena.*

To conclude, it is expected that supersymmetry may make itself manifest using *highly luminous* accelerators with centre-of-mass energy in excess of 1 TeV (e.g. LHC, SSC or an e^+e^- linear collider of 2 TeV). We could be fortunate and supersymmetry may manifest itself at lower energies as an *indirect* phenomenon, but that would be a great piece of luck. (Supersymmetry was indeed claimed in certain recent events detected at CERN, but the present background events are too many to draw unambiguous conclusions.)

Supersymmetry and N = 1 supergravity

If the fundamental laws of nature are supersymmetric we can draw some interesting conclusions:

(i) The $N = 1$ supersymmetrisation of the standard model of strong, weak and electromagnetic forces requires two multiplets of Higgs particles (plus, of course, their supersymmetric partners, the Higgsinos).

(ii) Supersymmetry imposes a new quantum number which is defined as $+1$ for all known particles and -1 for their supersymmetric partners. Thus (with beams of 'old' particles) these new partners must be produced in pairs to conserve this quantum number. Among the expected supersymmetry particles, therefore, there must be a lowest mass *stable* object which must be neutral in order to survive the big bang. Further, it must be weakly coupled to other particles otherwise it will be trapped by matter and concentrated in condensed form in the galaxies. The favourite candidates for this object are scalar neutrinos v, spin $\frac{1}{2}$ photinos $\tilde{\gamma}$, , or gravitinos – the spin $\frac{3}{2}$ partners of the (spin 2) gravitons.

(iii) If $N = 1$ supersymmetry is successful, the extension to include gravity (called $N = 1$ supergravity) cannot be too far behind. The relationship between supersymmetry and gravity comes about naturally. Gravity is treated in modern form as a manifestation of distortions in the geometry of spacetime (see Chapter 5 by Isham), and supersymmetry operations are related to geometry in a deep way. For example, two successive supersymmetry operations can generate a translation in space.

The argument for the inevitability of supergravity (if supersymmetry exists) goes as follows: the major theoretical problem regarding supersymmetry is supersymmetry breaking. The only decent known way to break supersymmetry is to break it spontaneously (i.e. like the onset of ferromagnetism). For this to work, one starts with a gauge theory of supersymmetry, i.e. a supergravity theory which (for the $N = 1$ case), would contain one spin $\frac{3}{2}$ gravitino for every spin 2 graviton. One would then postulate a 'super-Higgs' effect. (For an explanation of the Higgs effect see the chapter by Taylor.) This would require a spin $\frac{1}{2}$ and spin 0 multiplet of particles in addition to the graviton and gravitino, which interact with known particles only gravitationally (such particles would constitute 'shadow matter' because, on account of the weak coupling of the gravity, they would be almost undetectable).

The breaking of the supersymmetry is achieved by introducing a new superfield describing a supermultiplet of spin 0 and spin $\frac{1}{2}$ particles. This field gives rise to a new source of potential energy V, which causes the supersymmetric state to be unstable and hence drives the system into a state where supersymmetry is broken. As a result the spin $\frac{1}{2}$ member of the supersymmetry multiplet is absorbed by the spin $\frac{3}{2}$ gravitino, thereby endowing the gravitino with a rest-mass. The value of this mass is determined by the quantum expectation value of V, denoted $\langle V \rangle$, and by the Planck mass through the formula (gravitino mass)$^2 \simeq \langle V \rangle / m_{\mathrm{P}}^2$.

(iv) Interest therefore focuses on the precise value of $\langle V \rangle$. Some workers have estimated this by linking spontaneous supersymmetry breaking in supergravity to the spontaneous symmetry breaking that occurs in the electroweak unification theory. If one requires a zero value of the so-called cosmological constant in gravitational theory (see Chapter 3 by Guth and Steinhardt) then they find that the (spin $\frac{3}{2}$) gravitino mass is *roughly equal to* the mass of W, an important prediction. One may then estimate $\langle V \rangle^{1/4} \simeq 10^{10}$ GeV.

Right–left symmetry and preons

Quarks exist in both left and right states of chirality; so do electrically charged leptons (electrons, muons and tauons). Neutrinos seem to come only in left-hand form (i.e. they spin like left-handed corkscrews along their line of motion) – there is at present no evidence for the existence of right-handed neutrinos. Correspondingly, the weak force bosons, W^+, W^-, couple only with left-handed chiral currents. Is there a fundamental right–left symmetry in nature, spontaneously broken at an accessible energy – 1 TeV, say? Are there right-handed weak currents; are there right-chiral neutrinos? These are important questions for fundamental theory which may, or may not, be answered with the next generation of accelerators (LHC or SSC or $e^+ e^-$ linear collider).

A similar question mark hangs over the prospects of preons – a set of objects of which the present level of 'elementary' entities (quarks and leptons) – may be composed. The existence of preons would be the simplest answer to the 'Family Mystery', the twice-over replication (except for their masses) of apparently identical carbon-copies of quarks and leptons in the first family. Again, if preons exist, they would need accelerators with centre-of-mass energies beyond 1 TeV for their discovery, though the HERA accelerator at Hamburg will make an attempt at finding their signature indirectly, by measuring the quark form factor, below 1 TeV.

Unification of gravity with other forces

This brings me back to the supergravity story. So far we have considered ($N=1$) supergravity as following on the heels of ($N=1$) supersymmetry in order to provide for an orderly breaking of supersymmetry: there was no implication of a unification of gravity with other forces. Let us now discuss a true unification of gravity (or of supergravity) with the rest of particle physics.

The first physicist to conceive of unifying gravity with electromagnetism and to try to find experimental evidence for such a phenomenon was Michael Faraday. In a symbolic drawing – due to Alvaro de Rujula – one may see the equipmental set-up (figure 18.7). (The actual equipment which Faraday used is on exhibition at the Royal Institution in Piccadilly, London.) The failure of this attempt did not dismay Faraday. Fresh from his triumph with unifying electricity with magnetism, he wrote:

> If the hope should prove well founded, how great and mighty and sublime in its hitherto unchangeable character is the force I am trying to deal with, and how large may be the new domain of knowledge that may be opened to the mind of man.

The first semi-successful *theoretical* attempt (in the 1920s) to unify gravity with electromagnetism was that of Theodore Kaluza (and following him of Oskar Klein) who showed in a theory based on a *five-dimensional spacetime*, that the appropriate curvature component in the fifth dimension corresponds to electromagnetism. That is, if Einstein's gravitational field equations are generalised to a five-dimensional spacetime, they reproduce *both* ordinary gravitational physics (in ordinary four-dimensional spacetime) *and* Maxwell's theory of electromagnetism.

The problem of why we don't see the extra (fifth) dimension is solved ingeneously by supposing that space is 'rolled up' or, to use the technical term, compactified in the fifth dimension. The situation can be envisaged by analogy with a hose-pipe. Under low resolution the hose-pipe looks like a wiggly one-dimensional line, but on closer inspection each 'point' on the line is revealed as a little circle going around the hose. Similarly, each apparent point in space could be a tiny circle going around a fifth dimension.

If the circumference of these circles is R, and electrically charged matter is introduced into the theory, then a connection emerges between the value of the fundamental unit of electric charge, e, and Newton's gravitational constant, G. Specifically, the fine-structure constant $\alpha = e^2/\hbar c$ turns out to be approximately G/R^2. To obtain the observed value for $\alpha = 1/137$ requires $R \simeq 10^{-33}$ cm. Incredible audacity: first to conceive of a fifth dimension and secondly to suggest that, unlike the other four dimensions, the fifth must be compactified to a scale of length R as small as $(G/\alpha)^{\frac{1}{2}} \simeq 10^{-33}$ cm.

These ideas were beautifully generalised in an extended supergravity context, when Cremmer and Julia discovered in 1979 that the extended $N=8$ supergravity in four dimensions emerges as the zero-mass limit of the compactified $N=1$ supergravity in eleven dimensions. Thus, by grafting on to ordinary four-dimensional spacetime not *one*, but *seven*, extra

FIRST ATTEMPT TO UNIFY... ...ELECTRICITY & GRAVITY

FORCES OF THE WORLD, UNITE!

A: BALLS
B: CUSHION
C: COIL
D: GALVANOMETER
E: ROYAL INSTITUTION
F: M. FARADAY
S: STRING

Figure 18.5. Cartoon depicting Michael Faraday at the Royal Institution. Due to A. de Rujula.

dimensions, suitably compactified, gravity can be unified with all three other forces – electromagnetism, weak and strong. Technically, this too was an astounding achievement. Since 1979, all supergravitors have lived in higher dimensions.

At that time, this theory was hailed as the first T.O.E. (Theory of Everything). If it could be physically motivated as a spontaneously induced phase transition, the compactification of eleven-dimensional Kaluza–Klein supergravity down to four apparent dimensions would give, among its zero-mass particles, gravitons as well as gauge particles like the spin one photon Z as well as fifty-six fermions – all part of the unique multiplet of $N=8$ supergravity.

Unfortunately, the $N=8$ theory and this particular multiplet suffered from two fatal defects: the fermions were not chiral (i.e. did not have a definite handedness) and the theory did not have the content of the standard model so far as quarks, leptons or even the W^+, W^- were concerned. Also, in addition to the zero-mass particles, the theory predicts a collection of Planck mass

particles (with (mass)^2 = multiples of $1/R^2$), the so-called pyrgons, providing another embarrassment of riches.

Can one ever obtain direct evidence for the existence of higher dimensions? The answer is, possibly yes. If the extra dimensions happen to have been compactified to a small size through a spontaneous compactification mechanism – which, ideally, should be part of this theory – why should they remain small for ever? Why should these extra dimensions not share the universal cosmic expansion? Since α, G and R are expected to be related to each other we might be fortunate and find that α or G turn out to be time-dependent at the present experimental level if R is expanding. Such an effect might most simply be explained by postulating extra dimensions and their expansion at the present epoch. The experimental limits happen to be less than 1×10^{17} year^{-1} for $\dot{\alpha}$ while $\dot{G}/G$ is less than 1×10^{11} year^{-1}. A definite nonzero answer would be most welcome.

Anomaly-free supergravities

Where do we stand theoretically today as far as higher dimensions and extended supergravity theories are concerned? It would appear that the only theories which may combine *chiral* fermions and gravity are $N=1$ supergravity in $d=10$ dimensions or $N=2$ supergravity in six-dimensional (or in ten-dimensional) spacetimes. In order that such theories contain the known chiral quarks and leptons (as well as the W's, Z, photons and gluons) the most promising is the $N=1$, $d=10$ supergravity, but it would have to be supplemented with an additional supersymmetric multiplet of matter, known as a Yang–Mills multiplet (see the chapter by Isham for Yang–Mills fields), in addition to the supergravity multiplet. Likewise, the $N=2$ supergravity theory in six dimensions would need, not only an extra Yang–Mills field, but also some extra nonlinear matter fields. Thus a pure Kaluza–Klein supergravity theory will never be enough. Higher dimensions, yes; but to generate the known gauge theories of electroweak and strong forces, we need (higher-dimensional) super-Yang–Mills in addition.

As if this was not trouble enough, both $d=6$ or $d=10$ theories were shown to contain mathematical inconsistencies called anomalies, and to be replete with infinities. This impasse was broken only in Autumn 1984 by Green and Schwarz who showed that $N=1$ supergravity in ten dimensions with an added Yang–Mills characterised by the gauge group SO(32) (or alternatively $E8 \times E8$) could be made anomaly-free by an addition of certain numbers of new terms. They further showed that these additional terms were already present in a remarkable class of theories, invented some years earlier, that seek to replace particles by strings as the fundamental objects of study. Most likely, these string theories are also free of gravity infinities, though this is still an open question. So this brings us to the world of superstrings and the latest version of a Theory of Everything.

Supersymmetric strings

Consistent, relativistic string theories had already been written down in two, ten or twenty-six dimensions (the last being relevant only to bosonic strings), in the 1970s. A closed string is a loop which replaces a spacetime point. Its quantum oscillations correspond to particles of higher spins and higher masses, which may be arranged on a linear trajectory in a spin-versus-mass^2 (Regge) plot. If the slope parameter of this trajectory – the only parameter in the theory – is adjusted to equal the Newtonian gravitational constant, one can show, quite miraculously, that in the zeroth order of the closed bosonic string there emerges from the string theory Einstein's gravity in its fullness! (The higher orders give modifications to Einstein's theory, with corrections which have a range of Planck length $= 10^{-33}$ cm.)

Furthermore, the supersymmetric ten-dimensional string theory (descended from twenty-six dimensions) could exist in a 'heterotic' form, invented by Gross and his collaborators, with a built-in Yang–Mills gauge symmetry having a gauge group G of rank 16 which can either be $G = SO(32)/Z_2$ or $E_8 \times E_8$. This theory, though chiral, is anomaly-free. The descent from twenty-six to ten dimensions is accomplished by compactification on a sixteen-torus $(26 - 10 = 16)$, which, using the beautiful results of Frenkel and Kac, in fact reproduces a full tally of 496 Yang–Mills massless gauge particles associated with $SO(32)/Z_2$ or $E_8 \times E_8$ even though we started with only sixteen gauge particles (corresponding to the sixteen-torus). The remaining 480 gauge particles are solitons in the theory. The theory is free of both gravitational and chiral anomalies. The hope is that this theory may also be finite to all orders – perhaps the only finite theory of physics containing quantum gravity!

Can we proceed from ten down to four physical dimensions? Witten and his collaborators have attempted to show that the ten-dimensional theory can indeed be compactified to four-dimensional Minkowski spacetime plus an internal six-dimensional space (called a Calabi–Yau space) which preserves a residual $N = 1$ supersymmetry in four dimensions. A number of families emerge; their count is equal to one-half of what mathematicians call the Euler number of the compactified space. (This is a number that is determined by the topology of the space.) The coupling strengths between particles allowed by the theory are expected to be topologically determined. Another type of compactification is motivated using a construction which is toroidal in essence – the 'orbifold' construction. I shall not discuss this further.

It is all these remarkable features of superstring theories which make the string theorists 'purr' with deserved pride.

String theory as the 'Theory of Everything' (T.O.E.)

Could this be the long-awaited unified theory of all low energy phenomena in nature? The amazing part of the story is that the equivalence principle of Einstein's emerges from the theory, and does not have to be put in as Einstein did. But would such a theory be a T.O.E. – a Theory of Everything? The answer in my opinion is *No*. As remarked before, all theories which descend from higher to lower dimensions must contain massive particles in multiples of the Planck mass. Since no *direct* tests of the existence or interactions of such objects can ever be feasible, there will always remain the experimentally unexplored area of these higher masses and energies (in addition to the different level of mystery associated with the quantum of action, which still has to be incorporated within the structure of the theory). What we are saying is that before this can be called a T.O.E., one must prove, at the least, a *uniqueness* theorem which states that if a theory fits all known phenomena at low energies, it can have only *one* extrapolation to higher energies. From all past experience, this is unlikely – even as regards the basic framework. (Think of the framework of Newtonian gravity versus that of Einstein's gravity.)

There arises an important question at this stage of whether the string theories really do represent a wholly new attitude so far as fundamental theory is concerned, or whether we are in fact merely dealing with a conventional relativistic quantum field theory, but formulated in two dimensions.

The latter point of view has been (implicitly or explicitly) argued for by a number of authors: by Polyakov and those who have followed him, and latterly by Weinberg and the group at Trieste. According to this reasoning, the superstring theory can be regarded as equivalent to a theory of fields in a two-dimensional spacetime. In such a theory, gravitation takes a particularly simple form in which the *topology* of the spacetime determines the dynamics.

In addition, there are twenty-six fundamental, or 'preonic' fields living in this two-dimensional spacetime. The system of preonic fields should undergo a phase transition at Planck temperature and will acquire nonzero quantum expectation values for temperatures below this. As a result fields will display precisely the properties normally associated with twenty-six-dimensional spacetime. That is, what is normally regarded as twenty-six spacetime dimensions is mathematically equivalent to the expectation values of the twenty-six preonic fields. Spacetime was born at Planck time $(t \simeq 10^{-43} \text{ s})$.

The usual W, Z, etc. particles are recovered as composite structures built out of the preonic fields. To demonstrate that this complex assemblage conserves quantum probability (i.e. unitarity) is a major though familiar challenge, familiar in the sense that this problem is on par with the unitarity problem of theories of composite hadrons (protons and neutrons) in the context of a fundamental quark theory.

But the real and incandescent beauty of this theory manifests itself when one writes down the matrix elements. Many technical mathematical properties – conformality, modular invariance, holomorphy and the Riemann–Teichmuller variables (necessary for their description) – make for a unique specification of the amplitudes, never experienced in physics before. One can only hope nature is aware of our work on all this.

The other point of view favours the 'second quantised' version of the string theory. Here, unitarity problems for the likes of W's, Z's, photons (and also quarks and leptons in a spinning string version), present no difficulties. (It is as if one were writing the local field theory of hadrons.) This fact that 'local' field theories of *extended* objects – like strings – should exist at all, seems highly nontrivial at least within the mathematical framework used at present.

Perhaps the profoundest aspects of this second quantised formulation are represented by the work of Witten, who derives the basic interaction of strings in such a theory (at least for the purely bosonic open string) using a mathematical discipline called noncommutative geometry, which is *geometry freed of dependence on spacetime charts.* If this point of view succeeds in extensions to closed and spinning strings, the string theories would be ushering in a new era, like that of quantum theory in 1925 and 1926 when a new epistemology came into existence to replace the humbler point of view represented by the 'old' quantum theory.

Which of these two points of view will yield deeper insights, and in the end prevail, time will tell.

But apart from these matters of interpretation, the one crucial question which our experimental colleagues are entitled to ask, is this: what are the compelling experimental consequences of string theories?

The emergence of (necessarily a supersymmetric) standard model with the correct number of families may, of course, be a triumph, but will it establish the superiority of the string attitude? At present, there are few unambiguous *new* predictions. One of them concerns the existence of one or two new Z^0's. In figure 18.6 is shown the window for such Z^0's. Unfortunately, their masses (in fact, even their existence) are not firmly predicted by the theory. A possibly firmer and more spectacular prediction is the existence of fractionally charged dyons (particles with both electric and magnetic charges).

18.6 Passive and nonaccelerator experiments

Next we come to the passive experiments which are mainly concerned with testing *gauge* aspects of grand unification theories (unifying electroweak and strong nuclear interactions). These are the tests for (i) *magnetic monopoles* – though their formation during the hot early stages of the universe is predicted by gauge theories, one would not like too many monopoles to be around now; otherwise there will be problems with the magnitudes of cosmic magnetic fields. (In this context, note the claimed detection at Imperial College during 1985 of the 'South Kensington' monopole – the second monopole ever to be detected in laboratory conditions.) (ii) *Cosmological strings,*

which are good for galaxy seeding. (iii) *Domain walls,* which apparently would be a cosmological disaster.

Surely, this set of predictions presents a mixed bag of desirables and undesirables. In addition there is the question mark on varieties of remnant dark matter, endemic to most of our theories and whose ever-lengthening list is given in table 18.2. These topics are dealt with in more depth in Chapter 3 by Guth and Steinhardt.

Among the most celebrated passive and nonaccelerator experiments is proton decay. A limit on $p \rightarrow e^+ + \pi^0 \gtrsim 2.5 \times 10^{32}$ years partial decay time is suggested by current experiments. There are, however, claims for (nine) candidate events for $p \rightarrow e^+ + K^0$, $N \rightarrow \nu + \eta$ and $N \rightarrow \nu + K^0$ modes. (A firm detection of K's would signal supersymmetry and also explain the longer lifetime.) A worrisome background is due to atmospheric neutrinos which would make it difficult, on earth, to be sure of a real signal for proton decay if its life much exceeds 10^{33} years. Pati, Sreekantan and Salam have suggested experiments on the moon where even though the primary flux of cosmic rays is unhindered by the existence of an

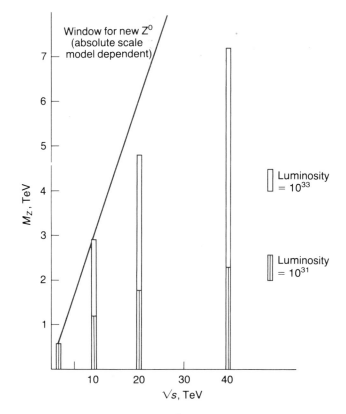

Figure 18.6. Window for new Z^0 particle (absolute scale model dependent).

Table 18.2. *Expected dark matter*

	Mass	Time of origin (s) (and corresponding temperature)
Invisible axion	10^{-5} eV	$10^{-30}(10^{12}\,\text{GeV})$
v	30 eV	1 (1 MeV)
Light $\tilde{\gamma}$, gravitino	keV	10^{-4} (100 MeV)
Heavy $\tilde{\gamma}$, gravitino, axiano, sneutrino, v	GeV	
monopole	10^{16} GeV	10^{-34} (10^{14} GeV)
Kaluza–Klein particles and shadow matter (maximons, pyrgons, etc.)	10^{18}–10^{19} GeV	10^{-43} (10^{15} GeV)
Quark nuggets	10^{15} g	10^{-5} (300 MeV)
Primordial black holes	$<10^{15}$ g	$<10^{12}$ (MeV)

atmosphere for magnetic fields, an experiment carried out in a tunnel or a cavern with 100 m of moon-rock surrounding it on all sides, would cut down the backgrounds – in particular of v_e neutrinos – to a figure less than 1/100 of the background on earth. If proton lifetime lies within the range 10^{34}–10^{35} years, experiments on the moon may become necessary for unambiguous detection.

The cost of such moon experiments consists in taking around some 100 tonnes of detecting material to the moon, plus the cost of the making of the cavern; it may come to around one billion dollars. Such outlays would become feasible if moon colonisation programmes are pursued seriously. (No doubt this will happen if there is a banning of nuclear weapons, since technological, advanced societies must spend funds on high technology projects, in order to keep the overall economy healthy.)

This concludes my survey of particle physics.

Glossary

absolute zero   the temperature at which thermal disorder completely disappears, and which is therefore the ultimate limit of 'coldness'. On the Kelvin (absolute) temperature scale it is by definition the zero of temperature: on the Celsius scale it lies at about -273 degrees.

angular momentum   a property of rotary motion analogous to the familiar concept of momentum in linear motion.

anisotropic superfluid (superconductor)   a system of fermions in which Cooper pairs form in a state of finite relative orbital motion and possibly finite total spin.

antiferromagnet   a solid in which the spins of neighbouring atoms are oppositely aligned. The lattice is composed of two equivalent sublattices, and on each sublattice the spins are magnetised, as in a ferromagnet, but the directions of the magnetisations are opposed so that there is no net magnetisation.

antimatter   for every variety of particle there exists an antiparticle with opposite properties such as sign of electrical charge. When a particle and its antiparticle meet they can mutually annihilate and produce energy. Thus, antiquark, antiproton, etc.

Argand diagram   a diagram in which the length and phase-angle of a complex quantity is displayed.

astrophysics   all applications of the laws of physics, chemistry and the other physical sciences to the understanding of astronomical phenomena.

asymptotic freedom   in general, there are strong, QCD forces between quarks due to the exchange of gluons. Asymptotic freedom is the principle which says that these forces become weaker for very close encounters between quarks, so that the quarks become 'free' of the forces at very short distances. In high-energy collisions, the quarks do sometimes become very close and, since they then behave almost like free particles, it is not too difficult to calculate their motion during the close collision.

attractor   a mechanical system may be such that its dynamical evolution causes it to approach a stable end-state. In the phase space representing the system, the representative point tends to a fixed set of points called an attractor. The attractor may be a point, a line, or a fractal.

autocatalysis   the ability of certain chemicals to enhance by their presence the rate of their own production in a sequence of chemical reactions. Part of the more general class of feedback processes.

β-decay   decay of a radioactive nucleus with production of an electron (β-particle). The underlying process is the transmutation of a neutron into a proton with electron and neutrino produced as a consequence. This process is controlled by the weak interaction and was its first known manifestation.

baryon   nuclear particle, e.g. proton, built from three quarks.

baryon number   a quantity assigned to elementary particles: quarks are assigned a baryon number $\frac{1}{3}$, and antiquarks $-\frac{1}{3}$; protons and neutrons, as well as a number of unstable particles, are each composed of three quarks and hence have baryon number 1, while antiprotons and antineutrons have baryon number -1; particles not composed of quarks, such as the electron or photon, have baryon number 0. Baryon number is conserved in all observations, but theories such as grand unified theories imply that it may not always be conserved, particularly at the very high energies achieved in the early universe.

Bell's inequality   one of a family of inequalities concerning the probabilities of joint occurrence of certain events in the two well separated parts of a composite system, implied by any hidden variables theory which satisfies an appropriate locality condition. The first example was derived by J. S. Bell in 1964.

Bell's theorem   the theorem that no hidden variables theory satisfying an appropriate locality condition can make statistical predictions in complete agreement with those of quantum mechanics. Specifically, there are situations in which quantum mechanics predicts a violation of Bell's inequality.

bifurcation   a phenomenon whereby the number of solutions of certain type presented by a dynamical system changes abruptly, as one of the parameters defining the dynamics crosses a critical value.

big bang theory   the theory, which has been generally accepted since the 1960s, that the universe began approximately 10 to 20 billion years ago in a state of enormous temperature and density and has since expanded and cooled to its present state.

binary stars   in binary stars, the two stars form a physically bound pair under their mutual gravitational attraction. The stars move in elliptical orbits about their common centre of mass. As many as 50% of all stars may be members of binary star systems.

black hole   a region of space-time which cannot be seen by distant observers because light is trapped by a strong gravitational field. The boundary of this region is called an event horizon because it separates events (i.e. those in the hole) that cannot be seen from events outside the hole, which can. Black holes might form, for example, from the gravitational collapse of a massive star. When the star shrinks inside an event horizon, it will collapse without known limit, leaving the surrounding space empty. Thus the designation 'hole'. Spherical black holes (without electric charge) are known as Schwarzschild black holes. Rotating holes are nonspherical; they are known as Kerr black holes.

BL-Lacertae object   an extreme form of active galactic nucleus in which the light is dominated by nonthermal continuum emission and there are no emission lines present in the spectrum, in contrast to the case of quasars. These active nuclei are highly variable in intensity and often are very highly polarised. They are named after the brightest prototype of this class, the variable object BL-Lacertae.

Bose condensation   a phenomenon occurring in a system of bosons whose total number cannot change, at a temperature of the order of the degeneracy temperature, in which a finite fraction of all the particles begin to occupy a single one-particle state.

Bose statistics   the form of statistics applicable to bosons. It favours configurations in which many particles occupy the same state.

boson   generic name (after Satyendra Bose) for a class of particles possessing whole number units of intrinsic spin, measured in terms of Planck's constant divided by 2π. All known particles that are not bosons possess half-integer units of spin, and are called fermions. The particles associated with the transmission of forces – photons, gravitons, gluons, W and Z – are bosons. Certain assemblages of fermions, such as the helium-4 nucleus, in which the combined spin is an integer, behave as bosons.

Brans–Dicke theory   alternative theory of gravity to general relativity devised by Carl H. Brans and Robert H. Dicke around 1960. A scalar gravitational field is postulated in addition to the curved space-time metric. Predictions for observable effects differ slightly from those of general relativity.

brown dwarf   A term used to describe very low mass stars, less than about one-tenth the mass of the Sun, in which the central temperature is too low for the nuclear burning of hydrogen into helium to take place. They may derive their internal energy from the initial energy of collapse. They are expected to be very cool stars and hence strong infrared emitters. They could make an important contribution to the hidden or dark matter in galaxies and clusters of galaxies.

canonical   the canonical approach to dynamics refers to the scheme in which the basic constituent is a space of states and the evolution of the system is described by a curve in this space parametrised by time. This approach to classical physics is in many respects the basic one to adopt when attempting to include quantum effects. In the case of a field theory, it has the disadvantage that space and time are treated on a different footing, and hence it is not always an easy matter to show that the formalism is compatible with the theory of relativity.

causality   pertaining to the time development of a system and the requirement of special relativity whereby energy cannot be propagated at a speed faster than that of light.

Chandrasekhar limit   the upper limit to the masses of stars which are held up by degeneracy pressure. The mass is more or less the same for white dwarfs and neutron stars, being about the mass of

the Sun. In the case of white dwarfs, the degeneracy pressure is due to the electrons and in the case of neutron stars to neutron degeneracy pressure.

chaos   originally used by the Greeks to describe the limitless void, it is now used to describe unpredictable and apparently random structures. The study of chaos using topology and computers has become a major part of modern mathematics, revealing universal and fundamental laws of a remarkably simple kind.

chaotic dynamics   time-dependent aperiodic regime in which individual histories corresponding to initially close states tend subsequently to diverge exponentially.

charm   one of the quark flavours.

chemical clock   an asymptotically stable regime of a chemical system in which the concentrations of the reagents are periodic functions of time. Both the period and the amplitude are determined solely by the system's intrinsic parameters.

coarse-graining   an operation implementing some form of spatial averaging which smoothes out relatively small length-scale configurational structure while preserving the larger length-scale structure.

colour   property possessed by quarks and gluons. A 'threefold type' of charge akin to electrical charge, believed to be the source of the strong force between quarks and described by the quantum chromodynamic theory of the strong interaction.

commutation relations   in quantum mechanics, if one has two operators A and B, then it is often the case that the action of the product operator AB is not the same as that of BA. The difference, $AB-BA$, is called the commutator of A and B. Specifying the value of the commutator is known as a commutation relation.

complexity   in information sciences, complexity measures the length of the shortest description of a given (finite) sequence of symbols. In the physical sciences, complexity is associated with the ability of a system to display long range coherence in space and time, and to undergo transitions between different states. These two alternative views might merge in chaotic dynamics.

Compton wavelength   in relativistic quantum mechanics the quantity h/mc represents the wavelength of the quantum wave associated with a particle of mass m. This is called the Compton wavelength (after Arthur Compton) because it appears in the theory of Compton scattering – the scattering of photons by electrons.

condensed matter physics   primarily the physics of solids and liquids. Analogous behaviour is expected in some stellar interiors, particularly in white dwarfs and neutron stars.

conduction band   in an insulator or semiconductor this is the lowest empty energy band, and electrons excited to the conduction band can carry current. In a metal the conduction band is the partially filled band in which the current-carrying electrons occur.

conservative system   a system in which total energy is conserved in time, and the evolution of the observable properties is indifferent with respect to the direction of time.

Cooper pair   a complex of two fermions in a degenerate Fermi system, which resembles a diatomic molecule. Cooper pairs differ from diatomic molecules in being strongly overlapping and automatically Bose-condensed.

correlation length   the correlation length ξ gives a measure of the typical distance over which the fluctuations of one microscopic variable are correlated with the fluctuations of another.

cosmic ray astronomy   the astronomy associated with the detection, propagation and origin of cosmic rays from their sources to the Earth.

cosmic rays high energy particles of extraterrestrial origin which can be detected in observations made above the Earth's atmosphere. If the particles are of very high energy, they can give rise to air-showers when they penetrate into the atmosphere; these showers can be detected at ground level. The cosmic ray particles are mostly protons, electrons and helium nuclei with a few per cent of heavy elements, and are present throughout the disc of our Galaxy.

cosmic strings thin, massive, thread-like objects that are predicted to exist by some, but not all, grand unified theories; they have a thickness of about 10^{-29} cm and a mass of about 10^{22} g cm^{-1}, or 10^7 solar masses per light-year; they could be produced copiously in a random arrangement in the early universe and might play an important role in the formation of galactic structure.

cosmological constant a contribution to the equations of general relativity, independent of space and time, which Einstein proposed in 1917 in order to allow a static model of the universe; in the nonrelativistic limit, a positive cosmological constant describes a force causing particles to fly apart from one another, with an acceleration proportional to their separation and independent of their masses; the cosmological constant is proportional to the energy density of the vacuum, and its value is known to be very small or perhaps zero.

cosmological principle a basic assumption made in the construction of cosmological models which is that the Earth is not located at any special position in the Universe. It is assumed that we are in a typical position and that the large-scale features of the Universe which we observe would also be observed by any other suitably chosen observer at the present day.

coupling constants the various forces of nature act on subatomic particles, each with a certain strength. The strength can be expressed as a constant that determines how strongly the particle couples to the field associated with the force concerned.

critical exponent near a critical region one physical quantity, such as the magnetisation, is often proportional to a power of another quantity, such as the difference between the temperature and the critical temperature. The power that occurs is known as a critical exponent.

critical phenomena the phenomena which occur in the neighbourhood of a continuous phase transition, characterised by very long correlation lengths.

critical point a point in a phase diagram identifying conditions in which the correlation length associated with some appropriate set of microscopic variables is, in principle, as large as the physical system.

critical temperature the temperature at which a continuous phase transition occurs.

degeneracy temperature the characteristic temperature, for a given system of indistinguishable particles, below which the effects of the Fermi or Bose statistics obeyed by them become apparent. At temperatures well above the degeneracy temperature indistinguishability has little effect.

degenerate an electron gas is said to be degenerate when the temperature is sufficiently low that almost all electron states up to a certain energy are occupied, and all those with higher energy are empty.

detailed balance cancellation of the effect of a process by a simultaneously operating 'inverse' process.

determinism the philosophy that all events are completely determined by prior events.

differential geometry the mathematical discipline that studies curved spaces.

diffraction the wave properties of light are manifested when an obstacle obstructs the light path. The optical waves spill around the obstacle and spread out in a process called diffraction.

diffraction grating a diffraction grating consists of a very large number of grooves ruled closely together. Light is reflected from the grating at angles which depend both on the wavelength of the light and its incident angle. Gratings are used in optics to discriminate between close frequencies.

Dirac equation the equation in quantum mechanics that describes the time evolution of the state associated with electrons and other spin $\frac{1}{2}$ particles. Unlike Schrödinger's equation that it replaces, Dirac's equation is consistent with the special theory of relativity.

disorder the departure of the atoms of a solid from their regular lattice positions.

dissipative structures states of matter arising through bifurcation when a system is driven away from the state of thermodynamic equilibrium by external constraints exceeding a critical value.

dissipative system a system satisfying the second law of thermodynamics. Such a system gives rise to irreversible processes, associated with a time-asymmetric evolution of observable quantities.

domain a ferromagnetic material is composed of domains, in each of which there is magnetisation in a definite direction. Neighbouring domains generally have opposed directions of magnetisation because that lowers the energy of the magnetic body, so that the magnetisation per unit volume of the whole body is generally very much less than that of single domains. A domain wall is the boundary between two domains.

Drell–Yan annihilation annihilation of a quark and an antiquark, thereby producing a high energy or high momentum photon, W or Z boson.

dye laser this is a widely used laser which has a solution of dye molecules as its active element, excited by photon pumping. The dye can emit over a wide range of frequencies making it an ideal source of radiation tunable across the whole visible spectrum. Its large bandwidth makes it a good source of short optical pulses.

eightfold way classification scheme for elementary particles established *c.* 1960. Forerunner of quark model.

Einstein equivalence principle foundation for curved space-time, it states that bodies fall with the same acceleration and that physics in freely falling reference frames is independent of the velocity and location of the frames.

Einstein's general theory of relativity the theory of gravity in which the gravitational force is described mathematically by a curvature in space or space-time.

electron gas a system of electrons whose mutual interactions are sufficiently weak that they can be regarded as moving independently, subject only to the effects of the exclusion principle.

electron–volt (eV)   unit of energy. Typically $1-10$ eV is the amount of energy per atom involved in chemical reactions. 1 eV is the energy gained when an electron is accelerated by a potential of one volt.

energy band   a continuous range of energies in a solid in which there are possible states for the electrons. Energy bands are separated from one another by energy gaps.

energy gap   a range of energies in a solid for which there are no quantum states of the electrons.

entanglement   the impossibility of expressing certain quantum mechanical states of a system with two or more parts as the conjunction of definite quantum states of the separate parts.

entropy   a thermodynamic property of a macroscopic body which corresponds intuitively to the degree of disorder.

EPR   abbreviation of A. Einstein, B. Podolsky and N. Rosen, who presented an argument in 1935 that the quantum mechanical description of certain composite physical systems cannot be complete.

equivalence principle   tested by Galileo in the famous experiments at Pisa, this principle asserts the identity of the inertial mass of an object with its ability to produce, or react to, a gravitational field. In Newtonian physics this has to be inserted 'by hand' whereas in general relativity it emerges from the formalism automatically.

event horizon   the surface surrounding a black hole with the property that any light ray emitted inside it cannot escape because of the strength of the gravitational field.

eventuality   a contingency concerning a system which is either true or false if it is definite, but which (in view of a fundamental conceptual innovation of quantum mechanics) may be indefinite. A near-synonym for this term is 'proposition'.

exchange interaction   the spin-dependent part of the interaction between particles with spin.

exclusion principle   Pauli's exclusion principle says that there could not be more than one electron in each quantum state.

existence theorems   these are the theorems that assert the existence of mathematical objects satisfying a specific set of axioms. In the case of differential equations describing the time evolution of a physical system, they guarantee the existence of solutions to the equations that become unique if enough initial conditions are specified.

false vacuum   a peculiar state of matter which has never been observed, but which is predicted to exist by many modern theories of elementary particles, including the grand unified theories; like the ordinary vacuum, a false vacuum has no structure, so motion through a false vacuum cannot be detected; unlike the ordinary vacuum, a false vacuum has a large (positive) energy density and a large negative pressure; a false vacuum is the driving force behind the rapid expansion in the inflationary universe model (*see* inflationary universe).

Fermi constant   the parameter that fixes the strength with which the weak force couples to particles of matter in Fermi's original theory of the weak interaction.

fermion   generic name (after Enrico Fermi) for a class of particles possessing half-integral units of intrinsic spin, measured in terms of Planck's constant divided by 2π. All the known elementary particles of matter (i.e. the quarks and leptons) are fermions. Collections of particles in which the combined spin is half-integral (e.g. the helium-3 nucleus) also behave as fermions.

Fermi statistics   the form of statistics applicable to fermions. It forbids two particles to occupy the same state.

ferromagnet   a material such as iron in which there may be a permanent magnetic moment. In a ferromagnet the spins of the atoms are aligned parallel to one another.

Feynman diagrams   pictorial representations of mathematical expressions for the quantum field theoretic predictions of the scattering of elementary particles. Broadly speaking, the lines in a diagram describe the path of a particle and the vertices correspond to particle interactions that are localised at a space-time point.

field   a physical quantity, like the electric or magnetic field, which varies from point to point in space.

fixed point   a scale-invariant limit point of the flow of configurations or coupling constants generated by a coarse-graining operation.

flavour   whimsical name given to the quality that distinguishes the six known varieties of quarks. These flavours have the names up, down, strange, charmed, top (or truth) and bottom (or beauty). The six different leptons are also sometimes described using the term flavour.

fluctuations   spontaneous deviations of the macroscopic variables from a certain 'reference' state, arising from the thermal motion and the interactions of the molecules.

flux   in magnetism, the total number of lines of magnetic force passing through a specified loop.

four-wave mixing   this is the combination of three optical waves to generate a fourth. It occurs in many optical media provided frequencies and intensities are carefully chosen. It is widely used in phase-conjugation.

fractal geometry   generalisation of Euclidean geometry suitable for describing irregular and fragmented patterns. A noninteger 'fractal dimension' can frequently (but not always) be associated with such patterns.

frequency   the frequency of a periodic or harmonic motion which repeats itself in equal time units is the number of oscillations or cycles per unit of time. Its unit is the hertz (Hz).

gamma-ray astronomy   astronomy carried out in the waveband of photon energies $\varepsilon > 100$ MeV. Except at the very highest energies at which cosmic gamma rays can be detected by the electron–photon cascades which they initiate in the atmosphere, these studies have to be conducted from above the Earth's atmosphere. Many different types of process have characteristic gamma-ray properties including gamma rays produced by π^0 decays following collisions between cosmic rays and nuclei of the interstellar gas, line emission such as the electron–positron annihilation line at 512 keV, inverse Compton scattering gamma rays and other high energy astrophysical processes.

gauge group   the mathematical group associated with a particular set of gauge transformations.

gauge invariance   this would be better called 'local phase-angle independence'. It is the property of electrically charged particles in quantum theory that says that the phase-angle has no physical significance, and can be chosen at will at each point of space. QCD possesses a generalised form of gauge invariance.

gauge symmetry   abstract mathematical symmetry of a field related to the freedom to re-gauge, or re-scale, certain quantities in the theory (potentials) without affecting the values of the observable field quantities.

gauge theory a field theory based on the use of a field that possesses one or more gauge symmetries.

general relativity specific theory of gravitation in terms of curved space-time developed by Einstein; provides field equations to determine the space-time metric for a given distribution of matter.

generation leptons and quarks come two by two. Two leptons (such as electron and neutrino) and two quarks (such as up and down) form a generation. The first generation is $(e^-, v_e; u, d)$, the second is $(\mu^-, v_\mu; c, s)$ and the third $(\tau^-, v_\tau; t, b)$. The top quark (t) in the third generation is predicted by some theories but not yet seen.

geodesic a path or line of shortest distance joining two points in space (or space-time). A geodesic is a straight line if the space is flat. Another familiar geodesic is a great circle on the surface of a sphere. In the general theory of relativity, freely falling particles follow geodesic paths in space-time.

GeV one thousand million electron-volts. Sometimes called BeV: B for billion (US variety).

globular cluster a gravitationally bound spheroidal system of old stars containing about 10^6 stars. These are amongst the oldest stellar systems known within our own Galaxy and they have very low abundances of the heavy elements relative to the average cosmic abundances. The globular cluster population forms part of the halo population of our Galaxy and it is presumed that these clusters formed early in the formation and evolution of our Galaxy.

glueball hypothesised form of matter consisting entirely of gluons.

gluon carrier of interquark force. Plays a role in QCD analogous to that played by the photon in QED.

grand unified theory mathematical scheme in which the electromagnetic, weak and strong nuclear forces are unified into a consistent description.

gravitational constant fundamental constant with units of $cm^3 g^{-1} s^{-2}$ that determines the gravitational force between two bodies at a given separation.

gravitational radiation propagating waves of gravitational tidal force that are emitted by dynamical systems such as collapsing stars or binary star systems, and move with the speed of light.

gravitational redshift generic name for the shift in the frequency or wavelength of a signal that travels up or down in a gravitational field; effect is a redshift if signal travels upward, a blueshift if it travels downward.

gravitational singularity a region where the gravitational field has become so strong that the curvature of space-time is infinite. The occurrence of such a situation signals the breakdown of the theory and is a central feature of the classical theory of relativity.

gravitino the fermion partner of the graviton predicted by the supergravity extension of Einstein's theory of general relativity.

graviton the quantum of the gravitational field. In the quantum theory of general relativity, this elementary particle is massless and has a spin value of two.

hadron the word applied to an object made of quarks and/or antiquarks. Thus protons, neutrons, antiprotons, antineutrons and pions are examples of hadrons. Since quarks have 'colour' they experience the strong QCD forces and so therefore do hadrons. Examples of particles which are *not* hadrons are leptons (electrons, neutrinos, muons, etc.) and photons and W and Z particles.

Hawking radiation the radiation produced by a black hole when quantum effects are taken into account. It can be viewed as a type of virtual pair production in which one of the particles falls through the event horizon of the black hole and hence cannot escape to rejoin its partner.

heavy-fermion systems a class of recently discovered materials, usually rare-earth or actinide compounds, in which the 'effective mass' of the electrons appears to be hundreds or thousands of times the real electron mass.

Heisenberg model a model of magnetic systems in which each magnetic atom has a spin which is free to point in any direction in space. Neighbouring atoms are coupled by a force which tends to align the spins in parallel (for a ferromagnet) or opposite (for an antiferromagnet) directions.

Heisenberg's uncertainty relation in quantum mechanics, the position, x, and the momentum, p, of a particle do not have well-defined values simultaneously. The uncertainty, or statistical spread, in their measured values satisfies the relation $\Delta x \Delta p \geqslant \hbar/2$. Similar inequalities apply to other pairs of dynamical variables.

Hertzsprung–Russell diagram the plot of luminosity against temperature for stars. Stars are only found in certain well-defined regions of this diagram. There are various different ways in which the diagram can be plotted. The most convenient presentation for the observer is to plot the colours of stars against their magnitudes; this is then known as a colour–magnitude diagram. This diagram has to be related to the theoretician's luminosity–temperature diagram through suitable models for stellar atmospheres which enable the relation between the temperature of the surface layers of the star and its observed colour to be determined.

hidden variables theory one of a class of physical theories which deny that the quantum state of a physical system is a complete specification. The hidden variables are those components of the hypothetical complete state which are not contained in the quantum state.

Higgs fields and Higgs particles Higgs fields constitute a set of fundamental fields, named after P. W. Higgs, which induce spontaneous symmetry breaking, e.g. in the Glashow–Weinberg–Salam model and in grand unified theories; a Higgs particle is associated with a Higgs field in the same way that a photon is associated with the electromagnetic field. The Glashow–Weinberg–Salam theory predicts a neutral Higgs particle with a mass-energy in the vicinity of $100 \, GeV$ (billion electron-volts); grand unified theories typically predict Higgs particles with mass energies of order $10^{14} \, GeV$ (*see* spontaneous symmetry breaking.)

horizon distance the maximum distance, at any given time, that a light signal could have travelled since the beginning of the Universe.

Hubble's law the observation, first made by E. P. Hubble in the 1920s, that distant galaxies are receding from us with a velocity proportional to their distance; one infers that any two galaxies are receding from each other with a velocity proportional to their separation.

hysteresis the ability to follow two different branches of states, as a parameter built in the system varies first in a monotonic fashion and subsequently comes back to its initial value by varying in the opposite direction.

indefiniteness   the suspension of an eventuality between truth and falsity, or of a physical variable among its possible definite values, which occurs, according to quantum mechanics, in certain states of a system. This suspension is not a matter of ignorance on the part of the observer but is rather an objective fact.

inertial frame   a frame of reference in which force-free bodies move along straight lines; postulates of special relativity are said to be valid in an inertial frame.

inflationary universe   a cosmological model in which the Universe underwent an epoch of extraordinarily rapid expansion within the first 10^{-30} s or so after the big bang; in a typical version, the diameter of the Universe increased by a factor at least 10^{25} times larger (or perhaps much larger still) than had been previously thought; the model was proposed by A. H. Guth in 1981, but the original formulation contained a crucial flaw which was remedied by the development of the new inflationary universe model in 1982 by A. D. Linde, and by A. Albrecht and P. Steinhardt (*see* new inflationary universe).

information   a measure of the delocalisation of the state of the system in the space of all possible events.

infrared astronomy   astronomy carried out at wavelengths between about $1\,\mu m$ and $300\,\mu m$. Ground-based observations are possible in a number of astronomical 'windows' in the $1-5\,\mu m$, $8-13\,\mu m$, $18-22\,\mu m$ and $30\,\mu m$ wavebands from high, dry observing sites. Observations in the wavelength ranges $5-8\,\mu m$, $13-18\,\mu m$ and $30-300\,\mu m$ can only be carried out from above the Earth's atmosphere because of atmospheric absorption. The wavelength region $100\,\mu m$ to $1\,mm$ is often referred to as the submillimetre waveband.

interference   waves interfere with each other to produce a new wave motion which depends on whether the interfering waves are in step (in which case a large wave motion is produced) or out of step (in which case a small wave motion is produced). Optical interference was discovered by Thomas Young in 1801.

intermediate vector boson   generic name for W and Z bosons – the carriers of the weak force.

interstellar dust   dust particles in the space between the stars. These are responsible for the dark patches of obscuration seen on astronomical photographs. The particles are composed of common heavy elements such as carbon and silicon but there is no agreement about the exact composition of the dust grains. Typically, the particles have size about $1\,\mu m$ but there must be a wide range of particle sizes present to explain the interstellar extinction curve. The dust plays a key role in giant molecular clouds in protecting the fragile molecules from intense interstellar ionising and dissociating radiation. The energy absorbed by the grains is emitted in the far-infrared waveband, and this form of dust emission is one of the most important energy loss mechanisms for regions of star formation.

interval   the quantity in Minkowski space-time which replaces length in ordinary space.

inversion layer   a very thin layer of electrons trapped on an interface between a semiconductor and an insulator, or between two different semiconductors.

irreversibility   time-asymmetric evolution of an observable quantity of a physical system. For an isolated system irreversibility implies the monotonic evolution in the future toward the state of thermodynamic equilibrium.

Ising model   a simplified version of the Heisenberg model in which the atomic spins must be aligned parallel or antiparallel to a given direction.

jet   spray of particles produced from the vacuum by the passage of a high momentum quark or gluon. The direction of the jet indicates the direction of the said quark or gluon.

J-meson   name for the J- or ψ-meson, mass 3 GeV, composed of a charmed quark and charmed antiquark. Its discovery in 1974 instigated a scientific revolution in the later half of the decade.

kaon (K-meson)   variety of strange meson

Kelvin scale   the 'natural' or 'absolute' scale of temperature, on which the value of temperature corresponds roughly to the typical thermal energy. The Kelvin temperature is approximately the Celsius temperature plus 273 degrees.

KeV   one thousand electron volts.

Lamb shift   minute correction to the energy levels of atoms (specifically the first excited state of the hydrogen atom) predicted by quantum electrodynamics, and confirmed to great accuracy by Willis Lamb.

laser   the word laser stands for Light Amplification by Stimulated Emission of Radiation. Proposed by A. Schawlow and C. Townes, the first operating laser was constructed by T. Maiman. Lasers generate intense directional beams of coherent radiation through stimulated emission provided sufficient energy is provided to maintain a large number of radiating atoms in the laser.

lattice   regular solids are characterised by the arrangement of the atoms on a set of regularly spaced points known as the lattice sites.

lattice gas   a model of a condensed system in which atoms may be present on or absent from the sites of a lattice, but no movement of the sites or distortion of the lattice is allowed.

lepton   the generic name for certain particles of matter. The presently known ones are electrons, muons, tau particles and neutrinos. They experience electromagnetic and weak forces, but unlike the quarks, not the strong QCD forces (leptons have no 'colour' quantum numbers). Leptons belong to the class of particles known as fermions.

light-cone   the history, in space-time, of a light flash.

limit cycle   the attractor describing a time-periodic regime of a dissipative dynamical system. In the phase space a limit cycle is represented by a closed curve.

liquid crystal   substances intermediate in their properties between liquids and crystals. There is considerable variety in the type of structure that they can have, but in general they have the anisotropy of a complex crystalline solid but no crystalline long range order, and they can flow like a liquid.

localisation   the wave-function of an electron is said to be localised if it is confined to a small region of a large system rather than being extended through the system.

logistic equation   models the growth of a population as a competition between self-reproduction on the one side and inhibition arising from density-dependent effects on the other side.

Lorentz invariance   principle that physics in an inertial frame is independent of the velocity of the frame relative to any other frame.

low-temperature physics   usually defined as the physics of matter below about 20 degrees absolute (-253 degrees Celsius).

Lyapounov exponent   measures the rate of exponential separation of initially nearby states of a dynamical system. In a system undergoing chaotic dynamics there is at least one positive Lyapounov exponent.

magnetic moment   the intrinsic spins of the electrons in an atom or ion, together with the motion of the electrons round the nucleus, give rise to a magnetic field around the atom. The magnitude of this field is determined by the magnetic moment of the atom or ion.

magnetic monopole   hypothetical particle that carries an isolated north or south magnetic pole. All known magnets are dipoles.

magnetic susceptibility   when a magnetic field is applied to a material, magnetisation is induced. The ratio of the magnetisation induced to the applied magnetic field is the magnetic susceptibility.

magnetosphere   the region about a star or planet in which the magnetic field of the body itself dominates the gas dynamics of the system. The Earth's magnetosphere is bounded by a shock wave associated with the flow of the solar wind past the magnetic cavity associated with the Earth's magnetic field. In the case of compact stars embedded in the winds from companion stars, magnetospheres similar to that of the Earth are set up. In the case of pulsars, the magnetosphere is bounded by the 'light cylinder' at which the corotation velocity of the magnetic field approaches the velocity of light.

manifold   a curved space described by the mathematical discipline of differential geometry.

Markovian process   a random process in which the probability of performing a transition to a certain state at a given time depends solely on the state in which the system is found at this time.

master equation   an equation describing the evolution of the probability of a state at a given time as the balance between transitions leading to this state, and transitions removing the system from this state.

matter fields   the fields whose quanta describe the elementary particles making up the material content of the Universe (as opposed to the gravitons and their supersymmetric partners).

Maxwell's equations   the equations that describe the variations in space and time of the electromagnetic field (after James Clerk Maxwell).

Meissner effect   the phenomenon in which a metal cooled through its superconducting transition temperature in the presence of a magnetic field completely expels the field.

meson   particle consisting of a quark and an antiquark.

metric   mathematical variable that describes the geometry of space-time; it is a tensor made up of ten functions or 'components'.

metric tensor   the mathematical object that describes the deviation of Pythagoras's theorem in a curved space.

microwave background radiation   thermal radiation with a temperature of about 3 K that is apparently uniformly distributed in the Universe; the radiation, discovered by A. A. Penzias and R. W. Wilson in 1964, is believed to be a redshifted remnant of the hot radiation that was in thermal equilibrium with matter during the first hundred thousand years after the big bang.

Minkowski metric   form of the metric that is valid in an inertial frame; underlying geometry of special relativity.

Minkowski space-time   space and time considered together, with special importance attached to the progress of a light flash, and to the light-cone and the 'interval'.

multiphoton process   in very intense radiation fields atoms or molecules can absorb several photons simultaneously in a multiphoton process.

muon   charged lepton. The analogue of electron in the second generation of particles.

neutral current   weak interaction where no change takes place in the charges of the participants.

neutrino   electrically neutral, massless lepton. There are three known varieties, one in each generation of particles, associated with electron, muon, and tau leptons. It only takes part in weak interactions.

neutron star   a compact star in which the internal pressure support is provided by neutron degeneracy pressure. Their masses cannot exceed roughly the mass of the Sun because of the Chandrasekhar limit. Their central densities are roughly nuclear densities and neutron stars may therefore be thought of as giant nuclei. Their radii are about 10 km and the corresponding central densities are about 10^{18} kg m^{-3}. Pulsars are magnetised rotating neutron stars and X-ray binaries consist of a neutron star as a member of the binary system, the X-ray emission being associated with mass transfer onto the neutron star.

new inflationary universe   a revised form of the inflationary universe model that provides a mechanism to avoid the gross inhomogeneities which result from the theory as originally proposed.

Newton's universal gravitational constant   the constant that sets the scale of gravitational forces. Its value is independent of the actual constitution of the matter producing the gravitational field.

nonlinear optics   this is concerned with the optical properties of matter in intense radiation fields. The induced electromagnetic polarisation does not depend linearly on the radiation strength but is severely distorted by the strong field. Optical harmonics, frequency mixing and intensity dependent refractive indices are all produced by this nonlinear response.

normal state (of a superconductor)   the state of a superconducting metal above its transition temperature, in which its behaviour is essentially indistinguishable from that of a metal which never becomes superconducting.

nucleon   generic name for neutrons and protons which are the constituents of a nucleus.

nucleosynthesis   the nuclear processes by which the chemical elements are synthesised. The principal sites for these processes are the central regions of stars where the temperatures are sufficiently high for the synthesis of, for example, helium from protons and neutrons or carbon from three helium nuclei to take place. Other forms of nucleosynthesis include explosive nucleosynthesis which takes place during supernova outbursts and primordial nucleosynthesis in which light elements are synthesised during the hot early phases of the hot big bang.

open system   a system communicating with the environment by the exchange of energy and matter.

optical chaos   in many nonlinear optical systems the output response varies in an unpredictable and uncontrollable fashion despite being governed by deterministic laws. Such optical chaos shows a rich and unexpected variety of phenomena which are only recently being discovered.

optical soliton   a soliton is a wave pulse which propagates without changing shape or dispersion. It maintains its shape by a balancing between linear dispersion and nonlinear compression and occurs in many areas of wave physics where nonlinearities are important. Optical solitons are generated in nonlinear media excited by strong laser pulses.

order parameter   a variable such as magnetisation used to describe the degree of order in a phase below its transition temperature. In a continuous phase transition the order parameter goes continuously to zero as the critical temperature is approached from below.

parity   the operation of studying a system or sequence of events reflected in a mirror.

particle physics   the physics of so-called 'elementary' particles, electrons, protons, pions, $W_{\pm}$ and Z^0 bosons, neutrinos etc.

partition-function   a weighted sum extending over all the possible microscopic arrangements of a macroscopic system, whose evaluation permits the determination of the observable properties of that system, in equilibrium.

parton   conjectured constituent of hadrons, these days normally identified with quarks.

Pauli exclusion principle   principle of quantum theory first enunciated by Wolfgang Pauli. It applies to spin-$\frac{1}{2}$ particles, such as quarks or leptons and states that at most one such can carry any given set of quantum numbers. It thus underwrites the electronic structure of atoms and the quark structure of hadrons.

period doubling   the motion of a particle under the influence of a force may settle down to a regular orbit with a definite period. If the force acts nonlinearly on the particle, and is increased, then the orbit period (the time taken to return to a previous position) may suddenly double when the motion changes to a more complex pattern. This doubling from a simple motion (called a one-cycle) to the more complex form (a two-cycle) is period doubling. The process may continue until an n-cycle is produced. Period doubling is a major phenomenon in nonlinear systems, especially in lasers where the particle motion is replaced by radiation fields.

perturbation theory   mathematical approximation in which a small disturbance added to an exactly soluble system is analysed by a series expansion in powers of the small disturbance.

phase-angle   a complex number has a phase-angle as well as a length.

phase conjugation   this novel form of nonlinear mixing of optical waves generates an output wave which retraces precisely the path taken by the input wave. The phase conjugation reverses the wave front variation in the incident field and can compensate exactly aberrations and distortions in the input. The technique is a kind of holography in real time.

phase space   a space whose coordinates are given by the set of independent variables characterising the state of a dynamical system.

phase transition   a change of state such as occurs in the boiling or freezing of a liquid, or in the change between ferromagnetic and paramagnetic states of a magnetic solid. An abrupt change, characterised by a jump in an order parameter (q.v.) is known as 'first-order'; a change in which the order parameter evolves smoothly to or from zero is called continuous.

photon   quantum particle of the electromagnetic field.

pion (π-meson)   the lightest meson. Predicted by Yukawa, to explain the force binding the nucleus. It comes in three varieties distinguished by their electrical charges $+1$, 0, -1 labelled π^+, π^0, π^-.

planar spin model   similar to the Heisenberg model, except that the spin of the atom is restricted to lie in a plane instead of being free to point in any direction in space.

Planck constant   the fundamental constant of nature $\hbar = 1.05 \times 10^{-34}$ kg m s^{-1}, which characterises quantum physics. Its simplest occurrence is in Heisenberg's uncertainty principle. For macroscopic systems, $\hbar$ is usually negligibly small.

Planck length   the fundamental quantity with the units of length that is expected to define the regime in which quantum gravity effects are of importance. It is defined as $L_P = (G\hbar/c^3)^{\frac{1}{2}}$ where $\hbar$ is Planck's constant, G is Newton's constant and c is the speed of light. The numerical value is $L_P \simeq 10^{-36}$ m.

Planck time   the time taken for a photon travelling at the speed of light to move a distance of the Planck length.

positron (antielectron)   carries positive electric charge. Annihilation with electron produces energy and new varieties of hadrons and quarks.

potentiality   a peculiarly quantum mechanical mode of reality, which is intermediate between full actuality and bare logical possibility. When an eventuality is characterised as a potentiality, it is neither true nor false but indefinite, but it has a definite probability of turning out to be true if the system is subjected to physical conditions which suffice to make it actual.

Potts model   a generalisation of the Ising model in which the two states for each lattice site of the Ising model are replaced by n equivalent states.

PPN parameters   denotes parametrised post-Newtonian parameters, dimensionless parameters that describe the first relativistic corrections beyond Newtonian gravity in the solar system; their values depend on the theory of gravity adopted.

predictability   the ability to predict the future behaviour of a dynamical system on the basis of the present knowledge available on this system.

principle of equivalence   a principle which states that all bodies should fall with the same acceleration; also denoted the 'weak equivalence principle'.

proton   positively charged constituent of the nucleus that gives it electrical charge. Built from (three) quarks.

psi   a name by which the J- or psi- (ψ-)meson is known.

quantum chromodynamics (QCD)   the modern theory of the strong forces between quarks, and hence of the forces between hadrons. It is a generalisation of quantum electrodynamics, with colour charge replacing electric charge and gluons replacing photons.

quantum electrodynamics (QED)   the theory of photons and electrons (or other electrically charged particles) and their interactions. It is called 'quantum' when the electromagnetic radiation (i.e. light etc.) is being treated by quantum theory, so that its discrete photon nature is important.

quantum electronics   this is the name used for those parts of quantum optics which have practical device applications.

quantum field theory   the theory that describes the quantum effects of a classical system of fields defined on space-time and satisfying various partial differential equations.

quantum Hall effect   in a two-dimensional electron system at sufficiently low temperature and in sufficiently high magnetic field the ratio of the current to the voltage applied in a direction perpendicular to the current is very accurately a multiple (integer or fraction with small odd denominator) of $e^2/\hbar$, where e is the electron charge and $\hbar$ is Planck's constant.

quantum liquid   a system of particles which are both sufficiently mobile and at sufficiently low temperature to display the effects of quantum-mechanical indistinguishability. Examples include the electrons in superconducting metals and the atoms in liquid helium.

quark   fundamental particle of all hadrons. There are six known varieties (or 'flavours') of quarks, and they combine in twos or threes to make up particles such as protons, neutrons and mesons.

quasar   an extreme form of active galactic nucleus in which the luminosity of the nucleus far exceeds the luminosity of the underlying galaxy. As a result, these objects have a stellar appearance on photographic plates. The first members of this class were discovered through the optical identification of extragalactic radio sources and hence the origin of the name 'quasi-stellar radio source' abbreviated to quasar. The quasars are the most luminous objects known in the Universe.

Rabi frequency   this is the frequency at which atomic population is coherently transferred from one state to another by a resonant radiation field; it is named after its discoverer I. Rabi. It plays a central role in atom–field interactions, showing up in emitted light spectra and in time dependence.

radio astronomy   the astronomy associated with radio observations of celestial objects. The waveband extends from low radio frequencies ($10\,\text{MHz}$, $\lambda = 30\,\text{m}$) to centimetre and millimetre wavelengths. At the low frequency end of the range, the limit is imposed by the Earth's ionosphere and at the upper end by water vapour absorption in the atmosphere. Within this waveband, many sophisticated radio telescope systems have been constructed, either using single dishes or combining them in arrays using the principles of aperture synthesis and interferometry to obtain high angular resolution.

relativistic plasma   a plasma consisting of particles which have relativistic energies, i.e. for typical particles, their kinetic energies exceed their rest-mass-energies mc^2. Relativistic plasmas are found in such astronomical objects as supernova remnants, radio galaxies, the interstellar medium and the nuclei of galaxies. In many of these cases, the relativistic plasma provides most of the pressure of the medium. The spectrum of the particle energies is generally of power-law rather than Maxwellian form so that many of the particles are ultrarelativistic, i.e. $E \gg mc^2$.

renormalisation   strictly, the rescaling of some parameter in a field theory. In practice, nearly all renormalisations involve an infinite rescaling, so the term has come to be identified with a mathematical procedure for circumventing otherwise nonsensical infinite terms in quantum field theory by absorbing them into observable constants in the theory, such as mass, charge, etc.

renormalisation group   the way in which coupling constants enter into field theory often involves certain simple scaling relations that are described by a group (in the mathematical sense). In statistical mechanics, the renormalisation group method systematically implements some form of coarse-graining operation to expose the character of the large-scale phenomena, in physical systems where many scales are important.

Rydberg atom   extremely highly excited atoms are called Rydberg atoms (Rydberg was an early systematiser of atomic spectra). These atoms are huge (the size of viruses for the largest) and fragile but interact very strongly with radiation. Quantum opticians use them to test basic ideas about atom–photon interactions.

scale-invariance   a physical system is said to exhibit scale-invariance if its appearance remains unchanged (in a statistical sense, and to within simple readjustments of the units of measurements) by a coarse-graining operation.

Schrödinger's equation   the equation that describes the propagation of the waves associated with subatomic particles. In a more general context it describes the time evolution of the state of a quantum system.

Schwarzschild radius   the effective radius of a spherically symmetric black hole, $r_g = (2GM/c^2)^{\frac{1}{2}} = 3(M/M_\odot)$ km. The gravitational redshift of radiation emitted at this radius is infinite to an external observer. Photons cannot escape from a black hole to the outside world from within this radius. The Schwarzschild radius is therefore the smallest physical size which an astronomical object of mass M can have. Matter falling within the radius r_g inevitably collapses into the singularity.

self-organisation   spontaneous emergence of order, arising when certain parameters built in a system reach critical values.

singularity   region of space-time where physical variables become infinite, such as density, tidal forces, pressure, and world line of observer terminates; a bad place to be.

space-time   in both the special and general theories of relativity it is necessary to treat space and time on an equal footing. The ensuing mathematical space is called 'space-time'.

special relativity   a theory of space, time and motion formulated by Einstein in 1905. The theory was generalised in 1915 to include gravity (*see* general theory of relativity).

spin   electrons, protons and neutrons have an intrinsic angular momentum, known as spin, of $\frac{1}{2}\hbar$ and a magnetic moment parallel (or antiparallel) to that angular momentum. When electrons are combined together to form an atom or ion there is a resultant angular momentum which is a combination of the intrinsic spin of the electrons and the angular momentum due to their motion about the nucleus, which is called the spin of the atom or ion; there is also a magnetic moment associated with this angular momentum (unless it is zero). Such atoms or ions with nonzero spin are therefore magnetic atoms or ions. The protons and neutrons in a nucleus also combine together in the same way to give a nuclear spin, but the magnetic moments associated with the nuclei are much smaller, and are only important for magnetic materials at exceedingly low temperatures.

spontaneous emission   an excited atom can shed its excitation by radiating a photon in a spontaneous emission. This emission is independent of external radiation and is entirely random and uncontrolled.

spontaneous symmetry breaking   in many physical systems the

actual state of the system does not reflect the underlying symmetries of the dynamics because the manifestly symmetric state is unstable. The system then trades stability for asymmetry. The symmetry breaking in this case is said to be spontaneous.

state space   the mathematical space whose points represent the states of a physical system.

stationary nonequilibrium state   time-independent state of a system subjected to fixed constraints.

stellar wind   a steady or unsteady outflow of material from the surface of a star. In many classes of star hot coronae are observed and these are believed to be due to heating by waves generated in the upper layers of the star. This results in the outflow of mass in the form of a stellar wind. For a star like the Sun, the mass outflow in the solar wind amounts to only about $10^{-13} M_\odot \, y^{-1}$ but in massive blue supergiant stars the mass loss in the form of stellar winds can amount to as much as 10^{-4} to $10^{-5} M_\odot \, y^{-1}$.

stimulated emission   incident radiation can induce an atom to radiate by stimulated emission at a rate which depends on the intensity of the incident light. Stimulated emission has a definite phase relationship with the incident light and is the driving force behind the laser. It was discovered by Einstein.

strange attractor   an attractor in the phase space of some dynamical system having fractional dimensionality. Strange attractors are associated with chaotic dynamics.

strangeness   property possessed by all matter containing a strange quark. This quark has charge $-\frac{1}{3}$ and partners the charmed quark in the second generation of particles.

string theory   the latest theory of fundamental physics in which the basic entity is a one-dimensional object rather than the 'zero-dimensional' point of conventional elementary particle physics.

strong equivalence principle   generalisation of the Einstein equivalence principle, stating that all bodies, including those with self-gravitational binding, fall with the same acceleration, and that physics in freely falling reference frames, including local gravitational physics, is independent of the velocity and location of the frame.

strong force (interaction)   the dominant force which acts between hadrons, for example, the force which binds neutrons and protons in nuclei. The inter-hadron force is now known to be a remnant of the more powerful force that acts between the quarks which make up the hadrons. This force, which is conveyed by the exchange of gluons, is described by the theory of quantum chromodynamics (QCD).

SU(N)   mathematical structure known as a 'group' that describes operations on N objects. Examples include SU(2) applied to the two quarks or two leptons in a generation and SU(3) applied to the three colours of quark. The three colours and two flavours have recently been combined to yield a set of five entities that can be described by a grand unified theory exploiting SU(5).

sublattice magnetisation   in an antiferromagnet the magnetic atoms can be divided into two equivalent classes, each magnetised in opposite directions. The total magnetisation of one of these classes is the sublattice magnetisation.

subspace   a subset of a vector space which is closed under the operations of vector addition and scalar multiplication.

superconductivity   a phenomenon occurring in some metals at very low temperatures, in which the resistance drops to zero and the metal shows many other anomalous properties.

superconductor   a piece of superconducting metal below the transition temperature at which superconductivity sets in.

superfluid   a liquid which undergoes the phenomenon of superfluidity, below the temperature at which this phenomenon sets in.

superfluidity   a phenomenon occurring in liquid helium-4 below about 2.17 degrees, in which the liquid flows through thin capillaries without apparent friction and displays many other anomalous properties. Liquid helium-3 is also thought to be superfluid below about 3×10^{-3} degrees.

supergravity   the theory that is obtained when the ideas of supersymmetry are applied to general relativity. It predicts that the spin-2 graviton will be accompanied by at least one spin-$\frac{3}{2}$ gravitino.

supernova   a stellar explosion in which a star may be completely disrupted, leaving a compact stellar remnant such as a neutron star or black hole. At maximum light, the supernova can have luminosity about 10^8 or 10^9 times that of the Sun. The luminosity decays after the initial outburst, in certain classes of supernova, the decline being exponential with a half-life of about 80 days. In massive stars, the supernova occurs when the star has used up all its available nuclear fuel and it reaches a lower energy state through gravitational collapse to form a more compact star. In white dwarf stars in binary systems, accretion of mass onto the surface of a neutron star can be sufficient to take the star over the upper mass limit for stability as a white dwarf and it collapses to form a neutron star resulting in a supernova explosion.

superposition principle   a quantum mechanical principle according to which any two states can be combined (actually in infinitely many ways) to form states which have characteristics intermediate between those of the two which are combined. In particular, if an eventuality is true in one of the states and false in the other, then it is indefinite in a superposition of the two states.

superstring theory   a version of string theory which incorporates the ideas of supersymmetry.

supersymmetry   an invariance principle that aspires to place fermions and bosons on an equal footing.

symmetry   if a theory or process does not change when certain operations are performed on it, then we say that it possesses a symmetry with respect to those operations. For example a circle remains unchanged under rotation or reflection. It therefore has rotational and reflection symmetry.

symmetry braking   spontaneous emergence of a state of lower symmetry than the symmetry of the hitherto prevailing state.

synchrotron   modern form of particle accelerator.

system flow   the evolution of the spectrum of configurations, or the associated effective coupling constants, under the action of repeated coarse-graining.

tau lepton   negatively charged lepton in the third generation of particles. A heavier analogue of the electron and muon.

thermal convection   energy transfer in a fluid by a mechanism of bulk hydrodynamic movement.

thermodynamic equilibrium   the state reached ultimately by an isolated system.

thermodynamic potential   a function of the state of a system which takes its extreme value on the asymptotically stable state reached by the system in the course of time.

topology the branch of mathematics that treats the 'large-scale' structure of curved spaces. Topological properties of a geometrical space are those properties that are unchanged by continuous distortion of the space. An example is that in two-dimensional space a simple closed curve divides the space into two regions, one inside the curve and the other outside.

transition metal one of the metals such as iron, manganese or platinum in the centre of the periodic table.

turbulence a hydrodynamic flow characterised by an irregular space and time dependence. Chaotic dynamics and fractal geometry constitute the natural models capturing the essence of turbulence.

U(1) the symmetry group associated with electromagnetic gauge invariance.

ultraviolet astronomy astonomy carried out in the waveband 300 nm to about 10 nm. At these wavelengths, the atmosphere is opaque to radiation and hence these astronomies have to be conducted from above the Earth's atmosphere. The wavelength band 300 to 120 nm can be successfully explored using telescopes which form a natural extension of optical techniques. At shorter wavelengths, new approaches have to be taken because the mirror materials become nonreflecting. In addition at short ultraviolet wavelengths, $\lambda < 91.2$ nm, the interstellar medium is likely to become opaque because of Lyman-continuum absorption.

uncertainty principle *see* Heisenberg's uncertainty relation

unified theories the attempts to unite the theories of the strong, electromagnetic, and weak forces of nature. Ultimately it is hoped that gravity will also be incorporated in this scheme.

universality the phenomenon whereby many microscopically quite different physical systems exhibit critical point behaviour with quantitatively identical features such as critical indices.

universality class this is a way of classifying the behaviour of systems near the critical points of continuous phase transitions. Systems in the same universality class have the same behaviour in the critical region, when an appropriate matching is made between the physical variables, and have the same critical exponents.

upsilon very massive (9 GeV) meson built from a bottom quark and bottom antiquark. Discovered in 1977, it is a member of the most massive family of particles known at present.

vacancy a site on a lattice on which there is no atom present.

valence band the highest completely filled energy band of a solid. In an insulator or semiconductor empty states in the valence band can carry an electric current as positively charged 'holes'.

vector a quantity which has both magnitude and direction, such as the spin of a magnetic atom in the Heisenberg model.

vector space a set of elements (called vectors) for which a binary operation of vector addition is defined, such that $u_1 + u_2$ is a vector if u_1 and u_2 are vectors; and a binary operation scalar multiplication is defined, such that cu is a vector if u is a vector and c is a scalar (a real number or a complex number, according to specification of the kind of vector space); and a standard collection of conditions governing these two operations is satisfied.

virtual particle a quantum particle that exists only temporarily, for example while being exchanged between other particles. Because of Heisenberg's uncertainty relation a virtual particle need not satisfy the usual relationship between energy, momentum and mass.

vortex in a planar spin model a vortex is a pattern of spins in which the spin direction rotates by 360 degrees along any path which surrounds the centre of the vortex. The name is taken from hydrodynamics, where it denotes the kind of circular flow pattern that can be observed in water flowing out of a bathtub. The vortices in superfluid helium films are of this sort.

wave-function the mathematical object in quantum theory which determines probabilities of different results of experiments. It is a complex quantity, so it has an amplitude (whose square gives the probability) and a phase-angle. The phase-angle has no direct physical interpretation, but is important in interference effects, where two wave-functions are added together.

wavelength the distance between adjacent peaks in a wave-train is the wavelength.

weak interaction (force) one of the fundamental forces of nature. Its most famous manifestation is in β-decay; it is also involved in some radioactive decays of nuclei, and neutrino interactions.

Weber bar gravitational-wave detector pioneered by Joseph Weber, consisting of a solid aluminium cylinder, suspended and isolated from surrounding vibrations. The cylinder can weigh several tonnes.

Weinberg angle (θ_W) parameter in the electroweak theory. Relates properties, such as mass, of the W and Z bosons and their interactions.

Weinberg–Salam model a name for the electroweak theory.

white dwarf compact star with mass less than about 1.4 solar masses, typical radius of 1000 km; supported against gravity by quantum-mechanical degeneracy pressure of electrons.

world-line in space-time, the history of a particle is represented by a world-line. The position of the particle in space at any particular time t is found by slicing space-time at time t and seeing where the slice cuts the world-line.

W particles very massive charged ($+$ or $-$) particles that convey part of the weak force between leptons and hadrons.

X particle hypothetical exceedingly massive particle predicted by grand unified theories to convey a very short-ranged interaction between quarks and leptons. An X particle would be able to change a quark into a lepton or an antiquark.

X-ray astronomy astronomy carried out in the waveband roughly 0.1–100 keV. The atmosphere is opaque to radiation at these wavelengths and so observations have to be carried out from above the Earth's atmosphere. It has been found that many classes of object are X-ray emitters, including stars, supernovae and active galaxies.

zero-point energy in quantum mechanics nothing of interest can have zero energy and quantum fluctuations lead to a kind of jittering motion even at lowest energies. This minimum energy due to these quantum fluctuations is termed zero-point energy.

Z particle a particle that is identical to the photon in all respects except mass. It conveys part of the weak force between hadrons and leptons, and its existence was a distinctive prediction of the Glashow–Salam–Weinberg theory of the amalgamated weak–electromagnetic force.

Index